Fluid Mechanics—Volume I

Fluid Mechanics—Volume I

Editors

Vasily Novozhilov
Cunlu Zhao

MDPI

Basel • Beijing • Wuhan • Barcelona • Belgrade • Novi Sad • Cluj • Manchester

Editors
Vasily Novozhilov
Victoria University
Melbourne, VIC
Australia

Cunlu Zhao
Xi'an Jiaotong University
Xi'an
China

Editorial Office
MDPI AG
Grosspeteranlage 5
4052 Basel, Switzerland

This is a reprint of articles from the Topic published online in the open access journals *Energies* (ISSN 1996-1073), *Mathematics* (ISSN 2227-7390), *Fluids* (ISSN 2311-5521), *Applied Sciences* (ISSN 2076-3417), and *Materials* (ISSN 1996-1944) (available at: https://www.mdpi.com/topics/fluid).

For citation purposes, cite each article independently as indicated on the article page online and as indicated below:

Lastname, A.A.; Lastname, B.B. Article Title. *Journal Name* **Year**, *Volume Number*, Page Range.

Volume I
ISBN 978-3-7258-1463-3 (Hbk)
ISBN 978-3-7258-1464-0 (PDF)
doi.org/10.3390/books978-3-7258-1464-0

Set
ISBN 978-3-7258-1461-9 (Hbk)
ISBN 978-3-7258-1462-6 (PDF)

Contents

energies

MDPI

Article

Mathematical Modeling of Sintering Air Leakage through Holes

Jin Cai [1], Xiangwei Kong [1,*] and Mingzhu Yu [1,2]

1 School of Mechanical Engineering and Automation, Northeastern University, Shenyang 110819, China; 15140662469@126.com (J.C.); niunianjixiangruyi@126.com (M.Y.)
2 Anshan Iron & Steel Corp. (AISC), Anshan 114051, China
* Correspondence: xwkong@me.neu.edu.cn or shawnkongneu@163.com; Tel.: +86-138-4027-0169

Abstract: The air leakage in sintering machines affects the technological and economic indexes of the sintering process. It is of great significance to monitor and estimate the key areas. Mathematical models of sintering air leakage through holes in the steady-state process are given based on the fluid mechanics to predict the flow rate and effect on the key area. It was found that the hole model is the application of constant orifice outflow in the computation of sintering air leakage. The counter-flow bed model is suitable for predicting the flow rate through a complete break in sintering wind boxes. Furthermore, This paper proposes a new hole–bed generalized model to cover all the possible hole diameters for further high-precision application. The model connects the leakage hole diameter with the sintering process for the first time and establishes their coupling relationship. The pressure state in the sintering system depends on the ratio of the leakage hole area to the sintering bed area. The proposed fast estimation models are a step forward in developing more precise and powerful calculation tools to foresee the effects and consequences of sintering air leakage. It has a good prospect for reducing and replacing complex manual measurement and bringing some insight into the state of the art that could be improved in the future.

Keywords: the flow rate; mathematical models; sintering air leakage; steady-state

Citation: Cai, J.; Kong, X.; Yu, M. Mathematical Modeling of Sintering Air Leakage through Holes. *Energies* **2022**, *15*, 4224. https://doi.org/10.3390/en15124224

Academic Editors: Vasily Novozhilov and Cunlu Zhao

Received: 29 April 2022
Accepted: 6 June 2022
Published: 8 June 2022

1. Introduction

Air leakage is a worldwide problem in sintering plants [1]. It directly leads to the reduction in the sinter output and quality. It is also the waste of a large amount of energy. Mastering and evaluating the air leakage state of sintering is the first step to solving the problem. The air leakage rate in sintering machines is generally 30~60%. Leakage between pallets and wind boxes accounts for the greatest proportion of the total, including head and tail position gaps, slide-way sealing, abrasion, corrosion, connecting flange, etc. They usually exist in the form of gaps or holes. Due to the large structure, production environment and instrument limitations, the development of sintering air leakage detection and energy-saving technology remains sluggish. At present, the gas analysis method [2], airflow distribution on bed surface method [3], calorimetry method [4], etc., are the commonly used methods to detect the air leakage rate. For local air leakage points, the appropriate instruments can be used to detect the air leakage directly. The thermal anemometer is used to measure the air leakage of bed cracks and holes, pallets gap, etc., at the No. 3 sintering plant in Nagoya Works of Nippon Steel [5]. Reference [6] introduced the air leakage detection system based on the acoustic principle. The microphone was used to test and compare the sound pressure of the old pallet and the new one. It was found that the sound pressure of the pallet with large damage is significantly higher, which proves the feasibility of using the sound pressure to evaluate the air leakage. However, most off-line detection methods require complicated, time-consuming, low-efficiency or high-cost processes. Moreover, the result tends to have bad repeatability and accuracy. The impact of air leakage on the sintering bed remains in the virtue of experience to judge.

On the other side, the research of generalized gas leakage detection in pipelines has gradually become a hot research field. Many research results provide a variety of detection and quantitative means for the leakage problem. Therefore, those results significantly benefit the research and quantitative evaluation of sintering air leakage.

The study on the leakage model for gas pipelines enables engineers to estimate the leakage without instruments, which become the basis for consequence analysis and simulation of leakage diffusion of gas pipeline. Some studies on the gas leakage model have been carried out in recent years. Mathematical models of gas leakage are given according to the fluid mechanics.

The "hole model" [7,8] is applicable for the cases where the leakage hole is small enough, while the "pipe model" [9,10] is for the cases where the pipe section is a complete break. The hole-pipe model [11,12] is the synthesis of the above two. It is a universal leakage model considering the coupling effect of leakage hole size and pipeline pressure drop. Geoff Hankinson [13] presented a program of full-scale experiments. The results provided essential data for validating mathematical models and are used in developing risk assessment methodologies for gas pipelines. The calculation results of the pipe model of Picard [14] proved that the ideal gas assumption makes the gas leakage rate too small. Therefore, the real gas state equation must be added to the model to calculate the accurate leakage. Yuhua Dong et al. [15,16] added the real gas state equation to the medium and low-pressure hole-pipe model. The compressibility coefficient was introduced, so the hole-pipe model was suitable for high-pressure gas. Mahgerefteh et al. [17–19] combined the characteristic line, nested grid and fast mathematical algorithm to put forward a two-phase hydrocarbon pipeline's fracture and leakage model. Young [20–22] assumed that the initial pressure in the pipe is constant and the leakage is an adiabatic process and obtained a simplified calculation model of the small hole leakage rate in a high-pressure pipeline. Kostowski et al. [23,24] presented a method of accounting for leakage utilizing a reference flow equation with a flow coefficient, and the dependency of the flow coefficient on pressure was demonstrated both with the literature data and the authors' experimental results. It was eventually shown that the impact of the flow coefficient on the predicted outflow rate is of lesser importance than that of the applied flow model. The literature [25] proposed a prediction model of leakage intensity of leaked oil and seepage–diffusion range in the soil. The characteristics of dynamic seepage–diffusion of leaked oil in porous soil are investigated, which provides an essential basis and technical support for pipeline accident treatment.

Recently, computational fluid dynamics also promoted the development of gas leakage and diffusion models. A. Ebrahimi-Moghadam et al. [26,27] used the numerical method to investigate leakage estimation of compressible gas in above-ground and buried urban distribution natural gas pipelines. Mahmood Farzaneh-Gord et al. [28] developed an accurate equation for predicting methane emission into the environment during the natural gas purging process with the assistance of the CFD simulation. Xinhong Li et al. [29] presented a computational fluid dynamics (CFD) based approach to describe the behavior of underwater gas release and dispersion from subsea gas pipelines leak. Juliane Fiates et al. [30] developed a series of CFD simulations and a computational tool for handling the gas dispersion of heavy gases, such as LNG and CO2. Meanwhile, gas leakage detection technology is developing rapidly. Many research results provide a variety of detection and quantitative means for the leakage problem, such as the ultrasound method [31,32], acoustic emission method [33–36], infrared thermography technology [37–40], etc.

Compared with the leakage problems in the gas transmission pipeline, the study and detection methods for sintering air leakage still have a long way to go. Therefore, developing a fast estimation model for sintering air leakage is essential to reduce and replace complex manual measurements. Additionally, it can evaluate the impact of air leakage on sintering beds quantitatively. It is also a beneficial supplement to the systematization and standardization of air leakage monitoring technology in sintering machines.

Unlike gas pipelines, air leakage in the sintering machine is negative pressure leakage. The steady sintering process can be regarded as a high-temperature granular bed, and there is a huge temperature difference along the running direction of the pallets. Thus, its fast estimation model has unique characteristics. This paper gives mathematical models of air leakage in sintering machines by using fluid mechanics. They are the hole model, counter-flow bed model and hole–bed model. Application and validation examples of the models are offered. The proposed fast estimation models are a step forward in developing more precise and powerful calculation tools to foresee the effects and consequences of sintering air leakage.

2. Description of Simplified Model of Sintering System

2.1. System Description

Figure 1 shows that the pallets carry the sintering material across the ignition furnace during normal production to start the sintering process. Vacuum suction is formed through the main exhaust fans to promote the airflow in the sintering bed. The air is sent to the material layer to accelerate the sintering process. The gas is heated after passing through the bed with some chemical reaction processes.

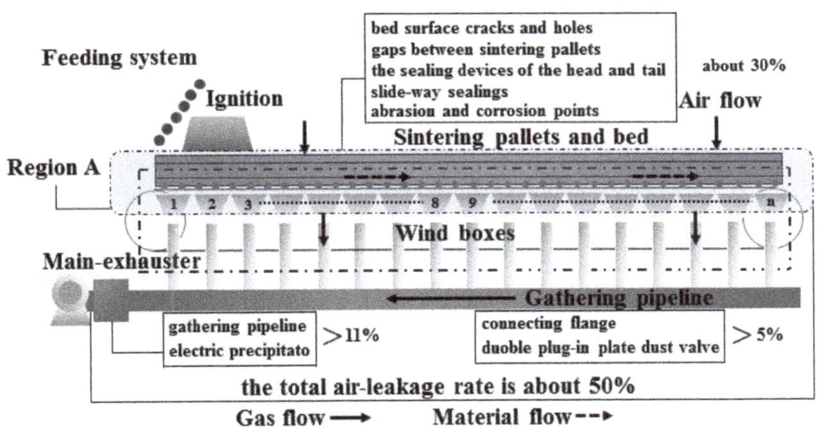

Figure 1. Air leakage distribution in the sintering machine.

There are many air leakage areas in the sintering machine. The leakage between pallets and wind boxes (cf. Region A in Figure 1) accounts for the greatest proportion of the total. These air leakage points are close to the sintering bed. Therefore, region A is the key air leakage monitoring and evaluation area in sintering plants. Figure 2 shows some typical leakage in region A.

Figure 2. Some typical leakage in region A.

The simplified model diagram of the sintering system is shown in Figure 3, and the equivalent leakage occurs in region A. Several assumptions were made:

(1) All leakage points (cf. Figure 3a) occurring in region A are equivalent to one hole for convenience;
(2) The sintering process is steady, so the thermal state of each wind box is stable;
(3) The cross-section area of wind boxes is tapered along the vertical direction (cf. Figure 3a). It is reasonable to treat Region A with the same equivalent diameter;
(4) Because the packed bed has a vast interfacial area, its frictional loss is usually much greater than that of the rest of the sintering system. Therefore, the local and on-way resistance inside region A can be ignored;
(5) The absolute pressure variation is usually less than 10% in the sintering bed. The working temperature along the length of the sintering machine varies greatly. Thus, the state equation for the incompressible ideal gas is used for the study (cf. Equation (1)). It can convert the different isothermal beds of each wind box position into the same temperature state.

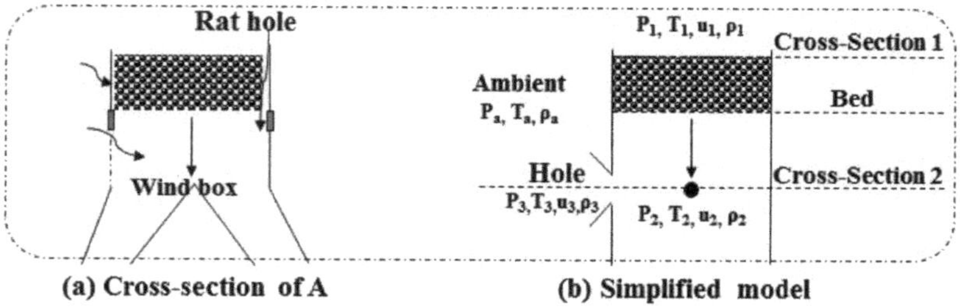

Figure 3. Cross-section of region A and simplified model.

Accordingly, the sintering system (Region A) can be simplified as an isothermal bed in Figure 3b. Various cross-section positions are considered: cross-section 1 on the surface of the sintering bed; cross-section 2 inside wind boxes and on a level with the hole; cross-section 3 is the equivalent leakage hole; cross-section a is the ambient environment; cross-section bed is the sintering bed position.

2.2. Main Equations of the Simplified Model

By considering the flow system of Figure 3b, from cross-section 1 to cross-section 2, the mechanical energy balance equation with no chemicals, the state equation and the continuity equation give

$$\left.\begin{array}{l}\Delta\left(\frac{u^2}{2}\right)+\int\frac{dP}{\rho_{bed}}+F_{total}=0\Rightarrow\frac{P_1-P_2}{\rho_{bed}}=F_{packed}\\ Q_m=\rho_1u_1A_{bed}=\rho_2u_2A_{bed}=\rho_{bed}u_{bed}A_{bed}\\ \rho_1=\frac{P_aM}{RT_1},\rho_2=\frac{P_aM}{RT_2},\rho_{bed}=\frac{P_aM}{RT_{bed}}\\ T_1=T_{bed}=T_2,P_1=P_a=101{,}325\text{ Pa},P_2=P_{bed}=known\end{array}\right\}\quad(1)$$

Here, subscripts 1, 2, a and bed represent cross-section 1, cross-section 2, the ambient environment and the sintering bed. P is the absolute pressure, Pa; T is the temperature, K. ρ is the density, kg/m^3; u is the velocity, m/s; A_{bed} is the area of the sintering bed, m^2; Q_m is the flow rate of the sintering system (Region A), kg/s. M is the molecular weight of the air, 29 kg/kmol; R is the constant of gas, 8314 J/kmol·K; P_a = 101,325 Pa. $\Delta(u^2/2)$ is the kinetic energy term, and this is generally negligible for gases because one rarely reaches very high velocities in packed beds. $\int(dP/\rho_{bed})$ is the workflow term. When the fluid density does not vary much as it passes through the packed bed, one can use an average fluid density. Thus, based on the direction of the airflow in Figure 3b, $\int\frac{dP}{\rho_{bed}}=\int_1^2\frac{dP}{\rho_{bed}}=\frac{\Delta P_{bed}}{\rho_{bed}}=\frac{P_1-P_2}{\rho_{bed}}$. ΔP_{bed} is the negative pressure of the sintering bed, Pa. F_{total} is the total friction. F_{total} = F_{packed}, F_{packed}, the frictional loss of packed beds. Additionally, because the system is interconnected, the pressure in wind boxes is equal everywhere. "*known*" represents known values in sintering plants.

The frictional loss for flow through packed beds can be expressed as

$$F_{packed} = \underbrace{k_1 \frac{\mu(1-\varepsilon)^2 H}{d_p^2 \cdot \varepsilon^3 \rho_{bed}} \cdot u_{bed}}_{\text{Viscoustic loss term}} + \underbrace{k_2 \frac{(1-\varepsilon)H}{d_p \cdot \varepsilon^3} \cdot u_{bed}^2}_{\text{Turbulent loss term}} \tag{2}$$

Here, H is the bed height, m; ε is the bed porosity. d_p is the particle size, m; μ is the air viscosity, kg/m·s. When k_1 = 150, k_2 = 1.75, Equation (2) is an expression proposed by Ergun [41]. The particular research results by Hinkley [42,43], which are more suitable for the sintering bed are cited: k_1 = 323 ± 15, k_2 = 3.78 ± 0.15. This two-term expression of Equation (2) fits the frictional loss well; when the characteristic Reynolds number $R_{ep} < 20$, the viscous loss term dominates and can be used alone with negligible error. On the other hand, when $R_{ep} > 1000$, only the turbulent loss term needs to be used. According to the actual R_{ep} in the sintering process, only the turbulent loss term of Equation (2) needs to be used.

The flow rate Q_m is derived by Equations (1) and (2)

$$Q_m = \rho_{bed} A_{bed} \sqrt{\frac{P_1 - P_2}{H\rho_{bed}} \cdot \frac{\varepsilon^3 d_p}{3.78(1-\varepsilon)}} = \rho_{bed} A_{bed} \sqrt{\frac{P_a - P_{bed}}{H\rho_{bed}} \cdot \frac{\varepsilon^3 d_p}{3.78(1-\varepsilon)}} \tag{3}$$

For Equation (3), there is still a variable T_{bed} unsolved.

T_{bed} is the mixed temperature of the whole sintering bed and can be obtained by the Equation (4) as follows:

$$\left.\begin{array}{l} \rho_{bed} = \frac{P_a M}{R T_{bed}}, Q_m = \sum\limits_{i=1}^{n} q_{m,i} = \rho_{bed} A_{bed} u_{bed} \\ \sum\limits_{i=1}^{n} q_{m,i} = \sum\limits_{i=1}^{n} \rho_{bed,i} A_{bed,i} \sqrt{\frac{P_{1,i} - P_{2,i}}{H\rho_{bed,i}} \cdot \frac{\varepsilon^3 d_p}{3.78(1-\varepsilon)}} \\ \rho_{bed,i} = \frac{P_a M}{R T_{bed,i}}, T_{1,i} = T_{bed,i} = T_{2,i} = known \\ P_{1,i} = P_a = 101,325 \text{ Pa}, P_{bed,i} = P_{2,i} = P_{bed} = known \end{array}\right\} \tag{4}$$

Thus, the analytic formula of T_{bed} is shown by Equation (5)

$$T_{bed} = \frac{P_a M}{R\left(\sum\limits_{i=1}^{24} q_{m,i}\right)^2} \frac{A_{bed}^2 (P_a - P_{bed})}{H} \frac{\varepsilon^3 d_p}{3.78(1-\varepsilon)} \tag{5}$$

i is the number of wind boxes (cf. Figure 1: 1, 2, . . . , n); $q_{m,i}$ is the flow rate through the sintering bed at each wind box location; $P_{bed,i}$, $T_{bed,i}$, $u_{bed,i}$ and $\rho_{bed,i}$ are the absolute pressure, temperature, velocity and density of the sintering bed at each wind box location; $P_{1,i}$, $T_{1,i}$, $u_{1,i}$ and $\rho_{1,i}$ are the absolute pressure, temperature, velocity and density of cross-section 1 at each wind box location; $P_{2,i}$, $T_{2,i}$, $u_{2,i}$ and $\rho_{2,i}$ are the absolute pressure, temperature, velocity and density of cross-section 2 at each wind box location. n is the total amount of wind boxes.

Equations (1)–(5) are the simplified equations of sintering airflow through the bed under a steady-state state, among which Equation (3) is the flow rate of the sintering system. Based on the general simplified model of the sintering system, in what follows, three air leakage models are described. The leakage flow rate could be solved by Equation (6).

$$Q_3 = C_d A_3 \sqrt{2(P_a - P_2) \cdot \rho_a} \quad \begin{cases} P_2 = P_{bed} = known, \ (d_3 \in \text{small hole}) \\ P_2 = \text{Equation (12)}, \ (d_3 \leq D_e) \end{cases} \tag{6}$$

Here subscript 3 represents the hole, as shown in Figure 3b. d_3 is the equivalent diameter of the hole. A_3 is the leakage hole area; Q_3 is the leakage flow rate of the hole, kg/s. D_e is the equivalent diameter of the sintering bed. C_d is the flow coefficient.

2.3. Hole Model

When the hole diameter is relatively small, air leakage would be calculated by the hole model. The assumptions are as follows: The ambient environment around Region A is regarded as an infinite tank with atmospheric pressure. Air leakage does not affect the pressure inside the sintering system. Therefore, the leakage can be considered a constant orifice outflow in fluid mechanics. As shown in Figure 3b, the Bernoulli equation and the state equation are used from the ambient environment tank (cross-section a) to the hole (cross-section 3).

$$\left.\begin{array}{l} \frac{P_a}{\rho_a} + \frac{u_a^2}{2} = \frac{P_3}{\rho_3} + \frac{u_3^2}{2} + \zeta_3 \frac{u_3^2}{2}, \rho_a = \frac{P_a M}{R T_a}, \rho_3 = \frac{P_a M}{R T_3} \\ T_a = T_3 = known, P_a = 101{,}325 \text{ Pa}, P_3 = P_2 = P_{bed} = known \end{array}\right\} \quad (7)$$

In the infinite tank $u_a << u_3$, thus,

$$Q_3 = C_d \rho_3 u_3 A_3 = C_d A_3 \sqrt{2(p_a - p_2) \cdot \rho_a}, \ (d_3 \in \text{small hole}) \quad (8)$$

Here, ζ_3 is the local resistance coefficient of the hole. C_d is the flow coefficient. The necking phenomenon appears when the air flows from the area with higher pressure to lower pressure through the hole. There must be differences between the flow rate calculated by the hole model and the actual flow. Thus, C_d is the ratio of the real flow rate to the theoretical flow rate. This paper mainly focuses on the estimation models of sintering air leakage, so the value is set first to be 1.0 to study and discuss.

Therefore, Equation (8) is the analytic formula of the hole model for fast estimation of the air leakage. The hole model is the application of constant orifice outflow in the computation of sintering air leakage.

2.4. Counter-Flow Bed Model

A complete break usually does not exist for the sintering system, but the case analysis helps propose a further model. As shown in Figure 4a, when a complete break of the sintering system appears at the cross-section 2 location, no effective air passes through the sintering bed. The normal sintering process is completely interrupted. The state of point 2 is the same as that of point 3. Therefore, $P_2 = P_3 = P_a$, $A_3 = A_{bed}$.

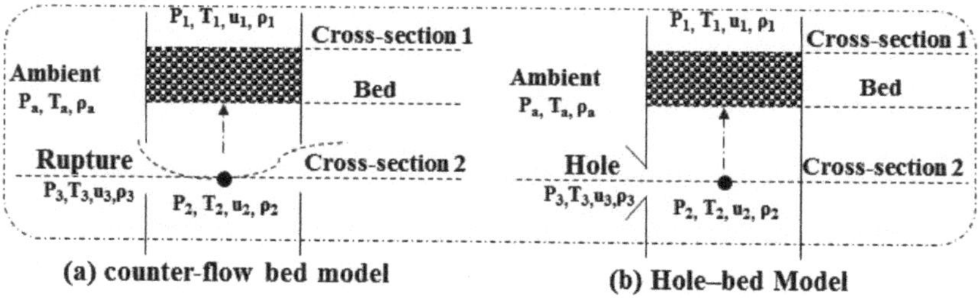

Figure 4. The principles of the counter-flow bed and the hole–bed models.

In order to study the effect of leakage on the sintering bed, we can make the flow rate through the bed in regular operation be the maximum leakage flow limit, $Q_3 = Q_m$. Accordingly, that means the gas needs counter-flow through the sintering bed to unify the mathematical relationship of both flow and pressure in the complete break, as shown in Figure 4a.

It is the application of the general simplified model (cf. Section 2.2) according to the counter-flow direction actually. In particular, note that the direction of the pressure integral is from cross-section 2 to cross-section 1 in this model. The flow rate Q_m is described by a mechanical energy balance equation, continuity equation, the state equation and the frictional loss equation.

Replace the pressure relationships of Equation (4) with Equation (9), and then T_{bed} is obtained by Equations (4) and (5),

$$P_3 = P_2 = P_{2,i} = P_a = 101,325 \text{ Pa}, P_1 = P_{1,i} = P_{bed,i} = P_{bed} = known \tag{9}$$

The leakage flow rate Q_3 is derived by Equation (3)

$$Q_3 = Q_m = \rho_{bed} A_{bed} \sqrt{\frac{P_2 - P_1}{H\rho_{bed}} \cdot \frac{\varepsilon^3 d_p}{3.78(1-\varepsilon)}} = \rho_{bed} A_{bed} \sqrt{\frac{P_a - P_{bed}}{H\rho_{bed}} \cdot \frac{\varepsilon^3 d_p}{3.78(1-\varepsilon)}} \tag{10}$$

The final expression is precisely the same as Equation (3), but note that the pressure relationship at point 2 is the counter-flow system. This model is suitable for a complete break of the sintering system, which is the counter-flow application of the sintering permeability equation.

2.5. Hole–Bed Model

The "hole model" and "counter-flow bed model" actually reflect two extreme cases of sintering air leakage. The hole size in the hole model is small enough to keep the pressure of the sintering system constant and unaffected. In the counter-flow bed model, the area of the leakage hole is equal to the cross-sectional area of the sintering bed. This model considers the system pressure change caused by a complete break, but the influence of variation in leakage hole size on pressure is ignored. According to the high leakage rate characteristics in sintering, the equivalent hole is always a large one. Neither the hole model nor the counter-flow bed model would be suitable. Therefore, a new model bridging this gap would benefit sintering air leakage measurement. This section describes a model that considers all these aspects, thus covering all the possible hole diameters from small to complete break.

Figure 4b shows that the air leakage from the hole flows through the counter-flow bed system. The leakage flow rate Q_3 from the hole is expressed by the hole model (Equation (8)), and the flow rate Q_m in the counter-flow bed system is obtained by Equation (10). According to the continuity equation, the state equation, the hole model Equation (8) and the counter-flow bed model Equation (10):

$$\left.\begin{array}{l} Q_m = Q_3 = A_3\sqrt{2(P_a - P_2)\cdot\rho_a} = \rho_{bed}A_{bed}\sqrt{\frac{P_2-P_1}{H\rho_{bed}} \cdot \frac{\varepsilon^3 d_p}{3.78(1-\varepsilon)}} \\ \rho_a = \frac{P_a M}{RT_a}, \rho_{bed} = \frac{P_a M}{RT_{bed}} \\ T_3 = T_a = known, T_2 = T_{bed} = T_1 \\ P_1 = P_{bed} = known, P_3 = P_a = 101,325 \text{ Pa} \end{array}\right\} \tag{11}$$

There are three unknown variables: Q_m, P_2 and T_{bed} in Equation (11). Firstly, the analytical formula of P_2 is derived by Equation (11).

$$P_2 = \frac{2\left(\frac{A_3}{A_{bed}}\right)^2 \frac{T_{bed}}{T_3} P_a + \frac{\varepsilon^3 d_p}{3.78(1-\varepsilon)H} P_{bed}}{\left[2\left(\frac{A_3}{A_{bed}}\right)^2 \frac{T_{bed}}{T_3} + \frac{\varepsilon^3 d_p}{3.78(1-\varepsilon)H}\right]} \tag{12}$$

Replace the pressure relationships of Equation (4) with Equation (13), and then T_{bed} is obtained by Equations (4) and (5),

$$P_{2,i} = P_a = 101,325, P_{bed,i} = P_{1,i} = P_{bed} = known \tag{13}$$

Then, P_2 can be calculated according to Equation (12), and the leakage flow rate of the hole Q_3 can be obtained by Equation (14).

$$\begin{aligned} Q_3 &= Q_m = C_d A_3 \sqrt{2(P_a - P_2) \cdot \rho_a} \\ P_2 &= \text{Equation (12)}, \ (d_3 \le D_e) \end{aligned} \quad (14)$$

From Equation (12), the pressure state in the sintering system depends both on bed parameters (T_{bed}, d_p, ε, H) and leakage hole parameters (A_3, T_3). Therefore, it can reflect the influence of the size of the leakage hole on the pressure drop of the sintering bed and the coupling relationship between leakage and process parameters.

Equation (14) is the hole–bed generalized model equation covering all the possible hole diameters. The model is preliminarily coupled with the parameters of the sintering granular bed, so it has obvious advantages in replacing manual repeated measurements.

3. Application Example of Models

Take a 360 m^2 sintering machine as an example. The sintering machine is arranged on both sides. Each side is connected with the corresponding gas gathering pipeline and main exhaust fan, as shown in Figure 1. The state parameters of the sintering system can obtain easily in the HMI (Human–Machine Interface) system. The parameters of the sintering machine and wind boxes are given in Tables 1 and 2. Therefore, fast estimation models can be applied to leakage holes found by any means (the image processing technology, manual inspection, statistics in capital repair, etc.) in region A. It dramatically benefits energy-saving and fine production to master the sintering machine's real-time and accurate leakage state.

Table 1. The known parameters of the sintering machine.

Item	Parameters	Unit	Value
Sintering area of oneside	A_{bed}	m^2	180
Area of the sintering bed at each wind box location	$A_{bed,i}$ (i = 1~6)	m^2	3 × 2
	$A_{bed,i}$ (i = 7~24)	m^2	4 × 2
Bed height	H	m	0.75
Bed porosity	ε		0.3
Molecular weight of the air	M	kg/kmol	29
Constant of gas	R	J/kmol·K	8314
Particle size	d_p	m	0.004
Number of wind boxes	n	group	24
Ambient environment temperature	T_a	K	300
Ambient environment pressure	P_a	Pa	101,325
Pressure of the sintering bed	P_{bed}	Pa	84,325
Negative pressure of gathering pipeline	ΔP_{bed}	Pa	17,000
Temperature of gathering pipeline	T_{bed}	K	390~470

Table 2. The parameters of each wind box's location.

Num. of Wind Boxes	$T_{bed,i}$ = *Known* (K)	$P_{bed,i}$ = *Known* (Pa)	$\Delta P_{bed,i}$ = *Known* (Pa)	$q_{m,i}$ Solved (Nm^3/h)
1	345.99	84,325	17,000	16,228.62311
2	370.87	84,325	17,000	15,674.8212
3	347.15	84,325	17,000	16,201.48649
4	349.15	84,325	17,000	16,155.01719
5	346.15	84,325	17,000	16,224.87202
6	341.15	84,325	17,000	16,343.3379
7	340.15	84,325	17,000	21,823.12532
8	339.15	84,325	17,000	21,855.27491
9	339.15	84,325	17,000	21,855.27491
10	340.15	84,325	17,000	21,823.12532

Table 2. *Cont.*

Num. of Wind Boxes	$T_{bed,i} = Known$ (K)	$P_{bed,i} = Known$ (Pa)	$\Delta P_{bed,i} = Known$ (Pa)	$q_{m,i}$ Solved (Nm^3/h)
11	341.15	84,325	17,000	21,791.1172
12	341.15	84,325	17,000	21,791.1172
13	348.15	84,325	17,000	21,570.93572
14	352.15	84,325	17,000	21,448.07591
15	358.15	84,325	17,000	21,267.65989
16	363.15	84,325	17,000	21,120.74145
17	382.15	84,325	17,000	20,588.99994
18	402.15	84,325	17,000	20,070.49797
19	423.15	84,325	17,000	19,566.13347
20	441.47	84,325	17,000	19,155.85699
21	503.15	84,325	17,000	17,943.34659
22	587.65	84,325	17,000	16,603.23879
23	647.78	84,325	17,000	15,813.87919
24	622.85	84,325	17,000	16,127.25489

As shown in Tables 1 and 2, the pressure of the gas-gathering pipeline is a constant negative pressure of 17,000 Pa. The negative pressure of each wind box is slightly different due to the non-uniform distribution of bed porosity and the local resistance of the gas pipeline. $P_{bed} = P_{bed,I} = 84{,}325$ Pa is fundamentally reasonable anywhere in the system. Table 2 also shows the temperature of each wind box $T_{bed,i} = known$. The ambient environment condition gives $T_a = 300$ K, $P_a = 101{,}325$ Pa.

According to the data source above, the leakage flow rate from holes with various diameters can be calculated using fast estimation models.

3.1. Hole Model Results

The results of the hole are shown in Figure 5. It can be seen that when the hole diameter increases, so do the flow rate of holes. The leakage flow rate exceeds the maximum limit $Q_m = 459{,}039$ Nm^3/h (cf. Figure 6a) when the hole diameter is about $d_3 > 1.0$. It indicates that the hole model overestimates the leakage flow rate. However, the velocity of the hole is always constant. The phenomenon's cause is that the hole model only considers variation in hole diameter but ignores the influence of leakage on bed pressure drop and temperature inside the wind boxes. Because the pressure difference between the ambient environment and inside the wind box keeps constant, the velocity keeps stable. The flow rate of the hole then rises with its diameter increasing.

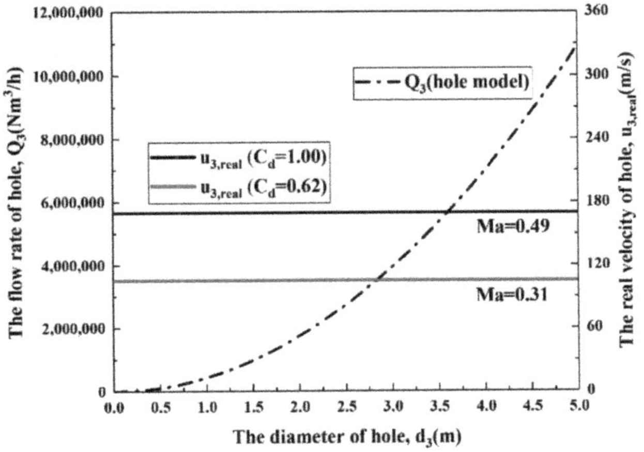

Figure 5. The flow rate and velocity of holes as a function of hole diameter, according to the hole model.

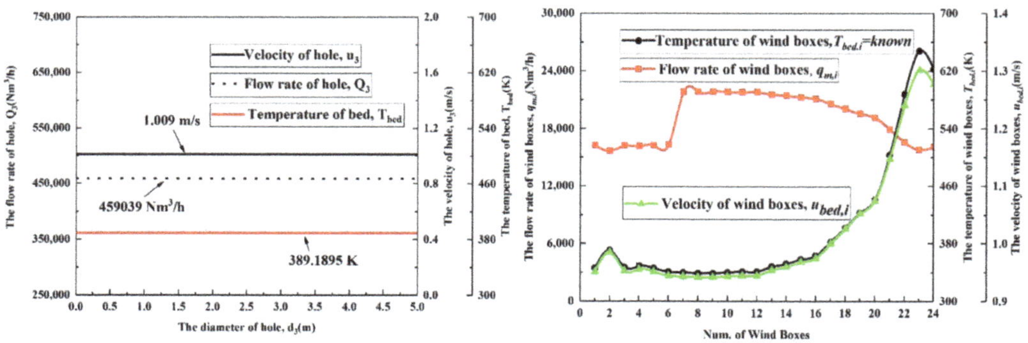

(a) (b)

Figure 6. The results of the counter-flow bed model. (**a**) The flow rate, velocity and temperature of holes. (**b**) The flow rate, velocity and temperature of the sintering bed at each wind box location.

Due to the necking phenomenon, the real average velocity of the hole can be defined as follows:

$$u_{3,real} = \frac{Q_3}{\rho_3 A_3} = C_d \sqrt{\frac{2(P_a - P_2)}{\rho_3}} \quad (15)$$

Ma stands for Mach number in Figure 5, $Ma = u_{3,real}/340$. when $C_d = 1.0$, $u_{3,real} = 169$ m/s, $Ma = 0.49$, the fluid belongs to the weakly compressible range. According to the following text, $C_d = 0.62$ is closer to reality than $C_d = 1.0$. At this time, $u_{3,real} = 105$ m/s, $Ma = 0.31$, the fluid belongs to the incompressible range. Due to the normal working negative pressure limit (0–17,000 Pa), our assumption of incompressibility of gas is correct.

3.2. Counter-Flow Bed Model Results

The counter-flow bed model was applied for the cases in the complete breaking. The leakage flow rate is always equal to the maximum limit of the system. As shown in Figure 6a, no matter how the hole diameter changes, the hole's flow rate, velocity and temperature remain unchanged. The counter-flow model considers the pressure state in a complete break, but the diameter variation is not considered.

Because of a significant temperature difference among the wind boxes, the temperature represents the leak location (wind boxes). It can be seen from Figure 6b that in the case of complete breaking, the flow rate is not evenly distributed. The leakage flow rate and velocity are different with the different temperature locations. From Figure 6b and Equation (4), the variation tendency of the $u_{bed,i}$ is consistent with the $T_{bed,i}$. Due to the different sizes of wind box inlets (cf. Table 1: $A_{bed,i}$), the flow rate of the first six (No. 1~6) wind boxes is lower than the others (No. 7–24) and rapidly increases between No. 6 and 7 wind boxes. The leakage flow rate has a slow downward trend from No. 7 to 24 wind boxes. It is mainly related to the temperature trend. Thus, $T_{bed,i} = known$ must offer accurate and timely feedback. Any changes in $T_{bed,i}$ lead to the changes in $q_{m,i}$, T_{bed} and Q_m, especially at high-temperature locations. It indicates that the counter-flow bed model allows the existence of the unsteady state, and timely feedback of process temperature changes can improve the accuracy of the bed model.

Since the sizes of wind box inlets (No. 7–24) are the same, the effect of temperature difference on the estimation accuracy can be evaluated in the counter-flow model. The estimation deviation E_T between different temperature locations can be defined:

$$E_T = \frac{q_{m,j} - \left(\sum_{i=1}^{n} q_{m,i}/n\right)}{\left(\sum_{i=1}^{n} q_{m,i}/n\right)} \cdot 100\% \quad (16)$$

Thus, from $q_{m,i}$ in Table 2, the maximum estimated deviation at different temperature locations is −17.32% under the same breaking size. It is regarded as a non-existent limit deviation in practice.

3.3. Hole–Bed Model Results

3.3.1. Discussion of Flow and Pressure Results

The hole–bed results are shown in Figure 7. As shown in Figure 7a, when the hole is relatively small, air leakage calculated by the hole–bed model is similar to that by the hole model. While the hole diameter approximates the bed equivalent diameter, here, it gives the same result as the bed model. It is closer to the facts. The hole–bed model, which combines the other two models mentioned above, considers their advantages, thus covering all the possible hole diameters and quantifying the effect on the pressure of the sintering bed. The calculation results show that the local leakage flow rate with various diameters is entirely mastered at one time based on the steady process.

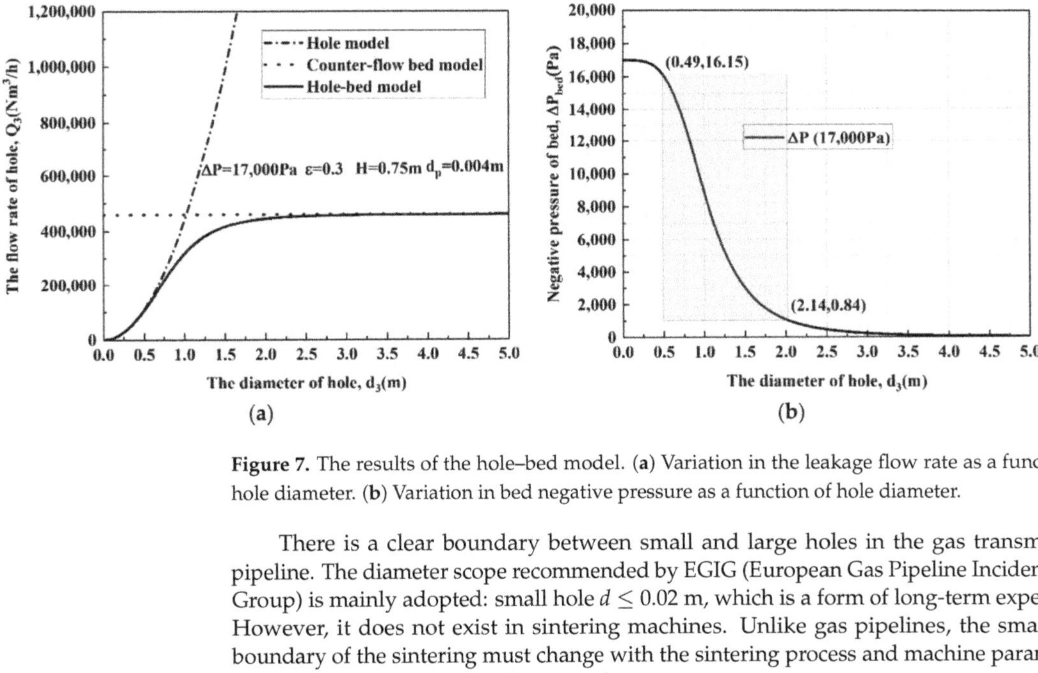

Figure 7. The results of the hole–bed model. (**a**) Variation in the leakage flow rate as a function of hole diameter. (**b**) Variation in bed negative pressure as a function of hole diameter.

There is a clear boundary between small and large holes in the gas transmission pipeline. The diameter scope recommended by EGIG (European Gas Pipeline Incident Data Group) is mainly adopted: small hole $d \leq 0.02$ m, which is a form of long-term experience. However, it does not exist in sintering machines. Unlike gas pipelines, the small hole boundary of the sintering must change with the sintering process and machine parameters, which is beyond the scope of this article.

The estimation deviation E_m between the two models can be defined:

$$E_m = \frac{Q_{3,hole} - Q_{3,hole-bed}}{Q_{3,hole-bed}} \cdot 100\% \quad (17)$$

$Q_{3,hole}$ and $Q_{3,hole\text{-}bed}$ are the leakage flow rate calculated by the hole and hole–bed models. According to the results of fast estimation models, when the hole diameter is 0.58 m, the estimation deviation between the hole model and the hole–bed model reaches 5.00%.

It can be seen from Figure 7b that with the increase in the hole diameter, the negative pressure of the sintering bed decreases. Therefore, air leakage is highly unfavorable to normal sintering production.

The change rate of negative pressure can be defined in the hole–bed model:

$$\frac{P_{bed} - P_2}{\Delta P_{bed}} \cdot 100\% \quad (18)$$

When the hole diameter reaches 0.49 m, the change rate of negative pressure comes to 5%. There is a rapid decreasing stage between about d_3 = 0.49 and d_3 = 2.14. When the d_3 = 2.14, the change rate reaches 95%, and then the negative pressure decreases very slowly. In this case, normal sintering is almost completely interrupted, and only a small amount of effective air flows through the sintering bed. Therefore, leakage must be strictly controlled to ensure normal production.

Thus, comprehensively considering the calculation results of flow rate and pressure drop, 0.49~0.58 m can be preliminarily regarded as a noteworthy diameter area in this machine and sintering process.

3.3.2. Discussion of Different Process Parameters

The leakage is preliminarily combined with the sintering process through the bed equation representing the sintering process's permeability in the hole–bed model. Figure 8 shows the variation in the leakage flow rate as a function of hole diameter under different process conditions. It found that even if the hole diameter is the same, the leakage rate is different under different process conditions.

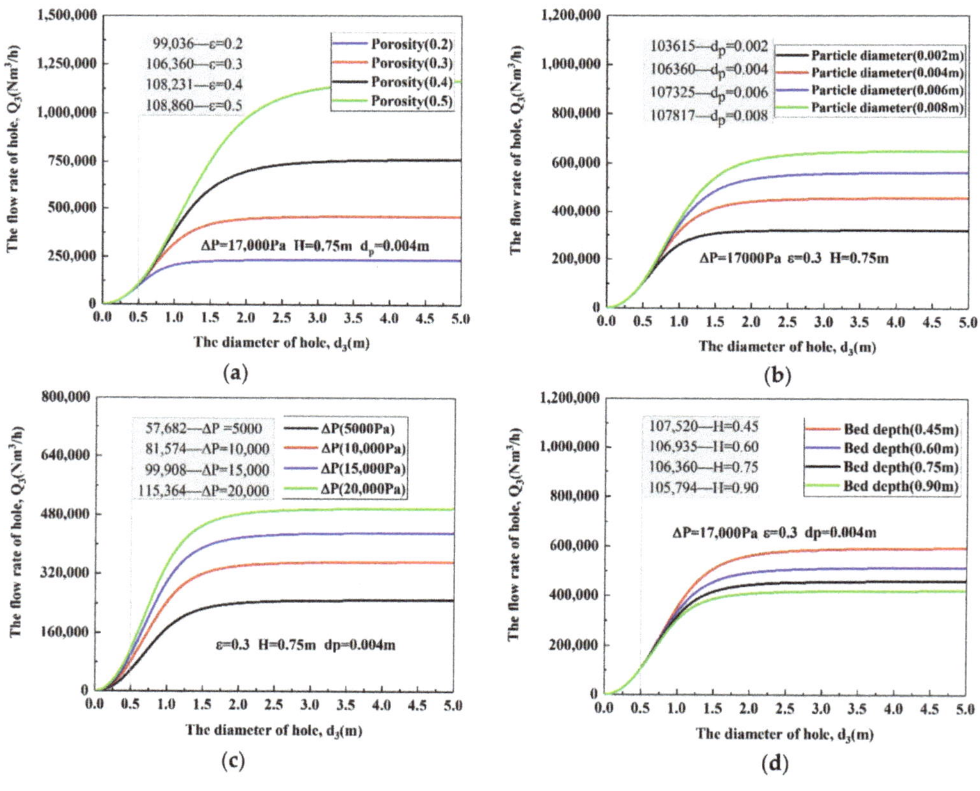

Figure 8. The variation in air leakage flow rate as a function of hole diameter under different process conditions: (**a**) bed porosity; (**b**) particle diameters; (**c**) initial negative pressure; (**d**) bed depth.

There is little difference among processes for small holes (cf. Figure 7: d_3 = 0.5 m), but the difference is apparent when the hole is large (about d_3 = 0.5~2.14 m). The difference gradually and completely stabilized near the fixed values (about d_3 > 2.14 m). That is because the variations in process parameters mean the change in bed permeability. The maximum flow rate in the hole–bed model change according to the permeability equation. The results of the large hole are closer to the leakage trend calculated by the bed model. The flow rate of each small hole also changes a little along with the maximum distribution. Thus, the process parameters affected the measurement repeatability. Air leakage measurement needs to be compared under the same process. It is also an important cause for the long-time test but poor repeatability in air leakage measurement.

More specifically, the estimation deviation E_Q under different processes can be defined:

$$E_Q = \frac{Q_{3,other} - Q_{3,based}}{Q_{3,based}} \cdot 100\% \tag{19}$$

$Q_{3,based}$ are the based flow rate in the hole–bed model in Figure 7. When d_3 = 0.5, $Q_{3,based}$ = 106,360.

$Q_{3,other}$ are the flow rate under other processes in Figure 8. When d_3 = 0.5, each $Q_{3,other}$ are given in the figures.

When d_3 = 0.5 m, the estimation deviation E_Q under different processes is offered as follows:

Figure 8a: ε = (0.2, 0.3, 0.4, 0.5), E_Q = (−6.88%, 0%, 1.76%, 2.09%);

Figure 8b: d_p = (0.002, 0.004, 0.006, 0.008), E_Q = (−2.58%, 0%, 0.91%, 1.37%);

Figure 8c: ΔP = (5, 10, 15, 20), E_Q = (−45.77%, −23.30%, −6.07%, 8.47%);

Figure 8d: H = (0.45, 0.60, 0.75, 0.90), E_Q = (1.09%, 0.54%, 0%, −0.53%).

According to the E_Q above, among the four factors, negative pressure and bed porosity have the greater relative impact on the leakage flow rate apparently. Bed height has the least impact on the prediction of the small hole.

When d_3 = 2.14 m and $Q_{3,based}$ = 447,442 Nm^3/h, the estimation deviation E_Q under different bed heights is offered as follows:

Figure 8d: H = (0.45, 0.60, 0.75, 0.90), E_Q = (27.12%, 11.11%, 0%, −8.33%).

Obviously, for large holes, the influence of bed height on leakage flow rate is vast. Therefore, any change in process parameters can not be ignored for large holes.

The hole–bed model, coupled with the permeability equation, can avoid some repeated measurements at the same location. Its prediction accuracy partly depends on the related expression of sintering permeability in the hole–bed model. We did not elaborate on the effect of each parameter on permeability, as they are in line with the general studies of the sintering permeability.

4. Model Validations and Flow Coefficient

4.1. Bed Model Validation

According to the result of the bed model, Qm = 459,043 Nm^3/h and T_{bed} = 390 K, which are almost the same as the measured average values (463,060 Nm^3/h, 390~450 K) of the gas gathering pipeline on one side and belong to the parameter in the reasonable range. This indicates that the estimation effect of the bed model is reliable.

4.2. Hole Model Validation

According to the results of the hole model, the air velocity of the hole is relatively high. We tried to use the thermal anemometer and pitot tube to measure the flow of local air leakage points at the site. All the results exceed the range of the instrument (45 m/s) at the center of the leakage point. Therefore, the field flow test did not complete the evaluation of the prediction effect of the hole and hole–bed models, and the accurate flow rate could not be determined.

A test device for sintering air leakage is established based on the principle of similarity in the laboratory. The modular design was used for easily realizing different test functions. Figure 9 shows the composition of the test device and its instruments. In the relatively low negative pressure range (0–2 kPa), the effect of the hole and hole–bed models was evaluated.

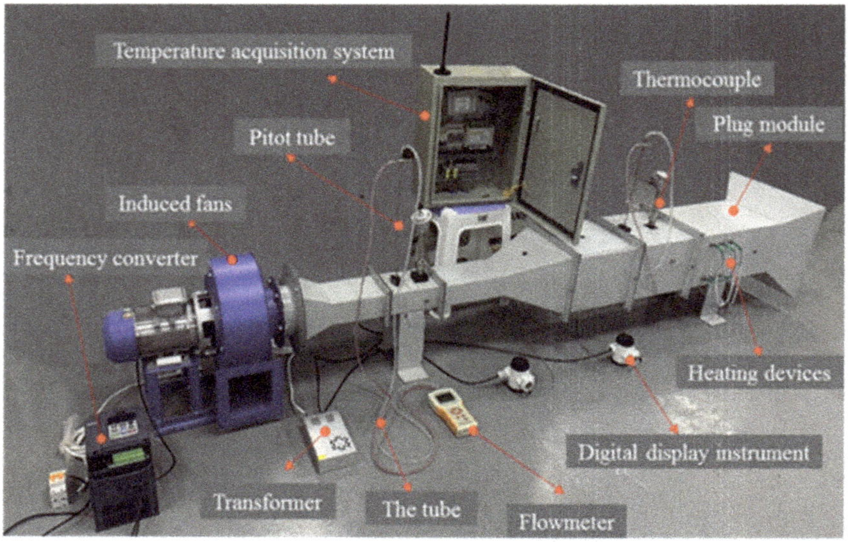

Figure 9. The composition of the test machine and its instruments.

Because there is no clear boundary between the small and large holes, it is difficult to measure the flow with a too-small hole. Therefore, the hole model can be verified: all processed holes are regarded as small, and the negative pressure and flow rate under each case are measured. Accordingly, the flow rate results of the hole model are given for comparison. The calculation of the hole–bed model should be based on the initial negative pressure without leakage. The experiment was carried out at room temperature and pressure.

The main experimental steps include the following: Keep the test device sealed without leakage holes. The entrance is plugged with a sinter module (H = 0.01 m, ε = 0.3). Start the frequency converter and fan to form steady-state initial negative pressures (ΔP_{bed} = 1290 Pa, 870 Pa, 720 Pa). Simulate the leakage from small to large with different circular leakage hole modules (d_3 = 0.02, 0.04, 0.06, 0.08 and 0.10 m). Hold the flow of the induced fan constant, and measure the wind velocity at the leakage point with a thermal anemometer. At the same time, the pitot tube is used to record the static pressure, dynamic pressure and flow rate before and after the leakage hole to ensure the reliability of the measured parameters.

The results of the experiment are shown in Table 3.

Table 3. Measurement results in the laboratory.

d_3, (m)	ΔP, (Pa)	$u_{3,mea}$, (m/s)	$Q_{3,mea}$, (m^3/h)	$Q_{3,hole}$, (m^3/h)	$C_{d,mea}$
0.00	720	-	-	-	-
	870	-	-	-	-
	1290	-	-	-	-
0.02	660	18.5	20.9124	36.1177	0.58
	820	23.0	25.9992	40.2582	0.64
	1200	27.5	31.0860	48.7011	0.64
0.04	560	17.5	79.1280	133.0767	0.59
	760	21.5	97.2144	155.0296	0.63
	960	24.5	110.7792	174.2382	0.64
0.06	410	15.5	157.6908	256.2019	0.62
	620	18.2	185.15952	315.0549	0.59
	820	22.4	227.8886	362.3242	0.63

Table 3. *Cont.*

d_3, (m)	ΔP, (Pa)	$u_{3,mea}$, (m/s)	$Q_{3,mea}$, (m^3/h)	$Q_{3,hole}$, (m^3/h)	$C_{d,mea}$
0.08	280	13.2	238.7405	376.3976	0.63
	430	15.6	282.1478	466.4468	0.60
	540	18.2	329.1725	522.7147	0.63
0.10	200	10.8	305.208	497.0532	0.61
	230	12.5	353.25	533.0302	0.66
	300	14.3	404.118	608.7633	0.66

According to Table 3, the hole model cannot ignore the flow coefficient. The flow coefficient C_d is the ratio of the measured flow rate $Q_{3,mea}$ to the theoretical one $Q_{3,hole}$, which is relatively stable (0.58~0.66) in laboratory research and entirely consistent with the general results of constant orifice outflow in fluid mechanics [44]. Therefore, we can consider that it is reasonable to use the average C_d = 0.62 in the hole model.

4.3. Hole–Bed Model Validation

It can be seen from Table 3 that the negative pressure of the system decreases gradually as the diameter of leakage holes increases. This trend is similar to that predicted by the hole–bed model. Based on the different initial negative pressure (720 Pa, 1290 Pa), Figure 10 shows the model validations of the hole–bed model in the laboratory (C_d = 1 and C_d = 0.62). Taking ±10% of the experimental test values as the reasonable prediction range, the flow rate (C_d = 1) calculated by the two models does not fall in this effective test region, which indicates that the factor of flow coefficient cannot be ignored in practical application. The overestimation error of the hole model is larger than the hole–bed model. When the hole diameter is 0.035 m, the estimation deviation E_m between the hole model and the hole–bed model reaches 5.00% (720 Pa and 1290 Pa). It indicates that the boundary between the small and large holes may be certain.

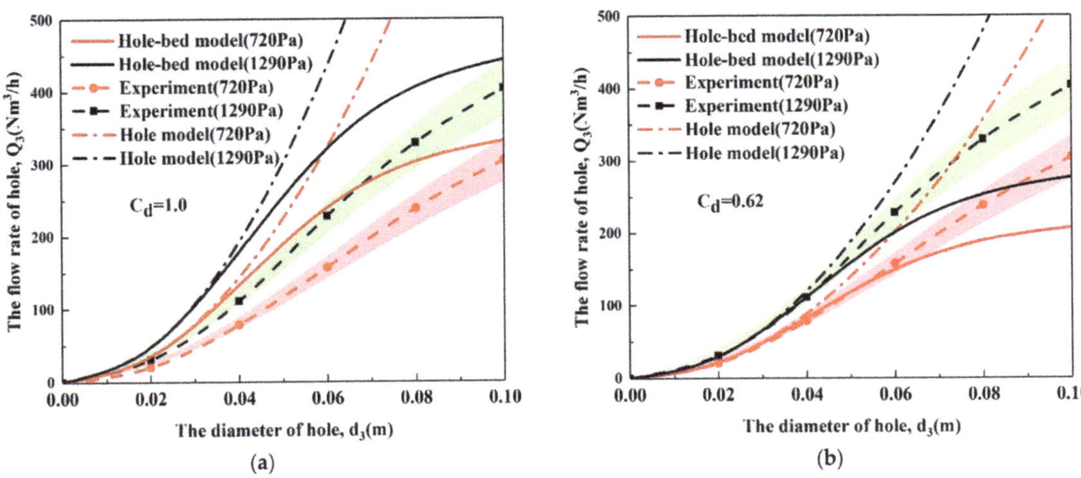

Figure 10. The model validations of the hole–bed model in the laboratory. (**a**) Case: the flow coefficient C_d = 1; (**b**) Case: the flow coefficient C_d = 0.62.

Based on the coupling relationship between the hole and hole–bed models, Figure 10b shows the effect of C_d = 0.62 is used in the hole–bed model. The application C_d = 0.62 enhances the prediction accuracy of the hole–bed model. The applicable diameter range for the flow coefficient is (0~0.06 m) and accounts for about 20% of the test device's whole diameter range (0~0.296 m). The diameter range of the hole model is much smaller than the hole–bed model. However, the flow coefficient is not applicable for all sizes of leakage holes. With the increase in hole diameter, the prediction deviation of the hole–bed model increases gradually. It indicates that the flow coefficient is affected by system pressure. Some research [24,45] also proves it.

According to the fluid mechanics, the recommended value of C_d is between 0.6 and 1.0. It is related to various factors such as the system's pressure, the shape and location of the hole, the forms of damage and corrosion, the material and roughness of the tubes, etc. Therefore, further systematic research is essential for a more accurate prediction but is not involved in this article. The reasonable diameter range on the sintering machine can be achieved at about (0~3 m) with similarity theory (1:9), which fully meets the precision requirement.

Within a specific diameter range of the hole, the fast estimation models with the flow coefficient are proved effective. It is suggested to use the generalized hole–bed model for programming calculation mainly. In contrast, the hole model can be used for mastering the flow rate by engineers directly because of its simple expression.

5. Conclusions

According to the fluid mechanics, three mathematical models of sintering air leakage are proposed in this paper. They are the hole model, counter-flow bed model and hole–bed model.

(1) The hole model conforms to the basic principle of constant orifice outflow. The hole model can predict the leakage flow rate when the hole is small. The counter-flow bed model can be used in a complete breaking. Although this case does not exist, it can estimate the flow rate of the total airflow; while the hole lies between the above two situations, the hole–bed model can be used;
(2) The hole bed model combines the characteristics of the other two models. The leakage is preliminarily combined with the sintering process through the bed equation. In this case, the model also allows the calculation of the decrease in pressure over the sintering bed. The pressure of the sintering bed is affected by leakage parameters;
(3) The calculation results show that the leakage flow rate of any hole is calculated and mastered at one time entirely in a steady process. It effectively reduces testing time, energy consumption and air leakage detection costs. With the model, real-time local leakage detection and a fast estimation system could be possible in the future and improve the detection means in the key area;
(4) Experiment results prove the validation of the models and give the flow coefficient for preliminary application. The flow coefficient can reduce the uncertainty of the estimation of air leakage.

However, there are still some limitations to these models in this paper. For example, it does not apply to other areas, the small hole's boundary is not clear, and the flow coefficient is fixed. Moreover, these models are still a step forward in the effort to develop more precise and powerful calculation tools to foresee the effects and consequences of sintering air leakage anyway. It has a good prospect for reducing and replacing complex manual measurement and bringing a deeper thinking direction to other research.

Author Contributions: J.C.: Conceptualization, validation, formal analysis, investigation, resources, data curation, writing—original draft preparation. X.K.: supervision, methodology. M.Y.: validation. All authors have read and agreed to the published version of the manuscript.

Funding: This research received no external funding.

Institutional Review Board Statement: Not applicable.

Informed Consent Statement: Not applicable.

Data Availability Statement: Not applicable.

Conflicts of Interest: The authors declare no conflict of interest.

Abbreviations

A_{bed}	The area of the sintering bed, (m^2)
$A_{bed,i}$	The area of the sintering bed at each wind box location, (m^2)
A_1	The area of cross-section 1, (m^2)
A_2	The area of cross-section 2, (m^2)
A_3	The leakage hole area, (m^2)
C_d	The flow coefficient
D_e	The equivalent diameter of the sintering bed, (m)
d_p	The particle size, (m)
d_3	The equivalent diameter of the hole, (m)
E_T	The estimation deviation between different temperature locations
E_m	The estimation deviation between the hole and hole–bed models
E_Q	The estimation deviation under different processes
F_{total}	The total friction.
F_{packed}	The frictional loss of packed beds.
H	The bed height, (m)
i	The number of wind boxes
M	The molecular weight of the air, (kg/kmol)
Ma	Mach number
n	The total amount of wind boxes.
P_a	The ambient environment pressure, (Pa)
P_1	The absolute pressure of cross-section 1, (Pa)
P_2	The absolute pressure of cross-section 2, (Pa)
P_3	The absolute pressure of the hole, (Pa)
$P_{1,i}$	The absolute pressure of cross-section 1 at each wind box location, (Pa)
$P_{2,i}$	The absolute pressure of cross-section 2 at each wind box location, (Pa)
$P_{bed,i}$	The absolute pressure at each wind box location, (Pa)
ΔP_{bed}	The negative pressure of the sintering bed, (Pa)
$\Delta P_{bed,i}$	The negative pressure of each wind box location, (Pa)
$q_{m,i}$	The flow rate through the sintering bed at each wind box location, (kg/s)
Q_3	The leakage flow rate of the hole, (kg/s)
Q_m	The flow rate of the sintering system(Region A), (kg/s)
$Q_{3,hole}$	The leakage flow rate calculated by the hole model in Figure 7, (Nm^3/h)
$Q_{3,hole\text{-}bed}$	The leakage flow rate calculated by the hole–bed model in Figure 7, (Nm^3/h)
$Q_{3,based}$	The based flow rate in the hole–bed model in Figure 7, (Nm^3/h)
$Q_{3,other}$	The flow rate under other processes in Figure 8, (Nm^3/h)
$Q_{3,mea}$	The measured value of the leakage flow rate, (Nm^3/h)
R	The constant of gas, ($J/kmol{\cdot}K$)
T_a	The ambient environment temperature, (K)
T_1	The temperature of cross-section 1, (K)
T_2	The temperature of cross-section 2, (K)
$T_{1,i}$	The temperature of cross-section 1 at each wind box location, (K)
$T_{2,i}$	The temperature of cross-section 2 at each wind box location, (K)
T_{bed}	The mixed temperature of the whole sintering bed, (K)
$T_{bed,i}$	The temperature at each wind box location, (K)
u	The velocity, (m/s)
ρ	The density, (kg/m^3)
$\rho_{bed,i}$	The density of the sintering bed at each wind box location, (kg/m^3)
$\rho_{1,i}$	The density of cross-section 1 at each wind box location, (kg/m^3)
$\rho_{2,i}$	The density of cross-section 2 at each wind box location, (kg/m^3)
$u_{1,i}$	The velocity of cross-section 1 at each wind box location, m/s
$u_{2,i}$	The velocity of cross-section 2 at each wind box location, m/s
$u_{bed,i}$	The velocity of the sintering bed at each wind box location, m/s
μ	The air viscosity, (kg/m/s)
ε	The bed porosity.
ζ_3	The local resistance coefficient of the hole.

HMI	Human Machine Interface
Known	The known values in sintering plants.
EGIG	European Gas Pipeline Incident Data Group

References

1. Wang, Y.Z.; Zhang, J.L.; Liu, Z.J.; Du, C.B. Recent Advances and Research Status in Energy Conservation of Iron Ore Sintering in China. *JOM* **2017**, *69*, 2404–2411. [CrossRef]
2. Fan, X.H.; Jiang, L.J.; Chen, X.L. Air Leakage Online Monitoring and Diagnosis Model for Sintering. In Proceedings of the 3rd International Symposium on High-Temperature Metallurgical Processing, TMS (The Minerals, Metals & Materials Society), Orlando, FL, USA, 11–15 March 2012; pp. 323–329.
3. Song, X.Y.; Li, W.H. On the Air Leakage Rate Measurement Technology of Sintering Machine System. *Energy Energy Conserv.* **2013**, *95*, 51–53. (In Chinese)
4. Jin, Y.L.; Xu, N.P.; Wu, S.Y.; Cheng, Z.F. Study and Application of a New Technology for Measuringthe Leakage Ratio of Sintering Machine. *J. Iron Steel Res.* **1999**, *11*, 67–70. (In Chinese)
5. Sakaue, H. Improvement of Production by Decreasing Air Leak at Nagoya No. 3 Sintering Plant. *Tetsu-to-Hagane-J. Iron Steel Inst. Jpn.* **2009**, *95*, 582–585. (In Japanese) [CrossRef]
6. Shiau, J.S.; Huang, T.Y.; Liu, S.H.; Hsieh, C.M.; Yeh, P.Y. Energy Saving Technology for Lowering Air Leakage of Sintering Pallets and Dust Collectors in Sinter Plant. *J. Mech. Eng. Autom.* **2018**, *8*, 233–249.
7. Woodard, J.L.; Mudan, K.S. Liquid and gas discharge rates through holes in process vessels. *J. Loss Prev. Process Ind.* **1991**, *4*, 161–165. [CrossRef]
8. Crowl, D.A.; Louvar, J.F. *Chemical Process Safety: Fundamentals with Application*, 3rd ed.; Pearson: Upper Saddle River, NJ, USA, 2012; pp. 122–146.
9. Turton, R. A new approach to non-choking adiabatic compressible flow of an ideal gas m pipes with friction. *Chem. Eng. J.* **1985**, *30*, 159–160. [CrossRef]
10. Levinspiel, O. *Engineering Flow and Heat Exchange*, 3rd ed.; Springer: New York, NY, USA, 2014; pp. 20–51.
11. Montiel, H.; Vilchez, J.A.; CasaL, J.; Arnaldos, J. Mathematical modelling of accidental gas releases. *J. Hazard. Mater.* **1998**, *59*, 211–233. [CrossRef]
12. Arnaldos, J.; Casal, J.; Montiel, H.; Sánchez-Carricondo, M.; Vílchez, J.A. Design of a computer tool for the evaluation of the consequences of accidental natural gas releases in distribution pipes. *J. Loss Prev. Process Ind.* **1998**, *11*, 135–148. [CrossRef]
13. Hankinson, G.; Lowesmith, B.J.; Genillon, P.; Hamaide, G. Experimental Studies of Releases of High Pressure Natural Gas from Punctures and Rips in Above-Ground Pipework. In Proceedings of the 3rd International Pipeline Conference (IPC2000), Calgary, AB, Canada, 1–5 October 2000.
14. Picard, D.J.; Bishnoi, P.R. The importance of real-fluid behavior in predicting release rates resulting from high-pressure sour-gas pipeline ruptures. *Can. J. Chem. Eng.* **1989**, *67*, 3–9. [CrossRef]
15. Dong, Y.H.; Gao, H.L.; Zhou, J.E.; Feng, Y.R. Evaluation of gas release rate through holes in pipelines. *J. Loss Prev. Process Ind.* **2002**, *15*, 423–428.
16. Dong, Y.H.; Gao, H.L.; Zhou, J.E.; Feng, Y.R. Mathematical modeling of gas release through holes in pipelines. *Chem. Eng. J.* **2003**, *92*, 237–241.
17. Mahgertfteh, H.; Saha, P.; Economou, I.G. A Study of the Dynamic Response of Emergency Shutdown Valves Following Full Bore Rupture of Gas Pipelines. *Process Saf. Environ. Protect.* **1997**, *75*, 201–209. [CrossRef]
18. Mahgertfteh, H.; Saha, P.; Economou, I.G. Fast numerical simulation for full bore rupture of pressurized pipelines. *AICHE J.* **1999**, *45*, 1191–1201. [CrossRef]
19. Mahgertfteh, H.; Oke, A.; Atti, O. Modelling outflow following rupture in pipeline networks. *Chem. Eng. Sci.* **2006**, *16*, 1811–1818. [CrossRef]
20. Jo, Y.D.; Ahn, B.J. A simple model for the release rate of hazardous gas from a hole on high-pressure pipelines. *J. Hazard. Mater.* **2003**, *97*, 31–46. [CrossRef]
21. Jo, Y.D.; Ahn, B.J. Analysis of hazard area associated with hydrogen gas transmission pipelines. *Int. J. Hydrog. Energy* **2006**, *31*, 2122–2130. [CrossRef]
22. Jo, Y.D.; Crowl, D.A. Individual risk analysis of high-pressure natural gas pipelines. *J. Loss Prev. Process Ind.* **2008**, *21*, 589–595. [CrossRef]
23. Kostowski, W.J.; Skorek, J. Application of experimental flow characteristics of pipeline ruptures in gas network simulation. *Inz. Chem. Proces* **2006**, *27*, 579–596.
24. Kostowski, W.J.; Skorek, J. Real gas flow simulation in damaged distribution pipelines. *Energy* **2012**, *45*, 481–488. [CrossRef]
25. He, G.X.; Lyu, X.D.; Liao, K.X.; Li, Y.S.; Sun, L.Y. A method for fast simulating the liquid seepage-diffusion process coupled with internal flow after leaking from buried pipelines. *J. Clean. Prod.* **2019**, *240*, 118167. [CrossRef]
26. Ebrahimi-Moghadam, A.; Farzaneh-Gord, M.; Deymi-Dashtebayaz, M. Correlations for estimating natural gas leakage from above-ground and buried urban distribution pipelines. *J. Nat. Gas Sci. Eng.* **2016**, *34*, 185–196. [CrossRef]
27. Ebrahimi-Moghadam, A.; Farzaneh-Gord, M.; Arabkoohsar, A.; Jabari-Moghadam, A. CFD analysis of natural gas emission from damaged pipelines: Correlation development for leakage estimation. *J. Clean. Prod.* **2018**, *199*, 257–271. [CrossRef]

28. Farzaneh-Gord, M.; Pahlevan-Zadeh, M.S.; Ebrahimi-Moghadam, A.; Rastgar, S. Measurement of methane emission into environment during natural gas purging process. *J. Environ. Pollut.* **2018**, *242*, 2014–2026. [CrossRef] [PubMed]
29. Li, X.H.; Chen, G.M.; Zhang, R.R.; Zhu, H.W.; Fu, J.M. Simulation and assessment of underwater gas release and dispersion from subsea gas pipelines leak. *Process Saf. Environ. Protect.* **2018**, *119*, 46–57.
30. Fiates, J.; Santos, R.R.C.; Neto, F.F.; Francesconi, A.Z.; Simoes, V.; Vianna, S.S.V. An alternative CFD tool for gas dispersion modelling of heavy gas. *J. Loss Prev. Process Ind.* **2016**, *44*, 583–593. [CrossRef]
31. Huseynov, J.; Baliga, S.; Dillencourt, M.; Bic, L.; Bagherzadeh, N. Gas-leak localization using distributed ultrasonic sensors. In Proceedings of the SPIE 7293, San Diego, CA, USA, 8–12 March 2009; Volume 7293.
32. Dudić, S.; Ignjatović, I.; Šešlija, D.; Blagojevićand, V.; Stojiljković, M. Leakage quantification of compressed air using ultrasound and infrared thermography. *Measurement* **2012**, *45*, 1689–1694. [CrossRef]
33. Liu, C.W.; Li, Y.X.; Fang, L.P.; Xu, M.H. New leak-localization approaches for gas pipelines using acoustic waves. *Measurement* **2019**, *134*, 54–65. [CrossRef]
34. Wang, S.; Yao, X.F. Aeroacoustics measurement of the gas leakage rate for single hole. *Rev. Sci. Instrum.* **2020**, *91*, 045102. [CrossRef]
35. Mostafapour, A.; Davoudi, S. Analysis of leakage in high pressure pipe using acoustic emission method. *Appl. Acoust.* **2013**, *74*, 335–342. [CrossRef]
36. Jin, H.; Zhang, L.B.; Liang, W.; Ding, Q.K. Integrated leakage detection and localization model for gas pipelines based on the acoustic wave method. *J. Loss Prev. Process Ind.* **2014**, *27*, 74–88. [CrossRef]
37. Li, J.K.; Jin, W.Q.; Wang, X.; Zhang, X. MRGC performance evaluation model of gas leak infrared imaging detection system. *Opt. Express* **2014**, *22*, A1701–A1712. [CrossRef]
38. Wang, S.; Yao, X.F.; Yang, H. The heat conduction model and leakage characterization of the sealing interface. *Int. J. Therm. Sci.* **2019**, *145*, 106027. [CrossRef]
39. Yang, H.; Yao, X.F.; Wang, S.; Yuan, L.; Ke, Y.C.; Liu, Y.H. Simultaneous determination of gas leakage location and leakage rate based on local temperature gradient. *Measurement* **2019**, *133*, 233–240. [CrossRef]
40. Wang, S.; Yao, X.F.; Yang, H.; Yuan, L.; Dong, Y.F. A new leakage measurement method for damaged seal material. *Meas. Sci. Technol.* **2018**, *29*, 075203. [CrossRef]
41. Ergun, S. Fluid flow through packed columns. *Chem. Eng. Prog.* **1952**, *48*, 89–94.
42. Hinkley, J.; Waters, A.G.; Litster, J.D. An investigation of pre-ignition air flow in ferrous sintering. *Int. J. Miner. Process.* **1994**, *42*, 37–52. [CrossRef]
43. Wang, G.; Wen, Z.; Lou, G.F.; Dou, R.F.; Li, X.W.; Liu, X.L.; Su, F.Y. Mathematical modeling and combustion characteristic evaluation of a flue gas recirculation iron ore sintering process. *Int. J. Heat Mass Transf.* **2016**, *97*, 964–974. [CrossRef]
44. Brower, W.B.; Eisler, E.; Filkorn, E.J.; Gonenc, J.; Plati, C.; Stagnitti, J. On the Compressible Flow Through an Orifice. *J. Fluids Eng.-Trans. ASME* **1993**, *115*, 660–664. [CrossRef]
45. Szente, V.; Vad, J. A semi-empirical model for characterisation of flow coefficient for pneumatic solenoid valves. *Period. Polytech-Mech. Eng.* **2003**, *47*, 131–142.

mathematics

MDPI

Article

Insight in Thermally Radiative Cilia-Driven Flow of Electrically Conducting Non-Newtonian Jeffrey Fluid under the Influence of Induced Magnetic Field

Fehid Ishtiaq [1], Rahmat Ellahi [1,2,*], Muhammad Mubashir Bhatti [3] and Sultan Z. Alamri [4]

1 Department of Mathematics and Statistics, International Islamic University, Islamabad 44000, Pakistan; fehid.phdma99@iiu.edu.pk
2 Fulbright Fellow, Department of Mechanical Engineering, University of California Riverside, Riverside, CA 92521, USA
3 College of Mathematics and Systems Science, Shandong University of Science and Technology, Qingdao 266590, China; mmbhatti@sdust.edu.cn
4 Faculty of Science, Taibah University Al-Madinah Al-Munawarah, Madinah 41411, Saudi Arabia; szamri@taibahu.edu.sa
* Correspondence: rellahi@alumni.ucr.edu or rahmatellahi@yahoo.com

Abstract: This paper investigates the mobility of cilia in a non-uniform tapered channel in the presence of an induced magnetic field and heat transfer. Thermal radiation effects are included in the heat transfer analysis. The Jeffrey model is a simpler linear model that uses time derivatives rather than convected derivatives as the Oldroyd-B model does; it depicts rheology other than Newtonian. The Jeffrey fluid model is used to investigate the rheology of a fluid with cilia motion. The proposed model examines the behavior of physiological fluids passing through non-uniform channels, which is responsible for symmetrical wave propagation and is commonly perceived between the contraction and expansion of concentric muscles. To formulate the mathematical modeling, the lubrication approach is used for momentum, energy, and magnetic field equations. The formulated linear but coupled differential equations have been solved analytically. Graphs for velocity profile, magnetic force function, induced magnetic field, current density, pressure rise, and heat profile are presented to describe the physical mechanisms of significant parameters. It is found that the eccentricity parameter of the cilia equations opposes the velocity and the magnetic force functions. The thermal radiation decreases the temperature profile while it increases for Prandtl and Eckert numbers. A promising impact of the magnetic Reynolds number and electric field on the current density profile is also observed.

Keywords: heat transfer; induced magnetic field; cilia motion; non-uniform tapered channel; non-Newtonian fluid; exact solutions

MSC: 80A05; 76W05; 76Z05; 76A10

Citation: Ishtiaq, F.; Ellahi, R.; Bhatti, M.M.; Alamri, S.Z. Insight in Thermally Radiative Cilia-Driven Flow of Electrically Conducting Non-Newtonian Jeffrey Fluid under the Influence of Induced Magnetic Field. *Mathematics* **2022**, *10*, 2007. https://doi.org/10.3390/math10122007

Academic Editors: Vasily Novozhilov and Cunlu Zhao

Received: 5 May 2022
Accepted: 7 June 2022
Published: 10 June 2022

1. Introduction

Cilia structures are highly complex, resembling flexible rods or fine hairs, and are found in all major animal species except nematodes. Immotile and motile cilia have two primary functions: detection and fluid flow generation. Sleigh [1] described the structure of flagella and cilia in 1962. He discussed the four main patterns of metachronal waves in detail, which are symplectic, antiplectic, dexioplectic, and laeoplectic metachronal waves. The combined movement of the cilia tips systematically degenerates in a wave-like manner causing a metachronal wave.

Each cilium has a diameter of around 250 nm and a typical length of 2 to 15 μm. They are often classified into two types: motile and immotile. The existence of a 9 + 0 arrangement of axonemes with nine doublet microtubules in immotile cilia and a 9 + 2 structure of axonemes with nine doublet microtubules with a center microtubule pair in motile cilia. Motile cilia travel back and forth in a concentrated way. They are particularly effective in creating flow, and due to their tiny dimensions, they function in a low Reynolds number hydrodynamic regime [2]. Motile cilia include ependymal cilia in the brain [3], which move cerebrospinal fluid in the brain ventricle system to nourish and support brain homeostasis, and respiratory cilia [4], which can also aid in the flow of mucus and dust particles breathed up and out of the lungs. Large embedded particles are picked up by the mucus layer and moved in an integrated ciliary action. Mucociliary clearance, based on the continuous flow of fluid through the airway lumens, prevents the buildup of inhaled pollutants and bacteria in the respiratory tract. Similarly, cilia are present in female fallopian tubes, which aid in transporting fertilized ovum from the ovary to the uterus [5]. Lardner and Shak [6] addressed the motions of cilia in general and carefully explained the role of cilia movement in fluid transportation through the efferent of the male reproductive tract. Sher Akbar [7] studied nanoparticles in the symmetric ciliated channel for the biomathematical analysis of carbon nanotubes. Several researchers have recently addressed cilia motion in various geometrical configurations, which are listed in the studies [8–10].

The most essential component regulating ciliary function is temperature. Temperature fluctuations affect the amounts of moisture and humidity required in ciliated cells to sustain normal physiological conditions. In particular, heat affects the ciliary circulation in a way that is comparable to its influence on many biological processes; as the temperature rises, the cilia beat faster until they are damaged by high temperatures [11,12]. Many human disorders are caused by ciliary motility defects, including male and female infertility, lung and renal disease, mental illness, and developmental abnormalities. These conditions are classified as motile ciliopathies [13,14]. As heat transport occurs due to temperature differences, the three primary mechanisms of heat transfer are conduction, convection, and radiation. Improving heat transport is important in a ciliated movement [15–19]. Akbar and Butt [20] investigated the viscoelastic fluid model under the impact of heat transfer to better understand the role of cilia in the respiratory system and its essential roles associated with fluid transport in the human body. Butt et al. [21] investigated heat transport effects in a ciliated circular tube using the PTT (Phan–Thien–Tanner) fluid model. Abdou Al-Zubaidi et al. [22] studied the heat analysis of blood flow by taking into account the cilia movement of Newtonian, dilatant, and pseudo-plastic fluids in a horizontal inclined channel in front of metachronal waves and achieved the analytical solution. McCash et al. [23] studied the effects of a heated viscous Newtonian fluid on peristaltic flow through a ciliated elliptic conduit. It was mentioned that elliptic cross-sections had a longer circular length, which allowed for better estimation of finer colling effects, heat transport analyses, etc.

The diverse functions of living cilia, on the other hand, have inspired many researchers. They have been studying artificial cilia's properties in microfluidic applications such as microsensors, microrobots, light, particle manipulation, droplet antifouling surfaces, self-cleaning microfluidic mixing, and microfluidic pumping. The microfluidic channel was used by Mayne and Toonder [24] to test the ability of magnetic cilia in fluid mixing and bacterium collection.

Magnetohydrodynamics (MHD) is a growing topic of study in medical sciences and advanced biomedical engineering. It has several medicinal uses, including cancer therapy, preventing bleeding in time surgery, cell departure, magnetic endoscopy, and drug delivery targeting using magnetic particles as drug movers. Many problems inside the human body are addressed using various diagnostic techniques such as magnetic resonance imaging. Rashidi et al. [25] presented a detailed analysis of the applications of an effective magnetic field in simple flow, pulsatile flow, peristaltic flow, and drug delivery. Farooq et al. [26] investigated the combined effects of magnetic fields and heat transfer on ciliary-generated flows in the human body. Akbar et al. [27] studied the effects of magnetic induction and electrical fields on the dissipative flow of ciliary-produced MHD copper-water nanofluid. Sadaf and Nadeem [28] investigated fluid flow in the presence of heat transfer and a magnetic field with cilia beating along a curved ciliated channel. Few core investigations relevant to the topic can be found in references [29–34].

In the current study, we used the Jeffrey fluid model to analyze cilia motion. The Jeffrey model is a simpler linear model that uses time derivatives rather than convected derivatives as the Oldroyd-B model does; it depicts rheology other than Newtonian. The proposed study focuses on the convected Jeffrey's model, which extends the original Maxwell convected viscoelastic model. The presented formulation (see Equation (10)) does not account for the stress time derivative (valid for steady flows). In comparison to other viscoelastic flow models, the proposed Jeffrey fluid is simpler (such as see the refs. [35–37]). Since its equations could be reduced to those of Newtonian models as a particular case, the Jeffrey model is regarded as a generalization of the frequently used Newtonian fluid model. The Jeffrey fluid model can characterize the stress relaxation feature of non-Newtonian fluids, whereas the standard viscous fluid model cannot. The Jeffrey fluid model accurately describes a class of non-Newtonian fluids with a distinctive memory time scale, also referred to as the relaxation time. Azaiez and Homsy [38] addressed other Jeffrey fluid models, such as the co-rotational variant. Similarly, the Jeffrey fluid model was addressed for polymeric flows [39,40], steady blood flows [41,42], steady boundary layer flows [43,44], etc.

Following the above literature review, it is seen that cilia-driven flow in the tapered channel is still unavailable. As a result, the Jeffrey fluid model was used to explain the cilia-induced flow through the asymmetric ciliated tapered channel. By using the Jeffrey fluid model, it is possible to analyze both Magnetohydrodynamic Newtonian and non-Newtonian phenomena with long wavelengths and low Reynolds numbers. The implications of induced magnetohydrodynamics on the heat transfer rate are also discussed. The governing equations are modeled and simplified using the lubrication hypothesis. The computational software Mathematica is used to find the analytical solution to the corresponding differential equations. Similarly, the impacts of different physical variables on the velocity profile, magnetic force function, induced magnetic field, current density, and temperature profile were graphically shown. In addition, streamlines are used to explain the trapping phenomena of cilia-driven flow.

2. Viscoelastic Fluid in a Ciliated Tapered Channel: Mathematical Modeling

We consider the Jeffrey fluid model under the effects of a magnetic field in a tapered channel whose inner wall is ciliated in the form of an asymmetric metachronal wave moving towards the right with wave speed c. The lower wall of the tapered channel is maintained at a temperature of T_0' and the upper wall at T_1'. A Cartesian coordinate system is taken for the tapered channel in which X' axis is across the axial direction and Y' axis lies along the transverse direction as shown in Figure 1.

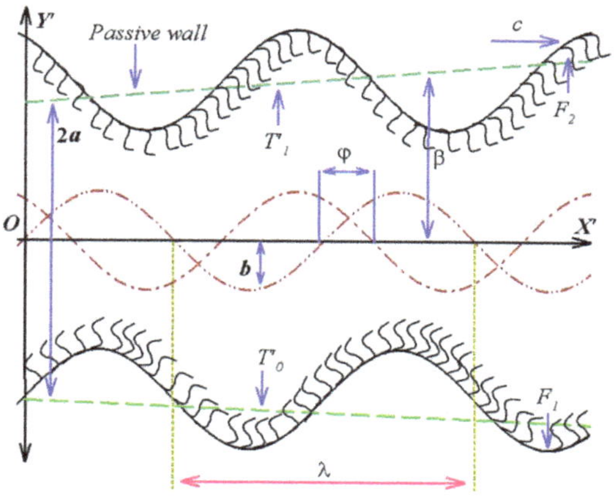

Figure 1. Geometrical structure of the non-uniform flow.

The envelope of cilia tips for the tapered channel is presumed to be written as [45]:

$$\left.\begin{array}{c} Y' = F_1(X',t') = -X'\tan\beta - a - a\epsilon\cos\left[-\frac{\varphi}{2} + \frac{2\pi}{\lambda}(X' - ct')\right] = H_1, \\ Y' = F_2(X',t') = X'\tan\beta + a + a\epsilon\cos\left[\frac{\varphi}{2} + \frac{2\pi}{\lambda}(X' - ct')\right] = H_2, \end{array}\right\} \quad (1)$$

where H_1 is the lower wall of the ciliated tapered channel, H_2 is the upper wall of the ciliated tapered channel, X' is the axial dimensional coordinate, t' is the dimensional time, β is the inclined angle of unperturbed walls, a is the channel half-width, λ is the wavelength, ϵ is a non-dimensional measure with respect to any of the cilia length, and φ is the phase difference, which may be in the range $0 \leq \varphi \leq \pi$ for symmetric contractions of wave $\varphi = \pi$. According to the sleigh, the cilia tips are assumed to move in elliptic paths so that the horizontal locus of the cilia tip is specified as

$$X' = F_3(X_0, X', t') = X_0 + \epsilon\omega a \sin\left[\frac{2\pi}{\lambda}(X' - ct')\right], \quad (2)$$

where X_0 is a reference position of the cilia and ω is a measure of the eccentricity of the elliptic path. To determine the horizontal and vertical position of the velocities of the cilia, we have

$$U' = \frac{\partial F_3}{\partial t'}|_{X_0} = \frac{\partial F_3}{\partial t'} + \frac{\partial F_3}{\partial X'}\frac{\partial X'}{\partial t'} = \frac{\partial F_3}{\partial t'} + \frac{\partial F_3}{\partial X'}U', \quad (3)$$

$$\left.\begin{array}{c} V' = \frac{\partial F_1}{\partial t'}|_{X_0} = \frac{\partial F_1}{\partial t'} + \frac{\partial F_1}{\partial X'}\frac{\partial X'}{\partial t'} = \frac{\partial F_1}{\partial t'} + \frac{\partial F_1}{\partial X'}U', \\ V' = \frac{\partial F_2}{\partial t'}|_{X_0} = \frac{\partial F_2}{\partial t'} + \frac{\partial F_2}{\partial X'}\frac{\partial X'}{\partial t'} = \frac{\partial F_2}{\partial t'} + \frac{\partial F_2}{\partial X'}U'. \end{array}\right\} \quad (4)$$

The alternate form of the velocity components is provided below by invoking Equation (2) in (3) and (1) in (4), respectively:

$$U' = \frac{2ac\omega\epsilon\pi\cos\left[\frac{2\pi}{\lambda}(X' - ct')\right]}{-\lambda + 2a\omega\epsilon\pi\cos\left[\frac{2\pi}{\lambda}(X' - ct')\right]}, \quad (5)$$

$$V' = \pm\frac{2ac\epsilon\pi\left(\sin\left[\frac{\varphi}{2} \pm \frac{2\pi}{\lambda}(X' - ct')\right] + \omega\cos\left[\frac{2\pi}{\lambda}(X' - ct')\right]\tan\beta\right)}{-\lambda + 2a\omega\epsilon\pi\cos\left[\frac{2\pi}{\lambda}(X' - ct')\right]}. \quad (6)$$

The equations governing the flow of a Jeffrey fluid in the presence of an induced magnetic field are as follows [46,47]:

1. The continuity equation

$$\nabla \cdot \boldsymbol{V}' = 0. \quad (7)$$

2. Maxwell's equation

$$\left.\begin{array}{c} \nabla \cdot \boldsymbol{B}' = 0, \nabla \times \boldsymbol{B}' = \mu'\boldsymbol{J}', \\ \nabla \times \boldsymbol{E}' = -\frac{\partial \boldsymbol{B}'}{\partial t'}, \boldsymbol{J}' = \sigma'(\boldsymbol{E}' + \boldsymbol{V}' \times \boldsymbol{B}'). \end{array}\right\} \quad (8)$$

3. The equation of motion

$$\rho\left(\frac{\partial \boldsymbol{V}'}{\partial t'} + (\boldsymbol{V}' \cdot \nabla)\boldsymbol{V}'\right) = \nabla \cdot \boldsymbol{T}' + \boldsymbol{J}' \times \boldsymbol{B}', \quad (9)$$

where $\boldsymbol{T}' = -p'\boldsymbol{I} + \boldsymbol{S}'$ in which the extra stress tensor $\boldsymbol{S}'$ for the Jeffrey fluid model is defined as [48,49]

$$\boldsymbol{S}' = \frac{\mu}{1 + \vartheta_1}\left(\dot{\Omega} + \vartheta_2\ddot{\Omega}\right), \quad (10)$$

4. The energy equation

$$\rho \xi' \frac{dT}{dt'} = \tau \cdot L - \nabla \cdot q - \nabla \cdot q'_r, \tag{11}$$

In the above equations, V' is the velocity vector, μ' is the magnetic permeability, σ' is the electric conductivity, ρ is the density of the fluid, p' is the pressure, ϑ_1 is the ratio of relaxation to retardation times, ϑ_2 is the retardation time, $\dot{\Omega}$ is the shear rate where dots show the differentiation with respect to time, $J' = J'_0 + J'_1$ is the current density in which J'_1 is the induced current density, $B' = B'_0 + B'_1$ is the magnetic field in which B'_1 is the induced magnetic field and B'_0 is the constant applied magnetic field, $E' = E'_0 + E'_1$ is the electric field (E'_0 is the constant applied electric field and E'_1 is the induced electric field), ξ' is the specific heat, $q = (-k' \frac{dT}{dt'}$, k' being the thermal conductivity) is the heat flux vector, and q'_r is the radiative heat flux.

Here we consider the effects of an induced magnetic field, for this low magnetic Reynolds number approximation must be dropped, and thus $E'_1 = B'_1 = J'_1 \neq 0$, and from Maxwell's Equation (8), we obtain the induction equation as follows [50]:

$$\nabla \times (V' \times B') + \zeta \nabla^2 B' = \frac{\partial B'}{\partial t'}, \tag{12}$$

where $\zeta = \frac{1}{\mu' \sigma'}$ is the magnetic diffusivity. A two-dimensional flow including an applied transverse constant magnetic field is represented by the equations:

$$V' = [U'(X', Y', t'), V'(X', Y', t'), 0], B'_0 = [0, B'_0, 0] \tag{13}$$

$$B'_1 = [B'_1(X', Y', t'), B'_2(X', Y', t'), 0], E'_0 = [0, 0, E'], \tag{14}$$

$$E'_1 = [0, 0, E'_3(X', Y', t')], \tag{15}$$

and thus, we received the following expression [39]:

$$J' \times B' = \begin{bmatrix} -\left\{\sigma'(E' + U'B'_0) + \frac{1}{\mu'}\left(\frac{\partial B'_2}{\partial X'} - \frac{\partial B'_1}{\partial Y'}\right)\right\}(B'_0 + B'_2), \\ +\left\{\sigma'(E' + U'B'_0) + \frac{1}{\mu'}\left(\frac{\partial B'_2}{\partial X'} - \frac{\partial B'_1}{\partial Y'}\right)\right\}B'_1 \end{bmatrix}. \tag{16}$$

In the above Equation (16) we utilized the value of J'_1 from Ampere's law instead of Ohm's law, i.e., ($J' = \sigma'(E'_0 + V' \times B'_0) + \frac{1}{\mu'}(\nabla \times B'_1)$). Also, we deduced the expression of Lorentz force for the applied magnetic field by ignoring the terms of the induced magnetic field, i.e., ($J' \times B' = [\sigma' U' B_0'^2, 0]$). The governing equations are now in a flowing form:

$$\frac{\partial U'}{\partial X'} + \frac{\partial V'}{\partial Y'} = 0, \tag{17}$$

$$\begin{aligned} &\rho\left(\frac{\partial U'}{\partial t'} + U'\frac{\partial U'}{\partial X'} + V'\frac{\partial V'}{\partial Y'}\right) \\ &= -\frac{\partial p'_m}{\partial X'} + \frac{\partial S'_{X'X'}}{\partial X'} + \frac{\partial S'_{X'Y'}}{\partial Y'} \\ &-\left\{\sigma'(E' + U'B'_0) + \frac{1}{\mu'}\left(\frac{\partial B'_2}{\partial X'} - \frac{\partial B'_1}{\partial Y'}\right)\right\}(B'_0 + B'_2), \end{aligned} \tag{18}$$

$$\begin{aligned} &\rho\left(\frac{\partial V'}{\partial t'} + U'\frac{\partial V'}{\partial X'} + V'\frac{\partial V'}{\partial Y'}\right) \\ &= -\frac{\partial p'_m}{\partial Y'} + \frac{\partial S'_{Y'X'}}{\partial X'} + \frac{\partial S'_{Y'Y'}}{\partial Y'} \\ &+\left\{\sigma'(E' + U'B'_0) + \frac{1}{\mu'}\left(\frac{\partial B'_2}{\partial X'} - \frac{\partial B'_1}{\partial Y'}\right)\right\}B'_1 \end{aligned} \tag{19}$$

$$\begin{aligned}
&\rho\xi'\left(\tfrac{\partial T}{\partial t'}+U'\tfrac{\partial T}{\partial X'}+V'\tfrac{\partial T}{\partial Y'}\right)\\
&\quad=k'\left(\tfrac{\partial^2 T}{\partial X'^2}+\tfrac{\partial^2 T}{\partial Y'^2}\right)\\
&\quad+\tfrac{\mu}{\vartheta_1+1}\left(1+\vartheta_2\left\{U'\tfrac{\partial}{\partial X'}+V'\tfrac{\partial}{\partial Y'}\right\}\right)\\
&\quad\times\left(2\left\{\left(\tfrac{\partial U'}{\partial X'}\right)^2+\left(\tfrac{\partial V'}{\partial Y'}\right)^2\right\}+\left(\tfrac{\partial U'}{\partial Y'}+\tfrac{\partial V'}{\partial X'}\right)^2\right)-\tfrac{\partial q_r'}{\partial Y'}
\end{aligned} \tag{20}$$

$$\frac{\partial B_1'}{\partial X'}+\frac{\partial B_2'}{\partial Y'}=0, \tag{21}$$

$$\frac{\partial B_1'}{\partial t'}=\zeta\left(\frac{\partial^2 B_1'}{\partial X'^2}+\frac{\partial^2 B_1'}{\partial Y'^2}\right)+\frac{\partial}{\partial Y'}(U'B_2'+U'B_0'-V'B_1'), \tag{22}$$

$$\frac{\partial B_2'}{\partial t'}=\zeta\left(\frac{\partial^2 B_2'}{\partial X'^2}+\frac{\partial^2 B_2'}{\partial Y'^2}\right)+\frac{\partial}{\partial X'}(U'B_2'+U'B_0'-V'B_1'), \tag{23}$$

$$\frac{\partial E_3'}{\partial Y'}=-\frac{\partial B_1'}{\partial t'}, \tag{24}$$

$$\frac{\partial E_3'}{\partial X'}=-\frac{\partial B_2'}{\partial t'}, \tag{25}$$

in which $p_m'=p'+\left\{(B_1')^2+(B_2')^2\right\}/2\mu'$. The associated boundary conditions are

$$\left.\begin{array}{c}
U'|_{Y'=H_1}=U'|_{Y'=H_2}=\frac{2ac\omega\epsilon\pi\cos\left[\frac{2\pi}{\lambda}(X'-ct')\right]}{-\lambda+2a\omega\epsilon\pi\cos\left[\frac{2\pi}{\lambda}(X'-ct')\right]},\\
V'|_{Y'=H_1}=\frac{\partial H_1}{\partial t'},V'|_{Y'=H_2}=\frac{\partial H_2}{\partial t'},\\
T|_{Y'=H_1}=T_0',T|_{Y'=H_2}=T_1'
\end{array}\right\}, \tag{26}$$

The q_r' is the radiative heat flux which is defined as [51]

$$q_r'=-\frac{4\sigma^*}{3k^*}\frac{\partial T^4}{\partial Y'}, \tag{27}$$

where k^* and σ^* are the mean absorption coefficient and the Stefan-Boltzmann constant. We assume the temperature differences within the flow are sufficiently small such that T^4 may be expressed as a linear function of temperature.

$$T^4\approx T^{*3}(4T-3T^*). \tag{28}$$

In view of above Equation (28), Equation (27) becomes

$$q_r'=-\frac{16\sigma^*T^{*3}}{3k^*}\frac{\partial T}{\partial Y'}, \tag{29}$$

Incorporating the non-dimensional quantities mentioned below:

$$\left.\begin{array}{c}
x=\frac{X'}{\lambda},y=\frac{Y'}{a},t=\frac{ct'}{\lambda},u=\frac{U'}{c},v=\frac{V'}{\delta c},h_1=\frac{H_1'}{a},\delta=\frac{a}{\lambda},Rd=\frac{16\sigma^*T^{*3}}{3k^*\mu\xi'}\\
h_x=\frac{B_1'}{B_0'},h_y=\frac{B_2'}{B_0'},E'=\frac{E'}{B_0'c},R_m=\sigma'\mu'ac,p_m=\frac{a^2p_m'}{\lambda\mu c},S=\frac{a}{\mu c}S',\\
Re=\frac{\rho ac}{\mu},M=aB_0'\sqrt{\frac{\sigma'}{\mu}},Pr=\frac{\mu\xi'}{k'},Ec=\frac{c^2}{\xi'(T_0'-T_S)},\theta=\frac{(T-T_S)}{(T_0'-T_S)},
\end{array}\right\}, \tag{30}$$

Now, using the aforesaid non-dimensional parameters, the dimensionless governing Equations (17)–(25) are described below:

$$\begin{aligned}&\delta Re\left(\tfrac{\partial u}{\partial t}+u\tfrac{\partial u}{\partial x}+v\tfrac{\partial u}{\partial y}\right)\\&=-\tfrac{\partial p_m}{\partial x}+\delta\tfrac{\partial}{\partial x}S_{xx}+\tfrac{\partial}{\partial y}S_{xy}-M^2(E+u)(1+h_y)\\&-\tfrac{M^2}{R_m}\left(\delta\tfrac{\partial h_y}{\partial x}-\tfrac{\partial h_x}{\partial y}\right)(1+h_y),\end{aligned} \tag{31}$$

$$\begin{aligned}&\delta^2 Re\left(\tfrac{\partial u}{\partial t}+\delta u\tfrac{\partial v}{\partial x}+\delta v\tfrac{\partial v}{\partial y}\right)\\&=-\tfrac{\partial p_m}{\partial y}+\delta^2\tfrac{\partial}{\partial x}S_{yx}+\delta\tfrac{\partial}{\partial y}S_{yy}+\delta M^2(E+u)h_x\\&+\delta\tfrac{M^2}{R_m}h_x\left(\delta\tfrac{\partial h_y}{\partial x}-\tfrac{\partial h_x}{\partial y}\right),\end{aligned} \tag{32}$$

$$\begin{aligned}&RePr\delta\left(\tfrac{\partial\theta}{\partial t}+u\tfrac{\partial\theta}{\partial x}+v\tfrac{\partial\theta}{\partial y}\right)\\&=\delta^2\tfrac{\partial^2\theta}{\partial x^2}+(1+PrRd)\tfrac{\partial^2\theta}{\partial y^2}\\&+\tfrac{1}{\vartheta_1+1}\,PrEcRePr\delta\left(\tfrac{\partial\theta}{\partial t}+u\tfrac{\partial\theta}{\partial x}+v\tfrac{\partial\theta}{\partial y}\right)\\&=\delta^2\tfrac{\partial^2\theta}{\partial x^2}+(1+PrRd)\tfrac{\partial^2\theta}{\partial y^2}\\&+\tfrac{1}{\vartheta_1+1}\,PrEc\\&\times\left[\left\{1+\vartheta_2\delta\left(u\tfrac{\partial}{\partial x}-v\tfrac{\partial}{\partial y}\right)\right\}\left\{2\delta^2\left(\left(\tfrac{\partial u}{\partial x}\right)^2+\left(\tfrac{\partial v}{\partial y}\right)^2\right)\right\}\right],\\&+\left(\tfrac{\partial u}{\partial y}+\delta^2\tfrac{\partial v}{\partial x}\right)^2\end{aligned} \tag{33}$$

$$\frac{\partial}{\partial y}(uh_y+u-\delta vh_x)+\frac{1}{R_m}\left(\delta^2\frac{\partial^2 h_x}{\partial x^2}+\frac{\partial^2 h_x}{\partial y^2}\right)=\delta\frac{\partial h_x}{\partial t}, \tag{34}$$

$$\frac{\partial}{\partial x}(uh_y+u-\delta vh_x)-\frac{1}{R_m}\left(\delta^2\frac{\partial^2 h_y}{\partial x^2}+\frac{\partial^2 h_y}{\partial y^2}\right)=-\frac{\partial h_y}{\partial t}, \tag{35}$$

wherein the above Equations (31)–(35) magnetic force functions are defined as: $h_x=\frac{\partial\phi}{\partial y}$, $h_y=-\delta\frac{\partial\phi}{\partial x}$, and the extra stress tensor in components form are defined as:

$$S_{xx}=\frac{2\delta}{\vartheta_1+1}\left(1+\vartheta_2\delta\left(\frac{\partial}{\partial t}+u\frac{\partial}{\partial x}+v\frac{\partial}{\partial y}\right)\right)\frac{\partial u}{\partial x}, \tag{36}$$

$$S_{xy}=S_{yx}=\frac{1}{\vartheta_1+1}\left(1+\vartheta_2\delta\left(\frac{\partial}{\partial t}+u\frac{\partial}{\partial x}+v\frac{\partial}{\partial y}\right)\right)\left(\delta^2\frac{\partial v}{\partial x}+\frac{\partial u}{\partial y}\right), \tag{37}$$

$$S_{yy}=\frac{2\delta}{\vartheta_1+1}\left(1+\vartheta_2\delta\left(\frac{\partial}{\partial t}+u\frac{\partial}{\partial x}+v\frac{\partial}{\partial y}\right)\right)\frac{\partial v}{\partial y}. \tag{38}$$

However, under the conditions of a long wavelength and a low Reynolds number, the dimensionless equations that describe the system have the following form:

$$-\frac{\partial p_m}{\partial x}+\frac{1}{\vartheta_1+1}\frac{\partial^2 u}{\partial y^2}-M^2(E+u)+\frac{M^2}{R_m}\frac{\partial h_x}{\partial y}=0, \tag{39}$$

$$\frac{\partial p_m}{\partial y}=0, \tag{40}$$

$$u+\frac{1}{R_m}\frac{\partial h_x}{\partial y}=0, \tag{41}$$

$$(1+PrRd)\frac{\partial^2\theta}{\partial y^2}+\frac{1}{1+\vartheta_1}PrEc\left(\frac{\partial u}{\partial y}\right)^2=0. \tag{42}$$

The corresponding boundary conditions are

$$\left.\begin{array}{c} u|_{y=h_1} = u|_{y=h_2} = \frac{2a\omega\epsilon\pi\delta\cos[2\pi(x-t)]}{-1+2a\omega\epsilon\pi\delta\cos[2\pi(x-t)]} \\ v|_{\,y=h_1} = \frac{\partial h_1}{\partial t}, v|_{\,y=h_2} = \frac{\partial h_2}{\partial t} \end{array}\right\}, \tag{43}$$

$$\theta|_{y=h_1} = 1, \theta|_{y=h_2} = m, \tag{44}$$

where m is the wall temperature ratio.

3. Analytical Solutions

With the help of computational software "Mathematica" the exact solution Equation (38) which satisfies the corresponding boundary conditions (Equations (43) and (44)) is obtained and we receive the axial velocity of the fluid in a ciliated tapered channel that is

$$u = \frac{\begin{array}{c}((EM^2 + \frac{\partial p_m}{\partial x})\cosh(\frac{(h_1-h_2)\sqrt{M^2(1+\vartheta_1)}}{\sqrt{2}}) - (EM^2 + 2M^2A + \frac{\partial p_m}{\partial x})\cosh(\frac{(h_1+h_2-2y)\sqrt{M^2(1+\vartheta_1)}}{\sqrt{2}})\times \\ (\cosh(\frac{(h_1+h_2)\sqrt{M^2(1+\vartheta_1)}}{\sqrt{2}}) + \sinh(\frac{(h_1+h_2)\sqrt{M^2(1+\vartheta_1)}}{\sqrt{2}})))\end{array}}{\begin{array}{c}M_2(\cosh(\sqrt{2}h_1\sqrt{M^2(1+\vartheta_1)}) + \cosh(\sqrt{2}h_2\sqrt{M^2(1+\vartheta_1)})) \\ +M_2(\sinh(\sqrt{2}h_1\sqrt{M^2(1+\vartheta_1)}) + \sinh(\sqrt{2}h_2\sqrt{M^2(1+\vartheta_1)}))\end{array}} \tag{45}$$

wherein above equation $A = \frac{2a\omega\epsilon\pi\delta\cos[2\pi(x-t)]}{-1+2a\omega\epsilon\pi\delta\cos[2\pi(x-t)]}$. The solution of Equation (42) can be written as:

$$\theta = \frac{\mathrm{sech}\left(\frac{(h_1-h_2)\sqrt{M^2(1+\vartheta_1)}}{\sqrt{2}}\right)^2}{32(h_1-h_2)M^4(1+PrRd)(1+\vartheta_1)} \times \left(\begin{array}{c} -4M^2\left[\begin{array}{c} Ec\{h_2\}^2\left(M^2(E+2A)+\frac{\partial p_m}{\partial x}\right)^2 Pr(h_1-y) + Ech_1M^4(E+2A)^2Pr(h_1-y)y+ \\ Ech_1\left(\frac{\partial p_m}{\partial x}\right)^2 Pr(h_1-y)y+ \\ h_2\left\{\begin{array}{c} -EcM^4(E+2A)^2Pr(h_1-y)(h_1+y)+ \\ Ec\left(\frac{\partial p_m}{\partial x}\right)^2 Pr\left(-\{h_1\}^2-y^2\right) + M^2\left(\begin{array}{c}4+4PrRd-2Ec \\ (E+2A)\frac{\partial p_m}{\partial x}Pr \\ (h_1-y)(h_1+y)\end{array}\right)\end{array}\right) - 2M^2 \\ \left(\left(2+2PrRd-2Ech_1(E+2A)\frac{\partial p_m}{\partial x}Pr(h_1-y)\right)y + 2(1+PrRd)(h_1-y)m\right)\end{array}\right) \\ (1+\vartheta_1) + \left\{\begin{array}{c} Ech_1\left(M^2(E+2A)+\frac{\partial p_m}{\partial x}\right)^2 Pr- \\ 16M^4(1+PrRd)(y(-1+m)-h_1m)(1+\vartheta_1) \\ -h_2\left(2EcM^2(E+2A)\frac{\partial p_m}{\partial x}Pr + \frac{\partial p_m}{\partial x}Pr + EcPr\left(\frac{\partial p_m}{\partial x}\right)^2 + M^4\left(\begin{array}{c}Ec(E+2A)_2Pr+16 \\ (1+PrRd)(1+\vartheta_1)\end{array}\right)\right)\end{array}\right) \\ \times\cosh\left((h_1-h_2)\sqrt{M^2(1+\vartheta_1)}\right) - Ec(h_1-h_2)\left(M^2(E+2A)+\frac{\partial p_m}{\partial x}\right)^2 Pr\times \\ \cosh\left(\sqrt{2}M(h_1+h_2-2y)\sqrt{1+\vartheta_1}\right)\end{array}\right), \tag{46}$$

In the fixed frame, the instantaneous flow rate is defined as

$$Q = \int_{h_1}^{h_2} u(x,y,t)dx. \tag{47}$$

The equation for pressure gradient is defined using Equations (45) and (47), and we have

$$\begin{aligned}
&\tfrac{\partial p_m}{\partial x} = -(M^2(\cosh(\tfrac{(h_1+h_2)\sqrt{M^2(1+\vartheta_1)}}{\sqrt{2}}) - \sinh(\tfrac{(h_1+h_2)\sqrt{M^2(1+\vartheta_1)}}{\sqrt{2}}))\times \\
&(M(-Eh_1 + Eh_2 + 2Q)\sqrt{(1+\vartheta_1)}\cosh(\sqrt{2}Mh_1\sqrt{1+\vartheta_1})+ \\
&M(-Eh_1 + Eh_2 + 2Q)\sqrt{1+\vartheta_1}\cosh(\sqrt{2}h_2Mh_1\sqrt{1+\vartheta_1}))- \\
&Eh_1\sqrt{M^2(1+\vartheta_1)}\sinh(\sqrt{2}h_1M\sqrt{1+\vartheta_1}) + Eh_2\sqrt{M^2(1+\vartheta_1)}\sinh(\sqrt{2}h_1M\sqrt{1+\vartheta_1})+ \\
&2Q\sqrt{M^2(1+\vartheta_1)}\sinh(\sqrt{2}h_1M\sqrt{1+\vartheta_1}) - Eh_1\sqrt{M^2(1+\vartheta_1)}\sinh(\sqrt{2}h_2M\sqrt{1+\vartheta_1})+ \\
&Eh_2\sqrt{M^2(1+\vartheta_1)}\sinh(\sqrt{2}h_2M\sqrt{1+\vartheta_1}) + 2Q\sqrt{M^2(1+\vartheta_1)}\sinh(\sqrt{2}h_2M\sqrt{1+\vartheta_1})+ \\
&2\sqrt{2}(E+2A)\sinh(\tfrac{(h_1+h_2)\sqrt{M^2(1+\vartheta_1)}}{\sqrt{2}})(\cosh(\tfrac{(h_1+h_2)\sqrt{M^2(1+\vartheta_1)}}{\sqrt{2}}) + \sinh(\tfrac{(h_1+h_2)\sqrt{M^2(1+\vartheta_1)}}{\sqrt{2}}))) \\
&/((2(-h_1M\sqrt{(1+\vartheta_1)}\cosh(\tfrac{(h_1+h_2)\sqrt{M^2(1+\vartheta_1)}}{\sqrt{2}}) + h_2M(1+\vartheta_1)\cosh(\tfrac{(h_1+h_2)\sqrt{M^2(1+\vartheta_1)}}{\sqrt{2}})+ \\
&\sqrt{2}\sinh(\tfrac{(h_1+h_2)\sqrt{M^2(1+\vartheta_1)}}{\sqrt{2}})))).
\end{aligned} \tag{48}$$

The expression for magnetic force function from Equation (41) can be written as

$$u + \frac{1}{R_m}\frac{\partial^2 \phi}{\partial y^2} = 0, \tag{49}$$

and the associated boundary conditions are

$$\phi|_{y=h_1} = 0, \phi|_{y=h_2} = 0, \tag{50}$$

Now, the solution of Equation (49) will take the form

$$\phi = \frac{\begin{aligned}R_m\Big(\tfrac{\partial p_m}{\partial x} + M^2\Big(2A + \tfrac{\partial p_m}{\partial x}(h_1-y)(h_2-y)(1+\vartheta_1) + E\big(1 + M^2(h_1-y)(h_2-y)(1+\vartheta_1)\big)\Big) \\ -M^2(E+2A) + \tfrac{\partial p_m}{\partial x}\Big(\cosh\Big(\tfrac{M(h_1+h_2-2y)\sqrt{(1+\vartheta_1)}}{\sqrt{2}}\Big)\operatorname{sech}\Big(\tfrac{M(h_1+h_2)\sqrt{(1+\vartheta_1)}}{\sqrt{2}}\Big)\Big)\Big)\end{aligned}}{4M^2(1+\vartheta_1)}, \tag{51}$$

and the axial induced magnetic field $h_x(x, y)$ is obtained by taking the derivative of Equation (51) w.r.t "y", and the expression of non-dimensional current density in the fixed frame is given as

$$J_z(x, y) = R_m\left(E + u - \frac{\partial^2 \phi}{\partial y^2}\right). \tag{52}$$

The expression for pressure rise can be written as

$$\Delta p_m = \int_0^1 \frac{dp_m}{dx}dx. \tag{53}$$

4. Graphical Results and Discussion

This section comprises seven subsections. The effects of various parameters on the velocity component (u), magnetic force function (ϕ), induced magnetic field (h_x), current density (J_z), pressure rise (Δp), temperature profile (θ) and trapping mechanism are presented.

4.1. Velocity Profiles

Figure 2a–e shows the behavior of axial velocity versus the different physical parameters. In Figure 2a, we observe that by increasing the values of the Hartmann number (M), the velocity of fluid gradually decreases in the center of the ciliated tapered channel because the Hartmann number is the ratio of electromagnetic forces to viscous forces. Therefore, large values of the Hartmann number create a resistance in the opposite direction of fluid movement as compared to the very small values of the Hartmann number, which cause a decline in the velocity field. Additionally, we can see the Lorentz force is less dominating closer to the walls of the channel. However, in Figure 2b, a similar type of behavior is noticed for the Jeffrey parameter (ϑ_1) on the velocity profile. Further, when the Jeffrey fluid parameter $\vartheta_1 \to 0$ the behavior of the fluid becomes Newtonian, which ensures that present results are valid for both Newtonian and non-Newtonian models. In Figure 2c, we observe that the velocity of the fluid is maximum at the center and gradually decreases at the wall for large values of eccentricity parameter (ω). Figure 2d depicts that as we increase the cilia length parameter (ϵ) the velocity of fluid is maximal at the center of the channel but behaves differently at the walls of the channel. From Figure 2e it is noticed that when the values of inclined angle (β) of the unperturbed walls increases, then the velocity profile significantly increases closer to the wall but decreases in the middle of the channel. This shows that the inclination angle is also responsible and plays an efficient role to control the flow.

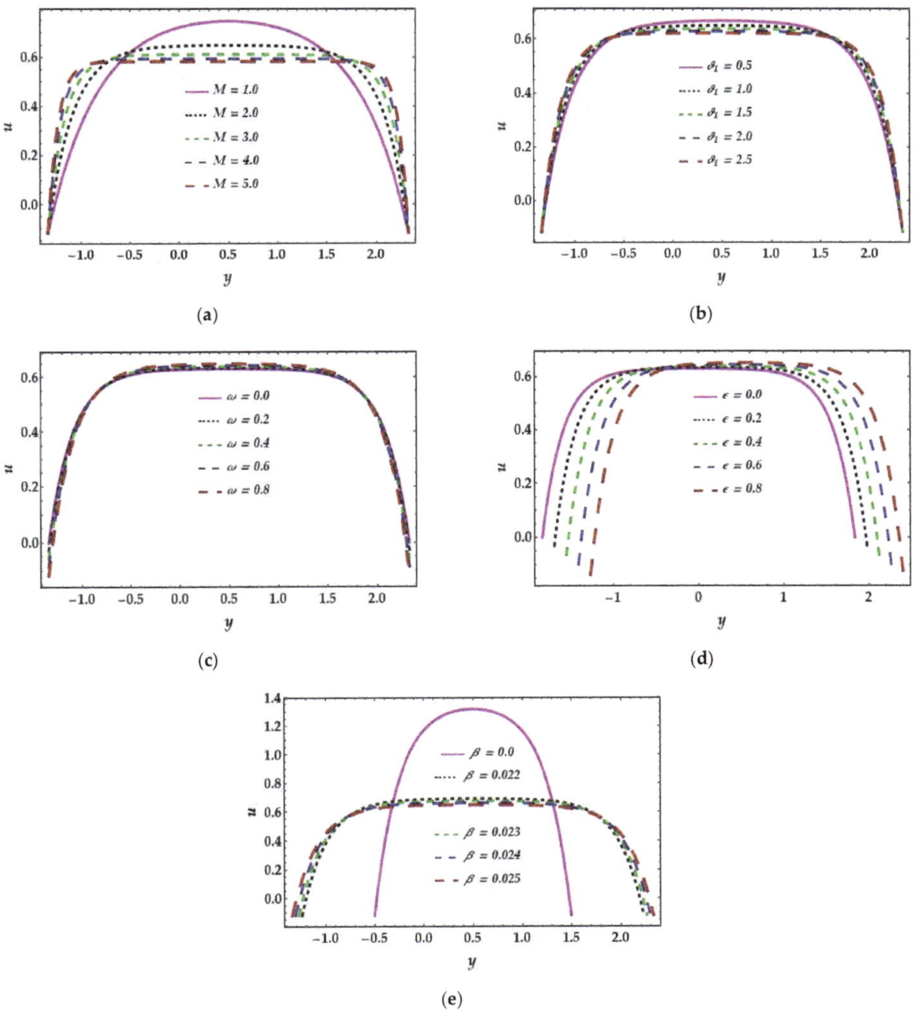

Figure 2. Effects of different physical parameters on axial velocity u.

4.2. Magnetic Force Function

The magnetic force function for different values of the Hartmann number (M) and Jeffrey fluid parameter (ϑ_1) is plotted in Figure 3a,b. It is noticed that the magnitude of magnetic force function shows a significant decline when (M) and (ϑ_1) increases. The Lorentz force dominates the magnetic force function, as seen in Figure 3a. However, in

Figure 3b, it can be seen that the Jeffrey fluid parameter has little or no influence on the magnetic force function but is decreasing. In Figure 3c, we see that increasing the magnetic Reynolds number (R_m) strengthens the magnetic force function. The magnetic Reynolds number is directly proportional to magnetic permeability and electromagnetic forces, as shown by Equation (31). Higher magnetic Reynolds numbers increase magnetic permeability, which helps in improving the magnetic force function. In Figure 3d, it is seen that increasing the eccentricity of the elliptic path (ω) optimizes the magnetic force function, although the changes are minor. Figure 3e shows that a change in the inclination angle (β) influences the behavior of the magnetic force component over the whole domain. Higher inclination values indicate an increasing trend in the velocity characteristic along the entire channel. Figure 3f illustrates that the magnitude of the magnetic force function at the wall of the channel is opposite to each other for different values of cilia length parameter (ϵ).

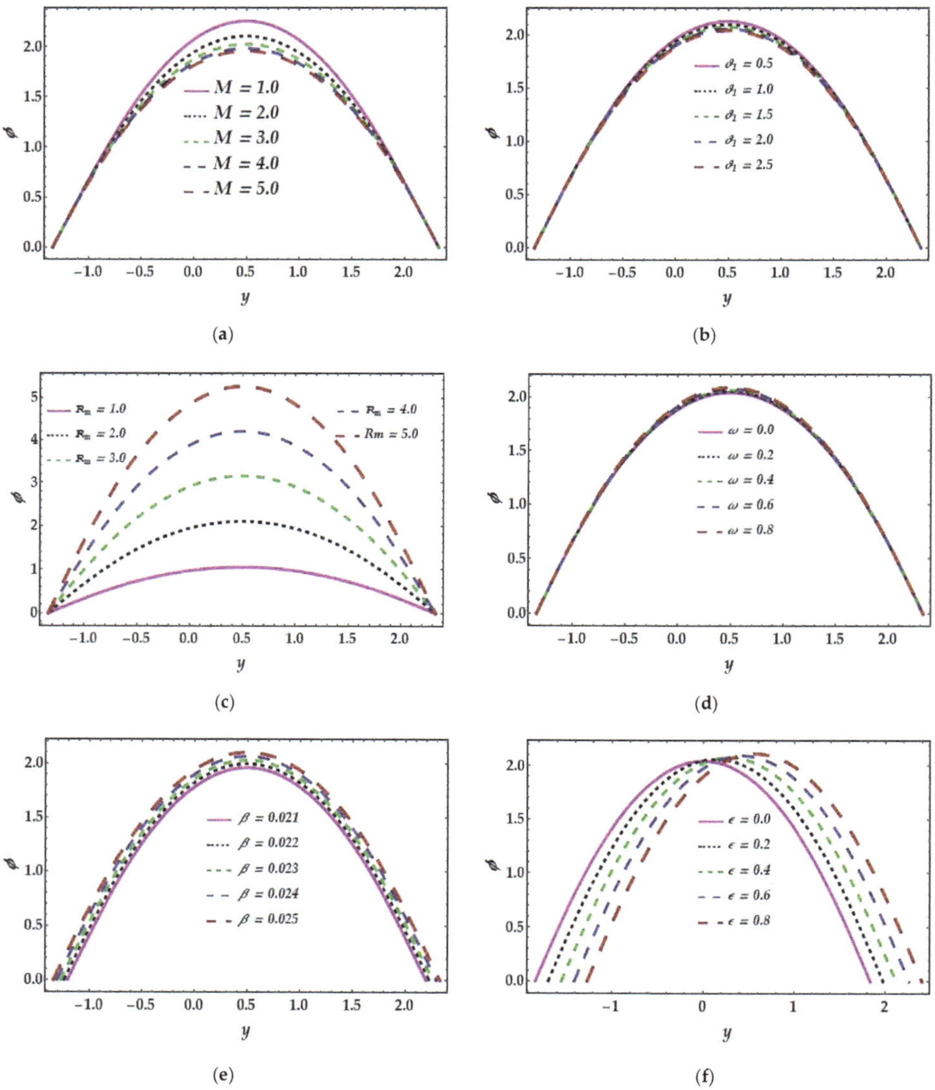

Figure 3. Effects of different physical parameters on magnetic force functions ϕ.

4.3. Magnetic Field Characteristics

Figure 4a–f depicts the findings for the induced magnetic field h_x across several important parameters. As seen in Figure 4a, when the Hartmann number (M) increases, the induced magnetic displays dual behavior, such as a reducing mechanism in the bottom part of the channel and a growing mechanism in the upper boundary. The Jeffrey fluid parameter (ϑ_1) and inclination angle (β) exhibit comparable effects on the induced magnetic field in Figure 4b,c; however, the effects from both variables on the induced magnetic field are insignificant. In the presence of a magnetic field Reynolds number (R_m), the induced magnetic field accumulates in the bottom half of the channel and decreases in the upper section, as illustrated in Figure 4d. In Figure 4e, a similar kind of pattern is observed for the eccentricity parameter (ω), although the impacts are essentially minor. The induced magnetic field increases over the entire domain under the significant effect of the cilia length parameter, as seen in Figure 4f.

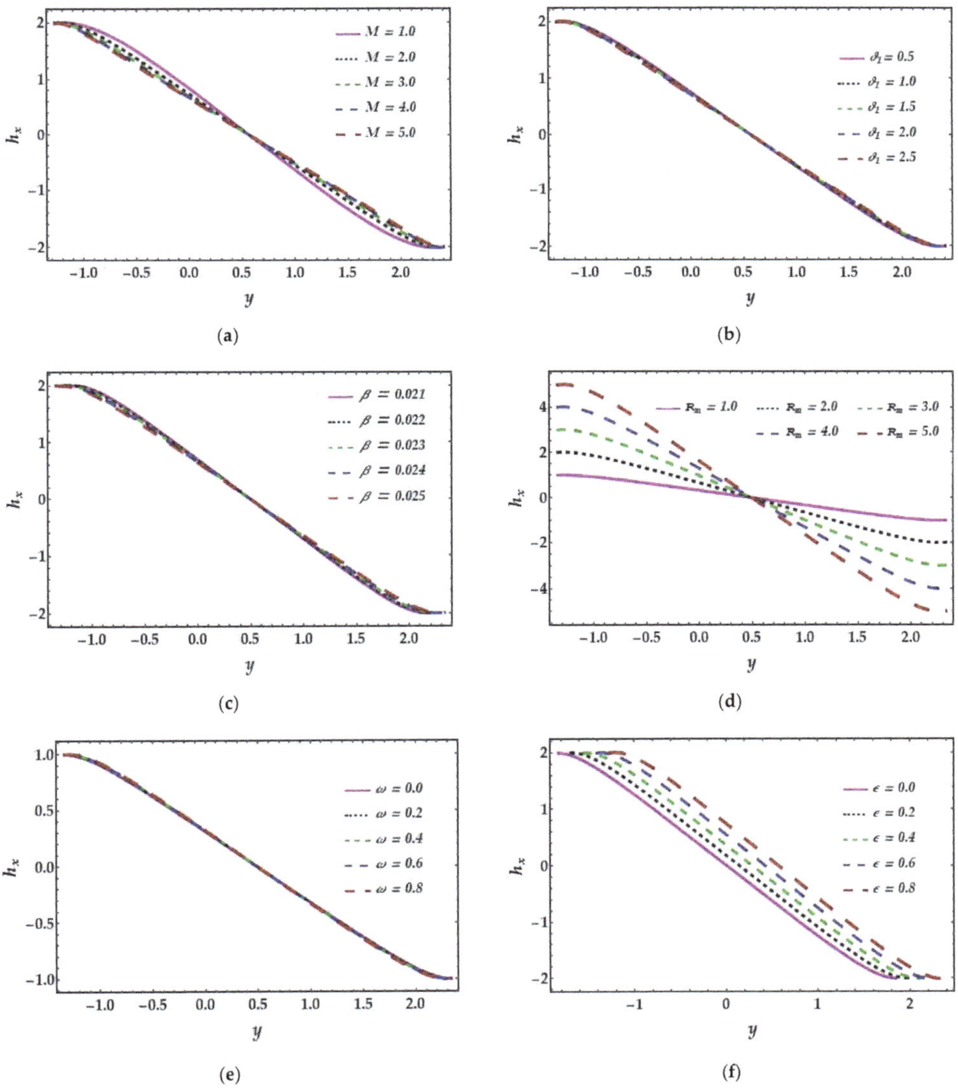

Figure 4. Effects of different physical parameters on induced magnetic fields h_x.

4.4. Current Density

Figure 5a–g depicts the variance in current density characteristics for all emerging parameters, and all graphs are parabolic in shape. It is also noticed that towards the middle of the channel and along the channel walls, the current density behaves in the opposite direction, as seen in Figure 5a–d for different values of the Hartmann number (M), Jeffrey parameter (ϑ_1), eccentricity parameter (ω), and inclined angle (β). Figure 5a,b shows that the Hartmann number and Jeffrey fluid parameter oppose the increase in current density in the streamwise direction, but adjacent to the boundaries, the response is entirely reversed. The eccentricity parameter shows negligible effects on the current density profile as shown in Figure 4c. The consequences of the inclined angle tend to reduce the current density profile in the channel's center, while it increases closer to the walls as illustrated in Figure 4d. The existence of a magnetic Reynolds number (R_m) and an electric field (E) helps in optimizing the current density profile dramatically, as seen in Figure 4e,f. As the cilia length rises, the current density grows in the upper portion of the channel while declining in the lower half. However, we identified a major point in this graph at $y \approx 0.2$.

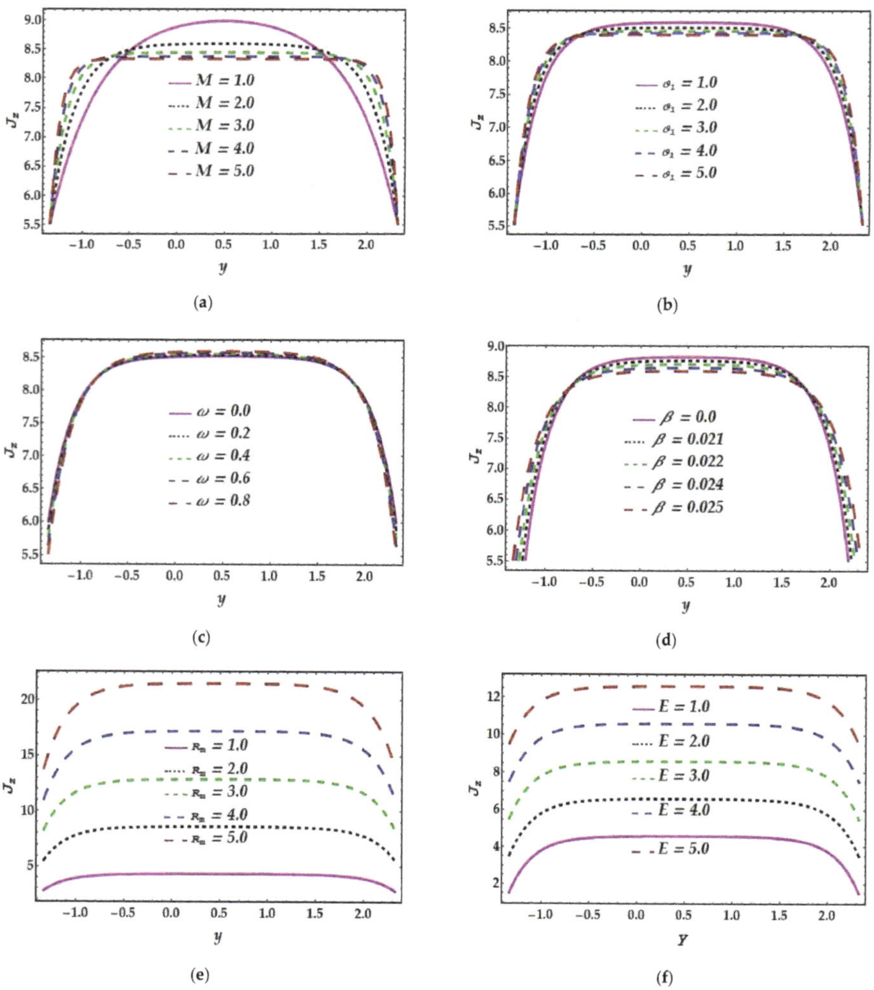

Figure 5. *Cont.*

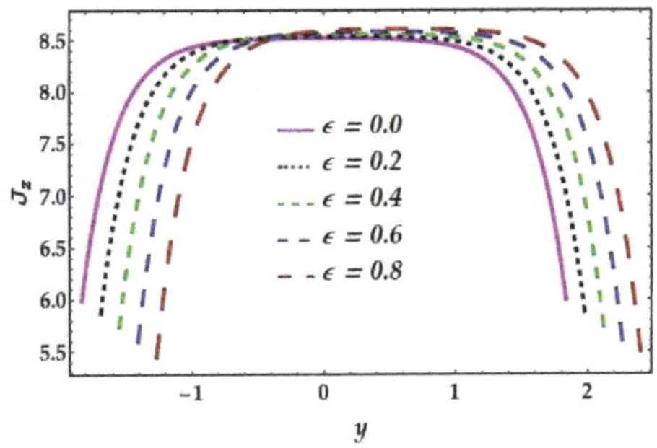

(g)

Figure 5. Effects of different physical parameters on current density J_z.

4.5. Pressure Rise

The main purpose of this part is to discuss the pumping characteristics along the whole flow region. Therefore, Figure 6a–e demonstrates the effects of the Hartmann number (M), Jeffrey parameter (ϑ_1), cilia length parameter (ϵ), the inclined angle of the unperturbed wall, and the eccentricity of an elliptical path (ω) on the pressure rise profile. It can be seen in Figure 6a that as the influence of Hartmann number (M) increases, the pressure rise phenomenon is observed to decrease equally throughout the pumping region. On the other hand, the large values of the cilia length parameter (ϵ) show that the pumping rate increases in the region ($-1 \leq Q < 0$) but decreases in the remaining region ($0 < Q \leq 1$) as illustrated in Figure 6b. Figure 6c,d shows the pumping rate rises in the region ($0 < Q \leq 1$) but decreases in the other region ($-1 \leq Q < 0$) for higher values of the Jeffrey parameter (ϑ_1) and inclined angle (β), respectively. Figure 6e represents the decreasing behavior of the pressure rise throughout the pumping region for larger values of eccentricity parameter (ω).

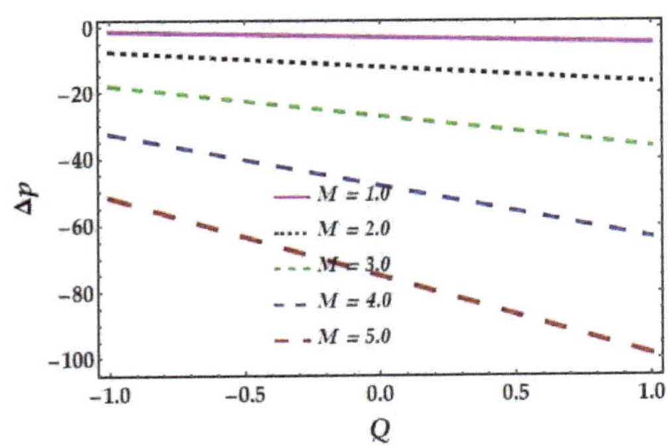

(a)

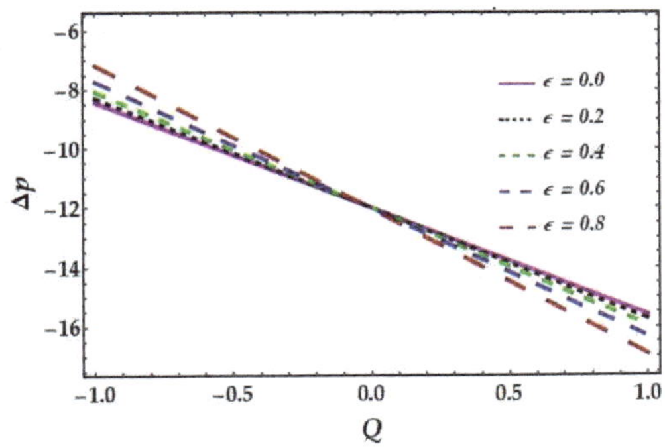

(b)

Figure 6. *Cont.*

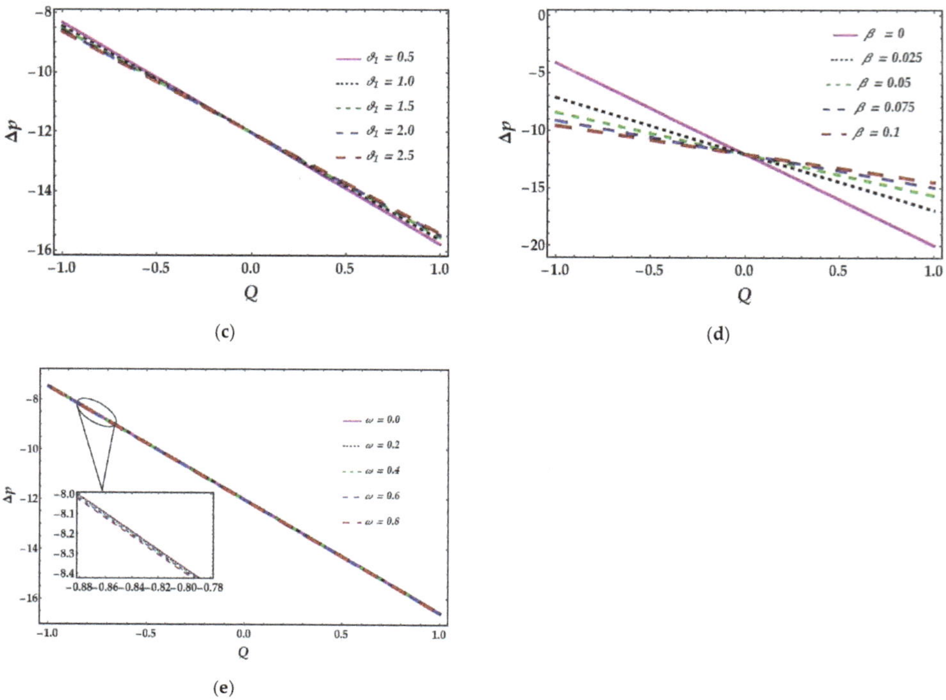

Figure 6. Effects of different physical parameters on pressure rise Δp.

4.6. Temperature Profile

Figure 7a,b demonstrates the effects of the Prandtl number (Pr) and the Eckert number (Ec). It is observed that the nature of temperature profiles exhibits an increasing tendency concerning both parameters. As a result, we may conclude that momentum diffusivity and advective transport contribute significantly to the enhancement of temperature profiles. Figure 7c shows the behavior of thermal radiation (Rd) on the temperature profile. Thermal radiation replicates the function of radiative heat transmission compared to thermal conduction heat transfer. When $Rd = 0$, the temperature profile reveals that radiation effects are absent. It can be seen that higher amounts of radiation reduce the temperature profile. Figure 7d illustrates that as the temperature ratio (m) values are increased, the temperature profile rises.

4.7. Trapping Phenomena

Trapping is another essential peristaltic transport mechanism. Trapping is the production of an inwardly flowing bolus of fluid by enclosed streamlines, which is propelled as a head together with the peristaltic wave. This physical phenomenon may be responsible for regulating blood thrombus creation and food bolus motility in the gastrointestinal system. Figures 8–13 show streamlines for all of the emerging parameters for the ciliated tapering channel. These images represent the trapping phenomena, in which a trapped bolus of fluid enveloped by streamlines moves along a metachronal wave. Figure 8 shows that as the Hartmann number (M) is increased, the number of trapped boluses diminishes. Figure 9 depicts similar results for the Jeffrey parameter (ϑ_1). The non-Newtonian effects inhibited the development of the trapping bolus, while in Figures 10 and 11, the opposite behavior is noticed for cilia length (ϵ) and inclined angle (β), respectively. Increased values in these parameters improve the volume and frequency of trapping boluses. Figures 12 and 13 show that the number of boluses reduces initially for smaller values of electric field (E) and eccentricity parameter (ω), whereas when $(E > 1)$ and $(\omega > 0)$ the effects are negligible, and no change occurs.

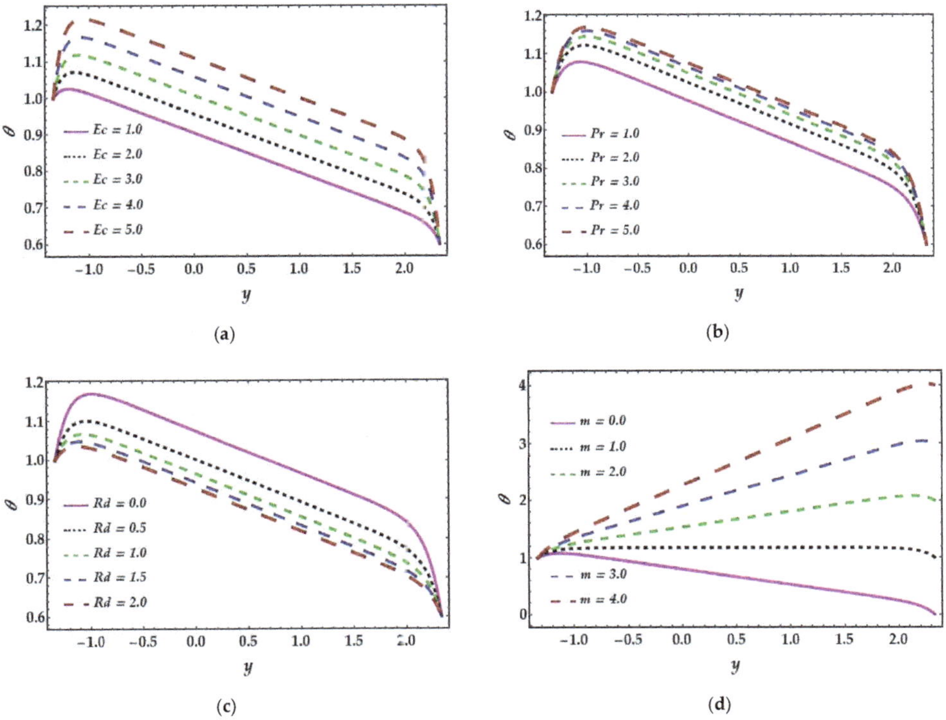

Figure 7. Effects of different physical parameters on temperature profile θ.

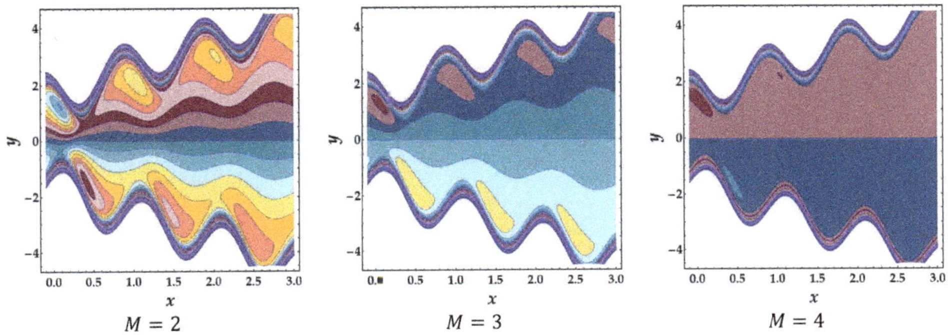

Figure 8. The behavior of Streamlines for different values of Hartmann number M.

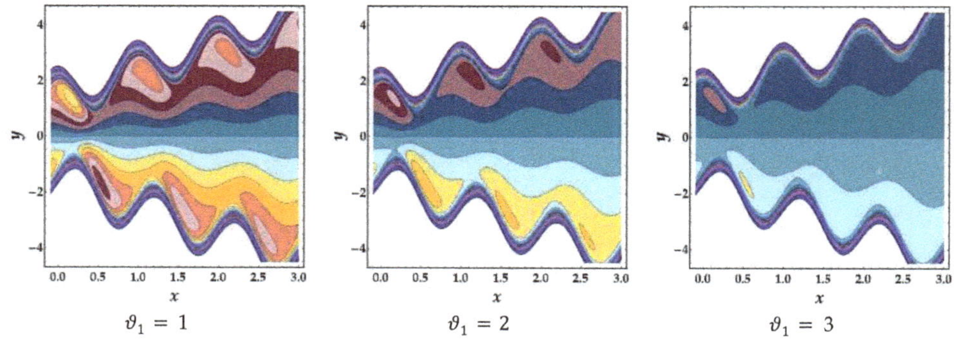

Figure 9. The behavior of Streamlines for different values of Jeffrey parameter ϑ_1.

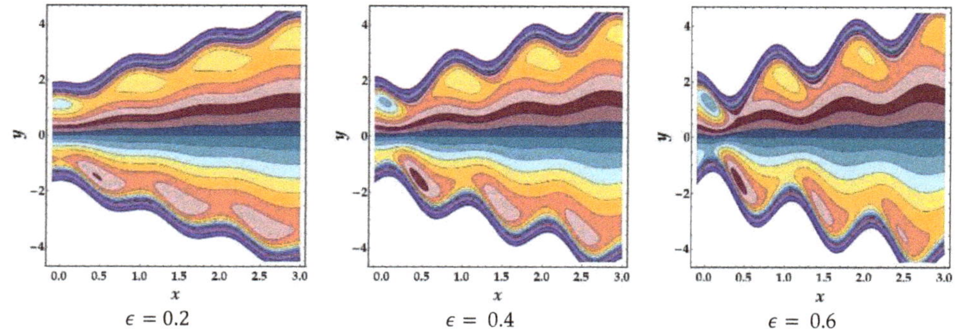

Figure 10. The behavior of Streamlines for different values of cilia length parameter ϵ.

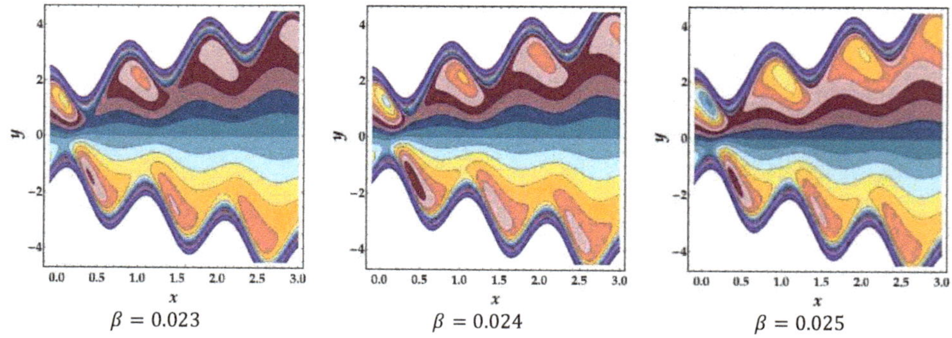

Figure 11. The behavior of Streamlines for different values of inclined angle β.

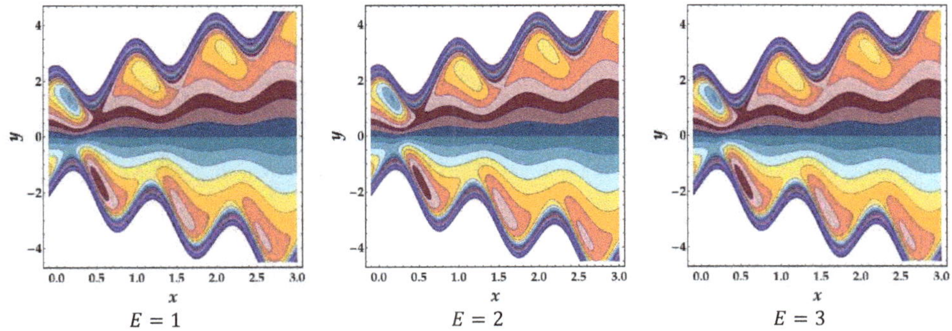

Figure 12. The behavior of Streamlines for different values of electric field E.

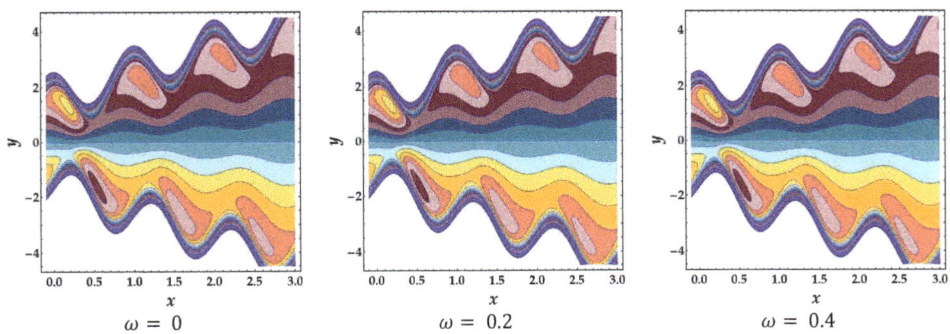

Figure 13. The behavior of Streamlines for different values of eccentricity parameter ω.

5. Concluding Remarks

In this study, the Jeffrey fluid model is used to evaluate cilia-driven flow through a ciliated, asymmetric tapering channel with thermally radiative heat transfer and magnetic phenomena. The suggested problem's mathematical formulation is simplified using lubrication theory. Since the finalized differential equations are coupled and linear, we employed computer tools to find analytical solutions. The most important findings are listed below:

i. The velocity profile decreases for large values of the Hartmann number and Jeffrey fluid parameter, while the eccentricity parameter exhibits the opposite trend.
ii. For the velocity profile and magnetic force function, large values of the cilia length parameter exhibit an opposing tendency closer to the walls.
iii. For a higher magnetic Reynolds number, inclined angle, and for higher eccentricity parameter values, the magnetic force function acts as an increasing function. However, for the Hartmann number and the Jeffrey parameter, the opposite result has been noticed.
iv. The effects of the Hartmann number, Jeffrey fluid, and inclined angle on induced magnetic exhibit similar behavior, although the effects of eccentricity parameter and magnetic Reynolds number are opposite.
v. The magnetic Reynolds number and electric field have had a considerable influence on current density, whereas the Hartmann number and Jeffrey fluid parameter have shown identical behavior.
vi. Temperature profiles reveal rising behavior for the Eckert and Prandtl numbers but decreasing behavior for the radiation parameter.

vii. The number of trapped boluses falls as the influence of the Hartmann number and Jeffrey fluid parameter improves but increases in the presence of the cilia length parameter and inclination angle.

viii. For large values of the electric field and eccentricity variable, the number of trapped boluses functions similarly.

Author Contributions: Investigation, Methodology and Writing—review and editing, F.I.; Supervision and Conceptualization, R.E.; validation, M.M.B.; Formal analysis, S.Z.A. All authors have read and agreed to the published version of the manuscript.

Funding: This research received no external funding.

Institutional Review Board Statement: Not applicable.

Informed Consent Statement: Not applicable.

Data Availability Statement: Not applicable.

Acknowledgments: Not applicable.

Conflicts of Interest: The authors declare no conflict of interest.

Nomenclature

$\boldsymbol{V}'$	Velocity vector
X', Y'	Cartesian coordinates system [m]
U', V'	Velocity components [m/s]
X_0	Reference position of the cilia
t'	Time [s]
ϵ	Cilia length parameter
ω	Eccentricity of elliptical path
a	Half-width of the channel [m]
c	Wave Speed [m/s]
λ	Wavelength [m]
φ	Phase difference
β	Inclined angle of unperturbed wall
μ'	Magnetic permeability [H/m]
ξ'	Specific heat [J·kg^{-1}·K^{-1}]
σ'	Electric conductivity [S/m]
p'	Pressure [N/m^2]
$\boldsymbol{q}$	Heat flux vector
q_r'	Radiative heat flux
k'	Thermal conductivity [W·m^{-1} K^{-1}]
$\boldsymbol{T}$	Temperature [K]
p_m'	Modified pressure [N/m^2]
ρ	Density [Kg/m^3]
k^*	Mean absorption coefficient [m^{-1}]
σ^*	Stefan-Boltzmann constant [W·m^{-2} K^{-4}]
δ	Wave number [m]
h_x	Dimensionless axial magnetic field
h_y	Dimensionless transverse magnetic field
Re	Reynolds number
R_m	Magnetic Reynolds number
M	Hartmann number
Pr	Prandtl number
Ec	Eckert Number
θ	Dimension less Temperature
Rd	Radiation parameter
Q	Instantaneous flow rate

ϕ	Magnetic force function
B'	Total magnetic field [T]
B'_0	Applied magnetic field
B'_1	Induced magnetic field
E'	Total electric field [V/m]
E'_0	Applied electric field
E'_1	Induced electric field
J'	Total current density [A/m^2]
J'_0	Applied current density
J'_1	Induced current density

References

1. Sleigh, M.A. *The Biology of Cilia and Flagella*; Pergamon Press: Oxford, UK, 1962.
2. Purcell, E.M. Life at Low Reynolds Number. *Am. J. Phys.* **1977**, *45*, 3–11. [CrossRef]
3. Breunig, J.J.; Arellano, J.I.; Rakic, P. Cilia in the Brain: Going with the Flow. *Nat. Neurosci.* **2010**, *13*, 654–655. [CrossRef] [PubMed]
4. Stannard, W.; O'Callaghan, C. Ciliary Function and the Role of Cilia in Clearance. *J. Aerosol Med.* **2006**, *19*, 110–115. [CrossRef] [PubMed]
5. Lyons, R.A.; Saridogan, E.; Djahanbakhch, O. The Reproductive Significance of Human Fallopian Tube Cilia. *Hum. Reprod. Update* **2006**, *12*, 363–372. [CrossRef]
6. Lardner, T.J.; Shack, W.J. Cilia Transport. *Bull. Math. Biophys.* **1972**, *34*, 325–335. [CrossRef]
7. Sher Akbar, N. Biomathematical Analysis of Carbon Nanotubes Due to Ciliary Motion. *Int. J. Biomath.* **2015**, *8*, 1550023. [CrossRef]
8. Nadeem, S.; Sadaf, H. Trapping Study of Nanofluids in an Annulus with Cilia. *AIP Adv.* **2015**, *5*, 127204. [CrossRef]
9. Javid, K.; Alqsair, U.F.; Hassan, M.; Bhatti, M.M.; Ahmad, T.; Bobescu, E. Cilia-Assisted Flow of Viscoelastic Fluid in a Divergent Channel under Porosity Effects. *Biomech. Model. Mechanobiol.* **2021**, *20*, 1399–1412. [CrossRef]
10. Khan, W.U.; Imran, A.; Raja, M.A.Z.; Shoaib, M.; Awan, S.E.; Kausar, K.; He, Y. A Novel Mathematical Modeling with Solution for Movement of Fluid through Ciliary Caused Metachronal Waves in a Channel. *Sci. Rep.* **2021**, *11*, 20601. [CrossRef]
11. Gray, J. The Mechanism of Ciliary Movement. III.—The Effect of Temperature. *Proc. R. Soc. Lond. B Biol. Sci.* **1923**, *95*, 6–15.
12. Baetjer, A.M. Effect of Ambient Temperature and Vapor Pressure on Cilia-Mucus Clearance Rate. *J. Appl. Physiol.* **1967**, *23*, 498–504. [CrossRef]
13. Bisgrove, B.W.; Yost, H.J. The Roles of Cilia in Developmental Disorders and Disease. *Development* **2006**, *133*, 4131–4143. [CrossRef]
14. Fliegauf, M.; Benzing, T.; Omran, H. When Cilia Go Bad: Cilia Defects and Ciliopathies. *Nat. Rev. Mol. Cell Biol.* **2007**, *8*, 880–893. [CrossRef]
15. Sahadevan, V.; Chen, C.Y. Microfluidic Applications of Artificial Cilia: Recent Progress, Demonstration, and Future Perspectives. *Micromachines* **2022**, *13*, 735. [CrossRef]
16. Zachariah, E.; Hale, Z.E.; Sadoshima, J. Primary Cilia and Their Role in Acquired Heart Disease. *Cells* **2022**, *11*, 960.
17. Riaz, A.; Bobescu, E.; Ramesh, K.; Ellahi, R. Entropy Analysis for Cilia-Generated Motion of Cu-Blood Flow of Nanofluid in an Annulus. *Symmetry* **2021**, *13*, 2358. [CrossRef]
18. Maqbool, K.; Shaheen, S.; Bobescu, E.; Ellahi, R. Thermal and Concentration Analysis of Phan-Thien-Tanner Fluid Flow Due to Ciliary Movement in a Peripheral Layer. *J. Cent. South Univ.* **2021**, *28*, 3327–3339. [CrossRef]
19. Alamri, S.Z.; Ellahi, R.; Shehzad, N.; Zeeshan, A. Convective Radiative Plane Poiseuille Flow of Nanofluid through Porous Medium with Slip: An Application of Stefan Blowing. *J. Mol. Liq.* **2019**, *273*, 292–304. [CrossRef]
20. Akbar, N.S.; Butt, A.W. Heat Transfer Analysis of Viscoelastic Fluid Flow Due to Metachronal Wave of Cilia. *Int. J. Biomath.* **2014**, *7*, 1450066. [CrossRef]
21. Butt, A.W.; Akbar, N.S.; Mir, N.A. Heat Transfer Analysis of Peristaltic Flow of a Phan-Thien-Tanner Fluid Model Due to Metachronal Wave of Cilia. *Biomech. Model. Mechanobiol.* **2020**, *19*, 1925–1933. [CrossRef]
22. Al-Zubaidi, A.; Nazeer, M.; Khalid, K.; Yaseen, S.; Saleem, S.; Hussain, F. Thermal Analysis of Blood Flow of Newtonian, Pseudo-Plastic, and Dilatant Fluids through an Inclined Wavy Channel Due to Metachronal Wave of Cilia. *Adv. Mech. Eng.* **2021**, *13*, 168781402110490. [CrossRef]
23. McCash, L.B.; Nadeem, S.; Akhtar, S.; Saleem, A.; Saleem, S.; Issakhov, A. Novel Idea about the Peristaltic Flow of Heated Newtonian Fluid in Elliptic Duct Having Ciliated Walls. *Alex. Eng. J.* **2022**, *61*, 2697–2707. [CrossRef]
24. Mayne, R.; den Toonder, J.M.J. (Eds.) *Atlas of Cilia Bioengineering and Biocomputing*; River: Gistrup, Denmark, 2018.
25. Rashidi, S.; Esfahani, J.A.; Maskaniyan, M. Applications of Magnetohydrodynamics in Biological Systems-a Review on the Numerical Studies. *J. Magn. Magn. Mater.* **2017**, *439*, 358–372. [CrossRef]
26. Ahmad Farooq, A.; Shah, Z.; Alzahrani, E.O. Heat Transfer Analysis of a Magneto-Bio-Fluid Transport with Variable Thermal Viscosity through a Vertical Ciliated Channel. *Symmetry* **2019**, *11*, 1240. [CrossRef]
27. Akbar, N.S.; Tripathi, D.; Khan, Z.H.; Bég, O.A. Mathematical Model for Ciliary-Induced Transport in MHD Flow of Cu-H_2O Nanofluids with Magnetic Induction. *Chin. J. Phys.* **2017**, *55*, 947–962. [CrossRef]
28. Sadaf, H.; Nadeem, S. Fluid Flow Analysis of Cilia Beating in a Curved Channel in the Presence of Magnetic Field and Heat Transfer. *Can. J. Phys.* **2020**, *98*, 191–197. [CrossRef]

29. Tripathi, D.; Bég, O.A. A Numerical Study of Oscillating Peristaltic Flow of Generalized Maxwell Viscoelastic Fluids through a Porous Medium. *Transp. Porous Media* **2012**, *95*, 337–348. [CrossRef]
30. Elelamy, A.F.; Elgazery, N.S.; Ellahi, R. Blood flow of MHD non-Newtonian Nanofluid with Heat Transfer and Sip Effects: Application of Bacterial Growth in Heart Valve. *Int. J. Numer. Methods Heat Fluid Flow* **2020**, *30*, 4883–4908. [CrossRef]
31. Zhu, J.; Xu, Y.; Han, X. A Non-Newtonian Magnetohydrodynamics (MHD) Nanofluid Flow and Heat Transfer with Nonlinear Slip and Temperature Jump. *Mathematics* **2019**, *7*, 1199. [CrossRef]
32. Turkyilmazoglu, M.; Pop, I. Heat and Mass Transfer of Unsteady Natural Convection Flow of Some Nanofluids Past a Vertical Infinite Flat Plate with Radiation Effect. *Int. J. Heat Mass Transf.* **2013**, *59*, 167–171. [CrossRef]
33. Othman, M.I.A.; Said, S.; Marin, M. A Novel Model of Plane Waves of Two-Temperature Fiber-Reinforced Thermoelastic Medium under the Effect of Gravity with Three-Phase-Lag Model. *Int. J. Numer. Methods Heat Fluid Flow* **2019**, *29*, 4788–4806. [CrossRef]
34. Goodarzi, M.; Tlili, I.; Tian, Z.; Safaei, M.R. Efficiency Assessment of Using Graphene Nanoplatelets-Silver/Water Nanofluids in Microchannel Heat Sinks with Different Cross-Sections for Electronics Cooling. *Int. J. Numer. Methods Heat Fluid Flow* **2019**, *30*, 347–372. [CrossRef]
35. Baranovskii, E.S. Mixed Initial–Boundary Value Problem for Equations of Motion of Kelvin–Voigt Fluids. *Comput. Math. Math. Phys.* **2016**, *56*, 1363–1371. [CrossRef]
36. Doubova, A.; Fernández-Cara, E. On the Control of Viscoelastic Jeffreys Fluids. *Syst. Control Lett.* **2012**, *61*, 573–579. [CrossRef]
37. Su, Z.-G.; Li, T.-F.; Luo, K.; Yi, H.-L. Nonlinear Behavior of Electrohydrodynamic Flow in Viscoelastic Fluids. *Phys. Rev. Fluids* **2021**, *6*, 093701. [CrossRef]
38. Azaiez, J.; Homsy, G.M. Linear Stability of Free Shear Flow of Viscoelastic Liquids. *J. Fluid Mech.* **1994**, *268*, 37–69. [CrossRef]
39. Baranovskii, E.S. Global Solutions for a Model of Polymeric Flows with Wall Slip. *Math. Methods Appl. Sci.* **2017**, *40*, 5035–5043. [CrossRef]
40. Baranovskii, E.S. Flows of a Polymer Fluid in Domain with Impermeable Boundaries. *Comput. Math. Math. Phys.* **2014**, *54*, 1589–1596. [CrossRef]
41. Hayat, T.; Khan, M.; Asghar, S.; Siddiqui, A.M. A Mathematical Model of Peristalsis in Tubes through a Porous Medium. *J. Porous Media* **2006**, *9*, 55–67. [CrossRef]
42. Kothandapani, M.; Srinivas, S. Peristaltic Transport of a Jeffrey Fluid under the Effect of Magnetic Field in an Asymmetric Channel. *Int. J. Non-Linear Mech.* **2008**, *43*, 915–924. [CrossRef]
43. Tripathi, D.; Pandey, S.K.; Bég, O.A. Mathematical Modelling of Heat Transfer Effects on Swallowing Dynamics of Viscoelastic Food Bolus through the Human Oesophagus. *Int. J. Therm. Sci.* **2013**, *70*, 41–53. [CrossRef]
44. Muzara, H.; Shateyi, S. MHD Laminar Boundary Layer Flow of a Jeffrey Fluid Past a Vertical Plate Influenced by Viscous Dissipation and a Heat Source/Sink. *Mathematics* **2021**, *9*, 1896. [CrossRef]
45. Ur Rehman, K.; Shatanawi, W.; Al-Mdallal, Q.M. A Comparative Remark on Heat Transfer in Thermally Stratified MHD Jeffrey Fluid Flow with Thermal Radiations Subject to Cylindrical/Plane Surfaces. *Case Stud. Therm. Eng.* **2022**, *32*, 101913. [CrossRef]
46. Zaher, A.Z.; Moawad, A.M.A.; Mekheimer, K.S.; Bhatti, M.M. Residual Time of Sinusoidal Metachronal Ciliary Flow of Non-Newtonian Fluid through Ciliated Walls: Fertilization and Implantation. *Biomech. Model. Mechanobiol.* **2021**, *20*, 609–630. [CrossRef]
47. Saleem, S.; Animasaun, I.L.; Yook, S.-J.; Al-Mdallal, Q.M.; Shah, N.A.; Faisal, M. Insight into the Motion of Water Conveying Three Kinds of Nanoparticles Shapes on a Horizontal Surface: Significance of Thermo-Migration and Brownian Motion. *Surf. Interfaces* **2022**, *30*, 101854. [CrossRef]
48. Jeffreys, H. *The Earth*; Cambridge University Press: Cambridge, UK, 1929; p. 265.
49. Ali, A.; Awais, M.; Al-Zubaidi, A.; Saleem, S.; Khan Marwat, D.N. Hartmann Boundary Layer in Peristaltic Flow for Viscoelastic Fluid: Existence. *Ain Shams Eng. J.* **2022**, *13*, 101555. [CrossRef]
50. Bhatti, M.M.; Abdelsalam, S.I. Bio-Inspired Peristaltic Propulsion of Hybrid Nanofluid Flow with Tantalum (Ta) and Gold (Au) Nanoparticles under Magnetic Effects. *Waves Random Complex Media* **2021**, 1–26. [CrossRef]
51. Kothandapani, M.; Prakash, J. Effects of Thermal Radiation Parameter and Magnetic Field on the Peristaltic Motion of Williamson Nanofluids in a Tapered Asymmetric Channel. *Int. J. Heat Mass Transf.* **2015**, *81*, 234–245. [CrossRef]

energies

MDPI

Article

Free Energy Changes during Spherical Droplet Deposition—Mechanistic Model

Jacek A. Michalski [1,*] and Slawomir Jakiela [2]

[1] Faculty of Civil Engineering, Mechanics and Petrochemistry, Institute of Chemistry, Warsaw University of Technology, Ignacego Łukasiewicza 17, 09-400 Płock, Poland

[2] Department of Physics and Biophysics, Institute of Biology, Warsaw University of Life Sciences, Nowoursynowska 159, Building 34, 02-776 Warsaw, Poland; slawomir_jakiela@sggw.edu.pl

* Correspondence: jacek.michalski@pw.edu.pl; Tel.: +48-24-367-2193

Abstract: On the basis of theoretical considerations (mechanistic model), an equation was determined that allowed to calculate the free energy (Helmholtz) of a spherical droplet deposited on a flat surface in a system without external forces. Assuming isochoric and isothermal transformation of the system and a very fast conversion of mechanical energy into heat, the obtained equation allows to determine the trajectory of thermodynamic transformation consisting of the spreading of the droplet on the surface of the substrate. The similarities and differences in the behaviour of spherical droplets described by the mechanistic model and Young's model, together with its improvements, were discussed. The trajectories of free energy changes during the spreading of droplets in a system in which the adhesive force acting perpendicular to the wetted surface was considered as well.

Keywords: wetting; droplet spreading; free energy of a droplet; contact angle

Citation: Michalski, J.A.; Jakiela, S. Free Energy Changes during Spherical Droplet Deposition—Mechanistic Model. *Energies* **2022**, *15*, 4725. https://doi.org/10.3390/en15134725

Academic Editors: Vasily Novozhilov and Cunlu Zhao

Received: 21 May 2022
Accepted: 23 June 2022
Published: 28 June 2022

1. Introduction

The phenomenon of wetting has accompanied mankind for thousands of years. Probably, at first, it was important mainly during washing, then indirectly during drying clothes. However, just over two hundred years ago, the first papers that attempted a mathematical description of this phenomenon appeared. At present, this phenomenon is important not only at home, but also during manufacturing processes (e.g., painting and varnishing), transport (e.g., lubrication), chemical and medical analysis (microfluidics), and many others.

The most cited paper is that written as an essay by Young in 1805 [1]. It contains a description (in words) of the balance of forces, or rather stresses, acting on a spherical droplet at the point where all phases meet—the droplet, its fluid surroundings, and the flat surface of the solid substrate. It is worth noting that this relationship was written in the form of an equation only over 100 years later [2]. This equation correlates the surface tension at the interface of liquid phases with the tangential stresses to the surface to be wetted. These stresses reflect the intermolecular interactions characterizing the interfacial surfaces and therefore have the properties of physicochemical parameters. The equation shows that for a given three—hase system, the contact angle should be independent of the droplet volume. However, based on the results of the experiments, it is known that the contact angle changes with the volume of the droplets [3]. This effect is especially visible with small, but still macroscopic, droplets.

Due to these observations, an additional term was introduced into Young's equation corresponding to the stresses acting along the three—phase contact line [4]. This hypothesis was verified based on the experimentally measured contact angle of gas bubbles deposited on a solid substrate. It is worth adding that these stresses may cause shrinking or stretching of this line, depending on the direction of action of this force described by its sign (positive or negative). It should be noted that the introduction of an additional term improved the accuracy of matching the experimental results to the modified Young's equation. Apart from the fact that the modified Young's equation is not the best representation of the experimental results, the determined values of the forces acting along the three—phase contact lines are greater by 5–6 orders of magnitude than their values calculated on the basis of molecular simulation [5].

Also in 1805, Laplace published his multi—volume work on mechanics [6]. In one of the chapters, he proved that the curvature of a flexible surface requires that a certain force acts perpendicular to it. If this flexible surface is an interphase surface, its curvature is due to the pressure difference on both sides of it. One can find out about it by observing the column of liquid in the thin capillaries immersed in it. However, even in 1883, when mathematically modeling the shape of the droplets was performed, a constant parameter was adopted instead of this pressure difference [7]. The current form of Laplace–Young's equation (taking into account pressures) is due to Neuman [8]. It is worth emphasizing here that a perfectly spherical shape of a drop is possible only when the pressure inside it is identical throughout its volume.

In the first half of the twentieth century, the phenomenon of spreading liquid droplets on porous or rough surfaces was mainly studied. Wenzel [9] noticed that the actual wetted surface on the surface of such materials is larger than the geometrically determined one. As a result of this observation, he introduced the roughness coefficient to the Young's equation, which is the ratio of the two surfaces mentioned. It is worth emphasizing that this factor only affects the tangential stress to the wetted surface of the solid substrate. Cassie and Baxter [10] took into account that as the liquid spreads over the porous surface, it closes a certain amount of gas contained in the pores. As a result of this approach to Young's equation, another factor was introduced, which modified the value of the contact angle on the flat substrate to the contact angle at the inlet to the pores. This coefficient did not take into account the dependence on the pore diameter, and the determined contact angle for the pores was not compared to the contact angle observed on a flat substrate. Also in this case, the introduced factor only influenced the tangential stress between the spreading liquid and the substrate.

The first relationship based on thermodynamic considerations was derived by Reiss [11] who determined the work of creating a spherical drop in a system not subjected to external forces. The derivation was made on the basis of the free energy (Helmholtz) definition for a two—phase liquid droplet—fluid environment. It is worth noting that during the derivation, no potential mechanism ensuring the isochoric nature of the system was indicated (despite the liquid evaporation), assuming only its isothermal nature.

In 1977, Boruvka and Neuman published a paper in which they derived equations for the internal energy of a macroscopic three—phase system based on the equations used to describe molecular systems [11]. In the obtained equation, apart from typical thermodynamic parameters, the internal energy depended on the interfacial areas, the length of the lines separating the three phases, and the force acting along them, as well as on point energy sources corresponding to local point forces acting in the system. In the case of a spherical droplet deposited on a solid substrate, determining its minimum internal energy requires maintaining the same entropy of all phases present in the system. On the other hand, further determination of the sufficient condition for this minimum is very complicated from the mathematical point of view and is not presented in this paper.

The need for entropy balancing in all phases has been eliminated by using free energy (Helmholtz) for thermodynamic analysis of a liquid droplet deposited on a solid substrate and assuming the isothermal nature of the system [12,13]. This approach made it possible to precisely define the necessary condition for the occurrence of the minimum free energy [14], but still it was not possible to determine the sufficient condition determined on the basis of equilibrium thermodynamics. Only the analysis based on non—equilibrium thermodynamics made it possible to determine the equation which was the condition of a sufficient minimum of free energy (Helmholtz) [15], and also allowed to determine the trajectories of free energy changes during droplet deposition on a solid substrate. On this basis, it was possible to determine whether the droplet spreading over the surface is spontaneous or forced. It is worth noting that the thermodynamic trick described in Appendix [15] ensured the isochoric nature of each of the fluid phases present in the isothermal system under consideration.

It should be emphasized that the equations describing thermodynamic functions (internal energy, thermodynamic potential, free energy) in multi—phase systems were formulated on the basis of ad hoc summation of the energy of interfacial surfaces, energy of phase contact lines, etc. However, the definitions of these state functions (among others in [16]) link their change during thermodynamic transformations with the work performed by the system or performed on the system. It is only as a result of these works that the interfacial surfaces, phase contact lines, etc. change. Each of the performed works results from the forces acting in the system and the change of interfacial surfaces, the position of the phase dividing lines, etc. Therefore, it is surprising that the equations obtained from the models [11–15] do not indicate the existence of adhesion forces perpendicular to the interface at the liquid–solid substrate boundary. After all, the change of molecules from the continuous phase to the droplet molecules should be associated with a change in the energy of the system due to the differentiation of forces acting between the molecules and the surface of the solid substrate. The Lenard–Jones model [17] shows such an impact.

In the available literature, one can find a mathematical model which takes into account all known forces acting on a droplet depositing on a solid substrate [18]. The balance of forces causing the droplet spreading was determined as the minimum of its mechanical energy. Apart from the necessary condition, a sufficient condition for the occurrence of such a minimum has also been established. So far, this model has not been associated with any of the thermodynamic state functions.

From the beginning of the second half of the 20th century until now, a number of mathematical models less related to thermodynamics, or rather to thermodynamic functions, were also formulated. In each of them, hypotheses of phenomena were formulated, which would allow for the balancing of force or stress components perpendicular to the substrate, acting along the three—phase contact line. The first hypothesis was put forward by Deriagin, assuming the existence of a jointing–disjoining pressure characteristic for a given system [19,20]. The result of the occurrence of such a phenomenon would have to be a deformation of the curvature of the interfacial surface between the liquid phases in the immediate vicinity of the three—phase contact line, i.e., the pressure occurring there that differs from the pressure in the remaining droplet volume. Unfortunately, this contradicts the Laplace condition [6] imposing a constant pressure in its entire volume, because in the event of a pressure difference, forces forcing the liquid to move would appear. The second hypothesis formulates the phenomenon of pinning occurring in the substrate along the three—phase contact line [21]. This phenomenon would be associated with a change in the structure of the substrate and the appearance of forces attracting or repelling the liquid forming the droplet in this place. Considering the large variety of wetted substrates, such a phenomenon would have to occur from the molecular to the macroscopic scale. So far, it has not been possible to illustrate it experimentally.

It is worth noting that in both models [19–21], there is no adhesion phenomenon consisting in the interaction with the substrate of all the liquid molecules located next to it, see the Lenard—Jones interaction [17]. On the other hand, the interactions between the droplet liquid and the substrate occur only along the three—phase contact line.

The droplet size that can be simulated in molecular modeling (10^5 molecules at most) is not yet in the range that can be studied experimentally. However, the numerical contact angles determined can be compared to those calculated from the Laplace—Young equation with the experimentally determined constants for macroscopic droplets. Unfortunately, the first attempts were unsuccessful [22] and it was necessary to apply appropriate computational schemes to obtain the compliance of both results [23,24]. It should be noted, however, that the simulations are performed for two—dimensional droplets and partially end when Rayleigh instabilities occur, which means that the simulated droplets are far from thermodynamic equilibrium.

A negative or positive sign of a change in free energy during a thermodynamic transformation indicates whether the transformation is spontaneous or forced (inter alia [16]). However, in order to be able to make such an assessment, it is necessary to know the equation describing this thermodynamic function, or rather the trajectory of its changes while spreading over a solid substrate. In the case of Laplace—Young's equation, such a relationship has been derived [15].

The aim of this work is to formulate an equation defining the free energy (Helmholtz) of a drop resting on a solid substrate based on a mechanistic model taking into account the forces of adhesion between a liquid and a solid [18]. With some simplifications, this relationship would allow the determination of the areas of spontaneous or forced spreading of the droplets on the substrate. It should be emphasized that formulating conclusions regarding the nature of the transformation (spontaneous or forced) is possible based on the sign of the free energy difference between the final and the initial state of the system, but it is not possible based on the magnitude of the forces acting in the system in each of these states.

2. Theory

The analysis of the behavior of a droplet spreading on the surface of the substrate should be started with writing the equation defining its free energy (Helmholtz):

$$dF = -S\,dT - p\,dV + dW \tag{1}$$

and description of the geometry of the system under consideration.

The system consists of a droplet whose volume corresponds to the volume of a fully spherical drop of radius R_π. The droplet spreading on the flat surface of the substrate forms a segment of a sphere with a radius R and a height z. Both these values characterize the contact angle φ. The relationships between the mentioned parameters are determined by the following equations:

$$z = R(1 - \cos\varphi) \tag{2a}$$

$$r = R\sin\varphi \tag{2b}$$

$$\lambda = 2\pi r \tag{2c}$$

$$R = R_\pi \left[\frac{4}{(1 - \cos\varphi)^2 (2 + \cos\varphi)}\right]^{\frac{1}{3}} \tag{2d}$$

Figure 1 schematically shows the adopted coordinate system and geometric parameters used in Equation (2a–d). It should be emphasized that the wetted surface is circular due to the spherical shape of the droplet segment.

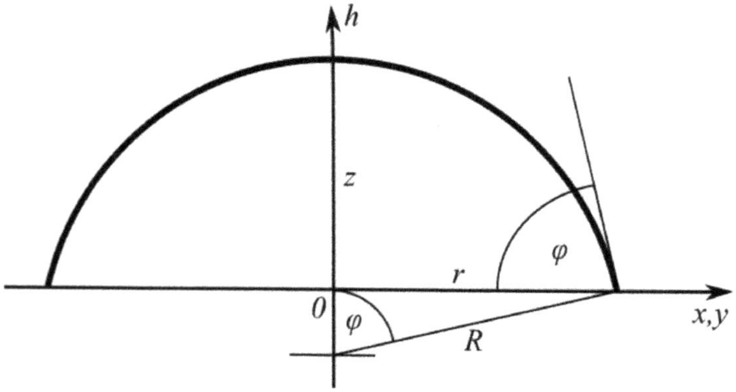

Figure 1. The geometric meaning of the parameters indicated in the figure and applied in the formulae.

In our previous paper [18], the equation describing the differential mechanical work performed by the droplet spreading on the substrate in a system not affected by external forces was derived:

$$dW = 2\pi\sigma_{CI}R_\pi^2 \left\{ \left[\frac{4}{(1-cos\varphi)^2(2+cos\varphi)}\right]^{\frac{2}{3}} \frac{(1-cos\varphi)(1+cos\varphi)}{2+cos\varphi} sin\varphi \right. \\ -2E\frac{(1+cos\varphi)}{(1-cos\varphi)(2+cos\varphi)^2} sin\varphi \\ -B\left[\frac{4}{(1-cos\varphi)^2(2+cos\varphi)}\right]^{\frac{2}{3}} \frac{sin\varphi}{2+cos\varphi} \\ \left. -D\left[\frac{4}{(1-cos\varphi)^2(2+cos\varphi)}\right]^{\frac{1}{3}} \frac{1}{2+cos\varphi} \right\} d\varphi \tag{3}$$

in which the constant parameters E, B, D are defined analogously to those given in our previous paper [18] as:

$$E = \frac{\varepsilon R_\pi}{\sigma_{CI}} \tag{4a}$$

$$B = \frac{\sigma_{FI}}{\sigma_{CI}} \tag{4b}$$

$$D = \frac{F_L}{\sigma_{CI}R_\pi} \tag{4c}$$

Before proceeding to further discussion, it is necessary to quote the conditions and limitations that were applied during the derivation of the above equation [18]:

1. In the vicinity of the state of equilibrium of forces acting on the droplet, it takes the shape of a sphere segment, and the fluid velocity and its changes become so small that the inertial forces of the moving fluid and its impact on the surface between fluids due to the so—called dynamic pressure become negligible.
2. During droplet deposition on the substrate, the temperature of any of the system components does not change.
3. The liquid forming the droplet does not change its volume—the entire system under nature of the system, it is possible to avoid the need to consider the issue of droplet evaporation [15]. On the other hand, the assumption of complete insolubility of the components of both fluid phases allows to ignore the influence of the Marangoni effect.

Due to the isothermal and isochoric nature of the system under consideration, Equation (1) is simplified to the following form:

$$dF = dW \tag{5}$$

Free energy is a function of state, therefore its change does not depend on the transformation trajectory, but only on the parameters characterizing the initial and final state of the system. This means that the droplet transformation from the moment it touches the substrate at one point until reaching the equilibrium state is described by the following integral:

$$f = \frac{dF}{2\pi\sigma_{CI}R_\pi^2} = \int_{\varphi=180^\circ}^{\varphi} \left\{ \left[\frac{4}{(1-cos\varphi)^2(2+cos\varphi)}\right]^{\frac{2}{3}} \frac{(1-cos\varphi)(1+cos\varphi)}{2+cos\varphi} sin\varphi \right. \\ -2E\frac{(1+cos\varphi)}{(1-cos\varphi)(2+cos\varphi)^2} sin\varphi - B\left[\frac{4}{(1-cos\varphi)^2(2+cos\varphi)}\right]^{\frac{2}{3}} \frac{sin\varphi}{2+cos\varphi} \\ \left. -D\left[\frac{4}{(1-cos\varphi)^2(2+cos\varphi)}\right]^{\frac{1}{3}} \frac{1}{2+cos\varphi} \right\} d\varphi \tag{6}$$

and as a result, we get the equation:

$$f = \frac{F}{2\pi\sigma_{CI}R_{\pi}^{2}} = \{ \quad -(1-\cos\varphi)^2\left[\frac{4}{(1-\cos\varphi)^2(2+\cos\varphi)}\right]^{\frac{2}{3}}\left[\sqrt[3]{3}(2+\cos\varphi)^{\frac{2}{3}}{}_2F_1\left(\frac{2}{3},\frac{2}{3};\frac{5}{3};\frac{1-\cos\varphi}{3}\right)-1\right]$$
$$-\frac{2}{9}E\left[\frac{3}{2+\cos\varphi}+\ln\frac{(2+\cos\varphi)^2}{(1-\cos\varphi)^2}\right]-B(1-\cos\varphi)(1+\cos\varphi)\left[\frac{4}{(1-\cos\varphi)^2(2+\cos\varphi)}\right]^{\frac{2}{3}} \quad (7)$$
$$+D\sin\varphi\left[\frac{4}{(1-\cos\varphi)^2(2+\cos\varphi)}\right]^{\frac{1}{3}}\}_{\varphi=180^\circ}^{\varphi}$$

in which the expression ${}_2F_1\left(\frac{2}{3},\frac{2}{3};\frac{5}{3};\frac{1-\cos\varphi}{3}\right)$ denotes the hypergeometric function described by Gauss back in the 19th century [25].

Before the drop spreads and reaches the state of equilibrium of forces, the liquid fragments inside it will obtain a certain kinetic energy not included in Equation (7). As a result of the viscosity of the fluid, this kinetic energy will be converted into heat, causing an increase in the entropy of the system, which, due to its isothermal nature, will not change the expression SdT in Equation (1). Moreover, in our considerations, we do not analyze the kinetics (hydrodynamics) of the liquid flow and, as a result, the appearance and disappearance of its kinetic energy. Therefore, we can assume that the kinetic energy that appears is transformed infinitely fast into heat, and according to the isothermal condition of the system, this heat is infinitely removed outside the system. This assumption corresponds exactly to the infinitely high viscosity of the liquid or the infinitely slow spreading of the droplet. It is also worth noting that such an assumption does not affect the form of Equations (6) and (7) because the viscosity of the liquid and time are not parameters of the model, but only affect the shape of the thermodynamic transformation trajectory, determining its asymptotic course. This, in turn, allows one for the determination of areas for which the droplet spreading is spontaneous or forced.

Based on the above observations, it is possible to interpret the relationship of the following terms in Equations (6) and (7) with specific sources (forces) in the system under consideration. The first term is related to the pressure changes inside the droplet caused by the change in the curvature of the interfacial surface between the fluid phases, and actually with the technical work (Vdp) performed in the system. The second term describes the change in free energy caused by the adhesive force acting between the liquid and the substrate—perpendicular to the substrate. The third one corresponds to the free energy changes due to the force acting on the three—phase contact line—tangent to the substrate. On the other hand, the fourth one corresponds to the free energy changes caused by the force acting along the three—phase contact line causing it to shrink or stretch.

The hypergeometric function is expressed as a series with an infinitely large number of terms. Therefore, instead of calculating it, it is much more convenient to perform a numerical calculation of the dimensionless free energy f using Equation (6).

3. Results and Discussion

Due to the large variety of behaviors of the analyzed system, a detailed analysis was carried out for only three cases identical to those described in [18].

3.1. Young's Solution

The classical Young's equation takes into account only surface tension stresses (forces) acting at the interface of liquid phases and tangential stresses to the substrate surface, stresses shrinking or stretching the wetted circumference [1,2]. In the case of the derived model [18], the balance of forces is described by the following equation:

$$cos\varphi = \pm\sqrt{1-B} \quad (8)$$

a stable solution is obtained only for $\varphi < 90^\circ$ [18], i.e., when the sufficient condition for the occurrence of the minimum free energy is fulfilled. It is also worth noting that real solutions can only be obtained for $0 < B < 1$.

It should be emphasized that the contact angle tends to the right angle ($\varphi \to 90^\circ$) while the value of the parameter B tends to unity ($B \to 1$), which means that the surface tension stress is balanced by the force stretching the wetted circuit. On the other hand, in Young's equation, the contact angle reaches this value when $B \to 0$, which means that the droplet is not affected by any forces, and yet it rests on the substrate.

In the case under consideration, the dimensionless free energy is described by the equation:

$$f = \frac{dF}{2\pi\sigma_{CI}R_\pi^2} = \int_{\varphi=180^\circ}^{\varphi} \frac{sin\varphi}{2+cos\varphi}\left[\frac{4}{(1-cos\varphi)^2(2+cos\varphi)}\right]^{\frac{2}{3}}[(1-cos\varphi)(1+cos\varphi)-B]\,d\varphi \quad (9)$$

Changes in the free energy of a spherical droplet during its spreading on a solid substrate (Young's case) are shown in Figure 2. After a droplet comes into contact with the substrate, it is necessary to supply it with some mechanical energy in order to initiate its spreading. Only after reaching a certain maximum value of free energy, corresponding to the unstable balance of forces, the drop will start to spread spontaneously. This maximum is higher the higher the B parameter value is. After this maximum is exceeded, the droplet spreads spontaneously on the substrate until it reaches the minimum free energy value. The droplet can spread further only as a result of supplying it with additional mechanical energy. It is worth noting that for $B \cong 0.74$ the level of this minimum corresponds to the free energy of the drop characterizing its state before contact with the substrate. For higher values of this parameter, the free energy at its minimum point is higher than the initial one. For the value of $B = 1$, the minimum of free energy disappears, and there is only the inflection point. It is also worth noting that for $B = 0$, there is no local free energy maximum and the droplet spreads spontaneously over the entire surface of the flat substrate. Moreover, the free energy of a fully spherical droplet is higher than that of a flat liquid surface. This means that this spherical droplet is characterized by a thermodynamically unstable being.

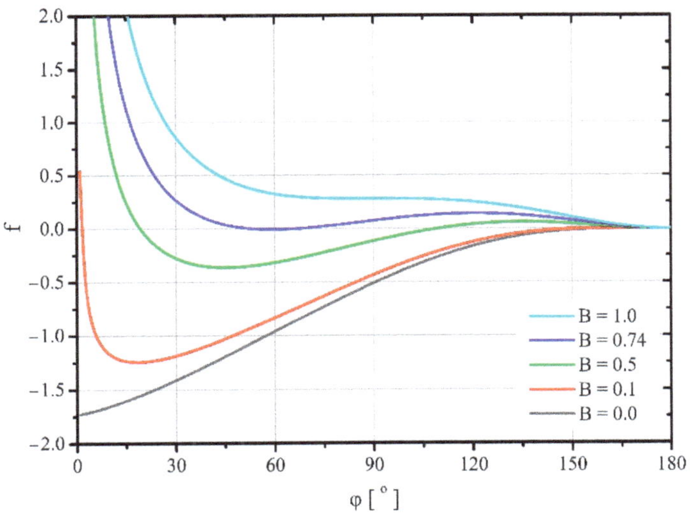

Figure 2. Changes in the dimensionless free energy of a spherical droplet during its spreading on a solid surface—the Young case.

In the traditional Young's model [1,2], for the value of the parameter $B = 0$, the contact angle is equal to $\varphi = 90^\circ$. For such a system, there is a minimum area of the spherical cup, which justifies the existence of the minimum free energy determined on the basis of the so—called surface energy of the drop. However, in such a system, the balance of forces acting in the direction perpendicular to the surface of the substrate is evidently not satisfied.

There is simply no stress along the three—phase contact line to balance the surface tension of the contacting interface between the fluid phases at this point.

3.2. Improved Young's Solution

When the droplet is subjected to a force tangential to the surface of the substrate and a force that stretches or shrinks the three—phase contact line, the balance of forces (minimum free energy) is described by the following equation [18]:

$$sin^2\varphi - B - D\frac{1}{sin\varphi}\left[\frac{4}{(1-cos\varphi)^2(2+cos\varphi)}\right]^{-\frac{1}{3}} = 0 \tag{10}$$

On the other hand, the equation describing the dimensionless free energy (6) simplifies to the form:

$$\begin{aligned} f = \frac{dF}{2\pi\sigma_{CI}R_\pi^2} = & \int_{\varphi=180^\circ}^{\varphi}\left\{\left[\frac{4}{(1-cos\varphi)^2(2+cos\varphi)}\right]^{\frac{2}{3}}\frac{(1-cos\varphi)(1+cos\varphi)}{2+cos\varphi}sin\varphi\right. \\ & -B\left[\frac{4}{(1-cos\varphi)^2(2+cos\varphi)}\right]^{\frac{2}{3}}\frac{sin\varphi}{2+cos\varphi} \\ & \left.-D\left[\frac{4}{(1-cos\varphi)^2(2+cos\varphi)}\right]^{\frac{1}{3}}\frac{1}{2+cos\varphi}\right\}d\varphi \end{aligned} \tag{11}$$

Due to the richness of the behavior of the considered system and the limited volume of the article, only three graphs shown in Figures 3–5 were prepared for illustration. A single, stable minimum of free energy (Helmholtz) occurs whenever one of the following conditions is met [18]: $B > 1.92$ or $D < -1.07$. In such cases, the droplets spread spontaneously on the substrate. However, due to the very significant values of parameter B, the analysis of this issue would not assess the impact of parameter D on the solutions of Young's Equation (8), but would only contribute to the discussion of an unrelated case.

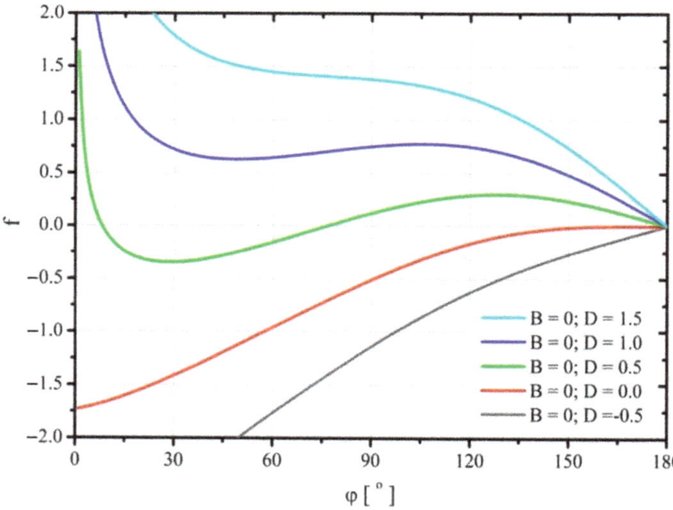

Figure 3. Change of the dimensionless free energy of the system during spreading of the droplet for the parameter $B = 0$.

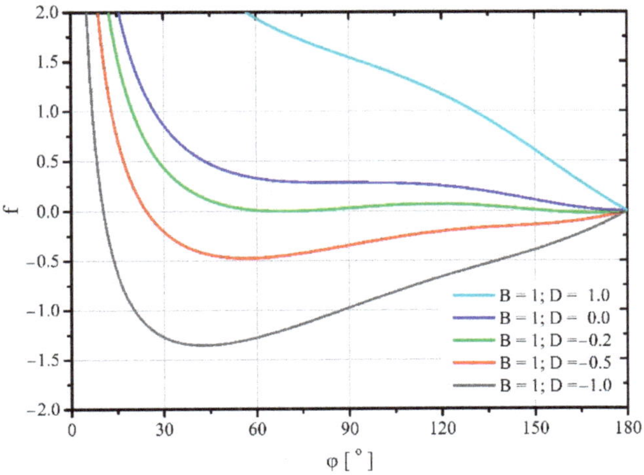

Figure 4. Change of the dimensionless free energy of the system during spreading of the droplet for the parameter $B = 1$.

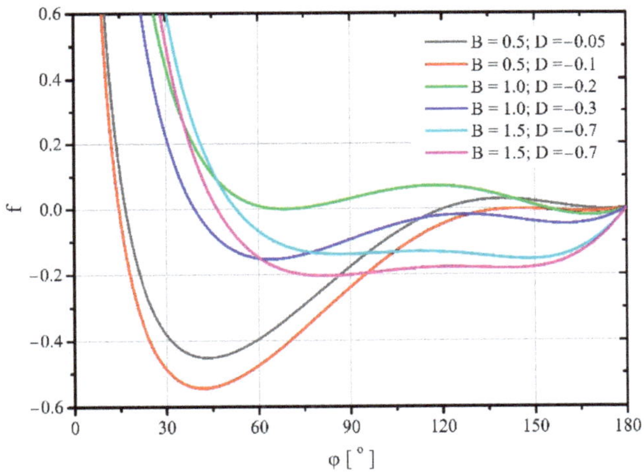

Figure 5. Changes of dimensionless free energy during droplet spreading for the values of parameters B and D for which there are two stable minima of this function.

In the considered case, a single and stable minimum of free energy also occurs when one of the following conditions is met: $-1.42 < B < 0$ and $D > 1$ or $0 < D < 3.5$ and $B < 1$ [18]. Thus, the analysis covered two extreme cases for which Young's solution should be met, i.e., for the parameters $B = 0$ and $B = 1$. The results are shown in Figures 3 and 4. However, in order to initiate the spontaneous spread of the droplet on the substrate, initially it is necessary to provide the droplet with a certain amount of mechanical energy so as to exceed a certain threshold value of free energy. The graphs also show example curves for which the conditions for the occurrence of a single stable minimum of free energy were not met. Thus, in Figure 3 it is the curve for $B = 0$ and $D = 1.5$, and in Figure 4 the curve for $B = 1$ and $D = 1$. In both cases, the spreading of the liquid on the substrate requires additional energy to be supplied to the system. Moreover, the curve for $B = 0$ and $D = -0.5$ was also plotted in Figure 3. For such parameter values, the droplet will spontaneously spread over the entire surface of the substrate. Figure 4 also shows the curve corresponding to the values of parameters $B = 1$ and $D = -0.2$. In this case, the spreading droplet can reach two free energy minima corresponding to two contact angle values.

The formulated model [18] predicts the occurrence of two minima of the free energy of a droplet deposited on a flat surface. These stable minima occur when one of the following conditions is met: $0 < B < 1.92$ and $D < 1$ or $-1.07 < D < 0$ and $B > 0$ [18]. For such parameter values, the droplets spread spontaneously after contact with the substrate reaching the first minimum of free energy. Moving to the next minimum requires a certain amount of mechanical energy to be supplied. Therefore, it is important to determine this energy, or rather its minimum value, which is a measure of the system durability. For this purpose, Figure 5 was prepared, showing exemplary free energy trajectories of spreading droplets on the substrate.

With the increase in the value of the B parameter, the trajectories of free energy changes during the spreading of the droplet, shown in Figure 5, become increasingly flat. This means that a smaller and smaller portion of mechanical energy has to be delivered to the droplet so that it passes from one state to another one that meet the conditions of a stable balance of forces.

It is worth noting that the thermodynamic description of the phenomenon related to the improved Young's equation [15] does not indicate the existence of two stable minima of free energy. These minima represent the balance of forces acting on the droplet and can potentially correspond to the advancing and receding contact angles observed during the experiments [5].

3.3. Influence of Adhesion Force on Free Energy Variation

The cases described so far were related to the Young's equation [1,2] because they took into account only the same forces; more specifically, forces tangential to the surface of the substrate stretching or shrinking the wetted surface, stretching or shrinking the three—phase contact line, and the surface tension acting at the interface of fluid phases. On the other hand, the Young's equation does not take into account the adhesive force acting between the molecules of the liquid forming the drop and the substrate material and directed perpendicularly to it. This in turn means that the zero value of the parameter E appearing in Equations (3), (6) and (7).

Let us consider a typical case of Young, in which, apart from the forces tangential to the substrate, stretching or shrinking the wetted surface [1,2], we also take into account the adhesion force acting perpendicular to the surface of the substrate. Then, the equation describing the balance of forces is described by the following equation [18]:

$$sin^2\varphi - \frac{1}{2}E\left[\frac{4}{(1-cos\varphi)^2(2+cos\varphi)}\right]^{\frac{1}{3}} sin^2\varphi - B = 0 \tag{12}$$

In such a case, the Equation (6) describing the dimensionless free energy is simplified to the form:

$$\begin{aligned} f = \frac{dF}{2\pi\sigma_{CI}R_\pi^2} = & \int_{\varphi=180^\circ}^{\varphi}\left\{\left[\frac{4}{(1-cos\varphi)^2(2+cos\varphi)}\right]^{\frac{2}{3}}\frac{(1-cos\varphi)(1+cos\varphi)}{2+cos\varphi}sin\varphi\right. \\ & -2E\frac{(1+cos\varphi)}{(1-cos\varphi)(2+cos\varphi)^2}sin\varphi \\ & \left.-B\left[\frac{4}{(1-cos\varphi)^2(2+cos\varphi)}\right]^{\frac{2}{3}}\frac{sin\varphi}{2+cos\varphi}\right\}d\varphi \end{aligned} \tag{13}$$

The characteristic point of Young's equation, mentioned in Section 3.1, is determined by the contact angle $\varphi = 0$, for which there is evidently no equilibrium of forces acting perpendicular to the substrate to be wetted. On the other hand, transforming the Equation (12) in the absence of forces acting tangentially to the wetted substrate ($B = 0$), the following equation can be obtained:

$$\frac{2\sigma_{CI}}{R} = \varepsilon \quad (14)$$

which proves that the adhesion forces balance the surface tension forces acting on the wetted perimeter, and more precisely, the adhesion forces balance the overpressure inside the droplet caused by the curvature of the interfacial surface between the fluid phases. It is also worth noting that this balance of forces will be stable, which can be determined based on the sufficient condition of presence of the minimum free energy [18], $\frac{df}{d\varphi} = 0$ and $\frac{d^2f}{d\varphi^2} > 0$.

The results shown in [18] imply that for a surface adhesion force greater than or equal to zero ($E \geq 0$ or $\varepsilon \geq 0$), the equilibrium of forces in the analyzed system can only occur when $B \leq 1$. In addition, as shown in the previous paper [18], the solutions to the equilibrium equation of forces acting on the droplet for the value of the parameter $B = 0$ divide the entire area of solutions to the equilibrium equation of forces into two parts. In the first region where the parameters can vary within the range $0 < B < 1$ and $0 < E < 2$, a stable equilibrium of forces cannot exist for large values of the contact angle φ. On the other hand, in the second, for $0 < B < -\infty$ and $0 < E < +\infty$, this stable equilibrium of forces cannot occur for small values of this angle. Since the relationship between the free energy and the forces acting on the droplet (3) and (5) is the same for both areas, the shapes of the free energy changes as a function of the contact angle should have a topologically similar course. To explore this, Figure 6 was created, on which several trajectories for each of the areas were plotted.

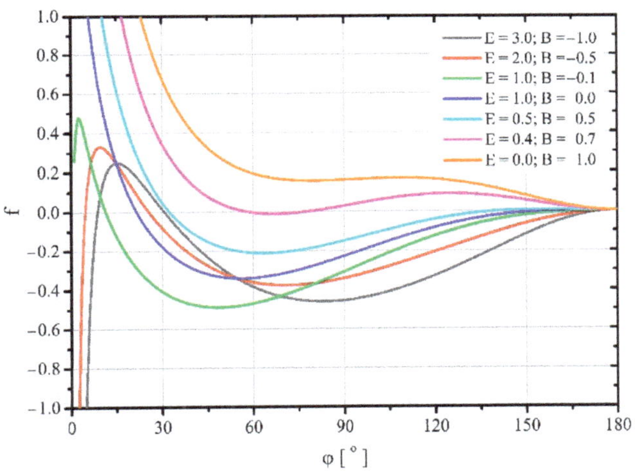

Figure 6. Changes in dimensionless free energy in a system in which there is no force acting along the three—phase contact line causing its stretching or shrinking ($D = 0$).

For negative values of parameter B, for which the forces tangential to the surface of the substrate cause shrinkage of the wetted area, after contact with the substrate, the droplets spread out spontaneously reaching the minimum free energy. However, further spreading of the droplets requires the application of mechanical energy to them until free energy reaches a certain maximum value. After this maximum is exceeded, further spreading of the droplets occurs spontaneously again until the surface of the substrate is completely covered with the liquid. On the other hand, for non—negative values of this parameter ($B \geq 0$), after the droplet comes into contact with the substrate, it is initially necessary to supply the droplet with a certain amount of mechanical energy to drive it to spread. Only after exceeding a certain maximum free energy of the system, the spontaneous spread of the droplets begins until the minimum of this energy is reached. Further spreading of the droplet requires the supply of mechanical energy to it. A certain exception is the course of the trajectory for $B = 0$, for which the initial maximum of energy does not occur and the droplet spreads spontaneously from the moment it touches the substrate.

It is worth noting that for positive values of parameter pairs B and E, the free energy characterizing its minimum may be greater than the initial free energy of the droplet. This means that achieving a stable balance of forces requires the supply of a certain amount of mechanical energy to a droplet.

Although the derivation of the equations shown in the paper concern the spherical shape of the droplet, the results of the experiments show that droplets of other shapes can also form on solid substrates [5]. Looking from the point of view of thermodynamics, such cases are very rarely analyzed mainly due to obvious mathematical difficulties. Currently published papers [26] concern the description of surface energy rather than the description of the total or free energy of such systems. The basic problem seems to be the combination of all types of energy (e.g., surface and adhesion) as a function of the state.

4. Conclusions

This work is an extension of the mechanistic model of liquid distribution on the surface of a solid substrate [18]. It presents the analytical derivation of the equation of the dependence of free energy (Helmholtz) on the contact angle, which allows to determine the nature of the spherical droplet spreading (spontaneous or forced) on a solid substrate. Appropriate calculations can be performed based on the knowledge of the unit forces (physicochemical parameters) acting parallel and perpendicular to the surface to be wetted and on the interface between the fluids.

In the case of solutions corresponding to the Young model [1,2], the minimum free energy (stable force equilibrium) may occur only when the stretching force of the wetted perimeter acts parallel to the substrate and perpendicular to the three—phase contact line. However, the free energy of a droplet sessile on the substrate will be lower than the energy of the spherical droplet floating above the substrate when the ratio of the stretching force of the wetted perimeter to the surface tension at the interface of fluid phases will be less than $B < 0.74$. This means that for higher values of this ratio, the droplet spreading will be forced.

The solutions corresponding to the improved Young's equation indicate that only in certain ranges of variability of physicochemical parameters (surface tension between fluid phases, tension stretching or shrinking the wetted area, unit force acting along the three—phase contact line), there is at least one stable minimum of free energy. The most important, however, is that in the case of tangential stress to the wetted surface stretching the wetted perimeter and the unitary force shrinking the wetted perimeter acting along the line of contact of three phases, there are two stable minima of the free energy of the system. This behavior confirms the results of experiments in which two contact angles advancing and receding are observed [5]. It is also important that the energy barrier (free energy change) necessary to overcome when passing from one minimum to another may be relatively small. It should be emphasized that such a theoretical explanation is provided only by a formulated mechanistic model.

In the case of very high adhesion forces, the free energy reaches a stable minimum, practically regardless of whether there is a stretching or shrinking tension of the wetted perimeter. Only after the droplet comes into contact with the substrate and the stretching tension on the wetted perimeter occurs is it necessary to provide a certain amount of work to the droplet so that it starts to spread spontaneously. For the shrinking tension acting on this perimeter, such spreading occurs spontaneously from the moment the droplet contacts the substrate.

If, during the spreading of the droplet, the mechanical energy is infinitely fast converted into heat, the derived dependence of the free energy on the contact angle can be treated as a trajectory of such isochoric and isothermal thermodynamic transformation.

Author Contributions: Conceptualization, J.A.M. and S.J.; methodology, J.A.M. and S.J.; validation, J.A.M. and S.J.; formal analysis, J.A.M. and S.J.; investigation, J.A.M. and S.J.; writing—original draft preparation, J.A.M. and S.J.; writing—review and editing, J.A.M. and S.J.; visualization, S.J. and J.A.M.; supervision, J.A.M. and S.J.; project administration, J.A.M. and S.J. All authors have read and agreed to the published version of the manuscript.

Funding: This research received no external funding. The APC was funded by Warsaw University of Technology.

Institutional Review Board Statement: Not applicable.

Informed Consent Statement: Not applicable.

Data Availability Statement: Not applicable.

Conflicts of Interest: The authors declare no conflict of interest.

Abbreviations

The following abbreviations are used in this manuscript:

Nomenclature

B	constant parameter
D	constant parameter
f	dimensionless free energy
E	constant parameter
F	free energy (Helmholtz)
F_L	force acting along three-phase contact line
p	pressure
r	radius of the wetted surface
R	radius of the droplet bowl
R_π	radius of the sphere with a volume equal to the volume of the droplet
S	entropy
T	temperature
V	droplet volume
W	work
z	droplet height

Greek Letters

ε	unitary adhesion force perpendicular to solid surface (droplet–substrate interface)
λ	perimeter of the wetted surface
φ	contact angle
σ_{FI}	tension parallel to the substrate surface, perpendicular to the three-phase contact line, stretching or shrinking wetted area
σ_{CI}	surface tension (droplet—fluid surrounding interface)

References

1. Young, T. An Essay on the Cohesion of the Fluids. Philos. *Trans. R. Soc. Lond.* **1805**, *95*, 65–87.
2. Maxwell, J.C.; Strut, J.W. Capillary Action. In *Encyclopædia Britannica*, 11th ed.; Encyclopædia Britannica Inc.: London, UK; New York, NY, USA, 1911; pp. 256–277.
3. Good, R.J.; Koo, M.N. The Effect of Drop Size on Contact Angle. *J. Colloid Interface Sci.* **1979**, *71*, 283–292. [CrossRef]
4. Vesselovsky, V.C.; Pertzov, V.N. Adhesion of Air Bubbles to the Solid Surface. *Zhurnal Fiz. Khimii* **1936**, *8*, 245–259.
5. Erbil, H.Y. The debate on the dependence of apparent contact angles on drop contact area or three-phase contact line: A review. *Surf. Sci. Rep.* **2014**, *69*, 325–365. [CrossRef]
6. Laplace, P.S. Traite de Mecanique Celeste. In *Chez Courcier*; Imprimeur-Libraire pour les Mathematiques: Paris, France, 1805; Volume 4.
7. Bashforth, F.; Adams, J.C. An Attempt to Test. In *The Theories of Capillary Action*; University Press Warehouse: Cambridge, UK, 1883.
8. Neuman, F. *Theorie der Capillarität: Leipzig, Verlag von B. G. Teubner*; University of California Libraries: Los Angeles, CA, USA, 1894.
9. Wenzel, R.N. Resistance of solid surfaces to wetting by water. *Ind. Eng. Chem.* **1936**, *28*, 988–994. [CrossRef]
10. Cassie, A.B.; Baxter, S. Wettability of porous surfaces. *Trans. Faraday Soc.* **1944**, *40*, 546–551. [CrossRef]
11. Boruvka, L.; Neuman, A.W. Generalization of the classical theory of capillarity. *J. Chem. Phys.* **1977**, *66*, 5464–5476. [CrossRef]

12. Drelich, J.; Miller, J.D. The Line/Pseudo-Line Tension in Three Phase System. *Part. Sci. Technol.* **1992**, *10*, 1–20. [CrossRef]
13. Drelich, J. The Significance and Magnitude of the Line Tension in Three-Phase (Solid-Liquid-Fluid) Systems. *Colloid Surf. A* **1996**, *116*, 43–54. [CrossRef]
14. Widom, B. Line Tension and the Shape of a Sessile Drop. *J. Phys. Chem.* **1995**, *99*, 2803–2806. [CrossRef]
15. Torbus, S.; Dolata, M.; Jakiela, S.; Michalski, J.A. Analysis of Existing Thermodynamic Models of the Liquid Drop Deposited on the Substrate—A Sufficient Condition of the Minimum Free Energy of the System. *Coatings* **2019**, *9*, 791. [CrossRef]
16. Pohorecki, R.; Wronski, S. *Kinetyka I Termodynamika Procesów Inżynierii Chemicznej*; Wydawnictwo Naukowo-Techniczne: Warsaw, Poland, 1979.
17. Lennard-Jones, J.E. On the Determination of Molecular Fields. *Proc. R. Soc. Lond. A* **1924**, *106*, 463–477.
18. Michalski, J.A.; Jakiela, S. Spherical Droplet Deposition—Mechanistic Model. *Coatings* **2021**, *11*, 248. [CrossRef]
19. De Gennes, P.-G.; Brochard-Wyart, F.; Quéré, D. *Capillarity and Wetting Phenomena*; Springer: New York, NY, USA, 2004.
20. Kuchin, I.; Starov, V. Hysteresis of Contact Angle of Sessile Droplets on Smooth homogeneous Solid Substrates via Disjoining/Conjoining Pressure. *Langmuir* **2015**, *31*, 5345–5352. [CrossRef] [PubMed]
21. Tadmor, R. Open Problems in Wetting Phenomena: Pinning Retention Forces. *Langmuir* **2021**, *37*, 6357–6372. [CrossRef] [PubMed]
22. Nijmeijer, M.J.P.; Bruin, C.; Bakker, F.A. Wetting and drying of an inert wall by a fluid in a molecular-dynamic simulation. *Phys. Rev. A* **1990**, *42*, 6052–6059. [CrossRef] [PubMed]
23. Schmelzer, J. The curvature dependence of surface tension of small droplets. *J. Chem. Soc. Faraday Trans.* **1986**, *82*, 1421–1428. [CrossRef]
24. Kanduc, M. Going beyond the standard line tension: Size-dependent contact angles of water nanodroplets. *J. Chem. Phys.* **2017**, *147*, 174701-1–174701-8. [CrossRef] [PubMed]
25. Gauss, C.F. *Disquisitiones Generales Circa Seriem Infinitam*; Typis Dieterichianis: Gottingae, Germany, 1878.
26. Siqveland, L.M.; Skjaeveland, S.M. Derivations of the Young-Laplace equation. *Capillarity* **2021**, *4*, 23–30. [CrossRef]

materials

MDPI

Article

Effect of Leaching Behavior on the Geometric and Hydraulic Characteristics of Concrete Fracture

Yuan Wang [1], Mengmeng Tao [2,*], Di Feng [2], Yu Jiao [2], Yulong Niu [3] and Zhikui Wang [2]

1 Department of Water Conservancy and Hydropower Engineering, Hohai University, Nanjing 210098, China; wangyuan@hhu.edu.cn
2 Department of Civil and Transportation Engineering, Hohai University, Nanjing 210098, China; fengdi@126.com (D.F.); jiaoyu0125@163.com (Y.J.); wangzhikui23@163.com (Z.W.)
3 China Three Gorges Corporation, Beijing 100038, China; niuyl89@163.com
* Correspondence: taomm@hhu.edu.cn

Abstract: The leaching of material from concrete fracture surfaces has an impact on the structural concrete in service, but the number of studies that consider the effect of the coupling of the leaching, fracture geometry and hydraulic processes on concrete fractures is insufficient. In this study, a series of experiments was conducted, and a leaching model proposed, to investigate the mechanism of leaching behavior on the geometric and hydraulic characteristics of concrete fractures. Following the leaching experiment, the evolution of fracture geometric characteristics was observed by a three-dimensional (3D) laser scanning technique, finding that the fracture produces residual leached depth and local uneven leaching, which results in a decrease in roughness. The hydraulic characteristics were then investigated by permeability tests, and it was found that the fracture hydraulic aperture and permeability increase monotonically with leaching time. A simulation of fluid flow in a numerical fracture revealed the effect of residual leached depth and a decrease in roughness on the hydraulic characteristics. Finally, based on the analysis of the chemical composition of the leaching solution, a leaching model of concrete rough fracture surface is proposed and the mechanism of leaching behavior is discussed. These new findings are useful for the understanding of the development of leaching, local to concrete fracture surfaces.

Keywords: underground engineering; leaching of concrete; fracture geometric characteristics; hydraulic characteristics

Citation: Wang, Y.; Tao, M.; Feng, D.; Jiao, Y.; Niu, Y.; Wang, Z. Effect of Leaching Behavior on the Geometric and Hydraulic Characteristics of Concrete Fracture. *Materials* **2022**, *15*, 4584. https://doi.org/10.3390/ma15134584

Academic Editors: Vasily Novozhilov and Cunlu Zhao

Received: 19 May 2022
Accepted: 27 June 2022
Published: 29 June 2022

1. Introduction

Concrete is an excellent building material with high strength and reliable water resistance; it is widely used in underground engineering [1], such as in tunnel lining [2] and coal mine shafts [3]. The phenomenon of leaching consists in the dissolution of solid calcium in cement hydrates when concrete is exposed to any aggressive solution (most of the time, pure water or water with a very low calcium concentration) [4–6]. This leaching process involves the dissolution of the most soluble phase of cement hydrates, calcium hydroxide ($Ca(OH)_2$), and the subsequent transport of dissolved ions out to the environment [7–9]. The long-term effect of this leaching phenomenon is to weaken the material's solid matrix and the concrete's durability [6,7,10,11], thereby causing the degradation of concrete structures in aggressive environments [12]. At the same time, structural concretes in service develop fracturing from different causes, including early-age thermal shrinkage or long-term mechanical loadings [2]. Fractures provide preferential transport pathways for the ingress of water, which contributes greatly to the effect of leaching of hydration products from fractured regions in concrete [13,14]. Therefore, studying the leaching of hydration products from fractured regions in concrete has more important engineering significance than concrete itself [7,15] (Figure 1).

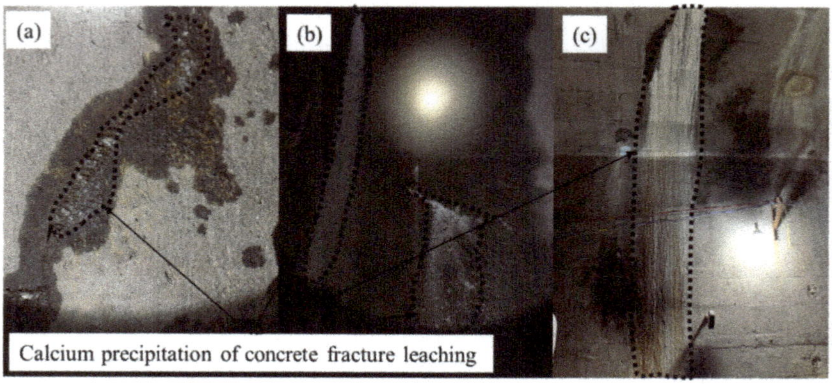

Figure 1. Different leaching degrees of hydration products from fractured regions in concrete (tunnel lining). (**a**) mild, (**b**) moderate and (**c**) severe.

Compared with the relatively flat surface of concrete, the surface of fractures in concrete is rough, and the geometric characteristics are complicated [16,17]. The roughness and complicated geometric characteristics affect the hydraulic characteristic of fractures (hydraulic aperture) [18–21]; the hydraulic characteristic controls the flow state of water in fractures [22] and the flow state, in turn, affects the leaching characteristics [7]. Leaching also alters the fracture geometric characteristics [23,24], and the evolution of fracture geometric characteristics then affect the characteristics of fracture space [25]; this determines the hydraulic characteristics of fractures [26], and the evolution of hydraulic characteristics will further affect the leaching characteristics on the surface of concrete fractures [13]. This is a coupling of the leaching, fracture geometry and hydraulic processes, shown in Figure 2. The effects of fracture geometric characteristics on hydraulic characteristics are many. Brown et al. [25] considered the roughness of fracture surface and modified the cubic law to obtain the hydraulic aperture. Zoorabadi et al. [27] established a new equation between mechanical and hydraulic aperture for different roughnesses. Chen et al. [28] studied the effect of fracture geometric characteristics on the permeability in deformable rough-walled fractures. In contrast, the effect of leaching on the fracture surface geometric characteristics was only explored by a few [29]. Recently, however, it has gradually attracted the attention of several scholars. For example, Wang et al. [30] and Duan et al. [31] studied the effect of leaching on a series of geometric characteristics of limestone fractures. Compared with limestone, concrete is an artificial material composed of fine and coarse aggregates and additives [6], and its physical and chemical properties are quite different. The evolution of fracture geometric and hydraulic characteristics after concrete fracture leaching was not well investigated in previous studies.

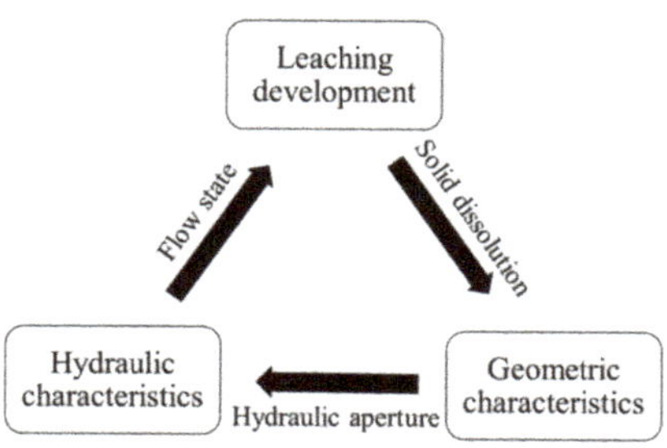

Figure 2. Coupling of the leaching, fracture geometry and hydraulic processes.

In summary, it is necessary to study the evolution law of fracture geometric characteristics after the leaching of concrete, and its effect on hydraulic characteristics. In this paper, the direct leaching method was used to carry out leaching experiments on concrete fractures; several three-dimensional (3D) laser scanning tests and permeability tests in different time periods were then conducted to observe the evolution of fracture geometric characteristics and their effects on the hydraulic characteristics. Numerical simulations were compared to reveal the effect of variation in fracture surface geometric characteristics on hydraulic characteristics. Finally, the chemical composition of the leaching solution was analyzed and a leaching model of rough fracture surface is proposed to discuss the mechanism of leaching on concrete rough fracture surface. The results of this research paper will contribute, to some extent, to a better understanding of the development and mechanism of leaching on concrete fractures.

2. Materials and Methods

In an attempt to approach a real rough fracture, the Brazilian splitting test was used to make a random fracture surface. A series of leaching experiments were carried out on concrete samples with single fractures. In the experiments, the fracture surface geometric characteristics were obtained by 3D laser scanning technology and the fracture permeability was tested using a permeability experimental setup every 120 h to observe the evolution law of leaching on the fracture surface geometric characteristics and the influence of this evolution on fracture permeability, respectively. To validate universal conclusions, two groups of reproducibility experiments were added.

2.1. Preparation of Fracture Samples

First, concrete samples with rough single fractures were prepared before the experiments. We referred to the method of preparing concrete samples with rough single fractures introduced by Anwar [14]. Firstly, the concrete was configured and poured into concrete cylinders. The Brazil splitting test was then used to produce rough single fractures. The main steps were as follows:

(1) Configuring the concrete. In order to meet the strength and water resistance requirements of general underground engineering [32], composite cement P.C.32.5 and medium sand (fineness modulus between 2.3 and 3.0) were selected and mixed with clean water. The mixing ratio of cement, sand and water was 1:3:0.55. A waterproof agent (SJM-1500, Suzhou Institute of Building Science Group Co.,Ltd., Suzhou, China) was added to reduce the permeability of the concrete (2% of cement dosage).

(2) Pouring the concrete cylinders. A release agent was applied on the inner wall of each cylindrical mold (height 50 mm and outer diameter 50 mm) to facilitate demolding after initial setting. The prepared concrete was poured into the molds three separate times. After each pouring, each mold was placed on the shaking table and shaken for 30 s. The purpose of shaking was to eliminate the bubbles in the concrete to make the filler denser. The shaking time was controlled to prevent the separation of solids and liquid in the concrete. After all the concrete was poured, the concrete-filled molds were put into the curing room at a relative humidity of 100% and temperature of 20 °C for the initial setting of the concrete. When the initial setting of the concrete was completed (about 8 h), each mold was disassembled and the concrete cylinder removed.

(3) Curing the concrete. The intact cement cylinders were placed in the curing room for further curing and left for 28 days to complete the curing process.

(4) Making the rough single fracture. A specially designed Brazilian splitting test machine was used to split each cylinder into two half-cylinders along the long axis, creating an artificial fracture by tensile stress within the sample.

Six rough fracture surfaces on three pairs of samples were processed in the same way. Under the same experimental conditions, the three pairs of samples were considered as reproducibility experiments to obtain the universal law. The three pairs of samples were named S1, S2 and S3, respectively, and the suffixes A- and B-side were added to the two halves of each pair of samples, respectively (see Figure 3). For example, S1A indicated the A side of sample S1.

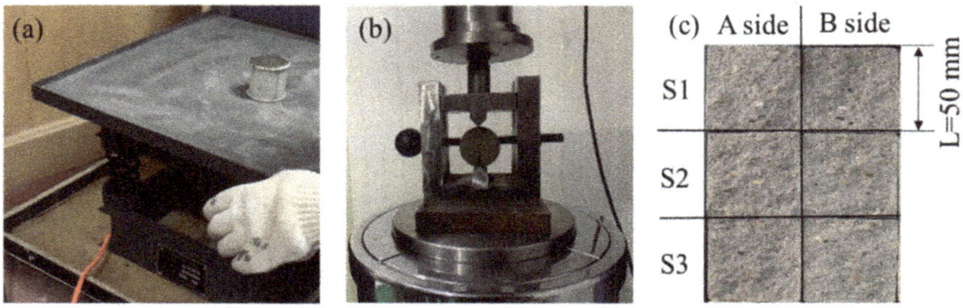

Figure 3. Process of samples preparation. (**a**) Pouring cement samples, (**b**) Brazilian splitting test and (**c**) numbering of fracture surfaces.

2.2. Leaching Experiments

The direct leaching method enables the placing of the cement-based material samples (cement, mortar, concrete, etc.) in an aggressive solution (deionized water, mineral water, sulfate, etc.) [33,34]. If the actions of water head and mechanical stress are not to be considered, the direct leaching method is the most widely used contact leaching method because of its simplicity [33]. In this paper, the effect of water head and mechanical stress were not considered, so the direct leaching method was selected. The six rough fracture surfaces from three pairs of samples were put in a constant temperature water tank (25 °C). Deionized water was selected as the aggressive solution because deionized water, itself, has good leaching ability, and was often used as the aggressive solution in leaching experiment by other scholars [5,6,10]. Based on the experimental experience of Duan et al. [31], the deionized water was replaced every 24 h in order to simulate the flow of the groundwater environment and maintain a high concentration gradient.

2.3. Evolution of Fracture Surface

2.3.1. Obtainment of Fracture Surface

3D laser scanning technology is a non-contact three-dimensional measurement technology which can quickly obtain a wide range of high-precision fracture surface geometries without damaging the fracture surface [28]. The scanning interval used in this paper was 0.05 mm and the scanning accuracy was 0.01 mm, which satisfied the requirements of obtaining the geometric characteristics of a fracture surface [35,36]. In order to control the same fracture surface scanned at different times in a unified coordinate system, identifiable coordinate marks were marked on the outer surface of each sample (see Figure 4).

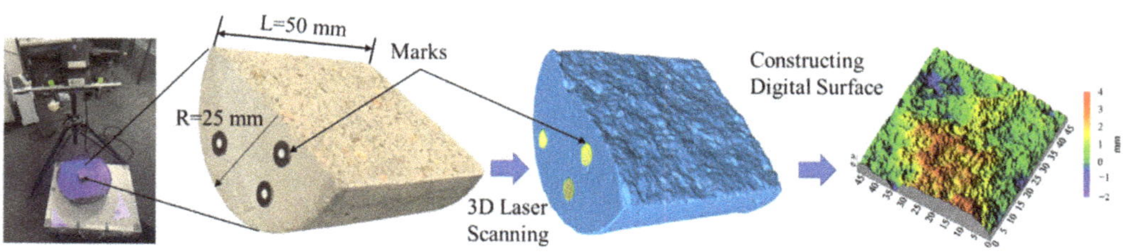

Figure 4. Obtainment process of fracture surfaces by 3D laser scanning.

As shown in Figure 5, red indicates high elevation and blue indicates low elevation. It can be seen that the fracture surface has a complicated geometry, from which key geometric characteristics must be extracted and characterized for further study.

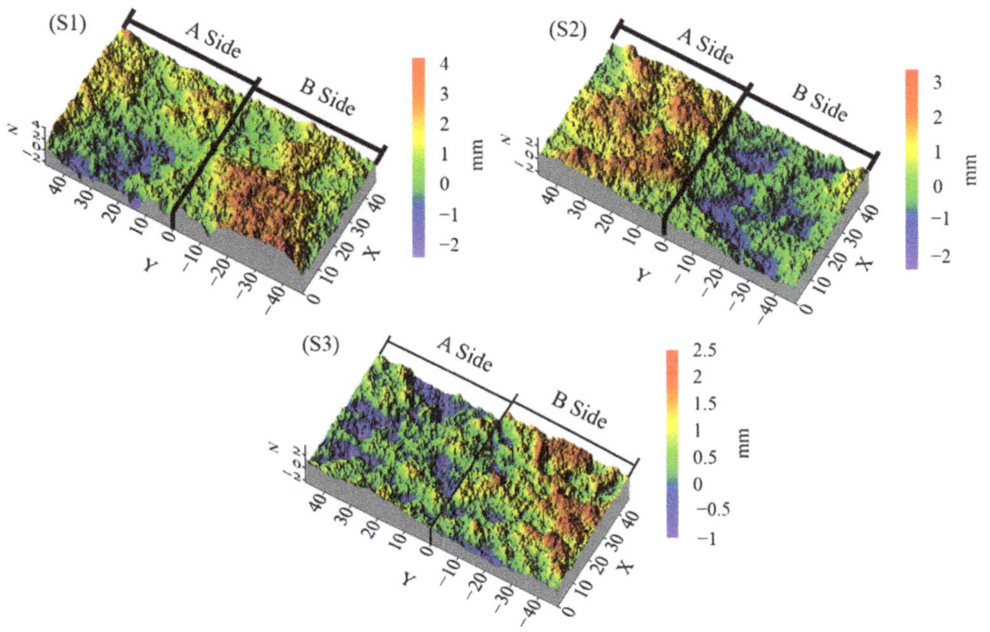

Figure 5. Geometry of concrete samples used in this study.

2.3.2. Analysis of Geometric Characteristics of Fracture Surface

After obtaining the fracture surface geometry in Section 2.3.1, the geometric characteristics were analyzed. Based on previous studies [20,25] on the relationship between fracture geometric characteristics and hydraulic characteristics, roughness is an important indicator affecting hydraulic characteristics [20]; it was, therefore, first necessary to accurately identify the surface roughness [37]. The joint roughness coefficient (JRC) is a basic parameter proposed by Barton [38] to describe the roughness fluctuation of a fracture surface and is widely used in engineering [39]. The JRC values can be calculated with a dimensionless parameter Z_2 [40], widely discussed, and defined as follows [40,41]:

$$Z_2 = \frac{1}{L}\int_{x=0}^{x=L}\left(\frac{dz}{dx}\right)^2 = \sqrt{\frac{1}{n}\sum_{i=1}^{n}\left(\frac{Z_{i+1}-Z_i}{X_{i+1}-X_i}\right)^2} \tag{1}$$

Subsequently, the JRC values can be evaluated by [40]:

$$JRC = 32.2 + 32.47 lg Z_2 \tag{2}$$

where, L is the nominal length of the profile, X_i is the ith segment of L, Z_i is the amplitude of the roughness of the profile. According to the definition of the JRC, it is initially the quantization parameter of the profile (2D). The JRC values of all profiles on the same fracture surface by scanning interval can be calculated and averaged to characterize the JRC value of the fracture surface (3D).

It can be seen from Table 1 that the JRC values of the fracture surfaces of the A- and B-side from the same pair of samples are almost the same, but not exactly consistent. This is because the fracturing process of concrete is often accompanied by particle disintegration, so the two surfaces of the A- and B-side cannot be completely consistent, and the JRC values are not exactly the same.

Table 1. Initial JRC values for sample surfaces.

Fracture Surface Number	JRC (0 h)
S1A	22.0
S1B	20.5
S2A	20.4
S2B	20.0
S3A	20.4
S3B	21.0

According to Equations (1) and (2) and the calculation method of the 3D fracture surface JRC, the JRC value of the 3D fracture surface represents the macro roughness of the fracture surface. In order to obtain a detailed description of the fracture surface at the meso level, additional parameters describing the geometric characteristics were needed. Therefore, the distribution of slope [42] was added to the description of the fracture surfaces at the meso level. The slope was defined as follows [42]:

$$Slope = \frac{dz}{dx} = \frac{Z_{i+1} - Z_i}{X_{i+1} - X_i} \quad (3)$$

It can be seen that the slope is one segment unit in the JRC calculation, which represents the rough fluctuation of a segment unit in the fracture surface. Correspondingly, the JRC is the averaging of the slopes of all segment units on the fracture surface. Therefore, the JRC and slope distribution can describe the roughness characteristics of the fracture surface more comprehensively at the macro and meso levels, respectively. Figure 6 shows the slope distribution of the fracture surfaces, where black represents the A-sides and red represents the B-sides.

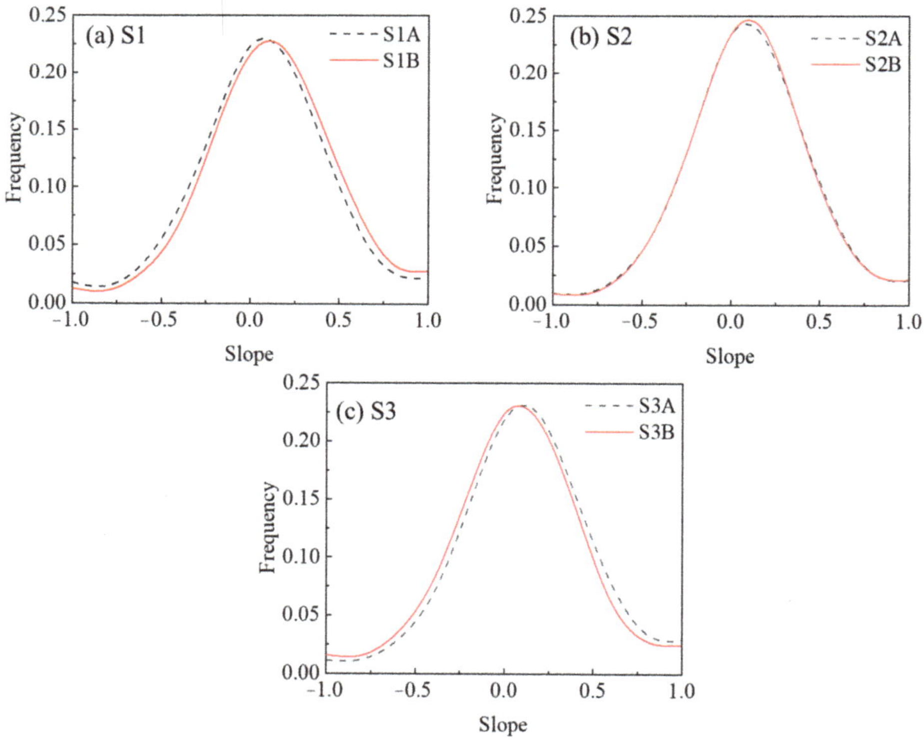

Figure 6. Distribution of surface slope.

Figure 6 shows that the slope distribution of the fracture surfaces exhibit Gaussian distribution. The slope of S1 is −0.037 ± 0.485 (for S1A) and 0.019 ± 0.436 (for S1B). The slope of S2 is −0.002 ± 0.433 (for S2A) and −0.002 ± 0.422 (for S2B). The slope of S3 is 0.020 ± 0.434 (for S3A) and −0.028 ± 0.450 (for S3B).

2.4. Permeability Test of Single Fracture

For the purpose of achieving the hydraulic characteristics of the fractures, we used a self-developed permeability experimental setup to test permeability of fractures (see Figure 7). The cell pressure (i.e., confining pressure) range was 0–80 MPa and the accuracy was 0.01 MPa. The water pressure difference (i.e., pressure difference between inlet- and outlet- pressure) range was 0~16 MPa and the accuracy was 0.01 MPa. These pressures were controlled by computer. The concrete sample was saturated according to the operational standard in order to remove air from the samples. After gently washing the fracture surfaces with deionized water, the two half-concrete samples were fitted carefully together to avoid introducing small debris into the fracture. Samples were marked to ensure that the two half-concrete samples matched equally between different experiments. The fitted sample was then confined within the rubber membrane by iron hoops. Filter paper was sandwiched between the sample and the cushion block to prevent small debris from blocking the outlet pipe, and the sample was then placed into the pressure cell (see Figure 8). Each concrete sample with a single fracture was subjected to a confining pressure of 2 MPa and a water pressure difference of 0.2 MPa, and monitored by computer to accurately maintain the specified pressure. The measurable flow rate was obtained in the process. The only liquid used in the permeability experiments was also deionized water.

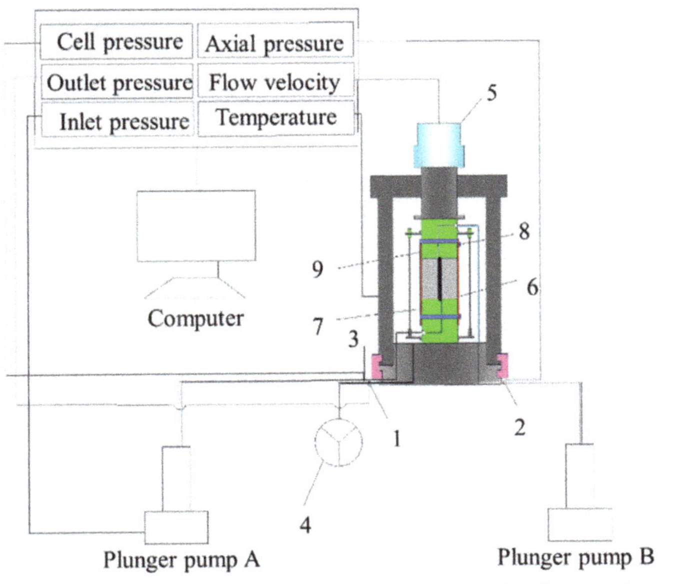

1-Cell pressure transducer; 4-Hydraulic pump; 7-Rubber membrane;
2-Outlet pressure transducer; 5-Top piston; 8-Iron hoop;
3-Inlet pressure transducer; 6-Filter paper; 9-Cushion block.

Figure 7. Schematic of experimental setup for permeability experiment.

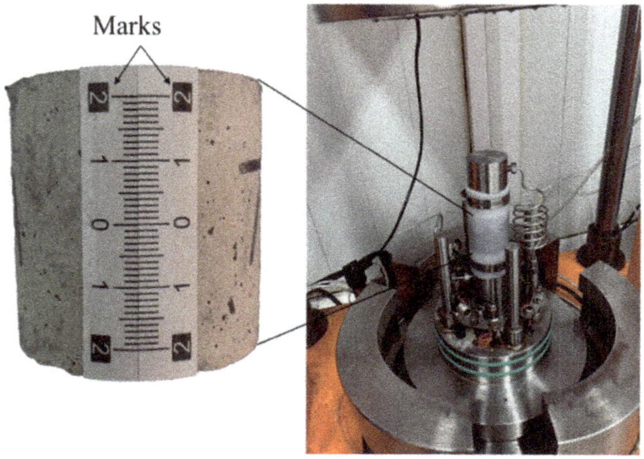

Figure 8. Sample loading.

The linear Darcy's law was used to describe the laminar fracture flow at low velocity, given by [43]:

$$k = \frac{\mu QL}{A\nabla P} \tag{4}$$

where k is the intrinsic permeability, μ is the fluid viscosity, Q is the flow rate, A is the cross-sectional area, and ∇P is the differential pressure. Similarly, the fracture permeability was evaluated via the hydraulic aperture obtained using a parallel plate approximation, namely [44]:

$$b_n = \left(\frac{12\mu LQ}{W\nabla P}\right)^{1/3} \tag{5}$$

$$k = \frac{b_n^2}{12} \tag{6}$$

where b_n is the hydraulic aperture and W is the sample width.

3. Results and Analyses

3.1. Variations in the Fracture Geometric Characteristics

Through the techniques and methods introduced in Section 2.3, the evolution results of the geometric characteristics of concrete fracture surfaces after the leaching experiments were summarized in universal laws and analyzed for causes.

3.1.1. Evolution of Fracture Surface Geometry

First, since the evolution laws are consistent, the S2 fracture surface was selected as an example to show its variation in geometry.

In each group surface variation shown in Figure 9, the upper half is the initial fracture surface before the leaching experiment (0 h), and the lower half is the fracture surface after the experiment (480 h). The middle part shows the intact fracture surface, and the left and right parts are enlarged images of dotted boxes to facilitate the observation of detailed geometry. The legend indicates the elevation of the fracture surface. The red wireframe in the figure shows the local areas before the experiment, and the white (or black) wireframe shows the changed area after the experiment. It can be seen that the red areas (several areas with high elevation) decrease in size and fade in color after leaching, while the blue areas (several areas with low elevation) increase in size and deepen in color after leaching. This shows the degraded depth of the fracture surface elevation, the surface solid components of the fracture surface (such as calcium hydroxide) being removed by the leaching experiment; that is, the leaching phenomenon was confirmed in the fracture surface geometry.

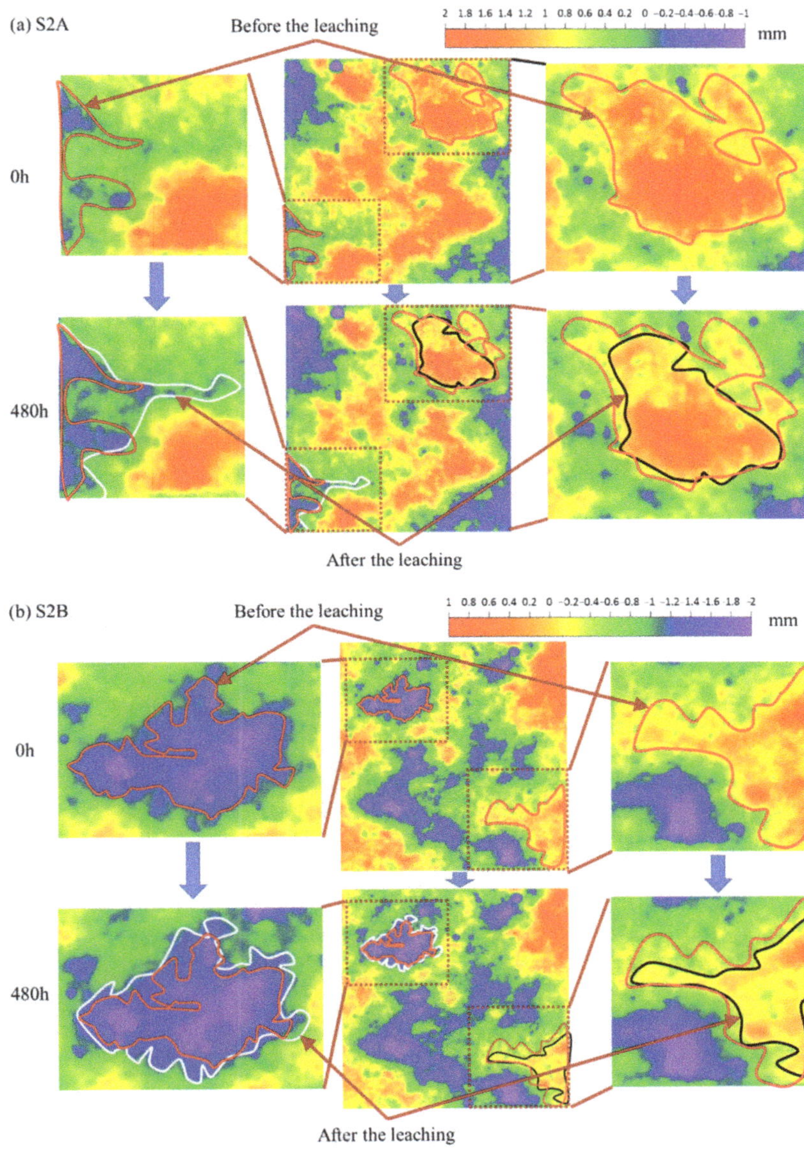

Figure 9. Temporal evolution in fracture surface geometry variation.

Second, in order to further obtain the degraded depth of the fracture surface elevation and summarize the universal law, the elevation distributions of all fracture surfaces are shown in Figure 10, where, the black line indicates the distribution of elevation before the leaching experiment and the red line indicates the distribution after the leaching experiment.

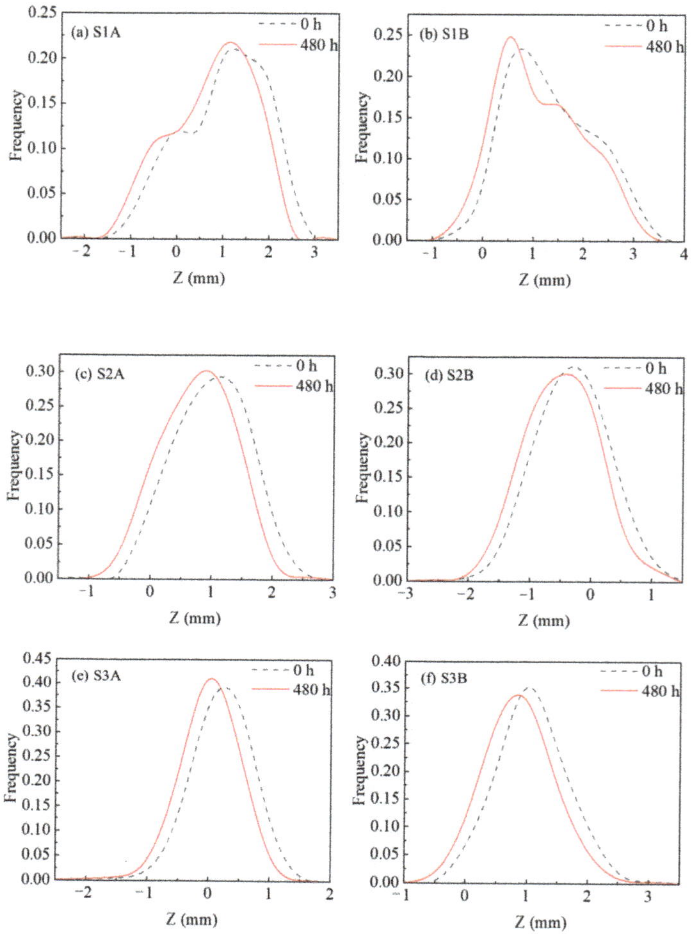

Figure 10. Temporal evolutions in elevation distribution variation.

It can be seen from Figure 10 that the peak value of the elevation distribution curve of all fracture samples moves to the left; that is, the macroscopic elevation of the fracture surface shows a declining law, which proves the results shown in the 2D cloud above (see Figure 9), and also shows that the influence of leaching on the morphology of the fracture surface has a universal law. The degraded depth of the fracture surface is called the leached depth [5,10].

3.1.2. Evolutions of Fracture Surface Macro Roughness

In order to study the evolution of fracture surface roughness, the JRC values of three pairs of samples at 0 h, 120 h, 240 h, 360 h and 480 h are listed in Table 2.

Table 2. Fracture surface JRC values at different leaching time.

Fracture Surface Number	JRC				
	0 h	120 h	240 h	360 h	480 h
S1A	22.0	21.5	21.3	21.3	21.2
S1B	20.5	20.5	20.0	19.9	19.5
S2A	20.4	20.0	19.3	19.2	19.0
S2B	20.0	19.7	19.4	19.2	18.9
S3A	20.4	20.2	19.8	19.6	19.6
S3B	21.0	20.9	20.8	20.5	20.4

It can be seen from Table 2 that the JRC value of the fracture surface in each test decreases with the increase in leaching time. It is not appropriate to compare the JRC between different tests in Table 2, because the concrete samples are different in each test. Therefore, in order to compare the sensitivity of leaching time to roughness, the ratios of the JRC relative to initial conditions (0 h) (i.e., normalized JRC) are calculated and shown in Figure 11.

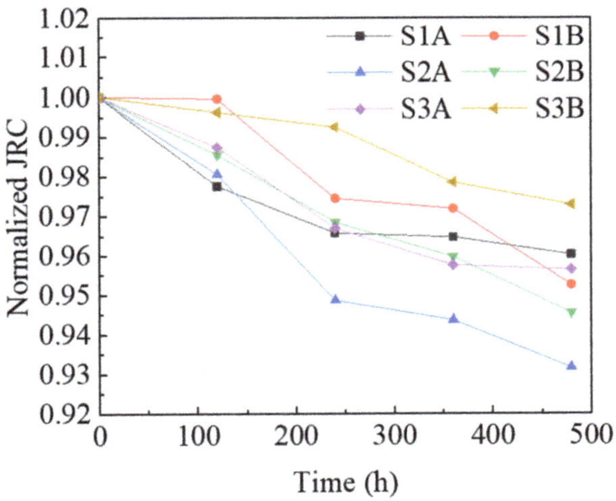

Figure 11. Temporal evolutions in fracture surface normalized JRC variation.

Figure 11 shows that S1A decreased by 3.97%, S1B decreased by 4.74%, S2A decreased by 6.83%, S2B decreased by 5.45%, S3A decreased by 4.35% and S3B decreased by 2.71% from 0 h to 480 h. These findings are consistent with those of Duan et al. [31] in that the roughness of the limestone fracture shows a decreasing trend after long-term leaching in deionized water.

It can be seen from Figure 10 that although there are exceptions (S2B and S3B), the elevation of fracture surfaces decreases unevenly and shows a more centralized trend. The peak value of fracture surface elevation distribution after the leaching experiment moves to a lower elevation than that before the experiment. The peak value is larger, that is, the frequency is larger. The centralization of elevation indicates that the fracture surface elevations decrease unevenly, but some of the higher elevations decrease more and some of the lower elevations decrease less. Therefore, the fracture surface tends to be flattened, with mainly the raised part of the fracture being flattened. According to the meaning of roughness (JRC), when the fracture surface geometric characteristics tend to be flat, its roughness decreases. This explains very well the continuous decrease in JRC seen in Figure 11. The evolution in roughness (JRC) is still at the macro level, and we need to continue to investigate the mechanism of the effect of leaching on the fracture surface characteristics at the meso level of the detailed part of the fracture surface.

3.1.3. Evolutions of Fracture Surface Local Details

In order to further study the evolution of the fracture surface details at the meso level and investigate the mechanism of the uneven variation of the fracture surface elevation, we first studied the evolution of the slope distribution of the fracture surface, as shown in Figure 12.

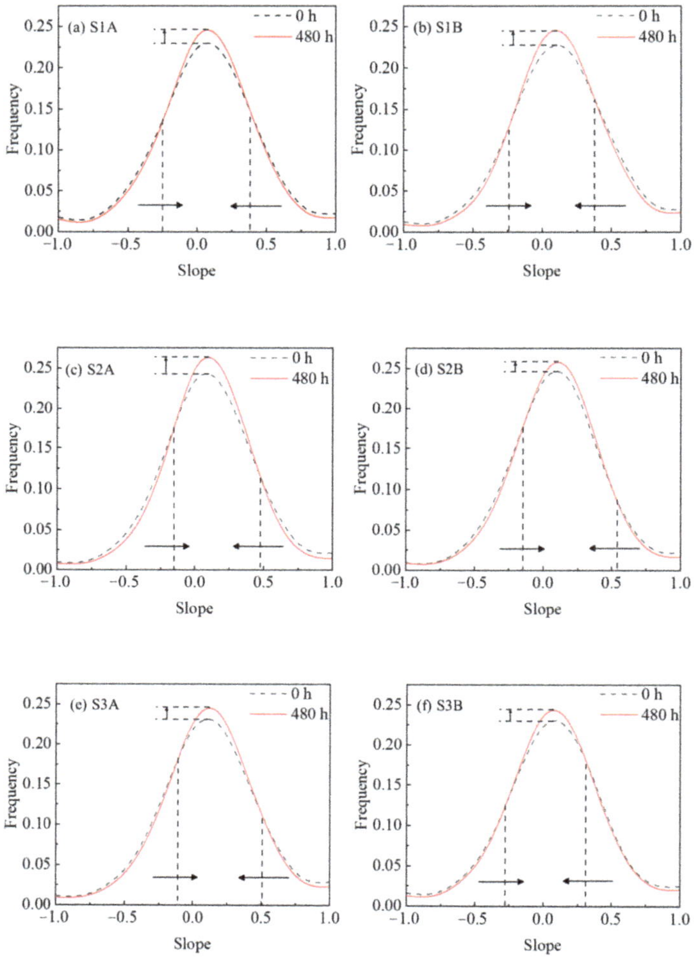

Figure 12. Temporal evolutions in slope distribution variation.

Where, black indicates the slope distribution of the fracture surface before the experiment, red indicates the slope distribution after the experiment, and the arrows point to the evolution trend in the slope distribution. It can be seen in Figure 12 that the area with a larger absolute value of slope on the fracture surface of all samples decreases and the area with a smaller absolute value of slope increases; furthermore, the slope distribution after the leaching experiment is more centralized towards the slope with a lower absolute value. Further proof can be obtained from the standard deviation of the slope distribution of the fracture surface from the start to the end of experiment. The standard deviation of S1A decreases from 0.485 to 0.455, that of S1B decreases from 0.436 to 0.407, that of

S2A decreases from 0.433 to 0.392, that of S2B decreases from 0.422 to 0.391, that of S3A decreases from 0.434 to 0.407, and that of S3B decreases from 0.450 to 0.432. The lower the standard deviation, the more centralized the distribution. From the centralization of slope distribution, it can be inferred that asperities with higher slopes were flattened to lower slopes by the leaching effect.

Finally, to verify this mechanism more visually, we fit the two fracture surface models from same pair of samples. According to the triangular stability principle, ignoring the deformation of fracture surface, it is assumed that there are only three contact points between the two fracture surfaces which intercept the profile along the seepage direction (see Figure 13).

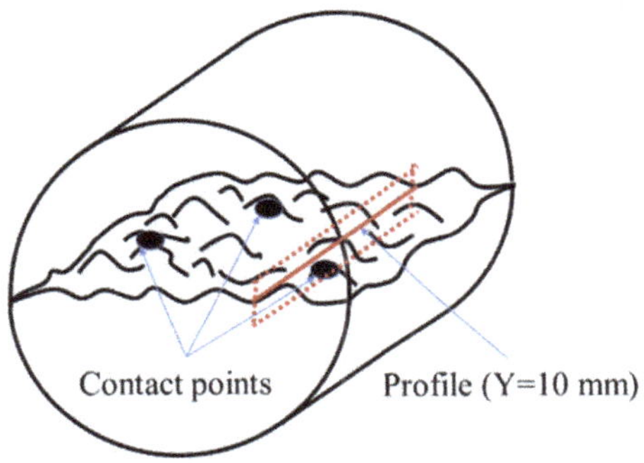

Figure 13. Illustration of fitting the two fracture surfaces and intercepting the profile.

The profile of S2 at Y = 10 mm was selected as an example for presentation and analysis.

It can be seen from Figure 14 that after the fracture surfaces of the same sample were fitted together, the initial profiles of A- and B-side surfaces are not exactly coincident (as outlined by the solid line in the figure), which is also consistent with the phenomenon that the JRC values of the A- and B-side surfaces are not same (see Table 1); this proves the existence of an initial aperture in the fracture, providing a channel for fluid flow. The profiles are shifted backward after being subjected to the leaching experiment (with the A-side surface at the bottom, the profile is shifted downward; with the B-side surface at the top, the profile is shifted upward), which is consistent with the evolution law of the elevation decrease of the fracture surface (leached depth) in Figure 10. However, the average value of the backward shift of the fracture profile is 63.8% of the average of leached depth. The leached depth is produced by the leaching of the cement component in the concrete, which also leads to the phenomenon of the profile on the same side shifting backward (solid to dashed line). It should be noted that when the fracture surface is fitted, especially when the permeability is being measured (the confining pressure is applied), the two surfaces of the A- and B-side will be in contact with each other. Therefore, the degraded depth of the fracture surface evolution (leached depth) is not completely consistent with the backward distance of the fracture profile after fitting. However, the backward distance will not be completely cancelled because concrete is a special material with a combination of many components. As mentioned in Section 2.1, regarding the preparation of samples, concrete mainly includes cement, water and sand particles, so sand particles and gravel will be embedded into the fracture surface of concrete (as shown in Figure 15). The leaching property of a sand particle is different from that of cement. The property of a sand particle is more stable, the leaching rate is low, and it is not easily dissolved. Therefore, when fitting fracture surfaces, these sand particles will become the support of the aperture and prevent the fracture surface from getting bigger. In terms of the profile, although there is backward distance, the amount of backward distance is not as large as the leached depth. We, therefore, refer to this backward distance as the residual leached depth after the sample is fitted.

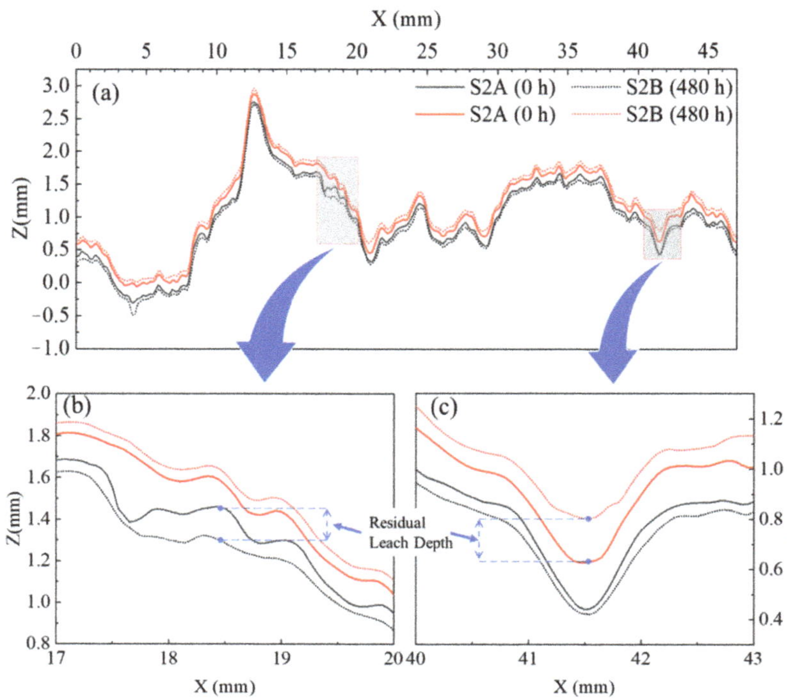

Figure 14. Temporal evolutions in profile variation. (**a**) profile of Y = 10 mm, (**b**) enlarged view of X = 17–20 mm and (**c**) enlarged view of X = 40–43 mm.

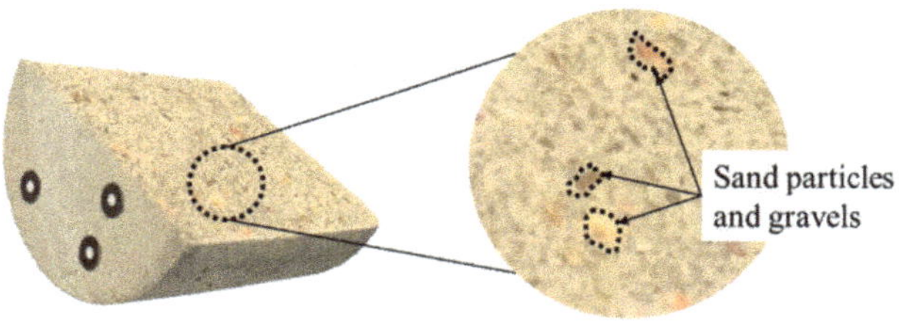

Figure 15. Illustration of sand particles and gravel embedded into the fracture surface.

It can be seen from the enlarged view (see Figure 14b,c) that, in addition to the residual leached depth of the profile, the profile local asperity with a large slope becomes smoother and the slope decreases after the leaching experiment. This directly proves the inference that the large slope of the asperity is flattened into a low slope, and this is caused by the leaching experiment; it also reveals the direct reason for the decline law of the JRC value of fracture surface. Moreover, it intuitively shows the evolution of the slope distribution in Figure 12, and further verifies the analysis of the evolution law of fracture surface elevation.

So far, the evolution law of fracture surface geometric characteristics has been observed and analyzed at the macro level (JRC) and the meso level (slope distribution and profile).

3.2. Variations in Fracture Permeability

In the permeability experiment, the samples were tested for different leaching times (0 h, 120 h, 240 h, 360 h and 480 h). In order to obtain the universal law, three groups of tests with the same conditions were carried out, specifically, with a confining pressure of 2 MPa and a seepage pressure difference of 0.2 MPa. The direct results of the experiment, the steady flow rate, are recorded in Table 3.

Table 3. Experimental results of flow rate.

Sample Number	Flow Rate (mL/min)				
	0 h	120 h	240 h	360 h	480 h
S1	3.74	4.96	7.60	8.41	9.22
S2	3.28	3.86	4.17	5.08	6.51
S3	3.00	5.42	5.94	6.46	7.22

It is seen in Table 3 that, under the same conditions (confining pressure 2 MPa, seepage pressure difference 0.2 MPa) after the leaching experiment, the flow rate in the fracture increases. S1 increases by 146.4%, S2 increases by 98.6% and S3 increases by 140.7%. The hydraulic aperture (b_n) and permeability (k) of each group of samples, according to Equations (1)–(3) introduced in Section 2.4, are shown in Tables 4 and 5, respectively.

Table 4. Calculated hydraulic aperture at different leaching times.

Sample Number	b_n (m)				
	0 h	120 h	240 h	360 h	480 h
S1	1.55×10^{-5}	1.7×10^{-5}	1.97×10^{-5}	2.03×10^{-5}	2.1×10^{-5}
S2	1.49×10^{-5}	1.57×10^{-5}	1.61×10^{-5}	1.72×10^{-5}	1.87×10^{-5}
S3	1.44×10^{-5}	1.76×10^{-5}	1.81×10^{-5}	1.86×10^{-5}	1.93×10^{-5}

Table 5. Calculated permeability at different leaching times.

Sample Number	k (m^2)				
	0 h	120 h	240 h	360 h	480 h
S1	2.01×10^{-11}	2.42×10^{-11}	3.22×10^{-11}	3.45×10^{-11}	3.66×10^{-11}
S2	1.84×10^{-11}	2.05×10^{-11}	2.16×10^{-11}	2.46×10^{-11}	2.9×10^{-11}
S3	1.73×10^{-11}	2.57×10^{-11}	2.73×10^{-11}	2.89×10^{-11}	3.11×10^{-11}

Similar to the normalized JRC in Section 3.1.2, it is not appropriate to compare the hydraulic characteristics between different tests in Tables 4 and 5, because the concrete samples have different initial mechanical apertures. Therefore, in order to compare the sensitivity of leaching time to the hydraulic characteristics, the ratios of hydraulic aperture and permeability, relative to initial conditions (0 h), i.e., normalized hydraulic aperture and normalized permeability, were calculated and are shown in Figure 16.

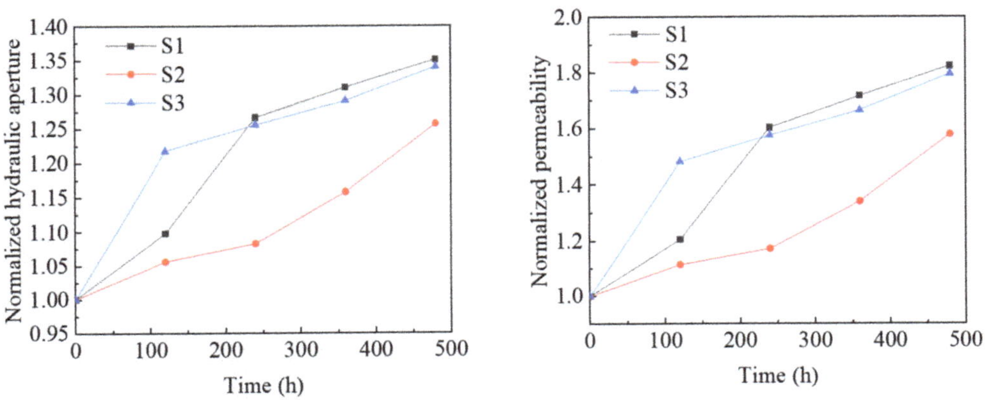

Figure 16. Temporal evolutions in fracture normalized hydraulic aperture (**left**) and permeability (**right**) variation.

It can be seen from Figure 16 that the curves for the evolutions of hydraulic aperture and permeability present the same shape, but the increase rate of permeability is much higher than that of hydraulic aperture. As the leaching time increases from 0 to 480 h, the hydraulic apertures of S1, S2 and S3 increase by 35%, 26%, 34%, while the permeability of S1, S2 and S3 increases by 82%, 58%, 80%, respectively. Since fracture is the main channel of fluid flow, combined with the evolution of fracture characteristics in the Section 3.1, it can be considered that the evolution of fracture hydraulic characteristics is closely related to the evolution of fracture geometric characteristics. The evolution of the geometric characteristics of the fracture surface after the leaching experiment was summarized and analyzed in Section 3.1, and the concept of residual leached depth was presented. The generation and development of the residual leached depth widens the seepage channel of the fracture and increases the hydraulic aperture. Meanwhile, the roughness and slope of the fracture surface also changes (the JRC value decreases and the slope distribution concentrates to a low absolute value) due to the local asperity of the fracture surface being flattened, caused by the leaching experiment. This will weaken the resistance effect of a rough sidewall on the fluid flow in the fracture, further improving the permeability of the fracture, and increasing the hydraulic characteristics of the fracture.

In order to validate this analysis, the fluid in the fracture was simulated on the profile (S2, Y = 10 mm), shown in Figure 14.

Incompressible Newtonian flow is governed by the well-known Navier–Stokes equations [43,45,46]:

$$\rho\left(\frac{\partial U}{\partial t}+U\cdot\nabla U\right)=-\nabla P+\mu\nabla^2 U+F \tag{7}$$

where ρ is the fluid density, U is the velocity vector of flow particle and F is the body force vector. Because the Navier–Stokes equation cannot be solved directly, the numerical method is widely used to solve it [47–49], and to simulate and study the flow characteristics of a fluid in rough fractures [50]. In this paper, Fluent 16.0 was used to simulate and analyze the seepage in rough fractures. Many scholars [46,51,52] verified the reliability of the calculation software. The inlet on the left-hand side of the fracture model was set as the velocity inflow boundary: the velocity magnitude was 1 m/s and the free outlet boundary was on the right-hand side. The upper and lower boundaries of the fracture model were set as the wall boundary conditions without fluid flow or slip.

Figure 17a,b shows the velocity distribution in the X direction of the fluid in the fracture before and after the leaching experiment. The seepage channel has been widened, and the hydraulic aperture and permeability have increased. The simulation results show that the hydraulic aperture increased by 82% and the permeability increased by 231%. Figure 17c,d is a local enlarged view, showing the seepage at the asperity where the local slope decreases at X = 17−20 mm. It can be seen that the resistance of the sidewall to the fluid is weakened.

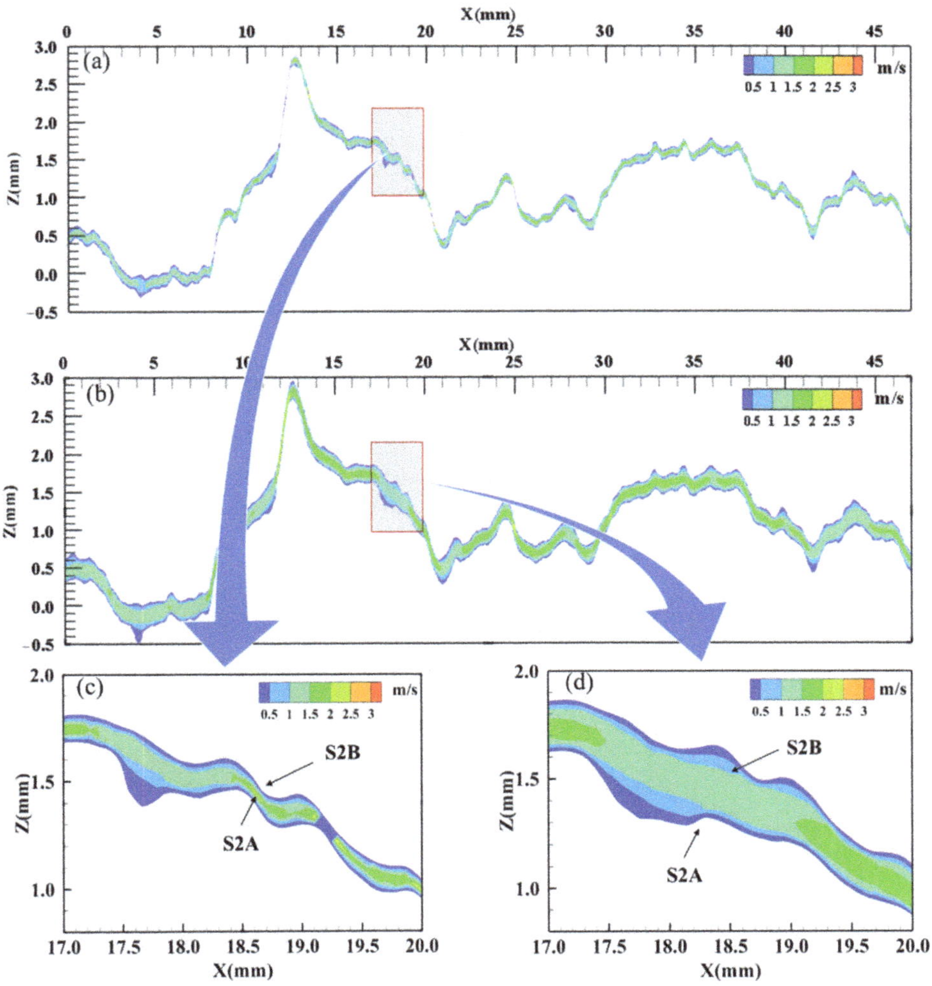

Figure 17. Numerical simulation of fluid flow in actual fracture model by Fluent 16.0. (**a**) Fracture before leaching, (**b**) fracture after leaching, (**c**) enlarged view of X = 17−20 mm before leaching and (**d**) enlarged view of X = 17−20 mm after leaching.

Since Figure 17 shows the actual profile of fracture before and after leaching experiment, both the mechanical aperture and roughness are variables. Therefore, in order to further analyze the influence of the evolution of roughness on the fluid in the fracture, the profile at Y = 10 mm and X = 17–20 mm on S2A, before and after leaching, were selected as the wall boundaries with different roughnesses, respectively; the profiles were horizontally translated by the same 0.2 mm aperture to establish an idealized fracture model. In this model, the influence of varying roughness on fracture seepage can be studied.

It can be seen from Figure 18 that the streamline in Figure 18a is significantly more tortuous than that in Figure 18b, and the streamline at the center of the aperture in Figure 18a is 3.65% longer than that in Figure 18b. The simulation results show that the hydraulic aperture of Figure 18b increases by 46% and the permeability increases by 112% compared with Figure 18a. This shows that when the mechanical aperture is the same, the decrease in fracture surface roughness leads to the increase in fracture permeability. The numerical simulation results are consistent with the experimental results, and validate the analysis of the evolution mechanism of fracture hydraulic characteristics well. Moreover, the results of the experiment and the numerical simulation agree well with the empirical equation according to a large number of experiments presented by Barton et al. [20,53]:

$$b_n = \frac{b_m^2}{JRC^{2.5}} \tag{8}$$

where b_m is mechanical aperture. Equation (8) shows that in the case of a constant mechanical aperture, as the JRC increases, the hydraulic aperture also increases.

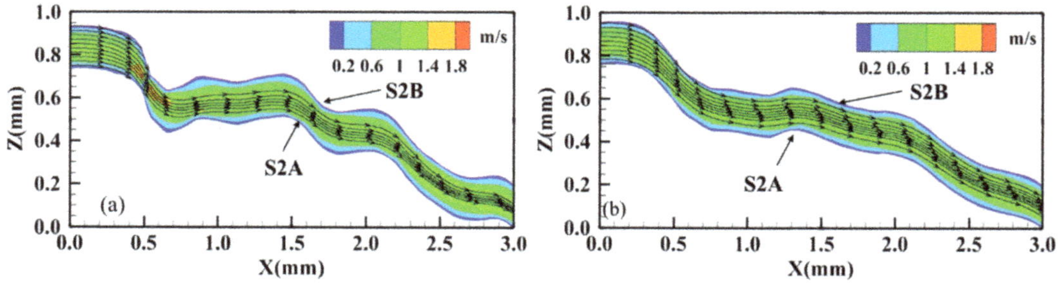

Figure 18. Numerical simulation of fluid flow in idealized fracture model (equal mechanical aperture) by Fluent 16.0. (**a**) fracture before leaching and (**b**) fracture after leaching.

4. Discussion

4.1. Mechanism of Rough Fracture Surface Leaching

To further investigate the mechanism of the effect of leaching on the geometric characteristics of the fracture surface, we measured the elements and concentrations of the deionized water, soaked during the leaching experiment, by inductively coupled plasma mass spectrometry (ICP-MS). Since the initial deionized water contained no other element, the measured element concentrations should originate from the concrete samples only. Figure 19 shows the chemical composition analysis results and pH values for a period of time after the replacement of the deionized water.

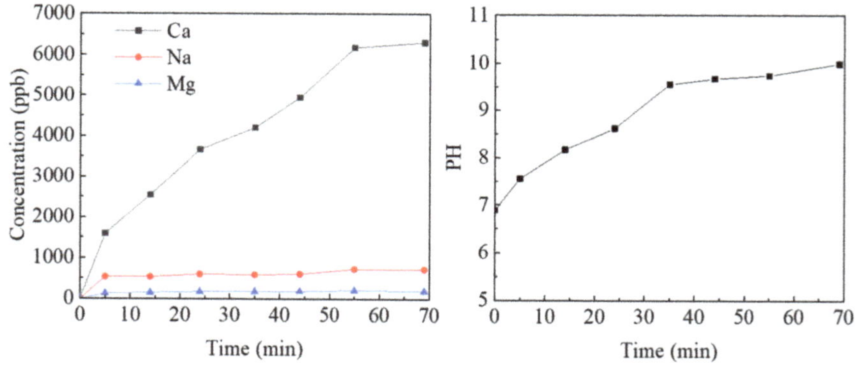

Figure 19. The analysis of chemical composition (**left**) and PH value (**right**) in leaching solution.

From the graph, it can be seen that the calcium ion concentration changes to the greatest extent and is consistent with the pH trend (R = 0.957). Referring to previous studies [4,11,54], it can be assumed that the main chemical reaction occurring in concrete leaching experiment is:

$$Ca(OH)_2 = Ca^{2+} + 2(OH)^- \tag{9}$$

That is, mainly the cement components in concrete undergo a leaching reaction, so the following calcium concentration is the main object of study. Figure 19 shows that the gradient of the calcium concentration curve is initially large, then small and, finally, stabilizes when it reaches 60–70 min. It indicates that the leaching reaction rate is from fast to slow and reaches the near equilibrium state at the later stage. The mechanism of the leaching reaction can be inferred from this. The initial deionized water has a low calcium concentration, so the gradient in calcium concentration between the water and the fracture solid surface is large, and the rate of leaching is fast. As the leaching reaction proceeds, the calcium concentration continues to rise, and the calcium concentration reaches a higher level at the later stage (after 60 min) and remains stable: the leaching reaction is basically balanced, and the PH value also tends to be stable. When the calcium concentration in deionized water is very low, the calcium concentration gradient is large, which is conducive to driving the rapid precipitation reaction of calcium hydroxide components on the concrete surface. When the calcium concentration in the leaching solution increases and the gradient of calcium concentration decreases, the leaching reaction rate of calcium hydroxide decreases and the reaction between solid and liquid reaches equilibrium, inhibiting leaching from proceeding. Thus, the existence of the calcium concentration gradient between the concrete fracture solid surface and the leaching solution can be considered as the mechanism of leaching reaction, which agrees well with the mainstream view [13,33].

Based on this theory, the mechanism by which the geometric characteristics of rough fracture surfaces is affected by leaching is further discussed. A rough fracture surface leaching model is proposed based on the profile (see Figure 14), as shown in Figure 20.

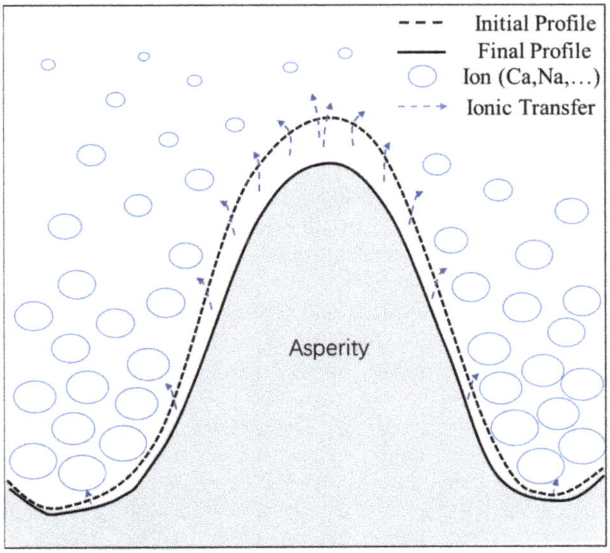

Figure 20. Illustration of concrete rough fracture leaching model.

In Figure 20, the dashed line is the initial profile; the solid line is the final profile after the leaching experiment; the blue circles represent ions or ion clusters (mainly calcium); the size and density indicate ion mass and concentration; the dashed line with arrows indicates the direction of ion transport; and the length of the dashed line indicates the rate of ion transport. Figure 20 shows the asperity in the slope of the raised part of the wide surrounding area; the nearby deionized water flow is great so the leached calcium ions are easily diffused and carried away by the water flow. Therefore, the concentration of calcium ions in the deionized water around the raised part will maintain a low value, and as the ion mass and concentration around the raised part in the model is small, this region maintains a higher gradient in calcium concentration. This, in turn, promotes the raised asperity part to leach further, and the ion transport rate becomes greater, as can be seen in the model. While the asperity in the slope of the depressed part is limited by the surrounding raised part, mobility is poor. Moreover, early by the higher initial calcium concentration gradient effect results in the leaching of calcium ions. Calcium ions tend to aggregate, so the concentration of calcium ions in deionized water around the depressed part will maintain a higher value. As the model around the depressed part of the ion mass and concentration is larger, and this region maintains a lower calcium concentration gradient, it inhibits the depressed asperity part from further leaching, so the ion transport rate in the model is smaller. In summary, the leaching in the raised part is promoted and the leaching in the depressed part is inhibited, so the leaching in the raised part is faster and greater than the leaching in the depressed part. After the leaching experiment, the profile of the raised part moved back more than that of the depressed part (see Figure 14b,c), resulting in the decline of the local slope, that is, the evolution law shown in the slope (see Figure 11). When the whole rough fracture surface is affected by this leaching mechanism, it leads to the reduction of roughness, that is, the evolution law shown in the roughness (see Figure 10). This model reveals effectively the mechanism of the geometric characteristics of the rough fracture surface under the action of leaching at the micro level.

4.2. Mechanism of the Effect of Leaching on Fracture Hydraulic Characteristics

In this paper, we found that fracture permeability increases gradually with the development of leaching. In the previous sections, we investigated the impact of leaching on the geometric characteristics of fracture surfaces, which leads to the evolution of fracture hydraulic characteristics. That is, the rough fracture surface is leached, and the surface elevation decreases; the residual leached depth is generated after the samples were fitted together, resulting in an increase in aperture. Moreover, the roughness of the fracture surface (JRC) decreases and the effect of inhibiting flow also decreases. These actions together lead to the increase in the hydraulic aperture and the permeability of fractures.

However, the law that the decrease in fracture surface roughness leads to the increase in permeability cannot be simply understood as the lower the roughness, the higher the permeability, especially between different samples. We observed that the initial average JRC of S1, S2 and S3 on both surfaces of the A- and B-sides are 21.3, 20.2, 20.7, but the hydraulic aperture is 1.55×10^{-5} m, 1.49×10^{-5} m, 1.44×10^{-5} m, respectively, and the permeability is 2.01×10^{-11} m^2, 1.84×10^{-11} m^2,1.73×10^{-11} m^2, respectively. The permeability of S2, which has the smallest JRC, is instead smaller than that of S1, which has the largest JRC. This is because the initial mechanical aperture between them is different (see Equation (8)). Furthermore, the JRC values of the three groups of samples are very close to each other, which makes the influence caused by other chance factors very large, so it is difficult to establish an accurate relationship between the JRC value of a fracture surface and its hydraulic characteristics (hydraulic aperture and permeability). In this study, the main purpose of designing three pairs of samples was to conduct a repeated experiment from which to derive a universal law governing the effect of leaching on the fracture surface. The relationship between the roughness JRC and the hydraulic characteristics obtained is, therefore, the only law presented by the same sample under the effect of leaching. The inability to establish an exact equation for the relationship between the JRC and hydraulic characteristics does not prevent the correctness of the conclusion that the decrease in fracture surface roughness caused by leaching leads to an increase in permeability. This is because it is based on the law observed for the same pair of fracture surfaces at different leaching times and is verified by repeatability experiments and analyzed by numerical simulations.

5. Conclusions

In this study, leaching behavior acting on the fracture was verified through laboratory work. The main conclusions are summarized as follows:

(1) The series of 3D laser scanning test results show that the elevation of fracture surfaces decreases unevenly, the JRC decreases monotonically with leaching time and the slope distribution centralizes to a low absolute value. The solid element of the concrete fracture surface was leached by an aggressive solution (deionized water) and the degree of this leaching is uneven in different areas of the fracture surface. The leached degree of the high slope area is greater than that of the low slope area, and this analysis was verified from the profile.
(2) The hydraulic characteristics and the fluid flow state in the fractures were investigated by using the permeability test combined with the fracture numerical model. The existence of residual leached depth widens the hydraulic aperture of the fracture, and the flattened asperity weakens the resistance to the fluid. This variation in geometric characteristics, caused by leaching, increases the hydraulic aperture and permeability of fractures.
(3) The chemical analysis of the soaked solution in the leaching experiment showed an inevitable linkage between the leached degree and the calcium concentration gradient in the solution. Therefore, a leaching model of concrete rough fracture surface was proposed to describe the mechanism of leaching on the fracture characteristics. The rough surface of the fracture leads to an uneven calcium concentration gradient near the asperity, resulting in a greater degree of leaching in the raised part than that in the depressed part.

This study confirms that the leaching behavior acting on a natural rough concrete fracture exposed to aggressive solution results in a long-term evolution of geometric and hydraulic characteristics. This conclusion is useful for a better detection of the variation in fracture of structural concrete in service from the standpoint of practical applications and for further illustration of hydraulic performance in the long term.

The leaching method used was relatively simple and the proposed model is straightforward because the interactions of normal pressure in the leaching processes are omitted; these are, however, applicable to open fracture without normal pressure in concrete structures. Generally, when considering a closed fracture under normal pressure, the phenomenon of pressure leaching may occur and be significant [44,55–57]. This phenomenon will make the concrete fracture leaching model more complicated but is a longstanding inspiration in the field of rock engineering [57–59]. In the near future, we will improve the current model by taking into account the normal pressure interaction to examine its effects on concrete fracture leaching.

Author Contributions: Conceptualization, Y.W. and M.T.; Data curation, M.T.; Formal analysis, Y.N.; Funding acquisition, Y.W.; Investigation, Y.W., M.T., D.F. and Z.W.; Methodology, Y.J.; Writing—original draft, M.T.; Writing—review & editing, Y.W. and M.T. All authors have read and agreed to the published version of the manuscript.

Funding: The research was funded by the National Natural Science Foundation of China (U1765204, 41772340).

Institutional Review Board Statement: Not applicable.

Informed Consent Statement: Informed consent was obtained from all subjects involved in the study.

Data Availability Statement: The data used to support the findings of this study are included within the article.

Conflicts of Interest: The authors wish to confirm that there are no known conflicts of interests associated with this publication and there has been no significant financial support for this work that could have influenced its outcome.

References

1. Kairong, H. Development and Prospects of Tunnels and Underground Works in China in Recent Two Years. *Tunn. Constr.* **2017**, *37*, 123–134.
2. Shi, L.; Zhou, H.; Gao, Y.; Lu, J.; Li, Q. Experimental study on the acoustic emission response and permeability evolution of tunnel lining concrete during deformation and failure. *Eur. J. Environ. Civ. Eng.* **2020**, *26*, 3398–3417. [CrossRef]
3. Behera, M.; Bhattacharyya, S.; Minocha, A.; Deoliya, R.; Maiti, S. Recycled aggregate from C&D waste & its use in concrete—A breakthrough towards sustainability in construction sector: A review. *Constr. Build. Mater.* **2014**, *68*, 501–516. [CrossRef]
4. De Larrard, T.; Benboudjema, F.; Colliat, J.-B.; Torrenti, J.M.; Deleruyelle, F. Concrete calcium leaching at variable temperature: Experimental data and numerical model inverse identification. *Comput. Mater. Sci.* **2010**, *49*, 35–45. [CrossRef]
5. Faucon, P.; Le Bescop, P.; Adenot, F.; Bonville, P.; Jacquinot, J.; Pineau, F.; Felix, B. Leaching of cement: Study of the surface layer. *Cem. Concr. Res.* **1996**, *26*, 1707–1715. [CrossRef]
6. Kamali, S.; Gérard, B.; Moranville, M. Modelling the leaching kinetics of cement-based materials—Influence of materials and environment. *Cem. Concr. Compos.* **2003**, *25*, 451–458. [CrossRef]
7. Li, K.; Li, L. Crack-altered durability properties and performance of structural concretes. *Cem. Concr. Res.* **2019**, *124*, 105811. [CrossRef]
8. Mainguy, M.; Tognazzi, C.; Torrenti, J.M.; Adenot, F. Modelling of leaching in pure cement paste and mortar. *Cem. Concr. Res.* **2000**, *30*, 83–90. [CrossRef]
9. Burlion, N.; Bernard, D.; Chen, D. X-ray microtomography: Application to microstructure analysis of a cementitious material during leaching process. *Cem. Concr. Res.* **2006**, *36*, 346–357. [CrossRef]
10. Kamali, S.; Moranville, M.; Leclercq, S. Material and environmental parameter effects on the leaching of cement pastes: Experiments and modelling. *Cem. Concr. Res.* **2008**, *38*, 575–585. [CrossRef]
11. Gérard, B.; Le Bellego, C.; Bernard, O. Simplified modelling of calcium leaching of concrete in various environments. *Mater. Struct.* **2002**, *35*, 632–640. [CrossRef]
12. Konečný, P.; Vořechovská, D.; Šomodíková, M.; Horňáková, M.; Rovnaníková, P. Extended evaluation of durability-related field inspection data from concrete bridges under service. *Arch. Metall. Mater.* **2020**, *65*, 81–89. [CrossRef]
13. Wang, J.; Huang, W.-H. Spectral distributions of the scattered photons within an acceptance angle in Thomson scattering. *J. Chin. Silic. Soc.* **2011**, *39*, 525–530. [CrossRef]
14. Anwar, I.; Hatambeigi, M.; Chojnicki, K.; Taha, M.R.; Stormont, J.C. Alteration in micro-mechanical characteristics of wellbore cement fracture surfaces due to fluid exposure. *J. Pet. Sci. Eng.* **2021**, *205*, 108935. [CrossRef]
15. Mengel, L.; Krauss, H.-W.; Lowke, D. Water transport through cracks in plain and reinforced concrete—Influencing factors and open questions. *Constr. Build. Mater.* **2020**, *254*, 118990. [CrossRef]
16. Akhavan, A.; Rajabipour, F. Quantifying Permeability, Electrical Conductivity, and Diffusion Coefficient of Rough Parallel Plates Simulating Cracks in Concrete. *J. Mater. Civ. Eng.* **2017**, *29*, 04017119. [CrossRef]
17. Wang, K.; Jansen, D.C.; Shah, S.P.; Karr, A.F. Permeability study of cracked concrete. *Cem. Concr. Res.* **1997**, *27*, 381–393. [CrossRef]
18. Dong, J.; Ju, Y. Quantitative characterization of single-phase flow through rough-walled fractures with variable apertures. *Géoméch. Geophys. Geo-Energy Geo-Resour.* **2020**, *6*, 1–14. [CrossRef]
19. Li, X.; Li, D.; Xu, Y. Modeling the effects of microcracks on water permeability of concrete using 3D discrete crack network. *Compos. Struct.* **2018**, *210*, 262–273. [CrossRef]
20. Wang, Y.; Su, B.Y. Research on the behavior of fluid flow in a single fracture and its equivalent hydraulic aperture. *Shuikexue Jinzhan Adv. Water Sci.* **2002**, *13*, 61–68.
21. Wang, Y. Coupling Characteristic of Stress and Fluid Flow Within a Single Fracture. *Chin. J. Rock Mech. Eng.* **2002**, *21*, 83–87. [CrossRef]
22. Chen, L.; Wang, Y.; Niu, Y. Influence of Roughness and Fracture Width on the Transition from Darcy to Non-Darcy Behavior for Flow through Rough Single Fracture. *Henan Sci.* **2017**, *35*, 1987–1994.
23. Yonghao, F.; Pubing, A.N.; Wei, Z.; Yikan, C.E.N. Permeating, leaching and autogenous healing of cracks of cementitious materials. *J. Chin. Silic. Soc.* **2008**, *36*, 451–456.
24. Chen, Y.; Cao, P.; Chen, R.; Teng, Y. Effect of water–rock interaction on the morphology of a rock surface. *Int. J. Rock Mech. Min. Sci.* **2010**, *47*, 816–822. [CrossRef]
25. Brown, S.R. Fluid flow through rock joints: The effect of surface roughness. *J. Geophys. Res. Earth Surf.* **1987**, *92*, 1337–1347. [CrossRef]
26. Im, K.; Elsworth, D.; Wang, C. Cyclic Permeability Evolution during Repose Then Reactivation of Fractures and Faults. *J. Geophys. Res. Solid Earth* **2019**, *124*, 4492–4506. [CrossRef]
27. Zoorabadi, M.; Saydam, S.; Timms, W.; Hebblewhite, B. Semi-Analytical Procedure for Considering Roughness Effect on Hydraulic Properties of Standard JRC Profiles. In Proceedings of the 47th U.S. Rock Mechanics/Geomechanics Symposium, San Francisco, CA, USA, 23–26 June 2013.
28. Chen, Y.; Liang, W.; Lian, H.; Yang, J.; Nguyen, V.P. Experimental study on the effect of fracture geometric characteristics on the permeability in deformable rough-walled fractures. *Int. J. Rock Mech. Min. Sci.* **2017**, *98*, 121–140. [CrossRef]
29. Brunet, J.-P.; Li, L.; Karpyn, Z.T.; Kutchko, B.G.; Strazisar, B.; Bromhal, G. Dynamic Evolution of Cement Composition and Transport Properties under Conditions Relevant to Geological Carbon Sequestration. *Energy Fuels* **2013**, *27*, 4208–4220. [CrossRef]

30. Wang, K.; Sheng, J.C.; Hui-Cai, G.; Xiao-Dan, T.; Luo, Y.L. Study on seepage characteristics of rough crack under coupling of stress-seepage erosion. *Yantu Lixue Rock Soil Mech.* **2020**, *41*, 30–40. [CrossRef]
31. Duan, L.; Deng, H.; Qi, Y.; Li, G.; Peng, M. Study on the evolution of seepage characteristics of single-fractured limestone under water-rock interaction. *Rock Soil Mech.* **2020**, *41*, 3671–3679.
32. Shiyong, W.U.; Xuhua, R.E.N.; Xiangrong, C.; Jixun, Z. Stability analysis and supporting design of sur-rounding rocks of diversion tunnel for jinping hydropower station. *Chin. J. Rock Mech. Eng.* **2005**, *24*, 3777–3782.
33. Zhang, K.; Shen, Z.; Gan, L. Advances in cement-based materials leaching test. *Adv. Sci. Technol. Water Resour.* **2018**, *38*, 86–94.
34. Yang, H.; Jiang, L.H.; Zhang, Y.; Yu, D.U.; Zhou, X.M. Summary of researches on chemical damage of concrete based on leaching process. *Adv. Sci. Technol. Water Resour.* **2011**, *55*, 371.
35. Ulusay, R. *The ISRM Suggested Methods for Rock Characterization, Testing and Monitoring: 2007–2014*; Springer International Publishing: Cham, Switzerland, 2014; Volume 15, pp. 47–48.
36. Xie, H.; Wang, J.-A.; Xie, W.-H. Photoelastic study of the contact mechanics of fractal joints. *Int. J. Rock Mech. Min. Sci.* **1997**, *34*, 865–874. [CrossRef]
37. Song, C.; Nakashima, S.; Kido, R.; Yasuhara, H.; Kishida, K. Short- and Long-Term Observations of Fracture Permeability in Granite by Flow-Through Tests and Comparative Observation by X-ray CT. *Int. J. Géoméch.* **2021**, *21*, 04021151. [CrossRef]
38. Barton, N. Review of a new shear-strength criterion for rock joints. *Eng. Geol.* **1973**, *7*, 287–332. [CrossRef]
39. Liu, X.; Zhu, W.; Yu, Q.; Chen, S.; Guan, K. Estimating the Joint Roughness Coefficient of Rock Joints from Translational Overlapping Statistical Parameters. *Rock Mech. Rock Eng.* **2018**, *52*, 753–769. [CrossRef]
40. Tse, R.; Cruden, D. Estimating joint roughness coefficients. *Int. J. Rock Mech. Min. Sci. Géoméch. Abstr.* **1979**, *16*, 303–307. [CrossRef]
41. Yong, R.; Ye, J.; Liang, Q.-F.; Huang, M.; Du, S.-G. Estimation of the joint roughness coefficient (JRC) of rock joints by vector similarity measures. *Bull. Eng. Geol. Environ.* **2017**, *77*, 735–749. [CrossRef]
42. Belem, T.; Homand-Etienne, F.; Souley, M. Quantitative Parameters for Rock Joint Surface Roughness. *Rock Mech. Rock Eng.* **2000**, *33*, 217–242. [CrossRef]
43. Zhang, Z.; Nemcik, J. Fluid flow regimes and nonlinear flow characteristics in deformable rock fractures. *J. Hydrol.* **2013**, *477*, 139–151. [CrossRef]
44. Yasuhara, H.; Kinoshita, N.; Ohfuji, H.; Takahashi, M.; Ito, K.; Kishida, K. Long-term observation of permeability in sedimentary rocks under high-temperature and stress conditions and its interpretation mediated by microstructural investigations. *Water Resour. Res.* **2015**, *51*, 5425–5449. [CrossRef]
45. Lee, S.H.; Yeo, I.W.; Lee, K.-K.; Detwiler, R.L. Tail shortening with developing eddies in a rough-walled rock fracture. *Geophys. Res. Lett.* **2015**, *42*, 6340–6347. [CrossRef]
46. Javadi, M.; Sharifzadeh, M.; Shahriar, K. A new geometrical model for non-linear fluid flow through rough fractures. *J. Hydrol.* **2010**, *389*, 18–30. [CrossRef]
47. Boutt, D.F.; Grasselli, G.; Fredrich, J.T.; Cook, B.K.; Williams, J.R. Trapping zones: The effect of fracture roughness on the directional anisotropy of fluid flow and colloid transport in a single fracture. *Geophys. Res. Lett.* **2006**, 33, L21402. [CrossRef]
48. Cardenas, M.B.; Slottke, D.T.; Ketcham, R.; Sharp, J.M. Effects of inertia and directionality on flow and transport in a rough asymmetric fracture. *J. Geophys. Res. Earth Surf.* **2009**, *114*, B06204. [CrossRef]
49. Koyama, T.; Neretnieks, I.; Jing, L. A numerical study on differences in using Navier–Stokes and Reynolds equations for modeling the fluid flow and particle transport in single rock fractures with shear. *Int. J. Rock Mech. Min. Sci.* **2008**, *45*, 1082–1101. [CrossRef]
50. Cardenas, M.B.; Slottke, D.T.; Ketcham, R.A.; Sharp, J.M. Navier-Stokes flow and transport simulations using real fractures shows heavy tailing due to eddies. *Geophys. Res. Lett.* **2007**, *34*, L14404. [CrossRef]
51. Liu, R.; Li, B.; Jiang, Y. Critical hydraulic gradient for nonlinear flow through rock fracture networks: The roles of aperture, surface roughness, and number of intersections. *Adv. Water Resour.* **2016**, *88*, 53–65. [CrossRef]
52. Crandall, D.; Bromhal, G.; Karpyn, Z.T. Numerical simulations examining the relationship between wall-roughness and fluid flow in rock fractures. *Int. J. Rock Mech. Min. Sci.* **2010**, *47*, 784–796. [CrossRef]
53. Barton, N.; Bandis, S.; Bakhtar, K. Strength, deformation and conductivity coupling of rock joints. *Int. J. Rock Mech. Min. Sci. Géoméch. Abstr.* **1985**, *22*, 121–140. [CrossRef]
54. Jung, M.S.; Hwang, J.P.; Kim, J.H.; Lee, C.K.; Ann, K.Y. Risk of environmental contamination arising from concrete structures, Part III: Alkali leaching. *KSCE J. Civ. Eng.* **2014**, *19*, 1597–1603. [CrossRef]
55. Ogata, S.; Yasuhara, H.; Kinoshita, N.; Kishida, K. Coupled thermal–hydraulic–mechanical–chemical modeling for permeability evolution of rocks through fracture generation and subsequent sealing. *Comput. Geosci.* **2020**, *24*, 1845–1864. [CrossRef]
56. Ogata, S.; Yasuhara, H.; Kinoshita, N.; Cheon, D.-S.; Kishida, K. Modeling of coupled thermal-hydraulic-mechanical-chemical processes for predicting the evolution in permeability and reactive transport behavior within single rock fractures. *Int. J. Rock Mech. Min. Sci.* **2018**, *107*, 271–281. [CrossRef]

57. Yasuhara, H.; Elsworth, D.; Polak, A. A mechanistic model for compaction of granular aggregates moderated by pressure solution. *J. Geophys. Res. Earth Surf.* **2003**, *108*, 2530. [CrossRef]
58. Beeler, N.M. Stress-induced, time-dependent fracture closure at hydrothermal conditions. *J. Geophys. Res. Earth Surf.* **2004**, *109*, B02211. [CrossRef]
59. Polak, A.; Elsworth, D.; Liu, J.; Grader, A.S. Spontaneous switching of permeability changes in a limestone fracture with net dissolution. *Water Resour. Res.* **2004**, *40*, 40. [CrossRef]

fluids

Article

Thermographic Observation and Hydrodynamic Patterns of Inclined Ethanol Droplet Train Impingement on a Non-Uniformly Heated Glass Surface

Baris Burak Kanbur [1,2], Sheng Quan Heng [1] and Fei Duan [1,*]

1 School of Mechanical and Aerospace Engineering, Nanyang Technological University (NTU), Singapore 639798, Singapore; bbkanbur@ntu.edu.sg (B.B.K.); sheng011@e.ntu.edu.sg (S.Q.H.)
2 Department of Mechanical Engineering, Technical University of Denmark (DTU), 2800 Kongens Lyngby, Denmark; babukan@dtu.dk (B.B.K.)
* Correspondence: feiduan@ntu.edu.sg

Abstract: Droplet train impingement is a fundamental approach to mimic the complicated interactions between the fluid and the substrate in advanced thermal engineering applications in industry. Differently from previous studies, the main original contribution of this study is to perform an inclined droplet train impingement on a non-uniformly heated surface. Ethanol was used as the liquid for droplet train impingement applications, while glass substrate was selected as the target surface. The inclined flow angle was 63 degrees. Both optical and thermographic observations were performed on the target surface by focusing on the droplet impact area. Three experimental sets were created with the Weber numbers 667.57, 841.90, and 998.01. A surface temperature range was selected between 85.00 °C and 200.00 °C, which was above the boiling point of the ethanol. The maximum spreading length was measured at 0.97 mm at the surface temperature of 82.00 °C for the experiment with the Weber number of 998.01, whilst the minimum spreading length was found at 0.18 mm at the highest surface temperature for the experiment with the Weber number of 667.57. A uniform splashing direction was observed above 170.00 °C for all experiments, which meant that the sign of the transition regime appeared.

Keywords: boiling; droplet evaporation; two-phase heat transfer; droplet spreading; thermal imaging; hydrodynamic patterns

Citation: Kanbur, B.B.; Heng, S.Q.; Duan, F. Thermographic Observation and Hydrodynamic Patterns of Inclined Ethanol Droplet Train Impingement on a Non-Uniformly Heated Glass Surface. *Fluids* **2022**, *7*, 229. https://doi.org/10.3390/fluids7070229

Academic Editors: Vasily Novozhilov and Cunlu Zhao

Received: 24 May 2022
Accepted: 5 July 2022
Published: 7 July 2022

1. Introduction

Droplet impingement onto a target surface is a fundamental approach to mimicking modern and sophisticated engineering applications in thermal-fluid engineering, such as thermal management of electronic packaging [1], data center cooling [2], spray cooling and jet impingement [3,4], fuel spraying [5], and drop-on-demand studies [6,7]. There are two types of droplet impingement techniques, which are single droplet impingement and droplet train impingement [8]. Single droplet impingement is the established way to observe rebounding, oscillation, wettability (on the target surface), the impact of coating, and the impact of the surface slope and elevation by focusing on the boiling behaviors related to the heat fluxes, evaporation rate, etc. [9–11]. However, the real interactions between the liquid and the target surface are more complicated than the represented cases of single droplet impingement as the liquid flow is continuous and multiple, which results in different scenarios, such as steady, transient, and pseudo-transient flows. Therefore, droplet train impingement has a wider perspective to consider the mentioned complex interactions because it provides multiple and continuous flows on the target surface. Thereby, more detailed investigations, such as observing the spreading diameter, crown rim formation, and splashing angle, can be performed [12–14]. Regarding this fact, there is a considerable amount of work related to droplet train impingement.

There are two main flow characteristics in droplet train impingement studies: the Weber number (We) and the Reynolds number (Re). The Weber number is the ratio of inertia to surface tension. The inertia is defined by the liquid density, characteristic velocity, and characteristic length. Using the We, the break-up and coalescence of the droplets can be quantified to help with the parametric settings of experimental studies. The Reynolds number is the ratio of the inertial forces to viscous forces for quantifying the turbulent flows. Both We and Re are defined in Equations (1) and (2), respectively.

$$\text{We} = \frac{\rho \cdot v^2 \cdot d}{\sigma} \quad (1)$$

$$\text{Re} = \frac{\rho \cdot v \cdot d}{\mu} \quad (2)$$

where ρ is the liquid density, v is the surface tension, d is the average diameter of the droplet in the air, σ is the surface tension of the liquid, and μ is the dynamic viscosity of the liquid.

One of the cornerstones in this area was presented by Yarin and Weiss [8], who described the experimental behavior of ethanol droplet train impingement on a hot surface while explaining the theoretical fundamentals of the boiling phenomenon. It was seen that the experimental results were in good agreement with the defined theory. The reason behind selecting the ethanol was its lower surface tension and boiling point compared to water, which made the boiling observation more observable at lower surface temperatures. However, water is also one of the most preferred liquids in droplet train impingement studies. Gradeck et al. [15] focused on the surface temperature range of 250.00–675.00 °C to observe the boiling behavior of water droplet train impingement above the Leidenfrost point. They preferred a very low We value: lower than 30.00. Li et al. [16] also preferred a low We study, below 200.00, by focusing on the impact of different droplet frequencies and We values. The impact of frequency was found to be negligible, but changes in the contact angle and the We values were found to be noteworthy in the boiling trends of water droplet train impingement. Another low We (<200.00) study was conducted by Dunand et al. [17] by impinging a water droplet train onto a surface that had a temperature value above the Leidenfrost point. Besides the hydrodynamic trends (e.g., splashing, crater formation, etc.), the temperature field was also observed via a thermographic camera within the impinging area and for the splashed droplets. That study was one of the first works aimed at the combined investigation of the thermal and hydrodynamic behaviors of a droplet train.

HFE-7100 is another preferred liquid in droplet train applications as it is widely used in the thermal management of electronic packaging and data center cooling, where the droplet train impingement method can closely mimic the heat transfer characteristics. Zhang et al. carried out single droplet train [18] and multiple droplet train [19] impingement studies to observe the crater and crown formation together. In a single droplet train impingement study with a We range of 270.00–600.00 [18], they were able to improve the given theory by Yarin and Weiss [8] by adding the velocity distribution and the liquid film thickness, which became more prominent at higher We values, into the theory. In the multiple droplet train impingement study [19], the surface temperature was lower than the boiling point, and the main focus was the hydrodynamic patterns of the multiple droplet train. Considering that each train impingement created a unique crater, a hump was observed between craters. The heat transfer rate increased when the spacing between the craters became greater. The heat transfer rate calculations were performed by analyzing the thermographic camera data. When the number of droplet trains increased from two to six, hydrodynamic pattern observation became more difficult as the supplied liquid amount significantly increased. The main outcome was that the global heat transfer rate increased by raising the spacing between the craters, whilst the local heat transfer rate showed the opposite trends in hexagonal-arranged droplet train studies. In the We range of 262.00–850.00, droplet train impingement studies were performed with high-frequency values [20,21]. It was deduced that the droplets produced radial momentum; hereby, the maximum heat transfer rate was observed in the crown propagation region. Compared to low-frequency and low-We studies, the high-frequency and high-We experiments have more complicated hydrodynamic trends and droplet–surface interactions. Thermal investigations were not only performed with thermographic cameras but also with numerical studies. Trujillo et al. [22,23] presented a numerical model for thermal boundary layer analysis during the droplet train impingement of HFE-7100. Computational simulations were designed for We values below 300.00, and the main focus was a better understanding of heat transfer trends. The numerical simulations showed that the crater walls behaved like barriers between the inside and outside regions of the crater, where the inside region had a higher heat transfer rate. Additionally, it was obtained that the heat transfer was a single-phase mechanism between the wall surface and the accumulated liquid (also called liquid film) at the current conditions. The increase in the surface temperature can provide a two-phase heat transfer mechanism between the wall surface and the liquid film.

In addition to the above-mentioned efforts, several studies were performed in our research group via water and ethanol droplet train impingement setups. Liu et al. [24–26] applied water droplet train impingement on a hot surface up to 246.00 °C for observing the spreading diameter, splashing angles, and jumps of the secondary droplets on the target surface. Another study by Liu et al. [27] developed a statistical analysis approach for the main droplet and the splashed droplets from the hot surface. Kanbur et al. [28,29] observed the transient hydrodynamic patterns of ethanol droplet train impingement at high-We values. They defined the spreading diameter and the splashing angle trends on a glass surface during the transient period. Shen et al. [30] also applied ethanol droplet train impingement, but the main focus was to see the impact of a nanotube coating on a titanium surface under steady conditions. The vast majority of those studies were vertical water droplet train impingement studies, except for a small section in Ref. [26] that applied the droplet train with three different impact angles. It was seen that there were no significant changes in the spreading diameter when the impact angles changed, but the decrease in the impact angle increased the splashing angle and made the splashing phenomenon observable even at high surface temperatures. Recently, Kanbur et al. [31,32] experimentally observed the hydrodynamic patterns and thermal gradients of inclined ethanol droplet train impingement on an aluminum target surface and then compared it with the gradients on a glass surface. It was inferred that the aluminum surface had a significantly better evaporation rate than the glass surface, so that the thermal gradients were more uniform on the aluminum surface as the thermal conductivity of aluminum is dramatically higher than the thermal conductivity of glass, which also provided smaller spreading lengths. In those studies [31,32], the aim was also to create a non-uniform temperature distribution on a hot aluminum surface before the impingement, but it was not very successful as the aluminum had a very high thermal conductivity and therefore the surface quickly became uniform.

Following recent trends and applications in the area, the main contribution of this study is to create a temperature non-uniformity on the target surface by investigating the hydrodynamic patterns and the thermal gradients when droplet train impingement is applied with an inclined flow at steady conditions. The following contributions can be given as follows: (i) inclined ethanol droplet train impingement is applied with three different We values at high frequency, which has not been performed previously as far as we know; (ii) hydrodynamic patterns and thermal gradients are observed at the same time using an optical camera and a thermographic camera together; (iii) a small temperature gradient is achieved on the glass surface before the impingement, which is preferred because many real applications have temperature non-uniformity on the target surface.

2. Experimental Setup

A simplified view of the experimental setup is shown in Figure 1. The ethanol liquid was stored in a stainless-steel tank. It was pressurized via a compressed air stream up to 10 bar. In the experiments, compressed air with a pressure of 2 bar was used to send the ethanol (purity $\geq$ 99.9%; density 789.00 kg/m^3; surface tension 0.022 N/m; dynamic viscosity 0.012 Pa·s; saturation temperature 78.40 °C, Sigma-Aldrich, St. Louis, MO, USA) from the tank to the droplet generator. A flow meter and a temperature sensor were located between the droplet generator and the tank to continuously measure the flow rate and the temperature. The ethanol temperature was kept constant at 26.00 °C, which was also the room temperature where the experiments were conducted. The droplet generator (FMP Technology GmbH, Erlangen, Germany) was coupled with a piezoelectric element that was connected to a function generator. The generated high frequencies in the generator produced successive ethanol droplets from the coupled droplet generator–piezoelectric element component. The nozzle part of the droplet generator was stainless-steel, and the size of the generated droplet was controlled by an orifice-like element called a pinhole. Different from previous studies in the literature, the droplet train impinged on the surface with an incline of 63.00 degrees. The inclined droplet train impingement process became steady after the first 100 ms. A glass substrate was selected as a target surface because there are a considerable amount of engineering cases where the ethanol and the glass are coupled as the liquid and the surface [28–30]. The inclined droplet train impinged on a single point, where it behaved like a point heat sink in the experiments. The droplet impact area and the inclined flow direction are also shown in Figure 1.

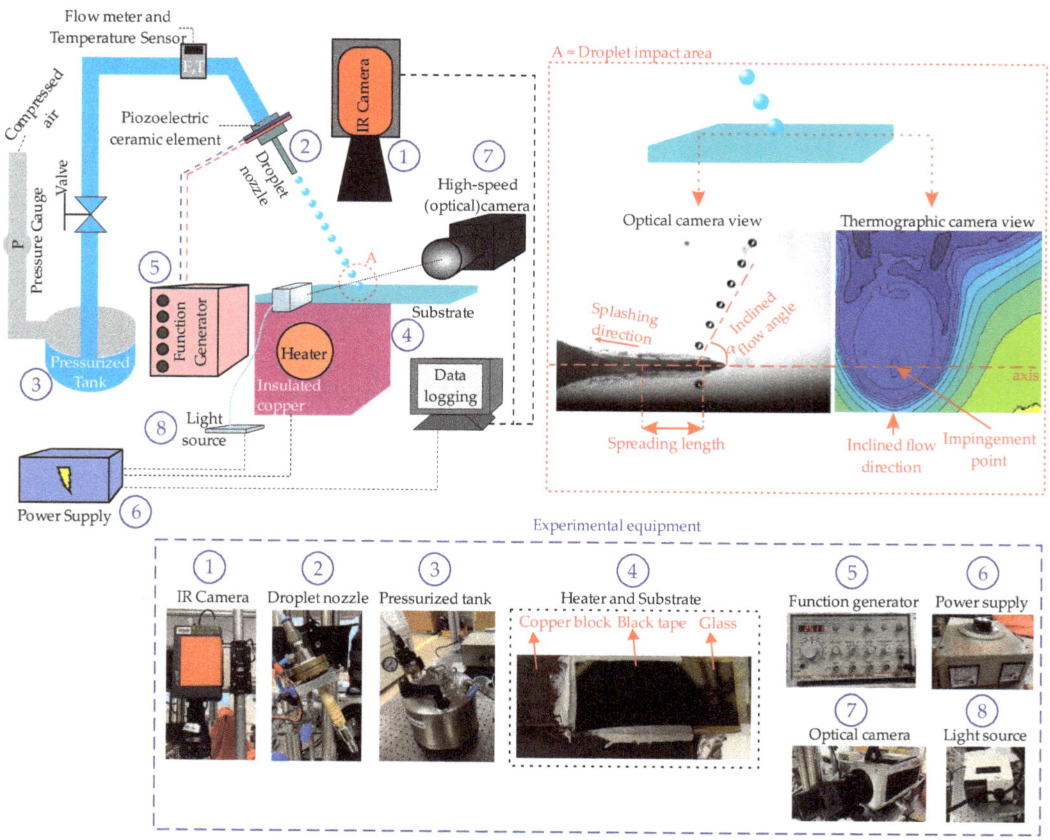

Figure 1. Simplified schematic of the experimental setup.

Furthermore, to create a temperature gradient on the target surface, the glass was heated from one side while the other side was open to the ambient environment. There was no uniform temperature on the glass surface, which was one of our main goals, as temperature gradients on the target surface are a reality in many engineering applications.

The heating of the glass was performed by using a copper block with a cartridge heater inside. The copper block was well insulated. The impingement was monitored via an optical camera (Phantom V711, 1024 × 512 resolution, images with 14,000 frames per second, 1 μs exposure time, and Canon MP-E 65 lens with 2× magnification) and a thermographic camera (ImageIR® 8300, Infratech, Dresden, Germany, 640 × 512 resolution) at the same time. It is worth noting that the thermographic camera was not a high-speed camera, unlike the optical camera, so the optical camera recorded the video while the thermographic camera captured images when the impingement point (on the surface) reached the target temperature. To collect accurate surface temperature data on the glass surface, the bottom side of the glass was covered with black tape, which minimized the reflection. Additionally, focusing on the droplet train impingement point, the temperature data were collected via the thermographic camera and a thermocouple; the temperature values were then compared to each other to avoid misreading and minimize uncertainty. The thermographic camera was calibrated before using it in the experiments. For the experimental investigation, the spreading length and the splashing angle were selected as the main focuses. When the droplet train impinged onto the target surface, an impact area (crater region) occurred. In vertical droplet train impingement studies, that area can be characterized by the spreading diameter as the impact area has a circular shape [28,29,33]. However, in the inclined flows, the impact area becomes more elliptical so that the spreading length is preferred to characterize the impact of the droplet train [31,32]. In addition, a splashing phenomena occurs when the surface energy levels become high enough; thus, splashing trends are also focused on in this study. Both the spreading length and the splashing can be seen in Figure 1. Following the experimental procedure, three different droplet trains were generated; each train type was titled as a unique experiment (Exp.), and their characteristics can be seen in Table 1.

Table 1. Droplet train characteristics.

Parameters	Experiment I	Experiment II	Experiment III
Pinhole diameter (D_0)	75.00 μm	100.00 μm	150.00 μm
Droplet diameter in air (d)	0.14 mm	0.17 mm	0.20 mm
Mean droplet spacing (l)	0.47 mm	0.47 mm	0.46 mm
Droplet velocity (ϑ)	11.52 m/s	11.85 m/s	11.70 m/s
Frequency (f)	24.77 kHz	25.12 kHz	25.46 kHz
Weber number (We)	667.57	841.90	998.01

Image and Data Analysis

The experimental setup provided two different media: (i) a video record of the droplet train impingement using an optical camera and (ii) a thermal image of the moment the droplet train impinged on the target surface at the desired surface temperature. In the study, an experimental investigation was performed with a surface temperature of 85.00 °C to 200.00 °C. Unlike the previous works performed at uniform surface temperatures, the current study aimed to create a thermal gradient on the target surface by heating the glass only from one side while the other side was open to the environment. Hence, the above-mentioned surface temperature term refers to the surface temperature at the droplet train impact point, which created an impact area (crater formation) around it. The procedure of image and data analysis is illustrated in Figure 2.

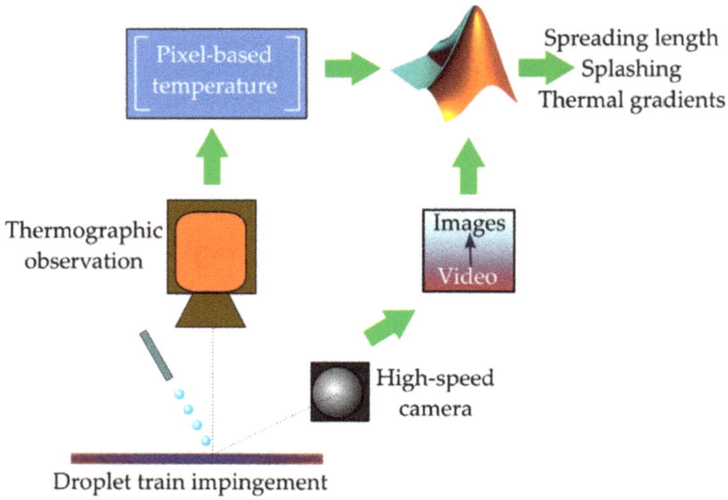

Figure 2. Image and data analysis steps of the study.

The main steps of the procedure are explained as follows: once the temperature at the impact point reached more than the target temperature (e.g., 85.00 °C, 110.00 °C), the thermographic camera captured a thermal image. In parallel to this, the corresponding optical camera view was captured from the recorded video at the same time. The optical camera images allowed us to measure the spreading length and observe the existence of the splashing phenomenon, whilst the thermal images presented the changes in the temperature gradient and the impact of the droplet train on the target surface. The thermographic camera defined the surface with 640 × 512 pixels, which was equal to the thermographic camera resolution. Each pixel had a temperature value, and all pixel data were transferred to the MATLAB environment for pixel-based thermal gradient analysis. In the thermographic analysis, we realized that the temperature value at each pixel point was not exactly the same as in the repeated experiment, although the experimental conditions were exactly the same. The deviation for each pixel-based temperature was found to be nearly (and sometimes above) 5% between the performed and the repeated experiments. In the current study, we aimed to observe the deviation of 1% for pixel-based temperature values, but surface temperatures did not provide the desired deviation value. To overcome this in future works, some potential solutions are given in the last paragraph of the Results and Discussion section. In the MATLAB environment, each pixel-based temperature corresponded to an element in a 640 × 512 matrix, therefore we were able to analyze, measure, and plot the thermographic data. Similar to the thermographic observations, the captured image from the recorded video was analyzed in the MATLAB environment for measuring the spreading length. In addition to the experimental procedure, it is worth noting that the observed surface temperature values were different from one another between Exp. I, II, and III due to the fact that the thermal image was manually captured (via hand), while the optical image was easily captured thanks to its high time frame value (14,000 frames per second). Although we tried to capture the thermal images at the same surface temperatures for each experimental set (e.g., 85.00 °C, 110.00 °C, 140.00 °C, 170.00 °C, 185.00 °C, and 200.00 °C), we could not catch the desired surface temperature due to the rapid changes in temperature values at the droplet impact point. Nevertheless, the obtained temperature values were sufficient to plot the spreading length trends for boiling investigation. Table 2 presents the obtained surface temperature values for the experimental sets. Optical and thermal images were collected for some of the obtained surface temperature values for observational analysis.

Table 2. The obtained surface temperature values for each experimental set.

Experiment I	Experiment II	Experiment III
87.00 °C	92.00 °C	95.00 °C
104.00 °C	96.00 °C	109.00 °C
120.00 °C	114.00 °C	120.00 °C
125.00 °C	118.00 °C	136.00 °C
132.00 °C	132.00 °C	155.00 °C
140.50 °C	142.50 °C	163.00 °C
149.00 °C	151.00 °C	174.50 °C
154.00 °C	160.00 °C	181.00 °C
158.00 °C	168.00 °C	185.00 °C
168.00 °C	177.00 °C	188.00 °C
172.50 °C	193.00 °C	194.00 °C
185.00 °C		200.00 °C

3. Results and Discussion

The optical images of droplet train impingement between the surface temperatures of 85.00 °C and 115.00 °C are shown for Experiments I–III in Figure 3. For all experimental sets, liquid pool accumulations were observed around the droplet impact area and were mostly opposite the inclined droplet train direction. The minimum accumulated liquid pool was seen for Experiment I, which had the lowest Weber number and the lowest droplet diameter and pinhole diameter (see Table 1). When the droplet diameter and Weber number increased, the accumulated liquid pool became more chaotic and difficult to observe, which was in agreement with the outcomes of Refs. [20,21]. Additionally, the crater wall became more visible at higher Weber numbers.

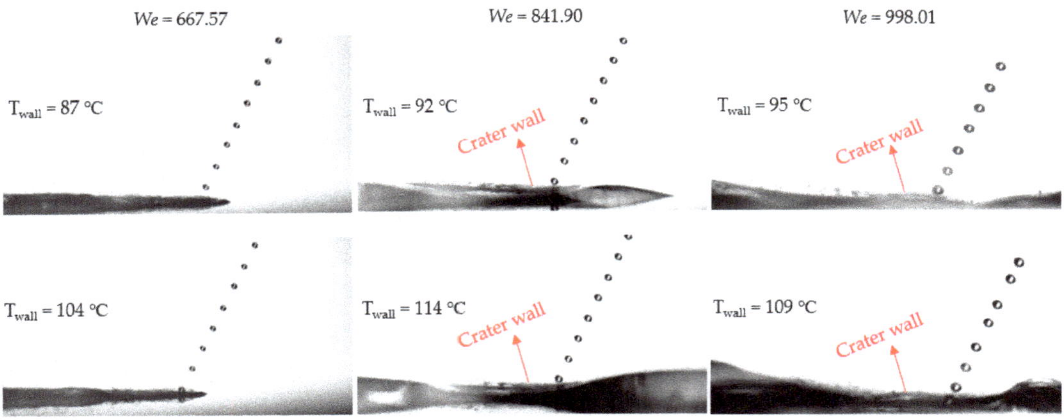

Figure 3. Steady-state hydrodynamic patterns of ethanol droplet train impingement on the target glass surface at different Weber numbers for the surface temperature values of 85.00 °C to 115.00 °C.

The first row of Figure 3 represents the lower surface temperature values when compared to the second row. Compared to the lower surface temperatures, the volume of the accumulated liquid pool was slightly smaller at higher surface temperature values. The reason behind that observation is the energy balance on the target glass surface. That is, if the droplet impinging on the surface had a lower temperature value than the ethanol boiling point, the mass of the supplied ethanol (via droplet impingement) would be equal to the mass of the accumulated liquid on the target surface. Because the surface temperature values were higher than the ethanol boiling point, there was a certain amount of ethanol evaporation from the surface. Following this, the mass and energy balance show that the supplied ethanol rate was equal to the sum of the evaporated ethanol and the accumulated liquid on the target surface. Nevertheless, Figure 3 infers that the evaporation rate was small at the observed surface temperature values; hence, the accumulated liquid pool was large for all experimental sets. The accumulated pool formation and the impact of the Weber number and droplet diameter (in the air) were also observed during the thermographic analysis, as shown in Figure 4.

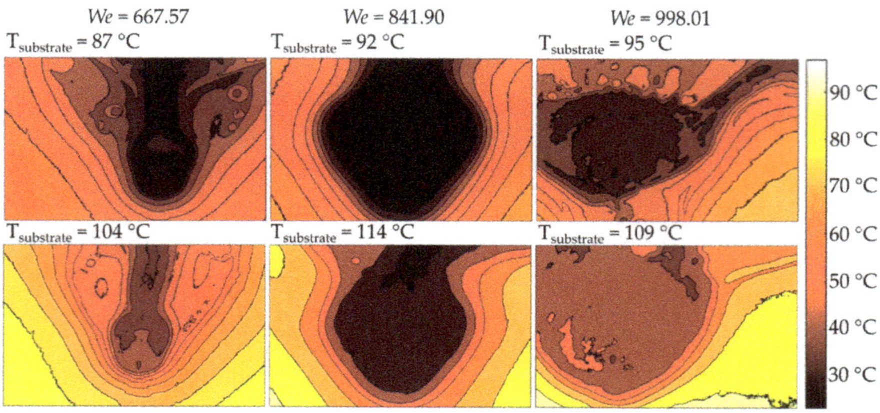

Figure 4. Steady-state thermographic images of ethanol droplet train impingement on the target glass surface at different Weber numbers for the surface temperature values of 85.00 °C to 115.00 °C.

Figure 4 shows that a smaller droplet diameter has a smaller impact area, which is in good agreement with the observed hydrodynamic patterns in Figure 3; therefore, it cooled down a limited area around and resulted in a higher thermal gradient on the surface. In real applications, to overcome high thermal gradients, the number of droplet trains should be increased, as also observed in Ref. [19]. In parallel to the discussion in Figure 3, the accumulated liquid pool became smaller at higher surface temperature values for all the experimental sets. Moreover, as also shown in Figure 3, the accumulated liquid flow was more chaotic at higher Weber numbers. In addition, similar to the outcomes in Refs. [22,23], thermographic images showed that the crater walls behaved like barriers between the inside and outside regions of the droplet impact area, which most probably resulted in different heat transfer coefficients in those regions. Figure 5 shows the steady-state hydrodynamic patterns of ethanol train impingements at higher surface temperature values: from 135.00 °C to 200.00 °C.

Compared to the lower surface temperature values in Figure 3, it can be clearly seen that the spreading length became smaller, and the volume of the accumulated ethanol liquid decreased at higher surface temperatures. The reason for this is related to the increment in the surface energy level of the glass, which resulted in a higher evaporation rate and therefore a smaller droplet impact area for all Weber number cases. Unlike (i) the vertical droplet train impingement studies that had Weber numbers higher than 1100 [28,29] and (ii) the inclined droplet train impingement studies at similar Weber numbers but on the aluminum surface that had significantly higher thermal conductivity than the glass, nucleate bubbles, which represents the nucleate boiling regime, were not observed near the droplet impact area. The reason for this might be the lower thermal conductivity value of the glass substrate, which was not able to provide sufficiently high surface energy levels, even though the surface temperatures were close to the inclined droplet train impingement on the aluminum surface [31]. The low thermal conductivity inherently affected the surface energy level so that the existing accumulated liquid prevented the formation of nucleate bubbles. Figure 5 also presents non-uniformly splashed microdroplets from the crown of the droplet impact area in the first row (surface temperatures of 136.00–142.00 °C). When the splashed droplets are observed, it can be assumed that the surface energy level was significantly high enough to break the main droplet into microdroplets from its crown. It must be noted that the nucleate bubbles were not observed in this range, unlike the vertical droplet train impingement studies [28,29]. The reason for this might be the above-mentioned facts: the glass substrate had lower thermal conductivity, and thus the accumulated ethanol droplet was in the opposite direction to the inclined flow.

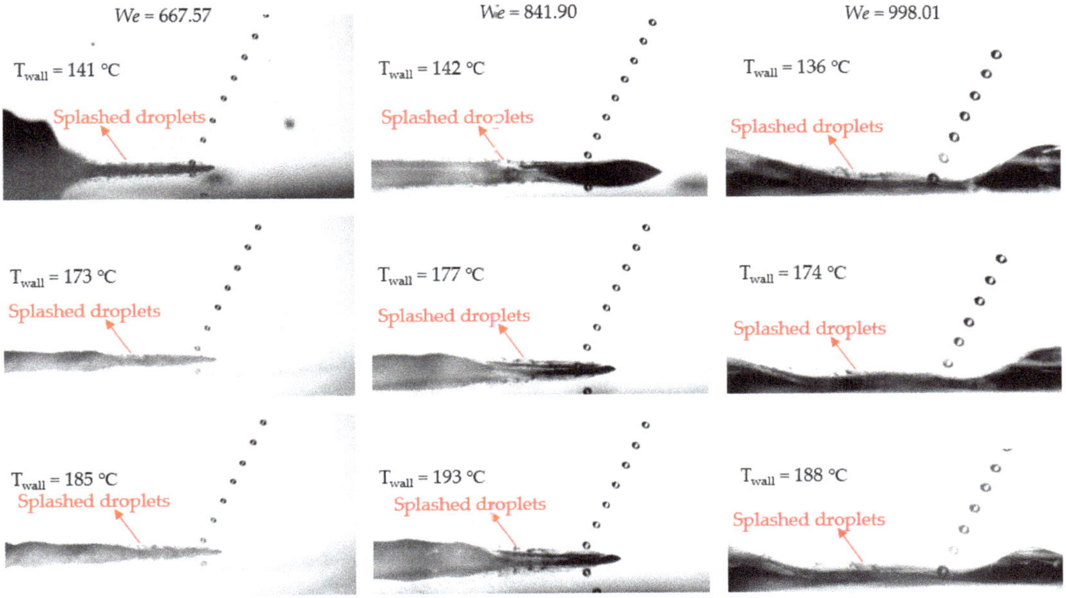

Figure 5. Steady-state hydrodynamic patterns of ethanol droplet train impingement on the target glass surface at different Weber numbers for the surface temperature values of 135.00 °C to 200.00 °C. A Supplementary Material (Video S1) can be seen as well.

By increasing the surface temperature above 170.00 °C, it was seen that the splashed microdroplets had a uniform direction opposite to the inclined flow direction. As already stated in Refs. [25,28,29,31], observing the splashing in a single direction can be assumed to be a sign of the transition regime. However, differently from previous ethanol droplet train impingement works [28,29,31], the splashing angle was slightly higher than zero and the increment/decrement in the angle value was negligible when the surface temperature continued to increase up to 200.00 °C; thus, we were not able to detect the exact initial moment of the transition regime. Therefore, it can be said that the nucleate boiling was in the surface temperature range of 100.00–135.00 °C, whereas the sign of the transition regime and the transition regime were observed in the temperature range of 170.00 °C–200.00 °C. To better visualize the flow trends, Figure 6 projects the thermographic camera measurements from the surface temperature of 135.00 °C to 200.00 °C.

Thermographic images support the observations performed in Figure 5, which stated that the spreading length became smaller by the rising of the surface temperature for all experiments. It was also seen that the crater wall continued to behave like a barrier between the inside and outside regions of the droplet impact area. Compared to the thermographic observations in Figure 4, the amount of accumulated liquid was notably decreased due to the high energy level and the high temperature of the surface, which provided a higher evaporation rate. At high temperatures, especially above 170.00 °C, which corresponded to the sign of transition regime referring to Figure 5, it was observed that Experiment I (minimum droplet diameter and Weber number) had a very local cooling impact on the plate, while greater Weber numbers (Experiments II and III) achieved a larger impact area that can be considered as a heat sink.

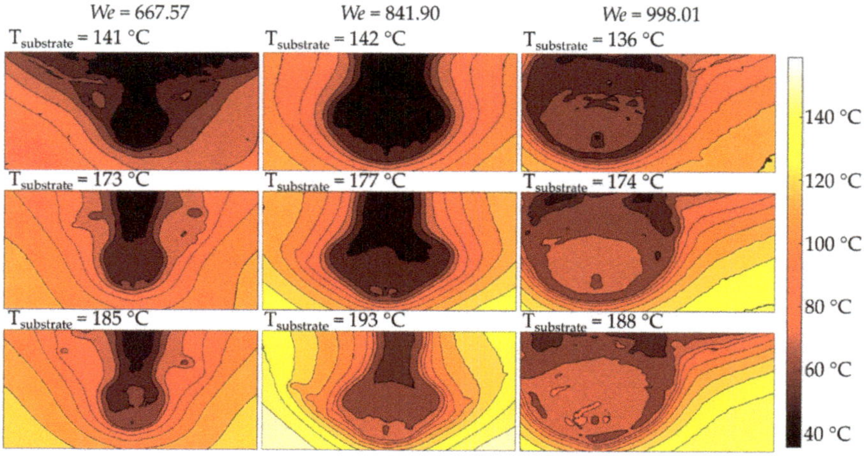

Figure 6. Steady-state thermographic images of ethanol droplet train impingement on the target glass surface at different Weber numbers for the surface temperature values of 135.00 °C to 200.00 °C.

Hydrodynamic pattern evaluation allowed us to measure the spreading length of the droplet impact areas via optical camera images. Figure 7a shows the measured spreading length values of Experiments I–III. Apart from the experiments above the boiling point of ethanol, an additional experiment was performed for surface temperature lower than the ethanol boiling point to see the spreading length before the boiling step. For Experiments I, II, and III, the lowest surface temperatures were 68.00 °C, 69.50 °C, and 76.00 °C, with the corresponding spreading length values of 0.38 mm, 0.59 mm, and 1.19 mm, respectively. When the surface temperature was increased above the ethanol boiling point, there was a notable decrement with the ratio of 15.67%, 10.77%, and 18.32% for Experiments I–III, respectively. The measured decrements showed the importance of substrate surface temperature when it was higher than the boiling point of the liquid compared to the operations below the boiling point. The decrease of the spreading length continued with the rising of the surface temperature, as observed in Figures 3–6, which supports the reason given due to the thermal energy balance between the supplied ethanol liquid and the surface energy level of the glass. From the surface temperature range of 85.00–90.00 °C to 140.00 °C, the decrease in the spreading length was found to be nearly 30.00% for all the experiments. From the surface temperature of 140.00 °C to 172.50 °C, the spreading length did not change significantly, but it fluctuated in Experiment I (Weber number of 667.57). Considering the fact that the increase in surface temperature provided a high surface energy level that meant a high evaporation rate, the splashed microdroplets might balance the thermal energy between the surface and the supplied liquid ethanol; therefore, the spreading length fluctuated in a specific range of 0.20–0.22 mm but did not decrease until a surface temperature of 172.50 °C. However, for the surface temperature of 172.50 °C to 185.00 °C, which was most probably in the transition regime regarding the discussion in Figure 5, the spreading length decreased from 0.20 mm to 0.18 mm.

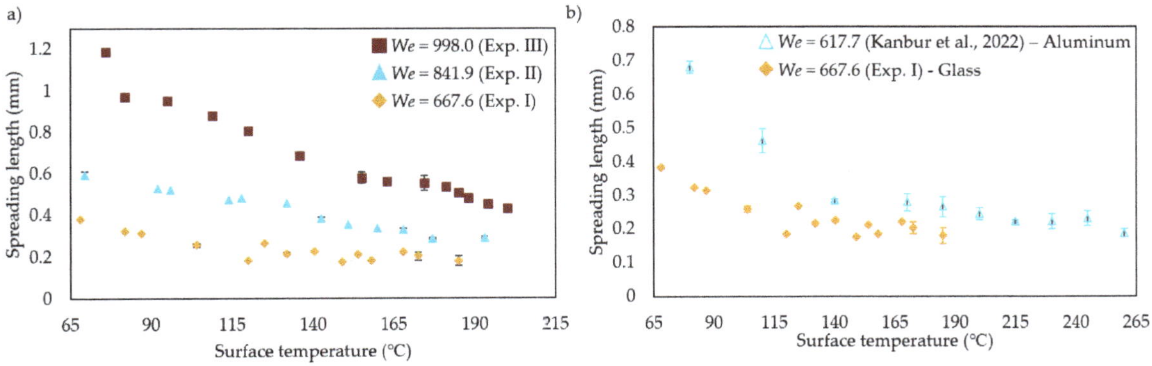

Figure 7. Spreading length measurements of (**a**) Experiments I–III and (**b**) comparative cases of Experiment I of this study and another study from Ref. [31].

For other experiments (Experiments I and II), the spreading length decreased continuously without any fluctuation, which indicated the micro-splashed droplets were not able to achieve the thermal balance between the supplied ethanol liquid and the glass surface. The minimum spreading lengths were 0.26 mm and 0.30 mm for Experiments II and III, respectively. The comparative spreading length analysis among different Weber numbers deduced that the glass surface temperature was critical for observing different boiling behaviors, but it was not the only critical factor because the supplied ethanol rate was also critical for manipulating the boiling behaviors at different (higher or lower) surface temperature values.

In addition to the current discussion, Figure 7b compares the spreading length values of two different experiments with very close Weber numbers to each other: (i) Experiment I of this study, and (ii) an inclined ethanol droplet train impingement (with the same angle) on an aluminum surface in Ref. [31]. The experiment in Ref. [31] was performed with a pinhole diameter of 100 µm, which is the same diameter as Experiment II, but the Weber number is close to Experiment I of the current study. When the spreading lengths of Experiments I and II are compared in Figure 7a, it can be seen that the greater pinhole diameter generated a higher ethanol liquid supply on the surface, which affected the evaporation behaviors and the accumulated liquid on the target surface; this can also be seen in Figures 3–6. However, Figure 7b shows that the spreading length values of the droplet train impingement in Ref. [31] were very close to the spreading length values of Experiment I of the current study, although the experiment in Ref. [31] had higher supplied ethanol liquid on the surface. To this end, it can be said that the higher thermal conductivity of the substrate significantly improves the evaporation rate and the boiling performance, which has also been discussed in Figure 5 above.

Following the given discussions above, the main limitations and potential future improvements can be mentioned. One of the main limitations is related to the experimental conditions. In the current work, we wanted to work with a non-uniformly heated surface, so the glass substrate was heated from a single side while the other side was open to the environment. A temperature gradient was achieved, but the repeated experiments (for each experimental set, Exp. I, II, or III) showed that the pixel-based temperature values were not the same in the repeated experiments, and the deviation between them was above 5%, whilst we preferred a deviation of 1% from the point of view of accuracy. To overcome this limitation, a better thermographic camera resolution could be used, or the target of thermal gradient creation could be sacrificed and the glass substrate uniformly heated by the heater block in Figure 1. Another limitation is the strong accumulated liquid flow in the opposite direction of the inclined ethanol flow, which made observation of the droplet impact area difficult, especially for higher Weber numbers. By decreasing the Weber number values, the boiling behaviors near the droplet impact area can be observed, or the inclined flow angle could be changed to higher values (closer to 90 degrees represents the vertical droplet train impingement). Alternatively, without changing the thermographic camera resolution or the inclined flow angle, a substrate with better thermal conductivity (such as aluminum in Ref. [32]) could be preferred; however, it must be noted that substrates with higher thermal conductivity can rapidly reach the uniform temperature distribution, and that can block the purpose of investigating the non-uniformly heated surface. Besides the limitations and their potential solutions, near-future insights could be related to the efficient and smart analysis of the camera data, particularly for thermographic camera images. Our recent works [7,33] showed that both thermal and optical images and videos can be used to analyze and predict the future steps of the flow (via extrapolation). For the current case, thermographic image-related extrapolation studies could be performed, but the hydrodynamic-related images will most probably need additional effort due to the moving pixels related to the droplet train movements before impinging the surface.

4. Conclusions

The presented study aimed to observe the hydrodynamic patterns of inclined ethanol droplet train impingement on a non-uniformly heated surface while a thermographic camera collected the thermal gradients of the moment the droplet impact area was formed in steady conditions. To create a non-uniform temperature distribution, a glass substrate was selected and heated on a single side while the other side was open to the environment. The inclined flow angle was adjusted to 63 degrees. Three different experimental sets were designed with the Weber numbers 667.57, 841.90, and 998.01. The glass substrate was heated up to 200.00 °C. The main conclusions are given as follows:

- Crater walls around the droplet impact area became more visible while the accumulated liquid pool was more chaotic when raising the Weber number.
- The spreading length decreased by the raising of the surface temperature for the Weber numbers of 841.90 and 998.01. However, a decrement was observed until a surface temperature of 172.50 °C, then the spreading length fluctuated until the surface temperature reached 180.00 °C for the experiment with the Weber number of 667.57. The splashed micro droplets from the crown rim of the droplet impact area might have achieved the thermal energy balance between the surface and the supplied ethanol liquid so that the spreading length fluctuated instead of there being a continuous decrement at the Weber number of 667.57. To highlight this speculative understanding, future works should focus on surface temperatures with smaller temperature intervals.
- Non-uniform splashing was observed at the surface temperature values of 135.00–142.00 °C, which showed that the nucleate bubbles were not able to be seen due to the strong flow of the accumulated liquid in the opposite direction to the inclined flow. In future works, experiments with lower Weber numbers could be carried out to observe the nucleate bubbles.
- Signs of the transition regime started to be observed from a surface temperature of 170.00 °C by observing the splashed microdroplets in a single direction. The splashing angle did not change significantly when the surface temperature increased to 200.00 °C; therefore, no exact temperature value was found to define the start of the transition regime. Smaller surface temperature intervals and/or experiments with low Weber numbers could be useful ways to detect this in the future.
- Thermographic images showed that the crater walls created a barrier between the inside and outside of the droplet impact area, which can affect the heat transfer coefficients in those regions. More detailed works could be carried out via better thermographic resolutions at lower Weber number values or numerical studies that can be performed under transient conditions.
- The greatest spreading length was found at 0.97 mm for the experiment with the Weber number of 998.01, while the minimum one was measured at 0.18 mm for the experiment with the Weber number of 667.57, when the surface temperatures ranged between 85.00 °C and 200.00 °C.

Future works could focus on the fundamental relationship between the droplet impact area and the target surface by considering the spreading length, splashed droplet, and crater wall formation by applying experimental and computational (computational fluid dynamics and/or molecular dynamic simulations) approaches. Additionally, parametric studies using different liquid types (water, HFE-7100, etc.), various flow angles, and surface materials could be performed.

Supplementary Materials: The following supporting information can be downloaded at: https://www.mdpi.com/article/10.3390/fluids7070229/s1, Video S1: Inclined droplet train impingement on the target surface at the surface temperature of 141 °C and the Weber number of 667.57.

Author Contributions: Conceptualization, B.B.K. and F.D.; methodology, B.B.K. and F.D.; software, B.B.K. and S.Q.H.; validation, B.B.K.; formal analysis, B.B.K.; investigation, B.B.K. and S.Q.H.; resources, F.D.; data curation, B.B.K.; writing—original draft preparation, B.B.K. and F.D.; writing—review and editing, B.B.K. and F.D.; visualization, B.B.K.; supervision, F.D.; project administration, F.D.; funding acquisition, F.D.; B.B.K. also declares that the formal analysis, data curation, writing—original draft preparation, and writing—review and editing were completed at his new affiliation, the Technical University of Denmark (DTU), Denmark, while the remaining efforts were performed at his previous affiliation, Nanyang Technological University (NTU), Singapore. All authors have read and agreed to the published version of the manuscript.

Funding: This research received no external funding. We received internal funding from the School of Mechanical and Aerospace Engineering, Nanyang Technological University (NTU).

Institutional Review Board Statement: Not applicable.

Informed Consent Statement: Not applicable.

Acknowledgments: The authors would like to thank the funding supports from the School of Mechanical and Aerospace Engineering, Nanyang Technological University.

Conflicts of Interest: The authors declare no conflict of interest. The funders had no role in the design of the study; in the collection, analyses, or interpretation of data; in the writing of the manuscript; or in the decision to publish the results.

References

1. Ma, C.; Tian, Y. Experimental investigation on two-phase two-component jet impingement heat transfer from simulated microelectronic heat sources. *Int. Commun. Heat Mass Transf.* **1990**, *17*, 399–408. [CrossRef]
2. Fan, S.; Tong, W.; Duan, F. Nucleate pool boiling heat transfer enhancement in saturated Novec 7100 using titanium dioxide nanotube arrays. *Int. Commun. Heat Mass Transf.* **2021**, *122*, 105166. [CrossRef]
3. Vontas, K.; Boscariol, C.; Andredaki, M.; Georgoulas, A.; Crua, C.; Walther, J.H.; Marengo, M. Droplet Impact on Suspended Metallic Meshes: Effects of Wettability, Reynolds and Weber Numbers. *Fluids* **2020**, *5*, 81. [CrossRef]
4. Qiu, L.; Dubey, S.; Choo, F.H.; Duan, F. Recent developments of jet impingement nucleate boiling. *Int. J. Heat Mass Transf.* **2015**, *89*, 42–58. [CrossRef]
5. Markt, D., Jr.; Raessi, M.; Lee, S.-Y.; Zhu, X. High-speed impact of micron-sized diesel drop trains—Splashing dynamics, secondary droplet formation, and effects of pre-existing film thickness. *Phys. Fluids* **2021**, *33*, 102120. [CrossRef]
6. Li, A.; Liu, X.; Wan, X.; Yang, Y. Thermal behaviors and fluid flow controlling the geometry of 7075 aluminum alloy single tracks during liquid metal flow rapid cooling additive manufacturing. *Int. Commun. Heat Mass Transf.* **2020**, *116*, 104664. [CrossRef]
7. Tanis-Kanbur, M.B.; Kumtepeli, V.; Kanbur, B.B.; Ren, J.; Duan, F. Transient Prediction of Nanoparticle-Laden Droplet Drying Patterns through Dynamic Mode Decomposition. *Langmuir* **2021**, *37*, 2787–2799. [CrossRef]
8. Yarin, A.L.; Weiss, D.A. Impact of drops on solid surfaces: Self-similar capillary waves, and splashing as a new type of kinematic discontinuity. *J. Fluid Mech.* **1995**, *283*, 141–173. [CrossRef]
9. Deng, T.; Varanasi, K.K.; Hsu, M.; Bhate, N.; Keimel, C.; Stein, J.; Blohm, M. Nonwetting of impinging droplets on textured surfaces. *Appl. Phys. Lett.* **2009**, *94*, 133109. [CrossRef]
10. Lee, J.B.; Lee, S.H. Dynamic Wetting and Spreading Characteristics of a Liquid Droplet Impinging on Hydrophobic Textured Surfaces. *Langmuir* **2011**, *27*, 6565–6573. [CrossRef]
11. Peng, C.; Xu, X.; Liang, X. Numerical investigation on crown behavior and energy evolution of droplet impinging onto thin film. *Int. Commun. Heat Mass Transf.* **2020**, *114*, 104532. [CrossRef]
12. Qian, Y.; Bian, S.; Zhao, P.; Wang, T.; Hua, Y.; Liu, Y.; Teng, Q.; Tao, C. The investigation of evaporation behavior of lubricating oil droplet at high ambient temperature. *Int. Commun. Heat Mass Transf.* **2020**, *117*, 104689. [CrossRef]

13. Zen, T.-S.; Chou, F.-C.; Ma, J.-L. Ethanol drop impact on an inclined moving surface. *Int. Commun. Heat Mass Transf.* **2010**, *37*, 1025–1030. [CrossRef]
14. Khavari, M.; Sun, C.; Lohse, D.; Tran, T. Fingering patterns during droplet impact on heated surfaces. *Soft Matter* **2015**, *11*, 3298–3303. [CrossRef]
15. Gradeck, M.; Seiler, N.; Ruyer, P.; Maillet, D. Heat transfer for Leidenfrost drops bouncing onto a hot surface. *Exp. Therm. Fluid Sci.* **2013**, *47*, 14–25. [CrossRef]
16. Li, J.; Zhang, H.; Liu, Q. Dynamics of a successive train of monodispersed millimetric-sized droplets impact on solid surfaces at low Weber number. *Exp. Therm. Fluid Sci.* **2018**, *102*, 81–93. [CrossRef]
17. Dunand, P.; Castanet, G.; Gradeck, M.; Lemoine, F.; Maillet, D. Heat transfer of droplets impinging onto a wall above the Leidenfrost temperature. *C. R. Mécanique* **2013**, *341*, 75–87. [CrossRef]
18. Zhang, T.; Muthusamy, J.; Alvarado, J.L.; Kanjirakat, A.; Sadr, R. Numerical and experimental investigations of crown propagation dynamics induced by droplet train impingement. *Int. J. Heat Fluid Flow* **2016**, *57*, 24–33. [CrossRef]
19. Zhang, T.; Alvarado, J.L.; Muthusamy, J.; Kanjirakat, A.; Sadr, R. Heat transfer characteristics of double, triple and hexagonally-arranged droplet train impingement arrays. *Int. J. Heat Mass Transf.* **2017**, *110*, 562–575. [CrossRef]
20. Zhang, T.; Alvarado, J.L.; Muthusamy, J.P.; Kanjirakat, A.; Sadr, R. Effects of High Frequency Droplet Train Impingement on Spreading-Splashing Transition, Film Hydrodynamics and Heat Transfer. *J. Heat Transf.* **2016**, *138*, 020902. [CrossRef]
21. Muthusamy, J.P.; Zhang, T.; Alvarado, J.L.; Kanjirakat, A.; Sadr, R. Effects of High Frequency Droplet Train Impingement on Crown Propagation Dynamics and Heat Transfer. *J. Heat Transf.* **2016**, *138*, 020903. [CrossRef]
22. Trujillo, M.F.; Lewis, S.R. Thermal boundary layer analysis corresponding to droplet train impingement. *Phys. Fluids* **2012**, *24*, 112102. [CrossRef]
23. Trujillo, M.F.; Alvarado, J.; Gehring, E.; Soriano, G.S. Numerical Simulations and Experimental Characterization of Heat Transfer From a Periodic Impingement of Droplets. *J. Heat Transf.* **2011**, *133*, 122201. [CrossRef]
24. Qiu, L.; Dubey, S.; Choo, F.H.; Duan, F. High jump of impinged droplets before Leidenfrost state. *Phys. Rev. E* **2019**, *99*, 033106. [CrossRef]
25. Qiu, L.; Dubey, S.; Choo, F.H.; Duan, F. Splashing of high speed droplet train impinging on a hot surface. *Appl. Phys. Lett.* **2015**, *107*, 164102. [CrossRef]
26. Qiu, L.; Dubey, S.; Choo, F.H.; Duan, F. The transitions of time-independent spreading diameter and splashing angle when a droplet train impinging onto a hot surface. *RSC Adv.* **2016**, *6*, 13644–13652. [CrossRef]
27. Qiu, L.; Dubey, S.; Choo, F.H.; Duan, F. The Statistical Analysis of Droplet Train Splashing After Impinging on a Superheated Surface. *J. Heat Transf.* **2017**, *139*, 052201. [CrossRef]
28. Kanbur, B.B.; Lee, M.Z.H.; Duan, F. Transient hydrodynamic patterns of high Weber number ethanol droplet train impingement on heated glass substrate. *Int. Commun. Heat Mass Transf.* **2021**, *126*, 105451. [CrossRef]
29. Kanbur, B.B.; Hui, M.L.Z.; Duan, F. Transient Transition of High Speed and High-Frequency Ethanol Droplet Train Impingement on the Heated Indium Tin Oxide (ITO) Surface. In Proceedings of the 5th International Conference on Smart and Sustainable Technologies (SpliTech), Split, Crotia, 23–26 September 2020. [CrossRef]
30. Shen, S.; Tong, W.; Duan, F. Hydrodynamic pattern transition of droplet train impinging onto heated titanium substrates with or without nanotube coating. *Int. J. Heat Mass Transf.* **2020**, *163*, 120409. [CrossRef]
31. Kanbur, B.B.; Quan, S.Q.; Duan, F. Hydrodynamic pattern investigation of ethanol droplet train impingement on heated aluminum surface. *Fluid Dyn. Mater. Process.* **2022**, *18*, 1711–1718. [CrossRef]
32. Kanbur, B.B.; Heng, S.Q.; Duan, F. Thermographic Observation of High-Frequency Ethanol Droplet Train Impingement on Heated Aluminum and Glass Surfaces. *Fluid Dyn. Mater. Process.* **2022**, *18*, 1331–1338. [CrossRef]
33. Kanbur, B.B.; Kumtepeli, V.; Duan, F. Thermal performance prediction of the battery surface via dynamic mode decomposition. *Energy* **2020**, *201*, 117642. [CrossRef]

fluids

MDPI

Article

Electrohydrodynamic Liquid Sheet Instability of Moving Viscoelastic Couple-Stress Dielectric Fluid Surrounded by an Inviscid Gas through Porous Medium

Mohamed Fahmy El-Sayed [1,2,*,†] and Agaeb Mahal Alanzi [3,†]

1 Department of Mathematics, College of Science, Qassim University, P.O. Box 6644, Buraidah 51452, Saudi Arabia
2 Department of Mathematics, Faculty of Education, Ain Shams University, Heliopolis (Roxy), Cairo P.O. Box 11341, Egypt
3 Department of Mathematics, College of Science and Arts in Al-Badaya, Qassim University, Buraidah P.O. Box 52571, Saudi Arabia; ag.alanazi@qu.edu.sa
* Correspondence: mo.elsayed@qu.edu.sa
† These authors contributed equally to this work.

Abstract: Viscoelastic liquid sheet of couple-stress type streaming with relative motion into an inviscid gas through porous molium is studied theoretically and quantitatively in this project. To derive the differential equations that describe liquids, gases, and the electric field, we linearized the governing equations of motion and continuity, Maxwell's equations in quasi-static approximation, and the appropriate boundary conditions at the two interfaces. Then we used the normal mode method. It was demonstrated analytically that the solutions to these differential equations can be found for both symmetric and antisymmetric disturbances, respectively. We could not obtain an explicit form of the growth rates since we could not solve the dispersion relations for both situations because they were obtained in highly complex forms. The Mathematica program is used to solve the dimensionless forms of the dispersion relations numerically using Gaster's theorem. Various influences on the stability analysis of the considered system have been studied in detail, and it is determined that the system in the presence of a porous material is more unstable than it would be otherwise. In a two-dimensional system, the antisymmetric disturbance case is found to be more unstable than the corresponding symmetric disturbance situation. Some characteristics, such as Wabe number, Ohnesorge number, and electric field, have destabilizing effects, whereas others, such as porosity, medium permeability, viscoelasticity parameter, gas-to-liquid viscosity ratio, and dielachic constants, have stabilizing effects. Finally, it is discovered that the gas-to-liquid velocity ratio plays a dual role in the stability condition depending on whether the gas-to-liquid velocity ratio $U \lesseqgtr 1$. In the past, we have only found evidence of very few previous studies.

Keywords: hydrodynamic stability; electrohydrodynamics; liquid sheet; couple-stress flunds; flows through porous medium

Citation: El-Sayed, M.F.; Alanzi, A.M. Electrohydrodynamic Liquid Sheet Instability of Moving Viscoelastic Couple-Stress Dielectric Fluid Surrounded by an Inviscid Gas through Porous Medium. *Fluids* **2022**, 7, 247. https://doi.org/10.3390/fluids7070247

Academic Editor: Vasily Novozhilov

Received: 28 May 2022
Accepted: 12 July 2022
Published: 18 July 2022

Publisher's Note: MDPI stays neutral with regard to jurisdictional claims in published maps and institutional affiliations.

1. Introduction

The instability of a thin liquid sheet has been extensively investigated in the past and is of great scientific and technological value. In relation to the process of atomization, it was studied by Squire [1], Hagerty and Shea [2], Dombrowski and Johns [3], and Li and Tankin [4] on the stability and breakup process of thin liquid sheets. An inviscid liquid sheet in a stationary inviscid gas medium was studied by Squire [1]. If the Weber number (We) is greater than one, he found that instability arises. Antisymmetric and symmetric waves are the only two types of waves that Hagerty and Shea [2] found to exist at any given frequency. Researchers discovered that antisymmetric waves have a higher growth rate than their symmetric counterparts. Dombrowski and Johns [3] were the first to investigate the instability of viscous liquid sheets and derive a drop-size relation that showed to be in good agreement with experimental results. They based some of their conclusions on a few educated guesses. Li and Tanken [4] found that the viscosity of a viscous liquid sheet moving in an inviscid gas is complicated at small wave numbers because the dispersion curve exhibits two local maxima, one corresponding to aerodynamic instability, and the other is viscosity enhanced.

Several practical applications rely on the ability of a liquid sheet to fragment into minute droplets when ejected into a gaseous medium, including spray painting and inkjet printing, as well as gas turbine and liquid rocket motors and oil burners [5]. Understanding how liquid sheets become unstable and break up is important both for science and industry. There are a wide variety of uses for this technology, and books about it tend to focus on a single application. Examples include Lefebvre [6], Yarin [7], and Lin [8], all of which deal with non-Newtonian fluids, as well as Lefebvre's book on internal combustion and Lin's book on liquid sheet and liquid jet breaking up. The review article by Dasgupta et al. [9] provides an excellent overview of the issue.

For practical reasons, electrohydrodynamics is an essential branch of fluid mechanics that studies the interactions between electric and hydrodynamic forces. Hydrodynamic motion and electric phenomena are linked in this manner. Consequently, the electrohydrodynamic equations of motion can be split into two groups: hydrodynamic equations and electrical fields-equation sets. The boundary conditions for these equations are also influenced by their connection. The electro-fluid dynamics of biological systems, liquid ejection in zero gravity environments, and liquid and gas insulation studies are all examples of electrohydrodynamics applications. Due to the complexity of the fluid-elastic interaction in the presence of an electric field, few results are known about the instability and breakup of electrohydrodynamic non-Newtonian liquid sheets, which is why new work on the fundamental electrohydrodynamic phenomena involving non-Newtonian fluids is increasingly needed [10–13]. Melcher [14] provides an overview of electrohydrodynamics that includes numerous references to recent advances in the subject.

Couple-stress fluids are becoming increasingly important in current technology and industry, making further study of these fluids desirable. Couple-stress fluids were first proposed by Stokes [15]. The study of the lubricating mechanics of synovial joints, a topic of current scientific inquiry, makes use of such fluids. Since the long chain hyaluronic acid molecules in synovial fluid are discovered as additive, this theory states that fluids with very large molecules exhibit noticeable magnitudes of Couple-stress [16–22]. In recent years, there has been a lot of interest in figuring out how non-Newtonian fluids flowing through porous media are affected by Couple-stress effects. The authors of [23–27] indicated an increased interest in the possibilities of boosting oil recovery efficiency from water flooding projects by mobility control with non-Newtonian displacing fluids in this area. It is also important to note that the flow through porous media is of interest to petroleum engineers and geophysical dynamicists; for example see refs. [28,29]. The present work hopes to provide a foundation for further investigations of the instability and breakup of viscoelastic liquid sheets in the presence of electric fields [30,31].

Electrohydrodynamic time instability of viscoelastic liquid sheet streaming with relative motion into an ambient inviscid gas through porous material is the focus of this paper. For both antisymmetric and symmetric disturbances, the three-dimensional dispersion relations in non-dimensional form have been established using complex forms. To the best of our knowledge, this subject has never been explored before, and certain limiting examples of prior efforts are found in the literature. Using a novel numerical technique to see the impacts of various parameters on the stability of the system, stability analysis and discussion are provided in this article's concluding remarks section.

2. Formulation and Perturbation Equations

We consider a viscoelastic dielectric liquid sheet of couple-stress type whose thickness is $2a$ that issues from a nozzle at a velocity U_{0l} and has a density ρ_l, pressure p_l, dielectric constant ϵ_l, kinematic viscosity $v_l(= \mu_l/\rho_l)$, and kinematic viscoelasticity $v_l'(= \mu_l'/\rho_l)$, where μ_l and μ_l' are the dynamic viscosity and dynamic viscoelasticity, respectively. This liquid sheet is surrounded by a moving inviscid dielectric gas whose velocity is U_{0g} and has a density ρ_g, pressure p_g, dielectric constant ϵ_g and kinematic viscosity ν_g, as shown in Figure 1. The whole system is influenced by the presence of an electric field E_0 parallel to the interfaces $y = \pm a$, and streaming through a porous medium whose porosity is ε, and whose permeability is k_1.

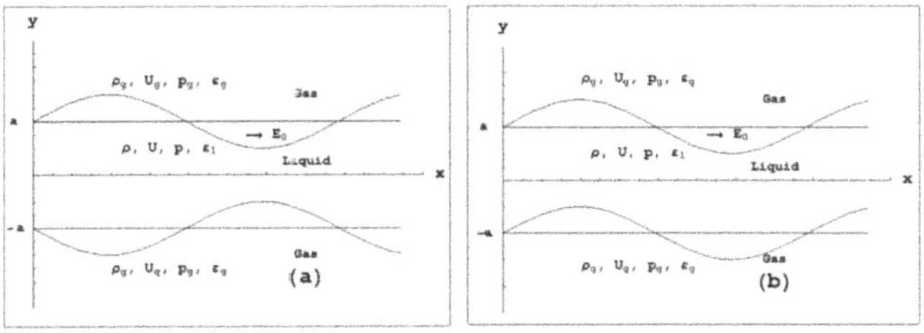

Figure 1. Description of the system for (**a**) Symmetric and (**b**) Antisymmetric disturbances.

The governing equations in three-dimensional Cartesian coordinates are the equation of continuity, Navier–Stokes equation of motion, and Maxwell's equations in both liquid and gas media defined by [32,33]

$$\frac{\partial u_l}{\partial x} + \frac{\partial v_l}{\partial y} + \frac{\partial w_l}{\partial z} = 0 \tag{1}$$

$$\frac{1}{\varepsilon}\frac{\partial u_l}{\partial t} + \frac{1}{\varepsilon^2}U_{0l}\frac{\partial u_l}{\partial x} = -\frac{1}{\rho_l}\frac{\partial p_l}{\partial x} - \frac{1}{k_1}\left(v_l - v_l'\nabla^2\right)u_l \tag{2}$$

$$\frac{1}{\varepsilon}\frac{\partial v_l}{\partial t} + \frac{1}{\varepsilon^2}U_{0l}\frac{\partial v_l}{\partial x} = -\frac{1}{\rho_l}\frac{\partial p_l}{\partial y} - \frac{1}{k_1}\left(v_l - v_l'\nabla^2\right)v_l \tag{3}$$

$$\frac{1}{\varepsilon}\frac{\partial w_l}{\partial t} + \frac{1}{\varepsilon^2}U_{0l}\frac{\partial w_l}{\partial x} = -\frac{1}{\rho_l}\frac{\partial p_l}{\partial z} - \frac{1}{k_1}\left(v_l - v_l'\nabla^2\right)w_l \tag{4}$$

and

$$\frac{\partial u_g}{\partial x} + \frac{\partial v_g}{\partial y} + \frac{\partial w_g}{\partial z} = 0 \tag{5}$$

$$\frac{1}{\varepsilon}\frac{\partial u_g}{\partial t} + \frac{1}{\varepsilon^2}U_{0g}\frac{\partial u_g}{\partial x} = -\frac{1}{\rho_g}\frac{\partial p_g}{\partial x} - \frac{v_g}{k_1}u_g \tag{6}$$

$$\frac{1}{\varepsilon}\frac{\partial v_g}{\partial t} + \frac{1}{\varepsilon^2}U_{0g}\frac{\partial v_g}{\partial x} = -\frac{1}{\rho_g}\frac{\partial p_g}{\partial y} - \frac{v_g}{k_1}v_g \tag{7}$$

$$\frac{1}{\varepsilon}\frac{\partial w_g}{\partial t} + \frac{1}{\varepsilon^2}U_{0g}\frac{\partial w_g}{\partial x} = \frac{1}{\rho_g}\frac{\partial p_g}{\partial z} - \frac{v_g}{k_1}w_g \tag{8}$$

$$\nabla \cdot \left(\epsilon_j \boldsymbol{E}_j\right) = 0 \tag{9}$$

$$\nabla \times \boldsymbol{E}_j = 0 \tag{10}$$

Maxwell's equations of motion are used to derive Equations (9) and (10) by assuming that the quasi-static approximation is applicable in this problem, and so the electric field $\boldsymbol{E}_j$ can be calculated from the gradient of a scalar electric potential ψ_j. The subscripts $(j = l, g)$ signify the liquid sheet and the gas.

The flow has been divided into a steady flow and a time dependent perturbation. The total electric field will be defined as $\boldsymbol{E}_j = E_0\boldsymbol{i} - \nabla\psi_j$, then Equation (9) indicates that electric potentials ψ_j in the two regions satisfy the following Laplace's equations

$$\nabla^2\psi_j = 0 \tag{11}$$

Let $u_{1j}, v_{1j}, w_{1j}, p_{1j}$, and ψ_{1j} where $(j = l, g)$, denote perturbations in the fluid velocity components u_{0j}, v_{0j}, w_{0j}, the pressure p_{0j}, and the electric potentials ψ_{0j}. Hence, we also can write

$$u_j = u_{0j} + u_{1j}, v_j = v_{1j}, w_j = w_{1j}, p_j = p_{0j} + p_{1j} \text{ and } \psi_j = \psi_{0j} + \psi_{1j}$$

Substituting for the above quantities into Equations (1)–(8), and (11), we obtain the following perturbed equations in the liquid sheet medium

$$\frac{\partial u_{1l}}{\partial x} + \frac{\partial v_{1l}}{\partial y} + \frac{\partial w_{1l}}{\partial z} = 0 \tag{12}$$

$$\frac{1}{\varepsilon^2}\left[\varepsilon\frac{\partial u_{1l}}{\partial t} + U_{0l}\frac{\partial u_{1l}}{\partial x} + \frac{\varepsilon^2}{k_1}\left(v_l - v_l'\nabla^2\right)u_{1l}\right] = -\frac{1}{\rho_l}\frac{\partial p_{1l}}{\partial x} \tag{13}$$

$$\frac{1}{\varepsilon^2}\left[\varepsilon\frac{\partial v_l}{\partial t} + U_{0l}\frac{\partial v_{1l}}{\partial x} + \frac{\varepsilon^2}{k_1}\left(v_l - v_l'\nabla^2\right)v_{1l}\right] = -\frac{1}{\rho_l}\frac{\partial p_{1l}}{\partial y} \tag{14}$$

$$\frac{1}{\varepsilon^2}\left[\varepsilon\frac{\partial w_{1l}}{\partial x} + U_{0l}\frac{\partial w_{1l}}{\partial x} + \frac{\varepsilon^2}{k_1}\left(v_l - v_l'\nabla^2\right)w_{1l}\right] = -\frac{1}{\rho_l}\frac{\partial p_{1l}}{\partial z} \tag{15}$$

and in the gas region as

$$\frac{\partial u_{1g}}{\partial x} + \frac{\partial v_{1g}}{\partial y} + \frac{\partial w_{1g}}{\partial z} = 0 \tag{16}$$

$$\frac{1}{\varepsilon^2}\left[\varepsilon\frac{\partial u_{1g}}{\partial t} + U_{0g}\frac{\partial u_{1g}}{\partial x} + \frac{\varepsilon^2 v_g}{k_1}u_{1g}\right] = -\frac{1}{\rho_g}\frac{\partial p_{1g}}{\partial x} \tag{17}$$

$$\frac{1}{\varepsilon^2}\left[\varepsilon\frac{\partial v_{1g}}{\partial t} + U_{0g}\frac{\partial v_{1g}}{\partial x} + \frac{\varepsilon^2 v_g}{k_1}v_{1g}\right] = -\frac{1}{\rho_g}\frac{\partial p_{1g}}{\partial y} \tag{18}$$

$$\frac{1}{\varepsilon^2}\left[\varepsilon\frac{\partial v_{1g}}{\partial t} + U_{0g}\frac{\partial v_{1g}}{\partial x} + \frac{\varepsilon^2 v_g}{k_1}w_{1g}\right] = -\frac{1}{\rho_g}\frac{\partial p_{1g}}{\partial z} \tag{19}$$

Together with the electric potential equations in both regions

$$\nabla^2\psi_{1l} = 0 \text{ and } \nabla^2\psi_{1g} = 0 \tag{20}$$

3. Normal Modes Analysis and Solutions

We seek solutions to Equations (13)–(20) in the liquid and gas medium whose dependency on x, y, z, and t is of the type [34] to analyze the disturbance into normal modes analysis

$$(u_{1l}, v_{1l}, w_{1l}, p_{1l}, \psi_{1l}) = (U_l, V_l, W_l, P_l, \Psi_l)\exp[i(kx + nz) + \omega t] \tag{21}$$

$$\left(u_{1g}, v_{1g}, w_{1g}, p_{1g}, \psi_{1g}\right) = \left(U_g, V_g, W_g, P_g, \Psi_g\right)\exp[i(kx + nz) + \omega t] \tag{22}$$

where $U_l, V_l, W_l, P_l, \Psi_l$ and $U_g, V_g, W_g, P_g, \Psi_g$ are functions of y only; k and n are the wave numbers in the x and z, respectively, and $\omega = \omega_r + i\omega_i$ is the complex frequency.

The interface displacement is given by

$$\eta = \eta_0\exp[i(kx + nz) + \omega t] \tag{23}$$

Substituting Equations (21) and (22) into Equations (13)–(20), we obtain in the liquid sheet medium the equation

$$ikU_l + DV_l + inW_l = 0 \tag{24}$$

$$\left\{(\varepsilon\omega + ikU_{0l}) + \frac{\varepsilon^2}{k_1}\left[v_l - v_l'\left(D^2 - m^2\right)\right]\right\}U_l = -\frac{\varepsilon^2}{\rho_l}(ikP_l) \tag{25}$$

$$\left\{(\varepsilon\omega + ikU_{0l}) + \frac{\varepsilon^2}{k_1}\left[v_l - v_l'\left(D^2 - m^2\right)\right]\right\}V_l = -\frac{\varepsilon^2}{\rho_l}(DP_l) \tag{26}$$

$$\left\{(\varepsilon\omega + ikU_{0l}) + \frac{\varepsilon^2}{k_1}\left[v_l - v_l'\left(D^2 - m^2\right)\right]\right\}W_l = -\frac{\varepsilon^2}{\rho_l}(inP_l) \tag{27}$$

$$\left(D^2 - m^2\right)\Psi_l = 0 \tag{28}$$

where $m^2 = k^2 + n^2$. Similarly, in the gas medium we obtain the following equations

$$ikU_g + DV_g + inW_g = 0 \tag{29}$$

$$\left[(\varepsilon\omega + ikU_{0g}) + \frac{\varepsilon^2 v_g}{k_1}\right]U_g = -\frac{\varepsilon^2}{\rho_l}(ikP_g) \tag{30}$$

$$\left[(\varepsilon\omega + ikU_{0g}) + \frac{\varepsilon^2 v_g}{k_1}\right]V_g = -\frac{\varepsilon^2}{\rho_l}(DP_g) \tag{31}$$

$$\left[(\varepsilon\omega + ikU_{0g}) + \frac{\varepsilon^2 v_g}{k_1}\right]W_g = -\frac{\varepsilon^2}{\rho_l}(inP_g) \tag{32}$$

$$\left(D^2 - m^2\right)\Psi_g = 0 \tag{33}$$

Multiplying Equation (25) by ik, operate by the operator D on Equation (26), multiply Equation (27) by in, and add the obtained three equations. Afterwards, using Equation (24), we obtain

$$\left(D^2 - m^2\right)P_l = 0 \tag{34}$$

The solution of Equation (34) is given by

$$p_{1l} = \left[C_1 e^{my} + C_2 e^{-my}\right]\exp[i(kx + nz) + \omega t] \tag{35}$$

where C_1 and C_2 are constants of integration to be determined. Then, we can write Equation (25) can be written in the form

$$\left(D^2 - s^2\right)U_l = \frac{ikk_1}{\rho_l v_l'}\left(C_1 e^{my} + C_2 e^{-my}\right) \tag{36}$$

where

$$s^2 = m^2 + \frac{k_1(\varepsilon\omega + ikU_{ol}) + \varepsilon^2 v_l}{\varepsilon^2 v_l'} \tag{37}$$

Hence, the general solution of nonhomogeneous differential Equation (36) is given by

$$u_{1l} = \left[C_3 e^{sy} + C_4 e^{-sy} + \frac{ikk_1}{\rho_l v_l'}\left(\frac{C_1 e^{my} + C_2 e^{-my}}{m^2 - s^2}\right)\right]\exp[i(kx + nz) + \omega t] \tag{38}$$

Similarly, Equations (26) and (27) can be written in the form

$$\left(D^2 - s^2\right)V_l = \frac{mk_1}{\rho_l v_l'}\left(C_1 e^{my} - C_2 e^{-my}\right) \tag{39}$$

$$\left(D^2 - s^2\right)W_l = \frac{ink_1}{\rho_l v'_l}\left(C_1 e^{my} + C_2 e^{-my}\right) \tag{40}$$

The solutions to nonhomogeneous differential Equations (39) and (40) are

$$v_{1l} = \left[C_5 e^{sy} + C_6 e^{-sy} + \frac{mk_1}{\rho_l v'_l}\left(\frac{C_1 e^{my} - C_2 e^{-my}}{m^2 - s^2}\right)\right]\exp[i(kx + nz) + \omega t] \tag{41}$$

$$w_{1l} = \left[C_7 e^{sy} + C_8 e^{-sy} + \frac{ink_1}{\rho_l v'_l}\left(\frac{C_1 e^{my} + C_2 e^{-my}}{m^2 - s^2}\right)\right]\exp[i(kx + nz) + \omega t] \tag{42}$$

where C_3–C_8 are constants to determined. Furthermore, from Equations (29)–(32), we obtain

$$\left(D^2 - m^2\right)P_g = 0 \tag{43}$$

The solution of Equations (28), (33) and (43) can be written in the forms
Hence, we can write

$$p_{1g} = \left[C_9 e^{my} + C_{10} e^{-my}\right]\exp[i(kx + nz) + \omega t] \tag{44}$$

$$\psi_{1l} = \left[A_1 e^{my} + A_2 e^{-my}\right]\exp[i(kx + nz) + \omega t] \tag{45}$$

$$\psi_{1g} = \left[B_1 e^{my} + B_2 e^{-my}\right]\exp[i(kx + nz) + \omega t] \tag{46}$$

where A_1, A_2, B_1, B_2, C_9 and C_{10} are constants of integration to be determined.

4. Boundary Conditions

If the disturbance is antisymmetric or symmetric, the required boundary conditions for the considered system are different. The upper and lower interfaces are used for antisymmetric disturbances, for example at $y = \pm a$ is described by $y = a + \eta$ and $y = -a + \eta$, respectively, while for symmetric disturbance, the two interfaces are described, respectively, by $y = a + \eta$ and $y = -a - \eta$, where η is defined by Equation (23).

The antisymmetric disturbance case boundary conditions at $y = \pm a$ are [8,13]

1. The kinematic boundary condition should be satisfied at the two interfaces, which states that the normal velocities of the liquid sheet at $y = \pm a$ are

$$v_{1l} = \varepsilon\frac{\partial\eta}{\partial t} + U_{0l}\frac{\partial\eta}{\partial x} \quad \text{at} \quad y = \pm a \tag{47}$$

and for the gas medium, this condition yields

$$v_{1g} = \varepsilon\frac{\partial\eta}{\partial t} + U_{0g}\frac{\partial\eta}{\partial x} \quad \text{at} \quad y = a \tag{48}$$

2. The perturbed velocity of the gas far away from the interface should be vanishes, i.e.,

$$v_{1g} = 0 \quad \text{at} \quad y \to \pm\infty \tag{49}$$

3. The stress tensor's tangential component must be continuous at the interfaces, i.e.,

$$\tau_{yx} = \frac{1}{k_1}\left(\mu_l - \mu'_l\nabla^2\right)\left(\frac{\partial u_{1l}}{\partial y} + \frac{\partial v_{1l}}{\partial x}\right) = 0 \quad \text{at} \quad y = \pm a \tag{50}$$

$$\tau_{yz} = \frac{1}{k_1}\left(\mu_l - \mu'_l\nabla^2\right)\left(\frac{\partial v_{1l}}{\partial z} + \frac{\partial w_{1l}}{\partial y}\right) = 0 \quad \text{at} \quad y = \pm a \tag{51}$$

4. At the interfaces, the electric field's tangential component is continuous.

$$\frac{\partial \psi_{1l}}{\partial x} = \frac{\partial \psi_{1g}}{\partial x} \quad \text{at} \quad y = \pm a \tag{52}$$

5. At interfaces, the electric displacement's normal component is continuous.

$$\epsilon_l \frac{\partial \psi_{1l}}{\partial y} - \epsilon_g \frac{\partial \psi_{1g}}{\partial y} = ik\eta E_0(\epsilon_l - \epsilon_g) \quad \text{at} \quad y = \pm a \tag{53}$$

6. The stress tensor normal component is broken up at the interface by the surface tension coefficient, i.e.,

$$-p_{1l} + \epsilon_l E_0 \frac{\partial \psi_{1l}}{\partial x} + 2\left(\mu_l - \mu_l' \nabla^2\right) \frac{\partial v_{1l}}{\partial y} = -p_{1g} + \epsilon_g E_0 \frac{\partial \psi_{1g}}{\partial x} + p_\sigma \quad \text{at} \quad y = \pm a \tag{54}$$

where p_σ is the pressure due to the surface tension σ.
Please note that for symmetric disturbances the boundary conditions that change forms are

7. The kinematic boundary condition (47) and the electric displacement boundary condition (53) at $y = -a$ take the forms

$$v_{1l} = -\left(\varepsilon \frac{\partial \eta}{\partial t} + U_{0l} \frac{\partial \eta}{\partial x}\right) \text{ at } y = -a \tag{55}$$

$$v_{1g} = -\left(\varepsilon \frac{\partial \eta}{\partial t} + U_{0g} \frac{\partial \eta}{\partial x}\right) \text{ at } y = -a \tag{56}$$

$$\epsilon_l \frac{\partial \psi_{1l}}{\partial y} - \epsilon_g \frac{\partial \psi_{1g}}{\partial y} = -ik\eta E_0(\epsilon_l - \epsilon_g) \text{ at } y = -a \tag{57}$$

5. The Antisymmetric Disturbance Case

For each of the fluid sheets and gas media (upper and lower phases), as well as the differential equation representing the influence of the electric field, we will try to find solutions in this section as follows:

5.1. Solutions in the Liquid Sheet Phase

The kinematic boundary condition (47), on using Equations (23) and (41), can be written at the two interfaces in the form

$$C_5 e^{sa} + C_6 e^{-sa} + \frac{mk_1}{\rho_l v_l'} \frac{(C_1 e^{ma} - C_2 e^{-ma})}{(m^2 - s^2)} = (\varepsilon\omega + ikU_{0l})\eta_0 \tag{58}$$

$$C_5 e^{-sa} + C_6 e^{sa} + \frac{mk_1}{\rho_l v_l'} \frac{(C_1 e^{-ma} - C_2 e^{ma})}{(m^2 - s^2)} = (\varepsilon\omega + ikU_{0l})\eta_0 \tag{59}$$

Solving Equations (58) and (59), we obtain

$$C_5 = \frac{(\varepsilon\omega + ikU_{0l})\eta_0}{2\cosh(sa)} - \frac{mk_1\{C_1 \sinh[(s+m)a] - C_2 \sinh[(s-m)a]\}}{\rho_l v_l'(m^2 - s^2)\sinh(2sa)} \tag{60}$$

$$C_6 = \frac{(\varepsilon\omega + ikU_{0l})\eta_0}{2\cosh(sa)} - \frac{mk_1\{C_1 \sinh[(s-m)a] - C_2 \sinh[(s+m)a]\}}{\rho_l v_l'(m^2 - s^2)\sinh(2sa)} \tag{61}$$

The tangential stress boundary condition (50), on using Equations (38), (41), (58) and (59), can be written at the two interfaces in the form

$$C_3 s e^{sa} - C_4 s e^{-sa} + \frac{ikk_1 m(C_1 e^{ma} - C_2 e^{-ma})}{\rho_l v_l'(m^2 - s^2)} + ik(\varepsilon\omega + ikU_{0l})\eta_0 = 0 \tag{62}$$

$$C_3 s e^{-sy} - C_4 s e^{sa} + \frac{ikk_1 m(C_1 e^{-ma} - C_2 e^{ma})}{\rho_l v_l'(m^2 - s^2)} + ik(\varepsilon\omega + ikU_{0l})\eta_0 = 0 \tag{63}$$

Solving Equations (62) and (63), we obtain

$$C_3 = -\frac{ikk_1 m\{C_1 \sinh[(s+m)a] - C_2 \sinh[(s-m)a]\}}{s\rho_l v_l'(m^2 - s^2)\sinh(2sa)} - \frac{ik(\varepsilon\omega + ikU_{0l})\eta_0}{2s\cosh(sa)} \tag{64}$$

$$C_4 = \frac{ikk_1 m\{C_1 \sinh[(s-m)a] - C_2 \sinh[(s+m)a]\}}{s\rho_l v_l'(m^2 - s^2)\sinh(2sa)} + \frac{ik(\varepsilon\omega + ikU_{0l})\eta_0}{2s\cosh(sa)} \tag{65}$$

The tangential stress boundary condition (51), on using Equations (41), (42), (58) and (59), can be written at the two interfaces in the form

$$s\left(C_7 e^{sa} - C_8 e^{-sa}\right) = -in(\varepsilon\omega + ikU_{0l})\eta_0 - \frac{inmk_1(C_1 e^{ma} - C_2 e^{-ma})}{\rho_l v_l'(m^2 - s^2)} \tag{66}$$

$$s\left(C_7 e^{-sa} - C_8 e^{sa}\right) = -in(\varepsilon\omega + ikU_{0l})\eta_0 - \frac{inmk_1(C_1 e^{-ma} - C_2 e^{ma})}{\rho_l v_l'(m^2 - s^2)} \tag{67}$$

Solving Equations (66) and (67), we obtain

$$C_7 = -\frac{in(\varepsilon\omega + ikU_{0l})\eta_0}{2s\cosh(sa)} - \frac{inmk_1\{C_1 \sinh[(s+m)a] - C_2 \sinh[(s-m)a]\}}{s\rho_l v_l'(m^2 - s^2)\sinh(2sa)} \tag{68}$$

$$C_8 = \frac{in(\varepsilon\omega + ikU_{0l})\eta_0}{2s\cosh(sa)} + \frac{inmk_1\{C_1 \sinh[(s-m)a] - C_2 \sinh[(s+m)a]]\}}{s\rho_l v_l'(m^2 - s^2)\sinh(2sa)} \tag{69}$$

The equation of continuity (12) at $y = \pm a$, on using Equations (38), (41), and (42) can be written in the form

$$\begin{aligned}\left[\sinh^2(sa)\cosh(ma) + \cosh^2(sa)\sinh(ma)\right]C_1 \\ - \left[\sinh^2(sa)\cosh(ma) - \cosh^2(sa)\sinh(ma)\right]C_2 \\ + \frac{\rho_l v_l'}{k_1 m}\left(m^2 + s^2\right)(\varepsilon\omega + ikU_{0l})\eta_0 \sinh^2(sa) = 0\end{aligned} \tag{70}$$

and

$$\begin{aligned}\left[\sinh^2(sa)\cosh(ma) - \cosh^2(sa)\sinh(ma)\right]C_1 \\ - \left[\sinh^2(sa)\cosh(ma) + \cosh^2(sa)\sinh(ma)\right]C_2 \\ + \frac{\rho_l v_l'}{k_1 m}\left(m^2 + s^2\right)(\varepsilon\omega + ikU_{0l})\eta_0 \sinh^2(sa) = 0\end{aligned} \tag{71}$$

Solving Equations (70) and (71), we obtain

$$C_1 = -C_2 = -\frac{\rho_l v_l'(m^2 + s^2)(\varepsilon\omega + ikU_{0l})\eta_0}{2k_1 m\cosh(ma)} \tag{72}$$

Substitute from Equations (72) into Equations (64), (65), (60), (61), (68), and (69), respectively, we obtain

$$C_3 = -C_4 = \frac{iks(\varepsilon\omega + ikU_{0l})\eta_0}{(m^2 - s^2)\cosh(sa)} \tag{73}$$

$$C_5 = C_6 = \frac{m^2(\varepsilon\omega + ikU_{0l})\eta_0}{(m^2 - s^2)\cosh(sa)} \tag{74}$$

$$C_7 = -C_8 = -\frac{ins(\varepsilon\omega + ikU_{0l})\eta_0}{(m^2 - s^2)\cosh(sa)} \tag{75}$$

Substitute from Equations (72) and (75) into Equations (38), (41) and (42) we obtain

$$\mathrm{p}_{1l} = -\frac{\rho_l v_l'(m^2 + s^2)(\varepsilon\omega + ikU_{0l})\eta_0 \sinh(my)}{k_1 m \cosh(ma)} \exp[i(kx + nz) + \omega t] \tag{76}$$

and

$$u_{1l} = \frac{ik(\varepsilon\omega + ikU_{0l})\eta_0}{(m^2 - s^2)}\left[\frac{2s\sinh(sy)}{\cosh(sa)} - \frac{(m^2 + s^2)\sinh(my)}{m\cosh(ma)}\right]\exp[i(kx + nz) + \omega t] \tag{77}$$

$$v_{1l} = \frac{(\varepsilon\omega + ikU_{0l})\eta_0}{(m^2 - s^2)}\left[\frac{2m^2\cosh(sy)}{\cosh(sa)} - \frac{(m^2 + s^2)\cosh(my)}{\cosh(ma)}\right]\exp[i(kx + nz) + \omega t] \tag{78}$$

$$w_{1l} = \frac{in(\varepsilon w + ikU_{0l})\eta_0}{(m^2 - s^2)}\left[\frac{2s\sinh(sy))}{\cosh(sa)} - \frac{(m^2 + s^2)\sinh(my))}{m\cosh(ma)}\right]\exp[i(kx + nz) + \omega t] \tag{79}$$

5.2. Solutions in the Gas Medium (in the Upper and Lower Phases)

Substitute form Equation (44) into Equations (30)–(32), we obtain

$$u_{1g} = -\frac{ik\varepsilon^2(C_9 e^{my} + C_{10}e^{-my})}{\rho_g\left[(\varepsilon\omega + ikU_{0g}) + \frac{\varepsilon^2 v_g}{k_1}\right]} \tag{80}$$

$$v_{1g} = -\frac{m\varepsilon^2(C_9 e^{my} - C_{10}e^{-my})}{\rho_g\left[(\varepsilon\omega + ikU_{0g}) + \frac{\varepsilon^2 v_g}{k_1}\right]} \tag{81}$$

$$w_{1g} = -\frac{in\varepsilon^2(C_9 e^{my} + C_{10}e^{-my})}{\rho_g\left[(\varepsilon\omega + ikU_{0g}) + \frac{\varepsilon^2 v_g}{k_1}\right]} \tag{82}$$

Using the kinematic boundary condition (48), and condition (49) in the two regions ($y \to \infty$) and ($y \to -\infty$), the in the upper gas region we obtain $C_9 = 0$, while in the lower gas region, we have $C_{10} = 0$. Then in the upper gas region ($y \geq a$), we have

$$C_{10} = \frac{\rho_g(\varepsilon\omega + ikU_{0g})\eta_0}{m\varepsilon^2}\left[(\varepsilon\omega + ikU_{0g}) + \frac{\varepsilon^2 v_g}{k_1}\right]e^{ma} \tag{83}$$

Therefore, in the upper gas medium, Equation (44) gives

$$p_{1g}^u = \frac{\rho_g(\varepsilon\omega + ikU_{0g})\eta_0}{m\varepsilon^2}\left[(\varepsilon\omega + ikU_{0g}) + \frac{\varepsilon^2 v_g}{k_1}\right]\exp[m(a - y)]\exp[i(kx + nz) + \omega t] \tag{84}$$

Furthermore, Equations (80)–(82) yield the velocity component for ($y \geq a$) as

$$u_{1g}^u = -\frac{ik(\varepsilon\omega + ikU_{0g})\eta_0}{m}\exp[m(a - y)]\exp[i(kx + nz) + \omega t] \tag{85}$$

$$v_{1g}^{u} = (\varepsilon\omega + ikU_{0g})\eta_0 \exp[m(a-y)] \exp[i(kx+nz)+\omega t] \tag{86}$$

$$w_{1g}^{u} = -\frac{ik(\varepsilon\omega + ikU_{0g})\eta_0}{m} \exp[m(a-y)] \exp[i(kx+nz)+\omega t] \tag{87}$$

Similarly, in the lower gas region $(y \leq -a)$, we have

$$C_9 = -\frac{\rho_g(\varepsilon\omega + ikU_{0g})\eta_0}{m\varepsilon^2}\left[(\varepsilon\omega + ikU_{0g}) + \frac{\varepsilon^2 v_g}{k_1}\right] e^{ma} \tag{88}$$

Therefore, in the lower gas medium, Equation (44) gives

$$p_{1g}^{l} = -\frac{\rho_l(\varepsilon\omega + ikU_{0g})\eta_0}{m\varepsilon^2}\left[(\varepsilon\omega + ikU_{0g}) + \frac{\varepsilon^2 v_g}{k_1}\right] \exp[m(a+y)] \exp[i(kx+nz)+\omega t] \tag{89}$$

Hence, Equations (80)–(82) yield

$$u_{1g}^{l} = \frac{ik(\varepsilon\omega + ikU_{0g})\eta_0}{m} \exp[m(a+y)] \exp[i(kx+nz)+\omega t] \tag{90}$$

$$v_{1g}^{l} = (\varepsilon\omega + ikU_{0g})\eta_0 \exp[m(a+y)] \exp[i(kx+nz)+\omega t] \tag{91}$$

$$w_{1g}^{l} = \frac{in(\varepsilon\omega + ikU_{0g})\eta_0}{m} \exp[m(a+y)] \exp[i(kx+nz)+\omega t] \tag{92}$$

Furthermore, the pressure due to the presence of surface tension is given by

$$p_\sigma = \sigma\left(\frac{\partial^2\eta}{\partial x^2} + \frac{\partial^2\eta}{\partial z^2}\right) = -\sigma m^2 \eta_0 \exp[i(kx+nz)+\omega t] \tag{93}$$

5.3. Solutions of the Electric Field (in the Upper and Lower Phases)

From Equation (46), since $\psi_{1g} \to 0$ as $y \to \pm\infty$, then in the upper gas region $(y \geq a)$, we take $B_1 = 0$, while in the lower gas region $(y \leq -a)$, we take $B_2 = 0$. Then substituting from Equation (45) of the liquid phase and Equation (46) for the gas phase (in the upper and lower regions) into the boundary conditions (52) and (53) at $y = \pm a$, we obtain

$$\left(A_1 e^{ma} + A_2 e^{-ma}\right) = B_2 e^{-ma} \tag{94}$$

$$\left(A_1 e^{-ma} + A_2 e^{ma}\right) = B_1 e^{-ma} \tag{95}$$

$$\epsilon_l m\left(A_1 e^{ma} - A_2 e^{-ma}\right) + \epsilon_g m B_2 e^{-ma} = ik\eta_0 E_0(\epsilon_l - \epsilon_g) \tag{96}$$

$$\epsilon_l m\left(A_1 e^{-ma} - A_2 e^{ma}\right) - \epsilon_g m B_1 e^{-ma} = ik\eta_0 E_0(\epsilon_l - \epsilon_g) \tag{97}$$

Now, solving Equations (94)–(97), we obtain

$$A_1 = -A_2 = \frac{ik\eta_0 E_0(\epsilon_l - \epsilon_g)}{2m\cosh(ma)[\epsilon_l + \epsilon_g \tanh(ma)]} \tag{98}$$

and

$$B_1 = -B_2 = -\frac{ik\eta_0 E_0(\epsilon_l - \epsilon_g)e^{ma}\tanh(ma)}{m[\epsilon_l + \epsilon_g \tanh(ma)]} \tag{99}$$

Therefore, the solutions (45) and (46) can be written in the form

$$\psi_{1l} = \frac{ik\eta_0 E_0(\epsilon_l - \epsilon_g)\sinh(my)}{m[\epsilon_l + \epsilon_g \tanh(ma)]\cosh(ma)} \exp[i(kx+nz)+\omega t] \tag{100}$$

$$\psi_{1g}^{u} = \frac{ik\eta_0 E_0(\epsilon_l - \epsilon_g)\tanh(ma)}{m[\epsilon_l + \epsilon_g \tanh(ma)]} \exp[m(a-y)] \exp[i(kx+nz)+\omega t] \tag{101}$$

$$\psi^l_{1g} = -\frac{ik\eta_0 E_0(\epsilon_l - \epsilon_g)\tanh(ma)}{m[\epsilon_l + \epsilon_g \tanh(ma)]}\exp[m(a+y)]\exp[i(kx+nz)+\omega t] \tag{102}$$

6. The Symmetric Disturbance Case

In this section, we will try to obtain the solutions for the above-mentioned differential equations describing each of the liquid sheet medium, and the gas medium (in the upper and lower phases), together with the solution of the differential equation of the electric field, in the corresponding case of symmetric disturbance. We note that the obtained solutions for the gas medium in this case is found to be similar to the solutions for the previous case of antisymmetric disturbances and will not be given here.

6.1. Solutions in the Liquid Sheet Phase

Following the same procedure shown in the previous case of antisymmetric disturbance together with the help of Equation (55) at $y = -a$, we obtain

$$C_1 = C_2 = -\frac{\rho_l v'_l(m^2+s^2)(\varepsilon\omega + ikU_{0l})\eta_0}{2k_1 m \sinh(ma)} \tag{103}$$

$$C_3 = C_4 = \frac{iks(\varepsilon\omega + ikU_{0l})\eta_0}{(m^2-s^2)\sinh(sa)} \tag{104}$$

$$C_5 = -C_6 = \frac{m^2(\varepsilon\omega + ikU_{0l})\eta_0}{(m^2-s^2)\sinh(sa)} \tag{105}$$

$$C_7 = C_8 = \frac{ins(\varepsilon\omega + ikU_{0l})\eta_0}{(m^2-s^2)\sinh(sa)} \tag{106}$$

Hence, we have the following solution

$$p_{1l} = -\frac{\rho_l v'_l(m^2+s^2)(\varepsilon\omega + ikU_{0l})\eta_0\cosh(my)}{k_1 m \sinh(ma)}\exp[i(kx+nz)+\omega t] \tag{107}$$

$$u_{1l} = \frac{ik(\varepsilon\omega + ikU_{0l})\eta_0}{(m^2-s^2)}\left[\frac{2s\cosh(sy)}{\sinh(sa)} - \frac{(m^2+s^2)\cosh(my)}{m\sinh(ma)}\right]\exp[i(kx+nz)+\omega t] \tag{108}$$

$$v_{1l} = \frac{(\varepsilon\omega + ikU_{0l})\eta_0}{(m^2-s^2)}\left[\frac{2m^2\sinh(sy)}{\sinh(sa)} - \frac{(m^2+s^2)\sinh(my)}{\sinh(ma)}\right]\exp[i(kx+nz)+\omega t] \tag{109}$$

$$w_{1l} = \frac{in(\varepsilon w + ikU_{0l})\eta_0}{(m^2-s^2)}\left[\frac{2s\cosh(sy))}{\sinh(sa)} - \frac{(m^2+s^2)\cosh(my))}{m\sinh(ma)}\right]\exp[i(kx+nz)+\omega t] \tag{110}$$

6.2. Solutions of the Electric Field (in the Upper and Lower Phases)

Following the same procedure given in the previous case of antisymmetric disturbance, together with the help of Equation (57) at $y = -a$, we obtain

$$A_1 = A_2 = \frac{ik\eta_0 E_0(\epsilon_l - \epsilon_g)}{2m\sinh(ma)[\epsilon_l + \epsilon_g\coth(ma)]} \tag{111}$$

and

$$B_1 = B_2 = \frac{ik\eta_0 E_0(\epsilon_l - \epsilon_g)e^{ma}\coth(ma)}{m[\epsilon_l + \epsilon_g\coth(ma)]} \tag{112}$$

Hence, we have the following solutions

$$\psi_{1l} = \frac{ik\eta_0 E_0(\epsilon_l - \epsilon_g)\cosh(my)}{m[\epsilon_l + \epsilon_g\coth(ma)]\sinh(ma)}\exp[i(kx+nz)+\omega t] \tag{113}$$

$$\psi_{1g}^{u} = \frac{ik\eta_0 E_0(\epsilon_l - \epsilon_g)\coth(ma)}{m[\epsilon_l + \epsilon_g \coth(ma)]} \exp[m(a-y)] \exp[i(kx+nz)+\omega t] \tag{114}$$

$$\psi_{1g}^{l} = \frac{ik\eta_0 E_0(\epsilon_l - \epsilon_g)\coth(ma)}{m[\epsilon_l + \epsilon_g \coth(ma)]} \exp[m(a+y)] \exp[i(kx+nz)+\omega t] \tag{115}$$

7. Non-Dimensional Dispersion Relations

Using the normal stress boundary condition (54) at $y = a$ we can now calculate the dispersion relation for the antisymmetric disturbance situation by substituting the following Equations (75), (78), (84), (85), (93), (100), and (101) into the normal stress boundary condition:

$$\begin{aligned} &\frac{\rho_l v_l'(m^2+s^2)(\varepsilon\omega + ikU_{0l})}{k_1}\tanh(ma) - \frac{k^2E_0^2(\epsilon_l-\epsilon_g)^2}{[\epsilon_l+\epsilon_g\tanh(ma)]}\tanh(ma) \\ &+\frac{2\mu m^2(\varepsilon\omega+ikU_{0l})}{k_1(m^2-s^2)}\left[2\,\mathrm{ms}\tanh(sa) - m^2\left(m^2+s^2\right)\tanh(ma)\right] \\ &+4\mu' m^3 s(\varepsilon\omega+ikU_{0l})\tanh(sa) + \frac{\rho_g(\varepsilon\omega+ikU_{0g})}{\varepsilon^2}\left[(\varepsilon\omega+ikU_{0g}) + \frac{\varepsilon^2 v_g}{k_1}\right] \\ &+\sigma m^3 = 0 \end{aligned} \tag{116}$$

To facilitate the analysis, we need to write the dispersion relation (7) in dimensionless form using the following non-dimensional quantities: Weber number $We = (\rho_l U_{0l}^2 a)/\sigma$, Reynolds liquid number $Re_l = (U_{0l}a)/v_l$, Ohnesorge number $Oh = \sqrt{We}/Re = v_l\sqrt{\rho_l/(a\sigma)}$ gas to liquid density ratio $\rho = \rho_g/\rho_l$, gas to liquid velocity ratio $U = U_{0g}/U_{0l}$, gas to liquid viscosity ratio $v = v_g/v_l$, Reynolds gas number $Re_g = (U/v)Re_l$, viscoelasticity parameter $V' = (v_l'/a^2)\sqrt{\rho_l/(a\sigma)}$, electric field parameter $E = E_0/\sqrt{\sigma}$, non-dimensional dielectric constants $\tilde{\epsilon}_l = \epsilon_l a, \tilde{\epsilon}_g = \epsilon_g a$, the non-dimensional medium permeability $\kappa_1 = k_1/a^2$ and let $K = ka, N = na$ represent the non-dimensional wave numbers. Furthermore, put $M = ma, S = sa$. We also define the quantities $\Omega = \Omega_r + i\Omega_i\sqrt{We}$, where $\Omega_r = \omega_r a^2\sqrt{\rho_l/(\sigma a)}, \Omega_i = (\omega_i a)/U_{0l}$. Hence the non-dimensional form of the dispersion relation (7) for antisymmetric disturbance case reduces to the form

$$\begin{aligned} &\left\{\frac{(\varepsilon\Omega+iK\sqrt{We})^2}{\varepsilon^2} + \frac{(2M^2V'+Oh)}{\kappa_1}(\varepsilon\Omega+iK\sqrt{We}) - \frac{K^2E^2(\tilde{\epsilon}_l-\tilde{\epsilon}_l)^2}{[\tilde{\epsilon}_l+\tilde{\epsilon}_g\tanh(M)]}\right\}\tanh(M) \\ &-\frac{4\varepsilon^2 V'(Oh)M^3(\varepsilon\Omega+iK\sqrt{We})}{\left[\kappa_1(\varepsilon\Omega+iK\sqrt{We})+\varepsilon^2(Oh)\right]}[S\tanh(S) - M\tanh(M)] \\ &+2(Oh)M^2(\varepsilon\Omega+iK\sqrt{We})\tanh(M) + 4V'M^3S(\varepsilon\Omega+iK\sqrt{We})\tanh(S) \\ &+\frac{\rho v(Oh)(\varepsilon\Omega+iKU\sqrt{We})}{\kappa_1} + \frac{\rho(\varepsilon\Omega+iKU\sqrt{We})^2}{\varepsilon^2} + M^3 = 0 \end{aligned} \tag{117}$$

where

$$S = \sqrt{M^2 + \frac{\left[\kappa_1(\varepsilon\Omega+iK\sqrt{We})+\varepsilon^2(Oh)\right]}{\varepsilon^2 V'}} \tag{118}$$

For the symmetric disturbance example, we use Equations (84), (86), (93), (107), (109), (113), and (114) in Equation (54) to derive the following non-dimensional dispersion relation in the form of a dispersion relation for the symmetric disturbance scenario.

$$\left\{\frac{(\varepsilon\Omega+iK\sqrt{We})^2}{\varepsilon^2}+\frac{(2M^2V'+Oh)}{\kappa_1}(\varepsilon\Omega+iK\sqrt{We})-\frac{K^2E^2(\tilde{\varepsilon}_l-\tilde{\varepsilon}_l)^2}{[\tilde{\varepsilon}_l+\tilde{\varepsilon}_g\coth(M)]}\right\}\coth(M)$$
$$-\frac{4\varepsilon^2V'(Oh)M^3(\varepsilon\Omega+iK\sqrt{We})}{\left[\kappa_1(\varepsilon\Omega+iK\sqrt{We})+\varepsilon^2(Oh)\right]}[S\coth(S)-M\coth(M)] \quad (119)$$
$$+2(Oh)M^2(\varepsilon\Omega+iK\sqrt{We})\coth(M)+4V'M^3S(\varepsilon\Omega+iK\sqrt{We})\coth(S)$$
$$+\frac{\rho v(oh)(\varepsilon\Omega+iKU\sqrt{We})}{\kappa_1}+\frac{\rho(\varepsilon\Omega+iKU\sqrt{We})^2}{\varepsilon^2}+M^3=0$$

The non-dimensional dispersion relations (7) and (7) for non-porous media ($\varepsilon = 1$ and $\kappa_1 \to \infty$), and absence of viscosity and viscoelasticity parameters ($Oh = 0, V' = 0$) and gas velocity ($U = 0$), reduce to the dispersion relations obtained by El- Sayed [35], which is a generalization of the work of Ibrahim and Akpan [36] in absence of electric field, and hence their results have been recovered. This limiting case in the absence of an electric field and presence of viscosity, reduces to the same results obtained by Dasgupta et al. [37], while in absence of viscosity and presence of gas velocity, it reduces to the same results of Nath et al. [38], respectively.

8. Stability Analysis and Discussion

The dispersion relations (7) and (7) are very complicated and cannot be solved analytically to express the real part of growth rates in terms of the wave numbers; otherwise, these two equations can be solved numerically using Mathematica software, to yield values of wave number K as a function of growth rate Ω for various values of the other physical parameters included in the analysis. In providing a guessed root of Ω_r, it is helpful to realize that the imaginary part $\Omega_i = -K$, in accordance with Gaster's theorem [39]. Please note that the three-dimensional results are obtained from the numerical solutions of the dispersion relations (7) and (7) with $N = 1, 2$ (the z-direction wave numbers), while the two-dimensional results corresponds to the case $N = 0$. Hence, by solving the dispersion relations (7) and (7) in the temporal mode of instability for both the antisymmetric and symmetric disturbances cases, the effects of various parameters on the stability of an electrified viscolastic couple-stress liquid sheet surrounded by an inviscid gas in porous medium can be examined. These dispersion relations are solved numerically using Mathematica via a new technique combined between Muller and Gaster methods, see refs. [40,41] to obtain relationships between the non-dimensional real part of growth rates Ω_r and the non-dimensional wave numbers K of the both disturbances. The effects of the other physical parameters including in this study are shown graphically in the following Figures 2–16 which exhibit the growth rate Ω_r as a function of the wave number K for the flow. Both symmetric and antisymmetric disturbances cause Ω_r to rise initially when K climbs to its maximum value, and thereafter it falls. The dominant growth rate and the related dominant wave number are both referred to as the dominant growth rate and the dominant wave number. The critical wave number K_c is the point where the growth rate curve intersects the wave number axis. Couple-stress liquid sheet becomes unstable with positive growth rate in the region beneath the growth rate curve whose wave number ranges from zero to the cutoff critical wave number, which is defined as the instability zone. With maximum dominant growth rates Ω_r and dominant wave number K in place, the liquid sheet becomes unstable. Figure 2 depicts the fluctuation of the non-dimensional growth rate Ω_r with the non-dimensional wave number K in the antisymmetric disturbance situation for two-dimensional configurations ($N = 0$) and three-dimensional ones ($N = 1, 2$) and constant physical parameter values. Two-dimensional disturbances ($N = 0$) have higher growth rates than three-dimensional disturbances ($N \geq 1$), and three-dimensional disturbances with ($N = 1$) have higher growth rates than those with ($N = 2$), as seen in the figure.

As a result, we can say that the system is more unstable in two dimensions than it is in three dimensions.

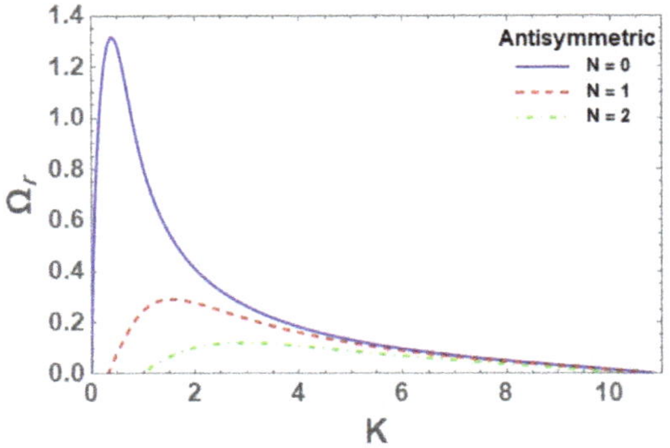

Figure 2. Variation in the non-dimensional growth rate Ω_r with different values of the non-dimensional wave number K the two- and three-dimensional of configuration $N = 0$ and $N = 1, 2$, respectively, when We = 1000; Oh = 0.1; ρ = 0.01; v = 0.2; κ_1 = 2; $\tilde{\varepsilon}_l$ = 0.3; $\tilde{\varepsilon}_g$ = 0.1; E = 3; U = 1.5; V' = 0.1; ε = 0.5, for axisymmetric disturbance case.

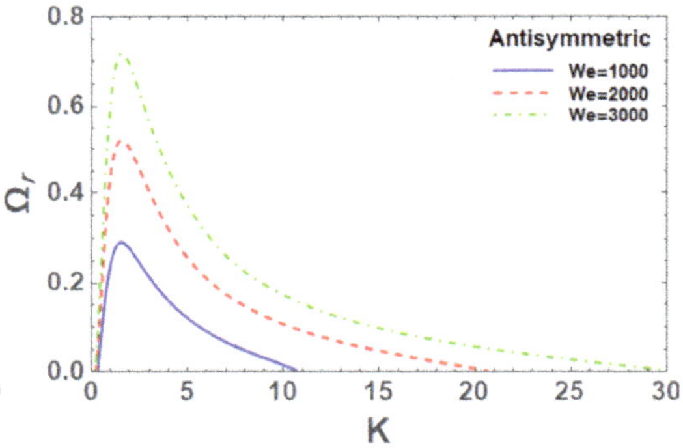

Figure 3. For various Weber numbers We, the variation in the non-dimensional growth rate Ω_r depends on the non-dimensional wave number K, We, when $N = 1; Oh = 0.1$; $\rho = 0.01$; linebreak $n = 1$; v = 0.2; $\kappa_1 = 2$; $\tilde{\varepsilon}_l = 0.3$; $\tilde{\varepsilon}_g = 0.1$; $E = 3$; $U = 1.5$; $V' = 0.1$; $\varepsilon = 0.5$, for axisymmetric disturbance case.

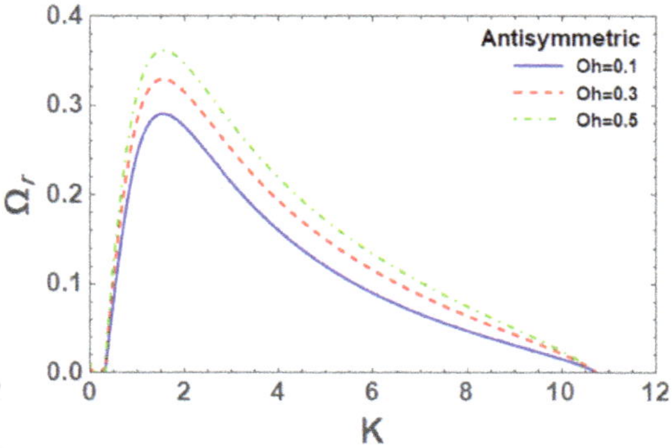

Figure 4. Non-dimensional growth rate Ω_r and non-dimensional wave number K for different values of Ohnesorge number Oh, when $We = 1000$; $N = 1$; $\rho = 0.01$; $v = 0.2$; $\kappa_1 = 2$; $\tilde{\varepsilon}_l = 0.3$; $\tilde{\varepsilon}_g = 0.1$; $E = 3$; $U = 1.5$; $V' = 0.1$; $\varepsilon = 0.5$, for axisymmetric disturbance case.

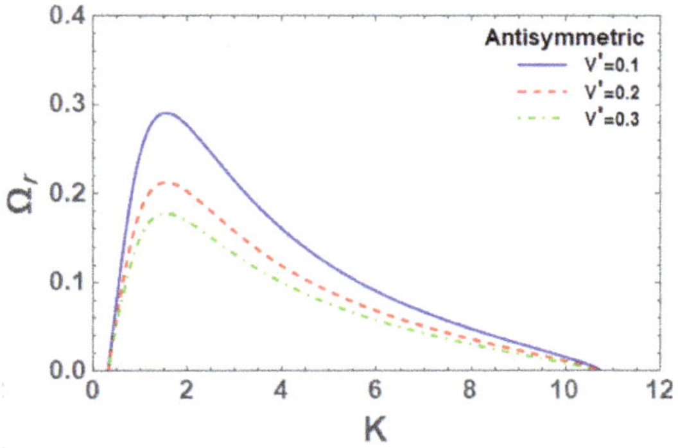

Figure 5. Non-dimensional growth rate Ω_r and non-dimensional wave number K for different values of viscoelastitily parameter V', when $We = 1000$; $Oh = 0.1$; $\rho = 0.01$; $N = 1$; $v = 0.2$; $\kappa_1 = 2$; $\tilde{\epsilon}_l = 0.3$; $\tilde{\epsilon}_g = 0.1$; $U = 1.5$; $\varepsilon = 0.5$, for axisymmetric disturbance case.

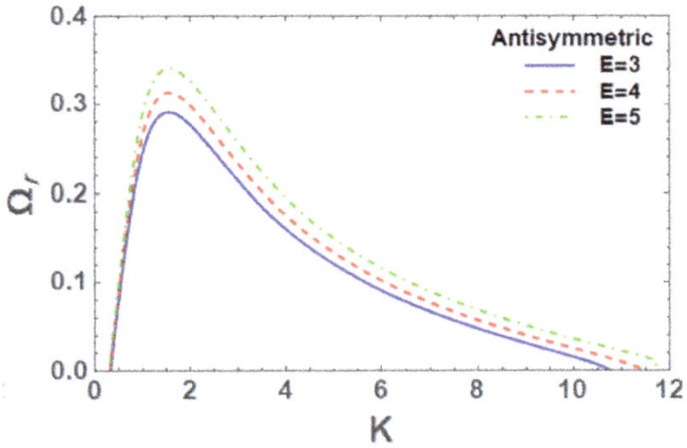

Figure 6. Analysis of non-dimensional growth rate Ω_r and non-dimensional wave number K for different electric field strengths E, when $We = 1000$; $Oh = 0.1$; $\rho = 0.01$; $N = 1$; $v = 0.2$; $\tilde{\epsilon}_l = 0.3$; $\tilde{\epsilon}_g = 0.1$; $E = 3$; $U = 1.5$; $V' = 0.1$; $\varepsilon = 0.5$, for axisymmetric disturbance case.

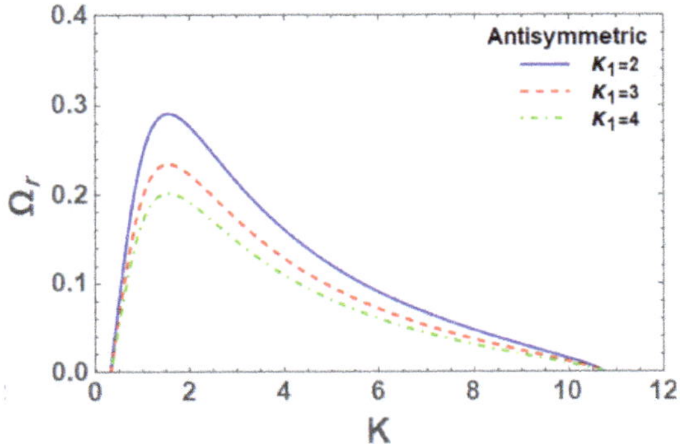

Figure 7. Variation of non-dimensional growth rate Ω_r with non-dimensional wave number K for various medium permeability values κ_1, when $We = 1000$; Oh $= 0.1$; $\rho = 0.01$; $n = 1$; $v = 0.2$; $\kappa_1 = 2$; $\tilde{\epsilon}_l = 0.3$; $\tilde{\epsilon}_g = 0.1$; $E = 3$; $U = 1.5$; $V' = 0.1$, for axisymmetric disturbance case.

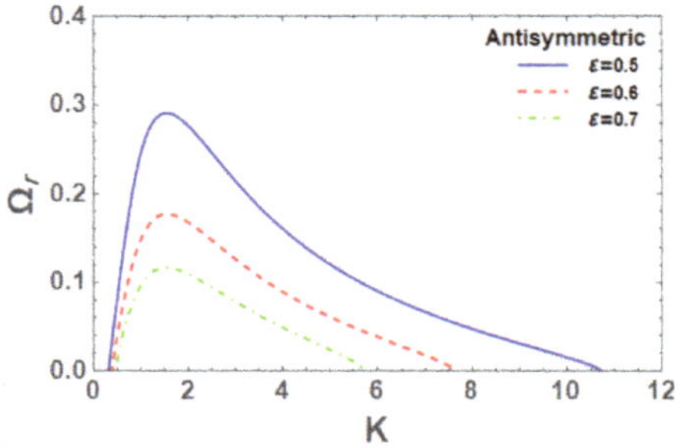

Figure 8. Non-dimensional growth rate Ω_r and non-dimensional wave number K for different values of porous medium porosity ε, when $We = 1000$; $Oh = 0.1$; $N = 1$; $v = 0.2$; $\kappa_1 = 2$; $\tilde{\varepsilon}_l = 0.3$; $\tilde{\varepsilon}_g = 0.1$; $E = 3$; $U = 1.5$; $V' = 0.1$; $\varepsilon = 0.5$, for axisymmetric disturbance case.

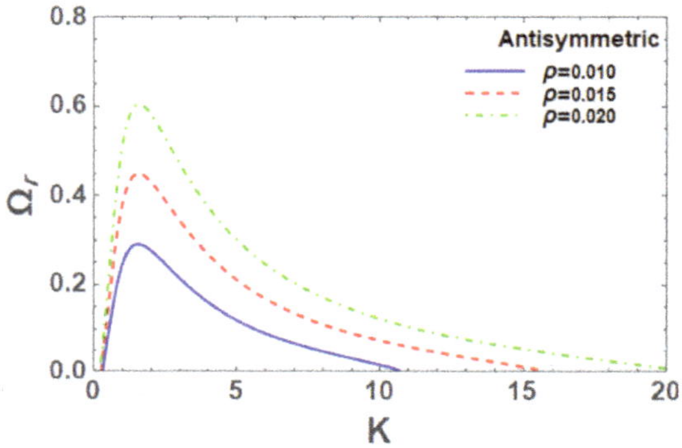

Figure 9. Non-dimensional growth rate Ω_r and non-dimensional wave number K for different gas to liquid densities ρ, when $We = 1000$; $Oh = 0.1$; $N = 1$; $\rho = 0.01$; $v = 0.2$; $\kappa_1 = 2$; $\tilde{\varepsilon}_l = 0.3$; $\tilde{\varepsilon}_g = 0.1$; $E = 3$; $V' = 0.1$; $\varepsilon = 0.5$, for axisymmetric disturbance case.

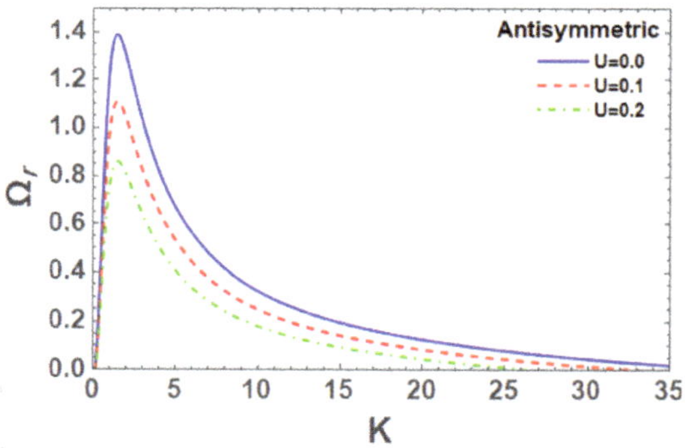

Figure 10. Change in non-dimensional growth rate Ω_r with non-dimensional wave number K for varied gas to liquid velocity values $U < 1$, when $We = 1000$; $Oh = 0.1$; $N = 1$; $\rho = 0.01$; $v = 0.2$; $\kappa_1 = 2$; $\tilde{\varepsilon}_l = 0.3$; $\tilde{\varepsilon}_g = 0.1$; $E = 3$; $V' = 0.1$; $\varepsilon = 0.5$, for axisymmetric disturbance case.

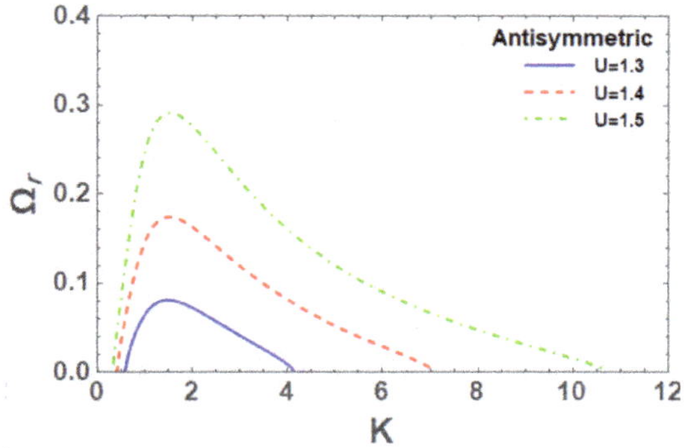

Figure 11. Variation of non-dimensional growth rate Ω_r with non-dimensional wave number K for varied gas to liquid velocities $U > 1$, when $We = 1000$; Oh $= 0.1$; $N = 1$; $\rho = 0.01; v = 0.2$; $\kappa_1 = 2$; $E = 3$; $V' = 0.1$; $\varepsilon = 0.5$, for axisymmetric disturbance case.

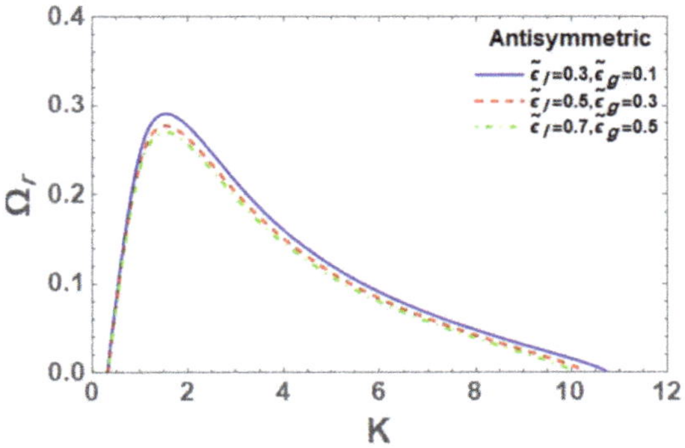

Figure 12. Change in non-dimensional growth rate Ω_r with non-dimensional wave number K for various dielectric constants values $\tilde{\epsilon}_l$ and $\tilde{\epsilon}_g$, when $We = 1000; Oh = 0.1; N = 1; \rho = 0.01$; $\kappa_1 = 2$; $\tilde{\epsilon}_l = 0.3$; $\tilde{\epsilon}_g = 0.1$; $E = 3; V' = 0.1$; $\varepsilon = 0.5$, for axisymmetric disturbance case.

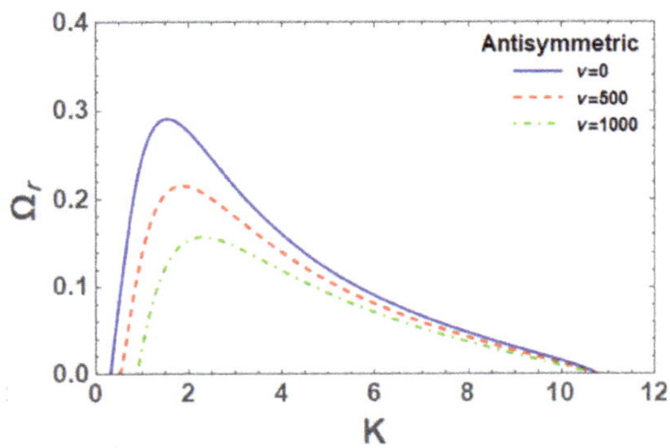

Figure 13. Non-dimensional growth rate Ω_r and non-dimensional wave number K for different gas to liquid viscosity values ν, when $We = 2000$; Oh $= 2$; $N = 1$; $v = 0.2$; $\rho = 0.1$; $U = 1.5$; $\tilde{\epsilon}_l = 0.3; \tilde{\epsilon}_g = 0.1; E = 10$; $V' = 5$; for axisymmetric disturbance case.

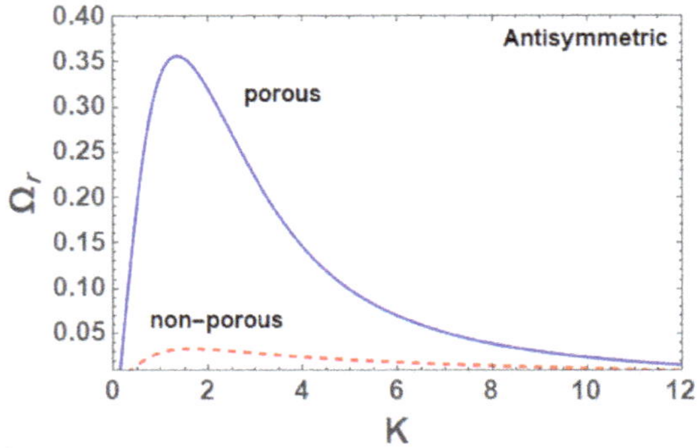

Figure 14. Non-dimensional growth rate Ω_r and non-dimensional wave number K for both case of porous medium ($\varepsilon = 0.5; \kappa_1 = 2$), and non-porous medium ($\varepsilon = 1;\ \kappa_1 \to \infty$), when $We = 2000$; Oh $= 2$; $N = 1$; $v = 0.2$; $\rho = 0.1; U = 1.5$; $\tilde{\varepsilon}_l = 0.3$; $\tilde{\varepsilon}_g = 0.1$; $E = 10$; $V' = 5$; for axisymmetric disturbance case.

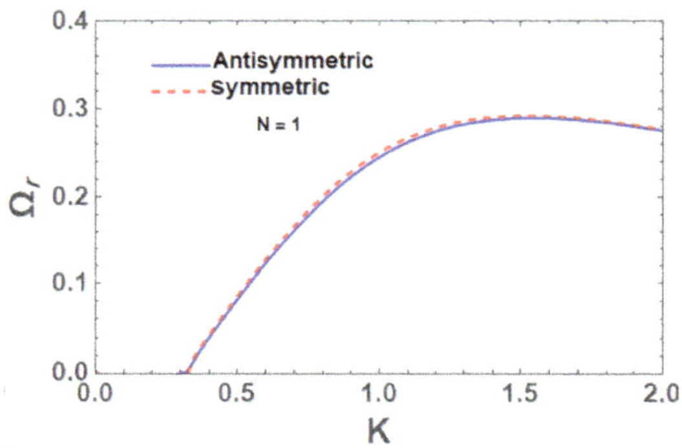

Figure 15. Variational of non-dimensional Ω_r with non-dimensional K in three-dimensional case ($N = 1$) for both anlisymmetric and symmetric disturbances with the same values of the parameters shown before.

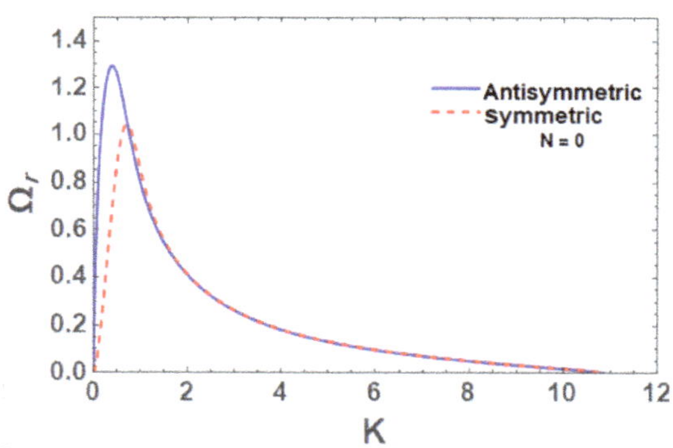

Figure 16. Change in in non-dimensional Ω_r with non-dimensional K in two-dimensional case ($N = 0$) for both antisymmetric and symmetric disturbances with the same values of the parameters used above.

8.1. Effect of the Weber Number

For various values of the Weber number and constant values of other parameters included in the analysis, the non-dimensional growth rate Ω_r is shown in Figure 3 for the antisymmetric disturbance example. This graph shows that when the Weber number rises, so do the dominant growth rates, dominant wave numbers, and critical wave numbers. Consequently, the instability zone expands, demonstrating that the Weber number destabilizes the system.

8.2. Effect of Ohnesorge Number

For antisymmetric disturbances with fixed physical parameters, the Ohnesorge number Oh affects the non-dimensional growth rate Ω_r against the non-dimensional wave number K in Figure 4. The figure shows that all curves have the same wave number at both the beginning and the conclusion, as can be seen. Increasing the Ohnesorge number Oh raises the maximum dominant wave numbers, and they are the same as the dominant wave number, while the critical wave numbers remain unchanged. As a result, we can infer that the Ohnesorge number Oh destabilizes the system under consideration as the unstable zone under the curves expands.

8.3. Effect of Viscolasticity Parameter

Viscoelasticity parameter and other physical constants remain constant in the antisymmetric disturbance situation, hence the growth rate Ω_r of the non-dimensional wave number K is shown in Figure 5. The graphic shows that the maximum dominating growth rates fall as the viscoelasticity parameter V' is increased, and that all curves have the same starting wave number value and critical wave number value. The instability zone shrinks when the viscoelasticity parameter V' is increased, indicating that the viscoelasticity parameter V' has a stabilizing effect.

8.4. Effect of Electric Field

According to the antisymmetric disturbance case, the relationship between non-dimensional growth rate Ω_r and non-dimensional wave number K can be seen in Figure 5 for various electric field E values and constant other physical parameters. Increasing the electric field E values increases both the maximum dominant wave numbers and critical wave numbers, but the dominant wave number remains constant. For this sort of disturbance, the electric field has a destabilizing effect because it increases the instability zones.

8.5. Effect of Medium Permeability

The antisymmetric disturbance mode is depicted in Figure 7, which shows how disturbance Ω_r growth rate changes with non-dimensional wave number K under various values of medium permeability κ_1 and constant values of other physical parameters. For example, we see in this figure that as permeability is increased, dominant and critical wave numbers remain at their predetermined values due to lower maximum dominant growth rates. Increased values of medium permeability κ_1, therefore, we suggest that medium permeability has stabilizing effects on this system of interest.

8.6. Effect of Porosity of Porous Medium

Using a non-dimensional growth rate Ω_r vs. a non-dimensional wave number K and the same values for the other parameters, Figure 8 depicts the effect of porous medium ε porosity on the instability of couple-stress viscoelastic electrified liquid sheets for antisymmetric disturbances. Increasing the porosity of porous medium decreases the maximum dominant growth rate Ω_r , but they retain at the same dominent wave number value. As a result, we may deduce that increasing the porosity of a porous medium ε stabilizes this system by raising the starting wave numbers and critical wave numbers, respectively, while decreasing the instability zones.

8.7. Effect of Gas to Liquid Density Ratio

Non-dimensional growth rate of disturbance Ω_r with non-dimensional wave number K for varying gas to liquid density ratio ρ and constant values of the other physical parameters is shown in Figure 9 for antisymmetric disturbance modes. When the gas to liquid density ratio grows, both the dominant growth rates and crucial wave numbers increase, while the dominant wave numbers occur at the same fixed wave number value and the starting points of the curves are coincidental. The instability zone expands as the gas-to-liquid density ratio increases, demonstrating the destabilizing effect of gas to liquid density on the examined structure.

8.8. Effect of Gas to Liquid Velocity Ratio

There is an antisymmetric form of disturbance when $U \lesseqgtr 1$, which is depicted in Figures 10 and 11 by the fluctuation of the non-dimensional growth rate Ω_r with the non-dimensional wave number K for various values of the gas to liquid velocity ratio U. With increasing gas to liquid velocity ratio U <1, both dominating growth rates and critical wave numbers drop, and thus the instability zone reduces as well, as seen in Figure 10. Therefore, we infer that the gas to liquid velocity ratio $0 \leq U < 1$ stabilizes the system in question, as seen in the following figure: As shown in Figure 11, the gas to liquid velocity ratio greater than one ($U > 1$) has a destabilizing effect on a similar system. Furthermore keep in mind that the dominant wave numbers fall and grow (slightly) as the gas to liquid velocity ratio increases in the cases when $U \lesseqgtr 1$. Therefore, we conclude that the ratio of gas to liquid velocity affects the stability of the system in two ways. Stabilization and destabilization of the system are both dependent on $U \lesseqgtr 1$, and vice versa.

8.9. Effect of Dielectric Constants

Non-dimensional growth rate Ω_r is shown in relation to the now-dimensional wave number K for different values of dielectric constant $\tilde{\epsilon}_l, \tilde{\epsilon}_g$ in liquid and gaseous media. As dielectric constants are increased, the resultant curves tend to converge and the instability zones narrow gradually, as seen by the curves' behavior. The dislectric constants $\tilde{\epsilon}_l, \tilde{\epsilon}_g$ in liquid and gas media have a small stabilizing influence on the system under consideration therefore we can draw this conclusion.

8.10. Effect of Gas to Liquid Viscosity Ratio

Figure 13 shows the non-dimensional growth rate Ω_r vs. the non-dimensional wave number K as a function of the gas to liquid viscosity ratio ν and various other factors. The figure clearly shows that the system is stable in the absence of gas viscosity, and we found that increasing the gas to liquid viscosity ratio ν caused all of the curves to nearly coincide, but the figure is not shown here. This result shows that the system is neutrally unstable for small values of the gas to liquid viscosity ratio. It is also worth noting that the maximum dominant growth rates and, thus, the instability zones shrink when the gas to liquid viscosity ratio rises significantly. For very low gas to liquid viscosity ratios the corresponding effect is only mildly stabilizing, but for high gas to liquid viscosity ratios the effect is stabilizing.

8.11. Effect of Porous Medium

As shown in Figure 14, for antisymmetric disturbance mode, the non-dimensional growth rate Ω_r and wave number K are related to each other in the presence (ε and κ_1) and absence ($\varepsilon = 1$ and $\kappa_1 \to \infty$) of porous media. The presence of porous media makes the system more unstable than it would be in the absence of porous medium, as can be seen from this image. All preceding figures that indicate the effect of other parameters, such as Figures 7 and 8, are based on this result. As a result, we have come to the conclusion that the system is more susceptible to breakup in the presence of porous medium than in the absence of such media.

8.12. Effect of Dimension

There is an inverse relationship between non-dimensional growth rates Ω_r and non-dimensional wave number K in the cases of a three-dimensional configuration $(N = 1)$ and a zero dimension $(N = 0)$, as shown in Figures 15 and 16. According to Figure 15, both symmetric and antisymmetric disturbances have nearly the same influence on the increasing rate of instability in the three-dimensional example. There is no difference between an antisymmetric or symmetric disturbance's impact on instability, because the growth rate curves are nearly identical. Figure 16 shows that for small wave numbers, the dominating growth rate for antisymmetric disturbance mode is higher than the equivalent symmetric disturbance mode, whereas for higher wave numbers, the growth rate curves for both antisymmetric and symmetric cases are coincident. To sum up, in the three-dimensional disturbance case, every parameter has an identical effect in both antisymmetric and symmetric mods, but in the two-dimensional disturbance case, the antisymmetric instability zone is larger than the symmetric instability zone, and as a result, the system is more unstable when configured in the antisymmetric mode than when configured in the symmetric mode.

9. Concluding Remarks

An incompressible viscoelastic dielectric liquid sheet of the couple-stress type streaming with relative velocity in an inviscid dielectric gas medium through a porous medium has been studied using fluid and electric field equations of motion with the corresponding appropriate boundary conditions in this work, resulting in an electrohydrodynamic instability analysis. It was found that there are two types of disturbances: antisymmetric and symmetrical. Analytically, we may derive the dispersion relations for both antisymmetric and symmetric disturbances, and by defining some non-dimensional qualities of the parameters included in our analysis, we can represent the derived dispersion equations in dimensionless forms. The influence of various factors on the instability of this type of liquid sheet problem can be explored by computationally resolving the non-dimensional dispersion relations using Mathematica software via the Gaster technique. Based on the preceding discussion and examination of the numbers, here are a few possible conclusions:

(1) When a system is placed in a two-dimensional configuration, it is more unstable when it is subjected to antisymmetric disturbance than when it is subjected to symmetric disturbance.
(2) A porous medium makes the system more unstable, and it breaks down more quickly, compared to the lack of a porous medium.
(3) Ohnesorge number, Weber number, and electric field all have a destabilizing effect on the system under consideration.
(4) The system is stabilized by the viscoelasticity parameter, the medium permeability, the porous medium porosity, and the gas to liquid viscosity ratio.
(5) We have found that the dielectric constants have a small stabilizing effect.
(6) The gas to liquid velocity ratio affects system stability in two ways: it stabilizes when $U < 1$ is less than one, and it destabilizes when $U > 1$ is more than one.

Author Contributions: Conceptualization, M.F.E.-S.; methodology, M.F.E.-S. and A.M.A; software, A.M.A.; validation, M.F.E.-S. and A.M.A.; formal analysis, M.F.E.-S.; investigation, A.M.A.; writing—original draft preparation, A.M.A.; writing—review and editing, M.F.E.-S.; supervision, M.F.E.-S. All authors have read and agreed to the published version of the manuscript.

Funding: This research received no external funding.

Institutional Review Board Statement: Not applicable.

Informed Consent Statement: Not applicable.

Data Availability Statement: Not applicable.

Conflicts of Interest: The authors declare no conflict of interest.

References

1. Squire, H.B. Investigation of the instability of a moving liquid Sheet. *Br. J. Appl. Phys.* **1953**, *4*, 167–169. [CrossRef]
2. Hagerty, W.W.; Shea, J.F. A study of the stability of plane fluid sheets. *J. Appl. Mech.* **1955**, *22*, 509–514. [CrossRef]
3. Dombrowski, N.; Johns, W.R. The aerodynamic instability and disintegration of viscous liquid sheets. *Chem. Eng. Sci.* **1963**, *18*, 203–214. [CrossRef]

4. Li, X.; Tankin, R.S. On the temporal instability of a two-dimensional viscous liquid sheet. *J. Fluid Mech.* **1991**, *225*, 425–443. [CrossRef]
5. Tong, M.-X.; Fu, Q.- F.; Yang, L.-J. Two-dimensional instability responce of an electrified viscoelastic planar liquid sheet subjected to unrelaxed axial elastic tension. *At. Sprays* **2015**, *25*, 99–121. [CrossRef]
6. Lefebvre, A.H. *Atomization and Sprays*; Hemisphere: New York, NY, USA , 1989.
7. Yarin, A.L. *Free Liquid Jets Films: Hydrodynamics and Rheology*; John Wiley & Sons: New York, NY, USA , 1993.
8. Lin, S.P. *Breakup of Liquid Sheets and Jets*, 2nd ed.; Cambridge University Press: New York, NY, USA , 2010.
9. Dasgupta, D.; Nath, S.; Makhopadhyay, A. Linear and nonlinear analysis of breakup of liquid sheats: A Review. *J. Indian Inst. Sci.* **2019**, *99*, 59–75. [CrossRef]
10. Liu, Z.; Braen, G.; Durst, F. Linear analysis of the instability of two-dimensional non-Newtonian liquid sheets. *J. Non–Newton. Fluid Mach.* **1998**, *78*, 133–166. [CrossRef]
11. Brenn, G.; Liu, Z.; Durst, F. Three dimensional temporal instability of non- Newtonian liquid sheets. *At. Sprays* **2001**, *11*, 49–84. [CrossRef]
12. Yang, L.-J.; Liu, Y.-X.; Fu, Q.-F.; Wang, C.; Ning, Y. Linear stability analysis of electrified viscoelastic liquid sheets. *At. Sprays* **2012**, *22*, 951–982. [CrossRef]
13. El-Sayed, M.F.; Moatimid, G.M.; Elsabaa, F.M.F.; Amer, M.F.E. Electrohydrodynamic instability of non-Newtonian dielectric liquid sheet issued into streaming dielectric gaseous environment. *Interfacial Phenom. Heat Transf.* **2015**, 3, 159–183. [CrossRef]
14. Melcher, J.R. *Continuum Electromechanics*; MIT Press: Cambridge, MA, USA , 1981.
15. Stokes, V.K. Couple stresses in fluids. *Phys. Fluids* **1966**, *9*, 1709–1715. [CrossRef]
16. Chavaraddi, K.B.; Awati, V.B.; Gouder, P.M. Effects of boundary roughness on Rayleigh-Taylor instability of a couple-stress fluid. *Gen. Math. Notes* **2013**, *17*, 66–75.
17. Chavaraddi, K.B.; Katagi, N.N.; Awati, V.B.; Gouder, P.M. Effect of boundary roughness on Kelvin-Helmholtz instability in couple stress fluid layer bounded above by a porous layer and below by rigid surface. *Int. J. Chem. Eng. Res.* **2014**, *4*, 35–43.
18. Chavaraddi, K.B.; Gouder, P.M.; Kudenatti, R.B. The influence of boundary roughness on Rayleigh-Taylor instability at the interface of superposed couple-stress fluids. *J. Adv. Res. Fluid Mech. Thermal Sci.* **2020**, *75*, 1–10. [CrossRef]
19. Rudrauah, R.; Chandrashekara, G. Effects of couple stress on the growth rate of Rayleigh-Taylor instability at the interface in a finite thickness couple stress fluid. *J. Appl. Fluid Mech.* **2010**, 3, 83–89.
20. Sharma, R.C.; Sunil; Sharma, Y.D.; Chandel, R.S. On couple-stress fluid permeated with suspended particles heated from below. *Arch. Mech.* **2002**, *54*, 287–298.
21. Kumar, P.; Lal, R.; Sharma, P. Effect of rotation on thermal instability in couple-stress elastic-viscous fluid. *Z. Naturforsh. A* **2004**, *59*, 407–4011. [CrossRef]
22. Kumar, P.; Sing, G.J. Analysis of stability in couple-stress magneto-fluid. *Nepal J. Math. Sci.* **2021**, *2*, 35–42. [CrossRef]
23. Nield, D.A.; Bejan, A. *Convection in Porous Medium*, 3rd ed.; Springer: New York, NY, USA , 2006.
24. Mathur, R.P.; Gupta, D. Effect of surface tension on the stability of superposed viscous-viscoelastic (couple-stress) fluids through porous medium. *Proc. Indian Natn. Sci. Acad.* **2011**, *77*, 335–342.
25. Shankar, B.M.; Shivakumara, I.S.; Ng, C.O. Stability of couple stress fluid flow through a horizontal porous layer. *J. Porous Med.* **2016**, *19*, 391–404. [CrossRef]
26. Rudraiah, N.; Shankar, B.M. Stability of Parallel couple strep viscous fluid flow in a channel. *Int. J. Appl. Math.* **2009**, *1*, 67–78.
27. Agoor, M.B.; Eldabe, N.T.M. Rayleigh-Taylor instability at the interface of superposed couple-stress casson fluids flow in porous medium under the effect of a magnetic field. *J. Appl. Fluid. Mech.* **2014**, *7*, 573–580.
28. Shirakumara, I.S.; Kumar, S.S.; Devaraju, N. Effect of non- uniform Temperature gradients on the onset of convection in couple stress fluid-saturated porous medium. *J. Appl. Fluid Mech.* **2012**, *5*, 49–55.
29. Sharma, R.C.; Sunil Pal, M. On superposed couple-stress fluid in porous medium. *Studia Geotech. Mech.* **2001**, *33*, 55–66.
30. Rana, G.C.; Saxena, H.; Gautam, P.K. The onset of electrohydrodynamic instability in a couple-stress nanofluid saturating a porous medium: Brinkman model. *Rev. Cubana Fis.* **2019**, *36*, 37–45. [CrossRef]
31. Rudrajah, N.; Shankar, B.M.; Ng, C.O. Electrohydrodynamic stability of couple stress fluid flow in a channel occupied by a porous medium. *Spec. Top. Rev. Porous Media* **2011**, *2*, 11–22. [CrossRef]
32. El-Sayed, M.F.; Eldabe, N.T.; Haroun, M.H.; Mastafa, D.M. Nonlinear electroviscous potential flow instability of two superposed couple-stress fluids streaming through porous medium. *J. Porous Med.* **2014**, *17*, 405–420. [CrossRef]
33. Chandrasekhar, S. *Hydrodynamic and Hydromagnetic Stability*; Dover Publications: New York, NY, USA , 1981.
34. Shivakumara, I.S.; Akkanagamma, M.; Ng, C.-O. Electrohydrodynamic instability of a rotating couple-stress dielectric fluid layer. *Int. J. Heat Mass Transfer.* **2013**, *62*, 761–771. [CrossRef]
35. El-Sayed, M.F. Three-dimensional Electrohydrodynamic temporal instability of a moving dielectric liquid sheet emanated into a gas medium. *Eur. Phys. J. E* **2004**, *15*, 443–455. [CrossRef]
36. Ibrahim, E.A.; Akpan, E.T. Liquid sheat instability. *Acta Mech.* **1998**, *131*, 153–167. [CrossRef]
37. Dasgupta, D.; Nath, S.; Bhanja, D. Linear instability analysis of viscous planar liquid sheet sandwiched between two moving gas streams. In *Advances in Mechanical Engineering*; Lecture Notes in Mechanical Engineering; Biswal, B., Sarkar, B., Mahanta, P., Eds.; Springer: Singapore, 2020; pp. 41–50.

38. Nath, S.; Mukhopadhyay, A.; Sen, S.; Tharakan, T.J. Influence of gas velocity on breakup of planar liquid sheets sandwiched between two gas streams. *At. Sprays* **2010**, *20*, 983–1003. [CrossRef]
39. Gaster, M. A note on the relation between temporally increasing and spatially-increasing disturbances in hydrodynamic stability. *J. Fluid Mech.* **1962**, *14*, 222–224. [CrossRef]
40. El-Sayed, M.F.; Moatimid, G.M.; Elsabaa, F.M.F.; Amer, M.F.E. Axlaymmotric and asymmetric instabilities of a non-Nowtonian liquid jet moving in an Inviscid streaming gas through porous media. *J. Porous Modia* **2016**, *19*, 751–769. [CrossRef]
41. El-Sayed, M.F.; Moatimid, G.M.; Elsabaa, F.M.F.; Amer, M.F.E. Electrohydrodynamic instability of a non-Newtonian dielectric liquid jet moving in a streaming dielectric gas with a surface tension gradient. *At. Sprays* **2016**, *26*, 349–376. [CrossRef]

applied sciences

MDPI

Brief Report

Model for Wall Shear Stress from Obliquely Impinging Planar Underexpanded Jets

Patrick Fillingham [1,*], Arjun Viswanathan [2] and Igor V. Novosselov [2]

[1] Department of Neurological Surgery, University of Washington, Seattle, WA 98105, USA
[2] Department of Mechanical Engineering, University of Washington, Seattle, WA 98105, USA; athreya.arjun@gmail.com (A.V.); ivn@uw.edu (I.V.N.)
* Correspondence: pfilling@uw.edu

Abstract: Though inclined under-expanded planar jets are used in many practical applications, the wall stress resulting from their impingement has not been adequately characterized. Reduced-order models for wall shear as a function of jet parameters have not been reported. This work uses computational fluid dynamics to determine wall shear stress as a function of the nozzle parameters and jet angle. The simulations of the impinging jet are validated against the experimental data and direct numerical simulation; then, the jet parameters are varied to formulate an empirical relationship for maximum wall shear stress as a function of a nozzle pressure ratio, standoff distance, jet Reynolds number, and impingement angle. The global expression for shear stress agrees with the numerical results within a mean deviation of 3%. The relationship can be used for applications where shear stress information is required to design or assess the performance of practical systems, such as surface cleaning, particle resuspension from the surface, and surface cooling.

Keywords: impinging jets; supersonic flow; wall shear stress; CFD

Citation: Fillingham, P.; Viswanathan, A.; Novosselov, I.V. Model for Wall Shear Stress from Obliquely Impinging Planar Underexpanded Jets. *Appl. Sci.* **2022**, *12*, 7311. https://doi.org/10.3390/app12147311

Academic Editors: Vasily Novozhilov and Cunlu Zhao

Received: 30 June 2022
Accepted: 19 July 2022
Published: 21 July 2022

1. Introduction

Impinging jets have a wide range of scientific and engineering applications and have been studied extensively. While most of these studies focus on heat and mass transfer [1–6], the focus of this work is related to aerodynamic particle resuspension from the surfaces such as in surface cleaning [7,8] and non-contact particle sampling from surfaces [9–11]. Resuspension rates were correlated to wall shear stress [12]; for example, Phares et al. [13] suggested that size-controlled microparticles' resuspension could be used to estimate the wall shear stress experimentally. Resuspension of small, deforming, or irregular shape particles, such as residues of energetic materials [9,14–16], microorganisms [17,18], and nanoparticles, require exposure to high velocities at the surface [19], which is associated with high wall shear stress. These levels of shear stresses are not typical for weather-induced conditions but can be generated by the impingement of supersonic or high-pressure under-expanded jets. Studies of under-expanded jets have typically been confined to axisymmetric jets with applications related to vertical take-off and landing aircraft [20–22].

Our earlier work parameterized the wall jet behavior from a normal impinging jet [23,24]. Compared with round or low aspect ratio nozzles, the high aspect ratio planar jets produce broader and more uniform regions of high shear stress. Obliquely impinging planar jets have several advantages for cooling [25–27], drying [28], and aerodynamic particle sampling. In non-contact surface sampling, the wall jet flow entrains and directs the resuspended particles for their subsequent collection and analysis in contrast to scattering the sample with axisymmetric jets [15]. Fillingham et al. [24] presented parameterization of the normal planar underexpanded jet, but the correlations developed in that work do not extend to oblique jets. Crafton et al. [29] and Ngyuen et al. [30] studied the flow field from round under-expanded jet impingement on an inclined surface. However, the authors did not consider planar jets or wall shear stress correlations.

The flow field of jets impinging normally and obliquely has been studied experimentally and by numerical methods to develop relationships with wall shear stress. Dorrepaal [31] found a similarity solution for subsonic 2-D incompressible non-orthogonal stagnation point flow. Beltaos [32] used Preston tube measurements to calculate the wall shear stress of the incompressible oblique planar jets. The behavior of compressible jets and characterization of wall shear stress was not considered. Chin et al. [33] studied the mass transfer of obliquely impinging planar jets but did not examine wall shear stress. Hwang et al. [3] conducted a computational study of obliquely impinging slot jets, but the $k - \epsilon$ turbulence model used is unreliable for wall shear stress measurement. Rajaratnam et al. [34] studied erosion from obliquely planar jets; however, their results do not directly apply to particle resuspension from a rigid surface.

Investigating inclined impinging jets can close a knowledge gap in microparticle resuspension applications where shear stress is closely correlated with particle resuspension [12,13,35,36]. This work presents a parametric study of obliquely impinging, under-expanded planar jets. Impingement angles from 0° to 60° are examined for jets standoff height to nozzle width ratios of 15–30 and nozzle pressure ratios from 1.0–3.4. The computational data is used to develop a simple relationship for maximum wall shear stress as a function of only the jet parameters: impingement angle, jet width, jet standoff height, and jet nozzle pressure. This reduced-order relationship allows evaluating wall shear stress without performing extensive CFD simulations or performing labor-intensive experiments.

2. Methodology

2.1. Problem Description

Obliquely impinging under-expanded jets can be characterized by four parameters: the standoff height (*H*), the jet width (*W*), the jet nozzle pressure ratio (*NPR*), and the jet impingement angle (θ). The schematic of the jet impingement and the definitions are shown in Figure 1. Table 1 summarizes the range of the values used in this study.

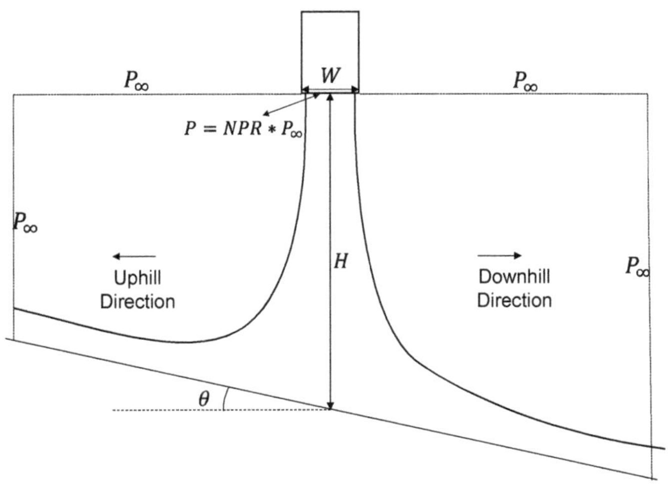

Figure 1. Schematic of oblique jet impingement on a flat plate.

Table 1. Range of geometric conditions used in the study.

Standoff Distance, H (mm)	15.0, 17.5, 30.0, 35.0, 60.0, 100.0
Width of the nozzle, W (mm)	0.5, 1.0, 2.0
Jet Angle, θ (°)	0, 15, 20, 30, 45, 60
Nozzle exit pressure ratio, *NPR*	1.0, 1.2, 1.4, 1.6, 1.8, 2.0, 2.2, 2.4, 2.6, 2.8, 3.0, 3.2, 3.4

The analysis is performed using non-dimensional parameters, such as the normalized maximum wall shear stress is defined as, $\tau^*_{max} = \frac{\tau_{max}}{\frac{1}{2}\rho_o U_o^2}$, where ρ_o and U_o are the density and velocity at the nozzle exit, respectively. The objective of this work is to develop a set of equations for the prediction of the maximum normalized wall shear stress as a function of the following non-dimensional parameters: jet height to width ratio—$\frac{H}{W}$, the nozzle pressure ratio—NPR, the jet Reynolds number—$Re = \frac{U_o W}{\nu_o}$, and the jet impingement angle—θ, a naught subscript describes the properties at the nozzle exit.

2.2. Computational Method

The scientific literature does not report the experimental data or DNS related to the wall jet developed from compressible impinging jets. To compute the flow properties needed for the estimation of shear stresses, the CFD simulations solve the steady-state Favre-Averaged Navier–Stokes equations:

$$\frac{\partial(\overline{\rho}\widetilde{u}_i)}{\partial x_i} = 0 \quad (1)$$

$$\frac{\partial(\overline{\rho}\widetilde{u}_i\widetilde{u}_j)}{\partial x_i} = -\frac{\partial\overline{p}}{\partial x_i} + \frac{\partial\overline{\tau_{ij}}}{\partial x_j} - \frac{\partial\left(\overline{\rho u_i'' u_j''}\right)}{\partial x_j} \quad (2)$$

$$\frac{\partial}{\partial x_j}\left(\overline{\rho}\widetilde{u}_j\left(\widetilde{h} + \frac{1}{2}\widetilde{u}_i\widetilde{u}_j\right) + \widetilde{u}_j\overline{\rho u_i'' u_j''}\right) = \frac{\partial}{\partial x_j}\left(\widetilde{u}_i\left(\overline{\tau_{ij}} - \overline{\rho u_i'' u_j''}\right) - \overline{q} - \overline{\rho u_j'' h''} + \overline{\tau_{ij} u_i''} - \frac{1}{2}\overline{\rho u_j'' u_i'' u_i''}\right) \quad (3)$$

Numerical simulations were performed using ANSYS FLUENT 17.2 software (Ansys, Canonsburg, PA, USA). A second-order scheme was used to solve the pressure. A third-order monotonic upstream-centered scheme was used for density, momentum, and turbulence, which was necessary to avoid the effects of numerical viscosity (associated with the low-order schemes) on the jet dissipation, as well as pressure–strain relationship. Since the flow contains non-negligible changes in temperature, the Sutherland model, based on the kinetic theory of ideal gases and an idealized intermolecular force potential, was used for viscosity calculation.

Turbulence closure models are challenged in modeling complex flow phenomena; however, Jaramillo et al. [37] demonstrated that $k-\omega$ models result in good agreement with DNS when calculating the mean flow of planar impinging jets. The $k-\omega$ shear stress transport (SST) model used in this work, uses $k-\epsilon$ away from the wall in the free stream and free jet portions of the flow while using $k-\omega$ near the wall to resolve the boundary layer. As demonstrated by Alvi et al. [38] and discussed by Fillingham et al. [23], the $k-\omega$ SST model [39] is a good choice for modeling underexpanded impinging jets while resolving the wall jet boundary layer. Shukla and Dewan. [40] also found $k-\omega$ SST to be superior to other closure models when considering planar impinging jets.

The computational grids contain ~500,000 quadrilateral elements. The grid is constructed so that, along the impingement surface, the first node in the wall-normal direction is located within a $y^+ = 1$ at the maximum shear stress location in every case. This is ensured by evaluating the maximum wall shear stress for the normal impingement case at the maximum NPR for each H/W ratio found in our previous work [24] and calculating the distance from the wall that yields $y^+ = 1$ for this maximum wall shear. The Δx of the mesh is generated via biasing from $\Delta x = 5\Delta y_{wall}$ at the impingement point to $\Delta x = 50\Delta y_{wall}$ at the edge of the domain. This grid resolution ensures that the viscous sublayer is resolved for the entire domain. The maximum x-direction spacing gives an element aspect ratio of less than 50:1. Mesh independence was confirmed by a simulation with the highest wall shear stress case and doubled number of elements in each direction, such that the first node was placed at $y^+ = 0.5$ still with a maximum aspect ratio of 50:1. This mesh consisted of approximately 2,000,000 elements; this further mesh refinement did not significantly affect the results, leading to only a 0.12% change in the maximum wall shear stress value.

The inlet boundary condition is defined as the exit of an isentropic nozzle where the flow is choked; thus, the boundary can be described by a total pressure, total temperature and a static pressure where the total pressure is necessarily (for an ideal diatomic gas) 1.893 times the static pressure. The walls are modeled as isothermal, no-slip boundaries. The outlets are defined as atmospheric pressure outlets. The outlets are located at 50 jet hydraulic diameters (100 jet slot widths) from the jet axis, corresponding to a minimum of 1.5 times the impingement height.

2.3. Model Validation

Validation of the CFD result is challenging in the absence of the experimental or DNS data for compressible planar impinging jets needed for direct comparison. We evaluated two flow regions: (i) impinging jet and (ii) wall jet region.

Impinging jet region validation: The implementation of $k-\omega$ SST jet in the impinging jet region was previously validated with underexpanded axisymmetric jets using Schlieren photography. The shape and the shock structures were shown to be in excellent agreement with experimental observations [23]. Additionally, the CFD simulations were compared with pressure profiles on the impingement surface from pressure-sensitive paint (PSP) experiments [24]. PSP utilizes the emission spectra of a luminophore by relating the emission intensity at specific wavelengths to the partial pressure of oxygen at the surface. Images were taken under wind-on and wind-off conditions, and the images' intensity ratios were related to pressure. The CFD approach was evaluated using oblique planar jet impingement against the PSP pressure measurements. The oblique jet produces an uphill shift in the impingement point from the geometric center [32]. Figure 2 shows the computed pressure profiles overlayed on the PSP measurements. The CFD simulations show good agreement in the shape and magnitude of the pressure profile.

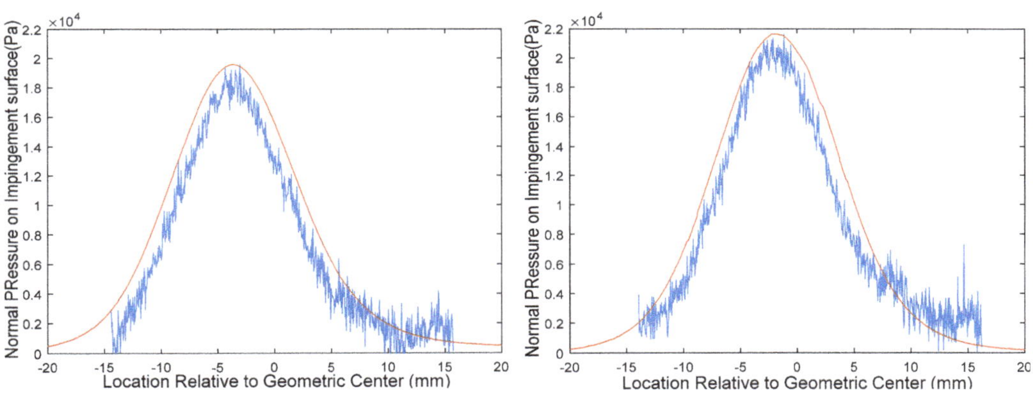

Figure 2. Normal pressure profiles from CFD (red) and pressure-sensitive paint (blue) experiments for h = 30 mm, d = 1 mm, NPR = 1.0 with the impingement angle of 30 degrees (**left**) and 15 degrees (**right**), as shown originally in Fillingham et al. [24].

Wall jet region validation: The $k-\omega$ SST simulations were compared to two separate DNS studies. First, the model was used to replicate DNS data from normal planar impinging jet conducted by Jaramillo et al. [37], reporting wall jet profiles downstream of the impingement point. Figure 3a shows the agreement in velocity profile from Jaramillo et al. and the $k-\omega$ SST computation. Second, the model was evaluated against the DNS of a classical wall jet conducted by Naqavi et al. [41]. Figure 3b–d compares the development of the wall jet thickness, $y_{1/2}$, the maximum velocity, U_m, and the wall shear stress.

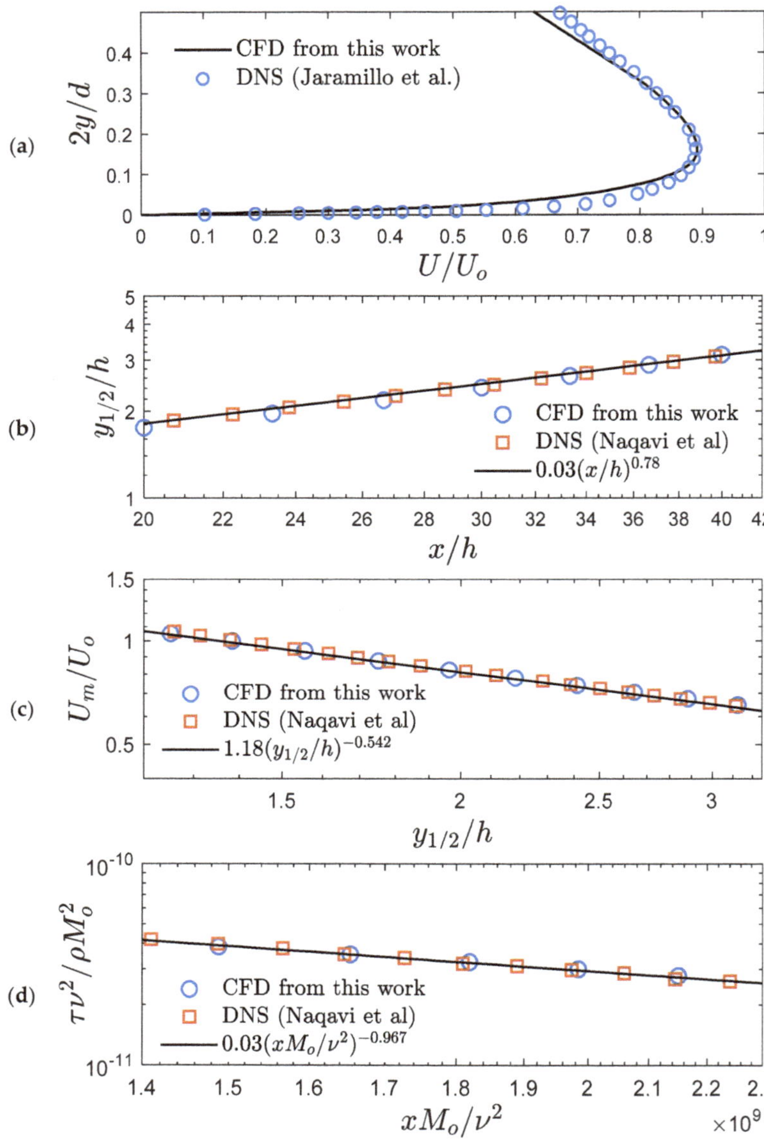

Figure 3. Validation of $k-\omega$ SST for impinging jet modeling as shown originally in Fillingham et al. [24]. (**a**) Comparison of the velocity profile at 8 jet widths downstream of impingement location against DNS from Jaramillo et al. [37]. Comparison with wall jet DNS simulation from Naqavi et al. [41]. (**b**) Decay of half maximum velocity location. (**c**) Maximum velocity. (**d**) Wall shear stress.

In summary, the $k-\omega$ SST was found to be an acceptable model for use in the parametric study because of the excellent agreement with the PSP measurements, validation from two DNS studies describing wall jet development, and previously reported Schlieren photography comparison for the axisymmetric impinging jet [23].

3. Results

The CFD data for the maximum wall shear stress was tabulated for each case and examined as a function of each jet parameter. As found in our previous work [24], supersonic flow is still present in the boundary layer for *H/W* values of less than 30. As such, we did not include these cases in the analysis, and the relations developed in this work should only be applied for *H/W* values above 30. Figure 4a plots the normalized maximum wall shear stress against the nozzle pressure ratio for various conditions. The angle is coded by colors, and the height-to-width ratio is represented by marker type. When each independent parameter is fixed, the maximum wall shear stress increases with approximately the square root of NPR for all cases. Figure 4b plots the normalized maximum wall shear stress against the height to width ratio. As expected, the maximum wall shear stress decreases with increasing height to width ratio. When all other parameters are fixed, the wall shear stress is proportional to the inverse of the height to width ratio. Figure 4c illustrates that maximum wall shear stress decreases with the impingement angle. This relationship was found to resemble a function of the form: $1 - \sin\theta$. After analyzing the maximum wall shear stress as a function of jet parameters, we propose an equation for normalized maximum wall shear stress of the form

$$\tau^*_{max} = \tau^*_{max,\,\theta=0}\left(1 - a\left(\frac{H}{W}\right)^b NPR^c \sin^d\theta\right). \quad (4)$$

(a)
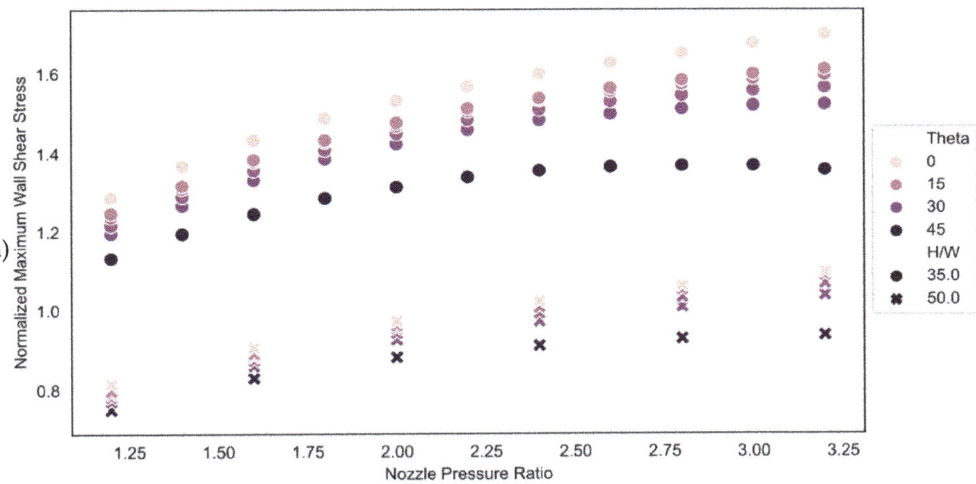

(b)
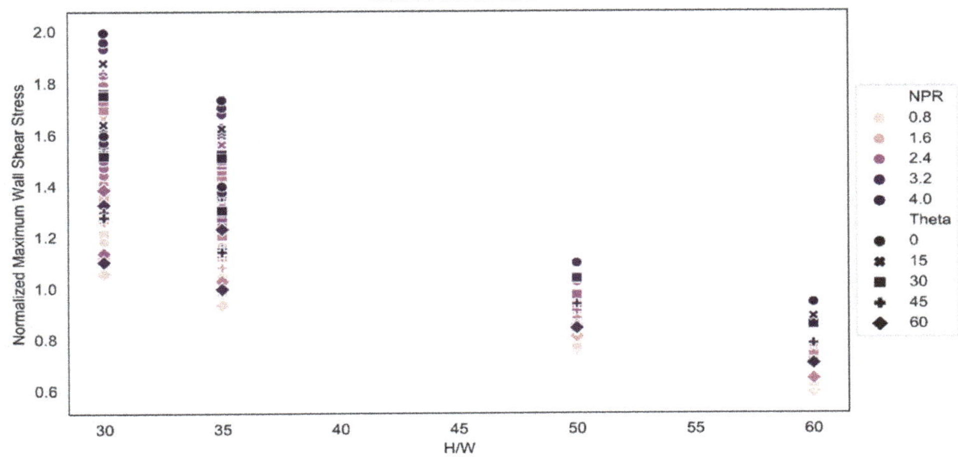

Figure 4. *Cont.*

(c)
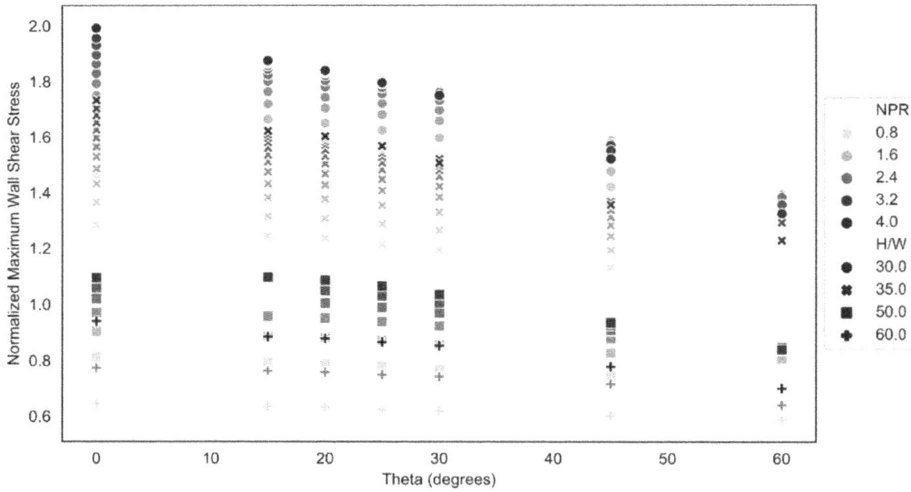

Figure 4. Normalized maximum wall shear stress plotted against Nozzle parameters with impingement angle represented by shade and height to width ratio designated by the symbol (**a**) nozzle pressure ratio, (**b**) height-to-width ratio, (**c**) impingement angle.

To obtain the global fit based on the form of Equation (1), we first determine the normalized maximum wall shear stress for a normal impingement angle $\theta = 0°$ as a function of height to width ratio, nozzle pressure ratio, and jet Reynolds number. Based on our previous work [23], we expect the Reynolds number term to be relevant as it accounts for the effective turbulent viscosity increase for jets with higher Reynolds numbers. This leads to increased energy dissipation in the free jet region and, thus, reduces the maximum wall shear stress after the impingement. Therefore, to describe the maximum wall shear stress at normal impingement, we propose the following equation:

$$\tau^*_{max,\,\theta=0} = \alpha\left(\frac{H}{W}\right)^{\beta} NPR^{\gamma} Re^{\lambda} \quad (5)$$

Least-squares fitting gives the following values for the coefficient and exponents:

$$\alpha = 9338.08$$

$$\beta = -1.146$$

$$\gamma = 0.589$$

$$\lambda = -0.301$$

The prediction of maximum wall shear stress from a normal impingement of underexpanded planar jets using Equation (5) agrees with the CFD data with a maximum deviation of 5.7% and a mean deviation of 2.1%. Figure 5 plots the predicted maximum wall shear stress against the calculated value from CFD.

After finding the normal impingement correlations, we calculated the coefficient and exponents for Equation (1). Least-squares fitting yields the following coefficients:

$$a = 0.208$$

$$b = -0.276$$

$$c = 1.213$$

$$d = 1.512$$

Inserting the coefficients into Equation (4) yields the final expression

$$\tau^*_{max} = 9338.08\left(\frac{H}{W}\right)^{-1.146} NPR^{0.589} Re^{-0.301}\left(1 - 0.208\left(\frac{H}{W}\right)^{-0.276} NPR^{1.213} \sin^{1.512}\theta\right) \quad (6)$$

Compared with the numerical data, the predicted normalized maximum wall shear stress has an average deviation of 3.3% and a maximum deviation from CFD of 10.3%. Figure 6 plots the predicted normalized maximum wall shear stress values against the values calculated via CFD.

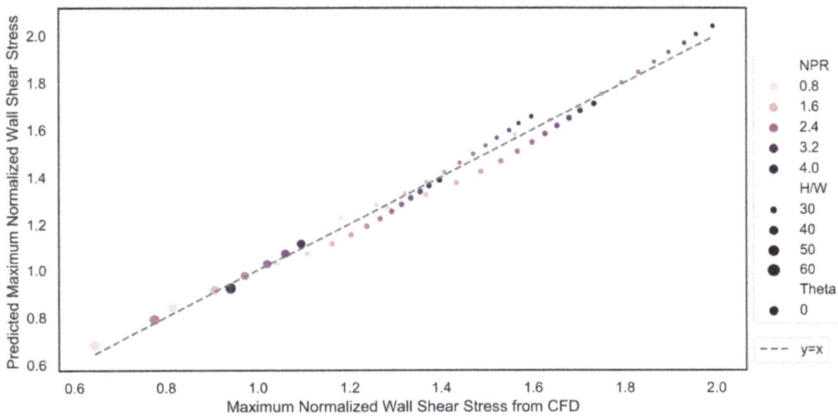

Figure 5. Predicted maximum normalized wall shear stress plotted against maximum normalized wall shear stress from CFD for normal impingement. Shade represents the nozzle pressure ratio, whereas the symbol represents the height-to-width ratio.

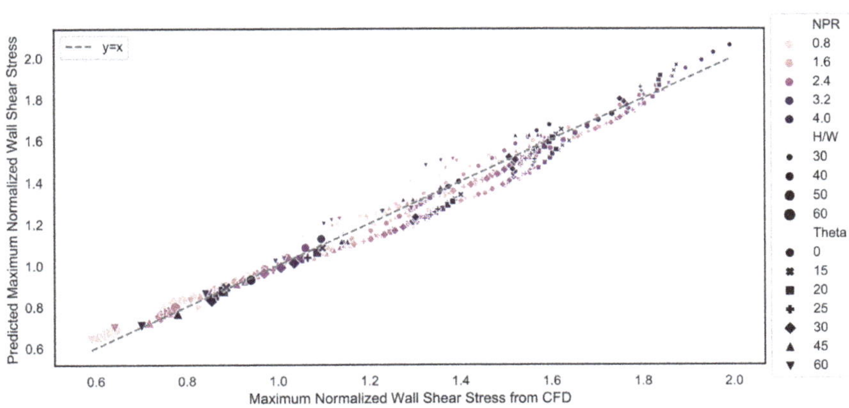

Figure 6. Predicted maximum normalized wall shear stress plotted against maximum normalized wall shear stress from CFD for all impingement angles. Shade represents the nozzle pressure ratio, the symbol represents the impingement angle, and the marker size represents the height-to-width ratio.

4. Conclusions

We report empirical formulation for predicting the maximum wall shear stress resulting from obliquely impinging under-expanded planar jets as a function of four jet parameters: the standoff height (**H**), the jet width (**W**), the jet nozzle pressure ratio (**NPR**), and the jet impingement angle (**θ**). The empirical relationship for the normal wall shear stress was developed in the form $\tau^*_{max,\ \theta=0} = \alpha\left(\frac{H}{W}\right)^{\beta} NPR^{\gamma} Re^{\lambda}$ with the coefficients (α = **9338.08**, β = **−1.146**, γ = **0.589**, λ = **−0.301**). The jet angle correction was implemented in the form $\tau^{*}_{max} = \tau^{*}_{max,\,\theta=0}\left(1 - a\left(\frac{H}{W}\right)^{b} NPR^{c} \sin^{d}\theta\right)$ with the coefficient (**$a = 0.208$, $b = -0.276$, $c = 1.213$, $d = 1.512$**) determined based on the least squire fit c the shear stresses calculated using scale resolved CFD simulations. The global expressio can predict wall shear stress within a mean error of 3.3%. The formulation can be use to estimate particle resuspension rates and heat transfer rates when correlated with wal shear stresses.

There are limitations to the relationships developed in this work. These expression will only hold for subsonic boundary layer flow and, as such, should not be extended t use for *H/W* of less than 30 or *NPR* above 3.4 without the expectation of error induced du to compressibility. Additionally, the expressions should not be extended to impingemen angles above 60° as the characteristics of the system change from an impinging jet to a wa jet, and the expressions will no longer apply.

Author Contributions: Conceptualization, P.F. and I.V.N.; methodology, P.F.; software, P.F. and A.V validation P.F.; formal analysis, P.F.; investigation, P.F. and A.V.; resources, I.V.N.; data curation, P.F writing—original draft preparation, P.F. and A.V.; writing—review and editing, I.V.N.; visualizatio P.F.; supervision, I.V.N.; project administration, I.V.N.; funding acquisition, I.V.N. All authors hav read and agreed to the published version of the manuscript.

Funding: This research was funded by DHS Science and Technology Directorate, and UK Hom Office; contract no. HSHQDC-15-C-B0033.

Institutional Review Board Statement: Not applicable.

Informed Consent Statement: Not applicable.

Data Availability Statement: The data used in this study will be made available upon reasonabl request to the corresponding author.

Acknowledgments: This work was facilitated by using advanced computational, storage, and the networking infrastructure provided by the Hyak supercomputer system at the University of Washington.

Conflicts of Interest: The authors declare no conflict of interests.

References

1. Coelho, P.J. Numerical simulation of radiative heat transfer from non-gray gases in three-dimensional enclosures. *J. Quant Spectrosc. Radiat. Transf.* **2002**, *74*, 307–328. [CrossRef]
2. Hofmann, H.M.; Kind, M.; Martin, H. Measurements on steady state heat transfer and flow structure and new correlations fo heat and mass transfer in submerged impinging jets. *Int. J. Heat Mass Transf.* **2007**, *50*, 3957–3965. [CrossRef]
3. Hwang, J.C.; Tsou, F.K.; Cho, W.C. K-EPSILON computations of flow and heat transfer in plane oblique impinging jets. In *Computational Methods and Experimental Measurements*; Springer: Berlin/Heidelberg, Germany, 1982.
4. Martin, H. Heat and mass transfer between impinging gas jets and solid surfaces. In *Advances in Heat Transfer*; Elsevier Amsterdam, The Netherlands, 1977; Volume 13, pp. 1–60.
5. Schauer, J. *The Flow Development And Heat Transfer Characteristics of Plane Turbulentimpinging Jets*; ProQuest Dissertations Publishing Ann Arbor, MI, USA, 1964.
6. Kim, D.; Lee, J. Influence of shock structure on heat transfer characteristics in supersonic under-expanded impinging jets. *Int. J Therm. Sci.* **2019**, *141*, 62–71. [CrossRef]
7. Kim, W.-J.; Karuppuchamy, V.; Heldman, D.R. Evaluation of maximum wall shear stress from air impingement to remove food deposits from stainless steel surfaces. *J. Food Eng.* **2021**, *316*, 110825. [CrossRef]
8. Maithani, R.; Sharma, S.; Kumar, A. Thermo-hydraulic and exergy analysis of inclined impinging jets on absorber plate of sola air heater. *Renew. Energy* **2021**, *179*, 84–95. [CrossRef]
9. Kottapalli, K.; Novosselov, I.V. Aerodynamic resuspension and contact removal of energetic particles from smooth, rough, and fibrous surfaces. *Talanta* **2021**, *231*, 122356. [CrossRef]
10. Novosselov, I.; SpecTree. Pulsed Jet Sampling of Particles and Vapors from Substrates. U.S. Patent 10274404, 30 April 2019.
11. Novosselov, I.V.; Ariessohn, P.C.; Dengler, E.D.; Hickner, M.; Inventors; Enertechnix, Assignee. Particle Interrogation Devices and Methods. U.S. Patent 8561486, 22 October 2013.
12. Fillingham, P.; Kottapalli, K.; Zhan, X.; Novosselov, I.V. Characterization of adhesion force in aerodynamic particle resuspension *J. Aerosol Sci.* **2019**, *128*, 89–98. [CrossRef]

13. Phares, D.J.; Smedley, G.T.; Flagan, R.C. The wall shear stress produced by the normal impingement of a jet on a flat surface. *J. Fluid Mech.* **2000**, *418*, 351–375. [CrossRef]
14. Keedy, R.; Dengler, E.; Ariessohn, P.; Novosselov, I.; Aliseda, A. Removal rates of explosive particles from a surface by impingement of a gas jet. *Aerosol Sci. Technol.* **2012**, *46*, 148–155. [CrossRef]
15. Kottapalli, K.; Novosselov, I.V. Experimental study of aerodynamic resuspension of RDX residue. *Aerosol Sci. Technol.* **2019**, *53*, 549–561. [CrossRef]
16. Fillingham, P.; Vaddi, R.S.; Bruning, A.; Israel, G.; Novosselov, I.V. Drag, lift, and torque on a prolate spheroid resting on a smooth surface in a linear shear flow. *Powder Technol.* **2021**, *377*, 958–965. [CrossRef]
17. Mercier-Bonin, M.; Dehouche, A.; Morchain, J.; Schmitz, P. Orientation and detachment dynamics of Bacillus spores from stainless steel under controlled shear flow: Modelling of the adhesion force. *Int. J. Food Microbiol.* **2011**, *146*, 182–191. [CrossRef]
18. Kesavan, J.S.; Humphreys, P.D.; Bottiger, J.R.; Valdes, E.R.; Rastogi, V.K.; Knox, C.K. Deposition method, relative humidity, and surface property effects of bacterial spore reaerosolization via pulsed air jet. *Aerosol Sci. Technol.* **2017**, *51*, 1027–1034. [CrossRef]
19. Henry, C.; Minier, J.-P. Progress in particle resuspension from rough surfaces by turbulent flows. *Prog. Energy Combust. Sci.* **2014**, *45*, 1–53. [CrossRef]
20. Kalghatgi, G.; Hunt, B. The occurrence of stagnation bubbles in supersonic jet impingement flows. *Aeronaut. Q.* **1976**, *27*, 169–185. [CrossRef]
21. Henderson, B.; Powell, A. Experiments concerning tones produced by an axisymmetric choked jet impinging on flat plates. *J. Sound Vib.* **1993**, *168*, 307–326. [CrossRef]
22. Krothapalli, A.; Rajkuperan, E.; Alvi, F.; Lourenco, L. Flow field and noise characteristics of a supersonic impinging jet. *J. Fluid Mech.* **1999**, *392*, 155–181. [CrossRef]
23. Fillingham, P.; Murali, H.; Novosselov, I.V. Nondimensional Parameter for Characterization of Wall Shear Stress From Underexpanded Axisymmetric Impinging Jets. *J. Fluids Eng.* **2017**, *139*, 111102. [CrossRef]
24. Fillingham, P.; Novosselov, I.V. Wall jet similarity of impinging planar underexpanded jets. *Int. J. Heat Fluid Flow* **2020**, *81*, 108516. [CrossRef]
25. Roy, S.; Patel, P. Study of heat transfer for a pair of rectangular jets impinging on an inclined surface. *Int. J. Heat Mass Transf.* **2003**, *46*, 411–425. [CrossRef]
26. Attalla, M.; Maghrabie, H.M.; Specht, E. Effect of inclination angle of a pair of air jets on heat transfer into the flat surface. *Exp. Therm. Fluid Sci.* **2017**, *85*, 85–94. [CrossRef]
27. Zhang, M.; Wang, N.; Han, J.-C. Internal heat transfer of film-cooled leading edge model with normal and tangential impinging jets. *Int. J. Heat Mass Transf.* **2019**, *139*, 193–204. [CrossRef]
28. Miguel-González, C.; García-Díaz, M.; Pereiras, B.; Vigil, M.; Rodríguez de Castro, A. Numerical model of a planar jet wiping system for continuous strip lines. *J. Mech. Sci. Technol.* **2021**, *35*, 2929–2938. [CrossRef]
29. Crafton, J.; Carter, C.; Elliott, G.; Sullivan, J. The impingement of sonic and sub-sonic jets onto a flat plate at inclined angles. *Exp. Methods Appl. Fluid Flow* **2006**, *41*, 699–710. [CrossRef]
30. Nguyen, T.; Blake, M. Flowfield Characteristics of a Supersonic Jet Impinging on an Inclined Surface. *Am. Inst. Aeronaut. Astronaut. AIAA J.* **2020**, *58*, 1240–1254. [CrossRef]
31. Dorrepaal, J.M. An exact solution of the Navier-Stokes equation which describes non-orthogonal stagnation-point flow in two dimensions. *J. Fluid Mech.* **1986**, *163*, 141–147. [CrossRef]
32. Beltaos, S. Oblique impingement of plane turbulent jets. *J. Hydraul. Div.* **1976**, *102*, 1177–1192. [CrossRef]
33. Chin, D.; Agarwal, M. Mass-Transfer from an Oblique Impinging Slot Jet. *J. Electrochem. Soc.* **1991**, *138*, 2643–2650. [CrossRef]
34. Mazurek, K.A.; Rajaratnam, N. Erosion of sand beds by obliquely impinging plane turbulent air jets. *J. Hydraul. Res.* **2005**, *43*, 567–573. [CrossRef]
35. Kesavan, J.; Humphreys, P.; Nasr, B.; Ahmadi, G.; Knox, C.K.; Valdes, E.; Rastogi, V.; Dhaniyala, S. Experimental and computational study of reaerosolization of 1 to 5 μm PSL microspheres using jet impingement. *Aerosol Sci. Technol.* **2017**, *51*, 377–387. [CrossRef]
36. Nasr, B.; Ahmadi, G.; Ferro, A.R.; Dhaniyala, S. A model for particle removal from surfaces with large-scale roughness in turbulent flows. *Aerosol Sci. Technol.* **2020**, *54*, 291–303. [CrossRef]
37. Jaramillo, J.; Trias, F.; Gorobets, A.; Pérez-Segarra, C.; Oliva, A. DNS and RANS modelling of a turbulent plane impinging jet. *Int. J. Heat Mass Transf.* **2012**, *55*, 789–801. [CrossRef]
38. Alvi, F.; Ladd, J.; Bower, W. Experimental and computational investigation of supersonic impinging jets. *Am. Inst. Aeronaut. Astronaut. AIAA J.* **2002**, *40*, 599–609. [CrossRef]
39. Menter, F. (Ed.) Zonal two equation kw turbulence models for aerodynamic flows. In Proceedings of the 23rd Fluid Dynamics, Plasmadynamics, and Lasers Conference, Orlando, FL, USA, 6–9 July 1993.
40. Shukla, A.K.; Dewan, A. Flow and thermal characteristics of jet impingement: Comprehensive review. *Int. J. Heat Technol.* **2017**, *35*, 153–166. [CrossRef]
41. Naqavi, I.Z.; Tyacke, J.C.; Tucker, P.G. Direct numerical simulation of a wall jet: Flow physics. *J. Fluid Mech.* **2018**, *852*, 507–542. [CrossRef]

mathematics MDPI

Article

Hydrodynamic Impacts of Short Laser Pulses on Plasmas

Gaetano Fiore [1,2,*], Monica De Angelis [1], Renato Fedele [2,3], Gabriele Guerriero [1] and Dušan Jovanović [4,5]

1 Department di Matematica e Applicazioni, Università di Napoli "Federico II", Complesso Universitario M. S. Angelo, Via Cintia, 80126 Napoli, Italy; modeange@unina.it (M.D.A.); gabriele.guerriero@unina.it (G.G.)
2 INFN, Sezione di Napoli, Complesso MSA, Via Cintia, 80126 Napoli, Italy; renato.fedele@na.infn.it
3 Department di Fisica, Università di Napoli "Federico II", Complesso Universitario M. S. Angelo, Via Cintia, 80126 Napoli, Italy
4 Institute of Physics, University of Belgrade, 11080 Belgrade, Serbia; dusan.jovanovic@ipb.ac.rs
5 Texas A & M University at Qatar, Doha 23874, Qatar
* Correspondence: gaetano.fiore@na.infn.it

Abstract: We determine conditions allowing for simplification of the description of the impact of a short and arbitrarily intense laser pulse onto a cold plasma at rest. If both the initial plasma density and pulse profile have plane symmetry, then suitable matched upper bounds on the maximum and the relative variations of the initial density, as well as on the intensity and duration of the pulse, ensure a strictly hydrodynamic evolution of the electron fluid without wave-breaking or vacuum-heating during its whole interaction with the pulse, while ions can be regarded as immobile. We use a recently developed fully relativistic plane model whereby the system of the Lorentz–Maxwell and continuity PDEs is reduced into a family of highly nonlinear but decoupled systems of non-autonomous Hamilton equations with one degree of freedom, the light-like coordinate $\xi = ct-z$ instead of time t as an independent variable, and new a priori estimates (eased by use of a Liapunov function) of the solutions in terms of the input data (i.e., the initial density and pulse profile). If the laser spot radius R is finite and is not too small, the same conclusions hold for the part of the plasma close to the axis $\vec{z}$ of cylindrical symmetry. These results may help in drastically simplifying the study of extreme acceleration mechanisms of electrons.

Keywords: plasma hydrodynamics; non-autonomous Hamilton equations; Liapunov function; relativistic electrodynamics; plasma wave; wave-breaking

MSC: 34C11; 34C99; 34C60; 76W05; 70H05; 70H40

Citation: Fiore, G.; De Angelis, M.; Fedele, R.; Guerriero, G.; Jovanović, D. Hydrodynamic Impacts of Short Laser Pulses on Plasmas. *Mathematics* **2022**, *10*, 2622. https://doi.org/10.3390/math10152622

Academic Editors: Vasily Novozhilov and Cunlu Zhao

Received: 17 June 2022
Accepted: 22 July 2022
Published: 27 July 2022

1. Introduction and Preliminaries

Laser–plasma interactions induced by ultra-intense laser pulses lead to a variety of very interesting phenomena [1–5], notably plasma compression for inertial fusion [6], laser wakefield acceleration (LWFA) [7–9] and other extremely compact acceleration mechanisms (e.g., hybrid laser-driven and particle-driven plasma wakefield acceleration [10]) of charged particles, which hopefully will be the basis of a generation of new table-top accelerators. This is paramount because accelerators have extremely important applications in particle physics, materials science, medicine, industry, environmental remediation, etc., and therefore huge investments (such as the EU-funded project *Eupraxia* [11–13]) are being developed all over the world to promote the development of such accelerators. Similar extreme conditions (huge electromagnetic fields and huge accelerations of charged particles in plasmas) occur in a number of violent astrophysical processes as well; see, e.g., [9] and references therein. In general, these phenomena are ruled by the equations of a kinetic theory coupled to Maxwell's equations, which can be only solved numerically via particle-in-cell (PIC) techniques. Unfortunately, PIC codes involve huge and expensive computations for each choice of the free parameters; even with the presently ever increasing computational power, exploring the parameter space blindly in order to single out interesting regions remains prohibitively expensive. Sometimes, good predictions can be obtained by treating the plasma as a multicomponent fluid (electrons and ions) and by numerically solving the simpler associated hydrodynamic equations via multifluid codes such as QFluid [14] or hybrid kinetic/fluid codes. In general, however, it is not known a priori in which conditions or spacetime regions this is possible. Therefore any analytical insights that can simplify the work, at least in special cases or in a limited space-time region, are welcome.

This applies in particular to studying the impact of a very short (and possibly very intense) laser pulse perpendicularly onto a cold diluted plasma at rest, as well as onto matter which is locally ionized into a plasma by the front of the pulse itself. As is well known, electrons start oscillating orthogonally to the direction $\vec{z}$ of pulse propagation and drifting in the positive z-direction, pushed respectively by the electric and magnetic parts of the Lorentz force induced by the pulse; thereafter, electrons start oscillating longitudinally as well (i.e., in $\vec{z}$-direction), pushed by the restoring electric force induced by charge separation. Readers can find such initial longitudinal motions in Figure 1c and from the electron worldlines reported in Figures 7 and 8. It turns out that the initial dynamics are simpler if the pulse is *essentially short*. We say here (see Definition 1) that the pulse is *essentially short* if it overcomes each electron before the z-displacement, Δ, of the latter reaches a negative minimum for the first time; an essentially short pulse is *strictly short* if it overcomes each electron before Δ becomes negative for the first time. In other words, we regard a pulse as strictly short (essentially short) if it overcomes each electron before it finishes the first half (three quarters) longitudinal oscillation. In the nonrelativistic (NR) regime, a pulse which is symmetric under inversion around its center is strictly short (essentially short) if its duration l/c does not respectively exceed half (1) times the NR plasma oscillation period $t_H^{nr} \equiv \sqrt{\pi m / n_b e^2}$ associated with the maximum, n_b, of the initial electron density; that is (see Proposition 1), if

$$G_b := \sqrt{\frac{n_b e^2}{\pi m c^2}}\, l \leq \begin{cases} 1/2 \\ 1 \end{cases} \tag{1}$$

where $-e, m$ are the electron charge and mass c is the speed of light. (If the pulse is a slowly modulated monochromatic wave (58) with wavelength $\lambda = 2\pi/k$, this implies a fortiori $\frac{4\pi e^2}{mc^2} n_b \lambda^2 \ll 1$, meaning that the plasma is underdense). The general *relativistic* plasma oscillation period is not independent of the oscillation amplitude, but grows with the latter, which in turn grows with the pulse intensity. Correspondingly, Equation (16) can be fulfilled with a larger G_b; in addition, it is compatible with maximizing the oscillation amplitude, and thus the energy transfer from the pulse to the plasma wave, as for a given n_b and pulse energy such a maximization can be achieved [3,15] through a suitable

$$l \sim \tilde{\xi}_2, \qquad \text{(i.e., when } G_b \sim 1/2 \text{ in the nonrelativistic regime).} \tag{2}$$

We believe that such impacts require a deeper understanding because, among other things, they may generate: (i) a plasma wave (PW) [16,17] or even an *ion bubble* (a region containing only ions, because all electrons have been expelled out of it) [18–24] producing the LWFA, i.e., accelerating a small bunch of (socalled *witness*) electrons trailing the pulse at very high energy in the forward direction; and (ii) the *slingshot effect* [15,25,26], i.e., the backward acceleration and expulsion of energetic electrons from the vacuum–plasma interface during or just after the impact. The present work is one out of a few papers [27–29] arguing that, with the help of the plane, fully relativistic Lagrangian model of [30,31], and very little computational power, we can obtain important information about such an impact, in particular the formation of a PW, its persistence before wave-breaking (WB), and the features of the latter. As is known, a small WB is not necessarily undesirable, and may be used to produce and inject the mentioned witness electrons into the PW (*self-injection*).

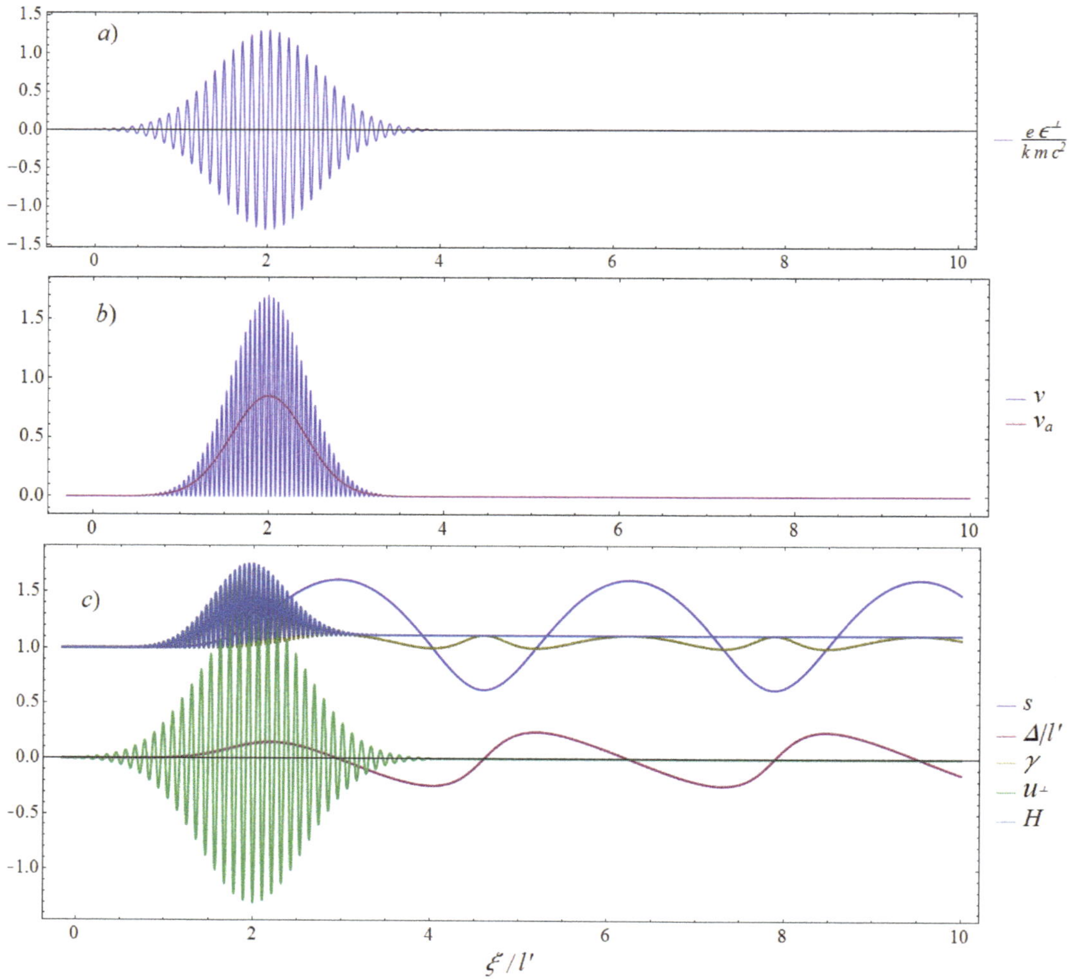

Figure 1. (**a**) Normalized amplitude of a linearly polarized (i.e., set $\psi = 0$ in (58)) monochromatic laser pulse slowly modulated by a Gaussian with *full width at half maximum* l' and peak amplitude $a_0 \equiv \lambda e E_M^{\perp}/mc^2 = 1.3$; this yields a moderately relativistic electron dynamics, and $\Delta_u \equiv \Delta^{(0)}(l) \simeq 0.45l'$. If $l' = 7.5$ µm (corresponding to a pulse duration of $\tau' = l'/c \simeq 2.5 \times 10^{-14}$ s), and the wavelength is $\lambda = 0.8$ µm, then the corresponding peak intensity must be $I = 7.25 \times 10^{18}$ W/cm^2; these are typical values obtainable by Ti:Sapphire lasers in LWFA experiments. (**b**) The corresponding forcing term $v(\xi)$ and average-over-cycle (60) $v_a(\xi)$ of the latter. (**c**) Corresponding solution of (8) and (9), or equivalently of (14) if $Kn_0 l'^2 \simeq 4$; this value is obtained if $l' = 7.5$ µm and $\widetilde{n_0}(Z) \equiv n_0 = 2 \times 10^{18}$ cm^{-3} (a typical value of the electron density used in LWFA experiments). As expected, s is insensitive to the rapid oscillations of $\epsilon^{\perp}$ for $\xi \in [0, l]$, while for $\xi \geq l$ the energy H is conserved and the solution is periodic. The length l is determined on physical grounds; if, e.g., the plasma is created locally by the impact of the pulse itself on a gas (e.g., hydrogen or helium), then $[0, l]$ has to contain all points ξ where the pulse intensity is sufficient to ionize the gas. Here, for simplicity and following convention, we have fixed it to be $l = 4l'$; the possible inaccuracy of such a cut is very small, because $\epsilon(0^+) = \epsilon(l^-)$ is 2^{-16} times the maximum $\epsilon(l/2)$ of the modulation, i.e., practically zero, which makes $G_b \equiv \sqrt{Kn_0}\, l \simeq 8$.

The plane model is as follows. We assume that the plasma is initially neutral, unmagnetized, and at rest, with zero densities in the region $z<0$. More precisely, the $t=0$ initial conditions for the electron fluid Eulerian density n_e and velocity $\boldsymbol{v}_e$ are of the type

$$\boldsymbol{v}_e(0,\boldsymbol{x}) = \boldsymbol{0}, \qquad n_e(0,\boldsymbol{x}) = \widetilde{n_0}(z), \tag{3}$$

where the initial electron (as well as proton) density $\widetilde{n_0}(z)$ fulfills

$$\widetilde{n_0}(z)=0 \text{ if } z\leq 0, \qquad 0<\widetilde{n_0}(z)\leq n_b \quad \text{if } z>0 \tag{4}$$

for some $n_b > 0$ (a few examples are reported in Figure 2). Assuming that prior to the impact the laser pulse is a free plane transverse wave travelling in the z-direction, i.e., that the electric and magnetic fields $\boldsymbol{E}, \boldsymbol{B}$ are of the form

$$\boldsymbol{E}(t,\boldsymbol{x}) = \boldsymbol{E}^{\perp}(t,\boldsymbol{x}) = \boldsymbol{\epsilon}^{\perp}(ct-z), \qquad \boldsymbol{B} = \boldsymbol{B}^{\perp} = \mathbf{k}\times\boldsymbol{E}^{\perp} \qquad \text{if } t\leq 0 \tag{5}$$

where $\boldsymbol{\epsilon}^{\perp}(\xi)$ has a bounded support with $\xi=0$ as the left extreme (i.e., the pulse reaches the plasma at $t=0$) and the superscript $\perp$ denotes vector components orthogonal to $\mathbf{k} \equiv \nabla z$. The input data of a specific problem are the functions $\widetilde{n_0}(z), \boldsymbol{\epsilon}^{\perp}(\xi)$; in addition, it is convenient to define the related functions

$$\boldsymbol{\alpha}^{\perp}(\xi) := -\int_{-\infty}^{\xi} d\zeta\, \boldsymbol{\epsilon}^{\perp}(\zeta), \qquad v(\xi) := \left[\frac{e\boldsymbol{\alpha}^{\perp}(\xi)}{mc^2}\right]^2, \tag{6}$$

$$\widetilde{N}(Z) := \int_0^Z d\zeta\, \widetilde{n_0}(\zeta), \qquad \mathcal{U}(\Delta;Z) := K\int_0^{\Delta} d\zeta\, (\Delta-\zeta)\, \widetilde{n_0}(Z+\zeta), \tag{7}$$

where $K := \frac{4\pi e^2}{mc^2}$. By definition, v is dimensionless and nonnegative, and $\widetilde{N}(Z)$ strictly grows with Z. We can describe the plasma as a fully relativistic collisionless fluid of electrons and a static fluid of ions; as usual, in the short time lapse of interest here, the motion of the much heavier ions is negligible, with $\boldsymbol{E}, \boldsymbol{B}$ and the plasma dynamical variables fulfilling the Lorentz–Maxwell and continuity equations. As at the impact time $t=0$ the plasma is made up of two static fluids, by continuity, such a hydrodynamical description (HD) is justified and we can neglect the depletion of the pulse at least for small $t>0$; the specific time lapse is determined a posteriori by self-consistency. This allows us (see [30,31], or [32–35] for shorter presentations) to reduce the system of the Lorentz–Maxwell and continuity partial differential equations (PDEs) into ordinary ones, more precisely into the following continuous family of *decoupled Hamilton equations for systems with one degree of freedom*. Each system determines the complete Lagrangian (in the sense of non-Eulerian) description of the motion of the electrons having a same initial longitudinal coordinate $Z > 0$ (the *Z-electrons*, for brevity), and reads

$$\Delta'(\xi,Z) = \frac{1+v(\xi)}{2s^2(\xi,Z)} - \frac{1}{2}, \qquad s'(\xi,Z) = K\Big\{\widetilde{N}[Z+\Delta(\xi,Z)] - \widetilde{N}(Z)\Big\}; \tag{8}$$

it is equipped with the initial conditions

$$\Delta(0,Z) = 0, \qquad s(0,Z) = 1. \tag{9}$$

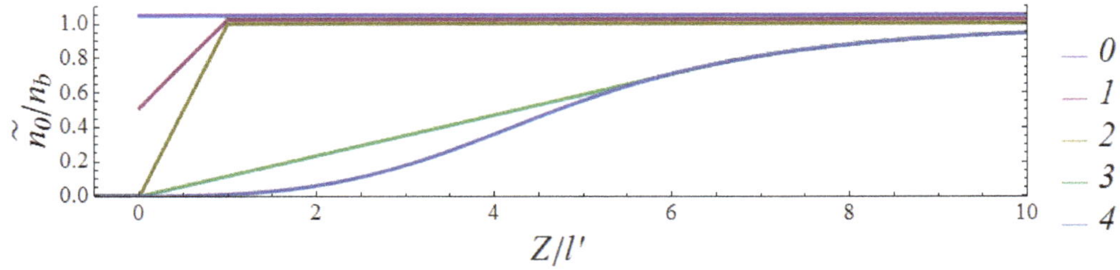

Figure 2. Plots of the ratios $\widetilde{n_0}/n_b$ for the following initial densities: (0) $\widetilde{n_0}(z) = n_b\,\theta(z)$. (1) $\widetilde{n_0}(z) = \frac{1}{2}n_b\,\theta(z)[1+\theta(l'-z)\,z/l' + \theta(z-l')]$. (2) $\widetilde{n_0}(z) = n_b\,[\theta(z)\,\theta(l'-z)\,z/l' + \theta(z-l')]$. (3) $\widetilde{n_0}(z) = n_b\left\{\frac{z}{\bar{z}}\,\frac{f(\bar{z})}{1+f(\bar{z})}\theta(z)\theta(\bar{z}-z) + \theta(z-\bar{z})\,\frac{f(z)}{1+f(z)}\right\}$, where $f(z) := (0.1\,z/l')^2 + (0.2\,z/l')^4$ and $\bar{z} = 6.5l'$; this grows as z for $z \le \bar{z}$ and coincides with the next one for $z > \bar{z}$. (4) $\widetilde{n_0}(z) = n_b\,\theta(z)\,\frac{f(z)}{1+f(z)}$.

Here, the unknown basic dynamical variables $\Delta(\xi, Z), s(\xi, Z)$ are respectivey the longitudinal displacement and s-factor of the Z-electrons espressed as functions of ξ, Z, while $z_e(\xi, Z) := Z+\Delta(\xi, Z)$ is the present longitudinal coordinate of the Z-electrons; s is the light-like component of the 4-velocity of the Z electrons, or equivalently is related to their 4-momentum by $p^0 - cp^z \equiv mc^2 s$; it is positive-definite. In the NR regime $|s-1| \ll 1$; in the present fully relativistic regime, it need only satisfy the inequality $s > 0$. We consider all dynamical variables f (in the Lagrangian description) as functions of ξ, Z instead of t, Z; in the cited papers [30–35], the two dependences are denoted as $\hat{f}(\xi, Z)$ and $f(t, Z)$, respectively, although here we use only the former and thus denote it simply as $f(\xi, Z)$ without the $\hat{}$. In the above, f' stands for the total derivative $df/d\xi := \partial f/\partial\xi + s'\partial f/\partial s + \Delta'\partial f/\partial\Delta$ and Z plays the role of the family parameter; all of the other electron dynamical variables can be expressed in terms of Δ, s and the initial coordinates $\mathbf{X} \equiv (X, Y, Z)$ of the generic electron fluid element. In particular, in the basic approximation the dimensionless variable $\boldsymbol{u}^\perp := \boldsymbol{p}^\perp/mc$, i.e., the electrons' transverse momentum in mc units, is given by $\boldsymbol{u}^\perp = \frac{e\boldsymbol{\alpha}^\perp}{mc^2}$; hence, $v = \boldsymbol{u}^{\perp 2}$. The light-like coordinate $\xi = ct - z$ in Minkowski spacetime can be adopted as an independent variable instead of time t, as all particles must travel at a speed lower than c; at the end, to express the solution as a function of t it is only necessary to replace ξ with the inverses $\tilde{\xi}(t, Z)$ of the strictly increasing (in ξ) functions $\hat{t}(\xi, Z) := (\xi + z_e(\xi, Z))/c$. Equation (8) consists of Hamilton equations, with $\xi, \Delta, -s$ playing the role of the usual t, q, p and the (dimensionless) Hamiltonian

$$H(\Delta, s, \xi; Z) \equiv \frac{s^2 + 1 + v(\xi)}{2s} + \mathcal{U}(\Delta; Z); \tag{10}$$

the first term provides the kinetic + rest mass energy, while $\mathcal{U}$ plays the role of potential energy due to the electric charges' mutual interaction. Consequently, along the solutions of (8) $H' = \partial H/\partial\xi = v'/2s$. Integrating the latter identity by parts and using the definition (10) of H, we find

$$\frac{(s-1)^2}{2s} + \mathcal{U}(\Delta; Z) = H(\xi, Z) - 1 - \frac{v(\xi)}{2s(\xi, Z)} = \int_0^\xi d\eta\, \frac{vs'}{2s^2}(\eta, Z) =: \nu(\xi, Z). \tag{11}$$

The Hamilton Equations in (8) are non-autonomous for $0 < \xi < l$, where $[0, l]$ is the smallest closed interval containing the support of $\boldsymbol{\epsilon}^\perp$; ultra-intense pulses are characterized by $\max_{\xi\in[0,l]}\{v(\xi)\} \gg 1$ and induce ultra-relativistic electron motions. For $\xi \ge l$, (8) can be solved by quadrature as well, using the energy integrals of motion $H(\xi, Z) = H(l, Z) =: h(Z) = \text{const}$.

Solving (8) and (9) yields the motions of the Z-electrons' fluid elements, which are fully represented through their worldlines in Minkowski space. In Figures 7 and 8 we have displayed the projections onto the z, ct plane of these worldlines for two specific sets of input data; as can be seen, the PW emerges from them as a collective effect. Mathematically, the PW features can be derived by reference to the Eulerian description of the electron fluid; the resulting flow is laminar, with xy plane symmetry. The Jacobian of the transformation $\boldsymbol{X} \mapsto \boldsymbol{x}_e \equiv (x_e, y_e, z_e)$ from the Lagrangian to the Eulerian coordinates reduces to $J(\xi, Z) = \partial z_e(\xi, Z)/\partial Z$, because $\boldsymbol{x}_e^{\perp} - \boldsymbol{X}^{\perp}$ does not depend on $\boldsymbol{X}^{\perp}$.

The HD breaks where worldlines intersect, leading to WB of the PW. No WB occurs as long as J remains positive. If the initial density is uniform, $\widetilde{n_0}(Z) = n_0 = \text{const}$, both Equation (8) and the initial conditions (9) become Z-independent, because (8 Right) takes the form $s' = M\Delta$, where $M := Kn_0$. Consequently, their solutions become Z-independent as well, and $J \equiv 1$ at all ξ. Otherwise, WB occurs after a sufficiently long time [36].

Our main goal here is to determine manageable sufficient conditions on $\widetilde{n_0}(z), \boldsymbol{\epsilon}^{\perp}(\xi)$ guaranteeing that $J(\xi, Z) > 0$ for all $Z > 0$ and $\xi \in [0, l]$, *without* solving the Cauchy problems (8) and (9). This ensures that there is no wave-breaking during the laser–plasma interaction (WBDLPI), i.e., while the Hamilton Equations (8) are non-autonomous (due to the dependence of v on ξ. We reach this goal by determining upper and lower bounds first on Δ, s, H (Section 2), then on J and $\partial s/\partial Z$ (Section 3), with the help of a suitable Liapunov function. These bounds provide useful approximations of these dynamical variables in the interval $0 \le \xi \le l$. As previously mentioned, the NR short-pulse conditions (1) are generalized by the ones (16) in the present fully relativistic regime. Inequalities (35) and (36) are respectively sufficient conditions for (16 Left) and (16 Right). Instead of (35), we can first check the stronger and more easily verifiable condition (40), or even the simplest and strongest one, $M_u l^2 \le 2$. In the case that (36) is satisfied, we can exclude WBDLPI in the NR regime if (48) is fulfilled; in the general case, if one of the three conditions of Proposition 1 is fulfilled, namely if the plasma is initially sufficiently diluted and/or the local relative variations of its density are sufficiently small (if $Q_0 < 1$, the strongest and easiest to compute, is not fulfilled, $Q_1 < 1$ or $Q_2 < 1$ can be used). In Section 4 we compare the dynamics of s, Δ, J, σ induced by the same pulse on five representative $\widetilde{n_0}(z)$ having the same upper bound and asymptotic value n_b, both by numerically solving the equations and by applying the mentioned inequalities. In particular, we find that the density profile at the very edge of the plasma is critical; for instance, if $\widetilde{n_0}(z) \sim z$ as $z \to 0^+$, then WB occurs earlier than if $\widetilde{n_0}(z) \sim z^2$ or if $\lim_{z\to 0^+} \widetilde{n_0}(z) > 0$ (discontinuous density at $z = 0$), albeit with the electrons colliding at very small relative velocities. To produce LWFA, the laser pulse is usually fired orthogonally to a supersonic gas jet (e.g., hydrogen or helium); outside the jet nozzle it is $\widetilde{n_0}(z) \sim z^2$, and thus our results imply that such impacts occur in the hydrodynamic regime under rather broad conditions.

For $\xi \ge l$, by using the conservation of energy we can show [28] that, while Δ and s are periodic with a suitable period ξ_H, J and σ are *linearly quasiperiodic*, that is, they are of the form

$$f(\xi) = a(\xi) + \xi\, b(\xi), \qquad \xi \ge l, \tag{12}$$

where a, b are periodic in ξ with period ξ_H and b has zero average over a period; $b(\xi)$ oscillates between positive and negative values, as does the second term, which dominates as $\xi \to \infty$, with ξ acting as a modulating amplitude. Therefore, the occurrence of WB after the laser–plasma interaction is best investigated by studying the dependence (12) [28].

The spacetime region where the present plane hydrodynamic model predicting a laminar and xy-symmetric flow is self-consistent is determined (Section 4) by the conditions $J > 0$ (no collisions) and (62) (undepleted pulse approximation). For typical LWFA experiments, the length and time sizes allowed by (62) are respectively on the order of several hundred microns and femtoseconds, respectively. The spacetime region where the predictions of the model can be trusted is further reduced (Section 4) by the finite transverse size R of the laser pulse. According to our model, other phenomenal characteristic of plasma physics, such as turbulent flows, diffusion, heating, heat exchange, as well as the very motion of ions, can be excluded inside the latter region, although of course they can and will occur outside.

Finally, we point out that recent advances in multi-timescale analysis have permitted (semi)-analytical studies of laser–plasma interaction, including aspects of WB, in the ultrarelativistic regime using Vlasov description [37].

2. A Priori Estimates of Δ, s, H for Small $\xi > 0$

The Cauchy problem (see (8) and (9)) is equivalent to the following integral one:

$$\Delta(\xi,Z) = \int_0^{\xi} d\eta\, \frac{1+v(\eta)}{2s^2(\eta,Z)} - \frac{\xi}{2}, \qquad s(\xi,Z) - 1 = \int_0^{\xi} d\eta \int_Z^{z_e(\eta,Z)} dZ'\, K\widetilde{n_0}(Z'). \tag{13}$$

If $\widetilde{n_0}(Z) \equiv n_0 = \text{const}$, then $\mathcal{U}(\Delta) = M\Delta^2/2$, $s' = M\Delta$, and (13) amounts to

$$s(\xi) = 1 + \frac{M}{2}\left[-\frac{\xi^2}{2} + \int_0^{\xi} d\eta\, (\xi-\eta)\frac{1+v}{s^2}(\eta)\right], \tag{14}$$

where $M := Kn_0$; once (14) is solved, we can obtain Δ from $\Delta = s'/M$. In Figure 1, we plot an example of a monochromatic laser pulse slowly modulated by a Gaussian along with the corresponding solution (s,Δ) in a constant density plasma; the qualitative behaviour of the solution remains the same if $\widetilde{n_0}(z) \neq \text{const}$. In the NR regime $v \ll 1$, whence $|\Delta/l| \ll 1$, $|\delta| \ll 1$, where $\delta := s-1$; at lowest order in δ, Δ (8) and (9) reduce to the equations $\delta' = M\Delta$, $\Delta' = v/2 - \delta$ of a forced NR harmonic oscillator with trivial initial conditions. The solution is

$$\Delta(\xi) = \int_0^{\xi} d\eta\, \frac{v(\eta)}{2} \cos\left[\sqrt{M}(\xi-\eta)\right], \quad \delta(\xi) = \int_0^{\xi} d\eta\, \frac{v(\eta)}{2} \sin\left[\sqrt{M}(\xi-\eta)\right]. \tag{15}$$

By (8 Right), the zeroes of $\Delta(\cdot,Z)$ are extrema of $s(\cdot,Z)$ and vice versa, because $\widetilde{N}(Z)$ grows with Z. Let us recall how Δ, s start evolving from their initial values (9). As previously mentioned, for $\xi > 0$ all electrons reached by the pulse start to oscillate transversely and drift forward; in fact, $v(\xi)$ becomes positive, implying in turn that the right-hand side (rhs) of (8 Left) and Δ does as well; the $Z=0$ electrons leave a layer of ions of finite thickness behind themselves that is completely evacuated of electrons. If the density vanished ($\widetilde{n_0} \equiv 0$), then we would obtain

$$s \equiv 1, \qquad \Delta(\xi,Z) = \int_0^{\xi} d\eta\, \frac{v(\eta)}{2} =: \Delta^{(0)}(\xi);$$

$\Delta^{(0)}(\xi)$ grows with ξ and is almost constant for $\xi > l$ if $v(\xi) \simeq 0$ for $\xi > l$ (which occurs if the pulse is slowly modulated (58)). On the contrary, as the density is positive, the growth of Δ implies the growth of the rhs of (8 Right), because the latter grows with Δ, as well as of $s(\xi,Z)-1$. Meanwhile, $\Delta(\xi,Z)$ continues to grow as long as $1+v(\xi) > s^2(\xi,Z)$, and reaches a maximum at $\tilde{\xi}_1(Z) \equiv$ the smallest $\xi > 0$ such that the rhs (8 Left) vanishes. $s(\xi,Z)$ keeps growing as long as $\Delta(\xi,Z) \geq 0$, reaches a maximum at the first zero $\tilde{\xi}_2 > \tilde{\xi}_1$ of $\Delta(\xi,Z)$, and decreases for $\xi > \tilde{\xi}_2$, while $\Delta(\xi,Z)$ is negative. $\Delta(\xi,Z)$ reaches a negative minimum at $\check{\xi}_3(Z) \equiv$ the smallest $\xi > \tilde{\xi}_2$, such that the rhs (8 Left) vanishes again. Here, we denote by $\tilde{\xi}_3(Z)$ the smallest $\xi > \check{\xi}_3$ such that $s(\xi,Z) = 1$ and $\mathcal{I} := [0,\tilde{\xi}_3]$. We encourage the reader to single out $\tilde{\xi}_1, \tilde{\xi}_2, \check{\xi}_3, \tilde{\xi}_3$ for the solution considered in Figure 1 from the graphs in 1c. If $\epsilon^{\perp}$ is slowly modulated, then $v(l) \simeq 0$ (see Appendix A.2 for details, and Appendix 5.4 in [31] for further information); then, $\check{\xi}_3 \simeq \tilde{\xi}_3$ if $l < \check{\xi}_3$.

Definition 1. *A pulse is strictly short, essentially short with respect to $\widetilde{n_0}$ if it respectively fulfills*

$$\begin{cases} \Delta(\xi,Z) \geq 0, \\ s(\xi,Z) \geq 1, \end{cases} \quad \forall \xi \in [0,l],\ Z \geq 0 \qquad \Leftrightarrow \qquad l \leq \begin{cases} \tilde{\xi}_2(Z) \\ \tilde{\xi}_3(Z) \end{cases} \quad \forall Z \geq 0. \tag{16}$$

In the NR regime, if $\widetilde{n_0}(Z) \equiv n_0$ =const, these respectively amount to requiring that the corresponding solution (15) fulfills $\Delta(l) \geq 0$, $\delta(l) \geq 0$; if $\widetilde{n_0}(Z) \neq$const, it is sufficient to replace n_0 with n_b to obtain sufficient conditions for the fulfillment of (16), said conditions being

$$\begin{cases} 2\Delta(\xi) = \int_0^\xi d\eta\, v(\eta)\, \cos\left[\sqrt{Kn_b}(\xi-\eta)\right] \geq 0\,, \\ 2\delta(\xi) = \int_0^\xi d\eta\, v(\eta)\, \sin\left[\sqrt{Kn_b}(\xi-\eta)\right] \geq 0\,, \end{cases} \qquad \forall \xi \in [0,l]. \tag{17}$$

Proposition 1. *If v is symmetric about $\xi = l/2$, i.e., $v(\xi) = v(l-\xi)$, then (17 Left) amounts to $G_b \leq 1/2$; if in addition v has a unique maximum in $\xi = l/2$, then (17 Right) amounts to $G_b \leq 1$.*

The proof is provided in Appendix A.1. The assumption that $v(\xi)$ be symmetric is satisfied with very good approximation if the pulse is a slowly modulated one (58) with a symmetric modulation $\epsilon(\xi)$ about $\xi = l/2$ (as in Figure 1b), per (59).

The following estimates hold for $\xi \in \mathcal{I} = [0, \tilde{\xi}_3]$. First, $s \geq 1$ and (13 Left) imply the bound

$$\Delta(\xi, Z) \leq \Delta^{(0)}(\xi). \tag{18}$$

2.1. Constant Density Case

If $\widetilde{n_0}(Z) \equiv n_0 > 0$, for $\xi \in \mathcal{I}$ we find by (18) $s(\xi)-1 = M\int_0^\xi d\eta\, \Delta(\eta) \leq M\int_0^\xi d\eta\, \Delta^{(0)}(\eta)$, i.e.,

$$s(\xi) \leq 1 + \frac{M}{2}\int_0^\xi d\eta\, (\xi - \eta)\, v(\eta) =: s^{(1)}(\xi). \tag{19}$$

As $s^{(1)}(\xi)$ grows strictly with ξ and is convex, Equation (19) and (13 Left) in turn imply

$$\Delta(\xi) \quad \geq \quad \int_0^\xi d\eta\, \frac{1+v(\eta)}{2[s^{(1)}(\eta)]^2} - \frac{\xi}{2} =: \Delta^{(1)}(\xi). \tag{20}$$

It is apparent that $\Delta^{(1)}(\xi)$ vanishes at $\xi = 0$ and grows with ξ for small $\xi > 0$ until its (unique) maximum point, while for larger ξ it decreases and becomes negative. Hence, a lower bound $\tilde{\xi}_2^{(1)}$ for $\tilde{\xi}_2$ is the smallest $\xi > 0$ such that $\Delta^{(1)}(\xi) = 0$. Therefore, $\Delta^{(1)}(l) \geq 0$ ensures that $\tilde{\xi}_2 \geq \tilde{\xi}_2^{(1)} \geq l$.

Equation (13 Right) implies that $s(\xi)-1 = M\int_0^\xi d\eta\, \Delta(\eta) \geq M\int_0^\xi d\eta\, \Delta^{(1)}(\eta)$, namely,

$$\frac{M}{2} f(\xi) =: s^{(2)}(\xi), \qquad f(\xi) := \int_0^\xi d\eta\, \frac{(\xi-\eta)[1+v(\eta)]}{[s^{(1)}(\eta)]^2} - \frac{\xi^2}{2}. \tag{21}$$

At least for small ξ, this is a more stringent lower bound for s in $\mathcal{I}$ than $s \geq 1$; $f(\xi)$ vanishes at $\xi = 0$, and grows with ξ for small $\xi > 0$ until its (unique) maximum point; for larger ξ, it decreases and becomes negative. Hence, a lower bound $\tilde{\xi}_3^{(1)}$ for $\tilde{\xi}_3$ is the smallest $\xi > 0$ such that $f(\xi) = 0$. Therefore, $f(l) \geq 0$ ensures that $\tilde{\xi}_3 \geq \tilde{\xi}_3^{(1)} \geq l$.

2.2. Generic Density Case

Let $\check{n}(\xi, Z) := \widetilde{n_0}\left[z_e(\xi, Z)\right]$, $\tilde{\xi}_2' := \min\{\tilde{\xi}_2, l\}$, $n_u, n_d > 0$ be some upper, lower bounds on $\check{n}$

$$n_d(Z) \leq \check{n}(\xi, Z) \leq n_u(Z) \tag{22}$$

for $0 \leq \xi \leq \tilde{\xi}_2'$. If $\widetilde{n_0}(Z) \equiv n_0$, then $\check{n} \equiv n_0$, and we can set $n_u = n_d = n_0$. In general, a Z-independent choice of n_u is $n_u = n_b$; see (4). More accurately, with $\Delta_u > 0$ such that

$$\Delta(\xi, Z) \leq \Delta_u \qquad \forall \xi \in [0, \tilde{\xi}_3], \tag{23}$$

then $0 \leq \Delta(\xi, Z) \leq \Delta_u(Z)$ for all $\xi \in [0, \tilde{\xi}_2]$ (by the definition of $\tilde{\xi}_2$), and (22) holds, choosing

$$n_u(Z) = \max_{Z' \in \left[Z, Z+\Delta_u\right]} \{\widetilde{n_0}(Z')\}, \qquad n_d(Z) = \min_{Z' \in \left[Z, Z+\Delta_u\right]} \{\widetilde{n_0}(Z')\}; \tag{24}$$

in general, (24 Left) is a lower (and therefore better) upper bound than $n_u = n_b$. Henceforth we abbreviate $M_u(Z) := Kn_u(Z)$, $M_d(Z) := Kn_d(Z)$. By (18), we can adopt the simple choice $\Delta_u := \Delta^{(0)}(l)$ (if $v(l) \simeq 0$, as occurs if the pulse is a slowly modulated one (58), then $\Delta^{(0)}(\xi) \simeq \Delta^{(0)}(l)$ if $\xi > l$, and (22) holds with $\Delta_u = \Delta^{(0)}(l)$ for all $0 \leq \xi \leq \tilde{\xi}_2$ even if $\tilde{\xi}_2 > l$).

Lemma 1. *For all $\xi \in [0, \tilde{\xi}_3]$, the rhs ν of (11) can be bound by*

$$\nu(\xi, Z) \leq \nu_u(\xi, Z) := \frac{M_u(Z)}{2} \left[\Delta^{(0)}(\xi)\right]^2 \tag{25}$$

Proof. For $\xi \in [0, \tilde{\xi}_2]$, the inequality is proven as follows:

$$\nu(\xi) = \int_0^{\xi} d\eta \, \frac{vs'}{2s^2}(\eta) \leq \int_0^{\xi} d\eta \, \frac{v}{2s^2}(\eta) \int_Z^{Z+\Delta(\eta)} dz \, M_u = M_u \int_0^{\xi} d\eta \, \frac{v\Delta}{2s^2}(\eta) \leq M_u \int_0^{\xi} d\eta \, \frac{v\Delta^{(0)}}{2}(\eta) = \nu_u(\xi)$$

using $\Delta^{(0)\prime} = v/2$; for brevity, we have not displayed the Z argument here. The inequality holds in $]\tilde{\xi}_2, \tilde{\xi}_3]$ as well, because there ν decreases, whereas ν_u grows. □

The maximum of $\nu(\xi, Z)$ in $[0, \tilde{\xi}_3]$ is in $\xi = \tilde{\xi}_2$ because $s' > 0$ in $]0, \tilde{\xi}_2[$ and $s' < 0$ in $]\tilde{\xi}_2, \tilde{\xi}_3[$. To obtain upper, lower bounds s_u, s_d for $s(\xi, Z)$ and a lower bound $\Delta_d(Z)$ for $\Delta(\xi, Z)$ in the longer interval $0 \leq \xi \leq \tilde{\xi}_3' := \min\{l, \tilde{\xi}_3\}$, we use (25) to majorize

$$\nu(\xi, Z) \leq \nu(\tilde{\xi}_2, Z) \leq \frac{M_u}{2} \left[\Delta^{(0)}(\tilde{\xi}_2)\right]^2 \leq \frac{M_u}{2} \Delta_u^2$$

(again, if $v(l) \simeq 0$, as occurs if the pulse is a slowly modulated one (58), then $\Delta^{(0)}(\xi) \simeq \Delta^{(0)}(l)$ if $\xi > l$, and these results remain valid if we replace $\tilde{\xi}_3'$ by $\tilde{\xi}_3$, even if $\tilde{\xi}_3 > l$).

When replaced in (11), this yields $(s-1)^2/2s \leq M_u \Delta_u^2/2$ and $\mathcal{U} \leq M_u \Delta_u^2/2$, whence

$$s_d \leq s(\xi, Z) \leq s_u, \quad \left.\begin{matrix} s_u \\ s_d \end{matrix}\right\} := 1 + \frac{M_u}{2}\Delta_u^2 \pm \sqrt{\left(1 + \frac{M_u}{2}\Delta_u^2\right)^2 - 1}, \qquad \Delta(\xi, Z) \geq \Delta_d(Z), \tag{26}$$

where Δ_d is the negative solution of the equation $\mathcal{U}(\Delta; Z) = M_u(Z)\Delta_u^2/2$ (as a first estimate, $\Delta_d = -\Delta_u$). Clearly, $1/s_d = s_u > 1$.

Proposition 2. *For all $\xi \in [0, \tilde{\xi}_3']$, the dynamical variables Δ, s are bounded as follows:*

$$\begin{aligned} &\Delta^{(0)}(\xi, Z) \geq \Delta(\xi, Z) \geq \Delta^{(1)}(\xi, Z), \\ &s^{(1)}(\xi, Z) \geq s(\xi, Z) \geq s^{(2)}(\xi, Z), \end{aligned} \tag{27}$$

where $\Delta^{(0)}(\xi) := \int_0^{\xi} d\eta\, v(\eta)/2$, *and*

$$s^{(1)}(\xi,Z) := \min\{s_u, 1+g(\xi,Z)\}, \quad g(\xi,Z) := \frac{M_u}{2}\int_0^{\xi} d\eta\,(\xi-\eta)\,v(\eta),$$

$$\Delta^{(1)}(\xi,Z) := \max\{\Delta_d, d(\xi,Z)\}, \quad d(\xi,Z) := \int_0^{\xi} d\eta\, \frac{1+v(\eta)}{2[s^{(1)}(\eta,Z)]^2} - \frac{\xi}{2},$$

$$s^{(2)}(\xi,Z) := \begin{cases} 1+\frac{M_d}{2} f(\xi,Z) & 0 \le \xi \le \tilde{\xi}_2^{(1)} \\ \max\left\{s_d, 1+\left[\frac{M_d}{2}-\frac{M_u'}{2}\right] f\left(\tilde{\xi}_2^{(1)},Z\right)+\frac{M_u'}{2} f(\xi,Z)\right\} & \tilde{\xi}_2^{(1)} < \xi \le \tilde{\xi}_3', \end{cases} \tag{28}$$

$$f(\xi,Z) := \int_0^{\xi} d\eta\,(\xi-\eta)\left\{\frac{1+v(\eta)}{[s^{(1)}(\eta,Z)]^2} - 1\right\},$$

where $\tilde{\xi}_2^{(1)}(Z) < \tilde{\xi}_2$ *is the zero of* $d(\xi,Z)$ *as well as the maximum point of* $f(\xi,Z)$, *and* $M_u'/K = n_u' := \max_{Z'\in[Z+\Delta_d,Z]}\{\widetilde{n_0}(Z')\}$. *Moreover, the value of the Hamiltonian is bounded by*

$$H_d := 1+\frac{v}{2s^{(1)}}+\nu_d \le H \le 1+\frac{v}{2s^{(2)}}+\frac{M_u}{2}\left[\Delta^{(0)}\right]^2 =: H_u, \tag{29}$$

where, dubbing the maximum of $d(\xi,Z)$ *by* $\tilde{\xi}_1^{(1)}(Z)$, *we have defined*

$$\nu_d(\xi,Z) := \begin{cases} \frac{M_d(Z)}{2}\left[\Delta^{(1)}(\xi,Z)\right]^2 & \xi \in [0,\tilde{\xi}_1^{(1)}], \\ \frac{M_d(Z)}{2}\left[\Delta^{(1)}(\tilde{\xi}_1^{(1)})\right]^2 & \xi \in [\tilde{\xi}_1^{(1)},\tilde{\xi}_2^{(1)}], \\ \frac{M_d(Z)}{2}\left[\Delta^{(1)}(\tilde{\xi}_1^{(1)},Z)\right]^2 + M_u'(Z)\int_{\tilde{\xi}_2^{(1)}}^{\xi} d\eta\,\frac{v\Delta^{(1)}}{2}(\eta,Z) & \xi \in [\tilde{\xi}_2^{(1)},\tilde{\xi}_3']. \end{cases} \tag{30}$$

Note that $\tilde{\xi}_1^{(1)}(Z) < \tilde{\xi}_1(Z)$. Equations (27) and (28) reduce to (18) and (21) if $\widetilde{n_0}(Z) \equiv n_0 = \text{const}$.

Proof. The left inequality in (27 Left) is the already proven (18). Equation (13) by (24), (18) implies $s(\xi,Z)-1 \le \int_0^{\xi} d\eta\, M_u\Delta(\eta,Z) \le M_u \int_0^{\xi} d\eta\,\Delta^{(0)}(\eta)$, which together with (26 Left) implies the left inequality in (27 Right); the latter, together with (13 Left), (26 Right), in turn implies the right inequality in (27 Left). First, $d(\xi,Z) = f'(\xi,Z)$ vanishes at $\xi = 0$, grows with ξ for small $\xi > 0$ until it reaches a maximum for sufficiently large ξ, then decreases to negative values. Hence, $\tilde{\xi}_2^{(1)}$, i.e., the smallest $\xi > 0$ such that $\Delta^{(1)}(\xi,Z) = 0$, is indeed a lower bound for $\tilde{\xi}_2$; meanwhile, $\tilde{\xi}_2^{(1)}$ is the maximum point of f and $s^{(2)}$. Equation (13) implies that for all $\xi \in [0,\tilde{\xi}_2]$

$$s(\xi,Z)-1 \ge M_d\int_0^{\xi} d\eta\,\Delta(\eta,Z) \ge M_d\int_0^{\xi} d\eta\, d(\eta,Z) = \frac{M_d}{2} f(\xi,Z). \tag{31}$$

If $\xi \in]\tilde{\xi}_2,\tilde{\xi}_3']$, integrating (8 Right) over $[\tilde{\xi}_2,\xi]$ and recalling that $\Delta_d \le \Delta < 0$ there, we find

$$\begin{aligned} s(\xi,Z)-s(\tilde{\xi}_2,Z) &= -K\int_{\tilde{\xi}_2}^{\xi} d\eta \int_{Z+\Delta(\eta Z)}^{Z} dZ'\,\widetilde{n_0}(Z') \ge M_u'\int_{\tilde{\xi}_2}^{\xi} d\eta\,\Delta(\eta,Z) \ge M_u'\int_{\tilde{\xi}_2}^{\xi} d\eta\,\Delta^{(1)}(\eta,Z) \\ &> M_u'\int_{\tilde{\xi}_2^{(1)}}^{\xi} d\eta\,\Delta^{(1)}(\eta,Z) \ge \frac{M_u'}{2}\left[f(\xi,Z)-f\left(\tilde{\xi}_2^{(1)},Z\right)\right] \end{aligned} \tag{32}$$

where the first inequality in the last line holds because $\Delta^{(1)}(\eta,Z) < 0$ if $\eta \in [\tilde{\xi}_2^{(1)}, \tilde{\xi}_2]$; as s has its maximum in $\tilde{\xi}_2$, (31) implies in particular that $s(\tilde{\xi}_2,Z) \geq s(\tilde{\xi}_2^{(1)},Z) \geq 1+\frac{M_d}{2}f(\tilde{\xi}_2^{(1)},Z)$, which when replaced in (32) provides

$$s(\xi,Z) \geq 1+\frac{M_d}{2}f(\tilde{\xi}_2^{(1)},Z) + \frac{M_u'}{2}\left[f(\xi,Z) - f\left(\tilde{\xi}_2^{(1)},Z\right)\right];$$

the latter inequality and (31) amount to the right inequality in (27 Right), which holds together with $s(\xi,Z) \geq 1$. The right inequality in (29) follows from (11) and (25). From $2\Delta^{(1)\prime} = (1+v)/s^{(1)2} - 1 \leq v/s^{(1)2}$, it follows for $\xi \in [0,\tilde{\xi}_2]$ that

$$\nu(\xi,Z) \geq M_d\int_0^\xi d\eta\, \frac{v\Delta}{2s^2}(\eta) \geq M_d\int_0^\xi d\eta\, \frac{v\Delta^{(1)}}{2s^{(1)2}}(\eta) \geq M_d\int_0^\xi d\eta\,[\Delta^{(1)}\Delta^{(1)\prime}](\eta) = \frac{M_d}{2}\left[\Delta^{(1)}(\xi)\right]^2,$$

where again, for brevity, we have not displayed the Z argument. As the rhs has its maximum in $\xi = \tilde{\xi}_1^{(1)}$, whereas $\nu(\xi,Z)$ can grow in $[\tilde{\xi}_1^{(1)},\tilde{\xi}_2]$, we obtain

$$\nu(\xi,Z) \geq \begin{cases} \frac{M_d}{2}\left[\Delta^{(1)}(\xi)\right]^2 & \xi \in [0,\tilde{\xi}_1^{(1)}] \\ \frac{M_d}{2}\left[\Delta^{(1)}\left(\tilde{\xi}_1^{(1)}\right)\right]^2 & \xi \in [\tilde{\xi}_1^{(1)},\tilde{\xi}_2] \end{cases} \tag{33}$$

If $\xi \in [\tilde{\xi}_2,\tilde{\xi}_3']$, then $s',\Delta < 0$ in $[\tilde{\xi}_2,\xi]$, and

$$\int_{\tilde{\xi}_2}^{\xi} d\eta\, \frac{vs'}{2s^2}(\eta) = -\int_{\tilde{\xi}_2}^{\xi} d\eta\, \frac{Kv}{2s^2}(\eta)\int_{Z+\Delta(\eta)}^{Z} dz\, \tilde{n}_0(z) \geq M_u'\int_{\tilde{\xi}_2}^{\xi} d\eta\, \frac{v\Delta}{2}(\eta) \geq M_u'\int_{\tilde{\xi}_2}^{\xi} d\eta\, \frac{v\Delta^{(1)}}{2}(\eta) > M_u'\int_{\tilde{\xi}_2^{(1)}}^{\xi} d\eta\, \frac{v\Delta^{(1)}}{2}(\eta)$$

where the last inequality holds because $\Delta^{(1)} < 0$ in $[\tilde{\xi}_2^{(1)},\tilde{\xi}_2]$; by summing this inequality and $\nu(\tilde{\xi}_2,Z) \geq M_d\left[\Delta^{(1)}(\tilde{\xi}_1^{(1)})\right]^2/2$ we obtain $\nu(\xi,Z) \geq M_d\left[\Delta^{(1)}(\tilde{\xi}_1^{(1)})\right]^2/2 - M_u'\int_{\tilde{\xi}_2^{(1)}}^{\xi} d\eta\, v\Delta^{(1)}(\eta)/2$, which actually holds for all $\xi \in]\tilde{\xi}_2^{(1)},\tilde{\xi}_3']$ by (33) and because the second term is negative if $\xi \in [\tilde{\xi}_2^{(1)},\tilde{\xi}_2]$. Summing up, we find $\nu(\xi,Z) \geq \nu_d(\xi,Z)$ in all the interval $[0,\tilde{\xi}_3']$; this, together with (11), implies the left inequality (29). □

By (26), inequalities (22) hold for all $\xi \in [0,\tilde{\xi}_3']$ if, rather than by (24), we define n_u, n_d by

$$n_u(Z) = \max_{Z' \in \left[Z+\Delta_d, Z+\Delta_u\right]}\{\tilde{n}_0(Z')\}, \qquad n_d(Z) = \min_{Z' \in \left[Z+\Delta_d, Z+\Delta_u\right]}\{\tilde{n}_0(Z')\}. \tag{34}$$

As mentioned previously, $\Delta^{(1)}(\xi,Z) = d(\xi,Z) = f'(\xi,Z)$ vanishes at $\xi = 0$, grows up to its unique positive maximum at $\tilde{\xi}_1^{(1)}$, then decreases to negative values; $\tilde{\xi}_2^{(1)}$ is the unique $\xi > \tilde{\xi}_1^{(1)}$ such that $\Delta^{(1)}(\xi,Z) = 0$. Hence, $\tilde{\xi}_2^{(1)}$ is a lower bound for $\tilde{\xi}_2$. Therefore, the condition

$$\Delta^{(1)}(l,Z) \geq 0 \tag{35}$$

ensures that $\tilde{\xi}_2(Z) \geq \tilde{\xi}_2^{(1)}(Z) \geq l$, namely, that the pulse is strictly short. Similarly, $s^{(2)} - 1$ vanishes at $\xi = 0$, grows up to its unique positive positive maximum at $\tilde{\xi}_2^{(1)}$, then decreases to negative values. Hence, a lower bound $\tilde{\xi}_3^{(1)}$ for $\tilde{\xi}_3$ is the unique $\xi > \tilde{\xi}_2^{(1)}$ such that $s^{(2)}(\xi,Z) = 1$, and the condition

$$s^{(2)}(l,Z) \geq 1 \tag{36}$$

ensures that $\tilde{\xi}_3(Z) \geq \tilde{\xi}_3^{(1)}(Z) \geq l \equiv \tilde{\xi}_3'$, namely, that the pulse is essentially short. Moreover, with this assumption we find by (29) the following upper and lower bounds on the final Z-electron energy $h(Z)$ after their interaction with the pulse:

$$H_d(l,Z) \leq h(Z) \leq H_u(l,Z). \tag{37}$$

In Figures 3 and 4, we plot $s(\cdot,Z), \Delta(\cdot,Z)$ for two values of Z and the associated upper and lower bounds corresponding to the densities of Figure 2, which have the same asymptotic value n_b. As can be seen, the bounds agree well. Moreover, a useful lower bound for $\Delta^{(1)}$ is provided by the following lemma.

Lemma 2.

$$\Delta^{(1)}(\xi,Z) \geq \Delta^{(0)}(\xi)\left[1 - \frac{M_u(Z)}{2}\xi^2 - M_u(Z)\,\xi\,\Delta^{(0)}(\xi)\right]. \tag{38}$$

Proof. $g \geq 0$ implies $1/(1+g) \geq 1-g$, $1/(1+g)^2 \geq (1-g)^2 \geq 1-2g$ such that

$$2\Delta^{(1)}(\xi,Z) \geq \int_0^\xi d\eta\,\{[1+v(\eta)][1-2g(\eta,Z)]-1\} = 2\Delta^{(0)}(\xi) - \int_0^\xi d\eta\,2g(\eta,Z)[1+v(\eta)]. \tag{39}$$

The definitions of $g, \Delta^{(0)}$ immediately imply the inequality $g(\xi,Z) \leq M_u\xi\Delta^{(0)}(\xi)$, whence

$$\int_0^\xi d\eta\,2g(\eta,Z) \leq M_u\int_0^\xi d\eta\,\eta 2\Delta^{(0)}(\eta) \leq M_u\int_0^\xi d\eta\,\eta 2\Delta^{(0)}(\xi) \leq M_u\xi^2\Delta^{(0)}(\xi),$$

$$\int_0^\xi d\eta\,2g(\eta,Z)v(\eta) \leq M_u\int_0^\xi d\eta\,\eta 2\Delta^{(0)}(\eta)v(\eta) \leq M_u\int_0^\xi d\eta\,\xi 4\left[\Delta^{(0)}\Delta^{(0)\prime}\right](\eta) = 2M_u\xi[\Delta^{(0)}(\xi)]^2$$

Replacing the latter in (39), we obtain (38). □

As a consequence, if the square bracket at the rhs (38) is nonnegative, then is as well $\Delta^{(1)}(\xi,Z)$, and therefore another condition ensuring that $\tilde{\xi}_2 > l$ (i.e., that the pulse is strictly short) is

$$M_u(Z)\,l^2\left[1+2\frac{\Delta_u}{l}\right] \leq 2, \tag{40}$$

which is more easily computable, although more difficult to satisfy, than (35).

More stringent (though less easily computable) bounds than (27) could be found by replacing them in (13) and reiterating the previous arguments. The first step is the new bound $\Delta(\xi,Z) \leq \Delta^{(2)}(\xi,Z) := \int_0^\xi d\eta\,\frac{1+v(\eta)}{2[s^{(2)}(\eta,Z)]^2} - \frac{\xi}{2}$, the second is setting $\Delta_u := \max_{\xi\in[0,l]}\{\Delta^{(2)}(\xi)\}$ instead of $\Delta_u := \Delta^{(0)}(l)$ in (24), etc. However, for the scope of the present work we content ourselves with these basic relations.

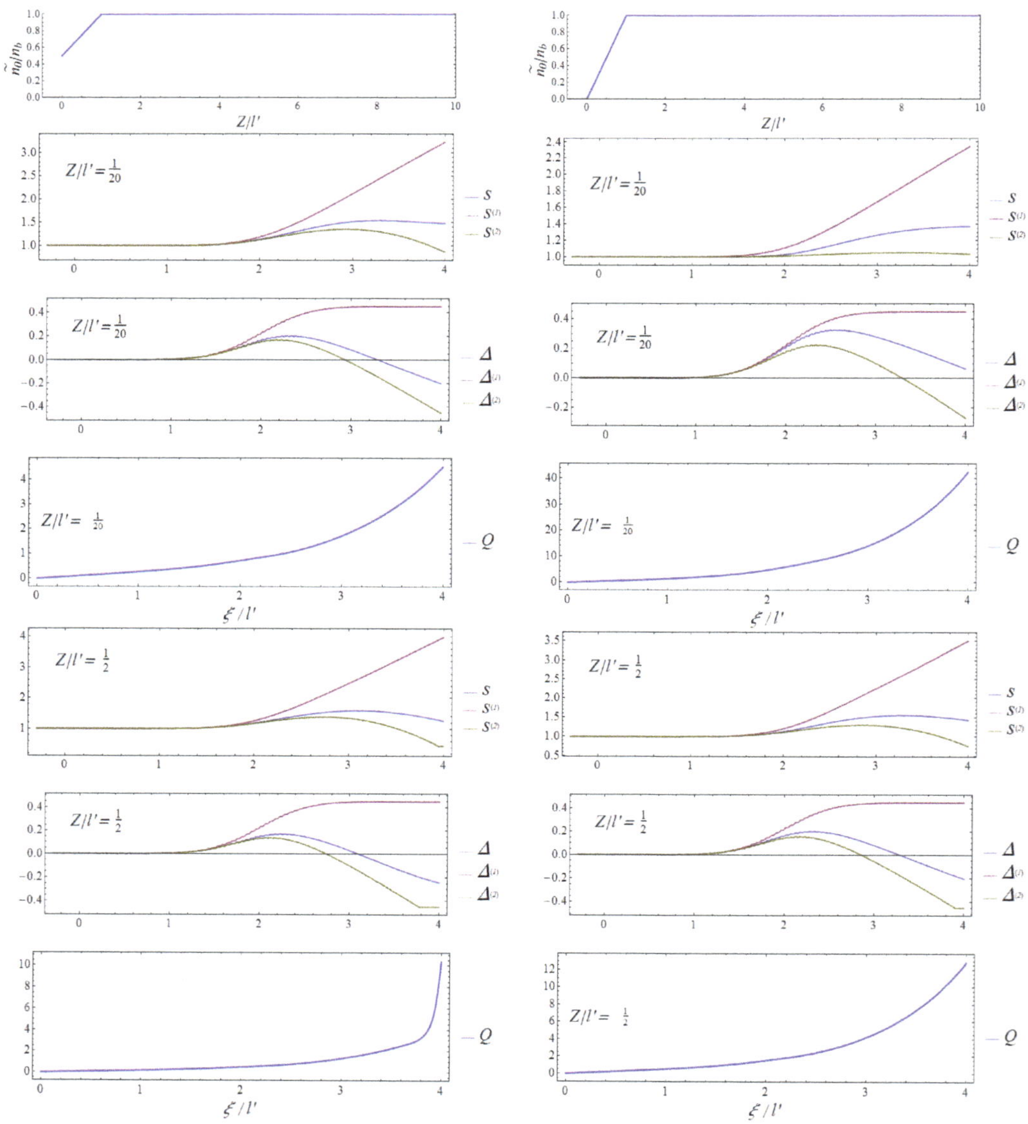

Figure 3. The initial electron densities (1), (2) of Figure 2 (first line; left and right, respectively). Below, assuming $n_b = 2 \times 10^{18}\text{cm}^{-3}$, we plot the corresponding s, Δ, their upper and lower bounds $s^{(1)}, s^{(2)}, \Delta^{(1)}, \Delta^{(2)}$, and the function Q, vs. ξ during interaction with the pulse in Figure 1 for the same sample values, $Z = l'/20$ and $Z = l'/2$ of Z. The values $Q_2(Z) := Q(l, Z)$ can be read off the plots. As can be seen, the bounds are much better for density (1); the values $Q_2(Z) \lesssim 1$ are consistent with all worldlines intersecting rather far from the laser-plasma interaction spacetime region. On the other hand, the large value of $Q_2(Z)$ for density (2) is an indication that worldlines intersect within or not far from the laser–plasma interaction spacetime region. Our computations lead to $Q_0(l'/20) = 10.64$, $Q_0(l'/2) =$ with density (1) and $Q_0(l'/20) = 88.53$, $Q_0(l'/2) = 32.35$ with density (2).

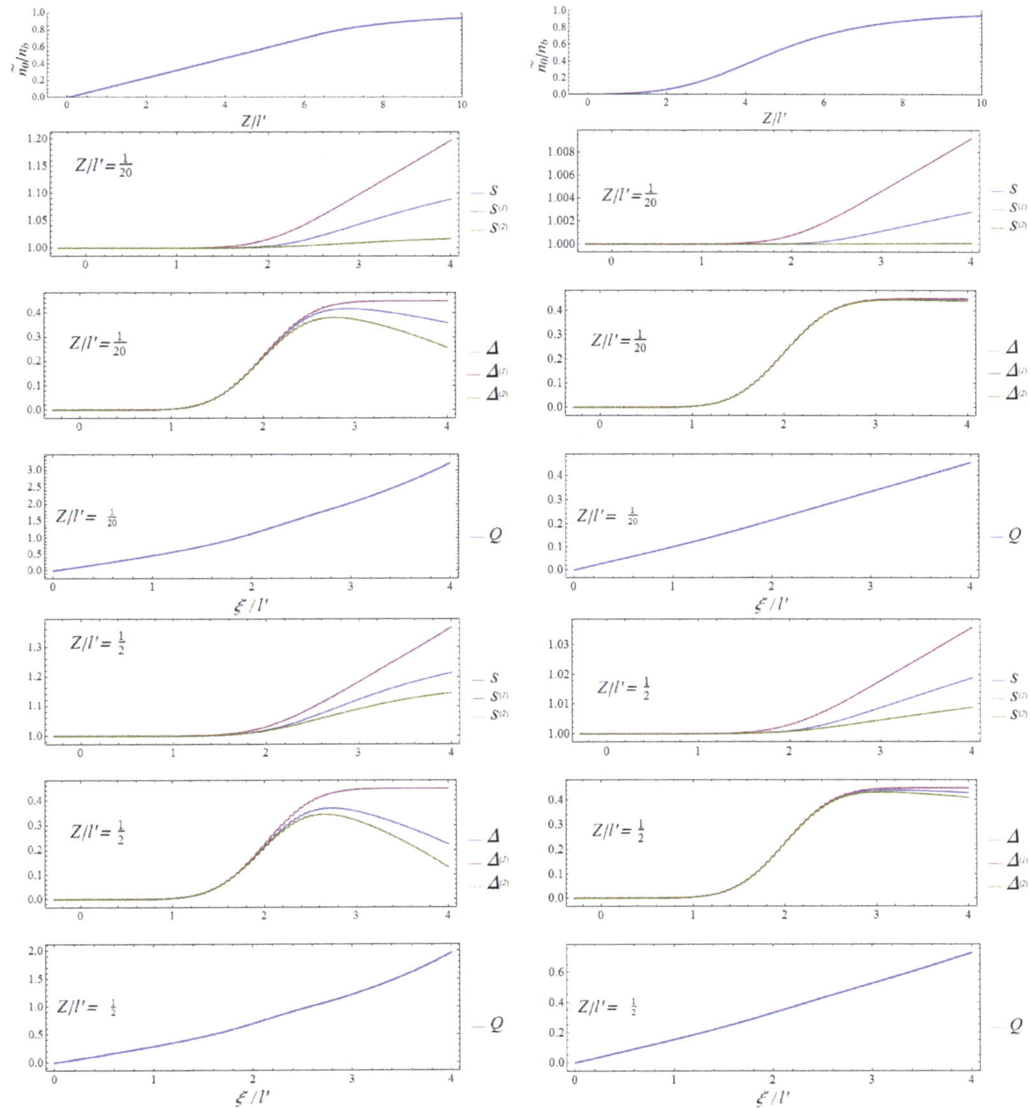

Figure 4. The initial electron densities (3), (4) of Figure 2 (left and right, respectively) with $n_b = 2 \times 10^{18}$ cm^{-3}, corresponding plots of s, Δ, their upper and lower bounds $s^{(1)}, s^{(2)}, \Delta^{(1)}, \Delta^{(2)}$, and the function $Q(\xi, Z)$, vs. ξ for the same sample values $Z = l'/20$ and $Z = l'/2$ of Z. The values $Q_2(Z) := Q(l, Z)$ can be read off the plots. As can be seen, the bounds are much better for density (4); the values $Q_2(Z) \leq 1$ are consistent with all worldlines intersecting rather far from the laser-plasma interaction spacetime region. On the other hand, the large value of $Q_2(Z)$ for density (3) is an indication that worldlines intersect not far from the laser–plasma interaction spacetime region. Our computations lead to $Q_0(l'/20) = 4.62$, $Q_0(l'/2) = 3.08$ with density (3) and $Q_0(l'/20) = 0.49$, $Q_0(l'/2) = 0.85$ with density (4).

3. Bounds on the Jacobian for Small $\xi > 0$

Differentiating (8), we find that the dimensionless variables

$$\varepsilon(\xi, Z) := J(\xi, Z) - 1 = \frac{\partial s(\xi, Z)}{\partial Z}, \qquad \sigma(\xi, Z) := l\frac{\partial s(\xi, Z)}{\partial Z} \tag{41}$$

fulfill the Cauchy problem

$$\begin{aligned} \varepsilon' &= -\kappa\sigma, \qquad \sigma' = Kl(\check{n} - \widetilde{n_0} + \check{n}\,\varepsilon), \\ \varepsilon(0,Z) &= 0, \quad \sigma(0,Z) = 0, \end{aligned} \tag{42}$$

where we have abbreviated $\kappa := \frac{1+v}{ls^3} \geq 0$. From σ, we can immediately obtain $\partial u^z/\partial Z$ via

$$l\frac{\partial u^z}{\partial Z} = -\frac{1+v}{2s^2}\sigma, \tag{43}$$

which is dimensionless, as well. To bound ε, σ for small ξ, we introduce the Liapunov function

$$V := \varepsilon^2 + \beta\sigma^2, \tag{44}$$

where $\beta(Z) \in \mathbb{R}^+$ is specified below. Clearly, $V(0,Z) = 0$. Equation (42) implies

$$V' = \varepsilon\sigma\, 2(\beta Kl\check{n} - \kappa) + \sigma\, 2\beta Kl(\check{n} - \widetilde{n_0}) \tag{45}$$

and as $2|\varepsilon\sigma| \leq V/\sqrt{\beta}$, $|\sigma| \leq \sqrt{V/\beta}$, we obtain

$$V' \leq 2A\sqrt{V} + 2BV \quad \Rightarrow \quad \left(\sqrt{V}\right)' \leq A + B\sqrt{V},$$

where we have abbreviated $B(\xi,Z) := |\beta Kl\check{n} - \kappa|/2\sqrt{\beta}$ and introduced some $A(Z)$ such that $A \geq Kl\sqrt{\beta}\max\{|\check{n} - \widetilde{n_0}|\}$. Per the comparison principle [38], $\sqrt{V} \leq R$, where $R(\xi, Z)$ is the solution of the Cauchy problem $R' = A + BR$, $R(0,Z) = 0$, which implies

$$|\varepsilon(\xi,Z)| \leq \sqrt{V(\xi,Z)} \leq R(\xi,Z) = A(Z)\int_0^\xi d\eta\, \exp\left[\int_\eta^\xi d\zeta\, B(\zeta,Z)\right] \tag{46}$$

and $\sqrt{b}|\sigma| \leq R$. As $R(\xi,Z)$ grows with ξ, if $R(l,Z) < 1$ then no WBDLPI may involve the Z-electrons. Choosing $\beta = 1/M_u l^2$, we find for all $\xi \in [0, \tilde{\xi}_3']$

$$2lB\,\beta = \left|\frac{1+v}{s^3} - \frac{\check{n}}{n_u}\right| \leq \left|\frac{1+v}{s^3} - 1\right| + \left|1 - \frac{\check{n}}{n_u}\right| \leq |D| + \delta,$$

$$\text{where} \qquad D(\xi,Z) := \frac{1+v(\xi)}{[s(\xi,Z)]^3} - 1, \quad \delta(Z) := 1 - \frac{n_d(Z)}{n_u(Z)}. \tag{47}$$

In the NR regime, which is characterized by $v \ll 1$, we have $s \simeq 1$, $\kappa \simeq 1/l$, $\Delta_u/l \ll 1$, $D \simeq 0$; setting $D = 0$ leads by a straightforward computation to

$$R(l,Z) \simeq R_{nr}(Z) := f[r(Z)], \qquad f(r) := 2\left(e^{r/2} - 1\right) \qquad r(Z) := \delta(Z)\sqrt{M_u(Z)}\,l.$$

As $f(r)$ grows with $r \geq 0$ and reaches the value 1 for $r \simeq 0.81$, we therefore find that the condition

$$r(Z) = \delta(Z)\sqrt{M_u(Z)}\,l < 0.81 \tag{48}$$

is sufficient to ensure that the Z-electrons are not involved in WBDLPI. This is automatically satisfied if $\sqrt{M_u(Z)}\,l < 0.81$, because by definition $\delta \leq 1$; otherwise, it is a very mild condition on the relative variation δ of the initial electron density across an interval of length $\Delta_u \ll l$ and in fact to violate (48) one needs a discontinuous (or a continuous and very steep) $\widetilde{n_0}(Z)$ with large relative variations around Z; see Section 4.

We now consider the general case. In the interval $[0, \tilde{\xi}_3']$, the inequalities $s^{(1)} \geq s \geq s^{(2)} \geq 1$ imply

$$\begin{aligned} |D(\xi, Z)| &\leq \mathcal{D}(\xi, Z) := \max\left\{ \frac{1+v(\xi)}{[s^{(2)}(\xi,Z)]^3} - 1 \,,\, 1 - \frac{1+v(\xi)}{[s^{(1)}(\xi,Z)]^3} \right\} \\ &\leq \tilde{v}(\xi) := \max\{v(\xi), 1\} \\ &\leq \max\{v_M, 1\} =: \tilde{v}_M, \end{aligned} \tag{49}$$

where $v_M, \tilde{v}_M$ are the maxima of $v, \tilde{v}$. Hence, we obtain the bounds

$$\begin{aligned} \int_\eta^\xi d\zeta\, B(\zeta) &\leq \frac{\sqrt{M_u}}{2}\left[(\xi-\eta)\,\delta + \int_\eta^\xi d\zeta\, \mathcal{D}(\zeta)\right] \\ &\leq \frac{\sqrt{M_u}}{2}\left[(\xi-\eta)\,\delta + \int_\eta^\xi d\zeta\, \tilde{v}(\zeta)\right] \\ &\leq \frac{\sqrt{M_u}}{2}(\tilde{v}_M+\delta)(\xi-\eta), \end{aligned}$$

which are replaced in (46) by choosing $A = \sqrt{M_u}\,\delta$, respectively implying

$$R(\xi, Z) \leq \delta(Z)\sqrt{M_u(Z)}\int_0^\xi d\eta\, \exp\left\{\frac{\sqrt{M_u(Z)}}{2}\left[(\xi-\eta)\,\delta(Z) + \int_\eta^\xi d\zeta\, \mathcal{D}(\zeta,Z)\right]\right\} =: Q(\xi,Z) \tag{50}$$

$$\begin{aligned} Q(l,Z) &\leq \delta(Z)\sqrt{M_u(Z)}\int_0^l d\eta\, \exp\left\{\frac{\sqrt{M_u(Z)}}{2}\left[(l-\eta)\,\delta(Z) + \int_\eta^l d\zeta\, \tilde{v}(\zeta)\right]\right\} =: Q_1(Z) \qquad (51) \\ &\leq \frac{2\delta(Z)}{\tilde{v}_M+\delta(Z)}\left\{\exp\left[\frac{\tilde{v}_M+\delta(Z)}{2}\sqrt{M_u(Z)}\, l\right] - 1\right\} =: Q_0(Z) \qquad (52) \end{aligned}$$

Here, we have redisplayed the Z-argument; $Q_2(Z) := Q(l,Z)$ is the most difficult to compute, while $Q_0(Z)$ is the easiest. We thus arrive at

Theorem 1. *Assume that condition (36) is fulfilled. Then, no WBDLPI involves the Z-electrons if in addition $Q_0(Z) < 1$, or at least $Q_1(Z) < 1$, or at least $Q_2(Z) < 1$. If one of these conditions is fulfilled for all Z, then WBDLPI occurs nowhere.*

Consequently, for any fixed pump there is no WBDLPI if n_b is sufficiently small. A simple sufficient condition is provided by the following corollary.

Corollary 1. *For any pulse (5) there is no WBDLPI if $Kn_b l^2 < 4\left[\log 2/(1+\tilde{v}_M)\right]^2$ and (36) are fulfilled. In particular, it suffices that $Kn_b l^2 < \min\{4\left[\log 2/(1+\tilde{v}_M)\right]^2, 2/(1+2\Delta_u/l)\}$.*

Proof. From $\delta \leq 1 \leq \tilde{v}_M$, it followss that $\delta\, l\sqrt{M_u} \leq L := \frac{\tilde{v}_M+\delta}{2}\, l\sqrt{M_b}$. Hence, $L < \log 2$ implies $e^L < 2$, whence

$$Q_0 = \delta\, l\sqrt{M_u}\,\frac{e^L - 1}{L} \leq e^L - 1 < 1,$$

which, together with (36), implies the first claim. The second follows as well, as (40) implies (36). □

4. Discussion and Conclusions

As we have seen, if inequality (35) is fulfilled, then $\tilde{\xi}_2 > l$, i.e., the pulse is strictly short (that is, it completely overcomes the Z-electrons while their longitudinal displacement remains nonnegative). If at least inequality (36) is fulfilled, then $\tilde{\xi}_3 > l$, andt the pulse is essentially short (that is, the pulse completely overcomes the Z-electrons before their longitudinal displacement reaches its first negative minimum), and the inequalities (27), (29), (50) and (52) apply. If in addition one of the conditions of Proposition 1 is satisfied, then WBDLPI can be excluded.

As seen above, the more easily computable (although more difficult to satisfy) condition (40) implies (35) and the inequality $M_u(Z)\,l^2 \le \frac{2}{1+2\Delta_u/l}$, which after substitution in (50) and (52) simplifies the computation of their rhs; in particular, (52) becomes

$$Q_0(Z) \le \frac{2\delta(Z)}{\tilde{v}_M+\delta(Z)}\left(e^C-1\right) =: \tilde{Q}_0(Z), \qquad C := \frac{\tilde{v}_M+\delta(Z)}{\sqrt{2[1+2\Delta_u/l]}}. \tag{53}$$

Therefore, (40) and $\tilde{Q}_0(Z) < 1$ provide a sufficient condition to exclude WBDLPI as well.

Note that in the above conditions, several dimensionless numbers characterizing the input data, viz. $\tilde{v}_M$, Δ_u/l, $G_b^2 = M_b l^2$, $M_u l^2$, $M_d l^2$, δ, and possibly s_u, $M_u' l^2$, play a key role in the main inequalities of the present paper. Therefore, their computation represents the first step in checking whether and where such conditions are fulfilled or violated.

In the NR regime (48) is equivalent to either inequality

$$\frac{n_d}{p} > \frac{n_u}{p} - \sqrt{\frac{2n_u}{p}} \quad \Leftrightarrow \quad \frac{n_u}{p} < 1+\frac{n_d}{p}+\sqrt{1+2\frac{n_d}{p}}, \qquad p := \frac{(0.81)^2}{2Kl^2}. \tag{54}$$

(In fact, inequality (48) amounts to $n_u - n_d < \sqrt{2p\,n_u}$, i.e., (54); taking the square one obtains the equivalent inequality $n_u^2 - 2n_u(n_d+p) + n_d^2 < 0$, which is fulfilled if $n_- < n_u < n_+$, where $n_\pm := n_d+p \pm \sqrt{(n_d+p)^2-n_d^2} = n_d+p\pm\sqrt{p^2+2pn_d}$ solve the equation $x^2-2x(n_d+p)+n_d^2 = 0$ in the unknown x; the left inequality is automatically satisfied because $n_u \ge n_d > n_d^2/n_+ = n_-$. Dividing the inequality $n_u < n_+$ by p, we obtain (54)).

If $\widetilde{n_0}$ grows in $[0, \bar{Z}+\Delta_u]$, then $q(z) := \widetilde{n_0}(z)/p$ does as well, and for all $\in [0, \bar{Z}]$ the previous conditions become

$$q(Z) > q(Z+\Delta_u) - \sqrt{2q(Z+\Delta_u)} \quad \Leftrightarrow \quad q(Z+\Delta_u) < 1+q(Z)+\sqrt{1+2q(Z)}. \tag{55}$$

This is fulfilled if, e.g., in $[0, \bar{Z}+\Delta_u]$ $\widetilde{n_0}$ is continuous (without excluding $\widetilde{n_0}(0^+) > 0$), at least piecewise C^1, and $0 \le \frac{dq(z)}{dz}\Delta_u < 1+\sqrt{1+2q(Z)}$ for all $Z \in [0, \bar{Z}]$ and $z \in [Z, Z+\Delta_u]$.

Now, we impose that $\widetilde{n_0}$ is continuous in $[-\infty, \bar{Z}+\Delta_u]$ and reaches a given value $\bar{n} > 0$ at $Z = \bar{Z}$ while respecting (55). We compare the minimum $\bar{Z}$ for a linear and a quadratic $\widetilde{n_0}$; note that dq/dZ for the former violates the above bound at $Z = 0$. We find

$$\widetilde{n_0}(Z) = n_1(Z) := \theta(Z)\bar{n}\frac{Z}{\bar{Z}} \quad \text{fulfills (55) if} \quad \frac{\bar{Z}}{\Delta_u} > \frac{\bar{n}}{2p} = \frac{K\bar{n}l^2}{(0.81)^2} =: \frac{\bar{Z}_1}{\Delta_u} \tag{56}$$

$$\widetilde{n_0}(Z) = n_2(Z) := \theta(Z)\bar{n}\frac{Z^2}{\bar{Z}^2} \quad \text{fulfills (55) if} \quad \frac{\bar{Z}}{\Delta_u} > \sqrt{\frac{\bar{n}}{2p}}+\sqrt{\frac{\bar{n}}{2p}+\frac{1}{4}}-\frac{1}{2} =: \frac{\bar{Z}_2}{\Delta_u} \tag{57}$$

where θ is the Heavisde step function. In fact, if $\widetilde{n_0} = n_1$, then (55 Right) becomes $\bar{n}\Delta_u/p\bar{Z} < 1+\sqrt{1+2\bar{n}Z/p\bar{Z}}$ for all Z; the rhs is lowest for $Z = 0$, whereby the inequality becomes (56 Right), as claimed. If $\widetilde{n_0} = n_2$, then (55 Left) becomes the condition

$$F(Z) := \sqrt{\frac{2p}{\bar{n}}}\bar{Z}(Z+\Delta_u) - \Delta_u^2 - 2\Delta_u Z > 0;$$

this is of first degree in Z, and is hence fulfilled for all $Z \in [0, \bar{Z}]$ if it is for $Z = 0, \bar{Z}$. The quadratic polynomial $F(\bar{Z})$ in $\bar{Z}$ is positive if $\bar{Z} > Z_2$, as claimed, because Z_2 is the positive solution of the equation $F(z) = 0$ in the unknown z, and $\bar{Z} > Z_2$ automatically makes $F(0) > 0$.

If $\sqrt{K\tilde{n}}\, l$ is considerably larger than 1, then $\bar{Z}_1$ is considerably larger than $\bar{Z}_2$; in particular, assuming (2) with $n_b = \tilde{n}$, i.e., $G_b \sim \pi$, yields $Z_1 \sim 15.04\Delta_u$, $Z_2 \sim 6.78\Delta_u$. Therefore, choosing $\bar{Z} \in]Z_2, Z_1[$, we can exclude WBDLPI by adopting $\widetilde{n_0}(Z) = n_2(Z)$, but not $\widetilde{n_0}(Z) = n_1(Z)$. Such a result is relevant for LWFA experiments, which usually fulfill (2). From the physical viewpoint, it allows us to exclude WBDLPI because: (i) the density $\widetilde{n_0}(Z)$ obtained just outside the nozzle of a supersonic gas jet (orthogonal to the $\vec{z}$) typically is $C^1(\mathbb{R})$, with $\widetilde{n_0}(0) = 0 = \frac{d\widetilde{n_0}}{dZ}(0)$ (see, e.g., Figure 2 in [39] or Figure 5 in [40]), and therefore is closer to type $n_2(Z)$ than to type $n_1(Z)$; and (ii) by causality, the effects of a pulse with a finite spot radius R near its symmetry axis $\vec{z}$ are the same as with a plane wave ($R = \infty$), at least for small ξ. From the viewpoint of mathematical modeling, this suggests that it makes a major difference whether we describe the edge of the plasma by n_1 or by n_2; in the first case, we can correctly predict the plasma evolution only by kinetic theory and PIC codes, while in the second we can do this by a hydrodynamic description and less computationally demanding multifluid codes.

If we allow a discontinuous (in $Z = 0$) linear Ansatz $\widetilde{n_0}(Z) = \theta(Z)\tilde{n}[a + (1-a)Z/\bar{Z}]$ ($0 < a \leq 1$), then (55) is fulfilled if $\bar{Z} > \tilde{n}(1-a)\Delta_u/p[1+\sqrt{1+2a\tilde{n}/p}]$, which is again smaller than Z_1.

Although our results apply to all $\boldsymbol{\epsilon}^{\perp}$ with support contained in $[0, l]$, regardless of their Fourier analysis, in most applications one deals with a modulated monochromatic wave

$$\boldsymbol{\epsilon}^{\perp}(\xi) = \underbrace{\epsilon(\xi)}_{\text{modulation}} \underbrace{[\mathbf{i}\cos\psi\, \sin(k\xi+\varphi_1)+\mathbf{j}\sin\psi \sin(k\xi+\varphi_2)]}_{\text{carrier wave } \boldsymbol{\epsilon}_o^{\perp}(\xi)}, \tag{58}$$

where $\mathbf{i} = \nabla x$, $\mathbf{j} = \nabla y$. The elliptic polarization in (58) is ruled by $\psi, \varphi_1, \varphi_2$; it reduces to a linear one in the direction of $\boldsymbol{a} := \mathbf{i}\cos\psi + \mathbf{j}\sin\psi$ if $\varphi_1 = \varphi_2$, to a circular one if $|\cos\psi| = |\sin\psi| = 1/\sqrt{2}$ and $\varphi_1 = \varphi_2 \pm \pi/2$. If, as follows from (1), $Kn_b\lambda^2 \ll 1$ ($\lambda = 2\pi/k$ is the wavelength), i.e., the plasma is underdense, then the relative variations of $\Delta(\xi, Z)$ (and $z_e(\xi, Z)$) in a ξ-interval of length $\leq \lambda$ are much smaller than those of $\boldsymbol{x}_e^{\perp}(\xi)$, and those of $s(\xi)$ even smaller; in fact, as $s > 0$, $v \geq 0$, the integral in (13 Left) averages the fast variations of v to yield much smaller relative variations of Δ, and the first integral in (13 Right) averages the residual small variations of $\widetilde{N}[z_e(\xi)]$ to yield an essentially smooth $s(\xi)$ (see, e.g., Figure 1). On the contrary, $\boldsymbol{\alpha}^{\perp}(\xi), \boldsymbol{x}_e^{\perp}(\xi)$ varies fast, as $\boldsymbol{\epsilon}^{\perp}(\xi)$. Under rather general assumptions (see Appendix 5.2) [31],

$$\boldsymbol{\alpha}^{\perp}(\xi) = -\frac{\epsilon(\xi)}{k}\, \boldsymbol{\epsilon}_p^{\perp}(\xi) + O\left(\frac{1}{k^2}\right) \simeq -\frac{\epsilon(\xi)}{k}\, \boldsymbol{\epsilon}_p^{\perp}(\xi), \tag{59}$$

where $\boldsymbol{\epsilon}_p^{\perp}(\xi) := \boldsymbol{\epsilon}_o^{\perp}(\xi + \lambda/4)$, and similarly for other integrals with modulated integrands. In the appendix, we recall upper bounds for the remainders $O(1/k^2)$. If $|\epsilon'| \ll |k\epsilon|$ (slow modulations); the right estimate is very good, and v can be approximated very well by $v \simeq \boldsymbol{\epsilon}_p^{\perp 2}(e\epsilon/kmc^2)^2$. For the reasons mentioned above, replacing v by its (approximated) average over a cycle,

$$v_a(\xi) := \frac{1}{2}\left(\frac{e\,\epsilon(\xi)}{kmc^2}\right)^2, \tag{60}$$

has only a small effect on Δ and almost no effect on s, V, and similarly on the functions $\Delta^{(0)}, s^{(1)}, \ldots$ introduced in Sections 2 and 3 to bound Δ, s, V; however, it simplifies their computation a great deal. As a consequence, the bounds (27), (29) and (37), as well as the short pulse conditions (35) and (36) and the no-WB conditions of proposition 1, remain essentially valid if in computing the bounds we replace v with v_a.

We illustrate the results obtained thus far considering a pulse (58) with a linear polarization (e.g., $\psi = 0$) and a modulation of Gaussian type, except that it is cut off outside of support $0 \le \xi \le l$:

$$\frac{e}{kmc^2}\,\epsilon(\xi) = a_0 \exp\left[-\frac{(\xi - l/2)^2}{l'^2} 2\log 2\right] \theta(\xi)\,\theta(l-\xi); \tag{61}$$

where l' is the *full width at half maximum* of the intensity I of the electromagnetic (EM) field. More precisely, we adopt the pulse plotted in Figure 1a, which has its maximum at $\xi = l/2$ and $a_0 = 1.3$; this yields a moderately relativistic electron dynamics and $\Delta_u \equiv \Delta^{(0)}(l) \simeq 0.45l'$. In Figure 1b, we plot the associated v and v_a. We performed all computations and plots running our specifically designed programs using an "off the shelf" general-purpose numerical package on a common notebook for an elapsed time bewteen several seconds and several minutes. Here, we compare the impact of such a pulse on the density profiles plotted in Figure 2.

The upper bound n_b plays the role of asymptotic value. Here, we choose $n_b = 2 \times 10^{18}$ cm^{-3}, which is the same as the n_0 of Figure 1, which yields $Kn_b l'^2 \simeq 4$; however, the results for the dimensionless variables remain the same if we change n_b, λ, l' while keeping λ/l' and $Kn_b l'^2$ constant. As already stated, in the case of the step-shaped density 0), if $\tilde{\xi}_2 > l$, i.e., if the pulse is strictly short (a sufficient condition for which is (35)) then $n_u = n_d = n_b$, $\delta = 0$, and by (50) and (52) there is no WBDLPI; more directly, this is a consequence of $J(\xi, Z) \equiv 1$ for $0 \le \xi \le l$, which follows from the Z-independence of Δ in such an interval. As for the other profiles, we have respectively plotted the following:

- In Figures 5 and 6, the solutions J, σ of Equation (42) and, by (43), the associated function $l'\partial u^z/\partial Z$ for $0 \le \xi \le 6l'$ and a few values of Z, assuming the initial electron density profiles (1), (2) and (3), and (4), respectively.
- In Figures 3 and 4, the solutions s, Δ of (8) and (9), their upper and lower bounds $s^{(1)}, s^{(2)}, \Delta^{(1)}, \Delta^{(2)}$ (Equation (28)), and the function Q of Equation (50) for $0 \le \xi \le 6l'$ and a few values of Z, assuming the initial electron density profiles (1)–(4), respectively.
- In Figures 7 and 8, the corresponding worldlines of the Z-electrons for $0 \le ct \le 20l'$ and $Z = n\,l'/20$, $n = 0, 1, ..., 200$ associated with the initial electron density profiles (3) and (4). The support of the EM pulse is coloured pink (the red part is the more intense part); the laser–plasma interaction takes place in the spacetime region which has nonempty intersection with worldlines of electrons or protons.

We compare the results for densities (1)–(2) and those for densities (3)–(4) side by side. In case (1), WBDLPI is avoided assuming that $\widetilde{n_0}(0) > 0$, although worldlines intersect and WB takes place not far from the laser–plasma interaction region. In case (2), WBDLPI takes place for $Z \simeq 0$ due to the steep growth of $\widetilde{n_0}(Z) = Z/l'n_b$ from the value $\widetilde{n_0}(0) = 0$. In case (3), although the growth $\widetilde{n_0}(Z) \propto Z$ is much less steep, again worldlines intersect and WB takes place not very far from the laser–plasma interaction region. Finally, in case (4) this occurs quite far from the latter, consistently with the results $|J - 1| \ll 1$, $Q_2 < 1$. Thus, we note that although such dynamics are moderately relativistic rather than nonrelativistic, switching from profile (1) to profile (2) or from profile (3) to profile (4) has the same qualitative effect of avoiding (or distancing from) WBDLPI. We additionally note that in cases (1) and (3) the $Z \simeq 0$ worldlines first intersect with very small angles, or equivalently that when the corresponding electrons collide their longitudinal momenta differ by only a very small amount. By Formula (43), $l\partial u^z/\partial Z$ is very small both because σ is as well and because $v \simeq 0$ and $s > 1$. Hence, we can expect that these collisions will lead only to very small momentum spreading.

Essentially the same results are reached when choosing a different pulse polarization because v_a is of the same type. In the case of circular polarization ($\varphi_1 - \varphi_2 = \pm\pi/2$, $\cos\psi = \pm\sin\psi$) and Gaussian modulation v, it will itself essentially coincide with v_a, thus displaying a single maximum (see Figure 1b).

As mentioned in the introduction, the equations of motion for the Z-electrons can be reduced to the form (8), and thus are *decoupled* from those of all Z'-electrons, $Z' \neq Z$, only in the idealization where the laser pulse is "undepleted", i.e., not affected by its interaction with the plasma. The latter is expected to be an acceptable approximation only for small $t, z > 0$. Actually, for a slowly modulated monochromatic wave we can show by self-consistency [28] that this is a good approximation in the spacetime region that is the intersection of the 'laser–plasma interaction' stripe $0 \leq ct - z \leq l$ with the orthogonal stripe

$$0 \leq \frac{e^2 n_b \lambda}{2mc^2}(ct + z) \ll 1. \tag{62}$$

In view of the inequalities $\lambda \ll l$ and (1) or (16), we can see that, to our satisfaction, this region is much longer than l in the $ct+z$ direction. Thus, if $n_b = 2 \times 10^{18}\text{cm}^{-3}$ and the pulse is as in Figure 1, we can consider the latter as undepleted and the electrons' motion determined above as accurate for time intervals $[0, t_d]$, where t_d is at least a few 10^{-13}s.

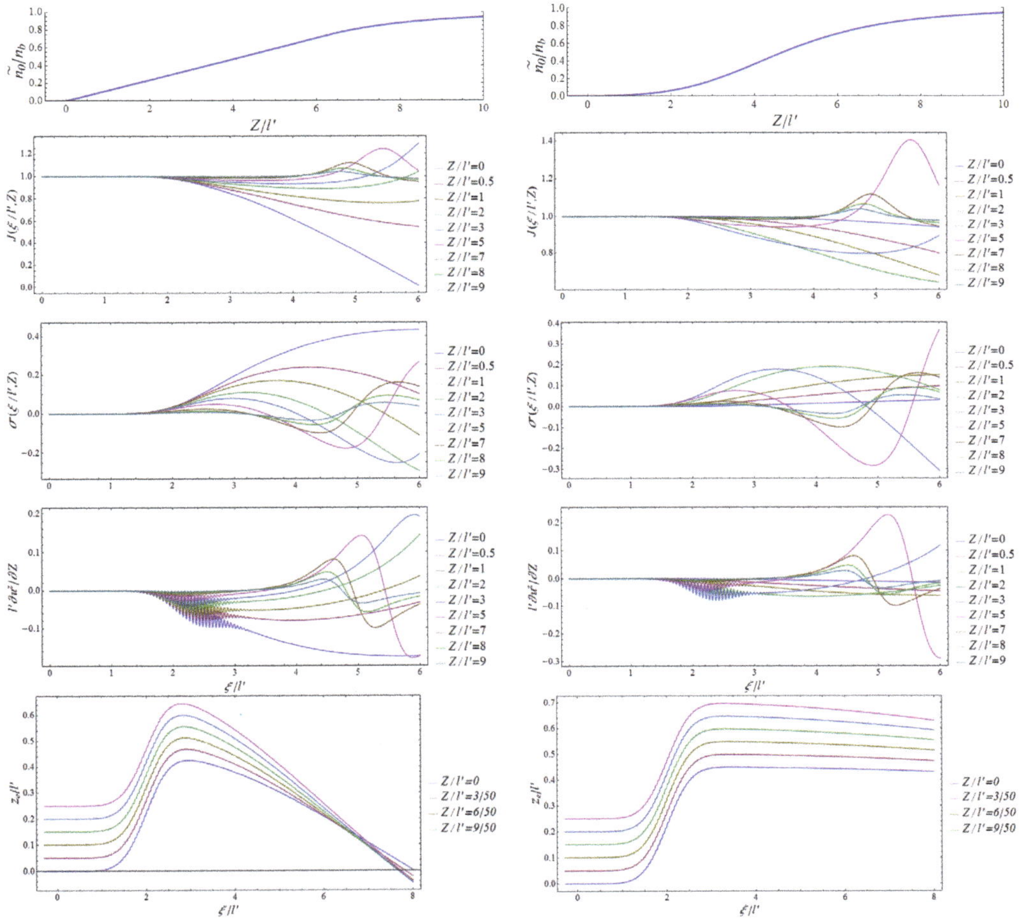

Figure 5. The initial electron densities (3), (4) of Figure 2 (left and right, respectively) with $n_b = 2 \times 10^{18}$ cm^{-3}, and below, the corresponding plots of $J, \sigma, l'\partial u^z/\partial z$ during interaction with the pulse in Figure 1 for a few sample values of Z. As can be seen, J remains positive at least for all $\xi \in [0, 2l]$ if the density is of type (4) (which grows as Z^2 for $Z \sim 0$), whereas it becomes negative for $\xi \sim 6.5l'$ and small Z if the density is of type (3) (which grows as Z for $Z \sim 0$). Correspondingly, the right-hand worldlines do not intersect, while the left-hand ones do (see the down z_e-graphs).

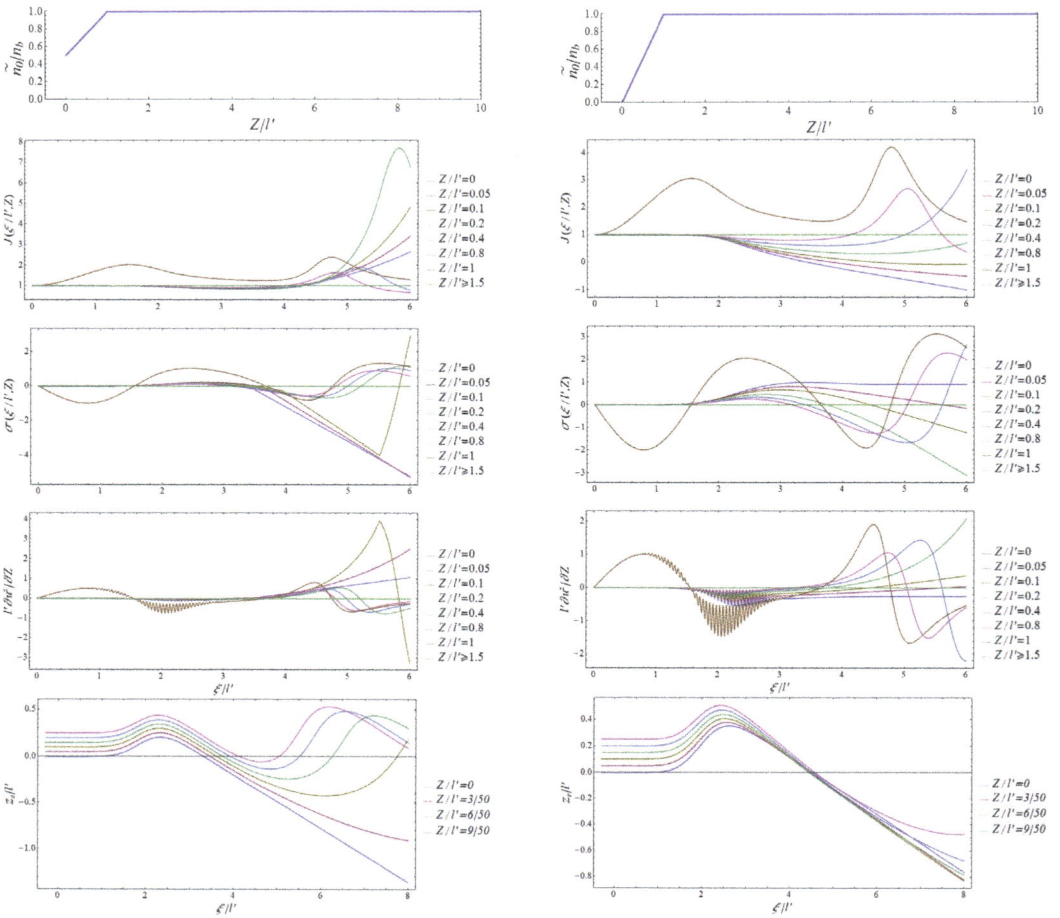

Figure 6. The initial electron densities (1), (2) of Figure 2 (first line; respectively left, right), and below, the corresponding plots of $J, \sigma, l'\partial u^z/\partial z$ vs. ξ during interaction with the pulse in Figure 1 for a few sample values of Z. As can be seen, the right J remains positive for $\xi < l$ and all Z, while the left J becomes negative for very small Z and $\xi \lesssim l$; correspondingly, the right worldlines do not intersect, while the right ones do (see the down z_e-graphs).

The above predictions are based on idealizing the initial laser pulse as a plane EM wave (5). In a more realistic picture, the $t = 0$ laser pulse is cylindrically symmetric around the $\vec{z}$-axis and has a *finite* spot radius R, namely, the $t = 0$ EM fields are of the form $\boldsymbol{E} = \boldsymbol{\epsilon}^{\perp}(-z)\,\chi(\rho)$, $\boldsymbol{B} = \mathbf{k}\times\boldsymbol{E}$, where $\rho^2 = x^2+y^2$, while $\chi(\rho) \geq 0$ is 1 for $\rho \leq 1$ and rapidly approaches zero for $\rho > R$. By causality, the motion of the electrons remains [25] strictly the same in the future Cauchy development $D^+(\mathcal{D})$ of $\mathcal{D} = \mathcal{D}_1 \cup \mathcal{D}_2$, where $\mathcal{D}_1 := \{(ct, \boldsymbol{x}) = (0, 0, 0, z > 0\}$ and $\mathcal{D}_2 := \{(0, \boldsymbol{x}) \,|\, \rho \leq R\}$, and almost the same in a neighbourhood of $D^+(\mathcal{D})$; therefore, the conditions described above remain sufficient to exclude WBDLPI at least in such a region. (Recall that the *future Cauchy development* $D^+(\mathcal{D})$ of a region $\mathcal{D}$ in Minkowski spacetime M^4 is defined as the set of all points $x \in M^4$ for which every past-directed causal, i.e., non-spacelike, line through x intersects $\mathcal{D}$.)

Finally, the conditions of Proposition 1 are very general in that they apply to discontinuous $\widetilde{n_0}$ or non-monotone $\widetilde{n_0}$; however, if $\widetilde{n_0}$ has a bounded derivative, it turns out that they are unnecessarily too strong for ensuring that no WBDLPI occurs. Weaker no-WBDLPI conditions under the latter assumptions are treated in [28].

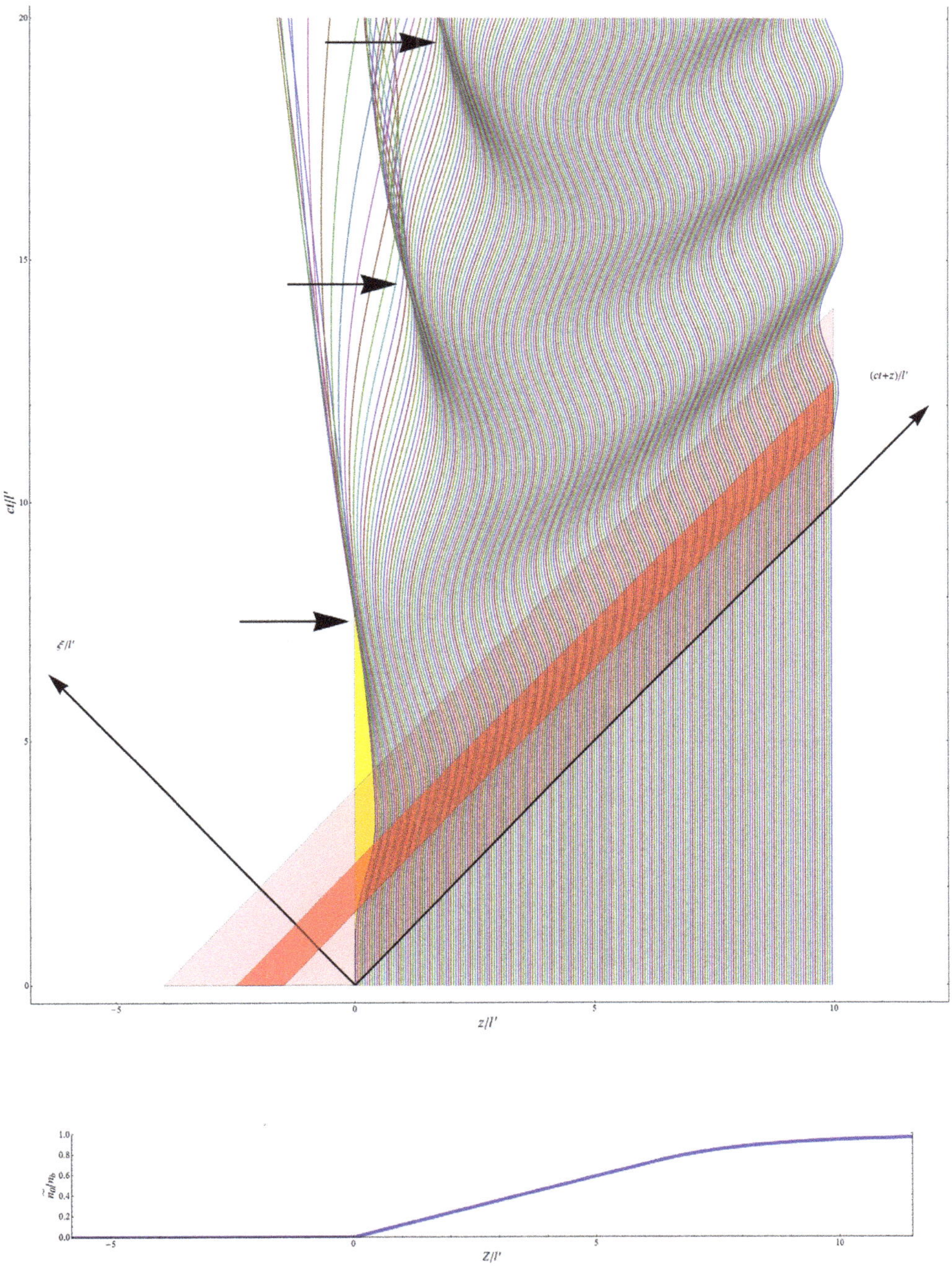

Figure 7. **Down**: the initial electron density (3) of Figure 2. **Up**: The worldlines of Z-electrons interacting with the pulse in Figure 1 for 200 equidistant values of Z; the support $0 \leq ct - z \leq l$ and the 'effective support' $(l - l')/2 \leq ct - z \leq (l + l')/2$ of the pulse are pink and red;, respectively, while the spacetime region of the pure-ion layer is yellow. Horizontal arrows pinpoint where particular subsets of worldlines first intersect.

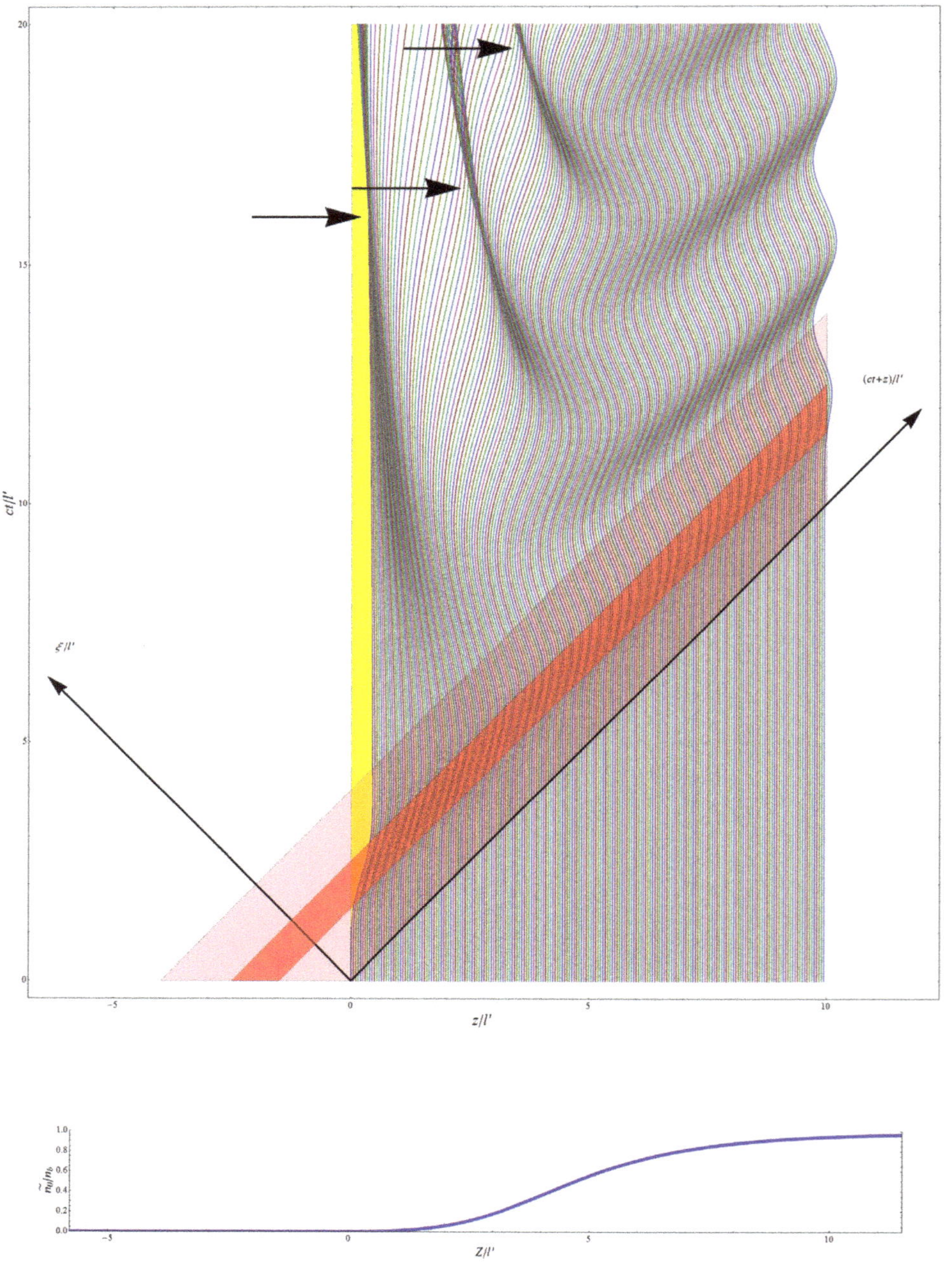

Figure 8. Down: The initial electron density (4) of Figure 2. **Up**: The worldlines of Z-electrons interacting with the pulse in Figure 1 for 200 equidistant values of Z; the support $0 \leq ct-z \leq l$ and the 'effective support' $(l-l')/2 \leq ct-z \leq (l+l')/2$ of the pulse are pink and red, respectively, while the spacetime region of the pure-ion layer is yellow. Horizontal arrows pinpoint where particular subsets of worldlines first intersect; as can be seen, small Z worldlines first intersect quite farther from the laser–plasma interaction spacetime region (shown in pink) than in the linear homogeneous case (3).

Author Contributions: Conceptualization, G.F., R.F. and D.J.; Formal analysis, G.F.; Investigation, G.F.; Methodology, M.D.A. and G.G.; Validation, D.J.; Visualization, G.F.; Writing, original draft, G.F.; Writing, review and editing, M.D.A., R.F., G.G. and D.J. All authors have read and agreed to the published version of the manuscript.

Funding: This research received no external funding.

Institutional Review Board Statement: Not applicable.

Informed Consent Statement: Not applicable.

Data Availability Statement: Not applicable.

Conflicts of Interest: The authors declare no conflict of interest.

Appendix A

Appendix A.1. Proof of Proposition 1

We abbreviate $\Omega := \sqrt{Kn_b}$, $\zeta := l - \eta$. Equation (17) yields

$$\begin{aligned}
2\Delta(l) &= \int_0^{l/2} d\eta\, v(\eta)\, \cos[\Omega(l-\eta)] + \int_{l/2}^{l} d\eta\, v(l-\eta)\, \cos[\Omega(l-\eta)] \\
&= \int_0^{l/2} d\eta\, v(\eta)\, \cos[\Omega(l-\eta)] - \int_{l/2}^{0} d\zeta\, v(\zeta)\, \cos(\Omega\zeta). \\
&= \int_0^{l/2} d\eta\, v(\eta)\, \cos[\Omega(l-\eta)] + \int_0^{l/2} d\eta\, v(\eta)\, \cos(\Omega\eta) \\
&= [1+\cos(\Omega l)] \int_0^{l/2} d\eta\, v(\eta) \cos(\Omega\eta) + \sin(\Omega l) \int_0^{l/2} d\eta\, v(\eta) \sin(\Omega\eta) \qquad \text{(A1)} \\
2\delta(l) &= \int_0^{l/2} d\eta\, v(\eta)\, \sin[\Omega(l-\eta)] + \int_{l/2}^{l} d\eta\, v(l-\eta)\, \sin[\Omega(l-\eta)] \\
&= \int_0^{l/2} d\eta\, v(\eta)\, \sin[\Omega(l-\eta)] - \int_{l/2}^{0} d\zeta\, v(\zeta)\, \sin(\Omega\zeta) \\
&= [1-\cos(\Omega l)] \int_0^{l/2} d\eta\, v(\eta) \sin(\Omega\eta) + \sin(\Omega l) \int_0^{l/2} d\eta\, v(\eta) \cos(\Omega\eta) \qquad \text{(A2)}
\end{aligned}$$

If $\Omega l \le \pi$, both integrands in (A1) are nonnegative, as are the factors of both integrals, and moreover the latter and $\Delta(l)$ itself are positive if $\Omega l < \pi$ and zero if $\Omega l = \pi$. In either case, $\Delta(\xi) > 0$ for all $\xi \in]0, l[$, because $\Delta'(l) = v(l) - \delta(l) < 0$ (as $v(l) \simeq 0$, $\delta(l) > 0$). Moreover, if $\varepsilon := \Omega l - \pi > 0$ is sufficiently small, then both integrals and $1+\cos(\Omega l) = 1-\cos\varepsilon \simeq \varepsilon^2/2$ are positive, whereas $\sin(\Omega l) = -\sin\varepsilon \simeq -\varepsilon$; the second negative term will dominate and make (A1) negative as well. Therefore, (17 Left) will be satisfied iff $\Omega l \le \pi$, i.e., if $G_b \le 1/2$.

Similarly, if $\Omega l = 2\pi$, the factors of both integrals in (A2) vanish, and $\delta l = 0$. If $\varepsilon := \Omega l - 2\pi \neq 0$ is sufficiently small, then $1-\cos(\Omega l) = 1-\cos\varepsilon \simeq \varepsilon^2/2$, whereas $\sin(\Omega l) = \sin\varepsilon \simeq \varepsilon$, and the second term dominates over the first. Under our assumptions, the second integral will be negative, because $v(\eta)$ is larger where $\cos(\Omega\eta) < 0$. Hence, (A2) will be negative if $\varepsilon > 0$ (and is sufficiently small); if $\varepsilon < 0$ (and is sufficiently small), then (A2) will be positive, and $\delta(\xi) > 0$ for $0 < \xi < l$, because $\delta'(l) = M\Delta(l) < 0$ (as in this case $\Delta(l) < 0$. Therefore, (17 Right) will be satisfied iff $\Omega l \le 2\pi$, i.e., if $G_b \le 1$.

Appendix A.2. Estimates of Oscillatory Integrals

Here, we recall useful estimates [31] of oscillatory integrals such as (6 Left) in case (58). Given a function $f \in C^2(\mathbb{R})$, integrating by parts we find that for all $n \in \mathbb{N}$

$$\int_{-\infty}^{\xi} d\zeta\, f(\zeta) e^{ik\zeta} = -\frac{i}{k} f(\xi) e^{ik\xi} + R_1^f(\xi), \tag{A3}$$

$$R_1^f(\xi) := \frac{i}{k} \int_{-\infty}^{\xi} d\zeta\, f'(\zeta)\, e^{ik\zeta} = \left(\frac{i}{k}\right)^2 \left[-f'(\xi)\, e^{ik\xi} + \int_{-\infty}^{\xi} d\zeta\, f''(\zeta)\, e^{ik\zeta}\right]. \tag{A4}$$

Hence, we find the following upper bounds for the remainder R_1^f:

$$\left|R_1^f(\xi)\right| \le \frac{1}{|k|^2} \left[|f'(\xi)| + \int_{-\infty}^{\xi} d\zeta\, |f''(\zeta)|\right] \le \frac{\|f'\|_\infty + \|f''\|_1}{|k|^2}, \tag{A5}$$

It follows that $R_1^f = O(1/k^2)$. All inequalities in (A5) are useful; the left inequalities are more stringent, while the right ones are ξ-independent.

Equations (A3) and (A5) and $R_1^f = O(1/k^2)$ hold if $f \in W^{2,1}(\mathbb{R})$ (a Sobolev space), and in particular if $f \in C^2(\mathbb{R})$ and $f, f', f'' \in L^1(\mathbb{R})$, because the previous steps can be performed under such assumptions. Equations (A3) will hold with a remainder $R_1^f = O(1/k^2)$ under weaker assumptions, e.g., if f' is bounded and piecewise continuous and $f, f', f'' \in L^1(\mathbb{R})$, although R_1^f will be a sum of contributions such as (A4) for every interval in which f' is continuous.

Letting $\xi \to \infty$ in (A3), (A5) we find for the Fourier transform $\tilde{f}(k) = \int_{-\infty}^{\infty} d\zeta\, f(\zeta) e^{-iky}$ of $f(\xi)$

$$|\tilde{f}(k)| \le \frac{\|f'\|_\infty + \|f''\|_1}{|k|^2}, \tag{A6}$$

hence, $\tilde{f}(k) = O(1/k^2)$ as well. Actually, if $f \in \mathcal{S}(\mathbb{R})$, then $\tilde{f}(k)$ decays much faster, as $|k| \to \infty$, because $\tilde{f} \in \mathcal{S}(\mathbb{R})$ as well. For instance, if $f(\xi) = \exp[-\xi^2/2\sigma]$, then $\tilde{f}(k) = \sqrt{\pi\sigma} \exp[-k^2\sigma/2]$.

To prove approximation (59), now we just need to choose $f = \epsilon$ and note that every component of $\boldsymbol{\alpha}^\perp$ will be a combination of (A3) and $(\text{A3})_{k \mapsto -k}$.

References

1. Kruer, W. *The Physics Of Laser Plasma Interactions*; CRC Press: Boca Raton, FL, USA, 2019; 200p.
2. Sprangle, P.; Esarey, E.; Ting, A. Nonlinear interaction of intense laser pulses in plasmas. *Phys. Rev.* **1990**, *A41*, 4463. [CrossRef] [PubMed]
3. Sprangle, P.; Esarey, E.; Ting, A. Nonlinear Theory of Intense Laser-Plasma Interactions, *Phys. Rev. Lett.* **1990**, *64*, 2011. [CrossRef] [PubMed]
4. Esarey, E.; Schroeder, C.B.; Leemans, W.P. Physics of laser-driven plasma-based electron accelerators. *Rev. Mod. Phys.* **2009**, *81*, 1229. [CrossRef]
5. Macchi, A. *A Superintense Laser-Plasma Interaction Theory Primer*; Springer: Berlin/Heidelberg, Germany, 2013.
6. Kuzenov, V.V.; Ryzhkov, S.V. Numerical simulation of the effect of laser radiation on matter in an external magnetic field. *J. Phys. Conf. Ser.* **2017**, *830*, 012124.
7. Tajima, T.; Dawson, J.M. Laser Electron Accelerator. *Phys. Rev. Lett.* **1979**, *43*, 267. [CrossRef]
8. Sprangle, P.; Esarey, E.; Ting, A.; Joyce, G. Laser wakefield acceleration and relativistic optical guiding. *Appl. Phys. Lett.* **1988**, *53*, 2146. [CrossRef]
9. Tajima, T.; Nakajima, K.; Mourou, G. Laser acceleration. *Riv. N. Cim.* **2017**, *40*, 34. [CrossRef]
10. Hidding, B.; Beaton, A.; Boulton, L.; Corde, S.; Doepp, A.; Habib, F.A.; Heinemann, T.; Irman, A.; Karsch, S.; Kirwan, G.; et al. Fundamentals and Applications of Hybrid LWFA-PWFA. *Appl. Sci.* **2019**, *9*, 2626. [CrossRef]
11. Weikum, M.K.; Akhter, T.; Alesini, P.D.; Alexandrova, A.S.; Anania, M.P.; Andreev, N.E.; Andriyash, I.; Aschikhin, A.; Assmann, R.W.; Audet, T.; et al. EuPRAXIA—A compact, cost-efficient particle and radiation source. *AIP Conf. Proc.* **2019**, *2160*, 040012. [CrossRef]
12. Weikum, M.K.; Akhter, T.; Alesini, D.; Alexandrova, A.S.; Anania, M.P.; Andreev, N.E.; Andriyash, I.A.; Aschikhin, A.; Assmann, R.W.; Audet, T.; et al. Status of the Horizon 2020 EuPRAXIA conceptual design study. *J. Phys. Conf. Ser.* **2019**, *1350*, 012059.

3. Assmann, R.W.; Weikum, M.K.; Akhter, T.; Alesini, D.; Alexandrova, A.S.; Anania, M.P.; Andreev, N.E.; Andriyash, I.; Artioli, M.; Aschikhin, A.; et al. EuPRAXIA Conceptual Design Report. *Eur. Phys. J. Spec. Top.* **2020**, *229*, 3675–4284; Erratum in *Eur. Phys. J. Spec. Top.* **2020**, *229*, 4285–4287. [CrossRef]
4. Tomassini, P.; Nicola, S.D.; Labate, L.; Londrillo, P.; Fedele, R.; Terzani, D.; Gizzi, L.A. The resonant multi-pulse ionization injection. *Phys. Plasmas* **2017**, *24*, 103120. [CrossRef]
5. Fiore, G.; Fedele, R.; de Angelis, U. The slingshot effect: A possible new laser-driven high energy acceleration mechanism for electrons. *Phys. Plasmas* **2014**, *21*, 113105. [CrossRef]
6. Akhiezer, A.I.; Polovin, R.V. Theory of wave motion of an electron plasma. *Sov. Phys. JETP* **1956**, *3*, 696. [CrossRef]
7. Gorbunov, L.M.; Kirsanov, V.I. Excitation of plasma waves by an electromagnetic wave packet. *Sov. Phys. JETP* **1987**, *66*, 290.
8. Rosenzweig, J.; Breizman, B.; Katsouleas, T.; Su, J. Acceleration and focusing of electrons in two-dimensional nonlinear plasma wake fields. *Phys. Rev.* **1991**, *A44*, R6189.
9. Mora, P.; Antonsen, T.M. Electron cavitation and acceleration in the wake of an ultraintense, self-focused laser pulse. *Phys. Rev.* **1996**, *E53*, R2068(R). [CrossRef]
10. Pukhov, A.; Meyer-ter-Vehn, J. Laser wake field acceleration: The highly non-linear broken-wave regime. *Appl. Phys.* **2002**, *B74*, 355–361. [CrossRef]
11. Kostyukov, I.; Pukhov, A.; Kiselev, S. Phenomenological theory of laser-plasma interaction in 'bubble' regime. *Phys. Plasmas* **2004**, *11*, 5256. [CrossRef]
12. Lu, W.; Huang, C.; Zhou, M.; Mori, W.; Katsouleas, T. Nonlinear theory for relativistic plasma wakefields in the blowout regime. *Phys. Rev. Lett.* **2006**, *96*, 165002. [CrossRef]
13. Lu, W.; Huang, C.; Zhou, M.; Tzoufras, M.; Tsung, F.S.; Mori, W.B.; Katsouleas, T. A nonlinear theory for multidimensional relativistic plasma wave wakefields. *Phys. Plasmas* **2006**, 13, 056709. [CrossRef]
14. Maslov, V.I.; Svystun, O.M.; Onishchenko, I.N.; Tkachenko, V.I. Dynamics of electron bunches at the laser–plasma interaction in the bubble regime. *Nucl. Instrum. Meth. Phys. Res.* **2016**, *A829*, 422. [CrossRef]
15. Fiore, G.; Nicola, S.D. A simple model of the slingshot effect. *Phys Rev. Acc. Beams* **2016**, *19*, 071302. [CrossRef]
16. Fiore, G.; Nicola, S.D. A "slingshot" laser-driven acceleration mechanism of plasma electrons. *Nucl. Instrum. Meth. Phys. Res.* **2016**, *A829*, 104–108. [CrossRef]
17. Fiore, G.; Catelan, P. On cold diluted plasmas hit by short laser pulses. *Nucl. Instrum. Meth. Phys. Res.* **2018**, *A 909*, 41–45. [CrossRef]
18. Fiore, G.; Akhter, T.; Nicola, S.D.; Fedele, R.; Jovanović, D. On the impact of short laser pulses on cold diluted plasmas. 2022, *in preparation.* [CrossRef]
19. Fiore, G. The time-dependent harmonic oscillator revisited. *arXiv* **2022**, arXiv:2205.01781.
20. Fiore, G. On plane-wave relativistic electrodynamics in plasmas and in vacuum. *J. Phys. A Math. Theory* **2014**, *47*, 225501.
21. Fiore, G. Travelling waves and a fruitful 'time' reparametrization in relativistic electrodynamics. *J. Phys. A Math. Theory* **2018**, *51*, 085203. [CrossRef]
22. Fiore, G. On plane waves in Diluted Relativistic Cold Plasmas. *Acta Appl. Math.* **2014**, *132*, 261. [CrossRef]
23. Fiore, G. On very short and intense laser-plasma interactions. *Ricerche Mat.* **2016**, *65*, 491–503. [CrossRef]
24. Fiore, G.; Catelan, P. Travelling waves and light-front approach in relativistic electrodynamics. *Ricerche Mat.* **2019**, *68* , 341–357. [CrossRef]
25. Fiore, G. Light-front approach to relativistic electrodynamics. *J. Phys. Conf. Ser.* **2021**, *1730*, 012106. [CrossRef]
26. Dawson, J.D. Nonlinear electron oscillations in a cold plasma. *Phys. Rev.* **1959**, *113*, 383. [CrossRef]
27. Jovanovic, D.; Fedele, R.; Belic, M.; Nicola, S.D. Adiabatic Vlasov theory of ultrastrong femtosecond laser pulse propagation in plasma. The scaling of ultrarelativistic quasi-stationary states: Spikes, peakons, and bubbles. *Phys. Plasmas* **2019**. *26*. 123104. [CrossRef]
28. Yoshizawa, T. *Stability Theory by Liapunov's Second Method*; Mathematical Society of Japan: Tokyo, Japan, 1966; 223p. [CrossRef]
29. Hosokai, T.; Kinoshita, K.; Watanabe, T.; Yoshii, K.; Ueda, T.; Zhidokov, A.; Uesaka, M. Supersonic gas jet target for generation of relativistic electrons with 12-TW 50-fs laser pulse. In Proceedings of the 8th European Conference, EPAC 2002, Paris, France, 3–7 June 2002; pp. 981–983.
30. Veisz, L.; Buck, A.; Nicolai, M.; Schmid, K.; Sears, C.M.S.; Sävert, A.; Mikhailova, J.M.; Krausz, F.; Kaluza, M.C. Complete characterization of laser wakefield acceleration. In Proceedings of the Volume 8079, Laser Acceleration of Electrons, Protons, and Ions; and Medical Applications of Laser-Generated Secondary Sources of Radiation and Particles, Prague, Czech Republic, 18–21 April 2011.

applied sciences

MDPI

Article

Experimental Investigation on Jet Vector Deflection Jumping Phenomenon of Coanda Effect Nozzle

Shaoqing Chi and Yunsong Gu *

Key Laboratory of Unsteady Aerodynamics and Flow Control, Ministry of Industry and Information Technolog
Nanjing University of Aeronautics and Astronautics, Nanjing 210016, China; shaoqing_chi@sina.com
* Correspondence: yunsonggu@nuaa.edu.cn

Abstract: The Coanda effect nozzle is a fluid thrust vectoring technology that uses the Coand effect to control jet vector deflection. The jumping phenomenon often occurs in the process o controlling jet vector deflection. This phenomenon leads to the nonlinearity of thrust vector contro It destroys the control performance of the aircraft and brings potential dangers to the safety of th aircraft. The jumping phenomenon occurs in an unsteady flow field different from the traditiona flow phenomenon. The flow structure in an unsteady flow field changes with time, so it is not easy t control by the traditional active flow control method. This paper explains the reasons for the jumpin phenomenon from two aspects: flow field stability and flow structure. Secondly, the unsteady flov field with the jumping phenomenon is studied and analyzed by a flow visualization experiment an dynamic force measurement. Furthermore, the dynamic modal decomposition (DMD) method i used to extract the characteristic frequencies of the critical vortices causing jets to jump in unstead flow fields. Finally, a pulsed jet with the same characteristic frequency is used to control the varyin vortices in the unsteady flow field. The experimental results show that the active flow control methoc which extracts the characteristic frequency of the critical flow field structure by DMD, effectivel suppresses the jumping phenomenon in the unsteady flow field. It also linearizes the process of je nonlinear vector deflection.

Keywords: Coanda effect; fluid thrust vectoring nozzle; active flow control; the jumping phenomenor DMD

Citation: Chi, S.; Gu, Y. Experimental Investigation on Jet Vector Deflection Jumping Phenomenon of Coanda Effect Nozzle. *Appl. Sci.* **2022**, *12*, 7567. https://doi.org/10.3390/app12157567

Academic Editors: Vasily Novozhilov and Cunlu Zhao

Received: 21 June 2022
Accepted: 25 July 2022
Published: 27 July 2022

1. Introduction

The jumping phenomenon of jet vector deflection is one of the critical problems of fluic thrust vectoring technology. The jumping phenomenon commonly exists in various form of fluid thrust vectoring technology. The sudden jump problem will cause the aircraft' attitude to change suddenly even if the pilot does not control it, which can very easil cause flight accidents. Therefore, the jump problem is the bottleneck and obstacle for th development and engineering application of passive thrust vector control technology basec on the Coanda effect.

The Coanda effect control method is derived from the aerial Coanda high−efficienc orienting−jet nozzle (ACHEON) program [1–4] funded by the European Union in 2013. Th program proposes to improve the deflection efficiency of the vectoring nozzle by using th Coanda effect of high−speed jet on the convex surface and the effect of plasma−acceleratin fluid delay separation. This control method adds a trailing edge plate at the outlet of th nozzle and uses the wall−attached effect of the fluid itself to control the deflection vecto This control method is also called trailing edge plate control. The sensitivity of plasm control to flow control is also significantly improved through this method. In 2016, Lu [5 et al. studied the specific application of the plasma control method, and further elaboratec on the application scope of an ion exciter in flow separation control. In the same yea Michel et al. [6–8] tried to produce a practical Coanda effect nozzle design guide, whic was designed to meet the actual needs of different projects. Michel et al. established th mathematical model of the Coanda effect nozzle based on the ACHEON project in the form of the integral equation, proposed a new aircraft architecture based on this model, and verified the model based on this nozzle through numerical simulation.

In 1960, Newman [9] proposed the inclined wall jet model, and suggested that the dimensionless reattachment distance is only a function of the wall deflection angle. Subsequently, the model has been widely studied. The position of the inclined wall and the outlet can be divided into two cases: no potential difference and potential difference. The main structural characteristics of the flow model are that the jet attaches at a certain distance downstream of the outlet and that there is a recirculation zone upstream of the reattachment point, with separation bubbles, reattachment points, and other flow structures. In 2000, Lai et al. [10] obtained consistent results of reattachment positions using pressure measurement and flow display, pointing out that the reattachment positions increase with the increase of wall deflection angle.

In 2013, Asghar, et al. [11] of the Iranian university of science and technology studied the vortex position, size, velocity, average turbulence intensity, Reynolds stress, and reattachment length (attachment point) in the recirculation zone under different wall inclination angles, with a potential difference using experiments and calculations, and obtained the near−wall velocity field and pressure field in the steady−state.

In 2014, Shantanu et al. [12] used a numerical simulation method to study the turbulent wall jet flow field structure under different deflection angles, including downstream velocity field, Reynolds stress, and wall static pressure distribution. The influence of the deflection angle on the flow field structure was studied in the range of small deflection angles. The above literature research results show that the main near−wall flow structure characteristics of the Coanda flow model in steady−state wall attachment include the separation bubble and reattachment structure. For the wall−attached deflection jet's flow structure, scholars' research mainly focuses on the near−wall flow structure and flow characteristics when the wall is stable. The transverse pressure gradient between the space environment and the separation area determines jet detachment and attachment. Zaitsev et al. [13] studied the flow characteristics of a supersonic jet on the inclined wall with a fixed deflection angle and, for the first time, obtained the pressure in the starting area, the range of the recirculation area, the jet trajectory, and pointed out that the jet separation and attachment are determined by the transverse pressure gradient developed between the space environment and the separation area. Allery et al. [14] used a hot wire anemometer to conduct experimental research on inclined wall jets with different incoming velocities at a fixed deflection angle of 30 degrees. They gave the deflection angle and Reynolds number boundary conditions for jet attachment and separation.

In 2015, Gillgrist et al. [15] used particle image velocimetry and conventional pressure sensors to obtain the results of the transient velocity field. The Reynolds stress field of the near−wall flow under the stable control state for the reverse flow vectoring nozzle model demonstrated that the lateral pressure gradient generated by the negative suction pressure and the reverse flow shear layer, under the combined action of the induced negative pressure on the wall, is the cause of the jet vector deflection. Allery et al. [16] experimentally conducted numerical simulation research on the phenomenon of wall−attached deflected jet due to the Coanda effect. Through practical means, they studied the effects of the inclination angle and Reynolds number on the wall−attached and separation phenomena, as well as hysteresis and jumping. They used the Galerkin projection of the Navier–Stokes equation on POD basis function to obtain a low−dimensional dynamic model, which qualitatively represented flow characteristics. Miozzi et al. [17] studied the phenomenon of the jet deviating from the straight direction due to the presence of the Coanda wall from the experimental point of view. The velocity field results clearly show that the inclination of the jet attached to the wall depends on the side wall distance itself. The self−similarity analysis along the inclined jet direction shows that the Coanda wall attachment effect will fail for wall distances more significant than five jet widths. Cornelius et al. [18] studied the physical mechanism of the Coanda jet wall separation process under a high−pressure ratio and qualitatively described the physical characteristics of wall−attached expansion and jet separation using an optical schlieren system.

For complex dynamic systems, commonly used dimensionality reduction methods include the inertial manifold/approximate inertial manifold method, the Eigen orthogonal decomposition POD method, and the primary manifold method [19], in which the POD method is widely used in the dimensionality reduction modeling of fluid dynamic systems. Because the POD method has a relatively small calculation, the numerical results or experimental data based on high−resolutions have clear physical significance. Lumley [20,21] first introduced the POD method into the turbulent field, and then Sirovich [22] introduced the snapshot method to study the dynamics of fluctuating flow and the dynamics of wave flow. Deane [23] et al. and Cao et al. [24] conducted a numerical simulation of the flow around a cylinder through the POD Galerkin method, and found that the evolution characteristics of the flow field could be accurately captured by using a less pod basis. The fluid thrust vectoring nozzle was an unsteady flow process in the process of vector deflection. As far as the usual structure analysis method of unsteady flow is concerned, the dynamic mode decomposition (DMD) method is commonly used. Schmid developed the DMD method from Koop in recent years; it is a low−dimensional system decomposition technology developed based on man analysis [25]. Dynamic mode decomposition (DMD) can solve or approximate the dynamic system according to the coherent growth structure, attenuation, and oscillation in time. This method has been widely used in the study of various unsteady flows, and derived the forms of optimal (opt−DMD) [26], optimal mode decomposition (OMD), and sparse improved DMD (SPDMD) [27], which has gradually became a new tool for hydrodynamics mechanism analysis. Using POD and DMD techniques, Sajadmanesh et al. [28,29] successfully identified separated bubbles in ultra−high lift turbine cascades and the flapping phenomenon in highly loaded low−pressure turbine cascades.

This paper uses the practical active flow control method to control the jet's transient vector deflection jumping phenomenon. The specific practical steps are as follows: firstly, the PIV experiment, oil flow experiment, dynamic pressure measurement, and dynamic force measurement experiment determine the control object and control position required by the active flow control method; secondly, the dynamic mode decomposition technique is used to determine the control frequency of the pulsed jet; finally, an experimental technique is used for verifying the effect of the active flow control experimental technique.

2. Experimental Setup and Methods

2.1. Coanda Effect Nozzle

The structure and mechanism of the Coanda effect nozzle are shown in Figure 1. The nozzle was comprised of a pair of Coanda walls with inclination, a secondary flow passage, a secondary flow control valve, and a central jet flow passage. As shown in Figure 1a, the secondary flow control valves at the upper and lower sides remained open when the jet remained neutral. At this time, two secondary flow fields were located at the upper and lower sides of the main jet and in the same direction as the main jet, respectively. The secondary flow was generated by the ejection effect of the main jet. This secondary flow was passive and did not need external energy injection. As shown in Figure 1b, when the jet deflected downward to the attached wall, the secondary flow control valve on the upper side of the nozzle remained open, and the passive secondary flow control valve on the lower side of the nozzle was closed. The Coanda effect was generated between the main jet and the lower side wall. At the same time, the secondary flow on the upper side of the jet generated a downward pressure difference, which eventually caused the jet to attach to the Coanda wall and complete the deflection attachment.

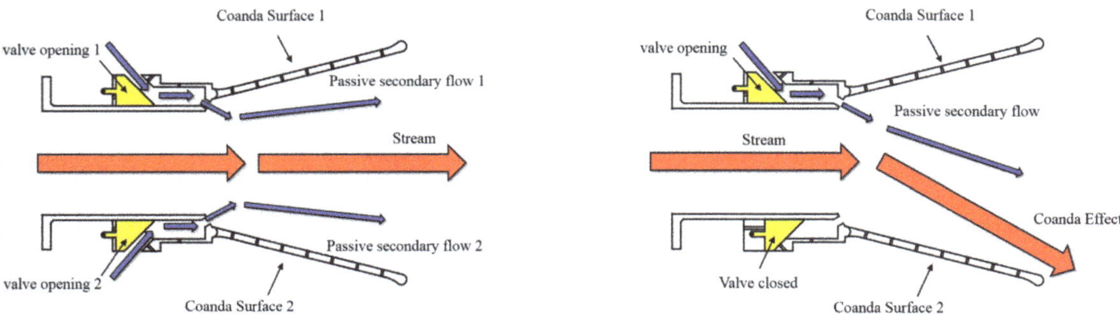

(**a**) (**b**)

Figure 1. Structure diagram of passive thrust vector. (**a**) Flow structure diagram of a jet under detachment condition; (**b**) flow structure diagram of jet deflection wall attachment condition.

2.2. Model Parameters and Experimental Setup

The Coanda effect nozzle is shown in Figure 2. The height of the main jet outlet was D = 40 mm, and the width was L = 100 mm. There ere passive secondary flow control joints with height h = 4 mm and width W = 100 mm, respectively, at the upper and lower parts of the main jet outlet area. The Coanda wall had an 18° deflection angle with the main jet direction, the plate length L was 100 mm, and the size of the whole plate was 100 mm × 100 mm. In this article, the fluid medium was air, and the incoming flow velocity was 30 m/s. At the upper and lower walls of the nozzle, four rows of pressure taps were arranged at S1 = 1.25D, S2 = 1D, S3 = 0.5D, and S4 = 0.1D, respectively. The pressure taps were connected with the dynamic pressure sensor to measure the dynamic pressure parameters of the wall. The vectoring nozzle controlled the injection flow of passive secondary flow by adjusting the opening of the control slot. The purpose of vector control was achieved by actively controlling the pulsed jet to control the main jet's flow.

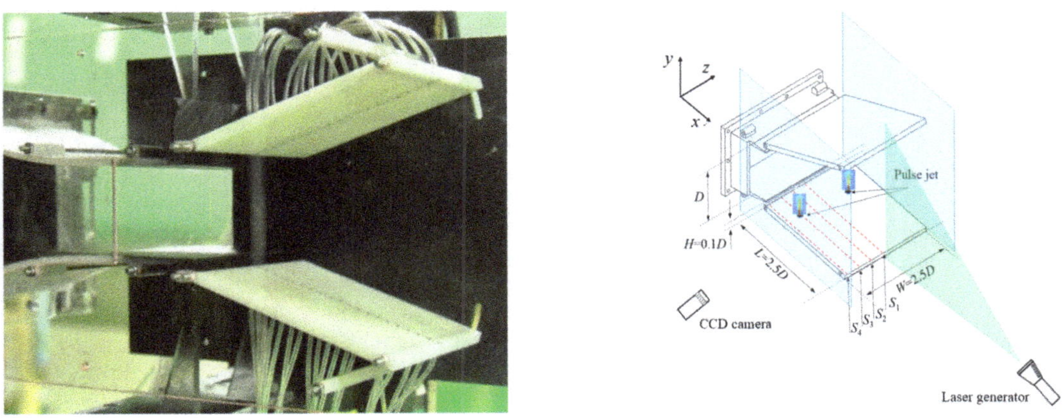

(**a**) (**b**)

Figure 2. The Coanda effect nozzle model and structural dimension schematics. (**a**) The Coanda effect nozzle model in PIV Layout; (**b**) structural dimension of model and section diagram of PIV shot.

The pulsed jet convection field was used for active flow control to improve the jet deflection rate under the condition of the jet attached to the wall. The structure of the pulsed jet generator is shown in Figure 3a. The pulsed jet exciter used the active compressed air source as the energy input of the pulsed jet. The pressure reducing and stabilizing valve was used to ensure the speed stability of the output jet. The signal generator was used to generate the frequency signal. The signal amplifier was used to input the frequency signal into the high−frequency solenoid valve. Finally, the high−frequency electromagnetic valv controlled the jet to output the constant frequency pulse jet. Among them, the model o a high−frequency solenoid valve was ZTO−45A−AA1−DDFA−1BA, which was drive by a 24VDC power supply, the maximum control frequency was 233 Hz, the power o response time was 6 ms, and the power off response time was 2 ms.

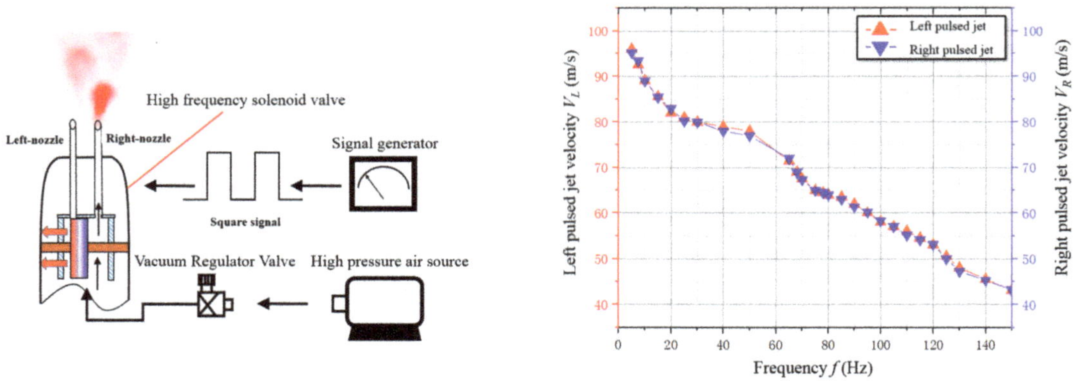

(**a**) (**b**)

Figure 3. The pulsed jet exciter. (**a**) Structure diagram of pulsed jet generator; (**b**) frequency chara teristic curve of the pulsed jet generator.

The pulsed jet adopted an active air source, and inputted a square−wave signal wit a duty cycle of 50% through the signal generator to control the high−frequency solenoi valve. By adjusting the frequency of the square wave signal, two jet holes could releas a pulsed jet with a specific frequency. The diameter of the two pulsed jet holes wa d = 2.5 mm, and the spacing was 1.5D = 60 mm. The curve of the velocity of pulsed je varying with frequency, is shown in Figure 3b. The velocity of the jet of the active sourc was V = 100 m/s, and the velocity of the pulsed jet decreased with the increase of th control frequency. The momentum ratio of the pulsed jet to the main jet was 1.47–0.736% The location of the jet hole was the lowest CP coefficient points (Z/W = 0.2, X/L = 0.215 and (Z/W = 0.8, X/L = 0.215) when the jet was in attachment condition.

Figure 4 shows the schematic diagram of the jet wind tunnel platform. In the experi ment, the charge−coupled device (CCD) camera was used to record the flow field data and its resolution was 2048 × 2048 pixels^2. The model of the CCD camera was Fast−CAM Mini ax50, and the camera's resolution was 1024 × 1024. The sensor adopted a CMO sensor. The frame rate could reach 2000 Hz, and the minimum exposure time was 1.05 μs The dynamic flow field structure in jet deflection and wall attachment was qualitativel displayed. The physical parameters in the dynamic flow field were measured. The pas sive secondary flow fluid thrust vectoring nozzle model was connected with the jet winc tunnel through the transition section. In the experiment, an Nd: YAG 200−MJ laser with a wavelength of 532 nm was used to illuminate the convection field. The shooting plane wa perpendicular to the horizontal plane in the PIV experiment, and was located at Z/W = 0.5 The experiment performed at a velocity of 30 m/s and a free stream turbulence intensit below 0.3%. In the experiment, the dynamic pressure measurement was synchronize with the PIV experiment, and the dynamic force measurement experiment was carrie out separately.

Figure 4. The layout of the dynamic pressure measurement on the active flow control experiment.

As shown in Figure 5a, an eight−channel pressure sensor module was used for dynamic pressure acquisition. The sensor used was SM5652−001−D−3−SR, compensated and calibrated by ceramic DIP under constant pressure excitation. Eight 8−channel pressure sensors with 64 channels and a pressure measuring the frequency of 1000 Hz were used in the experiment. Table 1 shows the technical parameters of the dynamic pressure sensor module. As shown in Figure 5b, the ATI−9610 six−component strain gauge balance made by ATI Corporation, USA, consisted of floating and fixed frames. The floating frame was fixed to the nozzle model, the bottom of the fixed frame was connected to the base, and the force, moment range, and total range measurement accuracy of the balance axes X, Y, and Z are shown in Table 2. The strain balance converted the acquired force signal into an electrical signal, amplified it through the back−end amplifier, and transmitted it to the NI data acquisition card for acquisition and storage. The acquisition frequency of ATI force balance was 200 Hz.

(a) (b)

Figure 5. Dynamic pressure and force measuring test equipment. (**a**) Eight−channel pressure sensor module; (**b**) the ATI−9610 six−component strain gauge balance.

Table 1. Technical parameters of the dynamic pressure sensor module.

Parameter	Numerical Value
Pressure measurement range	0.15 PSI
Acquisition accuracy	0.1 Pa
Acquisition resolution	0.3 Pa

Table 2. Measuring range and accuracy of each measuring axis of ATI balance.

Variable	F_x	F_y	F_z	M_x	M_y	M_z
Range	165 N	165 N	495 N	15 N·m	15 N·m	15 N·m
Precision (FS)	1.00%	1.00%	1.00%	1.00%	1.00%	1.00%

The dynamic pressure measurement baseline results are shown in Figure 6a. The random error of the measured pressure value within 1800 ms met the instrument standard. Furthermore, the pressure fluctuation had no evident periodicity within 1800 ms, so the pressure measurement error did not affect the experimental results. The dynamic torque measurement baseline results are shown in Figure 6b. The random error of the measured torque value within 700 ms met the instrument standard. Furthermore, the pressure fluctuation had no evident periodicity within 700 ms, so the dynamic torque measurement error did not affect the experimental results.

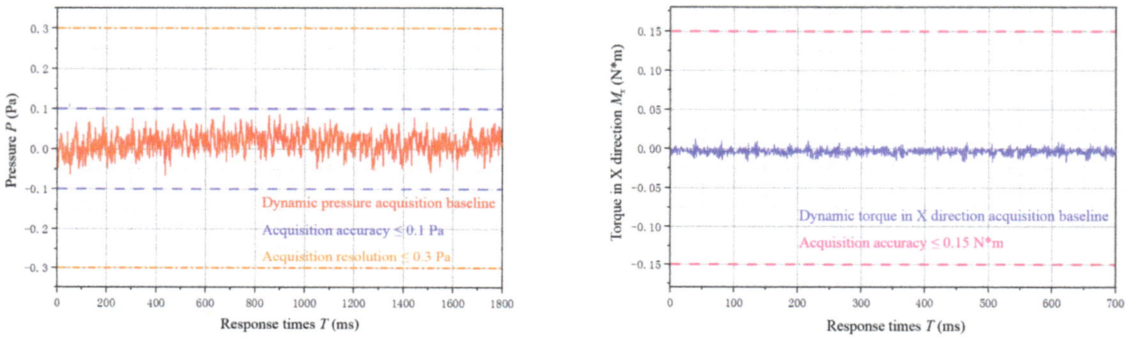

(a) (b)

Figure 6. The actual error results of dynamic pressure and force measurement experiments. (**a**) The dynamic pressure measurement baseline results; (**b**) the dynamic torque measurement baseline results.

The calculation formula of the force vector angle was:

$$\theta_E = \arctan(F_X/F_Y) \tag{1}$$

$$\theta = \theta_E - \theta_0 \tag{2}$$

The force measurement results selected the forces F_X and F_Y in X and Y directions and calculated the thrust vector angle under various working conditions through Formula (1) θ_C. Zero vector angle θ_0 was the thrust vector angle measured by the force balance when the jet remained horizontal. The thrust vector angle obtained through the measurement and calculation of each experimental state θ_E deducted the reference zero value of vector angle θ_0 to get the actual thrust vector deflection angle of the jet θ. Formulas (1) and (2) effectively eliminated the error caused by assembly with the model.

2.3. *Proper Orthogonal Decomposition*

A group of transient information of flow field $\{u_1, u_2, \ldots, u_N\}$ was described as:

$$u_i = \frac{1}{N}\sum_{j=1}^{N} u_j + v_i \tag{3}$$

where $\boldsymbol{u}_i$ was the transient flow field variable at time i, and $\boldsymbol{v}_i$ was the pulsation after subtracting the average value. The POD method used a linear combination of a set of optimal orthogonal basis functions to represent $\boldsymbol{v}_i$ [19], expressed as:

$$v_i = \sum_{j=1}^{N} a_j(t_i)\boldsymbol{p}_j \tag{4}$$

where $\boldsymbol{p}_j$ was the modal basis function of POD and $\boldsymbol{a}_j(\boldsymbol{t}_i)$ was the modal coefficient of mode p_j corresponding to time $\boldsymbol{t}_i$. The definition matrix $\boldsymbol{C} = \boldsymbol{V}^T\boldsymbol{V}$, where $\boldsymbol{V} = \{v_1, v_2, \ldots, v_N\}$. Then, the eigenvalues were solved as:

$$\boldsymbol{C}\boldsymbol{A}^j = \lambda_j \boldsymbol{A}^j \tag{5}$$

From Equation (5): λ_j was the eigenvalue; $\boldsymbol{A}^j$ was the corresponding eigenvector matrix, namely the modal coefficient matrix [20,21]; and $A^j = [\boldsymbol{a}_j(\boldsymbol{t}_1), \boldsymbol{a}_j(\boldsymbol{t}_2), \ldots, \boldsymbol{a}_j(\boldsymbol{t}_N)]^T$ rearranged the eigenvalues by size. The 1 order POD mode corresponded to the maximum eigenvalue. The pod mode could be solved in one step:

$$\boldsymbol{p}_j = \frac{1}{N\lambda_j}\sum_{i=1}^{N} \boldsymbol{A}_i^j v_i \tag{6}$$

POD could calculate the energy level of each mode in the flow field [22]. Energy was defined as:

$$E_i = \lambda_i / \sum_{j=1}^{N} \lambda_j \tag{7}$$

2.4. Dynamic Mode Decomposition

For the flow field change information $\{\boldsymbol{u}_1, \boldsymbol{u}_2, \ldots, \boldsymbol{u}_N\}$, it was assumed that there was a matrix $\boldsymbol{A}$, so that there was a linear transformation relationship between adjacent time layers:

$$\boldsymbol{u}_{i+1} = \boldsymbol{A}\boldsymbol{u}_i \tag{8}$$

Definitions $\boldsymbol{\psi}_0 = \{\boldsymbol{u}_1, \boldsymbol{u}_2, \ldots, \boldsymbol{u}_{N-1}\}$ and $\boldsymbol{\psi}_1 = \{\boldsymbol{u}_2, \boldsymbol{u}_3, \ldots, \boldsymbol{u}_N\}$ could give the following relation:

$$\boldsymbol{\psi}_1 = \boldsymbol{A}\boldsymbol{\psi}_0 = [\boldsymbol{A}\boldsymbol{u}_1, \boldsymbol{A}\boldsymbol{u}_2, \ldots, \boldsymbol{A}\boldsymbol{u}_{N-1}] \tag{9}$$

For finding matrix $\boldsymbol{A}$, DMD used a low−dimensional optimal approximation matrix $\tilde{\boldsymbol{A}}$ to replace $\boldsymbol{A}$. Solving $\tilde{\boldsymbol{A}}$ required a singular value decomposition of $\boldsymbol{\psi}_0$:

$$\boldsymbol{\psi}_0 = \boldsymbol{U}\boldsymbol{\Sigma}\boldsymbol{W}^H \tag{10}$$

where $\boldsymbol{U}$ was a left orthogonal matrix, $\boldsymbol{\Sigma}$ was a singular value diagonal matrix, $\boldsymbol{W}$ was a right orthogonal matrix, and H was a complex conjugate transpose [25]. The approximation matrix could be expressed as:

$$\widetilde{A} = \boldsymbol{U}^H\boldsymbol{\psi}_1\boldsymbol{W}\boldsymbol{\Sigma}^{-1} \tag{11}$$

The next step involved finding the eigenvalue $\tilde{A}\boldsymbol{\Lambda}_j = \lambda_j\boldsymbol{\Lambda}_j$ for $\tilde{A}$, where $\boldsymbol{\Lambda}_j$ was the eigenvector corresponding to the eigenvalue λ_j. Where DMD mode was $\boldsymbol{\Phi}_j = \boldsymbol{U}\boldsymbol{\Lambda}_j$, the mode amplitude was $\alpha_j = \boldsymbol{\Lambda}_j^{-1}\boldsymbol{U}^H\boldsymbol{u}_1$, and the mode growth rate was $g_j = \mathrm{Re}[\ln\lambda_j/\Delta t]$. The reconstructed flow field could be expressed as:

$$\boldsymbol{u}_j \approx \sum_{i=1}^{N} \boldsymbol{\phi}_i(\lambda_i)^{j-1}\alpha_i \tag{12}$$

where $(\lambda_i)^{j-1}\alpha_i$ was the modal coefficient.

The residual corresponding to the maximum difference between the sample data matrix $\boldsymbol{A}$ and the reconstructed data matrix $\boldsymbol{A'}$ was:

$$R = \max_{\substack{1 \le i \le n \\ 1 \le j \le m}} \left| A_{i,j} - A'_{i,j} \right| \tag{13}$$

In which $\boldsymbol{A}_{i,j}$ and $\boldsymbol{A'}_{i,j}$ were the elements of the $\boldsymbol{A}$ and $\boldsymbol{A'}$ matrices, respectively [28,29].

3. Results and Discussion

3.1. Jumping Phenomenon

Figure 7a shows the measured values at diffusion angle α = 12–22° in the experiment with the jet vector deflection angle θ and the secondary flow valve opening δ change curve. It is defined in the figure that when the jet is in the neutral condition, the jet vector deflection angle θ = 0°; when the jet is attached to the wall, $\theta = \alpha$. Defining the secondary flow valve opening as δ = 0, the secondary flow valve is fully opened, and the jet maintains a neutral working condition; when the opening of the secondary flow valve is 1, the secondary flow valve is completely closed, and the jet is attached to the Coanda wall at the closing side.

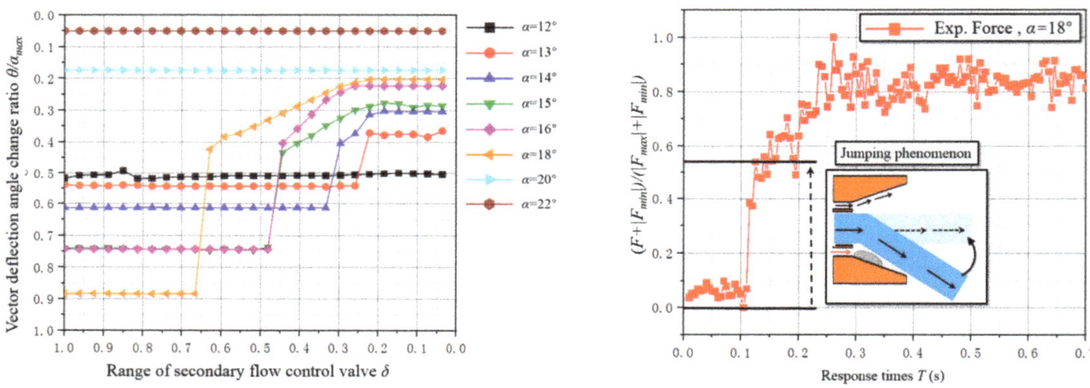

(a) (b)

Figure 7. The jumping phenomenon of jet vector deflection in the Coanda effect nozzle. (**a**) Jet vector deflection angle θ with secondary flow valve opening δ variation curve (diffusion angle α = 12–22°); (**b**) force variation curve of vectoring nozzle during jet dynamic detachment (diffusion angle α = 18°).

On the condition of a diffusion angle of α = 12°, the vectoring jet attaching to the wall is stable. This phenomenon shows that when the diffusion angle of the vectoring nozzle is too small, the Coanda effect will cause the jet to attach to the wall. When the diffusion angle is α = 13–18°, there is an evident jumping phenomenon in the change curve of the jet vector angle, and the vector angle does not have a linear relationship with the valve opening of secondary flow. When the secondary flow valve is gradually opened, the jet will remain attached to the wall for a while until the secondary flow valve is opened to a critical opening, and the jet will then suddenly leave the wall, resulting in a sudden jump. The vector change angle of the jumping phenomenon, $\Delta\theta$, with the diffusion angle within a specific range, α, gradually increases. When α > 20°, the jet remains neutral and will not produce vector deflection when the Coanda effect occurs. Therefore, in the Coanda effect nozzle, the Coanda effect has a specific effective range θ_C. When the distance between the vector jet and Coanda wall $\le \theta_C$, the jet will have the Coanda effect with the wall, resulting in the jet jumping to the Coanda wall; when the vector jet is attached to the Coanda wall, sufficient secondary flow will make the vector jet leave the Coanda wall and jump at the vector angle $\Delta\theta = \theta_C$.

Due to the spatial distance between the jet and the wall when the jet deflects towards the wall, it is not easy to install the pulsed jet exciter. Therefore, the phenomenon of the jet jumping off of the wall is studied in this paper. In order to better study the jump phenomenon in the Coanda effect nozzle, the expansion angle with the most apparent jump phenomenon is selected (α = 18°) as the working condition and is taken as the research object. Figure 7b shows the expansion angle and time−dependent force curve of the vectoring nozzle during jet dynamic wall separation at α = 18°. It can be seen from the figure that the nozzle will jump instantaneously after the force lags for a while, until the secondary flow reaches a certain level.

3.2. Analysis of the Attachment Flow Structure

The PIV image at Z/W = 0.5 inside the Coanda effect fluid thrust vectoring nozzle is shown in Figures 8 and 9. The region between the jet and the Coanda wall is region A. The velocity nephogram of the jet under neutral working conditions is shown in Figure 9a. Under this working condition, the secondary flow control valves open between the upper and lower sides of the jet. As shown in Figure 9b, when the main jet flow deflects downward, the upper control valve is open, and the lower control valve is closed. Only one passive secondary flow is injected in the same direction as the main jet into the nozzle flow field. The main jet remains separated from the side of the injection passive secondary flow and attached to the side of the closed valve. When the main jet flows to the lower wall, the internal flow field structure of the nozzle is divided into the main jet, the shear layer outside of the main jet, the separation bubble, and the return basin. The velocity difference between the main jet and the flow field in the nozzle causes the shear layer, and the shear layer leads to a large velocity gradient in the nozzle. The *K–H* instability caused by the high−speed shear sucks up many small vortex structures. There are apparent separation bubble structures on the wall side of the main jet, and several small and stable vortex structures in the separation bubble.

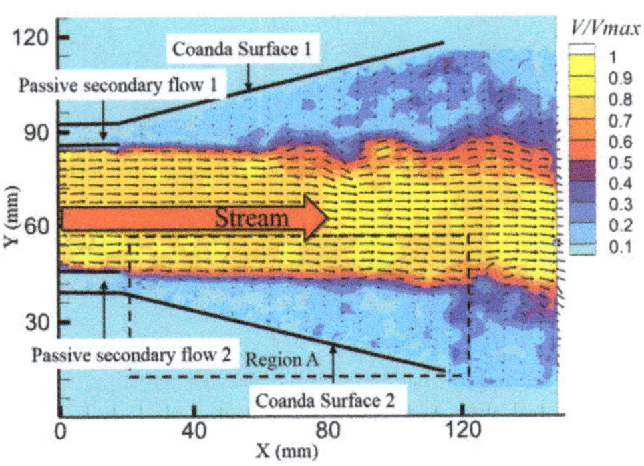

(a)

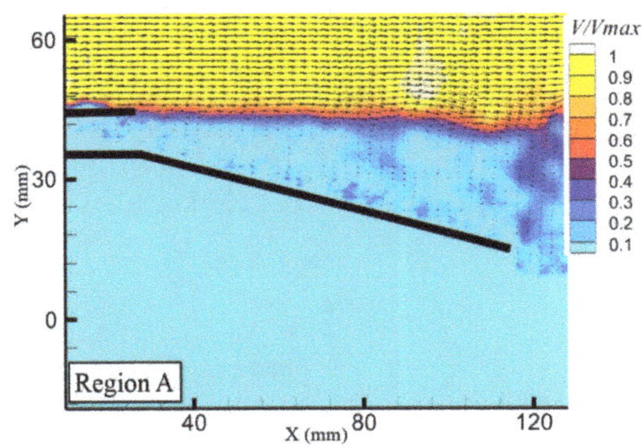

(b)

Figure 8. Detachment condition PIV visualization. (**a**) Velocity vector and velocity nephogram of the whole flow field under the jet detachment condition; (**b**) velocity vector and velocity nephogram of the Region A under the jet detachment condition.

In the POD method, the flow field is sorted according to the energy series, and the first−order mode is generally regarded as the time−average flow result of the flow field. The PIV results of jet attachment and jet detachment are usable to construct the snapshot set (snapshot = 100). As shown in Figure 10a, the first−order energy level in the wall−attached state is lower than in the neutral state. The first−order modal energy in the wall−attached state is 48.08%, while the first−order modal energy in the neutral state is 71.506%. It can be seen that the stability of the jet in a neutral state is higher than that in an attachment state. Therefore, when the jet deflects to the wall, the jumping phenomenon will occur after the jet vector deflects a certain angle, and when the jet deflects away from the wall, the jumping phenomenon will occur suddenly.

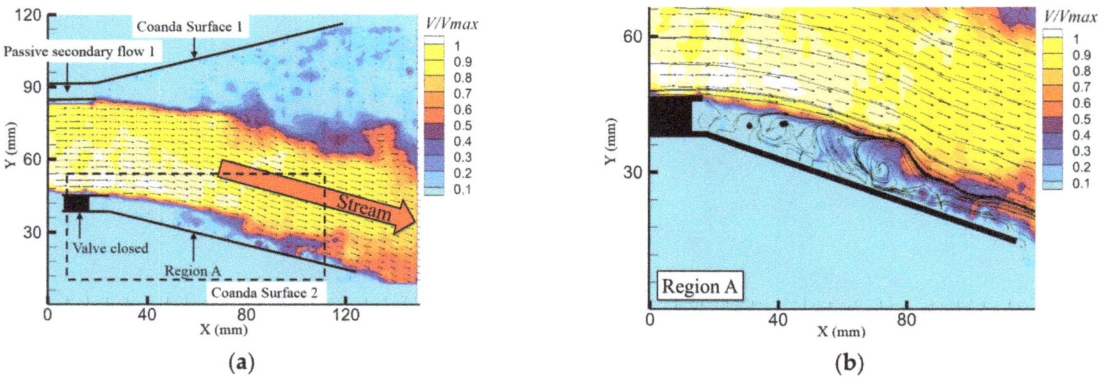

Figure 9. Attachment condition PIV visualization. (**a**) Velocity vector and velocity nephogram of the whole flow field under the jet attachment condition; (**b**) velocity vector, velocity streamline and velocity nephogram of the Region A under the jet attachment condition.

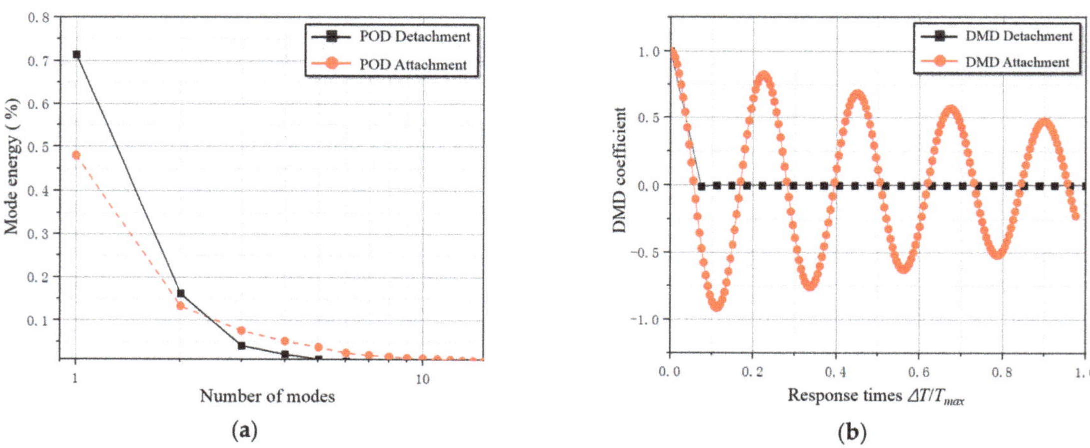

Figure 10. The POD results and DMD results of jet attachment and detachment conditions. (**a**) The modal energy distribution of the POD under attachment and detachment conditions; (**b**) the modal coefficient curve of the DMD under attachment and detachment conditions.

DMD is used to arrange the frequencies based on time series so that DMD can identify the stability of the flow, to a certain extent. When the modal curve is a straight line, the flow field is stable; when the modal curve is periodic, the flow field is periodic and stable; when the modal curve diverges, it indicates that the flow field is unstable; when the modal curve converges, it indicates that the flow field will be stable at a specific time. As shown in Figure 10b, the modal coefficient curve of the jet in the neutral state rapidly changes to a zero returning straight line, and the jet presents a completely stable flow state. This phenomenon indicates that the steady velocity of jet neutrality is very rapid. The modal coefficient curve of the jet attached to the wall is a pulsating curve with periodic frequency attenuation. It can be considered that the wall attachment flow is stable under steady flow and periodically stable under unsteady flow. The periodic frequency of the separation bubble structure affects the attachment flow, and changes the characteristic frequency of the flow structure under the wall−attached condition.

According to the above conclusion of jet flow stability, it can be inferred that the fundamental reason for the jet jump phenomenon is that the flow field structure inside the Coanda effect nozzle changes in the process of jet vector deflection. The stability of the neutral jet flow is steady, while the stability of the attachment flow is periodic. The vector deflection process is a process of mutual transformation between steady and periodic stable systems. The stability of the jet attached to the wall is worse than that of the jet in the neutral state, so the jet will jump when it leaves the Coanda wall.

Figure 11a shows the experimental results of oil flow on the Coanda wall at the side of the jet wall under the condition of jet wall attachment. The oil flow visualization experiment captured the shear stress distribution of the separation bubble structure on the Coanda wall. It can be seen that the separation bubble structure is a symmetrical flow structure along the Z−axis. The vortex−shaped structure of the shear stress line at both ends of the separation bubble indicates that there is a complex three−dimensional flow structure in the separation bubble. A pressure test on the location of the shear stress line was performed, and the test results are shown in Figure 11b. The pressure distribution on sections S1–S4 changes. At this time, the position of the lowest points (Z/W = 0.2, X/L = 0.215) and (Z/W = 0.8, X/L = 0.215) of the pressure coefficient in section S3 correspond to the core position of the vortex shear stress line. Therefore, the pulsed jet actuator is placed in the center of the vortex structure to change the flow structure in the field.

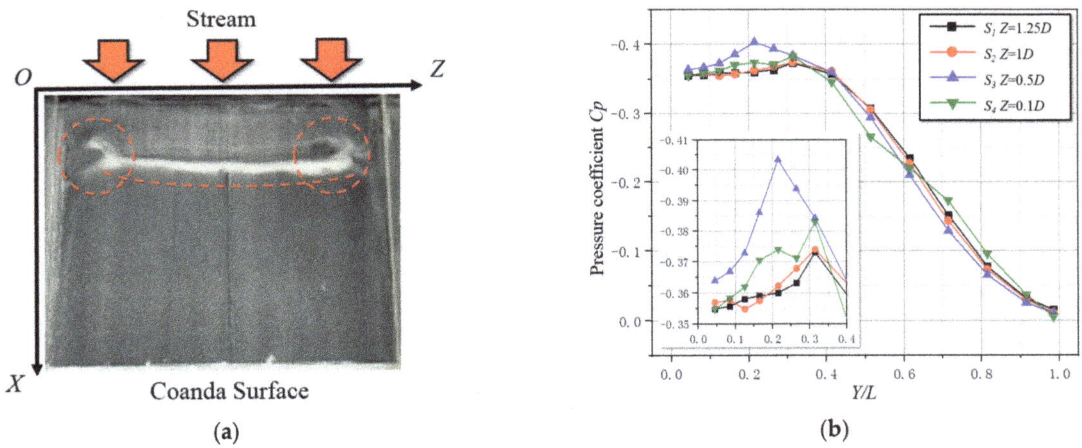

Figure 11. Experimental results of oil flow and pressure measurement under the condition of the jet attachment. (**a**) Results of oil flow experiments on the surface of the Coanda wall at the side of the jet attachment; (**b**) results of mean pressure measurements on the surface of the Coanda wall at the side of the jet attachment.

3.3. Active Flow Control Frequency Selection

The transient PIV flow field under the jet detachment process condition is shown in Figure 12; when the frame number = 10 and 30, the main jet attaches to the wall. The vortex structure produced by the outer shear layer near the side of the separation bubble structure is significant.

When the frame number = 50, the distance between the main jet and the wall increases. It is evident that the vorticity of the separation bubble structure begins to decrease. At this time, the separation bubble structure begins to move towards the trailing edge of the vectoring nozzle, and the tail of the main jet begins to leave the lower wall of the vectoring nozzle.

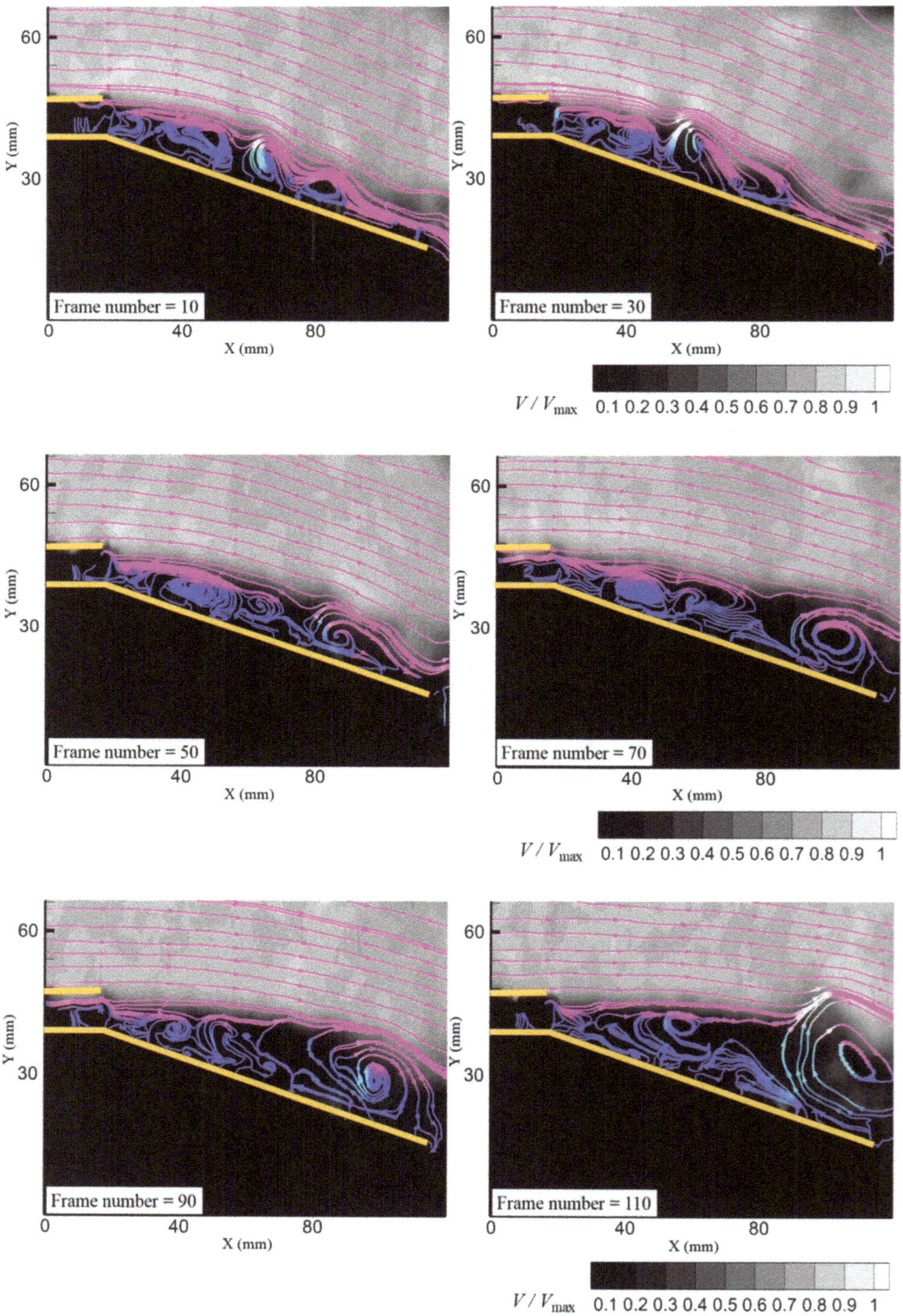

Figure 12. Transient PIV velocity nephogram flow field structure under the jet detachment condition (velocity streamline using RBG color).

When the frame number = 70, the vortex structure in the separation bubble area on the wall side has disappeared, the entrainment vortex begins to appear on the upper wall side of the vectoring nozzle, and the reverse suction vortex appears at the trailing edge of the lower wall of the vectoring nozzle.

When the frame number = 90 and 110, the vector jet is entirely neutral. It can be seen from the above description that when the separation bubble from before broke, the vector deflection angle of the jet is not apparent. When the separation bubble breaks, the jet begins to deflect rapidly. Therefore, maintaining the separation bubble structure can avoid the jumping phenomenon.

The PIV results of jet transient detachment process condition are used to construct the snapshot set (snapshot = 250). To determine the excitation frequency required for active flow control, dynamic mode decomposition was used to decompose the PIV snapshot set of the overall flow field and the separated bubble region (Region A). As shown in Figure 13a, the energy distribution of each mode of DMD is demonstrated. In the DMD of the global watershed, the energy proportion of the first five modes is 67.11%, 5.38%, 2.66%, 2.41%, and 2.09%, respectively. In the DMD of the separation bubble basin, the energy of the first five modes accounts for 75.48%, 9.07%, 2.59%, 1.52%, and 1.09%. Considering that the energy proportion of subsequent modes is less than 1%, the first five modes are further analyzed.

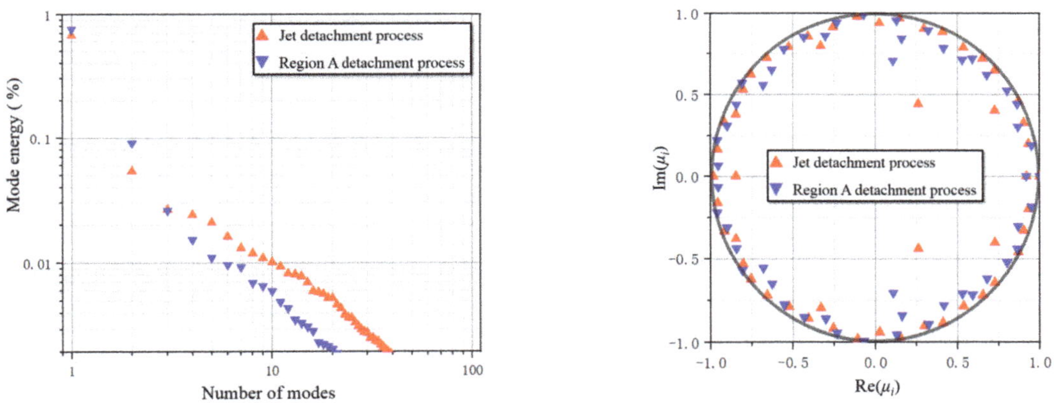

(a) (b)

Figure 13. DMD results of the PIV velocity nephogram under jet transient detachment process conditions. (**a**) The modal energy distribution of DMD under the conditions of the jet detachment process; (**b**) the distribution of the real and imaginary parts of the eigenvalues on the unit circle under the conditions of the jet detachment process.

Figure 13b shows the analytical solution of the dynamic mode decomposition μ_i The distribution diagram is of the real part and imaginary part of the μ_i. When the analytical solution of the DMD is distributed on the unit ring with radius one, it shows that the solved flow field is stable, and the flow field will not change with time, or has strong periodicity. When the analytical solution is outside of the unit ring, the solved flow field is divergent, and the flow in the flow field will become chaotic. When the analytical solution is in the unit ring, the solved flow field is concurrent, and the flow field will change into a more stable flow with time. The analytical solution of the transient off−wall condition of the jet is distributed in the unit ring, which shows that the off−wall process of the jet converges with time. In other words, when open, the passive secondary flow valve is on the wall side, and the injection of a passive secondary flow will not affect the flow stability of the transient jet deflection.

DMD decomposes the flow in the separation bubble region to accurately capture the required active flow control frequency. The DMD modal coefficient curve in the separation bubble area, and the modal coefficient curve, remain convergent. As such, the DMD in the separation bubble region captures the flow mode in the separation bubble rupture process.

As shown in Figure 14, the nephogram of the DMD velocity vector of the first−order mode is shown, and the PSD of the first−order mode is 6.2 Hz. The first−order mode is the time−average result of the separation bubble region during the transient wall separation of the jet.

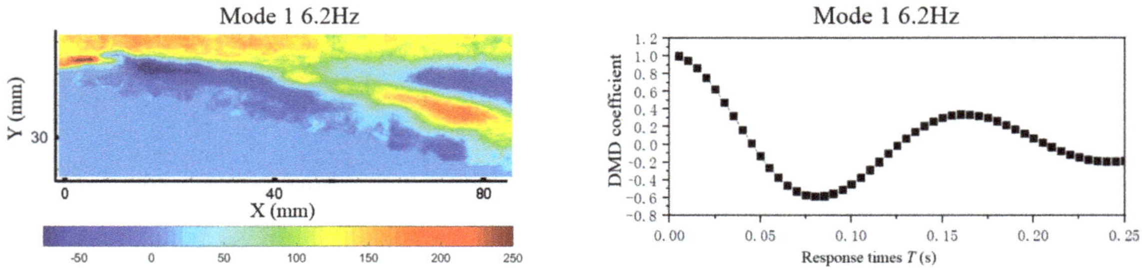

(a) (b)

Figure 14. First−order mode PSD = 6.2 Hz. (**a**) Decomposition results of the PIV velocity nephogram modal DMD in the Region A; (**b**) the DMD modal coefficient curve in the Region A.

In Figure 15a the nephogram of the DMD velocity vector of third−order mode is shown, and the PSD of third−order mode is 78.16 Hz. The high−value watershed of the third−order mode is similar to the separation shedding vortex structure. In Figure 15b, the attenuation degree of the modal coefficient curve of the third−order mode is significantly higher than that of other modes, indicating that the flow region represented by the third−order mode tends to be stable at the earliest. Therefore, the third−order mode is characterized as a separation shedding vortex structure.

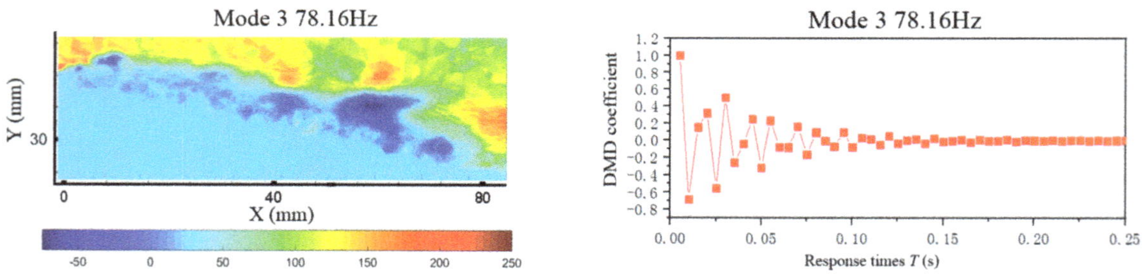

(a) (b)

Figure 15. Third−order mode PSD = 78.16 Hz. (**a**) Decomposition results of the PIV velocity nephogram modal DMD in the Region A; (**b**) the DMD modal coefficient curve in the Region A.

In Figure 16a, the nephogram of the DMD velocity vector of the fourth−order mode is shown, and the PSD of fourth mode is 89.45 Hz. The high−magnitude area of the fourth−order mode is the velocity region at the trailing edge of the lower wall of the nozzle. In Figure 16b, the modal coefficient curve of the fourth mode has a specific periodic frequency, indicating that the flow represented by the fourth−order mode still exists after the separation bubble breaks. Combined with the high magnitude location, the fourth−order mode characterizes the backward suction vortex at the trailing edge of the lower wall of the nozzle. The magnitude of the velocity of the inverted vortex is shallow, which does not affect the overall flow.

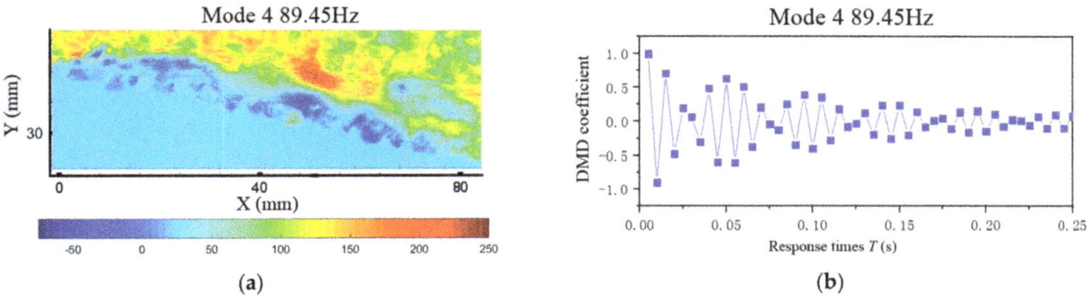

Figure 16. Fourth–order mode PSD = 89.45 Hz. (**a**) Decomposition results of the PIV velocity nephogram modal DMD in the Region A; (**b**) the DMD modal coefficient curve in the Region A.

As shown in Figure 17a, the nephogram of the DMD velocity vector of the fifth–order mode is shown, and the PSD of the fifth–order order mode is 97.83 Hz. The high–magnitude area of the fifth–order mode is the vortex region outside of the jet shear layer and inside the separation bubble. In Figure 17b, the attenuation degree of the modal coefficient curve of the fifth mode is linear. This linear attenuation trend is not consistent with the flow in the process of jet transient wall separation. Combined with the high magnitude location, the fifth–order mode characterizes the momentum transferred from the outer shear layer of the jet to the separation bubble. When the jet is attached to the wall in a steady state, the energy of the separation bubble structure is transferred from the momentum of the shear layer outside of the jet to the inside of the separation bubble structure. When the jet transiently leaves the wall, the momentum of the shear layer outside of the jet has no transmission medium, resulting in the gradual attenuation of the original momentum inside the separation bubble. Therefore, it can be judged that the fifth–order mode is characterized by the momentum of the shear layer outside of the jet. Combined with the above analysis, the control frequencies of the pulsed jet as 6.2 Hz and 78.16 Hz are selected. The two frequencies represent the global flow frequency and the separation bubble structure frequency in the jet separation process, respectively.

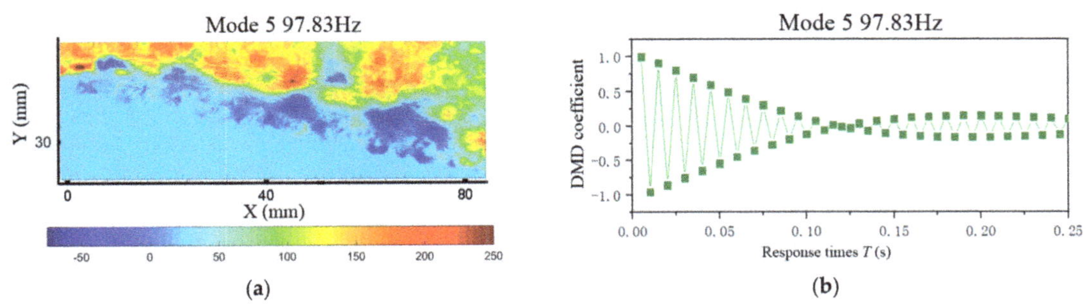

Figure 17. Fifth–order mode PSD = 97.83 Hz. (**a**) Decomposition results of the PIV velocity nephogram modal DMD in the Region A; (**b**) the DMD modal coefficient curve in the Region A.

3.4. Active Flow Control Results

Each mode frequency is used to control the jet transient vector deflection process to verify whether the control frequency of the pulsed jet is effective. The pulsed jet exciter characteristics determine the velocity of the pulsed jet.

Figure 18a shows the pressure change curve after adding the mode's PSD frequency pulsed jet in the separation bubble area. In the figure, the first–order mode 6.2 Hz and the third–order mode 78.16 Hz play the role of delay. The third–order mode characterizes the shedding vortex in the separation bubble, which shows that the characteristic frequency of the vortices in the separation bubble can inhibit the jumping phenomenon. The figure's pressure change curves of the third, fourth and fifth modes will decrease by a specific order of magnitude after the jet completes the detachment deflection process. The third, fourth, and fifth order excitation frequencies are the same as the flow frequency inside the separation bubble, so the flow structures with different frequencies in the separation bubble resonate and generate a certain negative pressure. The negative pressure causes the pressure to drop after the jet is neutral. Figure 18b shows the variation curve of the pressure differential. The figure shows that the pulsed excitation frequency of 78.16 Hz of the third−order mode can change the high amplitude single peak curve of the original working condition into a low amplitude three−peak curve. Furthermore, the curve amplitude of the third−order mode is lower than that of the first−order mode, indicating that the pressure change controlled by the third−order mode is more linear.

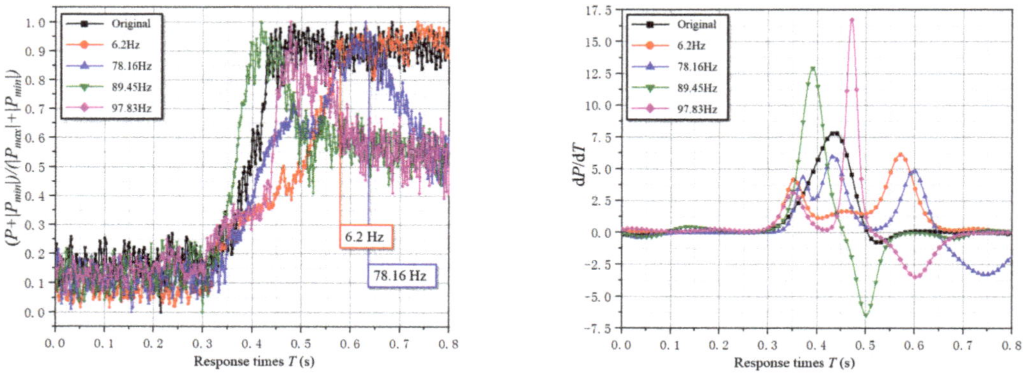

(a) (b)

Figure 18. Pressure change curve under the condition of the jet transient detachment process condition. (**a**) Adding mode PSD frequency pulse jet in the separation bubble area; (**b**) pressure differential change curve in the separation bubble area.

Figure 19a compares the vector force variation curves of the first−order modal 6.2 Hz condition and the third−order modal 78.16 Hz condition. The linearity analysis is carried out for the three working conditions. The results show that the original working condition R^2 = 0.7591, the first−order modal 6.2 Hz working condition R^2 = 0.9329, and the third−order modal 78.16 Hz working condition R^2 = 0.9637. The linear results show that the separation bubble structure excited by the *PSD* frequency pulsed jet with third−order mode has the best inhibition effect on the jump phenomenon. Figure 19b compares the differential variation curves of a vector force. Under the original condition, the high amplitude point appears in the early stage of the jet wall separation process. At this time, the separation bubble will break in the early stage of the jet dynamic wall separation. Adding the pulsed jet with the first−order modal frequency delays the high amplitude point to the later stage of the jet wall separation process. This phenomenon shows that although the pulse jet with the first−order modal frequency can delay the rupture time of the separation bubble structure, it cannot avoid the jumping phenomenon when the separation bubble structure breaks. The absolute amplitude of the vector force differential curve decreases significantly after adding the pulsed jet with third−order modal frequency. To further clarify the action mechanism of 78.16 Hz, the pressure changing with time in the S1 section is analyzed. According to the data results of the vortex jet experiment, the vector deflection angle control equation and linear correlation square R^2 of the nozzle fit. The calculation formula of phenomenon correlation coefficient R is:

$$R = \frac{\sum_{i=1}^{n}(x_i - \overline{x})(y_i - \overline{y})}{\sqrt{\sum_{i=1}^{n}(x_i - \overline{x})^2 \sum_{i=1}^{n}(y_i - \overline{y})^2}} \quad (14)$$

where: x_i and y_i are the coordinates of the appropriate point i, respectively, and $\overline{x}$ and $\overline{y}$ are the average of x_i and y_i, respectively.

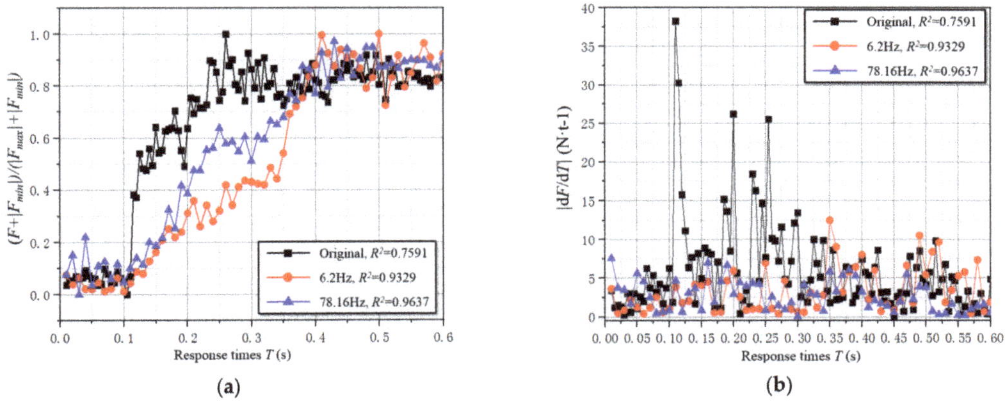

Figure 19. Vector force variation curve under the jet transient vector detachment process condition. (**a**) Comparison of original condition, 6.2 Hz condition, and 78.16 Hz condition; (**b**) vector force differential variation curve.

Figure 20a shows the dynamic pressure change cloud diagram in section S1 under the original working condition. When T = 0, the jet is completely attached to the wall. The blue low pressure represents the structural scale of the separation bubble. It can be observed from the figure that the separation bubble scale gradually grows with time, and disappears at a particular moment. This phenomenon describes the viscous tension generated when the main jet vector deflects the separation bubble structure. When the separation bubble bears enough tension, it will break instantly and cause a jump. Figure 20b shows the dynamic pressure change cloud diagram on S1 section under the 78.16 Hz working conditions. The figure shows that the blue low−pressure range decays linearly with time, and does not disappear instantaneously. This phenomenon indicates that the addition of a pulsed jet with the same frequency as the vortex structure in the separation bubble can enhance the structural strength of the separation bubble. After the separation bubble structure is subjected to the viscous tension generated by the vector deflection of the main jet, the separation bubble can avoid cracking, at the cost of reducing its volume.

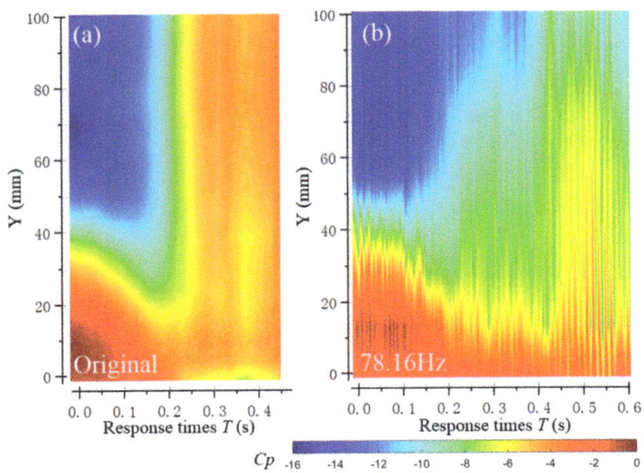

Figure 20. Nephogram of dynamic pressure change on S1 section. (**a**) Original working condition; (**b**) adding 78.16 Hz pulsed jet.

To sum up, the 78.16 Hz working condition has reached the hypothetical goal of this paper. The DMD method accurately obtains the characteristic frequency required in active flow control. The third−order modal results obtained by DMD in the separation bubble region characterize the vortex structure in the separation bubble structure. The active flow control method of a 78.16 Hz pulsed jet can effectively suppress the jump phenomenon and make the vector deflection angle velocity of the jet change linearly.

4. Conclusions

This paper investigates the vectoring deflection jumping phenomenon of the Coanda effect nozzle experimentally and numerically. The PIV technique is characterized the dynamic characteristics of jet deflection flow. According to the force measurement, pressure measurement, and DMD analysis of PIV measurement results, the phenomenon of jet deflection jump and the evolution law of vortex in a dynamic jet deflection detachment condition are investigated and discussed. Finally, according to the control object, control position, and control frequency obtained from the analysis results, the flow field's active flow control is carried out. According to the experimental model and conditions, the results are as follows:

The jumping phenomenon in the Coanda effect nozzle is caused by the different stability of the jet detachment condition and the jet attachment condition. The stability of detachment is steady, while the stability of the attachment flow is periodic. The vector deflection process is a process of mutual transformation between steady and periodic stable systems. The stability of the jet attachment is worse than that of the jet detachment state, so the jumping phenomenon will occur when the jet leaves the Coanda wall.

The DMD method can accurately analyze the jumping phenomenon in the unsteady flow field under the working conditions of this paper. The DMD method can extract the dominant frequency and the corresponding flow field modal structure in the unsteady flow field. The modal stability analysis of PIV results can accurately extract the shedding characteristic frequency of separated bubble structures. The DMD method provides a technical means for analyzing the flow field structure of the unsteady flow.

In the experimental conditions of this paper, by injecting a 78.16 Hz pulsed jet into the Coanda wall, the jumping phenomenon can be effectively suppressed, and the angular velocity of the jet vector changes linearly. The pulsed jet injection control position can effectively enhance the strength of the separation bubble structure and avoid the sudden rupture of the separation bubble structure. This active flow control method can provide a technical means for designing a new fluid thrust vectoring nozzle, by finding the characteristic frequency form of flow structure in the unsteady flow field.

Author Contributions: Funding acquisition, Y.G.; Writing—original draft, S.C. All authors have read and agreed to the published version of the manuscript.

Funding: This work was supported by A Project Funded by the Priority Academic Program Development of Jiangsu Higher Education Institutions, the National Natural Science Foundation of China, No. 11972017.

Conflicts of Interest: The authors declare no conflict of interest.

References

1. Hunter, C.A. Experimental, theoretical, and computational investigation of separated nozzle flows. In Proceedings of the 34th AIAA/ASME/SAE/ASEE Joint Propulsion Conference and Exhibit, Cleveland, OH, USA, 13–15 July 1998; p. 3107.
2. Flamm, J.D. Experimental study of a nozzle using fluidic counterflow for thrust vectoring. In Proceedings of the 34th AIAA/ASME/SAE/ASEE Joint Propulsion Conference and Exhibit, Cleveland, OH, USA, 13–15 July 1998; p. 3255.
3. Mason, M.S.; Crowther, W.J. Fluidic thrust vectoring for low observable air vehicles. In Proceedings of the 2nd AIAA flow control conference, Portland, OR, USA, 28 June–1 July 2004; p. 2210.
4. Banazadeh, A.; Saghafi, F.; Uhoreyshi, M.; Pilidis, P. Multi−directional co−flow fluidic thrust vectoring intended for a small gas turbine. In Proceedings of the AIAA Infotech@Aerospace 2007 Conference and Exhibit, Rohnert Park, CA, USA, 7–10 May 2007; p. 2940.

Wang, L.; Wong, C.W.; Lu, Z.; Wu, Z.; Zhou, Y. Novel sawtooth dielectric barrier discharge plasma actuator for flow separation control. *AIAA J.* **2017**, *55*, 1405–1416. [CrossRef]

Trancossi, M.; Maharshi, S.; Angeli, D. Mathematical modelling of a two streams Coanda effect nozzle. In *ASME International Mechanical Engineering Congress and Exposition*; American Society of Mechanical Engineers: New York, NY, USA, 2013; Volume 56178, p. V001T01A052.

Trancossi, M.; Dumas, A.; Das, S.S.; Pascoa, J. Design methods of Coanda effect nozzle with two streams. *Incas Bull.* **2014**, *6*, 83.

Trancossi, M.; Stewart, J.; Maharshi, S.; Angeli, D. Mathematical model of a constructal Coanda effect nozzle. *J. Appl. Fluid Mech.* **2016**, *9*, 2813–2822. [CrossRef]

Bourque, C.; Newman, B.G. Reattachment of a two−dimensional Incompressible jet to an Adjacent Flat Plate. *Aeronaut. Q.* **1960**, *11*, 201–232. [CrossRef]

0. Lai, J.C.S.; Lu, D. An inclined wall jet: Mean flow characteristics and effects of acoustic excitation. *Exp. Fluids* **2000**, *29*, 45–55. [CrossRef]

1. Nasr, A.; Lai, J.; Siah, O.; Young, J. Flow Characteristics of an Inclined Offset Jet. In Proceedings of the 36th AIAA Fluid Dynamics Conference, San Francisco, CA, USA, 5–8 June 2006.

2. Pramanik, S.; Kumar Das, M. Computational study of a turbulent wall jet flow on an oblique surface. *Int. J. Numer. Methods Heat Fluid Flow* **2014**, *24*, 290–324. [CrossRef]

3. Zaitsev, E.G. Investigation of the propagation of a two−dimensional wall jet over a step. *Fluid Dyn.* **1991**, *26*, 526–530. [CrossRef]

4. Allery, C.; Guerin, S.; Hamdouni, A.; Sakout, A. Experimental and numerical POD study of the Coanda effect used to reduce self−sustained tones. *Mech. Res. Commun.* **2004**, *31*, 105–120. [CrossRef]

5. Gillgrist, R.D.; Forliti, D.J.; Strykowski, P.J. On the Mechanisms Affecting Fluidic Vectoring Using Suction. *J. Fluids Eng.* **2015**, *129*, 91–99. [CrossRef]

6. Allery, C.; Beghein, C.; Hamdouni, A. On investigation of particle dispersion by a POD approach. *Int. Appl. Mech.* **2008**, *44*, 110–119. [CrossRef]

7. Miozzi, M.; Lalli, F.; Romano, G.P. Experimental investigation of a free−surface turbulent jet with Coanda effect. *Exp. Fluids* **2010**, *49*, 341–353. [CrossRef]

8. Cornelius, K.C.; Lucius, G.A. Physics of Coanda jet detachment at high−pressure ratio. *J. Aircr.* **1994**, *31*, 591–596. [CrossRef]

9. Rega, G.; Troger, H. Dimension reduction of dynamical systems: Methods, models, applications. *Nonlinear Dyn.* **2005**, *41*, 1–15. [CrossRef]

0. Lumley, J.L. The structure of inhomogeneous turbulent flows. *Atmos. Turbul. Radio Wave Propag.* **1967**, *6*, 166–178.

1. Berkooz, G.; Holmes, P.; Lumley, J.L. The proper orthogonal decomposition in the analysis of turbulent flows. *Annu. Rev. Fluid Mech.* **1993**, *25*, 539–575. [CrossRef]

2. Sirovich, L. Turbulence and the dynamics of coherent structures. I. Coherent structures. *Q. Appl. Math.* **1987**, *45*, 561–571. [CrossRef]

3. Deane, A.E.; Kevrekidis, I.G.; Karniadakis, G.E.; Orszag, S. Low−dimensional models for complex geometry flows: Application to grooved channels and circular cylinders. *Phys. Fluids A Fluid Dyn.* **1991**, *3*, 2337–2354. [CrossRef]

4. Cao, N.Z.; Aubry, N. *Numerical Simulation of a Wake Flow via a Reduced System*; ASME−PUBLICATIONS−FED: Montenegro, 1993; Volume 149, p. 53.

5. Schmid, P.J. Application of the dynamic mode decomposition to experimental data. *Exp. Fluids* **2011**, *50*, 1123–1130. [CrossRef]

6. Chen, K.K.; Tu, J.H.; Rowley, C.W. Variants of Dynamic Mode Decomposition: Boundary Condition, Koopman, and Fourier Analyses. *J. Nonlinear Sci.* **2012**, *22*, 887–915. [CrossRef]

7. Jovanovi, M.R.; Schmid, P.J.; Nichols, J.W. Sparsity−promoting dynamic mode decomposition. *Phys. Fluids* **2014**, *26*, 561–571. [CrossRef]

8. Sajadmanesh, S.M.; Mojaddam, M.; Mohseni, A.; Nikparto, A. Numerical identification of separation bubble in an ultra−high−lift turbine cascade using URANS simulation and proper orthogonal decomposition. *Aerosp. Sci. Technol.* **2019**, *93*, 105329. [CrossRef]

9. Sajadmanesh, S.M.; Mohseni, A.; Mojaddam, M. Vortex dynamics mechanisms of separated boundary layer in a highly loaded low pressure turbine cascade. *Int. J. Heat Fluid Flow* **2020**, *82*, 108540. [CrossRef]

mathematics MDPI

Article

Different Modes of Combustion Wave on a Lattice Burner

Vasily B. Novozhilov [1,*], Boris V. Lidskii [2] and Vladimir S. Posvyanskii [2]

1 Institute of Sustainable Industries and Liveable Cities, Victoria University, Melbourne, VIC 8001, Australia
2 N.N. Semenov Federal Research Centre for Chemical Physics, Russian Academy of Sciences, 4 Kosygina St., Building 1, Moscow 119991, Russia; fishgoit@mail.ru (B.V.L.); vsposv@yandex.ru (V.S.P.)
* Correspondence: vasily.novozhilov@vu.edu.au

Abstract: The stabilization of a planar premixed flame front on a lattice (porous) burner is considered. The developed model captures all the important features of the phenomenon, while also admitting qualitative analytical investigation. It has been rigorously mathematically proven that there exist two different stabilization regimes: one with flame front located nearby the surface of the burner and another with the flame front located inside the lattice. These two regimes result in qualitatively different gas temperature profiles along the flow that is monotonic and non-monotonic, respectively. The boundary between the two regimes is described in terms of dependence of the lattice solid material temperature on flow Peclet number. With similar temperature profiles, such dependencies may be both monotonic and non-monotonic. The transition between the two types of dependencies is controlled by the Arrhenius number. Conclusions of the study are supported by numerical analysis. They also compare favorably with the available experimental data. The novelty of the present approach is a fundamentally rigorous analytical analysis of the problem. The proposed analytical model, based on δ-function approximation of the chemical source term, agrees well (within 7% relative error) with the model based on the distributed description of the chemical reaction zone. The obtained results are important from both a theoretical and practical point of view. They demonstrate the existence of the two qualitatively different operating regimes for lattice burners, thus impacting design solutions for such devices. The results will be of great interest to the broader academic community, particularly in research areas where similar wave structures may emerge.

Keywords: premixed lattice burner; flame stabilization; analytical model; δ-function approximation; rigorous mathematical proof

MSC: 34B60

Citation: Novozhilov, V.B.; Lidskii, B.V.; Posvyanskii, V.S. Different Modes of Combustion Wave on a Lattice Burner. *Mathematics* **2022**, *10*, 2731. https://doi.org/10.3390/math10152731

Academic Editor: Yang Liu

Received: 23 June 2022
Accepted: 1 August 2022
Published: 2 August 2022

1. Introduction

Combustion waves are a natural way of flame spread. The most basic of such structures is the laminar pre-mixed flame propagating through a pipe or channel [1–4]. The most common dominant mechanism of such propagation is heat conduction from the flame front to the adjacent pre-heat zone, raising the temperature of the latter to values supporting a significant increase in the mixture reaction rate. This type of wave structure is known as a *thermal flame*. An alternative driving mechanism is the diffusion propagation of active radicals away from the reaction zone, which may support the flame being essentially driven by chain reactions, or the so-called *chain flame*.

A very specific standing combustion wave emerges around the burning fuel droplet [1]. More exotic examples are represented by *cold* or *isothermal* flames [3].

Flame fronts may also propagate through purely solid phase upon availability of both a solid-state reagent and oxidizer. The most important of such processes are the combustion of solid propellants [5], and the so-called *Self-Propagating High-Temperature Synthesis* (SPHTS) [4,6]. Combustion propagation in these cases is driven exclusively by heat conduction, with no convective effects (in the laboratory frame). Despite considerable simplification of the heat transfer mechanism, the flames of propellants, for example, are known to be capable of exhibiting very complicated dynamics, including chaotic dynamics [5].

A peculiar combustion wave is observed in fires, where propagation occurs in two-phase media, involving heat feedback from the gaseous diffusion flame to pyrolyzing solid material, which, in turn, supplies reactants back to flame [7,8].

The investigation of combustion waves with homogeneous multi-phase heat exchange within the wave structure has been somewhat more recent. One type of this process is convective combustion, where reacting gases interact with either the combustible or inert matrix of porous material [9–14].

Another is standing a combustion wave developing on a lattice (porous) burner (Figure 1). The latter device is of significant importance, both theoretically and practically.

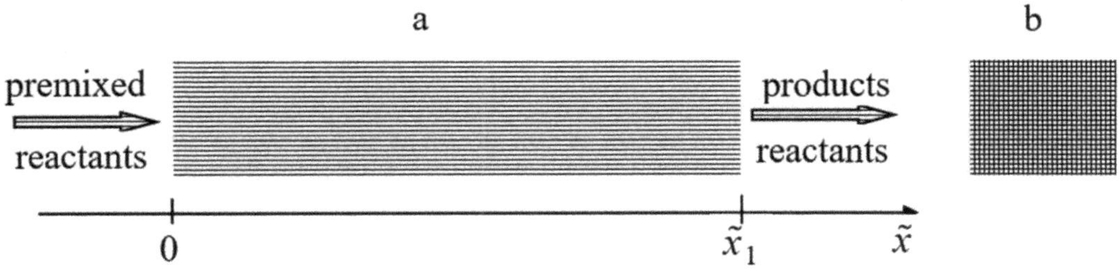

Figure 1. Schematic of the lattice burner considered: (**a**) side view (cross-section) and (**b**) front (upstream) view.

The combustion process in porous burners has been studied extensively [15–35], using both experimental and numerical modelling methods, with respect to different burner configurations, combustion regimes and employed fuels (for example, biofuels).

Barra et al. [31] investigated the effects of material properties on flame stabilization in porous burners. Significant influence of thermal conductivity, the volumetric heat transfer coefficient and the radiative extinction coefficient on stable operation limits were observed.

Barra and Ellzey [32] studied the heat recirculation process, due to solid matrix, for a range of equivalence ratios. They found that recirculation efficiency decreases with the increase of the equivalence ratio.

Djordjevic et al. [35] investigated flame stability in porous burners employing various ceramic sponge-like structures. They proposed a simple criterion for prediction of blow-off limits in combustion systems employing the porous burner concept.

Of particular interest are the results of Yakovlev et al. [15], Mital et al. [17] and Janvekar et al. [27], who observed (both numerically and experimentally) non-monotonic temperature profiles within the burner. Keshtkar et al. [34] also observed non-monotonic temperature profiles in the course of their numerical analysis of rectangular two-dimensional porous radiant burners.

Existence of both monotonic and non-monotonic temperature profiles is a key focus of the present study.

Recently, Arutyunov et al. [36], using numerical methods, investigated the nature of the upper limit of methane–air mixture combustion on a flat porous lattice. They demonstrated that the position of the flame front, relative to the lattice surface, depends strongly on the heat exchange rate between the gas and the solid lattice material. Within certain ranges of fuel preheating and injection rates unstable combustion regimes were observed. Further, Shmelev [37] established, numerically, the region of stable combustion for the same system. It was shown that this region covers a wide range of gas injection velocities and expands with increase of the width of the burner.

Despite the substantial volume of research referenced above, rigorous analytical studies of such a combustion system have not been conducted.

The present paper investigates combustion wave structure on a lattice (that is, on a laminar premixed porous burner).

The novelty and contribution of the study are as follows:

- Development of tractable (admitting analytical qualitative investigation) model of combustion wave stabilizing on a lattice (porous) burner;
- Development of approximated model, based on the description of chemical source term with δ-function, admitting an exact analytical solution, and the explicit construction of such a solution;
- Rigorous mathematical proof of existence of the two distinctive combustion wave modes, with monotonic and non-monotonic temperature profiles, respectively;
- Definition of critical curve in the space of control parameters, separating the two wave modes. Rigorous mathematical proof of existence of both monotonic and non-monotonic critical curves.

The existence of double-mode wave structure, rigorously proven in the present paper agrees qualitatively with experimental evidence, as discussed below. This finding impacts design solutions for industrial porous burners.

The paper is organized in the following way. The mathematical formulation of the problem is developed in Section 2. The results and discussion are presented in Section 3 (this section is separated into formal mathematical proof (Section 3.1), numerical illustration of the results (Section 3.2), analysis of the critical curve separating the two combustion regimes (Section 3.3) and comparison with experimental data (Section 3.4)). These sections are followed by the Conclusion Section, References and Appendix A.

2. Mathematical Formulation

Assuming one-dimensional flow structure and uniform material properties (of both the gas and the solid lattice), the following steady-state Heat Transfer Equation (HTE) is considered to model the combustion process (we remind that waved variables are dimensional; see subsection Ascents in the Notation table).

$$\widetilde{\lambda}\frac{d^2\widetilde{T}}{d\widetilde{x}^2}-\widetilde{\rho}\widetilde{c}_p\widetilde{u}\frac{d\widetilde{T}}{d\widetilde{x}}+\widetilde{Q}\widetilde{A}\widetilde{C}^2\left(\widetilde{T}-\widetilde{T}_0\right)^{1/2}\exp\left(-\frac{\widetilde{E}}{\widetilde{R}\widetilde{T}}\right)+\widetilde{h}\widetilde{S}\left(\widetilde{T}_s-\widetilde{T}\right)=0 \quad (1)$$

$$\widetilde{h}=\begin{cases}\widetilde{h}_s>0;\ \widetilde{x}\in[0,\widetilde{x}_1]\\ 0;\ \widetilde{x}\in(\widetilde{x}_1,\infty]\end{cases}$$

Flow is directed from left to right, and the lattice occupies the region from 0 to $\widetilde{x}_1$ (Figure 1).

The assumption of one-dimensional flow structure is standard in analysis of various types of combustion waves (see, for example, Zeldovich et al. [3]) and is known to lead to quantitatively correct and verifiable results. The major reason behind the assumption of uniform (e.g., temperature-independent) thermophysical properties is that is allows formal mathematical proof, presented below, to be conducted. Note, however, that this assumption is very reasonable everywhere in the flow, except for extremely narrow chemical reaction zone, as the temperature outside this zone does not deviate much from the inflow temperature.

The third term on the Left-Hand Side (LHS) of Equation (1) describes the contribution from a bimolecular reaction occurring in the premixed stream. The concentrations $\widetilde{C}_1$ and $\widetilde{C}_2$ of both reactants are assumed to be equal at all times, i.e., $\widetilde{C}_1(\widetilde{x})=\widetilde{C}_2(\widetilde{x})=\widetilde{C}(\widetilde{x})$. This means that combustion is being considered as stoichiometric. This assumption is made to allow for more concise mathematical proof but is not binding. The proof may be modified to remove this assumption. The fourth term describes heat transfer process between the gas and the solid lattice, with the heat transfer coefficient being assumed constant.

The chemical reaction rate is written in the form that makes it vanish at the initial temperature $\widetilde{T}_0$. This is a standard assumption in the combustion theory.

The kinetic equation is

$$-\tilde{u}\frac{d\tilde{C}}{d\tilde{x}} = \tilde{A}\tilde{C}^2\left(\tilde{T}-\tilde{T}_0\right)^{1/2}\exp\left(-\frac{\tilde{E}}{\tilde{R}\tilde{T}}\right) \tag{2}$$

Further simplification, which is being made is that variable concentration $\tilde{C}$, is replaced on the Right-Hand Side (RHS) of Equation (2) by its initial value $\tilde{C} \equiv \tilde{C}_0 \equiv$ const. This is reasonable in the view of much stronger dependence on reaction rate on temperature, compared to dependence on reactant concentrations.

The simplified kinetic equation takes the form

$$-\tilde{u}\frac{d\tilde{C}}{d\tilde{x}} = \tilde{A}\tilde{C}_0^2\left(\tilde{T}-\tilde{T}_0\right)^{1/2}\exp\left(-\frac{\tilde{E}}{\tilde{R}\tilde{T}}\right) \tag{3}$$

The following non-dimensional (scaling) variables are introduced

$$\xi = \frac{\tilde{x}}{\tilde{x}_1}\theta = \frac{\tilde{T}-\tilde{T}_0}{\left(\tilde{R}\tilde{T}_0^2/\tilde{E}\right)}\ Pe = \frac{\tilde{x}_1\tilde{u}}{\tilde{\kappa}}\ C = \frac{\tilde{C}}{\tilde{C}_0}$$

$$A = \tilde{Q}\tilde{A}\tilde{C}_0^2\frac{\tilde{x}_1^2}{\tilde{\lambda}}\left(\frac{\tilde{R}\tilde{T}_0^2}{\tilde{E}}\right)^{-1/2}\cdot\exp\left(-\frac{\tilde{E}}{\tilde{R}\tilde{T}_0}\right);\ h = \frac{\tilde{h}\tilde{S}\tilde{x}_1^2}{\tilde{\lambda}} \tag{4}$$

$$g = \frac{\tilde{x}_1^2}{\tilde{\kappa}}\tilde{A}\tilde{C}_0\left(\frac{\tilde{R}\tilde{T}_0^2}{\tilde{E}}\right)^{1/2}\cdot\exp\left(-\frac{\tilde{E}}{\tilde{R}\tilde{T}_0}\right)$$

The choice of the spatial scale (lattice width) and the reactant concentration scale (initial concentration) is natural. The temperature scale is chosen in the way that is commonly adopted (and reflecting most important scale) in combustion theory (see, for example, Merzhanov, and Khaikin [4]), as the ratio of excess temperature to characteristic temperature interval. Once these three scales are fixed, the other scales in Equation (4) are determined uniquely (to within an explicit inclusion of the reaction rate at initial temperature into the parameter A, which is also a common practice). Note that the emerging Peclet number Pe is a standard parameter in the analysis of convective heat transfer problems.

The problem formulation becomes

$$\frac{d^2\theta}{d\xi^2} - Pe\frac{d\theta}{d\xi} + Af(\theta)C^2(\theta)H(C) + h(\theta_s - \theta) = 0$$

$$\frac{dC}{d\xi} = -\frac{g}{Pe}f(\theta)H(C) \tag{5}$$

$$f(\theta) = \sqrt{\theta}\exp\left(\frac{\theta}{1+Ar\theta}\right)$$

$$0 \le \xi \le L$$

$$\theta_s > 0,\ h = \begin{cases} h_s > 0;\ \xi \in [0,1] \\ 0;\ \xi \in (1,L] \end{cases}$$

with the boundary conditions

$$\theta(0) = \theta_0,\ C(0) = 1,\ \frac{d\theta}{d\xi}(L) = 0 \tag{6}$$

where $L > 0$ is an arbitrarily large number.

Here, for the convenience of mathematical proofs, we assume that the temperatu at the left boundary is slightly different from zero, i.e., $\theta_0 > 0$ is an arbitrarily sma positive number.

3. Results and Discussion

3.1. Types of Solutions

First of all, we demonstrate existence of solutions of the set of Equation (5) with th boundary conditions in Equation (6).

This follows from the few Lemmas proved below.

Let us consider the set of Equation (5) with the boundary conditions

$$\theta(0) = \theta_0,\ C(0) = 1,\ \frac{d\theta}{d\xi}(0) = p \tag{7}$$

where p may assume any real value.

Define ξ_k in the following way: $\xi_k = L$ if $\theta(\xi) \neq 0$ on $[0, L)$; otherwise ξ_k is such tha $\theta(\xi_k) = 0$ and $\theta(\xi) \neq 0$ on $[0, \xi_k)$.

Lemma 1. *ξ_k, $\theta(\xi_k)$ and $\frac{d\theta}{d\xi}(\xi_k)$ are continuous functions of p, where p is any real value.*

Proof. The statement is correct for $\xi_k \leq 1$ since the solution of the Cauchy problem depend continuously on the initial conditions.

If $\xi_k > 1$, then $\theta(1)$ and $\frac{d\theta}{d\xi}(1)$ depend on p continuously. Therefore, ξ_k, $\theta(\xi_k)$ an $\frac{d\theta}{d\xi}(\xi_k)$ are also continuous functions of p. □

Lemma 2. *For any p_s there exists p such that for some $\xi_s \theta(\xi_s) = \theta_s$ and $\frac{d\theta}{d\xi}(\xi_s) > p_s$.*

Proof. Re-write the set of Equation (5) in the form

$$\frac{d^2\theta}{d\xi^2} - Pe\frac{d\theta}{d\xi} - A\frac{Pe}{g}\frac{dC}{d\xi}C^2(\theta) + h(\theta_s - \theta) = 0 \tag{8}$$

$$\frac{dC}{d\xi} = -\frac{g}{Pe}f(\theta)H(C)$$

and integrate the first equation on $[0, \xi]$.

We get

$$\frac{d\theta}{d\xi}(\xi) = p + Pe(\theta(\xi) - \theta_0) - \frac{APe}{3g}\left(1 - C^3H(C)\right) - \int_0^{\xi} h_s(\theta_s - \theta)d\xi \tag{9}$$

Therefore,

$$\frac{d\theta}{d\xi}(\xi) > p - \frac{APe}{3g} - \xi h_s\theta_s \tag{10}$$

Let us choose now

$$p > p_s + \frac{APe}{3g} + \frac{h_s\theta_s^2}{p_s} \tag{11}$$

Then, for $\xi \in \left[0, \frac{\theta_s}{p_s}\right]$,

$$\frac{d\theta}{d\xi}(\xi) > p_s;\ \theta\left(\frac{\theta_s}{p_s}\right) > \theta_s \tag{12}$$

□

Lemma 3. *$\theta(\xi)$increase monotonically if $p_s > \frac{APe}{3g}$.*

Proof. Integrating Equation (8) on $[\xi_s, \xi]$ we get

$$\frac{d\theta}{d\xi}(\xi) = p_s + Pe(\theta(\xi) - \theta_s) - \frac{APe}{3g}\left(1 - C^3H(C)\right) + \int_{\xi_s}^{\xi} h_s(\theta(\alpha) - \theta_s)d\alpha \qquad (13)$$

□

Lemma 4. *There exists p_0 such that $\xi_k = L$ and $\theta(\xi_k) = 0$.*

Proof. Let $q = \xi_k - L$ if $\xi_k < L$, and $q = \theta(\xi_k)$ if $\xi_k = L$. Then q is the continuous function of p, which assumes both positive and negative values. □

Theorem 1 (Existence of solution). *There exists a solution of the set of Equation (5) with the boundary conditions in Equation (6).*

Proof. $\frac{d\theta}{d\xi}(L)$ is a continuous function of p on $[p_0, +\infty)$. Since $\frac{d\theta}{d\xi}(L) \le 0$ at $p = p_0$, and there exists p such that $\frac{d\theta}{d\xi}(L) > 0$, then $\frac{d\theta}{d\xi}(L) = 0$ at some $p \in [p_0, +\infty)$. □

In the following analysis of the set of Equation (5), $\mu = \frac{A}{3g}$ will be assumed to be a large parameter.

Let us define ξ_0 in the following way:

$$C(\xi_0) = 0,\ C(\xi) \neq 0 \text{ on } [0, \xi_0) \qquad (14)$$

The following three cases are possible

(1) ξ_0 does not exist ($C(\xi) \neq 0$ on $[0, L)$);
(2) $1 < \xi_0 \le L$;
(3) $0 < \xi_0 \le 1$.

Consider first the case $1 < \xi_0 \le L$.

For $\xi_0 > 1$ the set of Equation (5) assumes the form

$$\frac{d^2\theta}{d\xi^2} - Pe\frac{d\theta}{d\xi} - 3\mu Pe\frac{dC}{d\xi}C^2(\theta)H(C) = 0$$

$$\frac{dC}{d\xi} = -\frac{g}{Pe}f(\theta)H(C) \qquad (15)$$

Lemma 5. *For $\xi \in [1, \xi_0]$, $\theta(\xi)$ does not decrease; for $\xi \in [\xi_0, L]$, $\theta(\xi) = \theta(\xi_0) = \theta^*$.*

Proof. This follows easily from Equation (15) and the condition $\frac{d\theta}{d\xi}(L) = 0$. □

Let now $\theta(\xi_0) = \theta^*$. Integrating Equation (15), we get

$$\frac{d\theta}{d\xi} - Pe\,\theta - \mu PeC^3(\xi)H(C) = -Pe\,\theta^* \qquad (16)$$

Let now, for $0 \le \hat{\xi} \le \xi_0$, $u(\hat{\xi}) = \theta^* - \theta(\xi_0 - \hat{\xi})$; $\hat{C}(\hat{\xi}) = C(\xi_0 - \hat{\xi})$.

The set of Equation (15) takes the form

$$\frac{du}{d\hat{\xi}} + Pe\,u = \mu Pe\hat{C}^3$$

$$\frac{d\hat{C}}{d\hat{\xi}} = \frac{g}{Pe}f(\theta^* - u) \qquad (17)$$

$$u(0) = \hat{C}(0) = 0$$

It follows from Equation (17) that $\frac{du}{d\hat{\xi}} < \mu Pe\hat{C}^3$.
Therefore

$$\mu Pe\frac{d\hat{C}}{d\hat{\xi}}\hat{C}^3 > \frac{g}{Pe}\frac{du}{d\hat{\xi}}f(\theta^* - u) \tag{18}$$

Integrating Equation (18) on $[0, \hat{\xi}]$, we obtain

$$\frac{\mu Pe}{4}\hat{C}^4 > \frac{g}{Pe}\int_{\theta^*-u}^{\theta^*} f(\theta)d\theta \tag{19}$$

Let $\alpha = \left.\frac{d\ln f(\theta)}{d\theta}\right|_{\theta=\theta^*}$.

Lemma 6.

$$f(\theta) > \alpha\int_0^{\theta} f(w)dw \tag{20}$$

Proof. Indeed, $\frac{d\ln f(\theta)}{d\theta} = \frac{1}{2\theta} + \frac{1}{(1+Ar\,\theta)^2}$ is monotonically decreasing function of θ. Therefore, for $w < \theta$ $\ln f(w) < \ln f(\theta) + \alpha(w - \theta)$, or $f(w) < f(\theta)\exp(\alpha(w-\theta))$. Integrating the later inequality, we prove Lemma 6. □

Further, define the constant $\hat{C}_0$ by

$$\frac{\mu Pe}{4}\left(\hat{C}_0\right)^4 = \frac{g}{Pe}\int_0^{\theta^*} f(\theta)d\theta \tag{21}$$

Lemma 7.

$$\frac{d\hat{C}}{d\hat{\xi}} > \frac{\alpha\mu Pe}{4}\left(\left(\hat{C}_0\right)^4 - \hat{C}^4\right) \tag{22}$$

Proof. It follows from Lemma 6 that

$$\frac{d\hat{C}}{d\hat{\xi}} > \frac{\alpha g}{Pe}\int_0^{\theta^*-u} f(w)dw = \frac{\alpha g}{Pe}\int_0^{\theta^*} f(w)dw - \frac{\alpha g}{Pe}\int_{\theta^*-u}^{\theta^*} f(w)dw \tag{23}$$

Since $\frac{\alpha g}{Pe}\int_0^{\theta^*} f(w)dw = \frac{\alpha\mu Pe}{4}\left(\hat{C}_0\right)^4$ and, according to Equation (19) $\frac{\alpha\mu Pe}{4}\hat{C}^4 > \frac{\alpha g}{Pe}\int_{\theta^*-u}^{\theta^*} f(w)dw$, then Lemma 7 is proved. □

Lemma 8.

$$\hat{C} > C_0\left(1 - 2\exp\left(\frac{\pi}{2}\right)\exp\left(-\alpha\mu PeC_0^3\hat{\xi}\right)\right) \tag{24}$$

Proof. Consider the following equation

$$\frac{dy}{d\varsigma} = \kappa\left(\left(\hat{C}_0\right)^4 - y^4\right), y(0) = 0 \tag{25}$$

for $\varsigma \geq 0, y < \hat{C}_0$.

Its exact solution is

$$\kappa\varsigma = \frac{1}{4(\hat{C}_0)^3}\ln\left(\frac{\hat{C}_0 + y}{\hat{C}_0 - y}\right) + \frac{1}{2(\hat{C}_0)^3}arctg\left(\frac{y}{\hat{C}_0}\right) \tag{26}$$

Since $y < \hat{C}_0$ $arctg\left(\frac{y}{\hat{C}_0}\right) < \frac{\pi}{4}$ and we obtain

$$4(\hat{C}_0)^3\kappa\varsigma - \frac{\pi}{2} < \ln\left(\frac{\hat{C}_0 + y}{\hat{C}_0 - y}\right) \tag{27}$$

or

$$y > \hat{C}_0\left(1 - \frac{2}{1 + \exp\left(4(\hat{C}_0)^3\kappa\varsigma - \frac{\pi}{2}\right)}\right) \tag{28}$$

Upon comparison with Equation (22) Lemma 8 is proved. □

We actually show that for $\alpha\mu \gg 1$ and $\xi_0 > 1$ (where $h \equiv 0$), the solution of the set of equations

$$\begin{gathered} \frac{d^2\theta}{d\xi^2} - Pe\frac{d\theta}{d\xi} + \mu Pe\delta(\xi - \xi_0) + h(\theta_s - \theta) = 0 \\ \theta(\xi_0) = \theta^* \\ \frac{\mu Pe}{4}C^4(1) = \frac{g}{Pe}\int_0^{\theta^*} f(w)dw \\ \theta(0) = 0\ \frac{d\theta}{d\xi}(L) = 0\ \xi \in [1, L] \end{gathered} \tag{29}$$

is a good approximation for solution of the set of Equation (5) on the interval $\xi \in [1, L]$.

It may be shown by similar arguments that for $0 < \xi_0 < 1$ (where $h > 0$), the solution of the set of Equation (5) is also well approximated by the solution of the set of Equation (29).

The requirements for μ given by the conditions $\alpha\mu \gg 1$ and the third of the conditions in the set of Equation (29) enforces a narrow reaction zone.

Analytical solution of the Boundary Value Problem (BVP) (29) is presented in Appendix A.

This analytical solution delivers two different types of temperature profiles, namely, non-monotonic and monotonic. These occur, for example, for the sets of parameters shown in the captions to Figures 2 and 3, respectively.

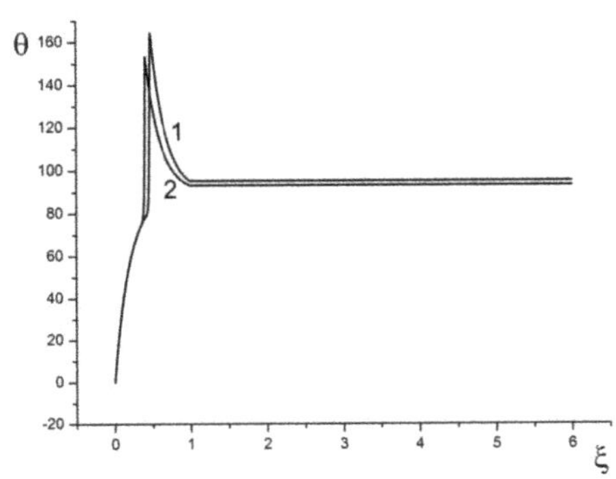

Figure 2. Non-monotonic solutions. 1—approximate model (29); 2—original model (5). Pe = 135; A = 2.75 × 10^{-18}; Ar = 0.012; h_s = 7.14 × 10^2; θ_s = 90; g = 1.0 × 10^{-20}; μ = 91.67; ξ_0 = 0.41.

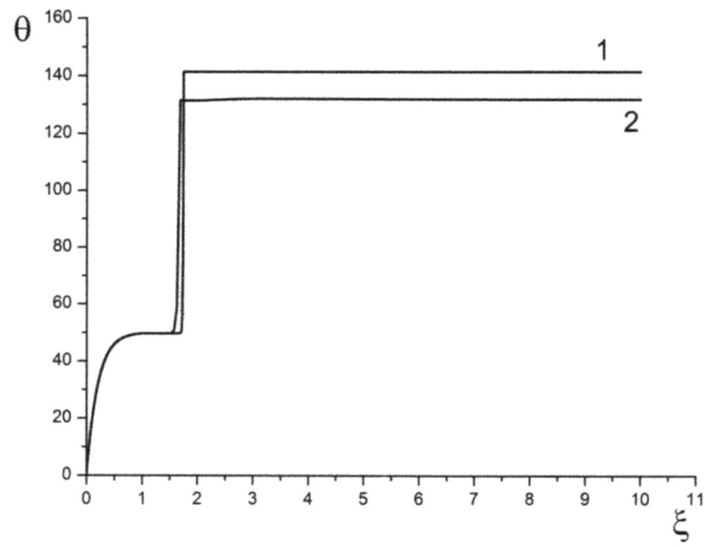

Figure 3. Monotonic solutions. 1—approximate model (29); 2—original model (5); Pe = 13 A = 2.75×10^{-18}; Ar = 0.012; $h_s = 7.14 \times 10^2$; $\theta_s = 50$; g = 1.0×10^{-20}; $\mu = 91.67$; $\xi_0 = 1.74$.

Since the Lemmas 5–8 prove approximation of the set Equation(5) by the set o Equation (29), and the two types of solutions can be demonstrated for the set o Equation (29), then the following Theorem below, the major result of the paper, is provec

Theorem 2 (Existence of the two types of solutions). *The set of Equations (5) and (6) admit two types of temperature profile solutions: monotonic and non-monotonic.*

3.2. Numerical Results

The Boundary Value Problem given by the Equations (5) and (6) is solved numericall using in-house computational code developed at the N.N. Semenov Federal Researcl Centre for Chemical Physics. The false transient (or establishment) method [38] is use with the following numerical scheme

$$\begin{aligned}\frac{\theta_k^{n+1}-\theta_k^n}{\tau} &= \frac{2}{(h_k+h_{k+1})}\left[\frac{\theta_{k+1}^{n+1}-\theta_k^{n+1}}{h_{k+1}} - \frac{\theta_k^{n+1}-\theta_{k-1}^{n+1}}{h_k}\right] - Pe\,\frac{\left(\theta_k^{n+1}-\theta_{k-1}^{n+1}\right)}{h_k} + \\ &+ Af\left(\theta_k^n\right)\left(C_k^n\right)^2 + h_s\left(\theta_s - \theta_k^{n+1}\right)\end{aligned} \quad (30)$$

$$C_k^n = \left(1 - \frac{g}{Pe}\int_0^{\xi_k} f(\theta^n(s))ds\right)H(C_k^n)$$

Iterations are performed until steady-state solution is achieved, with a certain a prior specified accuracy.

Based on existing data [39,40], physically meaningful ranges of the problem parameter are identified as follows:

$$Pe \in [4, 280],$$

$$A \in \left[4.7 \times 10^{-29}, 4.7 \times 10^{-12}\right],$$

$$Ar \in \left[1.2 \times 10^{-2}, 2.0 \times 10^{-2}\right],$$

$$g \in \left[9.7 \times 10^{-29}, 9.84 \times 10^{-12}\right],$$

$$\mu \in [75.0, 208.0],$$
$$\theta_s \in [50.0, 705.0],$$
$$h_s \in \left[7.14 \times 10^2, 1.14 \times 10^3\right]$$

It should be remembered that besides the above restrictions for the parameters A, g and μ, the relation $\mu = \frac{A}{3g}$ must always hold.

Numerical solutions plotted in Figures 2 and 3 confirm good approximation provided by the set of Equation (29) to the set of Equation (5), as was proven earlier in Section 3.1.

Analytical solutions of the approximating set of Equation (29) (Appendix A) show that this problem has non-monotonic solutions for $0 < \xi_0 < 1$ and monotonic for $\xi_0 > 1$.

It is instructive to identify, in the parameter space, the regions corresponding to the two types of solutions.

For this purpose, let us introduce the "critical curve", defined by the condition $\xi_0 = 1$ since the latter separates the two different types of solutions. From Equation (5)

$$1 - C(\xi) = \frac{g}{Pe}\int_0^{\xi} f(\theta(s))ds \tag{31}$$

Therefore, due to definition (14), the requirement $\xi_0 = 1$ translates into the following relationship between the governing parameters θ_s and Pe (with the parameter g being considered fixed)

$$\frac{g}{Pe}\int_0^{1} f(\theta(s))ds = 1 \tag{32}$$

The computed critical curve is presented in Figure 4.

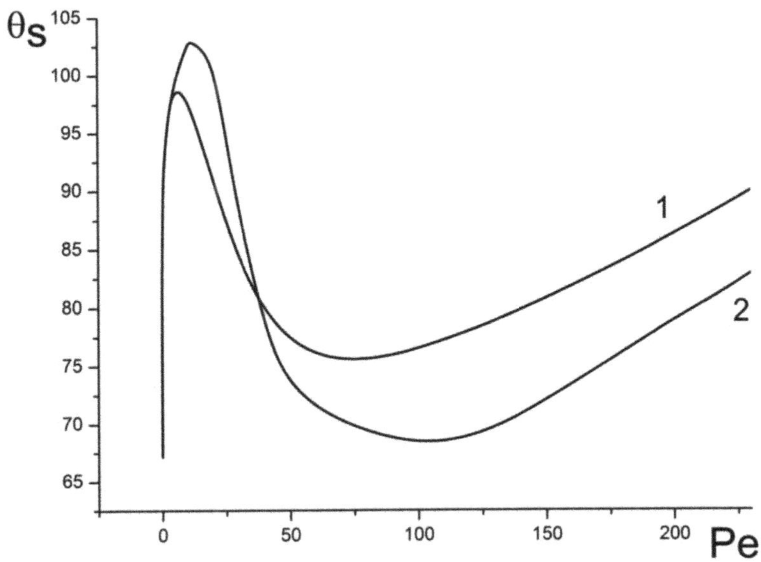

Figure 4. Critical curve $\theta_s = \theta_s(Pe)$. 1—approximate model (29); 2—original model (5); $A = 2.75 \times 10^{-18}$; $Ar = 0.012$; $h_s = 7.14 \times 10^2$; $g = 1.0 \times 10^{-20}$; $\mu = 91.67$.

3.3. Analysis of the Critical Curve Behavior

The behavior of critical curves and their dependence on parameters may be understood in more detail from Figures 5–9.

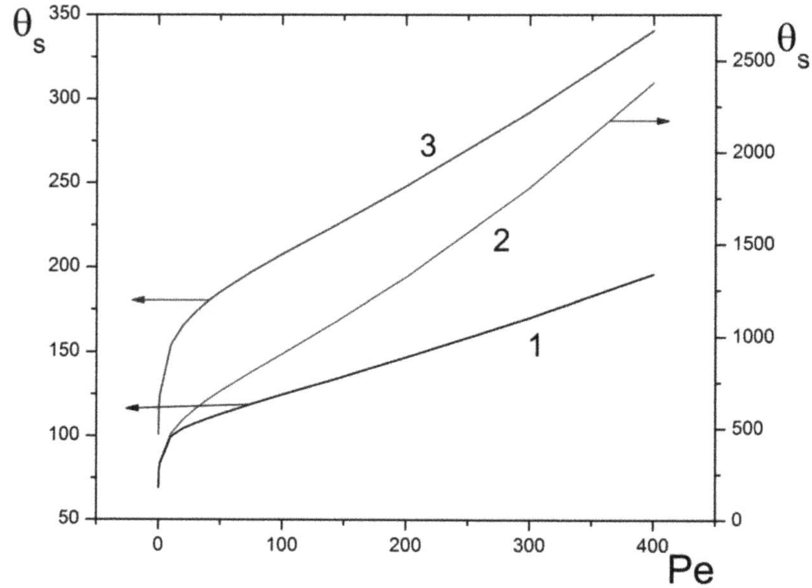

Figure 5. Monotonic critical curves. Approximate model (29). Ar = 0.02; $h_s = 7.14 \times 10^2$; $\mu = 75.0$; 1—$A = 2.75 \times 10^{-13}$; $g = 1.22 \times 10^{-15}$; 2—$A = 2.75 \times 10^{-15}$; $g = 1.22 \times 10^{-17}$; 3—$A = 2.75 \times 10^{-18}$; $g = 1.22 \times 10^{-20}$.

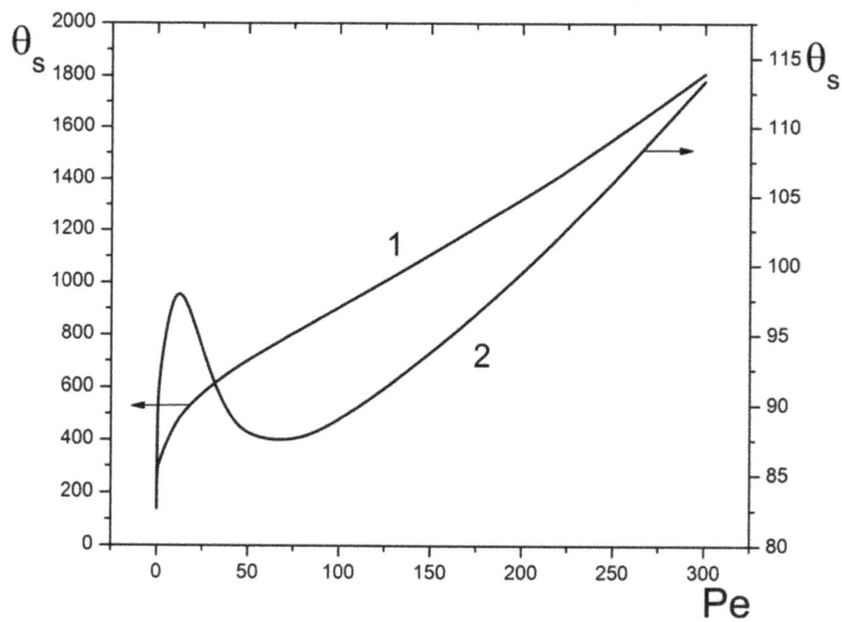

Figure 6. Transition from monotonic to non-monotonic critical curve. Approximate model (29). $A = 2.75 \times 10^{-18}$; $h_s = 7.14 \times 10^2$; $\mu = 75.0$; $g = 1.22 \times 10^{-20}$, 1—Ar = 0.02, 2—Ar = 0.012.

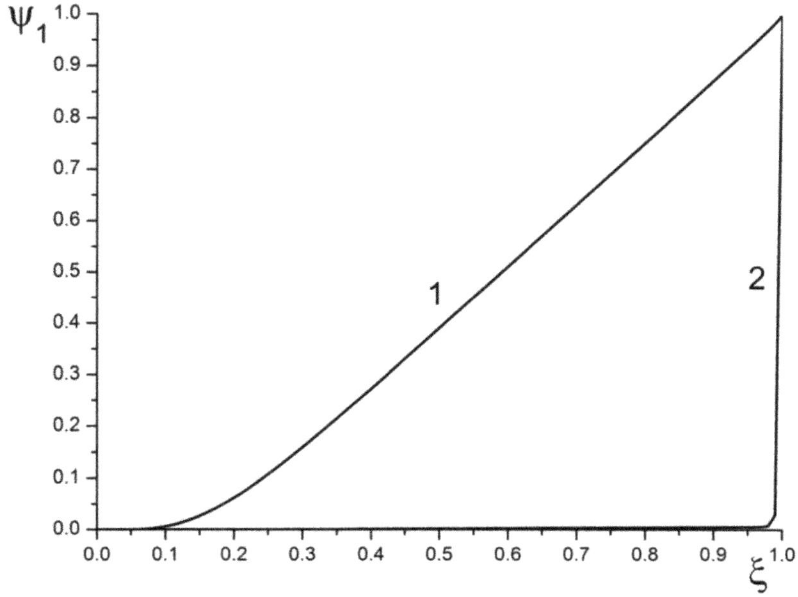

Figure 7. The integral $\Psi_1(\xi)$. Approximate model (29). A = 2.75 × 10^{-18}; h_s = 7.14 × 10^2; μ = 75.0; g = 1.22 × 10^{-20}. 1—monotonic critical curve. Ar = 0.02; Pe = 50.0; θ_s = 702.01. 2—non-monotonic critical curve. Ar = 0.012; Pe = 50.0; θ_s = 87.70.

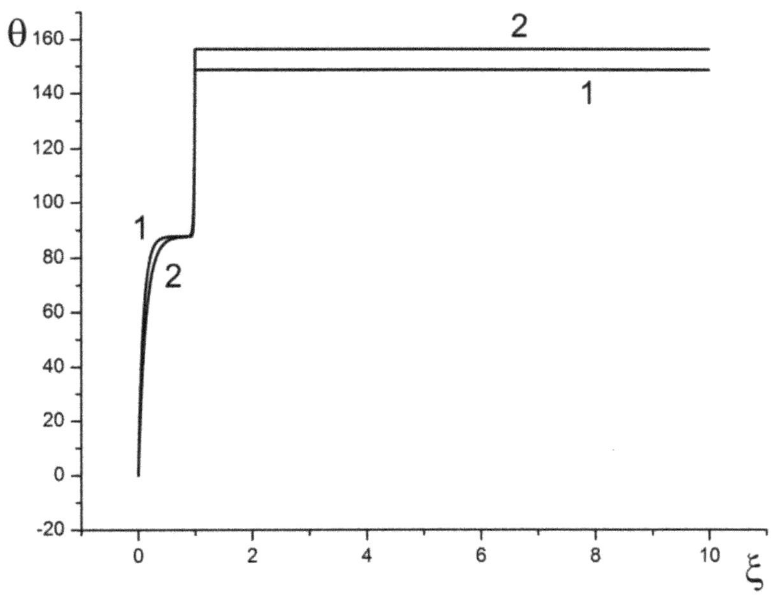

Figure 8. Solutions in the case of non-monotonic critical curve. Approximate model (29). A = 2.75 × 10^{-18}; Ar = 0.012; h_s = 7.14 × 10^2; μ = 75.0; g = 1.22 × 10^{-20}; θ_s = 87.70; 1—Pe = 50.0; 2—Pe = 84.35.

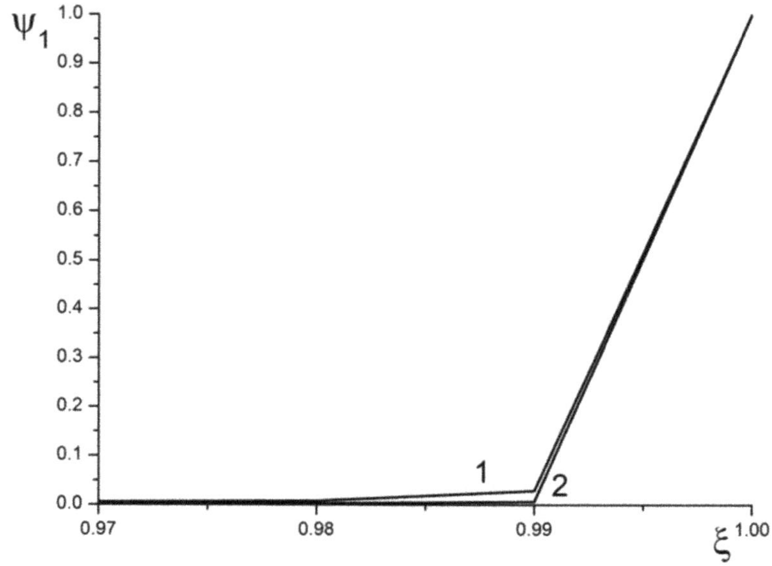

Figure 9. The integrals $\Psi_1(\xi)$ for the Figure 8 solutions. Approximate model (29). A = 2.75×10^{-18}; Ar = 0.012; $h_s = 7.14 \times 10^2$; $\mu = 75.0$; g = 1.22×10^{-20}; $\theta_s = 87.70$; 1—Pe = 50.0; 2—Pe = 84.35.

It turns out that both monotonic and non-monotonic critical curves exist.

The results presented in Figures 5–9 are obtained using the approximate model (29) with the delta function.

Figure 5 shows three curves drawn at different value of the parameters A and g. It is evident that all the three critical curves are monotonic.

Transition to non-monotonic behavior occurs upon variation of the Arrhenius number Ar (Figure 6), with all other parameters being kept constant. Non-monotonic behavior emerges upon decreasing the Arrhenius number.

The particular type (monotonic or non-monotonic) of the critical curve is controlled by behavior of the integral

$$\Psi_1(\xi) = \frac{g}{Pe}\int_0^{\xi} f(\theta(s))ds \tag{33}$$

which must be equal to unity at $\xi = 1$.

Figure 7 demonstrates these integrals for monotonic and non-monotonic curves. For the non-monotonic solution, the total value of the integral accumulates in the small vicinity of the point $\xi = 1$.

In the non-monotonic case (curve 2, Figure 6) the same value of $\theta_s = 87.70$ corresponds to the two values of the parameter Pe, namely $Pe = 50$ and $Pe = 84.35$.

The solutions $\theta(\xi)$ for these two values of Pe are presented in Figure 8, while Figure 9 presents the integrals $\Psi_1(\xi)$ for these solutions.

It is evident from Figure 9 that total values of both integrals accumulate in the small vicinity of the point $\xi = 1$.

The critical curve decreases at the point ($Pe = 50$, $\theta_s = 87.70$), and increases at the point ($Pe = 84.35$, $\theta_s = 87.70$).

It may be concluded that sharp behavior of the integral $\Psi_1(\xi)$ in the vicinity of the point $\xi = 1$ is the necessary condition for the critical curve to be non-monotonic.

The counterintuitive existence of non-monotonic critical curves may be explained in the following way.

Assuming the approximate δ- function model (29) and setting $\xi_0 = 1$ results in the following BVP for the temperature profile on $[0, 1]$

$$\frac{d^2\theta}{d\xi^2} - Pe\frac{d\theta}{d\xi} + h_s(\theta_s - \theta) = 0 \tag{34}$$

$$\theta(0) = 0; \; \frac{d\theta}{d\xi}(1) = \mu \cdot Pe$$

and, for a given Pe, the value of θ_s is determined from the condition

$$\frac{g}{Pe}\int_0^1 f(\theta(s))ds = 1 \tag{35}$$

where the source function $f(\theta)$ is given by Equation (5).

Let

$$\lambda_1 = \frac{Pe + \sqrt{Pe^2 + 4h_s}}{2}; \; \lambda_2 = -\frac{2h_s}{Pe + \sqrt{Pe^2 + 4h_s}} \tag{36}$$

Then λ_1 is sufficiently large for $h_s = 720$, $Pe < 50$, while at the same time, λ_2 is negative with sufficiently large absolute value, so that approximately

$$\theta(\xi) \approx \theta_s(1 - \exp(\lambda_2\xi)) + \frac{\mu Pe}{\lambda_1}\exp(\lambda_1(\xi - 1)) \tag{37}$$

Let us consider the case where the value $\frac{g}{Pe}f(\theta_s)$ is negligibly small. It is only in this case when the original problem is well approximated by the model with δ-function source. In this case we can assume

$$\theta(\xi) = \theta_s + \frac{\mu Pe}{\lambda_1}\exp(\lambda_1(\xi - 1)) \tag{38}$$

and then

$$\frac{d\theta}{d\xi} = \lambda_1(\theta - \theta_s) \tag{39}$$

We solve, under the above assumptions, the equation

$$F(\theta_s, Pe) = \frac{g}{Pe}\int_0^1 f(\theta(s))ds = 1 \tag{40}$$

with respect to θ_s at a fixed value of Pe.

It is easy to see that $F(\theta_s, Pe)$ is monotonic with respect to θ_s. Therefore, the solution $\theta_s(Pe)$ of the equation $F(\theta_s, Pe) = 1$ is unique with respect to θ_s.

If $\frac{\partial F}{\partial Pe} > 0$ at $F(\theta_s, Pe) = 1$, then $\frac{d\theta_s}{dPe} < 0$.

Let us calculate $\frac{\partial F}{\partial Pe}$.

Let $u = \theta - \theta_s$, then

$$F(\theta, Pe) = \frac{g}{Pe}\int_1^{\frac{\mu Pe}{\lambda_1}} \frac{f(\theta_s + u)}{\lambda_1 u}du \tag{41}$$

(We assumed that the value $\frac{g}{Pe}f(\theta_s)$ is small and the lower bound of the integral is irrelevant).

Further,

$$\begin{aligned}\frac{\partial F}{\partial Pe} &= \frac{2h_s}{\lambda_1^2\sqrt{Pe^2+4h_s}}\frac{\mu g}{Pe}\frac{f(\theta^*)}{\mu Pe} - \\ &- \frac{2g}{Pe^3\sqrt{Pe^2+4h_s}}\int\limits_1^{\frac{\mu Pe}{\lambda_1}}\frac{f(\theta_s+u)}{u}du = \\ &= \frac{2h_s g f(\theta^*)}{\lambda_1^2 Pe^2\sqrt{Pe^2+4h_s}} - \frac{2\lambda_1}{Pe^2\sqrt{Pe^2+4h_s}}\end{aligned} \quad (42)$$

Here $\theta^* = \theta(1) = \theta_s + \frac{\mu Pe}{\lambda_1}$, and we used the fact that

$$\int\limits_1^{\frac{\mu Pe}{\lambda_1}}\frac{f(\theta_s+u)}{u}du = \frac{\lambda_1 Pe}{g} \quad (43)$$

Therefore, under our assumptions, $\frac{d\theta_s}{dPe} < 0$ if $f(\theta^*) > \frac{\lambda_1^3}{gh_s}$.

It is easy to see that upon increasing μ or decreasing Ar (at fixed g, Pe, h_s), $\theta_s(Pe)$ decreases, while $f(\theta^*)$ increases.

Therefore, by increasing μ or decreasing Ar, we can obtain points on the critical curve where $\frac{d\theta_s}{dPe} < 0$.

3.4. Comparison with Experimental Data

Let us compare results of the present study with available experimental observations.

Yakovlev et al. [15] conducted a very detailed numerical investigation of flame stabilization process in thin-layered radial porous burner. It consisted of a mounting flange connected with a thin-layered porous shell. Flow configuration for the reported experiment may be considered as a flow through porous wall of the infinitely long hollow cylinder in the direction normal to the wall (i.e., in the plane perpendicular to the axial axis of symmetry of the cylinder). Flow direction is from the inner surface of the cylinder to the outer surface.

Within stable operating ranges, the flame front can be stabilized within the cavity of the burner (internal regime), within porous media (submerged flame) or above it (surface-stabilized flame), by adjusting the equivalence ratio and the flow rate.

The first of these regimes (stabilization upstream from the lattice) is not considered in the present study; however, the latter two are of most interest. Similar to the present study, Yakovlev et al. [15] observed two types of temperature distribution profiles, that is monotonic and non-monotonic. When the flame stabilized within porous media (submerged regime), then a non-monotonic profile was observed. When the flame stabilized downstream from the surface (surface-stabilized regime), then a monotonic profile was observed. This change of profile configuration upon flame front relocating from the porous to the gas region is quantitatively identical to the behavior reported in Figures 2 and 3 of the present study.

Mital et al. [17] and Janvekar et al. [27] considered porous burners of cylindrical shape, with the flow parallel to the to the axial axis of symmetry of the cylinder. This configuration would be asymptotically identical to the one used in the present study if the radius of the cylinder increases infinitely.

Both studies observed the flame stabilization regime inside the porous material (submerged reaction zone) with non-monotonic temperature profiles similar the one presented in Figure 2.

Thus, the studies of Yakovlev et al. [15], Mital et al. [17], Janvekar et al. [27] and Keshtkar et al. [34] provide indirect, but nevertheless very convincing support for the results of the present study.

4. Conclusions

The standing premixed combustion wave on a lattice burner has two qualitatively distinctive regimes: flame stabilizes either downstream from the burner surface or inside the solid lattice.

For the first time, this fact has been proved absolutely rigorously using a one-dimensional analytical model of reacting flow.

It has also been demonstrated using numerical simulations.

In the space of controlling parameters, the two stabilization regimes are separated by the specific dependence of the lattice temperature on the flow Peclet number (the critical curve), with all other parameters being fixed.

Transition from monotonic to non-monotonic regimes occurs as the Arrhenius number sufficiently decreases.

Results of the present study agree qualitatively with the available experimental data.

Novelties of the present study may be summarized as follows: a new model of the standing combustion wave stabilized on a lattice (porous) burner is proposed. The model allows for qualitative analytical investigation of the problem to be performed. An approximated model, based on description of chemical source term with δ-function, is also developed. The exact analytical solution is obtained for this model. Rigorous mathematical proof of existence of the two distinctive combustion wave modes, with monotonic and non-monotonic temperature profiles, respectively, is demonstrated for the first time. Definition of critical curve in the space of control parameters, separating the two combustion wave modes, is proposed. Rigorous mathematical proof of existence of both monotonic and non-monotonic critical curves is demonstrated.

In terms of quantitative results, proposed analytical model, based on δ-function approximation of chemical source term, agrees well (within 7% relative error) with the model based on distributed description of chemical reaction zone.

The proven existence of the two significantly different combustion regimes is of great practical importance and impacts design solutions for industrial porous burners. The results will be of great interest to the broader academic community, particularly in research areas where similar wave structures may emerge.

Author Contributions: Conceptualization, V.B.N., B.V.L. and V.S.P.; Formal analysis, V.B.N., B.V.L. and V.S.P.; Methodology, V.B.N., B.V.L. and V.S.P.; Writing—original draft, V.B.N., B.V.L. and V.S.P.; Writing—review & editing, V.B.N., B.V.L. and V.S.P. All the authors contributed equally to the study. All authors have read and agreed to the published version of the manuscript.

Funding: This research received no external funding.

Data Availability Statement: The data is available from the authors upon request.

Acknowledgments: We are grateful to Andrey Belyaev for helpful discussions of the current research on the subject. We are grateful to Inga Novozhilov for correcting English language in the manuscript.

Conflicts of Interest: The authors declare no conflict of interest.

Abbreviations

BVP	Boundary Value Problem
HTE	Heat Transfer Equation
LHS	Left-Hand Side
RHS	Right-Hand Side
Notation	
Ascents	
$\sim$	wave: dimensional, and only dimensional, variables
$\hat{}$	cap: auxiliary non-dimensional variables
Latin	
A	pre-exponential factor
Ar	Arrhenius number

C	reactant concentration
c_p	specific heat at constant pressure
E	activation energy
f	chemical source function
g	parameter
H	Heaviside function (unity for positive values of the argument; zero for non-positive values)
h	heat transfer coefficient
L	arbitrary large positive number
Pe	Peclet number
Q	heat of reaction
R	universal gas constant
S	lattice surface area per unit volume
T	temperature
u	gas mixture velocity
x	spatial coordinate
x_1	lattice thickness
Greek	
δ	Delta function
θ	temperature
κ	thermal diffusivity
λ	thermal conductivity
μ	parameter $(A/(3g))$
ξ	spatial coordinate
ρ	density
τ	time
Subscripts	
s	surface
0	ambient; initial

Appendix A

Here, an analytical solution for the approximate δ-function model (29) is presented. We solve the following problem

$$\frac{d^2\theta}{d\xi^2} - Pe\frac{d\theta}{d\xi} + \mu Pe\delta(\xi - \xi_0) + h(\theta_s - \theta) = 0 \quad \text{(A1)}$$

$$\theta(0) = 0,\ \frac{d\theta}{d\xi}(L) = 0,\ \xi \in (0, L)$$

The location ξ_0 is determined by the condition

$$\frac{g}{Pe}\int_0^{\xi_0} f(\theta(s))ds = 1 \quad \text{(A2)}$$

The source function $f(\theta)$ is given by Equation (5).
The parameters $\lambda_{1,2}$ used below are given by Equation (36).
Let us solve the problem (A1) at a fixed value of ξ_0.

Case 1 . $0 \le \xi_0 < 1$

In this case, the solution has the form

$$\Theta_1(\xi, \xi_0) = \begin{cases} \theta_s(1 - e^{\lambda_2\xi}) + C_{1,1}(\xi_0)(e^{\lambda_1\xi} - e^{\lambda_2\xi}); & \xi \in [0, \xi_0) \\ \theta_s + C_{3,1}(\xi_0)\left(e^{\lambda_1\xi} - \frac{\lambda_1}{\lambda_2}e^{(\lambda_1-\lambda_2)}e^{\lambda_2\xi}\right); & \xi \in (\xi_0, 1] \\ \theta_s + C_{3,1}(\xi_0)\left(e^{\lambda_1} - \frac{\lambda_1}{\lambda_2}e^{(\lambda_1-\lambda_2)}e^{\lambda_2}\right); & \xi \in (1, L] \end{cases} \quad \text{(A3)}$$

The matching conditions at $\xi = \xi_0$ are

$$\begin{cases} \theta_s(1-e^{\lambda_2\xi_0}) + C_{1,1}(e^{\lambda_1\xi_0} - e^{\lambda_2\xi_0}) = \theta_s + C_{3,1}\left(e^{\lambda_1\xi_0} - \frac{\lambda_1}{\lambda_2}e^{(\lambda_1-\lambda_2)}e^{\lambda_2\xi_0}\right) \\ C_{3,1}\lambda_1\left(e^{\lambda_1\xi_0} - e^{(\lambda_1-\lambda_2)}e^{\lambda_2\xi_0}\right) + \theta_s\lambda_2 e^{\lambda_2\xi_0} - C_{1,1}(\lambda_1 e^{\lambda_1\xi_0} - \lambda_2 e^{\lambda_2\xi_0}) = -\mu Pe \end{cases} \quad \text{(A4)}$$

from where

$$C_{1,1}(\xi_0) = \frac{\theta_s(\lambda_1-\lambda_2)\lambda_2 + \mu Pe\left(e^{(\lambda_1-\lambda_2)}e^{-\lambda_1\xi_0}\lambda_1 - e^{\lambda_2\xi_0}\lambda_2\right)}{(\lambda_1-\lambda_2)(e^{(\lambda_1-\lambda_2)}\lambda_1 - \lambda_2)} \quad \text{(A5)}$$

$$C_{3,1}(\xi_0) = \frac{\lambda_2(\theta_s(\lambda_1-\lambda_2) + \mu Pe(e^{-\lambda_1\xi_0} - e^{\lambda_2\xi_0}))}{(\lambda_1-\lambda_2)(e^{(\lambda_1-\lambda_2)}\lambda_1 - \lambda_2)}$$

Case 2. $\xi_0 = 1$

The solution has the form

$$\Theta_2(\xi,\xi_0) = \begin{cases} \theta_s(1-e^{\lambda_2\xi}) + C_{1,2}(\xi_0)(e^{\lambda_1\xi} - e^{\lambda_2\xi}); & \xi \in [0,1] \\ C_{3,2}(\xi_0); & \xi \in (1,L] \end{cases} \quad \text{(A6)}$$

The matching conditions are as follows

$$\begin{cases} \theta_s(1-e^{\lambda_2}) + C_{1,2}(e^{\lambda_1} - e^{\lambda_2}) = C_{3,2} \\ \theta_s\lambda_2 e^{\lambda_2} - C_{1,2}(\lambda_1 e^{\lambda_1} - \lambda_2 e^{\lambda_2}) = -\mu Pe \end{cases} \quad \text{(A7)}$$

$$C_{1,2}(\xi_0) = \frac{\theta_s\lambda_2 e^{\lambda_2} + \mu Pe}{(\lambda_1 e^{\lambda_1} - \lambda_2 e^{\lambda_2})} \quad \text{(A8)}$$

$$C_{3,2}(\xi_0) = \theta_s\left(1-e^{\lambda_2}\right) + \frac{(\theta_s\lambda_2 e^{\lambda_2} + \mu Pe)}{(\lambda_1 e^{\lambda_1} - \lambda_2 e^{\lambda_2})}\left(e^{\lambda_1} - e^{\lambda_2}\right)$$

Case 3. $\xi_0 > 1$

The solution has the form

$$\Theta_3(\xi,\xi_0) = \begin{cases} \theta_s(1-e^{\lambda_2\xi}) + C_{1,3}(\xi_0)(e^{\lambda_1\xi} - e^{\lambda_2\xi}); & \xi \in [0,1] \\ C_{3,3}(\xi_0) + C_{4,3}(\xi_0)e^{Pe\xi}; & \xi \in (1,\xi_0] \\ C_{3,3}(\xi_0) + C_{4,3}(\xi_0)e^{Pe\xi_0}; & \xi \in (\xi_0, L] \end{cases} \quad \text{(A9)}$$

The matching derivatives at $\xi = \xi_0$ give

$$C_{4,3}(\xi_0) = \mu e^{-Pe\xi_0} \quad \text{(A10)}$$

The matching conditions at $\xi = 1$ are

$$\begin{cases} \theta_s(1-e^{\lambda_2}) + C_{1,3}(e^{\lambda_1} - e^{\lambda_2}) = C_{3,3} + \mu e^{Pe(1-\xi_0)} \\ -\theta_s\lambda_2 e^{\lambda_2} + C_{1,3}(\lambda_1 e^{\lambda_1} - \lambda_2 e^{\lambda_2}) = \mu Pe e^{Pe(1-\xi_0)} \end{cases} \quad \text{(A11)}$$

$$C_{1,3}(\xi_0) = \frac{\theta_s\lambda_2 e^{\lambda_2} + \mu Pe e^{Pe(1-\xi_0)}}{(\lambda_1 e^{\lambda_1} - \lambda_2 e^{\lambda_2})} \quad \text{(A12)}$$

$$C_{3,3}(\xi_0) = \theta_s\left(1-e^{\lambda_2}\right) - \mu e^{Pe(1-\xi_0)} + \frac{\left(\theta_s\lambda_2 e^{\lambda_2} + \mu Pe e^{Pe(1-\xi_0)}\right)}{(\lambda_1 e^{\lambda_1} - \lambda_2 e^{\lambda_2})}\left(e^{\lambda_1} - e^{\lambda_2}\right)$$

The solution of the BVP (A1) and (A2) on the whole interval $[0, L]$ is

$$\Theta(\xi, \xi_0) = \begin{cases} \Theta_1(\xi, \xi_0); & \xi_0 \in [0, 1) \\ \Theta_2(\xi, \xi_0); & \xi_0 = 1 \\ \Theta_3(\xi, \xi_0); & \xi_0 \in (1, L) \end{cases} \tag{A13}$$

References

1. Williams, F.A. *Combustion Theory*; CRC Press: Boca Raton, FL, USA, 2018.
2. Kuo, K.K. *Principles of Combustion*; Wiley: Hoboken, NJ, USA, 2005.
3. Zeldovich, Y.B.; Barenblatt, G.I.; Librovich, V.B.; Makhviladze, G.M. *Mathematical Theory of Combustion and Explosion*; Nauka: Moscow, Russia, 1980. (In Russian)
4. Merzhanov, A.G.; Khaikin, B.I. Theory of Combustion Waves in Homogeneous Media. *Prog. Energy Combust. Sci.* **1988**, *14*, 1–98. (In Russian) [CrossRef]
5. Novozhilov, B.V.; Novozhilov, V.B. *Theory of Solid-Propellant Nonsteady Combustion*; Wiley-ASME Press: Hoboken, NJ, USA, 2021.
6. Merzhanov, A.G. Self-propagating high-temperature synthesis: Twenty years of search and findings. In *Combustion and Plasma Synthesis of High-Temperature Materials*; Munir, Z.A., Holt, J.B., Eds.; VCH: Vancouver, BC, Canada, 1990; pp. 1–53.
7. Hirano, T.; Saito, K. Fire Spread Phenomena: The Role of Observation in Experiment. *Prog. Energy Combust. Sci.* **1994**, *20*, 461–485. [CrossRef]
8. Fernandez-Pello, A.C.; Hirano, T. Controlling Mechanisms of Flame Spread. *Combust. Sci. Technol.* **1983**, *32*, 1–31. [CrossRef]
9. Smirnov, N.N.; Dimitrienko, I.D. Convective Combustion of Porous Compressible Propellants. *Combust. Flame* **1992**, *89*, 260–270. [CrossRef]
10. Aldushin, A.P.; Matkowsky, B.J.; Schult, D.A.; Shkadinskaya, G.V.; Shkadinsky, K.G.; Volpert, V.A. Porous medium combustion. In *Modeling in Combustion Science*; Lecture Notes in Physics; Buckmaster, J., Takeno, T., Eds.; Springer: Berlin/Heidelberg, Germany, 1995; Volume 449. [CrossRef]
11. Aldushin, A.P.; Seplyarsky, B.S. Propagation of waves of exothermal reaction in porous medium during gas blow-through. *Sov. Phys. Dokl.* **1978**, *23*, 483–485.
12. Shkadinsky, K.G.; Shkadinskaya, G.V.; Matkowsky, B.J.; Volpert, V.A. Two front traveling waves in filtration combustion. *SIAM J. Appl. Math.* **1993**, *53*, 128–140. [CrossRef]
13. Schult, D.A.; Matkowsky, B.J.; Volpert, V.A.; Fernandez-Pello, A.C. Propagation and Extinction of Forced Opposed Flow Smolder Waves. *Comb. Flame* **1995**, *101*, 471–490. [CrossRef]
14. Wahle, C.W.; Matkowsky, B.J. Rapid, Upward Buoyant Filtration Combustion Waves Driven by Convection. *Combust. Flame* **2001**, *124*, 14–34. [CrossRef]
15. Yakovlev, I.; Maznoy, A.; Zambalov, S. Pore-scale Study of Complex Flame Stabilization Phenomena in Thin-layered Radial Porous Burner. *Combust. Flame* **2021**, *231*, 111468. [CrossRef]
16. Sangjukta, D.; Sahoo, N.; Muthukumar, P. Experimental Studies on Biogas Combustion in a Novel Double Layer Inert Porous Radiant Burner. *Renew. Energy* **2020**, *149*, 1040–1052.
17. Mital, R.; Gore, J.P.; Viskanta, R. A Study of the Structure of Submerged Reaction Zone in Porous Ceramic Radiant Burners. *Combust. Flame* **1997**, *111*, 175–184. [CrossRef]
18. Fursenko, R.; Maznoy, A.; Odintsov, E.; Kirdyashkin, A.; Minaev, S.; Sudarshan, K. Temperature and Radiative Characteristics of Cylindrical Porous Ni–Al Burners. *Int. J. Heat Mass Transf.* **2016**, *98*, 277–284. [CrossRef]
19. Ferguson, J.C.; Sobhani, S.; Ihme, M. Pore-resolved Pimulations of porous Media Combustion with Conjugate Heat Transfer. *Proc. Combust. Inst.* **2021**, *38*, 2127–2134. [CrossRef]
20. Bedoya, C.; Dinkov, I.; Habisreuther Zarzalis, P.N.; Bockhorn, H.; Parthasarathy, P. Experimental study, 1D Volume-averaged Calculations and 3D Direct Pore Level Simulations of the Flame Stabilization in Porous Inert Media at Elevated Pressure. *Combust. Flame* **2015**, *162*, 3740–3754. [CrossRef]
21. Sirotkin, F.; Fursenko, R.; Kumar, S.; Minaev, S. Flame Anchoring Regime of Filtrational Gas Combustion: Theory and Experiment. *Proc. Combust. Inst.* **2017**, *36*, 4383–4389. [CrossRef]
22. Yakovlev, I.; Zambalov, S. Three-dimensional Pore-scale Numerical Simulation of Methane-air Combustion in Inert Porous Media under the Conditions of Upstream and Downstream Combustion Wave Propagation through the Media. *Combust. Flame* **2019**, *209*, 74–98. [CrossRef]
23. Billerot, P.-L.; Dufresne, L.; Lemaire, R.; Seers, P. 3D CFD Analysis of a Diamond Lattice-based Porous Burner. *Energy* **2020**, *207*, 118–160. [CrossRef]
24. Wood, S.; Harris, A.T. Porous Burners for Lean-burn Applications. *Prog. Energy Combust. Sci.* **2008**, *34*, 667–684. [CrossRef]
25. Wua, C.-Y.; Chen, K.-H.; Yang, S.Y. Experimental Study of Porous Metal Burners for Domestic Stove Applications. *Energy Convers. Manag.* **2014**, *77*, 380–388. [CrossRef]
26. Dehaj, M.S.; Ebrahimi, R.; Shams, M.; Farzaneh, M. Experimental Analysis of Natural Gas Combustion in a Porous Burner. *Exp. Therm. Fluid Sci.* **2017**, *84*, 134–143. [CrossRef]
27. Janvekar, A.A.; Miskam, M.A.; Abas, A.; Ahmad, Z.A.; Juntakan, T.; Abdullah, M.Z. Effects of the Preheat Layer Thickness on Surface/Submerged Flame during Porous Media Combustion of Micro Burner. *Energy* **2017**, *122*, 103–110. [CrossRef]

28. Gao, H.B.; Qu, Z.G.; Feng, X.B.; Tao, W.Q. Methane/Air Premixed Combustion in a Two-layer Porous Burner with Different Foam Materials. *Fuel* **2014**, *115*, 154–161. [CrossRef]
29. Al-attab, K.A.; Ho, J.C.; Zainal, Z.A. Experimental Investigation of Submerged Flame in Packed Bed Porous, Media Burner Fueled by Low Heating Value Producer Gas. *Exp. Therm. Fluid Sci.* **2015**, *62*, 1–8. [CrossRef]
30. Bouma, P.H.; De Goey, L.P.H. Premixed Combustion on Ceramic Foam Burners. *Combust. Flame* **1999**, *119*, 133–143. [CrossRef]
31. Barra, A.J.; Diepvens, G.; Ellzey, J.L.; Henneke, M.R. Numerical Study of the Effects of Material Properties on Flame Stabilization in a Porous Burner. *Combust. Flame* **2003**, *134*, 369–379. [CrossRef]
32. Barra, A.J.; Ellzey, J.L. Heat Recirculation and Heat Transfer in Porous Burners. *Combust. Flame* **2004**, *137*, 230–241. [CrossRef]
33. Lari, K.; Gandjalikhan Nassab, S.A. Transient Thermal Characteristics of Porous Radiant Burners. *Iran. J. Sci. Technol.* **2007**, *31*, 407–420.
34. Keshtkar, M.M.; Gandjalikhan Nassab, S.A. Theoretical Analysis of Porous Radiant Burners under 2-D Radiation Field Using Discrete Ordinates Method. *J. Quant. Spectrosc. Radiat. Transf.* **2009**, *110*, 1894–1907. [CrossRef]
35. Djordjevic, N.; Habisreuther, P.; Zarzalis, N. A Numerical Investigation of the Flame Stability in Porous Burners Employing Various Ceramic Sponge-like Structures. *Chem. Eng. Sci.* **2011**, *66*, 682–688. [CrossRef]
36. Arutyunov, V.S.; Belyaev, A.A.; Lidskii, B.V.; Nikitin, A.V.; Posvyanskii, V.S.; Shmelev, V.M. Numerical Solution of the Problem of Surface Combustion on Flat Porous Matrix. *AIP Conf. Proc.* **2018**, *2046*, 020075. [CrossRef]
37. Shmelev, V.M. Combustion of a Mixed Mixture in a Slot Matrix. *Russ. J. Phys. Chem. B* **2020**, *14*, 670–677. [CrossRef]
38. Godunov, S.K.; Ryaben'kii, V.S. *Difference Schemes*; North Holland: Amsterdam, The Netherlands, 1987.
39. Incropera, F.P.; DeWitt, D.P. *Fundamentals of Heat and Mass Transfer*; Wiley: Hoboken, NJ, USA, 2002.
40. Borman, G.L.; Ragland, K.W. *Combustion Technology*; McGraw-Hill: New York, NY, USA, 1998.

energies

MDPI

Article

Intake System Performance Stability as a Function of Flow Throttling

Adam Kozakiewicz, Stanisław Kachel *, Michał Frant and Maciej Majcher

Institute of Aviation Technology, Faculty of Mechatronics, Armament and Aerospace, Military University of Technology, 00-908 Warszawa, Poland
* Correspondence: stanislaw.kachel@wat.edu.pl; Tel.: +48-261-839-170

Abstract: This paper presents a numerical analysis of the stability of the flow parameters along the intake duct of an aircraft jet turbine engine. This problem has been investigated by many research teams and was included in the literature analysis. The unstable operation of a turbojet intake system can be the consequence of many adverse factors, including an intake vortex. The investigated intake system, due to its low location to the plane of the airport, is highly susceptible to the formation of an intake vortex. The phenomenon of an intake vortex can, in the worst-case scenario, result in the surging of the turbojet, and even engine stalling. This paper presents a developed model of the forward section of an aircraft, complete with its intake duct, and the method of its discretization. The intake-system model and numerical analysis were performed in Ansys Fluent. The flow parameters adopted for numerical simulations, under specific boundary conditions corresponded to the operating conditions of the engine cooperating with the investigated intake system. The numerical calculations were performed assuming an air-pressure rise in the end section of the engine-intake system, reflecting the reduction in the pitch angle of the inlet stator blades of the fan. As a result, the pressure distributions in a significant cross section in the intake system were obtained. The results were analyzed with the quantitative distribution of the pressure fields by applying a dimensionless potential-pressure ratio. The pressure ratio enabled a comparative analysis of the nonuniformity of the total-pressure distribution in selected cross sections of the intake system. The results were revealing in terms of growing unstable flows in the flow duct. A major conclusion drawn from the results, by testing the dimensionless potential-pressure ratio, was that, within certain limits, it was possible to improve the flow uniformity by increasing the throttling pressure.

Keywords: intake vortex; jet engine intake system; flow modeling; unstable engine operation; CFD

Citation: Kozakiewicz, A.; Kachel, S.; Frant, M.; Majcher, M. Intake System Performance Stability as a Function of Flow Throttling. *Energies* **2022**, *15*, 6291. https://doi.org/10.3390/en15176291

Academic Editors: Vasily Novozhilov and Cunlu Zhao

Received: 11 July 2022
Accepted: 25 August 2022
Published: 29 August 2022

1. Introduction

Investigations into the distribution of the flow parameters in the intake systems of turbojet engines remain important due to their impact on the operating efficiency of the entire system of a turbojet engine. The intake system is the first of the assemblies in the turbojet engine system. It has a significant effect on the engine's performance stability including changes to the engine operating parameters, as shown in Figure 1. Significant attention has been given to the problem of total-pressure disturbances in the AIP (aerodynamic interface plane) in front of the turbojet fan. A literature analysis concerning the heterogeneity of the flow fields was performed. One paper [1] concerned the optimization of an intake S-duct to reduce the total-pressure loss and limit disturbances. A combination of a genetic algorithm with CFD simulations was used. The Fluent CFD software was used to simulate the flow field in every tested configuration of the intake system, and the resulting calculations became the foundation for the entire optimization process.

This involved a thorough analysis of the 3D flow, with a focus on the flow fields in characteristic cross sections. An example of the results can be found in Figure 2, which illustrates the distribution of the pressure-rise deviation in the intake duct, and the distribution of the swirl angle of the stream.

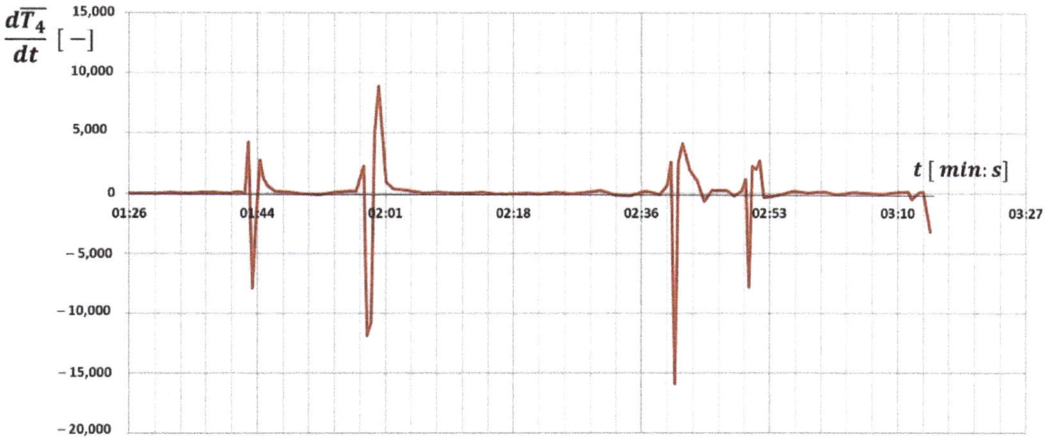

Figure 1. Temperature-change trend downstream of the jet engine turbine during an engine surge.

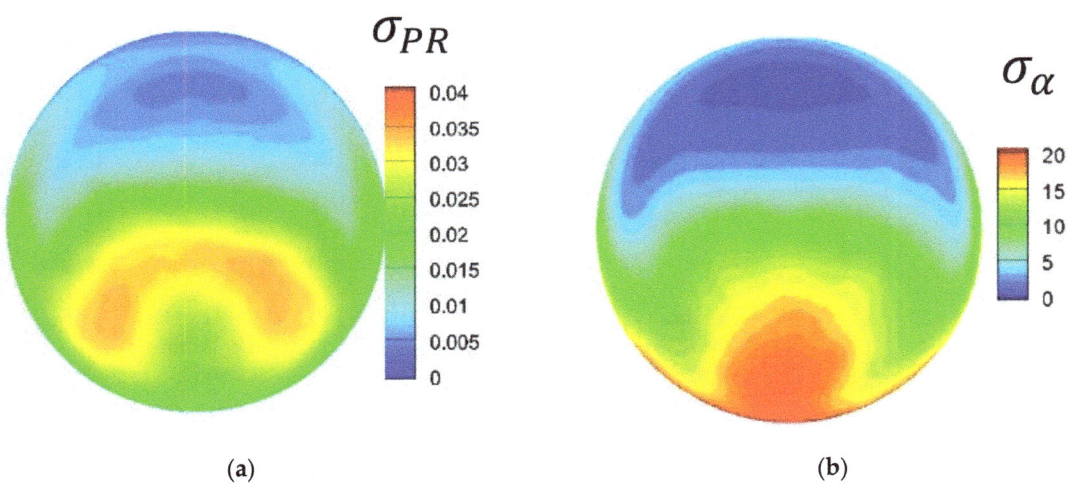

Figure 2. Standard-deviation distribution: (**a**) for the pressure-recovery factor; (**b**) for the flow-turbulence angle [1]. Adapted with permission from Ref. [1]. Copyright 2017, copyright the American Institute of Aeronautics and Astronautics AIAA.

The intake S-duct shape was optimized by free deformation combined with a genetic algorithm. At one extreme point, the total-pressure loss was reduced by 20% (using a Pareto front), while, at another extreme point, the average area turbulences were reduced by 10%, with both results deemed to be very good.

Another paper [2] investigated a modified DSI (diverterless supersonic inlet) intake system at various angles of attack and Mach numbers. The DSI, also known as a bump inlet, is an innovative design element for high-speed aircraft, offering considerable advantages in the design of stealth aircraft, and considered to be a state-of-the-art technology in some of the latest jet fighters. The modification of the geometry in the existing DSI was proposed to improve the performance of the aircraft's propulsion. A numerical analysis of the existing DSI was carried out with different angles of attack and Mach numbers for this purpose. The numerical model was validated with the available experimental data. The performance of the modified DSI was assessed with the criteria of the pressure-recovery factor, the pressure coefficient (being the ratio of a specified area to the average pressure within a cross section), and the drag coefficient. The analysis was carried out at subsonic Mach numbers with three different angles of attack at the same flight altitude. This provided data on the behavior of the parameters, including by testing the distribution of the pressure-field changes. An example of the results is shown in Figure 3.

Figure 3. Total-pressure distribution in the intake cross section at Ma = 0.8, and the angle of attack of $\alpha = 10°$ [2]. Adapted with permission from Ref. [2]. Copyright 2021, copyright the American Institute of Aeronautics and Astronautics AIAA.

Another paper [3] investigated the dependence of the total-pressure disturbances at the outlet of a high-diffusion-offset twin-intake S-duct system to determine whether a classic solution to the problem could be applied (Figure 4). The assessment of the combined characteristics of the time-dependent radial and circumferential disturbances demonstrated that local ring-based descriptors of flow distortions were more suitable for the characterization of maximum events. An EVT (extreme value theory) was applied to predict the maximum distortion levels that could emerge during the test duration that exceeded the available set of experimental data. It was found that, in comparison with the dynamic pressure, the total-pressure fluctuations were moderately dependent on the Mach number in all the investigated duct systems. However, changing the bend-offset ratio of the S-duct from 1.34 to 2.44 significantly changed the average total-pressure instability in the AIP (aerodynamic interface plane). It was discovered that higher secondary flows, caused by an increase in the final system, not only favored total-pressure loss, but also increased the overall flow-field instability within the AIP.

Another paper [4] presented the numerical (CFD) testing of a DSI intake design using Ansys. The CFD analysis was performed on a selection of DSI configurations, with a Mach number between 0.8 and 1.6. This provided results relevant to the performance of the tested design/configuration cases in the studied range of Mach numbers. They also identified the best geometric configuration that provided the most favorable pressure-recovery factor. Suggestions were given for future activities concerning the improvement in the pressure-recovery factor in the DSI-intake-system concept. The pressure-recovery factor is defined here as the ratio of the overall pressure in the inlet cross section (Figure 5) to the free-flow total pressure. The values of the factor cover the range between 0 and 1, with 1 being no energy loss. Higher pressure-recovery-factor values are more favorable, as the factor markedly affects the thrust force of turbojet engines. Based on the total-pressure distributions, it was found that using a tear-drop-shaped bump can produce better aerodynamic parameters than those provided by an ellipsoid bump.

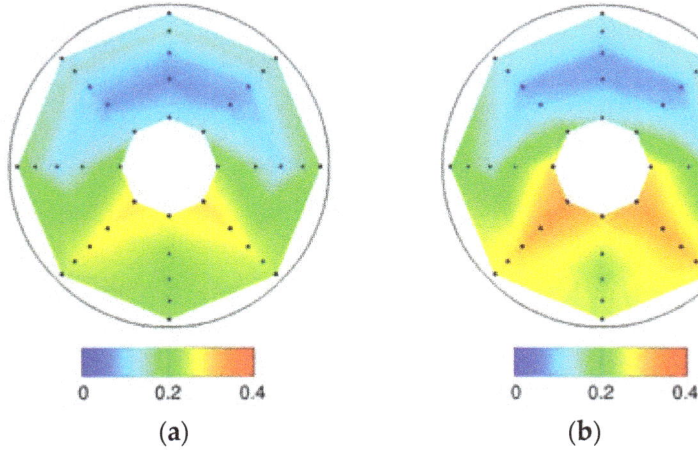

Figure 4. Distribution of the change in the dimensionless potential-pressure ratio in the AIP: (**a**) Duct A M_{AIP} = 0.2; (**b**) Duct A M_{AIP} = 0.36 [3]. Adapted with permission from Ref. [3]. Copyright 2018, copyright Elsevier: Amsterdam.

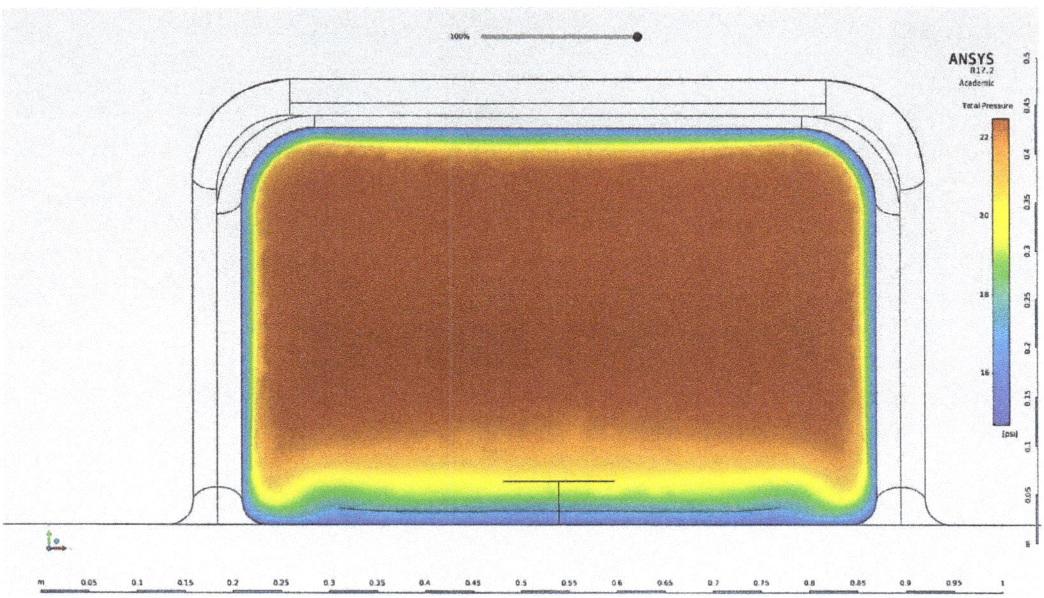

Figure 5. AIP total-pressure distribution [4]. Adapted with permission from Ref. [4]. Copyright 2019, copyright the American Institute of Aeronautics and Astronautics AIAA.

Another paper [5] contained an analysis of the changes in the intake-pressure-recovery factor and total-pressure distribution at different flight parameters of an F-16 jet fighter, which included the Mach number, angle of attack, and angle of roll. The intake parameters were investigated with reference to the AIP. The paper used a simplified computational model, which only included the intake tuck and the area under the aircraft fuselage (Figure 6).

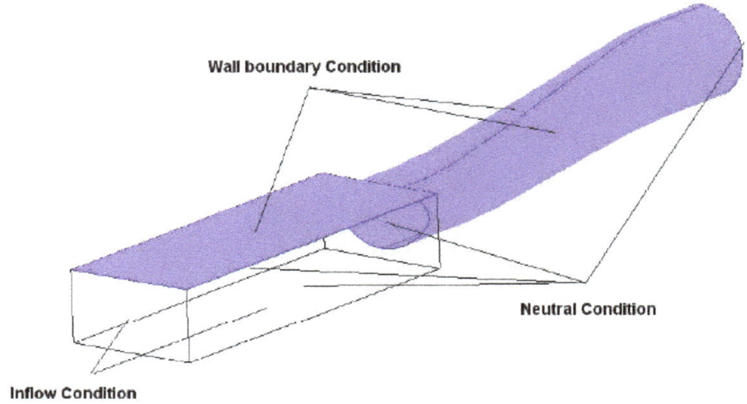

Figure 6. Mode of air-intake section of an F-16 jet fighter [5]. Adapted with permission from Ref. [5]. Copyright 2011, copyright Engineering Applications of Computational Fluid Mechanics.

Nonuniform-pressure-field distributions were achieved (Figure 7), which varied in intervals of 0.910–0.980 and 0.124–0.400. The flow stability through the intake was demonstrated to be affected by the aerodynamic factors of the whole flight of the aircraft.

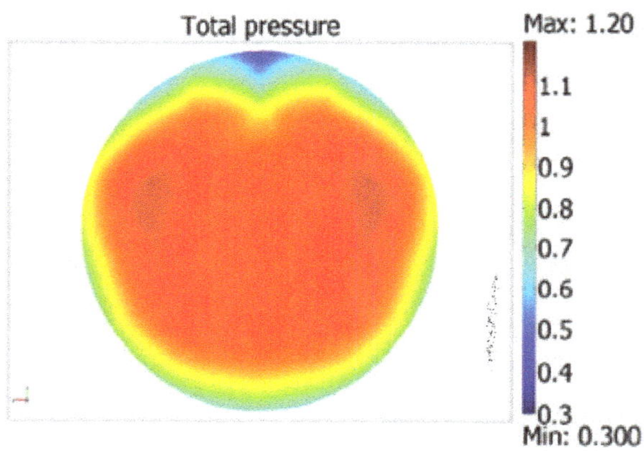

Figure 7. AIP cross-sectional pressure distribution in the air-intake section of an F-16 jet fighter [5]. Adapted with permission from Ref. [5]. Copyright 2011, copyright Engineering Applications of Computational Fluid Mechanics.

Another paper [6] involved working on improving the existing DSI intake sections based on an investigation into the distribution of the parameter fields. The examples included attempts to use a convex surface (a bump) ahead of the air intake, with early work already eliminating the risk of an epileptic form. An example of a simulated DSI performance is shown in Figure 8. The results show that there was the capacity for maintaining the structure of the shock waves during the operational mode, even at high angles of attack. The analysis of the shock-wave structures and the interaction of the shock-wave-boundary layer during supersonic maneuvers showed that the aerodynamic efficiency of a DSI with a thick boundary layer and high angles of attack was sufficient to ensure the performance of the entire supersonic flight envelope. The analysis proved that the DSI was sufficiently capable of ensuring continued operation during maneuvers at the design speed, and especially if the cowl structure was optimized for such maneuvers and flow-control technology was applied to improve the internal flow pattern.

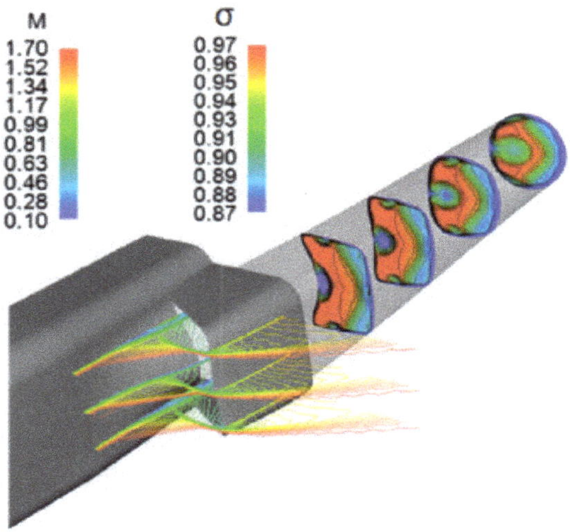

Figure 8. Distribution of changes in the Mach numbers and total pressure in a DSI intake system [6]. Adapted with permission from Ref. [6]. Copyright 2020, copyright Proc. Inst. Mech. Eng. Part G J. Mech. Eng.

In the literature analysis of this research area, the problem of studying the effect of a change in pressure in the inlet-fan zone (AIP (aerodynamic interface plane)) on inlet-flow disturbances was not encountered. The change in this pressure can be caused by various factors, the main one being a change in the setting angle of the inlet guide vanes. The control system causing a change in the angle of the inlet guide vanes due to its stable operation can cause instability in the operation of the inlet system. This article is part of the work carried out to study the cooperation of the inlet system with the aircraft engine fan.

2. Computational Model

The commercial computing package CFD Ansys Fluent was used to perform the numerical simulations. The package is based on an the FVM (finite volume method). An undisputed advantage of the method is the possibility of building nonorthogonal and nonuniform computational meshes, which can be critical in the computational tasks for objects with complex shapes [7], such as for a multirole aircraft (Figure 9)

First, a computational domain was prepared, and the shape of a cuboid with the dimensions 20 m × 10 m × 10 m was chosen. It was considered necessary to thoroughly investigate the zone directly ahead of the intake section, such as the zone of an intake vortex. Therefore, based on our own experience, it was decided to separate a subdomain in the form of a cuboid with the dimensions 7.16 m × 3.75 m × 4 m. The separated additional volume was required to compact the mesh in the immediate vicinity of the engine inlet without compacting it across the entire computational domain. Procedures for extracting additional volumes for the purpose of thickening the computational meshes were required in order to increase the accuracy of the calculations. Figure 10 shows the computational domain prepared for discretization, with a separated volume necessary to perform the mesh compaction.

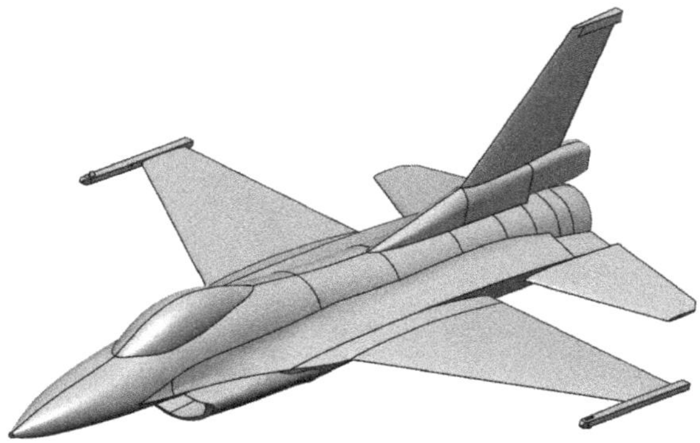

Figure 9. Surface model of the F-16 aircraft.

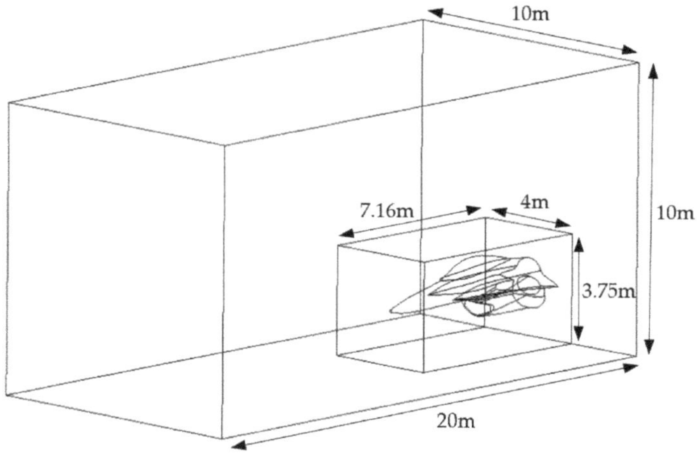

Figure 10. Computational domain prepared for discretization, with a separate additional volume for the compacting of the mesh.

Before starting the discretization process, the boundary conditions recognized by the Ansys Fluent software were assigned to the appropriate surfaces of the prepared calculation area. Thus, the "pressure-far-field" condition was assigned to the outer regions of the computational domain. This condition was used to simulate the free flow for a given Mach number, and the static-pressure and temperature values. A "pressure outlet" condition was set at the compressor inlet to establish a static-pressure value corresponding to the compressor operating conditions. The remaining surfaces were assigned "wall" conditions, including the fuselage surface of the airplane model and the surface representing the ground plane (under the fuselage surface). In the case of numerical flow analysis using viscous fluids, the "wall" condition allowed us to determine the nonslip on the surface, representing the fluid–solid boundary.

The model was adapted to carry out numerical three-dimensional flow simulations; hence, the discretization was made on the basis of discrete 3D elements with the use of structural and nonstructural meshes by building the so-called hybrid mesh. Triangular elements were modeled on the surfaces of the calculation area. In the area of the boundary layer (in the engine inlet), five layers of prismatic elements were modeled. The remaining volume of the computational domain was discretized with tetrahedral elements. Figure 11 shows the cross section of the mesh created in the aircraft plane of symmetry.

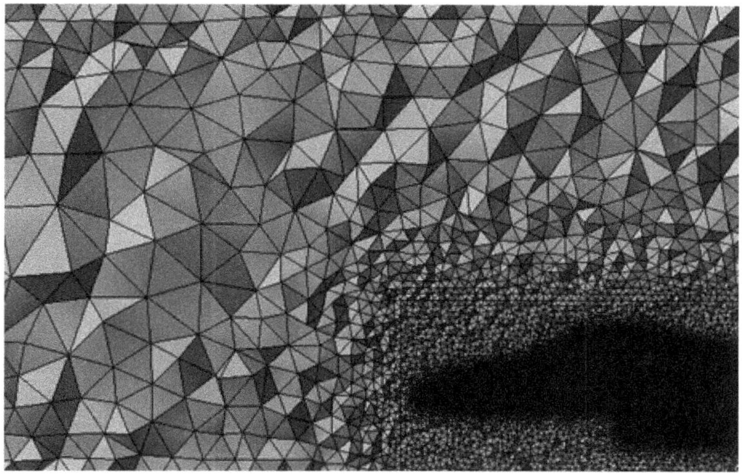

Figure 11. Cross section of the numerical mesh in the aircraft plane of symmetry.

As the FVM (finite volume method) was based on the direct digitization, in a physical space of equations that express the principles of conservation, the starting point involved equations of conservation formulated as integrals [7]. The equations and the method of their transformation are shown in [8,9].

Furthermore, the equations were averaged according to the formulas shown in [10–12], which produced Reynolds-averaged Navier–Stokes (RANS) equations. These nonlinear equations were used to solve the problem of interest. Note that the averaging of these equations resulted in the previously closed system of equations becoming an open system, as the six complementary relationships were missing that determined the components of a turbulent stress tensor [12], and hence, it was necessary to apply turbulence models. In this work, for the purpose of carrying out the numerical simulations, the Spalart–Allmaras turbulence model with its standard settings was used [13].

3. Analysis of the Results

Numerical flow simulations were performed for the 3D case, while the analysis of the obtained results was based on two-dimensional cross sections. An assessment of the flow field in an intake duct, which supplies air to the compressor in the longitudinal plane of the jet engine (Figure 12), does not provide a full picture of the processes that occur throughout the entire volume of the intake duct. This analysis was two-dimensional and shows that low-pressure zones are present in the inlet section within the boundary layer. This was confirmed by the results that show the total-pressure distribution in the cross section in Figure 13a. The color scale in Figure 13 and other figures maps the pressure scale shown in Figure 12.

An estimation of the parameters in a cross-sectional square to the intake-duct axis, and based on the distribution of the pressure parameters in the longitudinal section (Figure 12), is only possible up to the initial part of the intake duct, which is due to a certain regularity in the distribution of the parameters in the cross section (Figure 13a). For the final section of the intake duct, the outputs from the longitudinal section (Figure 12) can be encumbered with large errors (Figure 13b) in attempts to predict the total-pressure distribution. An initial irregularity in the distribution of the parameters was present at 0.5 m (Figure 14) from the inlet cross section A (Figure 13a).

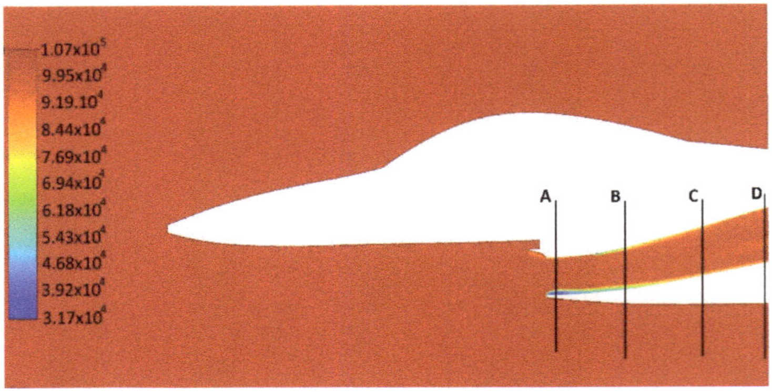

Figure 12. Overview of pressure-distribution changes, in pascals, along the intake duct of the turbojet engine, with cross sections A, B, C, and D shown.

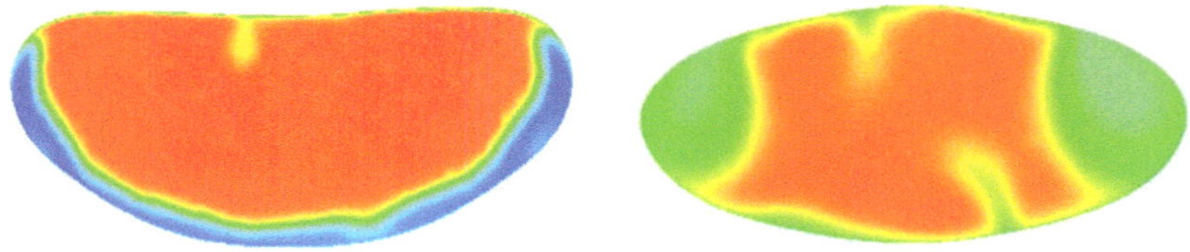

(**a**) (**b**)

Figure 13. Pressure distribution in the intake cross section at throttling pressure: $p = 65$ kPa: (**a**) cross section A; (**b**) cross section D.

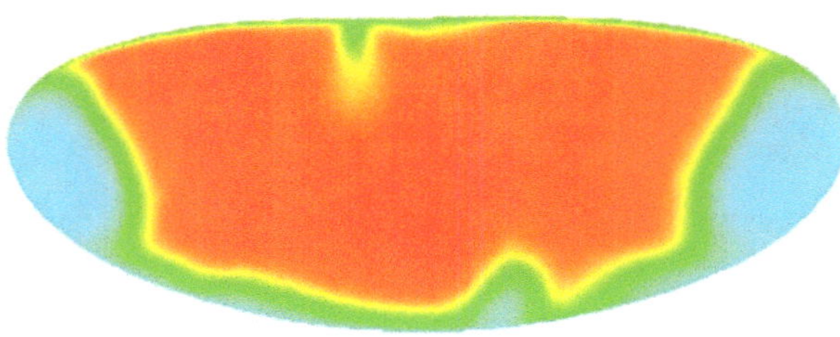

Figure 14. Pressure distribution in the intake cross section at throttling pressure $p = 65$ kPa for cross section B.

To eliminate the problems indicated above, a 3D numerical analysis was performed for a flow along the intake duct of the turbojet engine. Four cross sections (Figure 12) were qualified for further investigation to permit an assessment of the changes in the basic intake parameter, which was the total pressure along the intake duct. The cross sections selected were cross section A of the air inlet (4.5 m from the fuselage nose), cross section D, which connects the intake to the turbojet engine 6.0 m from the fuselage nose, and two intermediate cross sections (B and C, 5.0 m and 5.5 m, respectively, from the fuselage nose).

To analyze the distribution of the pressure fields (Figure 13), five main pressure intervals were defined. The pressure intervals are referred to as "series" in this paper. The tests were performed at a variable outlet pressure. For a pressure of 65 kPa at the intake-system outlet, which was the turbojet fan inlet, pressure intervals were assumed for each series, as listed in Table 1.

Table 1. Pressure intervals for each series.

Series No.	Pressure Interval (kPa)	Color
1	89.9–107.0	Red
2	81.6–89.9	Yellow
3	60.8–81.6	Green
4	48.3–60.8	Azure
5	35.9–48.3	Blue

Figure 15a shows the changes in the distribution of the pressure fields along the flow duct, produced in the air flow through the intake system at 65 kPa. Here, the pressure area grew outside of the main pressure zone vs. the duct length from 28.26% to 42.63% (Figure 15b), with the main pressure zone defined as the pressure values relevant to the zone designated Series 1 (Table 1).

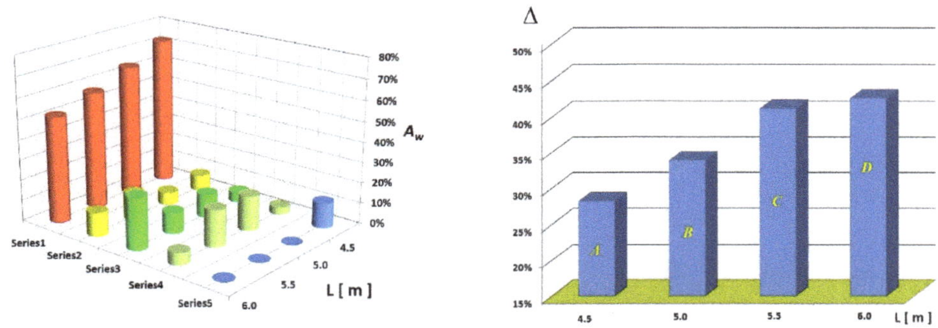

(**a**) (**b**)

Figure 15. Analysis of the distribution of pressure-field changes in the flow duct for $p = 65$ kPa: (**a**) A_w: relative surface area of specific total pressure (series) vs. the cross-section location (L); (**b**) Δ: relative total-pressure-area increase outside of the main zone.

The quantitative change in the distribution in each of the pressure zones is illustrated with the pie charts in Figure 16. There was a noticeable reduction in the main pressure-zone area (Series 1) from 72.58% to 51.99% (Figure 16). The relatively considerable zone of lowest pressure, Series 5, disappeared (Figures 15a and 16a) after occupying 12.03% in the first part of the intake. The Series 3 pressure zone had the highest growth, from 4.85% to 25.17% (Figures 15 and 16). The surface-area shape changed from a system that could be deemed a regular one (Figure 13a) into a rather irregular one (Figure 13b).

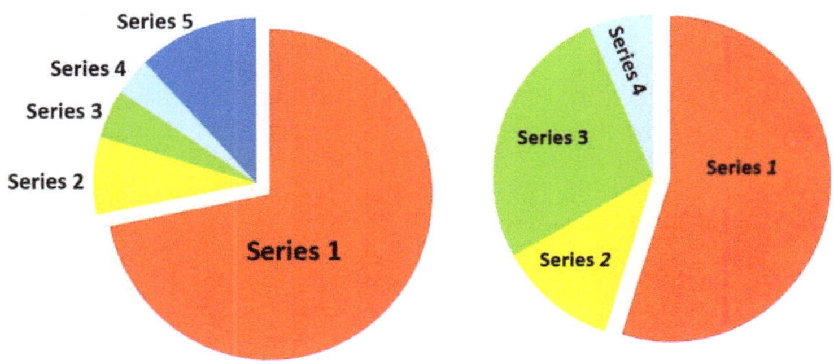

(**a**) (**b**)

Figure 16. Change in the ratios of pressure-distribution-series surface areas within the intake cross section for $p = 65$ kPa: (**a**) cross section A; (**b**) cross section D.

An analysis was carried out for selected operational options for the intake system, where each option was a change in the operating conditions of the turbojet fan. The changes involved throttling the pressure at the intake-system outlet, which could be achieved by changing the setting of the inlet-stator-blade pitch of the fan. This would change the pressure in the AIP cross section. The results of the numerical analyses are shown in Figures 17–21, with the throttling-pressure change within a range of $p = (75 \div 95)$ kPa. The results shown include the distributions of the total pressure in the cross section of the air-intake-duct inlet (plot lines upper-indexed "a") and the cross section of the air-intake-duct outlet ("b"), and the analysis of the distribution of the changes in the dimensionless potential-pressure ratio (1) in the flow duct at a specific throttling-pressure value ("c"):

$$\Delta p_{w,i} = \frac{p_i A_i}{\sum_{k=1}^{k=n} p_k A_k} \tag{1}$$

where i is the area number; n is the number of separated pressure areas; p_i is the average total pressure in the i-th area; A_i is the dimensionless surface area of the i-th area.

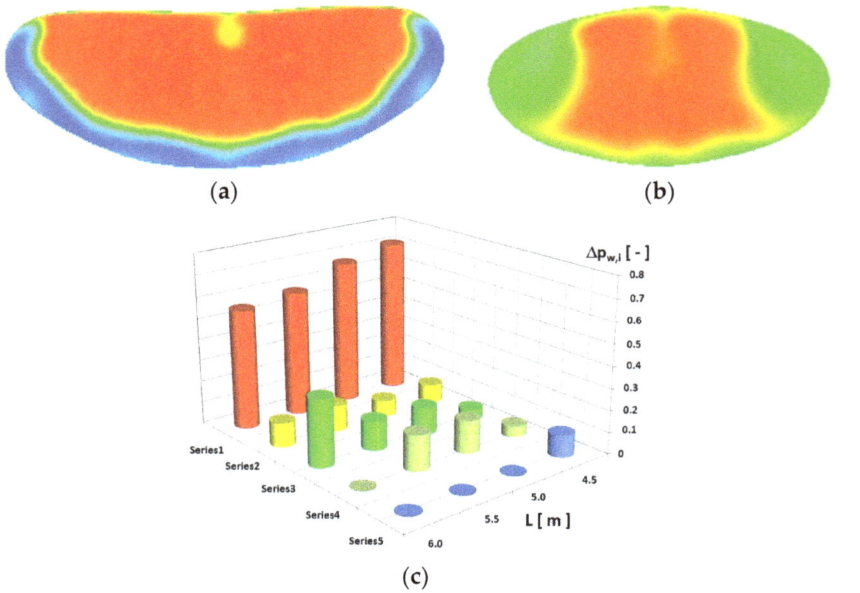

Figure 17. Total-pressure distribution in the inlet: (**a**) cross section A; (**b**) the outlet cross section D; (**c**) the distribution of the dimensionless potential-pressure ratio at Ma = 0.0075, beta 45°, p = 75 kPa.

The results concerning the performance of the inlet cross section at different throttling-pressure values provided a certain characteristic repeatability in the pressure-distribution zones, even with a partial change in the pressure range or area size. The flow-duct-outlet (final) cross section had an irregular pressure distribution, which changed with the throttling. The changes in the surface area in specific series of the individual cross sections are shown as plots in Figures 17–21, Part (c). The disappearance of the lowest-pressure zones was evident. This applied to Series 4 and 5, at the fuselage-boundary area, and it was a function of the cross-section location inside of the fuselage (increase in parameter L) and the throttling increase. A moderate-pressure area (Series 3) became evident, and very much so in Figure 22, or by comparing the results in Figure 16 to the results in Figure 22, and this area was a result of a throttling-pressure increase at the intake-duct end.

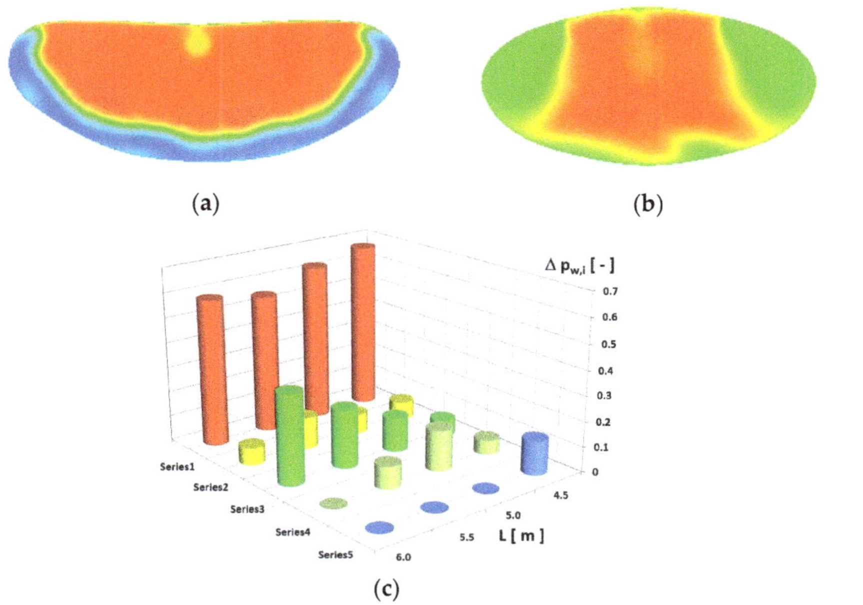

Figure 18. Total-pressure distribution in the inlet: (**a**) cross section A; (**b**) the outlet cross section D; (**c**) distribution of the dimensionless potential-pressure ratio at Ma = 0.0075, beta 45°, p = 80 kPa.

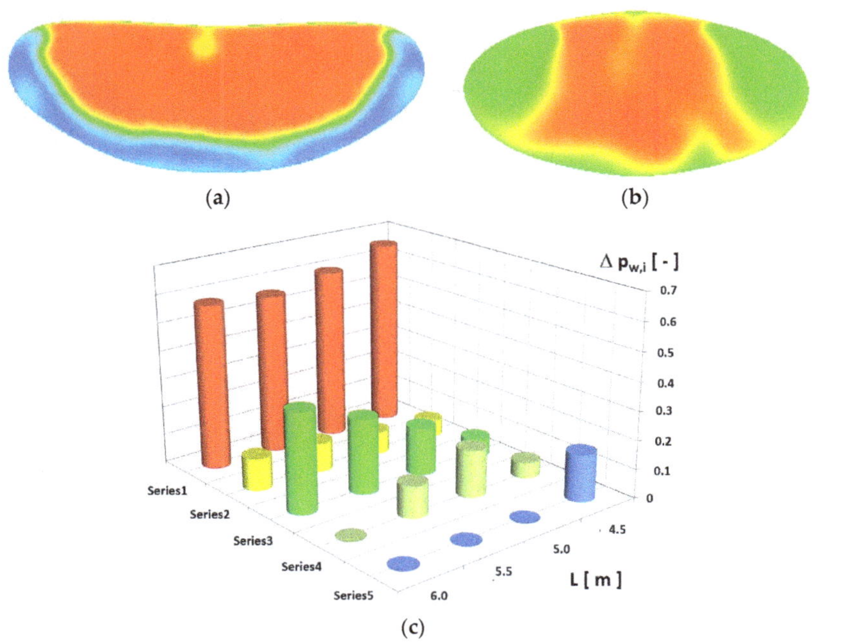

Figure 19. Total-pressure distribution in the inlet: (**a**) cross section A; (**b**) the outlet cross section D; (**c**) the distribution of the dimensionless potential-pressure ratio at Ma = 0.0075, beta 45°, p = 85 kPa.

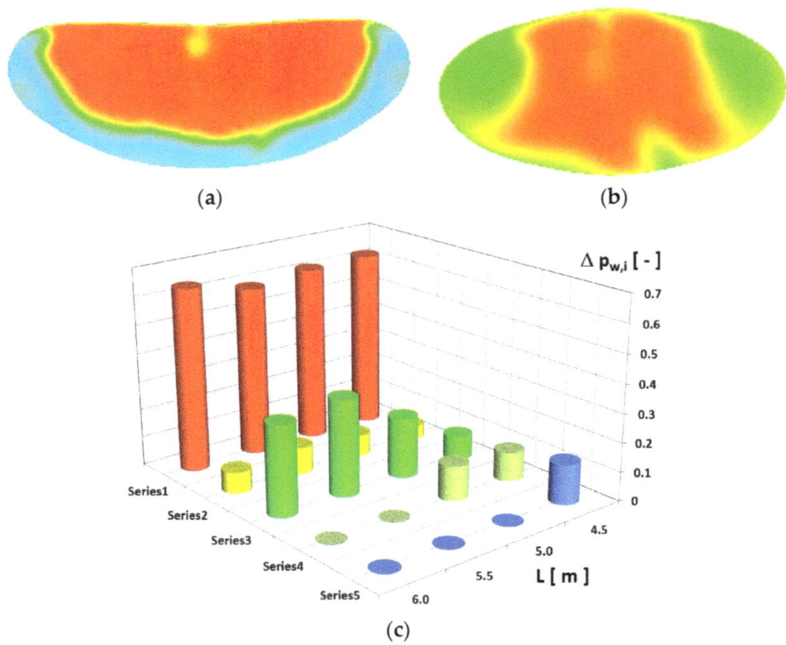

Figure 20. Total-pressure distribution in the inlet: (**a**) cross section A; (**b**) the outlet cross section D; (**c**) the distribution of the dimensionless potential-pressure ratio at Ma = 0.0075, beta 45°, p = 90 kPa.

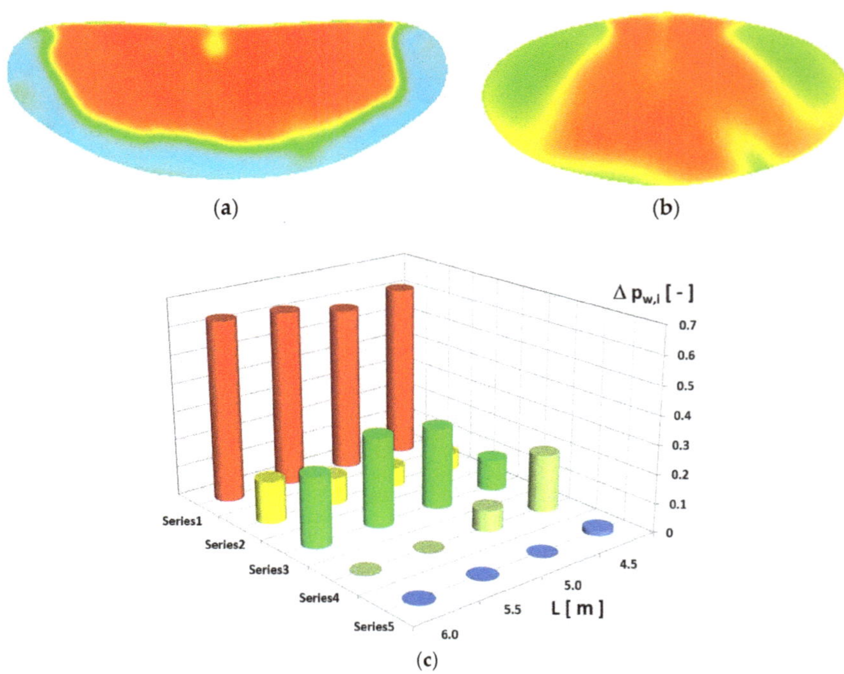

Figure 21. Total-pressure distribution in the inlet: (**a**) cross section A; (**b**) the outlet cross section D; (**c**) the distribution of the dimensionless potential-pressure ratio at Ma = 0.0075, beta 45°, p = 95 kPa.

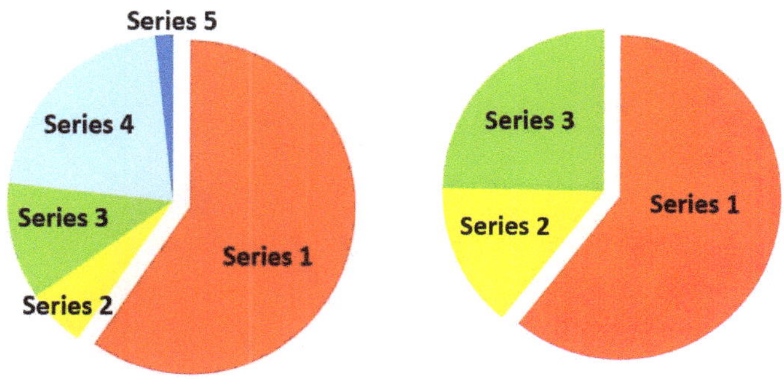

(a) (b)

Figure 22. Change in the ratios of pressure-distribution-series surface areas within the intake cross section: (**a**) cross section A; (**b**) cross section D (Ma = 0.0075, beta 45°, p = 95 kPa).

The moderate-pressure area (Series 3) in Figures 13, 17c, 18c, 19c, 20c and 21c, had its main part focused in the final sector of the starboard and port parts of the flow duct.

When analyzing the pressure distribution in the air-intake-inlet zone for an increase in the throttling pressure, as shown in Figures 16a, 17a, 18a, 19a, 20a, 21a and 22a, there was a distinct change in the areas occupied by each series. The areas beyond the main zone (series) increased at the inlet. The nature of the Series 1 pressure distribution also changed at the highest throttling pressure (Figure 22), and it did not change much down the duct.

To summarize the performance of the air-intake system with variations in the throttling pressure in a way that included the location of the tested cross section, a global dimensionless potential-pressure ratio was used: $\Delta p_i(p; L)$. The "Dimensionless potential pressure ratio" will be designated "DPPR" in this paper.

An analysis of the DPPR in the function of the cross-section location and changes in the air-intake-duct outlet pressure (which is the throttling pressure) is shown in Figure 23. The reference cross section for each throttling case was the inlet cross section (L = 4.5 m). The most significant changes were produced without the intake system being throttled (p = 55 kPa). This was characterized by a drop in the DPPR, evidenced by an increased nonuniformity of the total-pressure field. The DPPR fell by 12.0%, although it was improved by ≈1.0% at L = 5.0 m. The pressure throttling at the air-intake-duct outlet, with an increase in the pressure (p) to 95 kPa, caused increased nonuniformity in the DPPR. In the first step, for $p = 65$ kPa, the pressure parameter $\Delta p(L)$ increased in the end part of the intake duct by less than 1.0, with the trend dropping as a function of the L. A further increase in the pressure improved the DPPR, and the value increased above 1.0. The case with the most uniform values was $p = 80$ kPa, with the maximum pressure being 1.1% in the outlet (final) cross section of the air-intake duct. The highest DPPR value was produced in the system of parameters $p = 80$ kPa and $L = 6$ m, amounting to $\Delta p(p; L) = 1.03857$. The highest changes in the DPPR were within the outlet (final) cross section of the flow duct ($L = 6$ m) as a function of the throttling pressure, as is clearly shown in Figure 23. It can be argued that the process of increased throttling, related to an increase in pressure, improved parameter $\Delta p(p, L)$. In an aircraft turbojet engine, throttling is controlled by a system of variable-pitch stator rings. On the one hand, one purpose of the variable-pitch stators, including the inlet stator, such as the one used in the F-100-PW-229 engine, is to stabilize the fan performance by changing the inflow angle to the rotor ring. On the other hand, the results produced suggest a conclusion concerning the stabilization of the turbojet compressor performance by increasing the pressure in the AIP, which is achieved by changing the stator-ring-blade pitch.

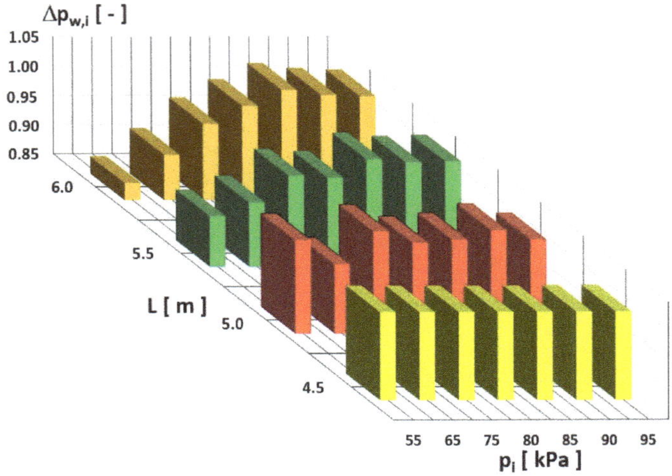

Figure 23. Distribution of changes in the potential pressure.

4. Conclusions

The research and testing results presented in this paper, which used a computational solver in Ansys Fluent that is capable of simulating flows, provide important information about the behavior of the turbojet engine air-intake system during an increase in the outlet pressure within the AIP, or flow throttling. The results illustrate the change in the nature of the distribution of the pressure fields, from relatively regular at the air-intake inlet, to zones that did not have any regularity at the end of the intake-flow channel. Those zones with the lowest pressure disappeared and were replaced with moderate-pressure zones along the intake duct. A major conclusion from the results of testing the DPPR (Figure 23) is that it is possible, within certain limits, to create an improvement in the flow uniformity by increasing the throttling pressure. The physical quantity of the DPPR is the relative value of p_iA_i, and Figure 23 shows the change in the pressure-potential coefficient as a function of the cross-sectional area tested and the change in the final pressure. The use of this coefficient allowed for a comparison of the pressure distributions in a tunnel of a changeable shape and area. The cross section $L = 4.5$ is taken as a reference. In some cross sections, the value of the resultant coefficient of the entire cross section changed a little. This information is of high utilitarian value for turbojet engine control and the programming of the engine's control system, which must protect the engine from unstable operation through control and adjustment.

For all the cases analyzed, an intake vortex was determined, the location of which changed with the point of stagnation (Figures 14 and 15). These results illustrate the significance of the influence of the intake vortex on the uneven distribution of such an important parameter as the total pressure, and thus there is the need to conduct further numerical analyses, as well as experimental research, in this area. To follow the mathematical approach of describing flow phenomena and numerical modeling, Ref. [14] introduces researchers to fluid-mechanics phenomena from the perspective of mathematical equilibrium.

Author Contributions: Conceptualization, A.K., S.K., M.F. and M.M.; methodology, A.K., M.F. and M.M.; software, A.K., S.K., M.F. and M.M.; validation, A.K., S.K., M.F. and M.M.; formal analysis, A.K., M.F. and M.M.; investigation, A.K., S.K., M.F. and M.M.; resources, A.K., S.K., M.F. and M.M.; data curation, A.K., M.F. and M.M.; writing—original draft preparation, A.K., S.K., M.F. and M.M.; writing—review and editing, A.K., S.K., M.F. and M.M.; visualization, M.F. and M.M.; supervision, A.K., M.F. and M.M.; project administration, A.K.; funding acquisition, A.K., S.K., M.F. and M.M. All authors have read and agreed to the published version of the manuscript.

Funding: This research was funded by the Military University of Technology, Warsaw, Poland, under research project No. UGB 781/2022.

Institutional Review Board Statement: Not applicable.

Informed Consent Statement: Not applicable.

Data Availability Statement: Not applicable.

Conflicts of Interest: The authors declare no conflict of interest.

References

1. Chiereghin, N.; Guglielmi, L.; Savill, M.; Kipouros, T.; Manca, E.; Rigobello, A.; Barison, M.; Benini, E. Shape Optimization of a Curved Duct with Free Form Deformations. In Proceedings of the American Institute of Aeronautics and Astronautics AIAA 2017-4114, Session: Optimization and Design Using CFD, Denver, CO, USA, 5–9 June 2017; Available online: https://www.researchgate.net/publication/317694229 (accessed on 2 June 2017). [CrossRef]
2. Afzal, A.A.; Safdar, M.M.; Javed, A. Performance Analysis of a Proposed Design of Diverterless Supersonic Inlet at Various Flight Conditions. In Proceedings of the AIAA Scitech 2021 Forum, Virtual Event, 11–15 and 19–21 January 2021. [CrossRef]
3. Tanguy, G.; Macmanus, D.; Garnier, E.; Martin, P. Characteristics of Unsteady Total Pressure Distortion for a Complex Aero-engine Intake Duct. In *Aerospace Science and Technology*; Elsevier: Amsterdam, The Netherlands, 2018; Volume 78, pp. 297–311. [CrossRef]
4. Montes, R.A.; Chandler, F.O. A CFD investigation of a diverterless supersonic inlet of ellipsoidal entrance shape. In Proceedings of the AIAA Propulsion and Energy Forum, Indianapolis, IN, USA, 19–22 August 2019; pp. 1–33. [CrossRef]
5. Ibrahim, I.H.; Ng, E.Y.K.; Wong, K. Flight Maneuverability Characteristics of the F-16 CFD and Correlation with its Intake Total Pressure Recovery and Distortion. *Eng. Appl. Comput. Fluid Mech.* **2011**, *5*, 223–234. [CrossRef]
6. Saheby, E.B.; Shen, X.; Hays, A.P. Design and performance study of a parametric diverterless supersonic inlet. *Proc. Inst. Mech. Eng. Part G J. Mech. Eng.* **2020**, *234*, 470–489. [CrossRef]
7. Chen, W.; Zheng, X.; Liu, S. Finite-Element-Mesh Based Method for Modeling and Optimization of Lattice Structures for Additive Manufacturing. *Materials* **2018**, *11*, 2073. [CrossRef] [PubMed]
8. Kozakiewicz, A.; Frant, M.; Majcher, M. Impact of the Intake Vortex on the Stability of the Turbine Jet Engine Intake System. *Int. Rev. Aerosp. Eng.* **2021**, *14*, 4. [CrossRef]
9. Frant, M.; Kozakiewicz, A.; Kachel, S. Analysis of Impact of Gust Angle and Velocity on the Position of Stagnation Point. *Adv. Sci. Technol. Res. J.* **2020**, *14*, 49–57. [CrossRef]
10. Lumley, J.L.; Yaglom, A.M. A Century of Turbulence. Flow. *Turbul. Combust.* **2001**, *66*, 241–286. [CrossRef]
11. Gry Liu, B.; Vanierschot, M. Numerical Study of the Hydrodynamic Characteristics Comparison between a Ducted Propeller and a Rim-Driven Thruster. *Appl. Sci.* **2021**, *11*, 4919. [CrossRef]
12. Elsner, J.W. *Turbulence of Flows*; State Scientific Publishing House: Kraków, Poland, 1987; pp. 27–30, 52–62, 379. (In Polish)
13. *ANSYS FLUENT Theory Guide, Release 14.0*; ANSYS, Inc.: Canonsburg, PA, USA, 2011.
14. Noii, N.; Khodadadian, A.; Wick, T. Bayesian inversion for anisotropic hydraulic phase-field fracture. *Comput. Methods Appl. Mech. Eng.* **2021**, *386*, 114118. [CrossRef]

fluids

MDPI

Article

Hydrodynamic Characteristics of Two Side-by-Side Cylinders at a Pitch Ratio of 2 at Low Subcritical Reynolds Numbers

Thiago Gomes, Jhon Goulart * and Carla Anflor

Group of Experimental and Computational Mechanics, University of Brasília, Gama 72.405-610, Brazil
* Correspondence: jvaz@unb.br

Abstract: Isothermal turbulent flow around circular cylinders arranged side-by-side was numerically simulated on a commercial finite-volumes platform, ANSYS® CFX, version 2020 R2. The turbulence was modeled by using k-ω shear stress transport (k-ω SST). Three different Reynolds numbers were computed, Re_d = 200, 1000, and 3000, which were based on the cylinder diameter, d, the free stream velocity, $U\infty$, and the kinematic viscosity of the fluid, ν. Sided cylinders were spaced apart from each other, forming a *p*/*d* ratio equal to 2, which was kept constant throughout the computations regardless of changes in the Reynolds number. The drag coefficient, C_d, as well as its time traces, was evaluated along with the different wake topologies experienced by the cylinders (wide wake *WW* and narrow wake *NW*). The simulations were able to predict the bistable flow over the cylinders and the Cd changes associated with the wakes. Whenever a new wake topology was identified, the shape drag changed in accordance with the instantaneous pressure distribution. A laminar simulation was carried out for the lowest Reynolds number case, showing that the adopted turbulence model did not affect the dynamic response of the flow. The Re_d = 3000 case was compared to Afgan's outcomes, whose simulations were carried out in a 3-D mesh using LES (Large Eddy Simulation), showing great agreement with their results.

Keywords: circular cylinders; side-by-side; detached angle; k-ω SST; bistability; turbulent flow

Citation: Gomes, T.; Goulart, J.; Anflor, C. Hydrodynamic Characteristics of Two Side-by-Side Cylinders at a Pitch Ratio of 2 at Low Subcritical Reynolds Numbers. *Fluids* **2022**, *7*, 287. https://doi.org/10.3390/fluids7090287

Academic Editors: Vasily Novozhilov and Mehrdad Massoudi

Received: 17 May 2022
Accepted: 3 August 2022
Published: 30 August 2022

1. Introduction

Due to vast applications in real life, the study of turbulent flow characteristics around cylinders has been the focus of attention for a long time. Knowledge of the flow field and its dynamics characteristics over such bodies is applied on a vast scale. Circular cylinders in pairs (side-by-side or in tandem) or even arranged in banks have been the subject of research since the very early twentieth century with the outcomes released by Grimison [1] and Wiemer [2]. A very thorough experimental campaign was carried out by Žukauskas [3] followed by Žukauskas et al. [4] and Žukauskas and Katinas [5]. In these works, the authors were very concerned with outlining the basis for heat transfer prediction in bank tubes, for instance in Žukauskas's work [3]. It is important to remind the reader that bank tubes or closed packet rods are the simplest geometries used for research to study the flow field and the features of its fluctuation over structures arranged in groups. We can easily cite the case of struts of a biplane wing or the flow past columns of a marine structure in offshore engineering, transmission lines, and heat exchanger tubes or bundles of risers [6,7]. In contrast to the high Reynolds numbers produced in such applications, low Reynolds numbers can be seen in the papermaking process [8]. According to the authors, the wood fibers can be modelled as sided cylindrical structures with dimensions of about 1 mm in length and 40 μm in diameter.

In recent times, the works [9–14], among others, were concerned with the wake interactions behind the cylinders when the gap between them changes. Furthermore, they also try to understand the flow changes in association with the Reynolds numbers, as was shown very well by Sumner [14].

Bearman and Wadcok [15] investigated experimentally the flow interaction in a pair of circular cylinders for a Reynolds number of 2.5×10^4. The work aimed to study the flow behavior as the *p/d* ratio changed. The authors observed that for sided circular cylinders separated from each other by a p/d ratio greater than 2, the wake formed downstream was similar to the one that takes place in a single cylinder. However, as the *p/d* ratio decreased, an asymmetric flow field appeared around the cylinder, also producing some effect on the vortex-shedding frequency. Furthermore, the authors, in 1973, also they pointed out that the shedding vortex mode could occur in phase and antiphase synchronization, but the second mode was seen more often. Years later, Meneghini and co-workers [16] also reached the same conclusion. At that time, the authors only associated this difference with the p/d ratio. Later, Zdravkovich and Pridden [7] investigated the wake formation and its relationship with the p/d ratio, reaching the same conclusions on the asymmetric flow field whenever the p/d ratio decreased below 2. Their experimental campaign conducted under subcritical Reynolds numbers (8×10^3 to 1.6×10^5) showed that different drags were assigned to the cylinders at the same time. Furthermore, the sum of the drags was less than twice the value of the drag for a single cylinder under the same Reynolds number.

Numerical and experimental works have aimed to study the bistability process in either a row of cylinders or in a pair of them [16–19]. According to Neumeister [19], bistable flow is only observed in the situation in which the wakes interact with each other, giving rise to stable wake topologies that can change randomly over time. Such a configuration leads to an asymmetric flow, forming dissimilar wakes (narrow wake *NW* and wide wake *WW*) downstream of the cylinders, whose main characteristics rule the aerodynamic forces on the cylinders' surfaces and the dynamics of the flow as well. Asymmetric wake formation was also shown by researchers [20–23]. Vila et al. [24] carried out an experimental campaign using two hot-wire probes and a single pressure transducer in a pair of circular cylinders for three different p/d ratios. They evaluated the turbulent signals of pressure and velocity acquired at the same time. The pressure time trace was gathered on the circular cylinder's surface, while the velocity signals were taken in the viscous wake. The authors performed this study for three different p/d ratios (1.26, 2.00, and 3.00) under a subcritical regime, $Re_d = 1.78 \times 10^4$. The authors identified that both the stagnation and boundary layer detachment points moved towards the tight gap as the p/d ratio decreased. Furthermore, with regard to the velocity and pressure time traces, the signals were seen to present long-term bistable behavior for the lowest p/d ratio. On the other hand, as the p/d ratio increased, this pattern tended to fade away. Spectral analysis using both *PSD* and *CWT* [18] tools showed energy peaks associated with the vortex shedding. The Strouhal number was seen to range from 0.22 to 0.24. These values were slightly higher than that identified in a single cylinder.

The outcomes for two-dimensional flow characteristics over circular cylinders arranged in pairs were reported by Kang [22]. In his numerical work, the author carried out simulations for various Reynolds numbers and T/D ratios, comprising $40 \leq \mathrm{Re} \leq 160$ and $0.2 \leq T/D \leq 5.0$, respectively. Unlike others, the author characterized the narrow gap space between cylinders as $T = p/d - 1$. The numerical results of Kang [22] identified up to six different topologies for wakes, depending on the distance between the centers of the cylinders and the Reynolds number. The author also stated that, for the studied Reynolds number range, the frequency of vortex shedding was influenced mainly by the spacing between the cylinders. For $0.5 < T/D < 1.5$, the vortex frequency dropped and was constantly synchronized with the movement of the wakes. On the other hand, the drag coefficients depended mainly on the spacing between the cylinders. Finally, the author concluded that as the T/D ratio increased, for values higher than 3, the flow characteristics again became significantly dependent on the Reynolds number.

In case of fluid–structure interaction (*FSI*), recently, Chen and co-authors [25,26] investigated the wake patterns of two-sided circular cylinders which were free to vibrate. The authors' investigations, in both works, aimed to provide an overview of the wake patterns for different p/d ratios and Reynolds numbers, whose values ranged from 60 up to

200. In the first paper by Chen [25], the authors simulated the fluid–structure interaction by using the immersed boundary (*IB*) method. The gap-spacing ratio (p/d) and the reduced velocity (U_r) were changed from 2 up to 5 and 0 up to 30, respectively. The authors identified up to eight wake flow patterns, whose existence was based on the gap spacing and the reduced velocity. For instance, the biased flow, which produces narrow and wide wakes, was only observed for a $2.3 < p/d < 2.5$ and U_r varied between 4.0 and 4.50. Later, in 2020, Chen [26] and co-authors furthered their numerical experiments by studying the effects of the p/d ratio and the reduced velocity U_r on the flow dynamics of two-sided cylinders free to vibrate. The authors concluded that the p/d ratio plays an important role in the dynamic response of the cylinders. The authors identified $St > 0.20$ for low U_r and gap ratios between 2.0 and 2.5.

The present work aimed to numerically investigate the hydrodynamic characteristics over two-sided cylinders. Three Reynolds numbers were simulated, Re_d = 200, 1000, and 3000, keeping the p/d ratio constant, equal to 2. In order to verify the quality of the computations, mean average values such as the stagnation and separation angles, (θ_{Est}) and (δ_{Sep}), and the drag forces, C_d, were compared to those reported by Afgan et al. [27] for Re_d = 3000 and a p/d ratio equal to 2. Finally, special attention was given to the bistability of the flow and its effect on the aerodynamic forces whenever a new topology was formed as well as the dynamic response of the flow behind the cylinders through the velocity time traces.

2. Materials and Methods

2.1. Governing Equations

For incompressible flow, the mass and the momentum conservation are ruled by:

$$\frac{\partial \overline{u_i}}{\partial x_i} = 0 \tag{1}$$

$$\frac{\partial \overline{u_i}}{\partial t} + \overline{u_j}\frac{\partial \overline{u_i}}{\partial x_j} = -\frac{1}{\rho}\frac{\partial \overline{P}}{\partial x_i} + \frac{\partial}{\partial x_j}\left[(\nu + \nu_t)\left(\frac{\partial \overline{u_i}}{\partial x_j} + \frac{\partial \overline{u_j}}{\partial x_i}\right)\right] \tag{2}$$

In Equations (1) and (2), $\overline{u_i}$ and $\overline{u_j}$ represent the velocity vector components, x_i is the spatial coordinates, $\overline{P}$ is the thermodynamic pressure, ρ is the fluid density, and ν and ν_t are the molecular and turbulent kinematic viscosity, respectively. The additional momentum diffusivity caused by the closure problem of the turbulence is represented by the turbulent viscosity, ν_t, which is approached through Boussinesq's idea, as follows:

$$\tau_{ij} = \nu_t\left(\frac{\partial \overline{u_i}}{\partial x_j} + \frac{\partial \overline{u_j}}{\partial x_i}\right) - \frac{2}{3}\delta_{ij}k \tag{3}$$

τ_{ij} is the Reynolds tensor, which comes from the decomposition of the nonlinear terms of the Navier–Stokes equation, and k represents the turbulent kinetic energy. So, additional equations are needed to model the turbulent kinematic viscosity, which is computed as a function of the turbulent kinetic energy field, k, and the specific rate of dissipation, ω.

The *k*-ω *SST* model is a two-equation turbulence model first introduced by Menter [28]. The model combines the advantages of the *k*-ε model and the *k*-ω model through a blending function that switches whenever it is possible. According to Menter [28], the two-equation model is ruled by the set of equations in Equation (4):

$$\begin{aligned} \frac{\partial k}{\partial t} + \overline{u_j}\frac{\partial k}{\partial x_j} &= \frac{\tau_{ij}}{\rho}\frac{\partial \overline{u_i}}{\partial x_j} - \beta * \omega k + \frac{\partial}{\partial x_j}\left[(\nu + \sigma_k \nu_t)\frac{\partial k}{\partial x_j}\right] \\ \frac{\partial \omega}{\partial t} + \overline{u_j}\frac{\partial \omega}{\partial x_j} &= \frac{\gamma \tau_{ij}}{\nu_t \rho}\frac{\partial \overline{u_i}}{\partial x_j} - \beta\omega^2 + \frac{\partial}{\partial x_j}\left[(\nu + \sigma_\omega \nu_t)\frac{\partial \omega}{\partial x_j}\right] + 2(1 - F_1)\frac{\sigma_{\omega 2}}{\omega}\frac{\partial k}{\partial x_j}\frac{\partial \omega}{\partial x_j} \end{aligned} \tag{4}$$

The bending function is F_1, which computes how far from the walls the problem is:

$$F_1 = \tanh\left(\arg^4\right)$$
$$\arg = \min\left[\max\left(\frac{\sqrt{k}}{\beta * \omega d}, \frac{500\upsilon}{d^2\omega}\right), \frac{4\rho\sigma_{\omega 2}k}{CD_{k\omega}d^2}\right] \quad (5)$$

Finally, the turbulent kinematic viscosity is calculated by

$$\nu_t = \frac{a_1 k}{\max(a_1\omega, \Omega F_2)} \quad (6)$$

where Ω is the absolute value of the vorticity and a_1 is a closure coefficient that is set to 0.30. For further information, see the complete description of the model in [28].

2.2. Computational Domain, Boundary Conditions, and Mesh Dependence

The rectangular computational domain is based on the work published by Afgan et al. [27]. Its dimensions were made dimensionless by using the diameter of the circular cylinder, d, its total length being $25d$, and height $22d$. From the inlet up to the cylinders' center, the computational domain is $10d$ long, whereas downstream of the cylinders, the flow travels $15d$ to reach the domain's outlet. Both cylinders are placed in the center of the domain. Furthermore, their centers are separated from each other by a distance, p. The dimensionless number that rules this distance is the p/d ratio, which was kept constant throughout this work, being equal to 2. The flow comes into the domain through the inlet with a free stream velocity, U_∞, oriented parallel to the x-axis. A prescribed velocity, $u = U_\infty$, $v = w = 0$, was imposed on the upper and lower faces of the domain (Figure 1a). The free stream turbulence intensity was set to 1% (in the work by Afgan et al. [27], the authors did not provide this information). No slip condition was applied to the walls, $u = v = 0$, and, finally, the outlet boundary. At the outlet boundary, the pressure difference was set to zero. A schematic view of the domain, its coordinates, boundaries, and mesh detail are depicted in Figure 1a. Figure 1b shows a schematic view of the circular cylinders and how they are oriented. The arrow over the cylinders indicates how the azimuthal positions are taken into account.

Based on the entrance velocity, three Reynolds numbers were simulated, Re_d = 200, 1000, and 3000, keeping the domain's dimensions, p/d ratio, and cylinder diameters unchanged. The commercial software *ANSYS CFX* is only able to perform analyses for 3-D domains. To ensure that the 2-D flow over the cylinders was symmetric, symmetry boundary conditions were applied to the x,y faces, which means that the spanwise derivative terms of the equations are set to zero.

Downstream of the upper and lower cylinders, two probes, i.e., points in the mesh where the temporal flow data were stored, were placed at a distance of $0.9d$ from the cylinder's center (Figure 1b). The probes were placed according to the work by De Paula et al. [18]. Velocity time traces were gathered by the virtual probes.

The mesh was built by splitting the domain into smaller ones, all of which were formed of smaller hexahedral volumes. Special care was taken near the walls (on the cylinders' surfaces), where $y+$ was carefully computed to ensure that the nondimensional distance from the wall would not exceed unity. The $y+$ was measured after each stationary run of a new mesh. Parallel to the z-axis, the mesh was built by splitting the third dimension into one volume (see the mesh detail in Figure 1a). The domain's thickness was 10 mm. Three different meshes were built and tested for Re_d = 3000. Afgan et al. [27] used the same configuration and Reynolds number for predicting the turbulent flow over side-by-side circular cylinders. Their outcomes were used as a benchmark for the present simulations. In Figure 2, one can see the $y+$ distribution on the cylinders' surface. It is possible to see that any generated mesh achieved the first imposition, that is $y+ \leq 1$. Actually, the coarsest one showed $y+ = 1$ at about 45° on the lower cylinder's surface. For the reader's guidance, in Figure 2, the flow reached the cylinders from left to right.

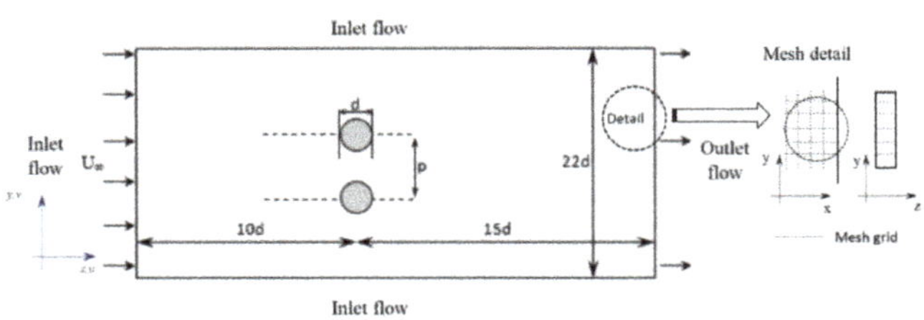

(**a**)

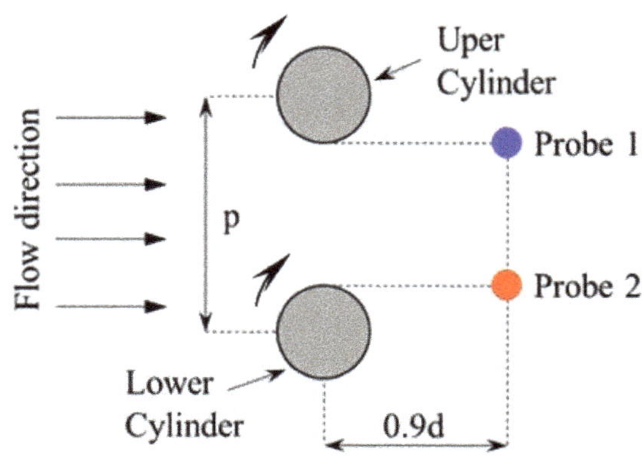

(**b**)

Figure 1. (**a**) Schematic view of the domain. (**b**) Probes' location.

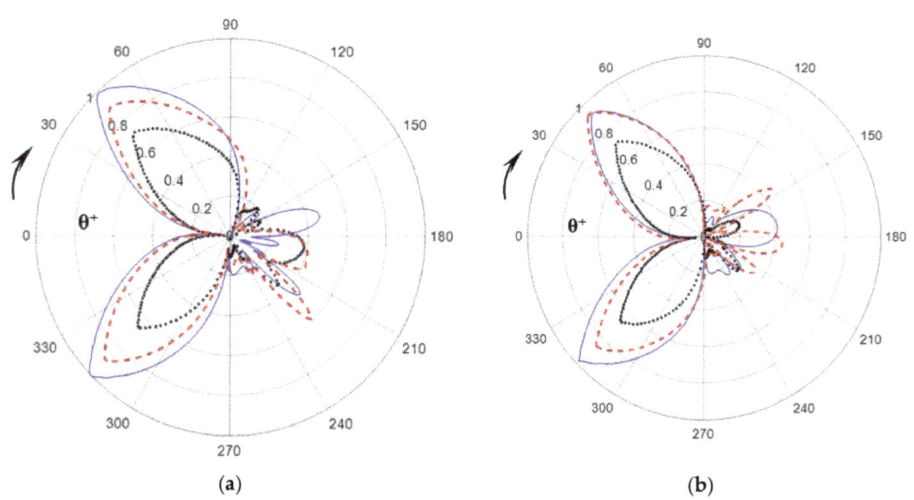

Figure 2. Nondimensional wall distance, y+, for cylinders' surfaces at Red = 3000. (**a**) Upper cylinder (**b**) Lower cylinder—Mesh 1 (coarse); Mesh 2 (medium); o Mesh 3 (fine).

As stated above, three different meshes were tested and compared with the results presented in the work by Afgan et al. [27]. The meshes were named from *M3*, the finest one to *M1*, the coarsest. The mesh characteristics and results are summarized in Table 1, along with the stagnation (θ_{Est}) and separation angles (δ_{Sep}) and the results published earlier by Afgan et al. [2]. From Table 1, it is possible to observe that the results from the meshes are quite close to those reported in Afgan's paper. The stagnation angles (θ_{Est}) were seen to be unchanged, regardless of the mesh.

Table 1. Mesh characteristic. The stagnation, θ_{Est}, and separation angles, δ_{Sep}, for Re_d = 3000 and p/d ratio 2.0.

	θ_{Est} *Upper*	θ_{Est} *Lower*	δ_{Sep} *Upper*		δ_{Sep} *Lower*		**Total Nodes**	**N° Divisions on Cylinders' Surfaces**
Afgan et al., 2011	353.7	8.1	83.8	262.1	99.8	276.3	——	——
Mesh 1	353.6	7.7	87.0	265.7	104.9	278.1	216,000	492
Mesh 2	353.6	7.7	85.5	263.5	101.7	277.3	235,000	504
Mesh 3	353.6	7.7	84.9	262.8	100.2	276.4	241,000	504

Only a very marginal difference of 0.4° is seen in comparison with Afgan's results. The highest variations were observed for the flow separation angle, δ_{Sep}. As the mesh becomes finer, the location where the flow detaches becomes closer to the value predicated by Afgan and his co-workers in their 3-D numerical simulation. The difference was found to range from 1% (M3) up to 4.70% (M1). Considering the results and the time-consuming simulation, the mesh M2 was chosen to carry out all computations. Each transient numerical simulation case took about 05 days on an i7 3.6 GHz computer with 06 cores and 32 GB of RAM.

Time-dependent computations were performed for 2900 s, which means seven flowthroughs for the highest Reynolds number. For the lowest Reynolds number, the total time was about 105 flowthroughs. The time-step for each Reynolds number was small enough to achieve a Courant number less than 1, as already carried out in the previous papers [29,30]. So, for Re_d = 200, 1000, and 3000, the time-step was set to 0.8, 0.14, and 0.03, respectively. During the numerical simulation, the temporal scheme was second-order backward Euler, the advective terms were discretized using an upwind second-order scheme, and the convergence criterion was set as at least 10^{-6} for each equation. The mean average data were averaged over the total time of the transient solution.

3. Results and Discussion

3.1. Stagnation (θ_{Est}) and Detached (δ_{Sep}) Angles

In Figure 3, mean average pressure (Figure 3a,b) and skin friction coefficients (Figure 3c,d) are shown as a function of the azimuthal position around the cylinders' surfaces. Both the pressure coefficient, C_p, and the skin friction coefficient, τ^*, were computed as shown by Achenbach [31] and Johansson [18] according to the expressions:

$$\begin{aligned} C_p &= \frac{P_\theta - P_0}{\frac{1}{2}\rho U_\infty^2} \\ \tau^* &= \mu \frac{\partial \overline{u}}{\partial y}\bigg|_{wall} \frac{\sqrt{\mathrm{Re}_d}}{\rho U_\infty^2} \end{aligned} \quad (7)$$

Focusing on the pressure distribution, one can see a similar distribution regardless of the Reynolds number. First, the stagnation point shifted towards the gap on both cylinders, taking place at about 354° and 7.70° in the upper and lower cylinders, respectively. According to the numerical work by Hensan [32], the movement of the stagnation angles is due to the repelling forces acting over the cylinders when they are close to each other. Furthermore, the repelling forces, according to the author, become stronger as the cylinders become closer. As stated before, despite the Reynolds number changing, the stagnation point location is not affected. Regarding the pressure values, the lowest coefficients occurred at different angular positions, depending on the cylinder. For the upper one (Figure 3a), the minima are placed at 75° and 285° for a Reynolds number of 3000. Moreover, regardless of the tube position, the pressure distribution fell at 180°, showing a valley for Re_d = 3000. In Figure 3c,d, the main purpose is to know where the boundary layer detaches. Here, the dissimilarities between the different Reynolds numbers are much more evident. In both tubes, the angle where the skin coefficient, τ^*, is maximal moves downstream as the

Reynolds number increases and the separation angle, δ_{Sep}, identified by $\tau^* = 0$, seems to be sensitive to Reynolds number in both cylinders. According to this criterion, the boundary layers were found to detach soon after 85°, first for the lowest Reynolds number, followed by $Re_d = 1000$ and 3000, in sequence. For Reynolds numbers of 1000 and 3000, the δ_{Sep} difference was found to be marginal. On the lower cylinder's surface, the τ^* distribution was found to be slightly different on the opposite side of the narrow gap. The skin friction coefficient was found to be zero at about 250° for $Re_d = 200$, followed by $Re_d = 1000$ and 3000, respectively. It is also interesting to notice that the separation angles are shifted in the narrow gap in comparison to the position where it takes place on the opposite side. The points where the boundary layer is detached are indicated by arrows in Figure 3c,d. Table 2 summarizes the stagnation and separation angles for each cylinder.

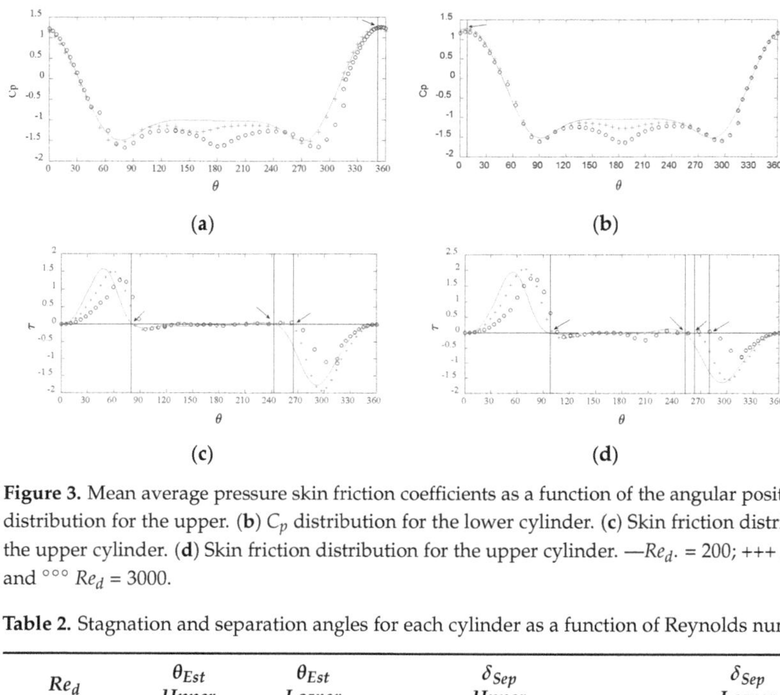

Figure 3. Mean average pressure skin friction coefficients as a function of the angular position. (**a**) C_p distribution for the upper. (**b**) C_p distribution for the lower cylinder. (**c**) Skin friction distribution for the upper cylinder. (**d**) Skin friction distribution for the upper cylinder. —Re_d. = 200; +++ Re_d = 1000 and °°° Re_d = 3000.

Table 2. Stagnation and separation angles for each cylinder as a function of Reynolds number.

Re_d	θ_{Est} *Upper*	θ_{Est} *Lower*	δ_{Sep} *Upper*		δ_{Sep} *Lower*	
200	351.20	9.30	82.20	246.50	97.90	252.60
1000	352.50	8.80	84.40	262.30	99.50	263.30
3000	353.60	7.70	85.50	263.50	101.70	277.30

3.2. *The Drag Coefficients and the Wake Interactions*

The instantaneous time history of the drag coefficients in both cylinders was gathered for every Reynolds number simulated. The drag coefficient is computed based on the following expression:

$$C_d = \frac{F_d}{\frac{1}{2}\rho U_\infty^2 dl} \quad (8)$$

where C_d is the drag coefficient and F_d is the total drag forces that are parallel to the x-axis. The circular cylinder diameter is d and l is the thickness of the domain (dimension parallel to the z-axis). The time, t^*, was made dimensionless by using the entrance velocity, U_∞, and the cylinder's diameter as $t* = \frac{tU_\infty}{d}$.

Figure 4a–c show the instantaneous drag time-trace, C_d, for each cylinder at Re_d = 200, 1000, and 3000, respectively. The mean average drag can be computed from each time history signal.

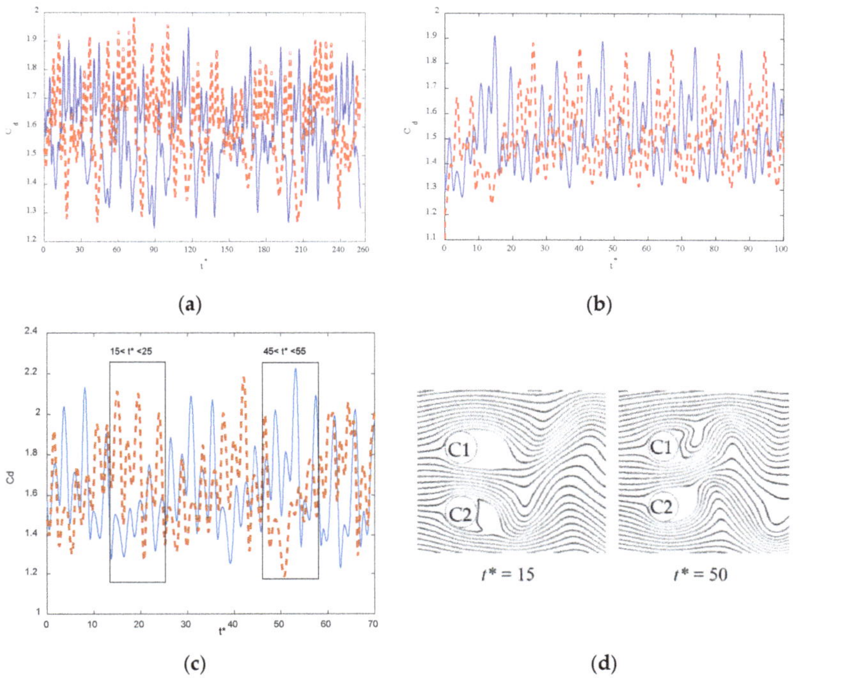

Figure 4. Instantaneous drag coefficients' time history. (**a**) Reynolds 200. (**b**) Reynolds 1000. (**c**) Reynolds 3000. (**d**) Stable modes of wakes—narrow and wide wakes formation for different t^*, under Re_d = 3000. — Upper cylinder. - - - Lower cylinder.

Vu et al. [33] suggested that the mean average drag coefficient should be computed as an equivalent drag, C_d^*.

$$C_d^* = \frac{\overline{C_d}|_{upper} + \overline{C_d}|_{lower}}{2\overline{C_{d0}}} \tag{9}$$

The computation would be based on each cylinder and compared to a single circular cylinder under the same Reynolds number, $\overline{C_{d0}}$, remembering that the time average processes were carried out over the total transient simulation.

Following Equation (9), the equivalent C_d^* was found to be 1.20, 1.33, and 2.16 for the Reynolds numbers 200, 1000, and 3000, respectively. These results are in good agreement with the results from Vu et al. [33], mainly for the cases at Re_d = 200 and 1000. In their paper, the authors simulated 2D flow around circular cylinders in pairs under the same Reynolds numbers and p/d ratio. The equivalent drag, C_d^*, was found to differ from ours by only 0.6% under Re_d = 200 and no difference was found under Re_d = 1000. Unfortunately, the same computation methodology could not be applied to Afgan's results in order to compare the outcomes for Re_d = 3000, since the authors evaluated the C_d based on the wakes' topologies. However, Afgan et al. [27] state that the sum of the drag coefficients of the two cylinders, separately, is slightly comparable to twice the drag found in a single circular cylinder for $1.25 \leq p/d \leq 5.0$. By carefully observing the data published by the authors, the total drag was found to be 2.80. Applying this methodology using $C_d = \overline{C_d}|_{upper} + \overline{C_d}|_{lower}$ yielded a value of 2.89, which differs by only 3.2% from Afgan's results [2].

The bistable flow process was also a target of our research. The bistability phenomenon takes place when the wakes are near enough to interact to each other. This interaction yields stable modes of wake topologies that change randomly over time. During the processes, a narrow and wide wake is formed behind each cylinder. In the first moment, a wide wake behind the lower cylinder moves out behind the upper cylinder, whose wake is narrow at first. This switching occurs over time and each mode lasts for a certain period of time [18,19]. Stable modes are observed in all cases. However, in the third case, Re_d = 3000 the observation is straightforward. Figure 4d shows the instantaneous streamlines over the cylinders corresponding to the times t^* = 15 and t^* = 50 for the case of Re_d = 3000. From the picture, it is possible to observe that the wake topology is stable for a certain period of time behind each cylinder (see the rectangles in Figure 4c). As previously stated, the direction of the central jet determines which cylinder will experience either the narrow wake (*NW*) or the wide one (*WW*), thus ruling the drag coefficient on each cylinder.

In fact, by observing the drag time-traces in each cylinder, higher and lower drag coefficients are assigned to different wakes' topologies. See the first stable mode at $15 \leq t^* \leq 25$ in Figure 4c. In this moment, we can identify that the lower cylinder, *C*2, experiences a higher value of drag and, at once, the viscous wake behind it is classified as a narrow wake (*NW*; Figure 4d). On the other hand, when the new topology takes place, from $t^* \sim 45$ to 55, the drag value in the upper cylinder is higher than that found in the lower one. Again Figure 4d identifies the narrow wake (*NW*) downstream of the upper cylinder, leading to a high drag value. The dependence of the drag force on the stable wake topologies has been pointed out by several authors [14,20,33]. During the time that the lower cylinder experiences the narrow wake, the mean average drag is computed differently; $\overline{C_d}\big|_{lower}$ is 1.60 and $\overline{C_d}\big|_{upper}$ is 1.30. According to Alam et al. [20], the drag coefficient difference between the narrow wake and the wider one is due to the pressure recovery behind each wake.

The instantaneous pressure distribution, $C_p{}'$, around each cylinder was also investigated. Figure 5a,b intend to show the instantaneous pressure distribution around the same cylinders in each stable mode. This information may help us to understand how the drag coefficient is associated with the wake topology, since in blunt bodies, the pressure (or shape) drag is expected to play a major role in the total drag [34].

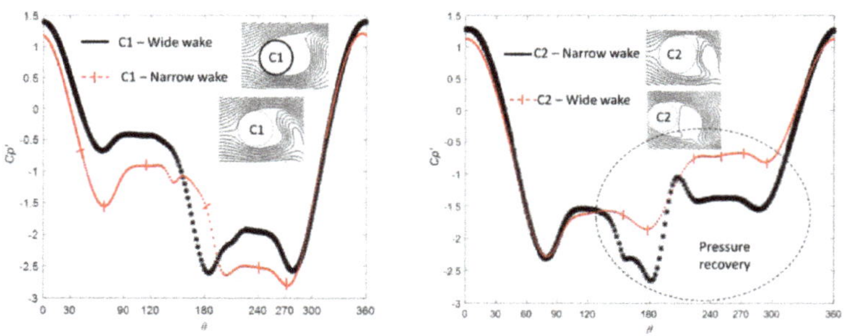

(a) (b)

Figure 5. Instantaneous pressure coefficient around the cylinders at Re_d = 3000. (**a**) Upper cylinder (C1) experiencing two different modes. (**b**) Lower cylinder (C2) experiencing two different modes.

In both pictures, the same cylinder is seen to experience almost the same pressure distribution at the very beginning, regardless of the wake topology behind them. After some angular positions, towards the rear part of the cylinders, the curves reveal different pressures. This behavior is seen to happen for the upper cylinder from $\theta \sim 40°$ to 270°, while on the lower cylinder's surface, the same behavior appears soon after θ = 120° up to 320° However, the most important part of the graph is the middle part. The reader can easily see that the pressure recovery is different for the same cylinder under the different modes (wake topologies). The upper cylinder, under the wide wake (*WW*), experiences higher levels of pressure in its front and rear part as well. On the other hand, when the mode switches and, therefore, the same cylinder is under the narrow wake (*NW*), the pressure is lowered in the same position in comparison with what would be under the other mode (Figure 5a). The same analogy can be employed for the lower cylinder, *C*2. In this case the differences between the levels of pressure are even larger. Furthermore, we can also easily observe the pressure recovery behind the cylinder subject to the wide wake (*WW*), indicating that the instantaneous shape drag should be lowered in comparison with the same cylinder in *NW* mode (Figure 5b).

Velocity time-traces were gathered at the monitoring point (Figure 1b), at once. The time, t^*, was made dimensionless, as mentioned before, and the velocity component received the same treatment by using the free stream velocity, U_∞, as follows:

$$u^* = \frac{u_{(t)}}{U_\infty}; v^* = \frac{v_{(t)}}{U_\infty} \tag{10}$$

where u and v are the axial and transversal instantaneous velocities. We also carried out numerical simulations for the lowest Re_d case without any turbulence model (laminar flow). In this case, we want to investigate whether the employed turbulence model affects somehow the spectral response of the flow field. The fast Fourier transform (FFT) employed here was performed using 2N data point for each velocity time-trace signal (N is the number of data on each signal). Before performing the FFT computations, the signals were windowed by a Hanning function. The FFT coefficients were scaled by the highest one. By observing the velocity signals in the wakes, one can see that the flow exhibits almost periodic patterns, mainly for higher Reynolds numbers. The Fourier transform coefficients of each signal are plotted along with the velocity time-traces (Figure 6b,d,f,h). The frequency was then made dimensionless through the Strouhal number as follows:

$$St = \frac{f\,d}{U_\infty} \tag{11}$$

where f is the main frequency in the spectrum in *Hz*.

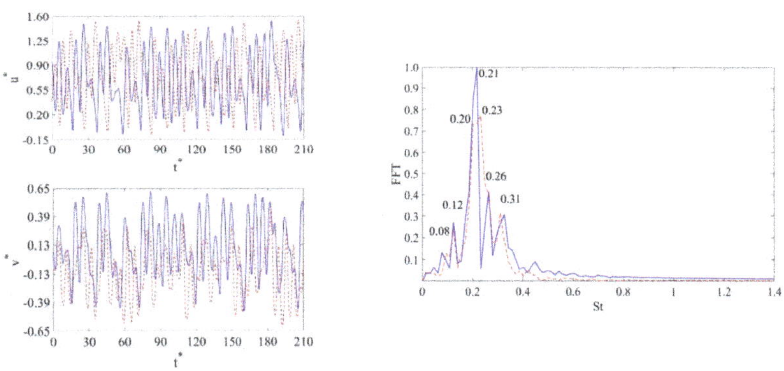

(**a**) $Re_{d.laminar} = 200$ (**b**) $Re_{d.laminar} = 200$

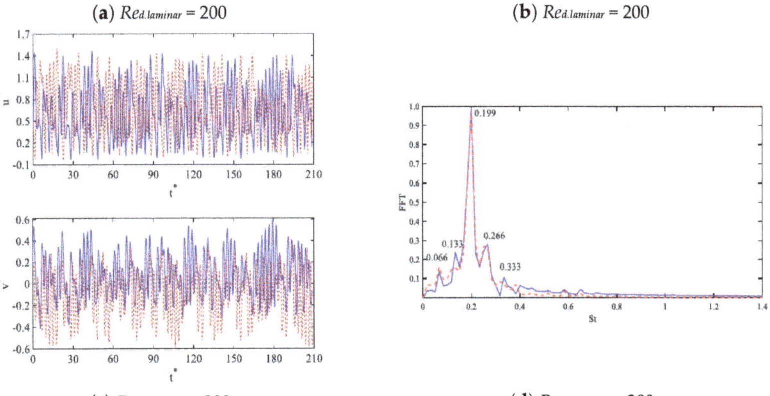

(**c**) $Re_{d.turbulent} = 200$ (**d**) $Re_{d.turbulent} = 200$

Figure 6. *Cont.*

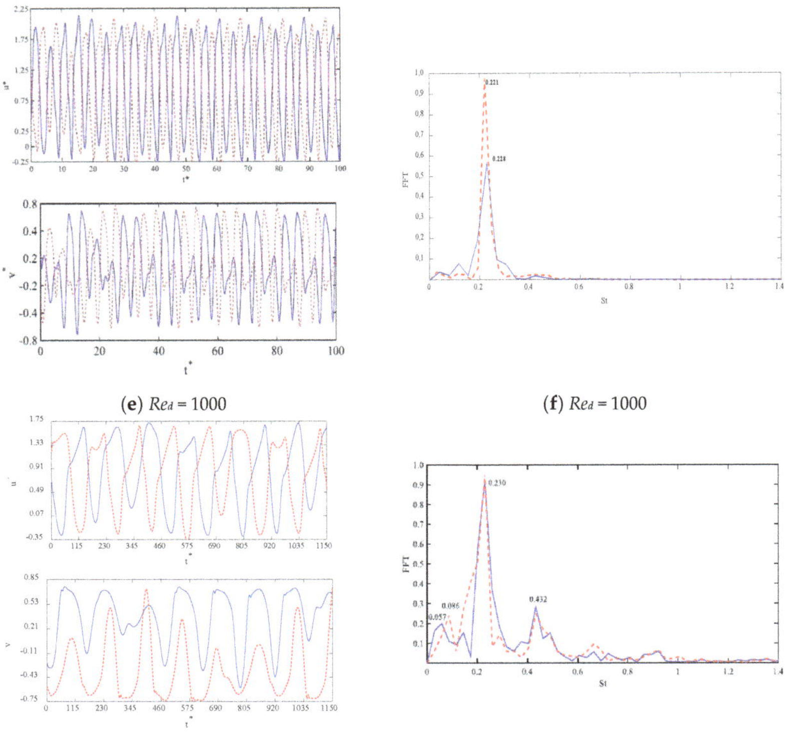

(**e**) $Re_d = 1000$ (**f**) $Re_d = 1000$

(**g**) $Re_d = 3000$ (**h**) $Re_d = 3000$

Figure 6. Time-traces of velocity components in the wake for upper and lower cylinders, along with the spectral computation of the flow velocity signal. — Upper cylinder. - - - Lower cylinder.

The dynamics of the flow were found to be in fair agreement with Bearman's findings [15]. As predicted by the authors [15], the cylinders are far enough apart to maintain a fundamental frequency very close to what one would expect in a single cylinder. In fact, the spectral analysis has shown that the fundamental frequency peak appears at about $St \sim 0.20$. A slight displacement towards higher frequencies is seen as the Reynolds number increases. Moreover, both velocity components presented the same frequencies. It should also be pointed out that the spectral response of the flow seems to be unaffected by the turbulence model. Both spectra for laminar and turbulent flow at $Re_d = 200$ showed almost the same fundamental peak at $St \sim 0.20$. Several peaks, besides the fundamental one, appear for each Reynolds number simulated, indicating that the biased flow deflection, i.e., the bistable flow, exists and possesses its very own signature in terms of spectral response. Indeed, such a feature was successfully seen in the C_d time-traces (Figure 4a–c).

Afgan and his co-workers [27] investigated the power spectral response of the flow dynamics for several p/d ratios under the same Reynolds number, 3000. For $p/d < 2.0$, the authors identified different peaks in the spectra; however, as the gap increased, the additional peaks vanished. In the same year, similar results were released by Verna and Verna and Govardhan [35], who investigated 2-D flow over side-by-side circular cylinders. Both works from 2011 [27,35] connect the secondary frequencies to different wake topologies, and lower frequency was assigned to the wide wake (*WW*); on the other hand, the higher one was related to the narrow wake (*NW*). Afgan and co-workers [27] associated the Strouhal numbers $St \sim 0.11$ and 0.39 to the wide wake (*WW*) and narrow wake (*NW*), respectively, for a p/d ratio 1.50; on the other hand, according to the authors, for $p/d = 2.0$, the peaks in the spectra were seen to be very close. Sided peaks, at distinguished frequencies, were also reported by Pang et al. [36] and Alam et al. [20]. In the former work, the authors studied the dynamical response of the flow field in a 2D domain under the Reynolds number 60,000 for several p/d ratios. The authors [36] associated a Strouhal number of 0.10 to the wide wake (*WW*) and the narrow wake (*NW*) was assigned to the Strouhal 0.3. Furthermore, intermediated frequencies, at *St* ~0.2, were found for $1.1 \leq p/d \leq 2.6$.

In Figure 6h, the reader can see that the Strouhal numbers are about 0.06 and 0.43, indicating dimensionless frequencies for different wake topologies. Higher frequency was associated with the narrow wake (*NW*) and vice versa.

In order to promote a better understanding of the wakes' characteristic frequency, the spectral response of the C_d signal, at Re_d = 3000, was analysed. Since we have very distinctive patterns of C_d, assigned to different wakes' topologies, the drag signal was split into C_d–narrow wake and C_d–wide wake. Figure 7a,b show both the C_d time-trace and the fast Fourier transform of those signals. From Figure 7b, one can observe that the spectral response of each wake topology is different. The wide wake (*WW*) stresses lower frequencies in comparison with the narrow wake (*NW*). The *WW* exhibits its main Fourier coefficients at *St* = 0.06 and 0.39, whereas for the *NW*, the most important coefficients are assigned to the fundamental frequency at about *St* = 0.47.

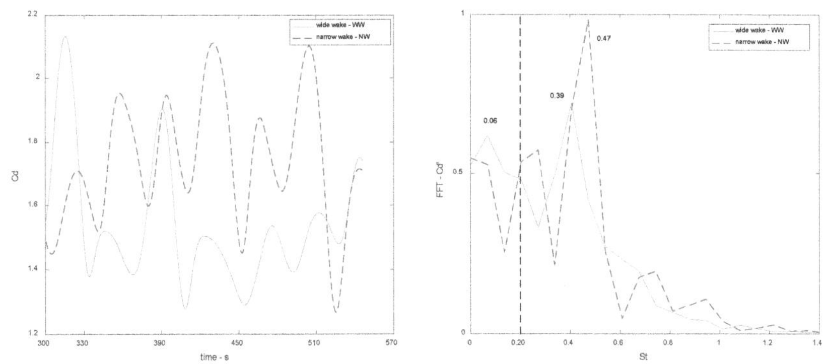

(a) (b)

Figure 7. (**a**) C_d time-traces for different wakes. (**b**) spectral computation of the flow C_d signal. — wide wake (*WW*). - - - narrow wake (*NW*).

4. Concluding Remarks

Isothermal, 2-D, and incompressible turbulent flow over a pair of circular cylinders arranged side-by-side was investigated by numerical simulation in this work. The main dimensionless number that qualified the simulation was the pitch–diameter ratio, *p/d*, which was kept constant at 2 throughout the work. In order to quantify the Reynolds effects, three Reynolds numbers were also simulated by varying the entrance velocity. The computations were carried out on a finite-volumes platform using Unsteady RANS-k-ω SST to overcome the closure problem of the turbulence.

The stagnation and detachment angles of the boundary layer were measured. The simulations were in quite good agreement when compared with the results reported by other authors. The Reynolds numbers were found to play a more important role in the stagnation angle, θ_{EST}, than the detached one, $\delta_{SE}P$, for any case simulated.

The mean and time history drag coefficients were also gathered. Asymmetric and irregular behavior of the drag coefficients was observed regardless of the Reynolds number. Furthermore, different wake topologies were formed behind the cylinders, causing the cylinders to experience different drag forces. The instantaneous drag force was seen as a function of the type of wake topology formed behind each one. Higher drag was assigned to narrow wakes (*NW*) and lower drags were seen to be related to wide wakes (*WW*).

Furthermore, the instantaneous $C_p{}'$ around the cylinder, under different wake topologies showed that the pressure recovers behind them when they are in the wide wake mode (*WW*); otherwise, the pressure is lowered, enhancing the drag force for the narrow wake mode (*NW*).

The Reynolds number seems to work in order to delay the changes between the wake topologies. As the Reynolds number increases, each wake topology lasts longer. In other words, as the Reynolds number increases, the number of changes experienced by each cylinder seems to decrease.

The spectral-flow response was also analyzed through the velocity time-traces behind the structures. A very well-distinguished peak was found in the spectrum at $St \sim 0.20$, which is in fair agreement with other research. A slight displacement towards higher frequencies could be seen as the Reynolds number increased, but it was marginal. Secondary peaks located on either of the main peaks in the spectrum were also observed, mainly for Reynolds numbers of 200 and 3000. The marginal peaks are associated with the different wake topologies behind the cylinders. The lower and higher frequency could be very well associated with the wakes through the spectral response of the C_d time-traces. In the narrow wake mode, which produced a higher C_d yield, high Fourier coefficients placed at higher frequencies, whereas for the wide wake, the higher Fourier coefficients were assigned to the lower frequencies.

The laminar simulations, $Re_d = 200$ (without any turbulence model), did not show any difference in the dynamics of the fluctuant flow field. This result shows that the employed turbulence model neither affects nor fosters the bistable flow mode in the pair of cylinders studied.

Author Contributions: Conceptualization, T.G. and J.G.; Funding acquisition, J.G.; Investigation, T.G. Methodology, T.G. and C.A.; Project administration, J.G.; Resources, C.A. and J.G.; Software and Data treatment, T.G. and J.G.; Validation, C.A. and J.G. Writing—original draft, T.G.; Writing—review and editing, C.A. and J.G. All authors have read and agreed to the published version of the manuscript.

Funding: This research received no external funding.

Informed Consent Statement: Not applicable.

Data Availability Statement: Not applicable.

Acknowledgments: The first author would like to thank the Coordenação de Aperfeiçoamento de Pessoal de Nível Superior-Brasil (CAPES)-Finance Code 001 for supporting him during this research with a fellowship. The second author thanks CNPQ and FAPDF for supporting him with financial resources through the projects n° 193.001.158/2015, 193.001356/2016 and 408869/2016-0.

Conflicts of Interest: The authors declare no conflict of interest.

References

1. Grimison, E.D. *Correlation and Utilization of New Data on Flow Resistence and Heat Transfer for Cross Flow of Gases Over Tube Banks* Transactions: Process Industries Division; American Society of Mechanical Engineers: New York, NY, USA, 1937; pp. 583–594.
2. Wiemer, P. *Untersuchung Über den Zugwiederstand Von Wasserrohrkesseln, Dissertation*; RWTH: Aachen, Germany, 1937.
3. Žukauskas, A.A. Heat transfer from tubes in crossflow. In *Advances in Heat Transfer*; Elsevier: Amsterdam, The Netherlands, 1972 Volume 8, pp. 93–160.
4. Žukauskas, A.A.; Katinas, V.J.; Perednis, E.E.; Sobolev, V.A. Viscous flow over inclined in-line tube bundles, and vibrations induced in the latter. *Fluid Mech. Sov. Res.* **1980**, *9*, 1–12.
5. Žukauskas, A.A.; Katinas, V.J. Fluid Dynamics forces on vibrating tubes of heat exchangers in cross flow. In *International Symposium on Flow-Induced Vibration and Noise*; ASME: Chicago, IL, USA, 1988; Volume 1, pp. 127–142.
6. Derakhshandeh, J.F.; Alam, M.M. A review of bluff body wakes. *Ocean Eng.* **2019**, *182*, 475–488. [CrossRef]
7. Zdravkocich, M.M.; Pridden, D.L. Interference between two circular cylinders; series of unexpected discontinuities. *J. Ind Aerodyn.* **1977**, *2*, 255–270. [CrossRef]
8. Young, E.W.K.; Martinex, D.M.; Olso, J.A. The sedimentation of papermaking fibers. *Am. Inst. Chem. Eng.* **2006**, *52*, 2697–2706 [CrossRef]
9. Ahmad, N.; Bihs, H.; Myrhaug, D.; Kamath, A.; Arntsen, O.A. Three-dimensional numerical modelling of wave-induced scour around piles in a side-by-side arrangement. *Coast. Eng.* **2018**, *138*, 132–151. [CrossRef]

10. Alam, M.; Zhou, Y. Flow around two side-by-side closely spaced circular cylinders. *J. Fluids Struct.* **2007**, *23*, 799–805. [CrossRef]
11. Alam, M.; Zheng, Q.; Hourigan, K. The wake and thrust by four side-by-side cylinders at a low Re. *J. Fluids Struct.* **2017**, *70*, 131–144. [CrossRef]
12. Kim, S.; Alam, M.M. Characteristics and suppression of flow-induced vibrations of two side-by-side circular cylinders. *J. Fluids Struct.* **2015**, *54*, 629–642. [CrossRef]
13. Shao, J.; Zhang, C. Large eddy simulations of the flow past two side-by-side circular cylinders. *Int. J. Comput. Fluid Dyn.* **2008**, *22*, 393–404. [CrossRef]
14. Sumner, D. Two circular cylinders in cross-flow: A review. *J. Fluids Struct.* **2010**, *26*, 849–899. [CrossRef]
15. Bearman, P.W.; Wadcok, A.J. The interaction between a pair for circular cylinders normal to a stream. *J. Fluid Mech.* **1973**, *61*, 499–511. [CrossRef]
16. Meneghini, J.R.; Saltara, F.; Siqueira, C.; Ferrari, J. Numerical Simulation of Flow Interference between Two Circular Cylinders in Tandem and Side-by-Side Arrangements. *J. Fluids Struct.* **2001**, *15*, 327–350. [CrossRef]
17. Olinto, C.R.; Indrusiak, M.L.S.; Endres, L.A.M.; Möller, S.V. Experimental study of the characteristics of the flow in the first rows of tube banks. *Nucl. Eng. Des.* **2009**, *239*, 2022–2034. [CrossRef]
18. De Paula, A.V.; Endres, L.A.M.; Möller, S.V. Experimental study of the bistability in the wake behind three cylinders in triangular arrangement. *J. Braz. Soc. Mech. Sci. Eng.* **2013**, *35*, 163–176. [CrossRef]
19. Neumeister, R.F.; Petry, A.P.; Möller, S.V. Characteristics of the wake formation and force distribution of the bistable flow on two cylinders side-by-side. *J. Braz. Soc. Mech. Sci. Eng.* **2018**, *40*, 564. [CrossRef]
20. Alam, M.; Moriya, M.; Sakamoto, H. Aerodynamic characteristics of two side-by-side circular cylinders and application of wavelet analysis on the switching phenomenon. *J. Fluids Struct.* **2003**, *18*, 325–346. [CrossRef]
21. Giacomello, M.V.; Rocha, L.A.; Schettini, E.B.; Silvestrini, J.H. Simulação numérica de escoamentos ao redor de cilindros com tranferência de calor. In *Anais da 5ª Escola de Primevera de Transição e Turbulência*; EPTT: Rio de Janeiro, Brasil, 2006; pp. 25–30.
22. Kang, S. Characteristics of flow over two circular cylinders in a side-by-side arrangement at low Reynolds numbers. *Phys. Fluid* **2003**, *15*, 712–714. [CrossRef]
23. Wang, Z.; Zhou, Y. Vortex interactions in a two side-by-side cylinder near-wake. *Int. J. Heat Fluid Flow* **2005**, *26*, 362–377. [CrossRef]
24. Vila, J.L.; Gomes, T.F.; De Melo, T.; Goulart, J.N.V. Experimental measurements of pressure and velocity fields around circular cylinders arranged in pair. In Proceedings of the 25th International Congress of Mechanical Engineering—COBEM, Uberlândia, Brazil, 20–25 October 2019.
25. Chen, W.; Ji, C.; Xu, D.; Srinil, N. Wake patterns of freely vibrating side-by-side circular cylinders in laminar flows. *J. Fluids Struct.* **2019**, *89*, 82–95. [CrossRef]
26. Chen, W.; Ji, C.; Xu, D.; An, H.; Zhang, Z. Flow-induced vibrations of two side-by-side circular cylinders at low Reynolds numbers. *Phys. Fluids* **2020**, *32*, 023601. [CrossRef]
27. Afgan, I.; Kahil, Y.; Benhamadouche, S.; Sagaut, P. Large eddy simulation of the flow around single and two side-by-side cylinders at subcritical Reynolds numbers. *Phys. Fluids* **2011**, *23*, 075101. [CrossRef]
28. Menter, F.R. Two-equation eddy-viscosity turbulence models for engineering applications. *AIAA J.* **1994**, *32*, 1598–1605. [CrossRef]
29. Candela, D.S.; Gomes, T.F.; Goulart, J.; Anflor, C.T.M. Numerical simulation of turbulent flow in an eccentric channel. *Eur. J. Mech. B/Fluids* **2020**, *83*, 86–98. [CrossRef]
30. Goulart, J.; Wissink, J.G.; Wrobel, L.C. Numerical simulation of turbulent flow in a channel containing a small slot. *Int. J. Heat Fluid Flow* **2016**, *61*, 343–354. [CrossRef]
31. Achenbach, E. Distribution of local pressure and skin friction around a circular cylinder in cross-flow up to Re = 5 × 106. *J. Fluid Mech.* **1968**, *34*, 625–639. [CrossRef]
32. Hensan, S.M.; Navid, N. Numerical simulation of flow over two side-by-side circular cylinders. *J. Hydrodyn.* **2011**, *23*, 792–805.
33. Vu, H.C.; Ahn, J.; Hwang, J.H. Numerical simulation of flow past two circular cylinders in tandem and side-by-side arrangement at low Reynolds numbers. *KSCE J. Civ. Eng.* **2015**, *20*, 1594–1604. [CrossRef]
34. White, F.M. *Fluid Mchanics*, 4th ed.; MacGraw-Hill: New York, NY, USA, 1999.
35. Verma, P.L.; Govardhan, M. Flow behind bluff bodies inf side-by-side arrangement. *J. Eng. Sci. Technol.* **2011**, *6*, 745–768.
36. Pang, J.H.; Zong, Z.; Zou, L.; Wang, Z. Numerical simulation of the flow around two side-by-side circular cylinders by IVCBC vortex method. *Ocean. Eng.* **2016**, *119*, 86–100. [CrossRef]

Σ mathematics

MDPI

Article

Two-Phase Flow of Eyring–Powell Fluid with Temperature Dependent Viscosity over a Vertical Stretching Sheet

Ahlam Aljabali [1], Abdul Rahman Mohd Kasim [1,*], Nur Syamilah Arifin [2], Noor Amalina Nisa Ariffin [3], Dennis Ling Chuan Ching [4], Iskandar Waini [5], Najiyah Safwa Khashi'ie [5] and Nurul Amira Zainal [5]

1 Centre for Mathematical Sciences, Universiti Malaysia Pahang, Gambang 26300, Pahang, Malaysia
2 Faculty of Computer and Mathematical Sciences, Universiti Teknologi Mara (UiTM) Cawangan Johor, Kampus Pasir Gudang, Masai 81750, Johor, Malaysia
3 Faculty of Computer and Mathematical Sciences, Universiti Teknologi Mara (UiTM) Cawangan Pahang, Kampus Jengka, Bandar Tun Abdul Razak, Jengka 26400, Pahang, Malaysia
4 Fundamental and Applied Science Department, Universiti Teknologi Petronas, Seri Iskandar 32610, Perak, Malaysia
5 Fakulti Teknologi Kejuruteraan Mekanikal dan Pembuatan, Universiti Teknikal Malaysia Melaka, Hang Tuah Jaya, Durian Tunggal 76100, Melaka, Malaysia
* Correspondence: rahmanmohd@ump.edu.my

Abstract: In this work, the mixed convection flow of non-Newtonian Eyring–Powell fluid with the effects of temperature dependent viscosity (TDV) were studied together with the interaction of dust particles under the influence of Newtonian Heating (NH) boundary condition, which assume to move over a vertical stretching sheet. Alternatively, the dusty fluid model was categorized as a two-phase flow that consists of phases of fluid and dust. Through the use of similarity transformations, governing equations of fluid and dust phases are reduced into ordinary differential equations (ODE), then solved by efficient numerical Keller–box method. Numerical solution and asymptotic results for limiting cases will be presented to investigate how the flow develops at the leading edge and its end behaviour. Comparison with the published outputs in literature evidence verified the precision of the present results. Graphical diagrams presenting velocity and temperature profiles (fluid and dust) were conversed for different influential parameters. The effects of skin friction and heat transfer rate were also evaluated. The discovery indicates that the presence of the dust particles have an effect on the fluid motion, which led to a deceleration in the fluid transference. The present flow model can match to the single phase fluid cases if the fluid particle interaction parameter is ignored. The fluid velocity and temperature distributions are always higher than dust particles, besides, the opposite trend between both phases is noticed with β. Meanwhile, both phases share the similar trend in conjunction with the rest factors. Almost all of the temperature profiles are not showing a significant change, since the viscosity of fluid is high, which can be perceived in the figures. Furthermore, the present study extends some theoretical knowledge of two-phase flow.

Keywords: dusty Eyring–Powell fluid; Newtonian heating; temperature dependent viscosity; vertical stretching sheet

MSC: 35Q30; 76D05; 35Q35; 34A45; 65Q10

Citation: Aljabali, A.; Mohd Kasim, A.R.; Arifin, N.S.; Ariffin, N.A.N.; Ling Chuan Ching, D.; Waini, I.; Khashi'ie, N.S.; Zainal, N.A. Two-Phase Flow of Eyring–Powell Fluid with Temperature Dependent Viscosity over a Vertical Stretching Sheet. *Mathematics* **2022**, *10*, 3111. https://doi.org/10.3390/math10173111

Academic Editors: Vasily Novozhilov and Cunlu Zhao

Received: 29 June 2022
Accepted: 28 July 2022
Published: 30 August 2022

1. Introduction

Research and studies in the area of heat and mass transport of fluids flow have discovered strategies for their development, as well as key problems. Nevertheless, it is important to choose the form of ideal fluid from the point of view of homogeneous or inhomogeneous, compressible or incompressible, Newtonian or non-Newtonian, and monophasic or polyphasic fluids, which have a significant role in determining suitable solutions for heat transfer and fluid flow enhancement. In the past few years, the new environment of accelerated technical progress has contributed to the emergence of creative approaches to analyse the suspension of fluid particles in a fluid flow that is also known as a two-phase flow model, which explains the actions of fluid dust characteristics. Industrial applications, such as petroleum transport, wastewater treatment, vehicle smoke emissions, power plant piping, and corrosive particulate matter in mining, generally involve fluid dust movement activities [1]. The movement of dust particles in a fluid leads to a two-phase cycle. Numerous pieces of research on the dynamics of solid particles in the fluid have been conducted as a result of recent advancements in the field of two-phase flow. The fluid and solid phases of this solid–liquid system are independently formulated using different continuum equations. This phenomenon includes micro-propulsion, aerosol filtration, powder transport, oil industry, and the flow of corpuscle in plasma (a liquid with suspended solids). Overall, it is very useful for modelling flow with a binary mixture of non-Newtonian fluid and solid particles linked to certain conditions. It can thus be suggested that this two-phase model could benefit in studying the dusty Eyring–Powell fluid that exhibits the binary characteristics of the Eyring–Powell fluid and spherical dust particles, such as undertaken here. Ref. [2] studied the process of radiative heat transfer in the flow of dusty liquid under the power generation aspects. The boundary layer flow of a dusty fluid with electrically conducting criteria in a porous medium has been studied by [3]. When the interaction of these phases is significant, the temperature of the fluid is always higher than that of the dust. In accordance with these applications, the literature includes a variety of works in corresponding flow for various contexts, such as multiple geometries, boundary conditions and fluid-based forms. The non-Newtonian Casson model with dust particles have been developed by [4,5]. In addition, [6,7] utilized the treating fluid–particle interaction with buoyancy forces on Jeffrey fluid with Newtonian heating indicates that the presence of the dust particles has an effect on the fluid motion, which led to decelerate the fluid transference. The natural convection flow caused by non-Newtonian fluid with dust nanoparticles has been addressed in [8,9]. Furthermore, [10] theoretically analysed along a vertical stretching sheet for magnetohydrodynamic (MHD) mixed convection of non-Newtonian tangent hyperbolic nanofluid flow with suspended dust particles. In [11], a detailed study has been done on a two-phase model implemented in the presence of hybrid nanoparticles on the dusty liquid flow through a stretching cylinder by employing the modified Fourier heat flux law. The study was conducted by considering the effect of viscous dissipation and non-linear thermal radiation, which demonstrated a two-phase dusty liquid movement across a permeable surface. Other contributions of flow models on dusty non-Newtonian fluid have drawn substantial interest among researchers under different conditions [12–15].

Consideration of the study on boundary layer phenomenon focussing on non-Newtonian heat-transported substances is essential for a deeper comprehension of engineering and industrial–technology issues, and the movement of these materials occurs extensively in various industrial processes, such as guided missiles, rain erosion, fluidisation, atmospheric failure, lunar ash fall, paint and aerosol spraying, as well as the cooling of nuclear reactors. Although the existence of such substances is greatly complicated and troublesome, a variety of constitutive models have been developed and studied to research the correct flow behaviour. Eyring–Powell fluid model is one of the subcategories of the non-Newtonian fluid model. It has a clear characteristic under other non-Newtonian models, conveniently derived from the kinetic theory of gases rather than empirical relations and comes baked from Newtonian behaviour for low and high shear rates. The rheological paradigm is known for its robustness and versatility in physical action. Ref. [12] addressed movement attributable to pulsatile pressure gradient of dusty non-Newtonian fluid with heat transfer in a channel. A preliminary analysis of the magnetohydrodynamic movement of the Eyring–Powell liquid under the suspension of nanoparticles and dust has been done by [13] in which this model has shown that the intensity of heat transfer in the aluminium oxide nanofluid was higher than that in the ferro oxide nanofluid with the current viscous variance parameter. In [14], by considering variable thermal conductivity and thermal radiation, analytical solutions of unstable flow Eyring–Powell and Carreau non-Newtonian fluids in the suspension of dust and nickel nanoparticles, a higher heat transfer rate was recorded in the nickel + Eyring–Powell mixture compared to the nickel + Carreau case. Ref. [15] attempted to examine the effects of heat and mass transfer in the presence of nonlinear convection and thermal radiation of MHD rheological Eyring–Powell fluid with dust and graphene nanoparticles in a mixture of ethylene glycol. In order to study the relationship between the fluid and dust phased, this analysis was conducted to further investigate the flow behaviour of dusty Powell–Eyring fluid in the vertical stretching sheet associated with temperature-dependent viscosity combined with NH as a thermal boundary conditions.

Over recent decades, the thermal boundary conditions have a great influence on the heat transfer in the laminar boundary layer flow problem. However, there is a situation in which the heat transfer rate is proportional to the local difference in temperature with ambient conditions or usually termed conjugate boundary conditions, which are driven by NH, should be considered as well. The case of NH has been mentioned by [16] during a study on the boundary layer flow over an upright plate. Ref. [17] made further headway during an investigation on free convection flow across horizontal surface. Ref. [18] proposed a series of solutions and numerical Eyring fluid flow with NH. Ref. [19] provided detailed nonlinear convective magneto nanofluid Eyring fluid with the effects of NH while Ref. [20] included the effects of the thermal radiation Eyring fluid subjected to NH boundary condition. Meanwhile, Refs. [21,22] highlighted the effects of MHD and NH on Powell–Eyring fluid over a stretching cylinder and inclined permeable surface, respectively. Ref. [23] reviewed the problem of temperature-dependent viscosity on mixed convection flow of Eyring–Powell fluid studied together with NH. A mathematical model of forced convective flow on non-Newtonian Eyring–Powell fluid under temperature-dependent viscosity circumstance is formulated by [24]. The effects of non-Newtonian magnetohydrodynamic nanofluid over a stretched plate with NH effect have been investigated by [25]. Other studies associated with particular impact for fluid–solid flow considering different fluid models were established and reported in [26–29].

Motivated by the impactful research as scrutinised above, this present study is dedicated to examining the two-phase boundary layer flow of Eyring–Powell fluid together with the temperature-dependent viscosity from a vertical stretching sheet, the temperature of which is higher than that of the ambient fluid. The simulation of mixed convection influence with NH was also implanted in this investigation. The mathematical formulations of the problems are constructed as mentioned in the study scope, which involves the derivation of governing equations for the proposed problem. The similarity transformation is used to transform the non-linear governing equations into ordinary differential equations (ODE). Then, the numerical solutions of the transformed equations are solved using the Keller-box method.

The step size of time and space can be arbitrary, since this method is implicit with second order accuracy, which makes it suitable to solve the parabolic partial differential equations efficiently [30]. However, the computation could be time consuming if the small step size of time and space is inserted. The algorithm of the Keller-box method is computed in Matlab software to generate the results and the figures for various non-dimensional parameters on the velocity and temperature profile. The comparison with the previous published result were tabulated to verify the present results by fixing several parameters. The output from the investigation are useful for the scientist and experimentalist in studying the behaviour of fluids, which have interactions with dust particles.

2. Mathematical Formulation

Flow suspended with particles affected by TDV over a vertical stretching sheet was introduced for the steady incompressible mixed convection of non-Newtonian Eyring fluid under NH condition. The term, T_∞, is related to temperature of ambient fluid. The x-axis was oriented to the vertical plane, and the y-axis to the plane was perpendicular. The sheet is stretched with the velocity, $u_w(x) = bx$ where, $b > 0$ is stretching rate. The flow was created by the stretching of the sheet due to the simultaneous application of two equal and opposite forces along the x-axis, holding the origin fixed and finding the flow to be limited to the area, $b > 0$. The configuration of a physical model is displayed in Figure 1.

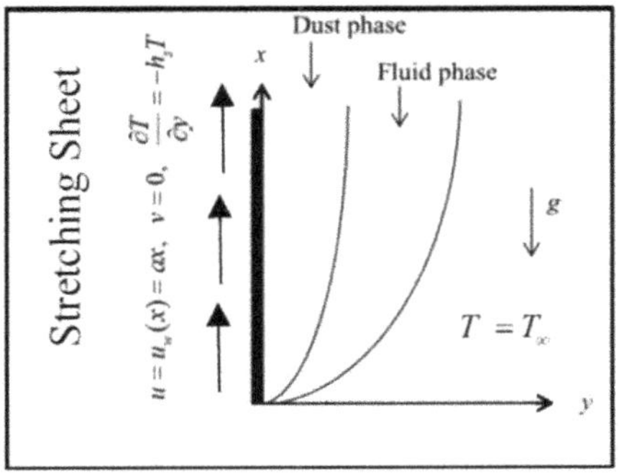

Figure 1. Physical sketch and coordinate system.

The solid particles are presumed to be spherical in shape and uniform size, where their density remains constant and the inter-particle collision may be neglected, since they are considered to be diluted throughout the flow. In terms of heat transfer from one atom to another, all of these were ignored: the volume fraction of dust particles, coagulation, phase transition and deposition. The fluid and sediment particle motions were linked only through drag-and-heat flow across them, the stokes linear drag theory was used for modelling the drag force. Based on the preceding assumptions, the basic two-dimensional boundary layer equations involving continuity, momentum and energy for both the fluid and particle phases with usual ratings can be written as [31,32] shown below:

For fluid phase:

$$\frac{\partial u}{\partial x} + \frac{\partial v}{\partial y} = 0, \tag{1}$$

$$u\frac{\partial u}{\partial x} + v\frac{\partial u}{\partial y} = \frac{1}{\rho}\frac{\partial}{\partial y}\left(\mu\frac{\partial u}{\partial y}\right) + \frac{1}{\rho\tilde{\beta}c}\left(\frac{\partial^2 u}{\partial y^2}\right) - \frac{1}{2\rho\tilde{\beta}c^3}\left(\frac{\partial u}{\partial y}\right)^2\frac{\partial^2 u}{\partial y^2} + \beta g(T - T_\infty) + \frac{\rho_p}{\rho\tau}(u_p - u), \tag{2}$$

$$\rho c_p\left(u\frac{\partial T}{\partial x} + v\frac{\partial T}{\partial y}\right) = k\left(\frac{\partial^2 T}{\partial y^2}\right) + \frac{\rho_p c_s}{\gamma_T}\left(T_p - T\right). \tag{3}$$

For dust phase:

$$\frac{\partial u_p}{\partial x} + \frac{\partial v_p}{\partial y} = 0, \tag{4}$$

$$\rho_p\left(u_p\frac{\partial u_p}{\partial x} + v_p\frac{\partial u_p}{\partial y}\right) = \frac{\rho_p}{\tau_v}(u - u_p), \tag{5}$$

$$\rho_p c_s\left(u_p\frac{\partial T_p}{\partial x} + v_p\frac{\partial T_p}{\partial y}\right) = -\frac{\rho_p c_s}{\gamma_T}\left(T_p - T\right). \tag{6}$$

Here, (u, v), T, ρ, c_p and μ represents the components of velocity in $(x,\ y)$ directions, temperature, density, specific heat at constant pressure and viscosity coefficient, respectively. Meanwhile, (u_p, v_p), T_p, ρ_p, c_s, τ_v and γ_T denote the velocity components in (x, y) directions, temperature, density, specific heat, velocity and thermal relaxation time for dust phase, respectively. The corresponding fluid and particle phase boundary conditions were given as.

$$\begin{array}{l} u = u_w(x) = ax,\ v = 0,\ \frac{\partial T}{\partial y} = -h_s T \text{ at } y = 0 \\ u \to 0,\ u_p \to 0,\ v_p \to v,\ T \to T_\infty,\ T_p \to T_\infty \text{ as } y \to \infty \end{array} \tag{7}$$

In (7), the parameter was corresponded to the velocity of the stretching surface with a being a positive constant of $u_w(x)$, thermal conductivity k, heat transfer coefficient h_s and ambient temperature T_∞. To obtain the set of similarity equations in the form of ordinary differential equations, the similarity transformations (8) were adopted and applied to the governing Equations (1)–(6).

$$\begin{aligned} &u = axf'(\eta),\ v = -(av)^{1/2}f(\eta),\ \eta = \left(\tfrac{a}{v}\right)^{1/2}y, \\ &\theta(\eta) = \tfrac{T-T_\infty}{T_\infty},\ u_p = axF'(\eta),\ v_p = -(av)^{1/2}F(\eta),\ \theta_p(\eta) = \tfrac{T_p-T_\infty}{T_\infty}, \end{aligned} \tag{8}$$

The Reynolds exponential viscosity model was used to predict temperature-dependent variability in viscosity that gave a detailed approach as:

$$\mu(\theta) = \mu_0 e^{-(\beta_1\theta)} = \mu_0\left[1-(\beta_1\theta)+O\left({\beta_1}^2\right)\right], \tag{9}$$

The Equations (1)–(6) are changeable from PDEs to ODEs, which can be represented as:

$$\begin{aligned} &(1+M)f'''(\eta) - (f'(\eta))^2 + f(\eta)f''(\eta) + \beta N(F'(\eta) - f'(\eta)) \\ &-BM(f''(\eta))^2 f'''(\eta) - \alpha f''(\eta)\theta'(\eta) - \alpha\theta(\eta)f'''(\eta) + \lambda\theta = 0, \end{aligned} \tag{10}$$

$$\theta''(\eta) + \mathrm{Pr}f(\eta)\theta'(\eta) + \frac{2}{3}\beta N(\theta_p(\eta) - \theta(\eta)) = 0, \tag{11}$$

$$(F'(\eta))^2 - F(\eta)F''(\eta) + \beta(F'(\eta) - f'(\eta)) = 0, \tag{12}$$

$${\theta_p}'(\eta)F(\eta) + \frac{2}{3}\frac{\beta}{\mathrm{Pr}\gamma}(\theta(\eta) - \theta_p(\eta)) = 0 \tag{13}$$

Subjected to boundary conditions:

$$\begin{aligned} &f(0) = 0, f'(0) = 1,\ \theta'(0) = -\gamma_1(1+\theta(0)) \text{ at } \eta = 0 \\ &f'(\eta) \to 0,\ F'(\eta) \to 0,\ F(\eta) \to f(\eta), \\ &\theta(\eta) \to 0,\ \theta_p(\eta) \to 0 \text{ as } \eta \to \infty \end{aligned} \tag{14}$$

In Equations (10)–(14), a notation prime ($'$) corresponds to the differentiation with respect to η. Additionally, the dimensionless numbers and parameters are as follows where $M = \frac{1}{\mu_0\tilde{\beta}c}$ and $B = \frac{a^3x^2}{2c^2\nu_f}$ are the fluid parameters, $\mathrm{Pr} = \nu_f/\alpha$ represents Prandtl number, $\alpha = k/\rho c_p$ represents viscosity parameter, γ_1 conjugate parameter of heat transfer, $\lambda = g_c\beta_T(T_f - T_\infty)/a^2x\lambda$ is mixed convection parameter, $\gamma = c_s/c_p$ specific heat ratio of mixture parameter, $N = \rho_p/\rho$ parameter of mass concentration of particle phase, fluid–particle interaction parameter $\beta = 1/a\tau_v$ and Reynolds number $\mathrm{Re}_x = (ax^2/\nu)$. A limiting case arising in this problem was without the presence of dust particles effect where the buoyancy force is negligible and can be obtained using the following expression [33]:

$$f(\eta) = S + A\eta + (1-A)(1-\exp(-\eta)), \theta(\eta) = \frac{\gamma_1}{1-\gamma_1}\exp(-\eta). \tag{15}$$

It is important to mention here that the comparison between the present results with the exact solution is necessary to claim the accuracy of the current model and its output. The primary physical quantity of importance is the dimensionless coefficient of skin friction and the local Nusselt number, which has been described by (16), where the shear stress and surface heat are compatible with those referred in [34].

$$C_{fx} = \frac{\tau_w}{\rho U_w^2(x)}, \quad Nu_x = \frac{xq_w}{k(T_w - T_\infty)} \tag{16}$$

where

$$\tau_w = \left(\mu_0 + \frac{1}{\tilde{\beta}C^*}\right)\frac{\partial u}{\partial y} - \frac{1}{6\tilde{\beta}}\left(\frac{1}{C^*}\frac{\partial u}{\partial y}\right)^3, \text{and } q_w = -k\left(\frac{\partial T}{\partial y}\right)_{y=0} \tag{17}$$

The shear stress and surface heat transfer is calculated using the following definition:

$$C_f \mathrm{Re}_x^{1/2} = ((1-\alpha\theta(0)) + M)f''(0) - \frac{B}{3}Mf''^3(0), \quad Nu_x \mathrm{Re}_x{}^{-1/2} = \gamma_1\left(\frac{1}{\theta(0)} + 1\right) \quad (18)$$

3. Results and Discussion

Equations (10)–(13) were solved numerically, along with boundary conditions (14) using the Keller-box approach as computed in the Matlab program. The Keller-box method comprise of four steps which are:

Step 1: The nonlinear partial differential equation are first transformed to first order system;
Step 2: The first order system is then approximated using central difference;
Step 3: The Newton's method is applied to linearize the system;
Step 4: The linearized system is solved by block elimination technique;

To initially integrate the procedure, the nonlinear structure of ordinary differential equations was converted into a structure of linear first-order equations. Our bulk calculations were viewed with $\eta_\infty = 8$ and identified as appropriate for all values of the parameters, which were considered asymptotically to achieve the far-field boundary conditions as seen in Figures 2–17. The interaction force among two phases is significant, where usually, both governing equations are coupled through the term of total fluid–particle interaction force per unit volume that is clearly different from single phase flow. Numerical computation are conducted for fluid parameters, namely M and B, Prandtl number Pr, viscosity parameter α, mixed convection parameter λ, γ_1 conjugate parameter of heat transfer, a parameter of mass concentration of particle phase N and fluid–particle interaction parameter β.

The local Nusselt number is one of the important characteristics in the heat transfer field, which indicates the ratio of convective heat transfer to conductive heat transfer. Hence, in order to check the accuracy of the numerical method used, the comparison in Nusselt number $Nu_x \mathrm{Re}_x{}^{-1/2}$ for a fixed value of Pr with the established results in [35–37] and it is revealed to be in strong agreement as displayed in Table 1. In addition, a direct comparative study was carried out with the exact Equation (15), as well as the existing study reported with the available published result by [38–41] shown in Table 2. From Tables 1 and 2, an excellent agreement is achieved, which indicates that the current model and its findings are acceptable. It is worth declaring here, even in the limiting cases, the present model does not exactly give the same solution, but the difference is very small. It is logical since the present model is more complex with multiple parameters. Table 3 demonstrates the variance of the skin friction coefficient and the Nusselt numbers for various parameters of the present analysis.

Figures 2–5 was plotted to understand the velocity and temperature distribution of the fluid and particle phase under variance of Pr and α. It was revealed that the velocity were decreased for both fluid and particle phases as Pr and α increases. The similar trend was noticed in temperature distribution for both phases in increasing Pr but contrary in the growing of α. At far from the surface, it is remarked the profile asymptotically reached the boundary conditions, and, therefore, the authors are confident on the correctness of present results.

Table 1. Comparative study on value $-\theta'(0)$.

Pr	[35]	[36]	[37]	Present
1	1.3333	1.3333	1.3333	1.3329
3	2.50970	2.50972	2.50972	2.50969
10	4.79690	4.79686	4.79687	4.79689
100	15.7120	15.7118	15.7120	15.7098

Table 2. Comparative study on $f''(0)$.

Existing Literature	Model of Problem	Boundary Condition	Limiting Cases	Value of $f''(0)$
Exact Solution (15)	$f'' = -e^{-\eta}$	-	-	-1.0000
[38]	$f''' - f'^2 + ff'' - A\left(f' + \frac{1}{2}\eta f''\right) + \lambda\theta = 0$	$f(0) = 0$ $f'(0) = 1$ $f'(\infty) = 0$	$A = \lambda = 0$	-1.0000
[39]	$f''' - f'^2 + ff'' - k_1\left(2f'f''' - f''^2 - ff'''\right) = 0$	$f(0) = 0$ $f'(0) = 1$ $f'(\infty) = 0$	$k_1 = 0$	-1.0000
[40]	$(1+M)(1+2\eta\gamma)f''' - \alpha M(1+2\eta\gamma)^2 f''^2 f'''$ $+2\gamma(1+M)f'' + ff'' - \frac{4}{3}\alpha M(\gamma + 2\eta\gamma^2)f''^3$ $-f'^2 + \lambda\theta\sin\varphi = 0$	$f(0) = 0$ $f'(0) = 1$ $f'(\infty) = 0$	$\alpha = \gamma = 0$ $\lambda = M = 0$	-1.0000
[41]	$(1+M)f''' - f'^2 + ff'' - MBf''^2 f''' - Hf' = 0$	$f(0) = 0$ $f'(0) = 1$ $f'(\infty) = 0$	$B = H = 0,$ $M = 0.0001$	-1.0000
Present study	$(1+M)f''' - f'^2 + ff'' + \beta N(F' - f')$ $-BMf''^2 f''' - \alpha f''\theta' - \alpha\theta f''' + \lambda\theta = 0,$	$f(0) = 0$ $f'(0) = 1$ $f'(\infty) = 0$	$B = M = 0$ $N = \beta = 0$ $\alpha = \lambda = 0$	-1.0015

Table 3. Numerical results of $C_f\mathrm{Re}_x^{1/2}$ and $Nu_x\mathrm{Re}_x^{-1/2}$ for various values of Pr, α, M, B, λ, γ_1, β and N.

Pr	α	M	B	λ	γ_1	β	N	$C_f\mathrm{Re}_x^{1/2}$	$Nu_x\mathrm{Re}_x^{-1/2}$
7	0.1	0.6	0.6	0.1	0.5	0.5	0.5	−1.076579	0.105283
9								−1.078272	0.104603
12								−1.079770	0.103939
10	0.2							−1.077788	0.104348
	0.4							−1.075612	0.104349
	0.7							−1.072307	0.104350
10	0.1	0.5						−1.048326	0.104357
		0.9						−1.173040	0.104324
		1.5						−1.351057	0.104292
10	0.1	0.6	0.1					−1.133436	0.104328
			0.5					−1.088912	0.104324
			0.9					−1.048210	0.104322
10	0.1	0.1	0.6	0.3				−1.754138	0.342542
				0.5				−1.741401	0.630439
				0.9				−1.706099	1.433335
10	0.1	0.6	0.6	0.1	0.1			−1.085300	0.104326
					0.6			−1.081074	0.104324
					1.2			−1.076098	0.104318
10	0.1	0.6	0.6	0.1	0.5	0.1		−1.180434	0.104421
						0.4		−1.114385	0.104359
						0.9		−1.044242	0.104282
10	0.1	0.6	0.6	0.1	0.5	0.5	0.1	−1.182364	0.104415
							0.4	−1.125083	0.104364
							0.9	−1.019745	0.104280

Figures 6–9 display the distribution on velocity and temperature for multiple values of M and B. It was found that with higher elasticity parameter (presence Eyring fluid), the magnitude of velocity for both fluid and particle were enhanced. The change in the velocity contributed to boosting the heat of the fluid (for increasing M) but against the heating development (for increasing B).

Figures 10 and 11 indicate that the amount of γ_1. Physically, the heat transfer rate with the γ_1 was reduced; this reduced the temperature and resulting thickness of the boundary layer. During the event with large values of the heat transfer equation, which implies a strong heat transfer rate as it slowly declines, the fluid became a small heat transfer rate. Through mixed convection and the effects of the fluid–particle interaction parameters β and λ, the velocity of all the phases and their associated boundary layer thicknesses increased. The response to the temperature profile was quite the opposite of the speed field with λ and β, as seen in Figures 12–15, further finding that the flow properties of the dusty fluid can be greatly regulated by changing the influence of the parameters of fluid–particle interaction β.

Figures 16 and 17 were plotted to evaluate all fluid and particle phase velocity and temperature components for the variance of N. Within the boundary layer, temperature profiles reduced with improvement within N. On the other hand, the velocity profile was improved by increasing the parameter N. This was because the fluid tends to raise the intensity of drag between the phases with the mass content of dust particles rising. The fluid movement was thus slowed down, resulting in reduced surface-phase energy, since the surface layer was pulled together with the liquid. By continuing to increase the mass content of the dust particles, more fluid-phase energy was converted into a larger number of particles, but less energy from the fluid phase was supplied to the individual particles. Therefore, it can be inferred that varying N will greatly affect the flow characteristics. Furthermore, the boundary momentum layer for ordinary Eyring fluid was observed thinner than that of the dusty Eyring fluid.

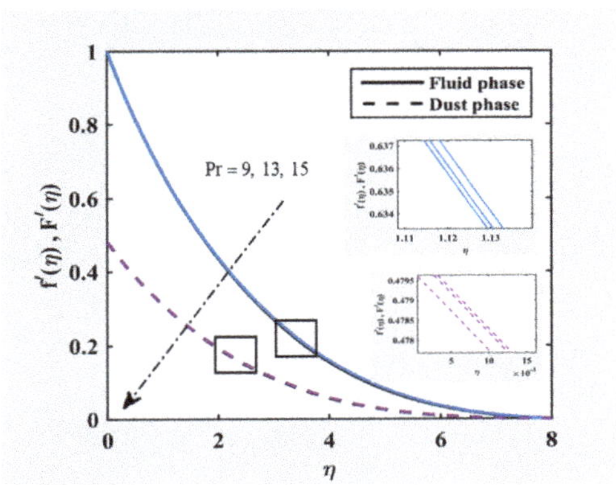

Figure 2. $f'(\eta)$ and $F'(\eta)$ at $M = B = \lambda = 0.5$, $\gamma_1 = \alpha = 0.1$ and $\beta = N = 0.6$ for various values of Pr.

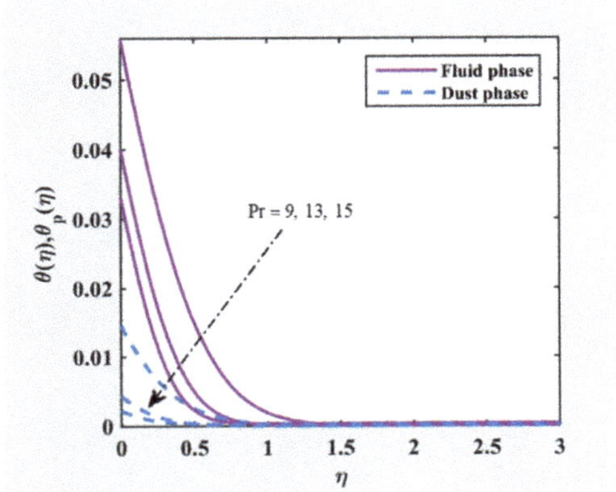

Figure 3. $\theta(\eta)$ and $\theta_p(\eta)$ at $M = B = \lambda = 0.5$, $\gamma_1 = \alpha = 0.1$ and $\beta = N = 0.6$ for various values of Pr.

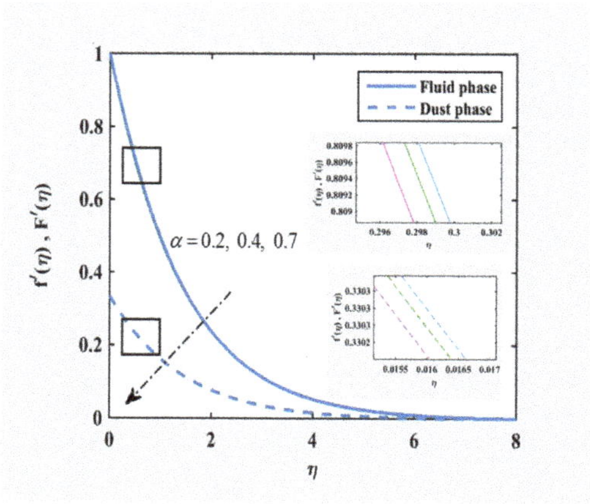

Figure 4. $f'(\eta)$ and $F'(\eta)$ at $M = B = \lambda = 0.5, \gamma_1 = 0.1, \beta = N = 0.6$ and $\Pr = 10$ for various values of α.

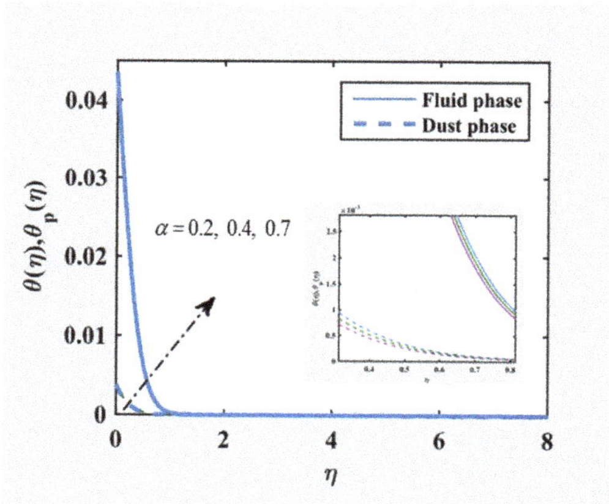

Figure 5. $\theta(\eta)$ and $\theta_p(\eta)$ at $M = B = \lambda = 0.5, \gamma_1 = 0.1, \beta = N = 0.6$ and $\Pr = 10$ for various values of α.

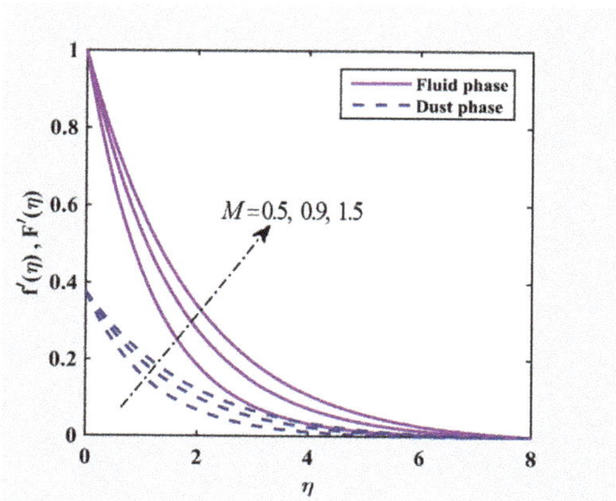

Figure 6. $f'(\eta)$ and $F'(\eta)$ at $B = \lambda = 0.5, \alpha = \gamma_1 = 0.1, \beta = N = 0.6$ and $\Pr = 10$ for various values of M.

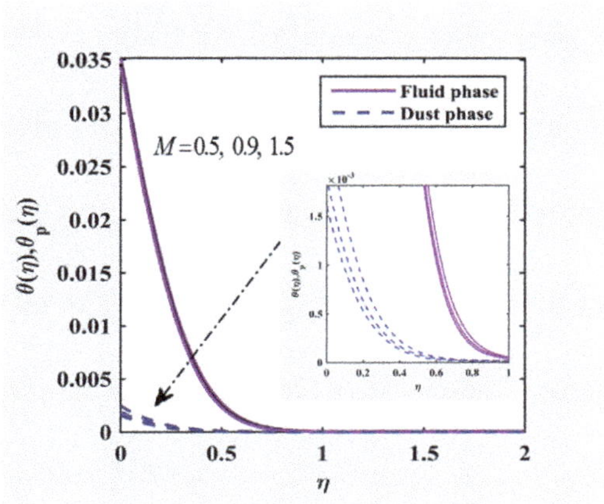

Figure 7. $\theta(\eta)$ and $\theta_p(\eta)$ at $B = \lambda = 0.5, \alpha = \gamma_1 = 0.1, \beta = N = 0.6$ and $\Pr = 10$ for various values of M.

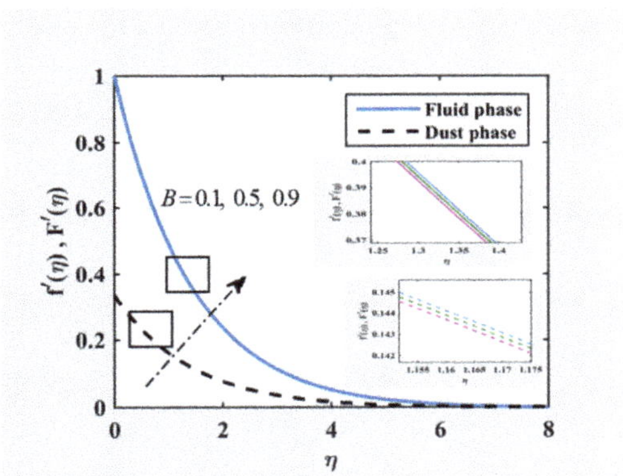

Figure 8. $f'(\eta)$ and $F'(\eta)$ at $M = \lambda = 0.5, \alpha = \gamma_1 = 0.1, \beta = N = 0.6$ and $\Pr = 10$ for various values of B.

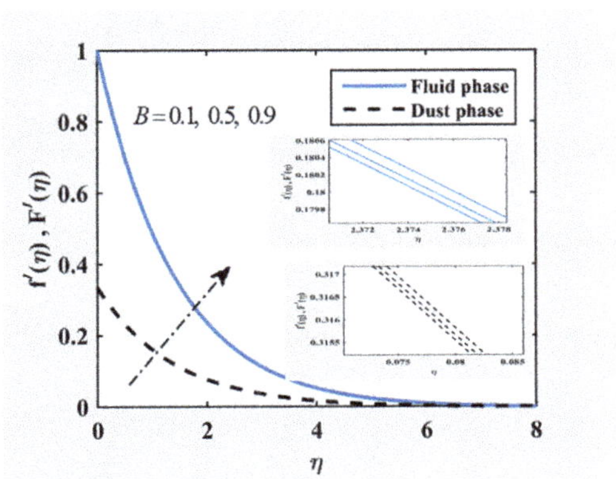

Figure 9. $\theta(\eta)$ and $\theta_p(\eta)$ at $M = \lambda = 0.5, \alpha = \gamma_1 = 0.1, \beta = N = 0.6$ and $\Pr = 10$ for various values of B.

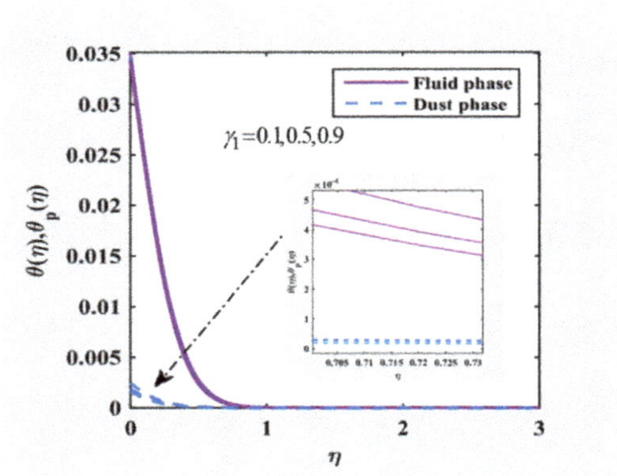

Figure 10. $f'(\eta)$ and $F'(\eta)$ at $M = B = \lambda = 0.5, \alpha = 0.1, \beta = N = 0.6$ and $\Pr = 10$ for various values of γ_1.

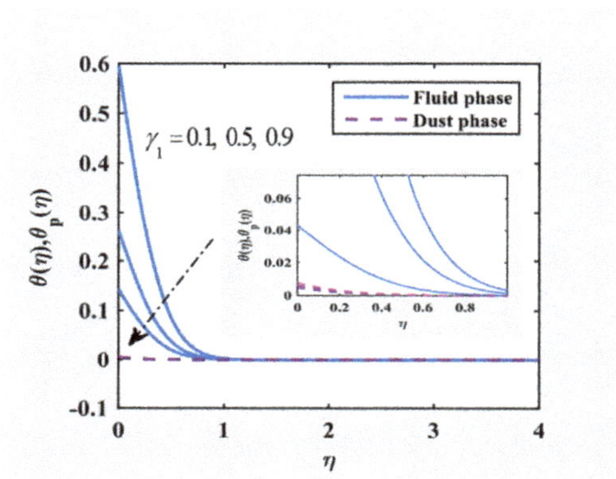

Figure 11. $\theta(\eta)$ and $\theta_p(\eta)$ at $M = B = \lambda = 0.5, \alpha = 0.1, \beta = N = 0.6$ and $\Pr = 10$ for various values of γ_1.

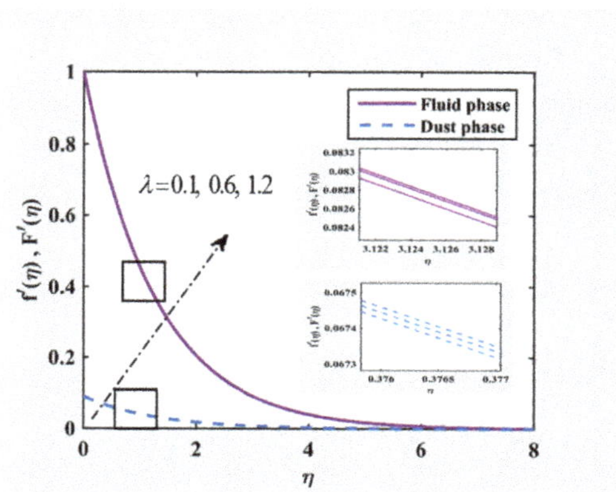

Figure 12. $f'(\eta)$ and $F'(\eta)$ at $M = B = 0.5, \gamma_1 = \alpha = 0.1, \beta = N = 0.6$ and $\Pr = 10$ for various values of λ.

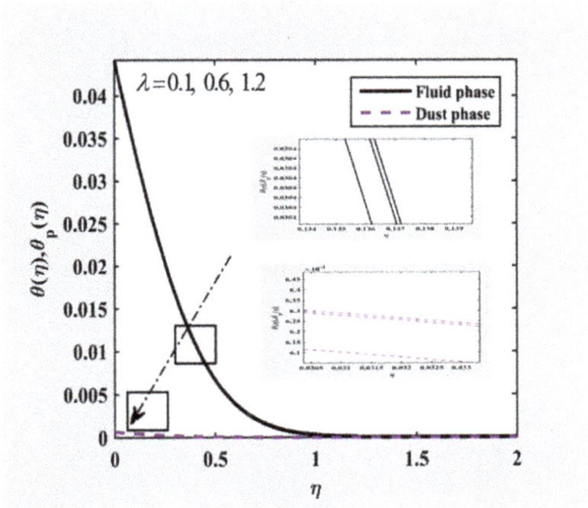

Figure 13. $\theta(\eta)$ and $\theta_p(\eta)$ at $M = B = 0.5, \gamma_1 = \alpha = 0.1, \beta = N = 0.6$ and $\Pr = 10$ for various values of λ.

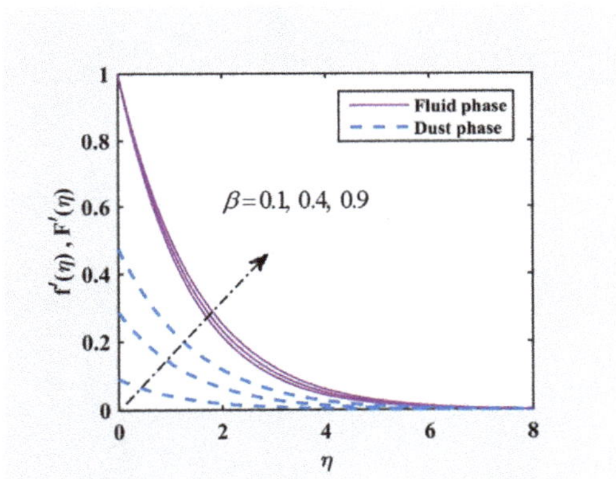

Figure 14. $f'(\eta)$ and $F'(\eta)$ at $M = B = \lambda = 0.5, N = 0.6, \gamma_1 = \alpha = 0.1$ and $\Pr = 10$ for various values of β.

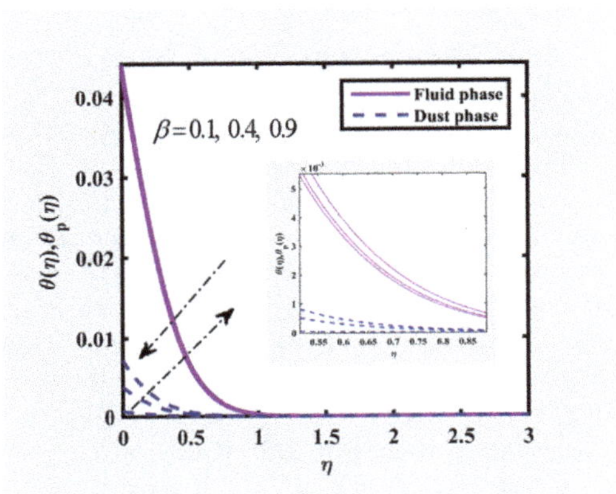

Figure 15. $\theta(\eta)$ and $\theta_p(\eta)$ at $M = B = \lambda = 0.5, N = 0.6, \gamma_1 = \alpha = 0.1$ and $\Pr = 10$ for various values of β.

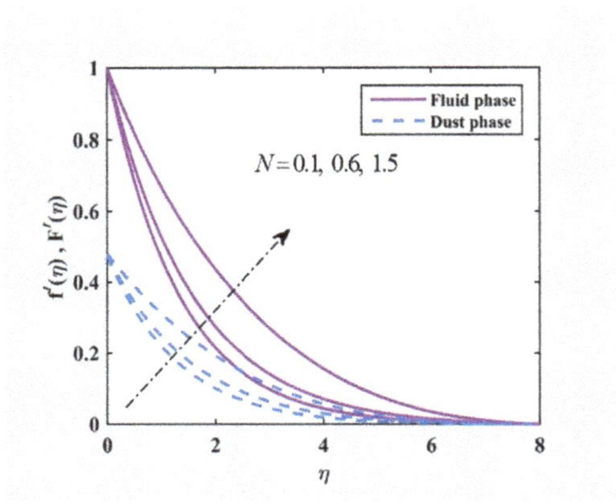

Figure 16. $f'(\eta)$ and $F'(\eta)$ at $M = B = \lambda = 0.5, \beta = 0.6, \gamma_1 = \alpha = 0.1$ and $\Pr = 10$ for various values of N.

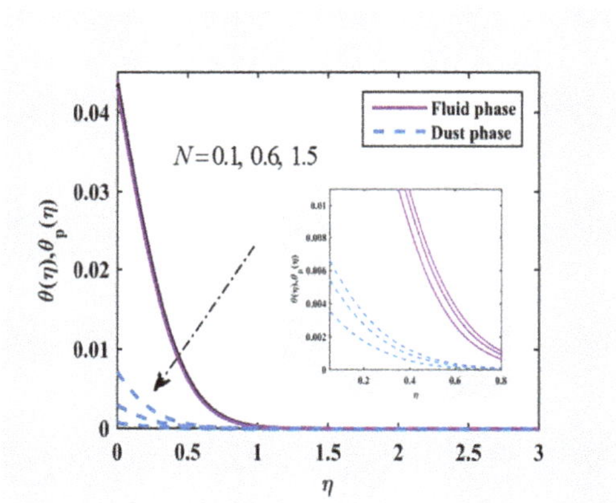

Figure 17. $\theta(\eta)$ and $\theta_p(\eta)$ at $M = B = \lambda = 0.5, \beta = 0.6, \gamma_1 = \alpha = 0.1$ and $\Pr = 10$ for various values of N.

4. Conclusions and Future Work

The results of temperature-dependent viscosity on the mixed convection flow of non-Newtonian Eyring–Powell fluid due to a vertical stretching surface interacting with dust particles under the control of NH boundary conditions have been numerically explained. Highly non-linear governing PDEs were first converted to ODEs, utilising similarity transformations and then numerically resolved using the Keller-box method. MATLAB program has been applied to conduct computational. Velocity and temperature profile for all fluid and particle phases were illustrated graphically. The value of sheer rate and heat transfer coefficient were calculated and have tabulated in table for various pertinent parameters. In addition, the current output was validated by comparative study with previously reported results and perceived a good agreement between them. In conjunction, a few important findings from the existing research are indicated as:

1. For certain applications, the fluid's flow and heat transfer can be regulated by embedding the particles of fine dust.
2. The process velocity (temperature) of fluid and dust have the opposite effects for buoyancy force parameter variability.

3. In the mixed convection regime, the local shear stress increases and the local rate of heat transfer decreases as the value of buoyancy parameter increases for all values of the Prandtl number and the viscosity variation parameter.
4. The fluid–particle interaction parameter variability is favourable for the thickness of the dust boundary layer. However, for the thickness of boundary layer of momentum, it is unfavourable.
5. The velocity distribution was suppressed with the Prandtl number compared to temperature distribution.
6. The velocity profiles increase and the viscosity of the fluid decrease near the surface of the plate owing to increase in the value of the viscosity variation parameter. The temperature profiles of both phases are enhanced by rising α.
7. The quantity of skin friction decreases with greater values of fluid parameters, mixed convection, conjugate parameter of heat transfer, mass concentration of particle phase and fluid–particle interaction parameter.
8. Increase in the value of the viscosity variation parameter leads to increase in the local shear stress and to decrease in the local rate of heat transfer. Its effect on the increase of the rate of heat transfer is less than that of the local shear stress.

Ultimately, it is worth concluding that the existence of dust particles has a substantial effect on the flow behaviour of Eyring fluid in the presence of TDV.

Author Contributions: Funding acquisition, A.R.M.K.; Investigation, A.A., N.A.N.A., I.W., N.S.K. and N.A.Z.; Methodology, A.A. and N.S.A.; Resources, A.R.M.K., N.S.A., N.A.N.A., D.L.C.C., I.W. and N.S.K.; Software, D.L.C.C. and N.A.Z.; Writing—original draft, A.A. All authors have read and agreed to the published version of the manuscript.

Funding: This research was funded by Universiti Malaysia Pahang under RDU213206 and the APC was funded by Universiti Malaysia Pahang through RDU213206 and Universiti Teknologi Petronas through 015MCO–030.

Institutional Review Board Statement: Not applicable.

Informed Consent Statement: Not applicable.

Data Availability Statement: Not applicable.

Acknowledgments: The authors wish to thank Universiti Malaysia Pahang for the financial support through RDU213206 and Universiti Teknologi Petronas through 015MCO–030. Our heartfelt thanks also go to Universiti Teknologi MARA Cawangan Johor and Cawangan Pahang as well as Universiti Teknikal Melaka for their assistance and support.

Conflicts of Interest: The authors of this manuscript declare no conflict of interest.

References

1. Ramesh, G.K.; Kumar, K.G.; Shehzad, S.; Gireesha, B. Enhancement of radiation on hydromagnetic Casson fluid flow towards a stretched cylinder with suspension of liquid-particles. *Can. J. Phys.* **2018**, *96*, 18–24. [CrossRef]
2. Manjunatha, P.T.; Gireesha, B.J.; Prasannakumara, B. Effect of Radiation on Flow and Heat Transfer of MHD Dusty Fluid Over a Stretching Cylinder Embedded in a Porous Medium in Presence of Heat Source. *Int. J. Appl. Comput. Math.* **2015**, *3*, 293–310. [CrossRef]
3. Kalpana, G.; Madhura, K.; Kudenatti, R.B. Impact of temperature-dependant viscosity and thermal conductivity on MHD boundary layer flow of two-phase dusty fluid through permeable medium. *Eng. Sci. Technol. Int. J.* **2018**, *22*, 416–427. [CrossRef]
4. Arifin, N.S.; Zokri, S.M.; Kasim, A.R.M.; Salleh, M.Z.; Mohammad, N.F.; Yusoff, W.N.S.W. Aligned magnetic field on dusty Casson fluid over a stretching sheet with Newtonian heating. *Malays. J. Fundam. Appl. Sci.* **2017**, *13*, 245–248. [CrossRef]
5. Kasim, A.R.M.; Arifin, N.S.; Zokri, S.M.; Salleh, M.Z.; Mohammad, N.F.; Ching, D.L.C.; Shafie, S.; Ariffin, N.A.N. Convective Transport of Fluid–Solid Interaction: A Study between Non-Newtonian Casson Model with Dust Particles. *Crystals* **2020**, *10*, 814. [CrossRef]
6. Kasim, A.R.M.; Arifin, N.S.; Zokri, S.M.; Salleh, M.Z. Fluid-particle interaction with buoyancy forces on Jeffrey fluid with Newtonian heating. *CFD Lett.* **2019**, *11*, 1–16.

7. Kasim, A.R.M.; Arifin, N.S.; Zokri, S.M.; Salleh, M.Z. The Investigation of a Fluid-Solid Interaction Mathematical Model under Combined Convective Jeffrey Flow and Radiation Effect Embedded Newtonian Heating as the Thermal Boundary Condition over a Vertical Stretching Sheet. In *Defect and Diffusion Forum*; Trans Tech Publications Ltd.: Kapellweg, Switzerland, 2020; Volume 399, pp. 65–75.
8. Siddiqa, S.; Begum, N.; Hossain, A.; Gorla, R.S.R. Natural convection flow of a two-phase dusty non-Newtonian fluid along a vertical surface. *Int. J. Heat Mass Transf.* **2017**, *113*, 482–489. [CrossRef]
9. Mohamed, R.A.; Hady, F.M.; Mahdy, A.; Abo-Zai, O.A. Laminar MHD natural convection flow due to non-Newtonian nanofluid with dust nanoparticles around an isothermal sphere: Non-similar solution. *Phys. Scr.* **2021**, *96*, 035215. [CrossRef]
10. Mahdy, A.; Hoshoudy, G.A. Two-phase mixed convection nanofluid flow of a dusty tangent hyperbolic past a nonlinearly stretching sheet. *J. Egypt. Math. Soc.* **2019**, *27*, 1–16. [CrossRef]
11. Kumar, R.S.V.; Gowda, R.J.P.; Radhika, M.; Prasannakumara, B.C. Two-phase flow of dusty fluid with suspended hybrid nanoparticles over a stretching cylinder with modified Fourier heat flux. *SN Appl. Sci.* **2021**, *3*, 1–9.
12. Shawky, H.M. Pulsatile flow with heat transfer of dusty magnetohydrodynamic Ree-Eyring fluid through a channel. *Heat Mass Transf.* **2009**, *45*, 1261–1269. [CrossRef]
13. Upadhya, S.M.; Mahesha; Raju, C.; Shehzad, S.; Abbasi, F. Flow of Eyring-Powell dusty fluid in a deferment of aluminum and ferrous oxide nanoparticles with Cattaneo-Christov heat flux. *Powder Technol.* **2018**, *340*, 68–76. [CrossRef]
14. Upadhya, S.M.; Mahesha; Raju, C.S.K. Comparative study of Eyring and Carreau fluids in a suspension of dust and nickel nanoparticles with variable conductivity. *Eur. Phys. J. Plus* **2018**, *133*, 156. [CrossRef]
15. Raju, C.S.K.; Saleem, S.; Al-Qarni, M.M.; Upadhya, S.M. Unsteady nonlinear convection on Eyring–Powell radiated flow with suspended graphene and dust particles. *Microsyst. Technol.* **2018**, *25*, 1321–1331. [CrossRef]
16. Merkin, J.H. Natural-convection boundary-layer flow on a vertical surface with Newtonian heating. *Int. J. Heat Fluid Flow* **1994**, *15*, 392–398. [CrossRef]
17. Lesnic, D.; Ingham, D.B.; Pop, I.; Storr, C. Free convection boundary-layer flow above a nearly horizontal surface in a porous medium with newtonian heating. *Heat Mass Transf.* **2003**, *40*, 665–672. [CrossRef]
18. Hayat, T.; Ali, S.; Farooq, M.A.; Alsaedi, A. On Comparison of Series and Numerical Solutions for Flow of Eyring-Powell Fluid with Newtonian Heating And Internal Heat Generation/Absorption. *PLoS ONE* **2015**, *10*, e0129613. [CrossRef]
19. Qayyum, S.; Hayat, T.; Shehzad, S.A.; Alsaedi, A. Nonlinear convective flow of Powell-Erying magneto nanofluid with Newtonian heating. *Results Phys.* **2017**, *7*, 2933–2940. [CrossRef]
20. Hayat, T.; Waqas, M.; Shehzad, S.A.; Alsaedi, A. Mixed convection stagnation-point flow of Powell-Eyring fluid with Newtonian heating, thermal radiation, and heat generation/absorption. *J. Aerosp. Eng.* **2017**, *30*, 04016077. [CrossRef]
21. Hayat, T.; Hussain, Z.; Farooq, M.; Alsaedi, A. Magnetohydrodynamic flow of Powell-Eyring fluid by a stretching cylinder with Newtonian heating. *Therm. Sci.* **2018**, *22*, 371–382. [CrossRef]
22. Parmar, A.; Jain, S. Unsteady convective flow for MHD Powell-Eyring fluid over inclined permeable surface. *J. Comput. Appl. Res. Mech. Eng.* **2019**, *9*, 297–312.
23. Aljabali, A.; Kasim, A.R.M.; Arifin, N.S.; Isa, S.M. Mixed Convection of Non-Newtonian Erying Powell Fluid with Temperature-Dependent Viscosity over a Vertically Stretched Surface. *Comput. Mater. Contin.* **2020**, *66*, 421–435. [CrossRef]
24. Aljabali, A.; Kasim, A.R.M.; Arifin, N.S.; Isa, S.M.; Ariffin, N.A.N. Analysis of Convective Transport of Temperature-Dependent Viscosity for Non-Newtonian Erying Powell Fluid: A Numerical Approach. *Comput. Mater. Contin.* **2020**, *66*, 675–689. [CrossRef]
25. Gangadhar, K.; Kumar, D.V.; Rao, M.V.S.; Kannan, T.; Sakthivel, G. Effects of Newtonian heating on the boundary layer flow of non-Newtonian magnetohydrodynamic nanofluid over a stretched plate using spectral relaxation method. *Int. J. Ambient Energy* **2019**, *43*, 1248–1261. [CrossRef]
26. Amin, W.N.Z.; Qasim, M.; Shafie, S. G-Jitter Induced Mixed Convection Flow between Two Parallel Plates with Newtonian Heating. *Sci. Proc. Ser.* **2019**, *1*, 107–110. [CrossRef]
27. Mohamed, M.K.A.; Ismail, N.A.; Hashim, N.; Shah, N.; Salleh, M. MHD Slip Flow and Heat Transfer on Stagnation Point of a Magnetite (Fe_3O_4) Ferrofluid towards a Stretching Sheet with Newtonian Heating. *CFD Lett.* **2019**, *11*, 17–27.
28. Ahmad, S.; Nadeem, S. Application of CNT-based micropolar hybrid nanofluid flow in the presence of Newtonian heating. *Appl. Nanosci.* **2020**, *10*, 5265–5277. [CrossRef]
29. Hasan, M.; Samad, A.; Hossain, M. Effects of Hall Current and Ohmic Heating on Non-Newtonian Fluid Flow in a Channel due to Peristaltic Wave. *Appl. Math.* **2020**, *11*, 292–306. [CrossRef]
30. Sarif, N.; Salleh, M.; Nazar, R. Numerical solution of flow and heat transfer over a stretching sheet with newtonian heating using the keller box method. *Procedia Eng.* **2013**, *53*, 542–544. [CrossRef]
31. Siddiqa, S.; Hossain, M.A.; Saha, S.C. Two-phase natural convection flow of a dusty fluid. *Int. J. Numer. Methods Heat Fluid Flow* **2015**, *25*, 1542–1556. [CrossRef]
32. Hayat, T.; Iqbal, Z.; Qasim, M.; Obaidat, S. Steady flow of an Eyring Powell fluid over a moving surface with convective boundary conditions. *Int. J. Heat Mass Transf.* **2012**, *55*, 1817–1822. [CrossRef]
33. Imtiaz, M.; Hayat, T.; Hussain, M.; Shehzad, S.A.; Chen, G.Q.; Ahmad, B. Mixed convection flow of nanofluid with Newtonian heating. *Eur. Phys. J. Plus* **2014**, *129*, 97. [CrossRef]
34. Javed, T.; Ali, N.; Abbas, Z.; Sajid, M. Flow of an Eyring-Powell non-Newtonian fluid over a stretching sheet. *Chem. Eng. Commun.* **2013**, *200*, 327–336. [CrossRef]

35. Grubka, L.; Bobba, K. Heat transfer characteristics of a continuous, stretching surface with variable temperature. *ASME J. Heat Transf.* **1985**, *107*, 248–250. [CrossRef]
36. El-Aziz, M.A. Unsteady mixed convection heat transfer along a vertical stretching surface with variable viscosity and viscous dissipation. *J. Egypt. Math. Soc.* **2014**, *22*, 529–537. [CrossRef]
37. Mamatha, S.; Mahesha; Raju, C.S.; Makinde, O.D. Effect of Convective Boundary Condition on MHD Carreau Dusty Fluid over a Stretching Sheet with Heat Source. *Defect Diffus. Forum* **2017**, *377*, 233–241. [CrossRef]
38. Ishak, A.; Nazar, R.; Pop, I. Boundary layer flow and heat transfer over an unsteady stretching vertical surface. *Meccanica* **2008**, *44*, 369–375. [CrossRef]
39. Arnold, J.C.; Asir, A.A.; Somasundaram, S.; Christopher, T. Heat transfer in a viscoelastic boundary layer flow over a stretching sheet. *Int. J. Heat Mass Transf.* **2010**, *53*, 1112–1118. [CrossRef]
40. Wahab, H.A.; Hussain, S.; Naeem, M. Mixed convection flow of Powell-Eyring fluid over a stretching cylinder with Newtonian heating. *Kuwait J. Sci.* **2016**, *43*, 1–13.
41. Khan, I.; Malik, M.Y.; Salahuddin, T.; Khan, M.; Rehman, K.U. Homogenous–heterogeneous reactions in MHD flow of Powell–Eyring fluid over a stretching sheet with Newtonian heating. *Neural Comput. Appl.* **2017**, *30*, 3581–3588. [CrossRef]

applied sciences

MDPI

Article

Numerical Study on the Effect of Deposit Layer on the Minimum Wall Thickness of Boiler Water Tube under Different Operating Conditions

Ahmed S. Aljohani [1,*], Khaled I. Ahmed [2], Saeed Asiri [3] and Mohamed H. Ahmed [4,5]

1 Mechanical Engineering Department, Shoaiba Power Plant, Saudi Electricity Company, Makkah 28211, Saudi Arabia
2 K.A. CARE Energy Research and Innovation Center, Mechanical Engineering Department, King Abdulaziz University, Jeddah 21589, Saudi Arabia
3 Department of Mechanical Engineering, King Abdulaziz University, Jeddah 21589, Saudi Arabia
4 Mechanical Engineering Department, King Abdulaziz University, Jeddah 21589, Saudi Arabia
5 Department of Mechanical Engineering, Assiut University, Assiut 71516, Egypt
* Correspondence: eng.asaj@gmail.com; Tel.: +966-56-984-0315

Abstract: Water wall tube temperature is a major parameter in the steam generator design which has a significant role in keeping the steam generator available. Thus, knowing the tube average temperature in different operating conditions is very important to avoid the causes of tube failures. High temperatures are a major cause of various types of failures, such as overheating, hydrogen damage, thermal stress, etc. Furthermore, deposits on the inner tube wall contribute to such failure by changing the thermal resistance of the tube wall, which causes a significant increase in the tube wall's average temperature, consequently lowering the allowable stress. Therefore, the model was created by using ANSYS FLUENT (Canonsburg, PA, USA) to determine the wall average water tube wall temperature considering the deposit layer thickness (magnetite). Furthermore, this model was verified. It was found that increasing tube thickness can increase the average tube temperature but combining it with increasing deposit thickness leads to higher temperatures. In other words, the effect of the deposit on the tube with higher thickness is higher than on the tube with lower thickness. By discussing the minimum thickness of the water wall tube, the suitable selection of the tube thickness and courses of action concerning the operating conditions that minimize the potential overheating of water tubes in the furnace section of the boiler can be determined.

Keywords: steam boiler; water tube; deposit layer; tube wall thickness; wall tube thermal conductivity; thermo-mechanical performance

Citation: Aljohani, A.S.; Ahmed, K.I.; Asiri, S.; Ahmed, M.H. Numerical Study on the Effect of Deposit Layer on the Minimum Wall Thickness of Boiler Water Tube under Different Operating Conditions. *Appl. Sci.* **2022**, *12*, 8838. https://doi.org/10.3390/app12178838

Academic Editors: Vasily Novozhilov and Cunlu Zhao

Received: 17 May 2022
Accepted: 15 August 2022
Published: 2 September 2022

1. Introduction

The increasing energy demand induces continuous development in creating new energy generation systems and enhances the existing ones, making them more robust and reliable. One such technology is steam thermal power plants, which possess a remarkable share in energy production worldwide and contribute to several applications. The boiler is an essential component in a thermal power plant to generate steam, driving the turbine to produce power or direct it to subsystems for subsequent use, such as in the desalination plant and fuel heating systems.

The tubes in water tube boilers are subjected to several types of failure [1] caused by stress rupture, waterside and fireside corrosion [2,3], fatigue [4], erosion [5] and lack of quality control. The main focus of the present work is the stress rupture, particularly in water tubes, caused by the overheating associated with being exposed to high temperatures and pressure that lead to tube failure. These conditions are induced in the short or long term. For example, the former is associated with decreasing cooling rate possibly attributed to complete or partial plugging, sudden shutdown of boiler circulating water pump, the rabid temperature increased during the start-up of the boiler evaporation of water in the liquid section, and the increase in the overall thermal resistance due to deposits accumulation inside the tube. The visual examination of the ruptured wall presents with fish mouth appearance, usually in overheating conditions [6]. Ref. [6] described that the plastic deformation in the rupture area shows deformation with elongated grains in the tube over a period during prolonged heat transfer. Refs. [7,8] further explained that the impact of the difference in temperature conditions on either side of the tube displays variation in the appearance. It has been reported that the different temperatures on either side of the wall, hot wall (fireside), and cool wall demonstrated microstructural changes, especially more toward the hotter side [7,8]. Concurring, ref. [9] stated that the fireside wall presented in the trials with partial degradation (spheroidization) of the lamellar structure mainly of iron carbide in pearlite colonies. The lamellar structure indicates the main driving force for change, excessive surface energy, which reduces the internal energy of the tube. Ref. [9] further pointed out that after the rupture of the tube, the rapid cooling causes a change in the structure and formation of bainite. However, Refs. [10,11] demonstrated that the substantial increase in the temperature lowers the yield stress of the metal. Thus, if the yield stress becomes equal to or less than the hoop stress of a metal tube at high temperature, the tube starts to deform, bulge, and thin [10,11]. Ref. [9] have studied the short-term overheating failure of a boiler water wall tube and examined the evolution of different microstructures during the failure through visual examination. They have concluded from the microstructure examination that the failure occurred due to overheating above the eutectoid temperature (lower critical temperature) of the tube material.

One solution to prevent overheating problems is enhancing the heat transfer through the tube wall to the water. Several researchers enhance the heat transfer by modifying the water tubes' inner surface area or inserting different shapes of obstructions inside tubes to induce more flow turbulence and mixings [12]. Either way, considerable concerns exist regarding these techniques for the possible increasing accumulation rate of deposits inside tubes.

Most of the literature paid attention to enhancing the heat transfer through the tubes without considering the adverse effect of the accumulated deposits inside the tube on the thermal-mechanical consequences. Therefore, the main objectives addressed in the present work are the numerical investigation of the impact of deposit thickness inside tubes on the minimum thickness of the water tube for safe operation under different working loads and the influence the operating conditions, such as heat flux and flow rate, on the minimum thickness of the furnace water tube during the boiler lifetime.

Ansys fluent software student version 2021 R2 provides a suite to cover the entire range of physics, which enables virtual access to any field of engineering simulation, thereby permitting designing models to identify problems and functional processes. Conventional gas boilers have been shown to dissipate a high heat loss. Thus, the engineers designed a new type of boiler with thermal efficiency [13,14]. Regardless, the efficient boiler engineers continued to develop to minimize complete heat loss and designed Ansys software that analyzes the internal water distribution within the heat exchanger [14,15]. Thus, ref. [14] pointed out that the engineers aim to improve the heat exchange coefficient within the tube to permit higher water flow velocity.

Furthermore, ref. [16] added that the system predicts the fluid flow behavior of water in the heat exchanger and throughout its distribution. In addition, ref. [16] showed that the software supported the designing of the simulation model for the boiler with boundary conditions of heat transfer and heat loss in the system. Thus, using the software, several simulations were garnered over many years for a substantial improvement in the flow and heat exchanging behavior of the boiler with the updated shape of the baffle.

This study concentrates on the CFD model, which provides easy simulation of different input parameters. The CFD is categorized into three types of systems: Reynolds-averaged Navier–Stokes (RANS) computations, large-eddy simulations (LES), and direct numerical simulations (DNS). The RANS technique applied is used mainly for resolving the mean values of each quantity that uses turbulent models for unclosed terms, whereas DNS solves the full instantaneous Navier–Stokes equations without the inclusion of turbulent motions models [17]. On the other hand, LES measures the turbulent large eddies and models the small-scale eddies that prove to be effective in establishing the mechanism required to stabilize the problem [17]. Ref. [17], integrated LES with Computational Fluid Dynamics to successfully simulate the effectiveness simulate the ultra-supercritical boiler at different operation load conditions, which concurred with the findings of [18]. Nevertheless, ref. [19] presented use of the DNS only to be feasible for calculating academic issues, such as heat flow due to the computational cost.

Refs. [20,21] studies have shown its application in terms of superheaters where the thermal efficiency of the furnace is based on the performance of the boiler.

CFD analysis is mostly employed at the designed stage, troubleshooting stage, and performance evaluation during the plant operation [22]. Ref. [22], stated that the outcome obtained from the CFD analysis had shown an effective approach to visualize the condition of the tubes and predict the further life of the tube depending upon the condition. Recently the implementation of the PRO-E design software enabled the authors to evaluate the thermal flow at different velocities in the 3D model simulation that provided a complex identification of the transfer of heat flux along with ANSYS analysis [22]. Moreover, it enabled the engineer to adjust the high-temperature zone to prevent tube erosion and refrain from tube leakage issues [23]. Ref. [23] demonstrated crude oil substitution with oil using the CFD approach. Thus, the advanced property of the system allowed several authors to study the operating parameters responsible for erosion without halting the process [23–25].

This article aims to study the impact of deposit thickness on the thickness of the water tube under different working loads and determine the minimum thickness of the tube wall by comparing the maximum allowable thickness with the hoop stress to assure safe operation. The tube thickness is the first line of defense to prevent the tube failure during regular operation. So, if the tube thickness is less than the required thickness, the boiler tube will fail. In other words, the tubes with thicknesses less than the required thickness will cost a lot during the life of the boiler.

2. Methodology

2.1. Physical Model

The present work focuses on the water tube boiler. Figure 1 shows a schematic of the different components of the water tube boiler. These are lower water drums where the water is introduced after passing through the economizer and upper steam drum where the steam is extracted and directed to the superheater section. Both drums are connected via water tubes, and the present work concentrates on the furnace section where the water remains in the liquid phase, as shown in Figure 1a. The considered tube has the current dimensions shown in Figure 1b in terms of tube length (L_p), outer pipe diameter (D_o), tube thickness (t_p), and deposit layer thickness (t_d). The water tubes are made from carbon steel seamless pipe (SA210 C) and the deposit layer (Magnetite) is considered in the present work using a generic material with a variable range of thickness. Table 1 shows the material properties of the deposit. The used material properties and driving dimensions for different parts are presented in Table 1. The outer diameter of the tubes and original thicknesses are 44.45 mm and 5.588 mm, respectively. The new thickness of the water tube is set between 1 mm to 6 mm, whereas the thicknesses are based on a deposit maximum length of 0.2 mm. The present work is focused on the water tubes in the furnace section and therefore is considered a single-phase with temperature-independent material properties at the average temperature of inlet water and outlet temperature at the working pressure. Note that the pressure is based on maximum operating pressure at 210 bar, however, which is expected to result in higher stresses on the water tubes.

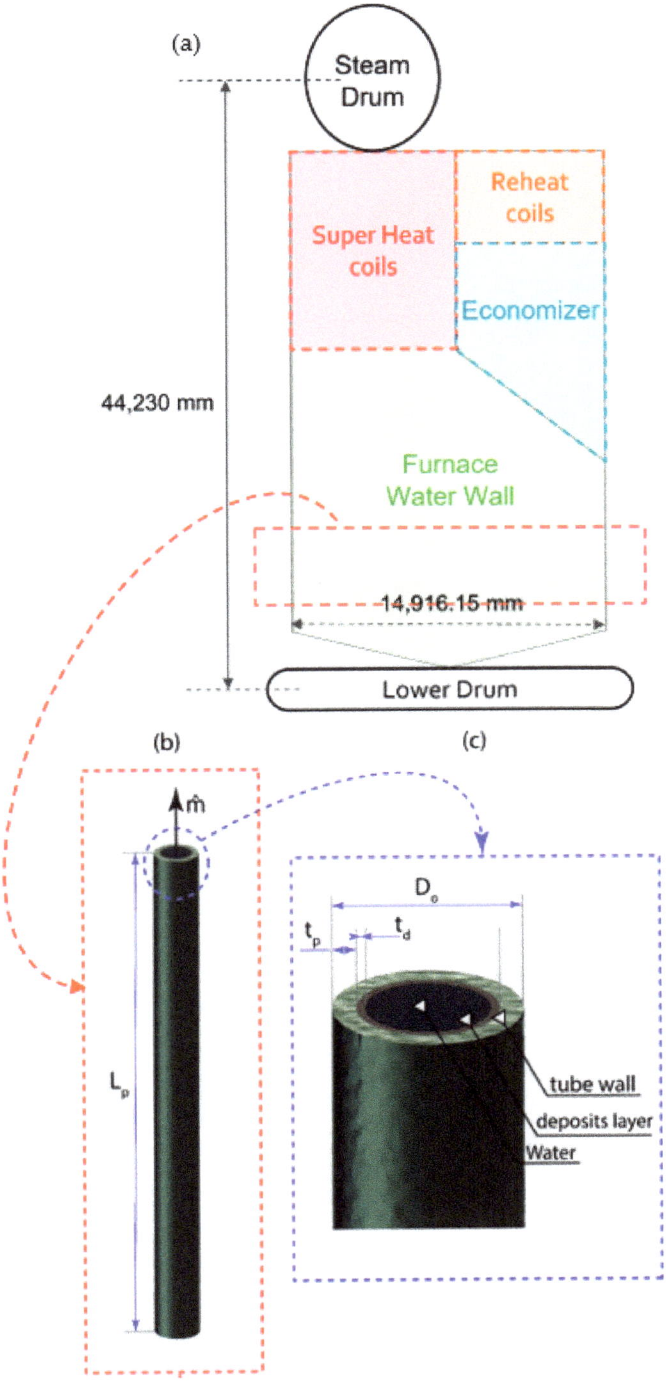

Figure 1. Schematic of (**a**) boiler, (**b**) the simulated water tube, and (**c**) magnification of the tube tip with written parametric dimensions.

Table 1. Material properties and related dimensions.

Domain	Deposit Material	Density [kg/m^3]	Specific Heat [j/kg·K]	Thermal Conductivity [mm]	Viscosity [kg/m·s]	Dimensions [mm]
Deposit	Magnetite	5175	586	0.625–1.39	NA *	t_d = from 0 to 0.2
Tube	Carbon Steel Seamless Pipe (SA210 C)	7830	465	53	NA *	t_p = from 1 to 6 L_p = 500 D_o = 44.45
Fluid	Water at P = 210 bar $T_{average}$ = 300 °C	736.44	5288.40	0.57314	9.04×10^5	D_i = from 42.45 to 32.28

* NA refers to not applicable.

2.2. Numerical Model

2.2.1. Governing Equations

A comprehensive three-dimensional thermal model conjugate with an incompressible turbulent flow model is developed to determine the thermo-fluid characteristics of the proposed heated water considering deposit layer. The governing equations include applying conservation of energy on solid parts and conservation of mass, momentum, and energy with the turbulence model on fluid parts.

Solid Regions: Tube Wall and Deposit Layer

The conservation of energy governs the thermal characteristics and energy transfer through the solid domain, including the tube wall and deposit layer, and can be written as follows:

- Conservation of Energy:

$$\nabla \cdot (k_s \nabla T) = 0 \tag{1}$$

where k_s is the thermal conductivity of the solid part (i.e., tube material or deposits material), and the term on the left-hand side refers to the heat flux by conduction.

Fluid Region

The flow inside the tube is considered to be single-phase, steady, Newtonian, incompressible, and turbulent. Therefore, the flow characteristics can be represented using the conservation of mass, momentum-based on Reynolds-averaged Navier–Stokes (RANS) equations coupled with turbulence model, and conservation of energy as indicated by Equations (2)–(7):

- Conservation of mass and momentum:

$$\frac{\partial \overline{u_i}}{\partial x_i} = 0 \tag{2}$$

$$\overline{u_i}\frac{\partial \overline{u_i}}{\partial x_i} = -\frac{1}{\rho_f}\frac{\partial \overline{P}}{\partial x_i} + \frac{\mu_f}{\rho_f}\frac{\partial^2 \overline{u_i}}{\partial {x_j}^2} - \frac{\partial}{\partial x_j}\left(\overline{u_i' u_i'}\right) \tag{3}$$

where $\overline{u_i}$ and x_i refer to the mean velocity of the flow and coordinate system, respectively, with $i = 1,\ 2,$ and 3. $\overline{P}$ is the mean pressure of the flow whereas ρ_f and μ_f represent water density and viscosity, respectively. $\overline{u_i' u_i'}$ is the turbulent Reynolds stress associated with the imposed turbulence over the mean flow, modelled and solved via turbulence models.

- Turbulence Model: Realizable k-ε turbulent model

the realizable k-ε turbulent model with enhanced wall function is used following [26,27] to account for the existing fully turbulent flow while obtaining an accurate solution at a reasonable computational time. The model is based on model transport equations for the turbulence kinetic energy (k) and its dissipation rate (ε). The modeled transport equations for Realizable k-ε model for steady, incompressible, fully turbulent flow can be written as follows

$$\frac{\partial}{\partial x_j}\left(\rho_f k u_j\right) = \frac{\partial}{\partial x_j}\left[\left(\mu + \frac{\mu_t}{\sigma_k}\right)\cdot\frac{\partial k}{\partial x_j}\right] + G_k + G_b - \rho\varepsilon \tag{4}$$

and

$$\frac{\partial}{\partial x_j}\left(\rho_f \varepsilon u_j\right) = \frac{\partial}{\partial x_j}\left[\left(\mu + \frac{\mu_t}{\sigma_\varepsilon}\right)\cdot\frac{\partial \varepsilon}{\partial x_j}\right] + \rho_f C_1 S\varepsilon - \rho_f C_2 \frac{\varepsilon^2}{k + \sqrt{\nu\varepsilon}} + C_{1\varepsilon}\frac{\varepsilon}{k} C_{3\varepsilon} G_b \tag{5}$$

where

$$C_1 = max\left[0.43, \frac{\eta}{\eta + 5}\right],\ \eta = S\frac{k}{\varepsilon},\ S = \sqrt{2S_{ij}S_{ij}} \tag{6}$$

where G_k and G_b refer to the generation of turbulence kinetic energy due to the mean velocity gradients and the generation of turbulence kinetic energy due to buoyancy, respectively. C_1 and $C_{1\varepsilon}$ are constants where σ_ε and σ_k are the turbulent Prandtl numbers for ε and k, respectively.

- Conservation of Energy:

$$\nabla\cdot\left(\vec{u}h\right) = \nabla\cdot(k_w \nabla T) \tag{7}$$

where h is the sensible enthalpy and equal to $\int_{T_{ref}}^{T} c_p dT$, and c_p is the specific heat at constant pressure.

2.2.2. Boundary Conditions

The applied boundary conditions are presented in Figure 2. To reduce the computational costs, the quartile of the domain (i.e., water tube) is considered, and therefore, a symmetry boundary condition is applied on the XZ plane, as indicated in Figure 2. Instead of applying constant heat flux on the outer surface of the water tube from one side and the other side being isolated, the heat transfer at the outer surface of the tube is as follows:

$$Q_s = \ddot{q}\cdot A_s = \frac{\dot{m}\cdot(h_O - h_i)}{\eta_{Boiler}} \tag{8}$$

where h_O and h_i are the enthalpy at outlet and enthalpy at inlet. A_s is the outer surface area of the water tube, η_{Boiler} is the boiler efficiency, $\dot{m}$ is the mass flow rate, and $\ddot{q}$ is the heat flux. The heat flux and mass flow rate are varied in the present work; the mass flow rate value is between 350, 231, and 156 kg/s and the heat flux value is between 263,703.57, 218,028.07, and 152,917.44 W/m^2 to represent different loads of 50%, 75%, and 100%. Furthermore, the heat flux is constant on the surface not affected by the outer surface shape.

Thermally coupled boundary conditions are used at the interfaces, such as the interface between the inner surface of the water tube and the outer surface of the deposit layer and the interface between the inner surface of the deposit layer with water volume circumferential surface in the case of the tube with deposit layer. Moreover, the inlet and outlet are set to mass flow rate inlet and pressure outlet boundary conditions, respectively. At the inlet, the constant mass flow rate is adopted equal to 0.368, 0.2432, and 0.1642, representing the total mass flow rate circulated between drums divided by the number of tubes. The Reynolds number for the presented cases beyond critical value and therefore turbulent flow model is considered. The inlet temperature of the water T_i is set to 290 °C and the minimum temperature recommended at the operating pressure is $P = 210$ bar. Nevertheless, the maximum value of the range, $P = 210$ bar, is considered in the present work, which leads to the minimum thickness attained under severe working conditions.

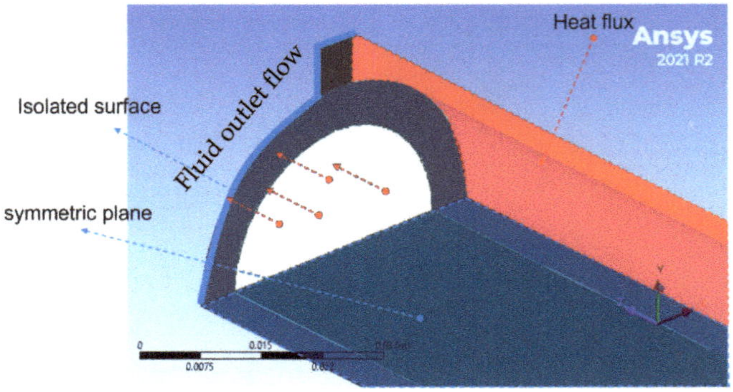

Figure 2. Boundary conditions on the simulated domain.

2.2.3. Minimum Tube Thickness Calculations (ASME)

The minimum thickness is determined using the following relation:

$$\frac{\sigma_{max}}{S_{hoop}} \geq 1 \quad (9)$$

$$S_{hoop} = \frac{P \cdot D_O}{2\, t_p} \quad (10)$$

where P is the operating pressure inside the tube, D_o is the outer diameter of the water tube, σ_{max} is the maximum allowable stress, t_p tube thickness which is varied between 1, 2, 3, 4, 5, and 6 mm. Note that both allowable stress and yield strength of carbon steel seamless pipe (SA210 C) are temperature-dependent, as shown in Figure 3.

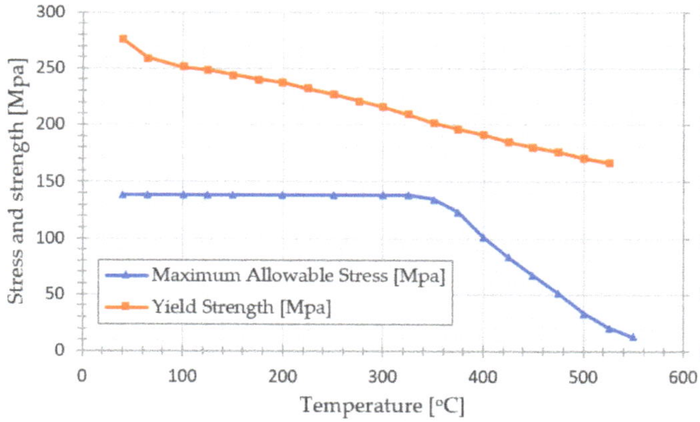

Figure 3. The maximum allowable stress and yield strength based on ASME code section II Part D for carbon steel seamless pipe (SA210 C) edited from [28].

2.3. Numerical Solution and Model Verification

The model was created in a 3D symmetric model by ANSYS. The model had three surface areas, and the axis of symmetry was on the XZ plane, as shown in Figure 2. These areas include water, scale, and tube. The water surface area had a dimension of 500 mm in length and 33.27 mm in width. Moreover, the scale and tube areas had a dimension of 500 mm in length and 1 mm in width. In addition, the mesh has been created in water, deposit layer, and tube wall domains via ANSYS meshing. A mesh independence test and validation are performed to exclude the influence of the mesh on the simulation calculations. The boundary conditions for the simulation are shown in Table 2.

Table 2. Boundary condition and geometry.

$L\ [m]$	5×10^{-1}	$T_i\ [K]$	5.63×10^{2}
D_i [m]	4.25×10^{-2}	$\dot{m}$ [kg/s]	3.6632×10^{-1}
D_O [m]	2.22×10^{-2}	$\ddot{q}$ [W/m^2]	3×10^{5}

Figure 4 shows the correlation Reynolds Number and Nusselt Number Curve with multiple element numbers of 476,370, 202,400, 137,410, and 92,516 mm (see Figure 5), and the 202,400 cells have been chosen to be accurately sufficient with the correlation of Reynolds Number and Nusselt number and lower computational time.

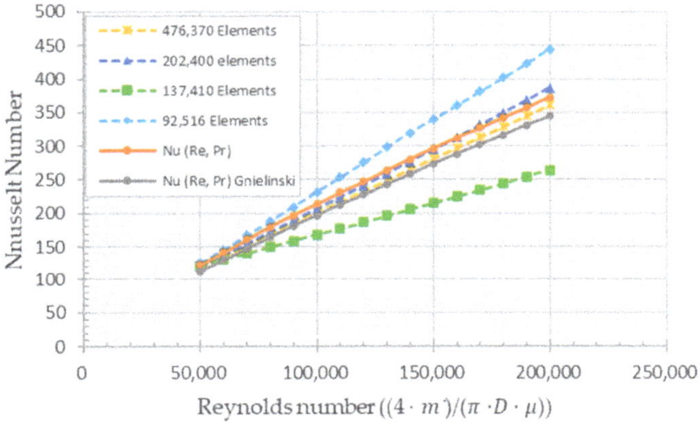

Figure 4. Correlation Reynolds Number and Nusselt Number Curve.

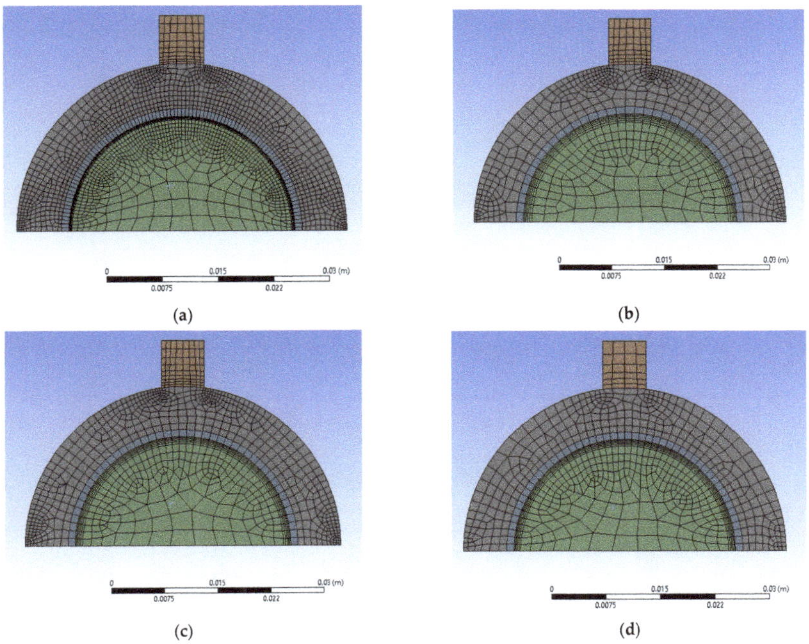

Figure 5. Mesh Dependency Test. (**a**) 476,370 Cells, (**b**) 202,400 Cells, (**c**) 137,410 Cells, and (**d**) 92,516 Cells.

The Reynolds number, convection heat transfer coefficient, and Nusselt number at a fully developed area were calculated as follows:

$$Q = \ddot{q} \cdot A_{OS} = \dot{m} \cdot C_p \cdot (T_O - T_i) \tag{11}$$

$$h_\infty = \frac{\ddot{q}}{T_s - T_O} \tag{12}$$

$$Nu_\infty = \frac{h_\infty \cdot D_i}{k} \tag{13}$$

$$Re = \frac{4 \cdot \dot{m}}{\pi \cdot D_i \cdot \mu} \tag{14}$$

$$PR = \frac{C_p \cdot \mu}{k} \tag{15}$$

$$Nu_\infty = 0.0214 \cdot \left(Re^{0.8} - 100\right) \cdot Pr^{0.4} \tag{16}$$

3. Results and Discussion

In the present section, the average water wall tube temperature was extracted at different loads, and the maximum allowable stress was calculated at the average tube temperature via linear interpolation that assures the safe operation of the water tubes in the furnace section of the boiler. Furthermore, the stress of working pressure $S_{\text{working Pressure}}$ was calculated via Equation (10). Increasing tube thickness can increase the average tube temperature but combining it with increasing deposit thickness leads to higher temperatures. In other words, the effect of the deposit on the tube with higher thickness is higher than on the tube with lower thickness. The following sections will analyze the impact of the deposits in full load, 75% of full load, and 50% of full load.

As shown in Figure 6, the high temperature concentrates on the region exposed to direct fire. So, this region is prone to overheating failure. Also, the temperature on the internal surface in this region is higher than in the opposite region. So, the Scale deposition in the area near the flame will be higher than in the other area.

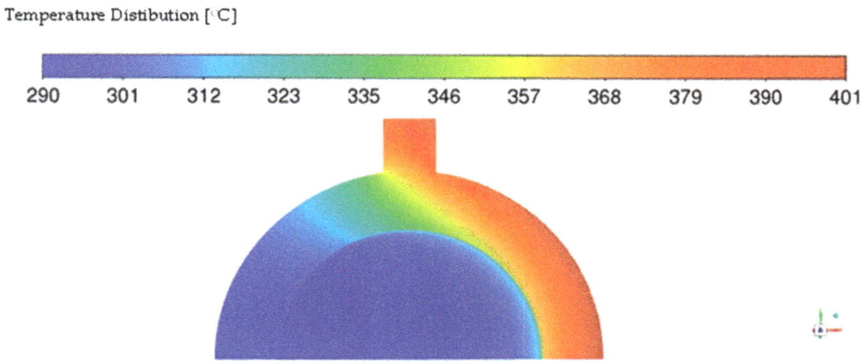

Figure 6. Average temperature distribution of water wall tube at full load.

As shown in Figure 7, the increase in the average temperature of the water wall tube is conjugated with accumulating deposit layer of Magnetite with low thermal conductivity. For example, the average temperature $T_{average}$ reaches 400.19, 395.56, 391.49, 387.90, 384.68, and 381.77 °C at $t_p = 6$, 5, 4, 3, 2, and 1 mm, respectively, in the case of $t_d = 0.2$ mm compared with 342.71, 341.48, 340.44, 339.51, 338.67, and 337.94 °C in the case of $t_d = 0.02$ mm at $t_p = 6$, 5, 4, 3, 2, and 1 mm, respectively. Therefore, an increase in the tube and deposit thickness leads to a significant reduction in yield strength (Figure 3) or allowable stress based on the ASME code (Figure 3). It can be noticed that in the deposit layer with $k = 0.625$ W/m K, the allowable stress divided by hoop stress Equation (9) at full load reaches 1.3, 0.72 in 6 mm and 3 mm tube thickness, respectively. See Figures 7 and 8.

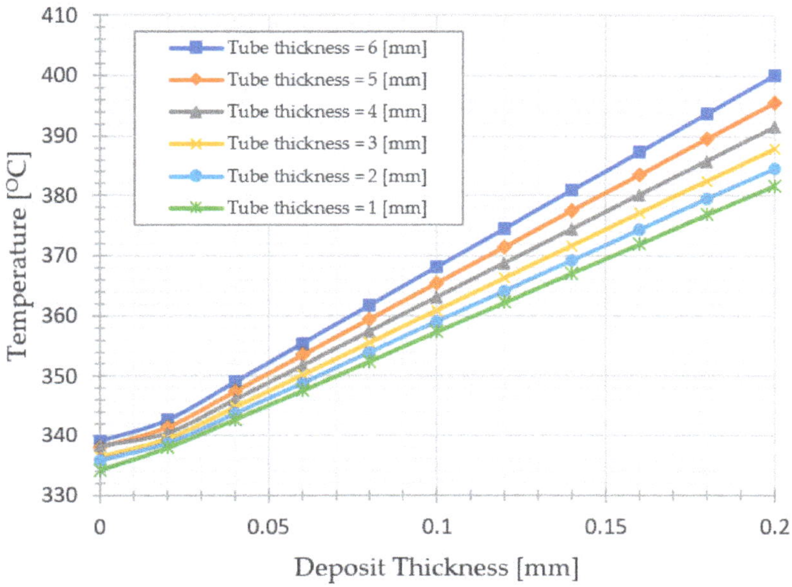

Figure 7. Thickness variation effects of deposit layer on average water wall tube temperature at full load.

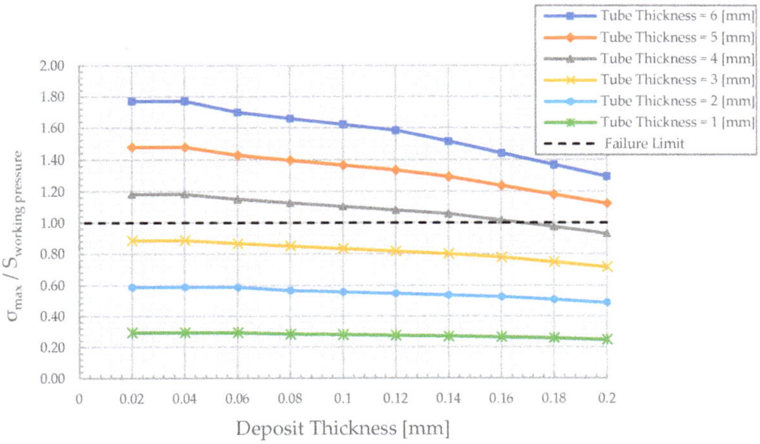

Figure 8. The effect of the deposits' thickness on the tube stress at full load.

Overall, the minimum value of the wall tube thickness is 4 mm, with magnetite thickness less than 0.16 mm, which is approximately equal to the maximum allowable stress in the ASME code at full load.

A similar analysis, as previously discussed, is conducted in the present section to determine the minimum tube thickness and shows the effect of magnetite deposit on the average tube temperature under 75% of full load. As shown in Figure 9, the average tube temperature was also affected by the deposits' thickness. However, it is noted that the temperatures here are lower than what was previously discussed. Furthermore, it can be noticed that the deposit layer with k = 0.625 W/m K, the allowable stress divided by hoop stress Equation (9) at 75% of full load, reaches 1.39, 0.75 in 6 mm and 3 mm tube thickness respectively; see Figure 10. The minimum thickness of the water wall tube is 4 mm, with magnetite thickness less than 0.18 mm.

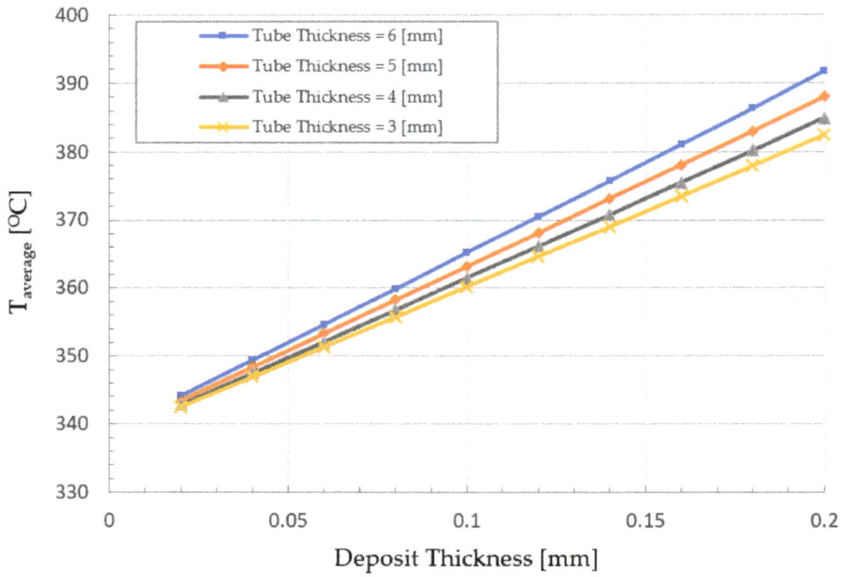

Figure 9. Thickness variation effects of deposit layer on average water wall tube temperature at 75% of full load.

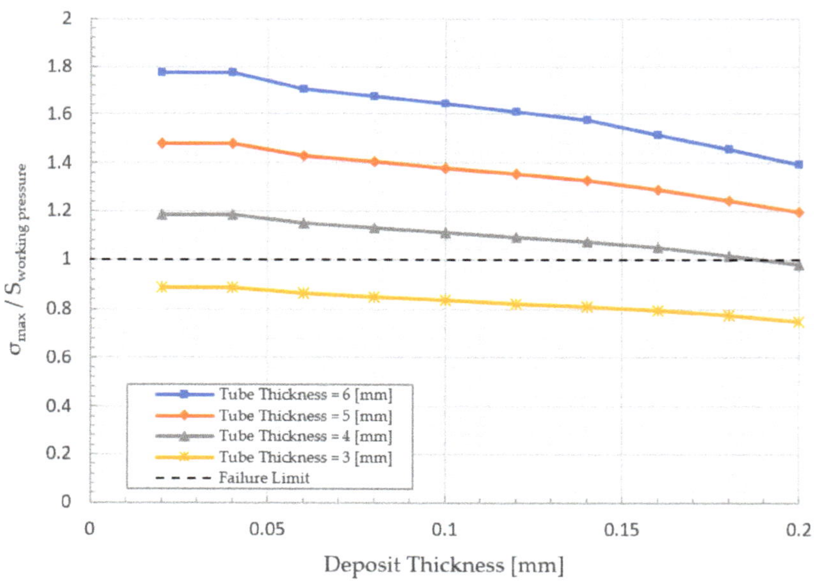

Figure 10. The effect of the deposits' thickness on the tube stress at 75% of full load.

In addition, the average temperature at 50% of full is discussed. As shown in Figure 11, the increase in the average temperature of the water wall tube is conjugated with accumulating deposit layer of Magnetite with low thermal conductivity. For example, the average temperature $T_{average}$ reaches 409.38, 405.25, 401.55, and 398.21 °C at t_p = 6, 5, 4, and 3 mm, respectively, in the case of t_d = 0.4 mm, compared with 338.89, 338.92, 339.00, and 338.82 °C in the case of t_d = 0.02 mm at t_p = 6, 5, 4, and 3 mm, respectively.

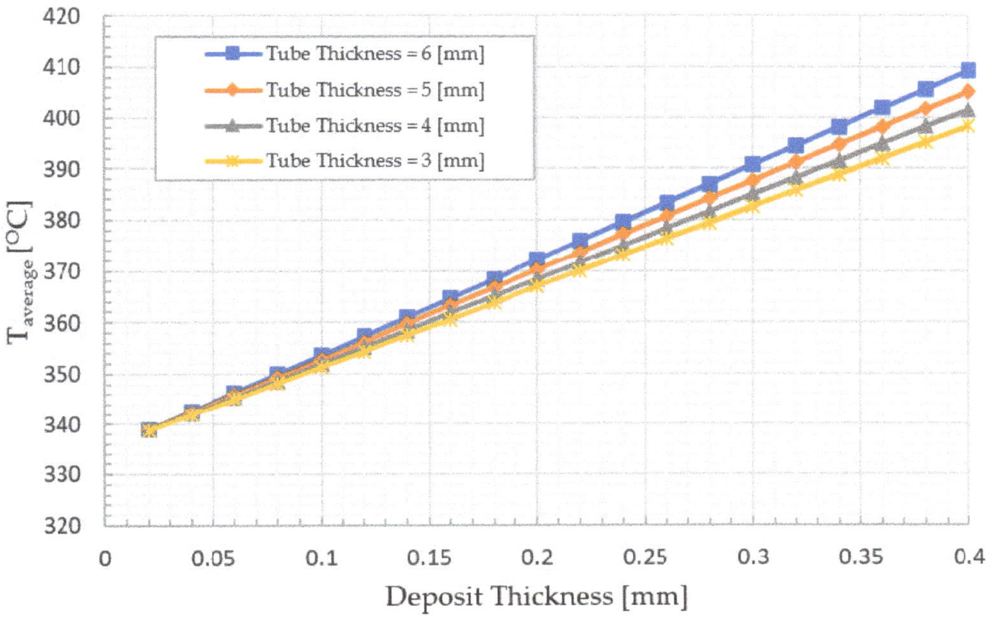

Figure 11. Thickness variation effects of deposit layer on average water wall tube temperature at 50% of full load.

Finally, as shown in Figures 7, 9 and 11, the change in average temperature in the tube with higher thickness is greater than the change in the tube with lower thickness. Chemical treatment and tube replacement are recommended for the water tubes in the case of the formation of a deposit layer with thermal conductivity less than 0.625 W/m K. Table 3 shows the relation between the maximum deposit thickness and the minimum tube thickness at different load. The table data was extracted from Figures 8, 10 and 12, the effect of changing of load can be observed. So, another temporal action is considering the operation should be at a lower load if it isn't feasible to conduct the chemical treatment.

Table 3. Summary.

Load [%]	Minimum Tube Thickness [mm]	Maximum Deposit Thickness [mm]
100%	4	0.16
75%	4	0.18
50%	4	0.28

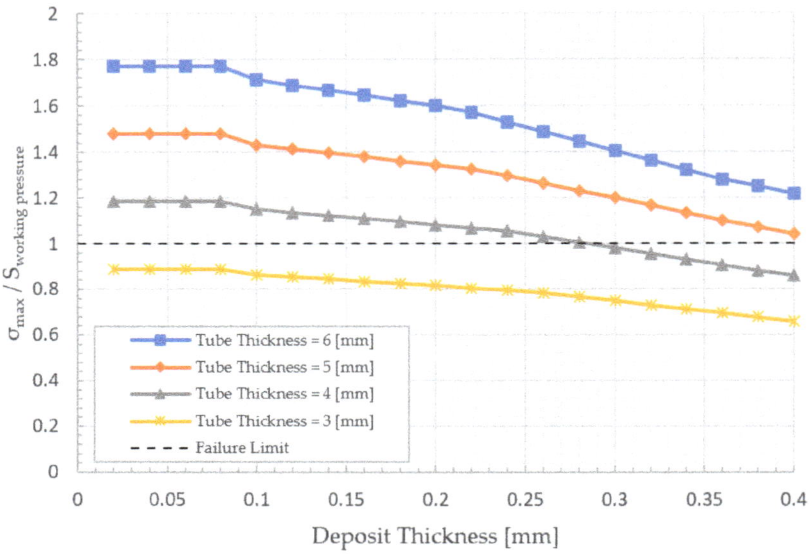

Figure 12. The effect of the deposit thickness on the tube stress at 50% of full load.

4. Conclusions

The present work was set to numerically determine the effect of the deposit layer thickness and thermal conductivity on the minimum thickness of the tube wall that attains safe operation, considering the different working loads of 50%, 75%, and 100% at the maximum operating working pressure of 210 bar. Based on the adopted parameters, the investigation of the minimum thickness of the water tube wall has revealed that:

1. The thermal conductivity of deposits affects the maximum temperature of the tubes. Consequently, the high potential for overheating, particularly at the deposit layer with thermal conductivity is 0.625 W/m^2 K. The maximum crown temperature exceeds 495 °C at 100% load in the 6 mm tube thickness with deposit thickness of 0.2 mm, which is near the maximum temperature of 539 °C of SA-210 C.
2. Increasing the thickness of the deposit layer leads to a linear increase in the tube wall average temperature. Consequently, it is essential to consider the consistency between the chosen wall tube thickness and the maintenance schedule considering the deposit accumulation rates and the operating conditions.
3. The deposit and tube thickness are mandatory inspection requirements. Furthermore, if the deposit thickness and tube thickness exceed the previous results, chemical cleaning and tube replacement are mandatory requirements.

The present study is limited to the water tube in the furnace section. The identified effect of the deposit layer of magnetite on the tube wall temperature and the corresponding minimum thickness assists our understanding of the role of the deposit layer attributes for suitable selection of the tube thickness. Furthermore, it presents courses of action concerning the operating conditions that minimize the potential overheating of water tubes in the furnace section of the boiler.

Author Contributions: Conceptualization, A.S.A. and K.I.A.; methodology, A.S.A. and K.I.A.; software, A.S.A. and K.I.A.; validation, A.S.A. and M.H.A.; formal analysis, A.S.A. and K.I.A.; investigation, A.S.A. and K.I.A.; resources, K.I.A. and M.H.A.; data curation, A.S.A.; writing—original draft preparation, A.S.A.; writing—review and editing, K.I.A. and M.H.A.; visualization, A.S.A., K.I.A. and M.H.A.; super-vision, K.I.A. and S.A.; project administration, K.I.A. All authors have read and agreed to the published version of the manuscript.

Funding: This research received no external funding.

Conflicts of Interest: The authors declare no conflict of interest.

References

1. Melville, J.; Foster, G.G. A pictorial review of failures in conventional boiler plant. *Press. Vessel. Pip.* **1975**, *3*, 1–25. [CrossRef]
2. Duarte, C.A.; Espejo, E.; Martinez, J.C. Failure analysis of the wall tubes of a water-tube boiler. *Eng. Fail. Anal.* **2017**, *79*, 704–713. [CrossRef]
3. Khajavi, M.R.; Abdolmaleki, A.R.; Adibi, N.; Mirfendereski, S. Failure analysis of bank front boiler tubes. *Eng. Fail. Anal.* **2007**, *14*, 731–738. [CrossRef]
4. Ghosh, D.; Roy, H.; Saha, A.; Subramanian, C. Failure Analysis of Boiler Water Wall Tube: A Case Study from Thermal Power Plant. *Fail. Anal. Prev.* **2022**, *22*, 203–208. [CrossRef]
5. Himarosa, R.A.; Hariyanto, S.D.; Hasan, W.H.; Muflikhun, M.A. Failure analysis of platen superheater tube, water wall tube, and sealpot plate: A case study from electricity power plant in indonesia. *Eng. Fail. Anal.* **2022**, *135*, 106–108. [CrossRef]
6. Taler, J.; Dzierwa, P.; Jaremkiewicz, M.; Taler, D.; Kaczmarski, K.; Trojan, M.; Sobota, T. Thermal stress monitoring in thick walled pressure components of steam boilers. *Energy* **2019**, *175*, 645–666. [CrossRef]
7. Sun, L.; Yan, W. Prediction of wall temperature and oxide scale thickness of ferritic–martensitic steel superheater tubes. *Appl. Therm. Eng.* **2018**, *134*, 171–181. [CrossRef]
8. Modliński, N.; Szczepanek, K.; Nabagło, D.; Madejski, P.; Modliński, Z. Mathematical procedure for predicting tube metal temperature in the second stage reheater of the operating flexibly steam boiler. *Appl. Therm. Eng.* **2019**, *146*, 854–865. [CrossRef]
9. Munda, P.; Husain, M.M.; Rajinikanth, V.; Metya, A.K. Evolution of Microstructure During Short-term Overheating Failure of a Boiler Water Wall Tube Made of Carbon Steel. *J. Fail. Anal. Prev.* **2018**, *18*, 199–211. [CrossRef]
10. Ardy, H.; Bangun, D.A. Failure Analysis of Superheater Boiler Tube SA 213 T12. In *IOP Conference Series*; Materials Science and Engineering: Bandung, Indonesia, 2019.
11. Dehnavi, F.; Eslami, A.; Ashrafizadeh, F. A case study on failure of superheater tubes in an industrial power plant. *Eng. Fail. Anal.* **2017**, *80*, 368–377. [CrossRef]
12. Kumar, B.; Srivastava, G.P.; Kumar, M.; Patil, A.K. A review of heat transfer and fluid flow mechanism in heat exchanger tube with inserts. *Chem. Eng. Processing—Process Intensif.* **2018**, *123*, 126–137. [CrossRef]
13. Barma, M.C.; Saidur, R.; Rahman, S.M.A.; Allouhi, A.; Akash, B.A.; Sait, S.M. A review on boilers energy use, energy savings, and emissions reductions. *Renew. Sustain. Energy Rev.* **2017**, *79*, 970–983. [CrossRef]
14. Kakaç, S.; Liu, H.; Pramuanjaroenkij, A. *Heat Exchangers: Selection, Rating, and Thermal Design*, 4th ed.; CRC Press: Boca Raton, FL, USA, 2020.
15. Mohite, N.T.; Benni, R.G. Optimization of Wall Thickness for Minimum Heat Losses for Induction Furnace. *Int. J. Eng. Res. Technol.* **2017**, *10*, 645–653.
16. Hu, Y.; Li, H.; Yan, J. Numerical investigation of heat transfer characteristics in utility boilers of oxy-coal combustion. *Appl. Energy* **2014**, *130*, 543–551. [CrossRef]
17. Haoshu, S.; Yuxin, W.; Minmin, Z.; Hai, Z.; Guangxi, Y.; Junfu, L. Large eddy simulation of a 660 MW utility boiler under variable load conditions. *Front. Energy* **2020**, *15*, 124–131.
18. Xueli, G.; Zhongxiao, Z.; Haojie, F.; Jian, Z.; Degui, B. Unsteady-state heat transfer characteristics of spiral water wall tube in advanced-ultra-supercritical boilers from experiments and distributed parameter model. *Energy* **2019**, *189*, 116158.
19. Paweł, M. Numerical study of a large-scale pulverized coal-fired boiler operation using CFD modeling based on the probability density function method. Applied Thermal Engineering. *Appl. Therm. Eng.* **2018**, *145*, 352–363.
20. Laubscher, R.; Rousseau, P. CFD study of pulverized coal-fired boiler evaporator and radiant superheaters at varying loads. *Appl. Therm. Eng.* **2019**, *160*, 114057. [CrossRef]
21. Huijun, F.; Zhuojun, X.; Chen, L.; Zhixiang, W.; Shaojun, X. Constructal design for supercharged boiler superheater. *Energy* **2020**, *191*, 116484.
22. Jalendar, M.; Kumar, G. Design and thermal analysis of steam boiler used in power plants. *Int. J. Recent Dev. Sci. Technol.* **2019**, *3*, 103–109.
23. Filkoski, R.; Petrovski, I. *Computational Fluid Dynamics in Function of Boilers' Revitalisation*; Academia: San Francisco, CA, USA, 2022.
24. Vandani, A.M.K.; Bidi, M.; Ahmadi, F. Exergy analysis and evolutionary optimization of boiler blowdown heat recovery in steam power plants. *Energy Convers. Manag.* **2015**, *106*, 1–9. [CrossRef]
25. Stifanese, R.; Belsanti, L.; Toselli, M.; Letardi, P.; Traverso, P. Corrosion investigation of a steam turbine after power generator failure onboard a vessel: A case study. *Eng. Fail. Anal.* **2016**, *64*, 58–66. [CrossRef]
26. Zhang, Z.; Yang, Z.; Nie, H.; Xu, L.; Yue, J.; Huang, Y. A thermal stress analysis of fluid–structure interaction applied to boiler water wall. *Asia-Pac. J. Chem. Eng.* **2020**, *15*, e2537. [CrossRef]
27. Li, Z.; Wu, Y.; Lu, J.; Zhang, D.; Zhang, H. Heat transfer to supercritical water in circular tubes with circumferentially non-uniform heating. *Appl. Therm. Eng.* **2014**, *70*, 190–200. [CrossRef]
28. ASME. *Pressure Vessel Code*; ASME: New York, NY, USA, 2004.

applied sciences

MDPI

Article

Performance of Textile Mask Materials in Varied Humidity: Filtration Efficiency, Breathability, and Quality Factor

Joelle M. Segovia [1], Ching-Hsuan Huang [2], Maxwell Mamishev [3], Nanhsun Yuan [1], Jiayang He [1] and Igor Novosselov [1,*]

1 Department of Mechanical Engineering, University of Washington, Seattle, WA 98195, USA
2 Department of Environmental and Occupational Health Sciences, School of Public Health, University of Washington, Seattle, WA 98195, USA
3 Nathan Hale High School, Seattle, WA 98125, USA
* Correspondence: ivn@uw.edu

Abstract: During the COVID-19 pandemic, reusable masks became ubiquitous; these masks were made from various fabrics without guidance from the research community or regulating agencies. Though reusable masks reduce the waste stream associated with disposable masks and promote the use of masks by the population, their efficacy in preventing the transmission of infectious agents has not been evaluated sufficiently. Among the unknowns is the effect of relative humidity (RH) on fabrics' filtration efficiency (FE) and breathability. This study evaluates the FE and breathability of several readily accessible mask materials in an aerosol chamber. Sodium chloride aerosols were used as the challenge aerosol with aerodynamic particle diameter in the 0.5 to 2.5 μm range. To mimic the variability in RH in the environment and the exhaled-breath condition, the chamber was operated at RH of 30% to 70%. The face velocity was varied between 0.05 m/s and 0.19 m/s to simulate different breathing rates. The FE and pressure drop were used to determine the quality factor of the materials. Among the tested materials, the 3M P100 filter has the highest pressure drop of 140 Pa; the N95 mask and the 3M P100 have almost 100% FE for all sizes of particles and tested face velocities; the surgical mask has nearly 90% FE for all the particles and the lowest pressure drop among the certified materials, which ranks it the second to the N95 mask in the quality factor. Other material performance data are presented as a function of relative humidity and aerosol size. The quality factor for each material was compared against reference filtration media and surgical masks. Multiple layers of selected materials are also tested. While the additional layers improve FE, the pressure drop increases linearly. Additionally, the certified materials performed approximately three times better than the highest performing non-certified material.

Keywords: mask; aerosol; filtration efficiency; breathability; quality factor

Citation: Segovia, J.M.; Huang, C.-H.; Mamishev, M.; Yuan, N.; He, J.; Novosselov, I. Performance of Textile Mask Materials in Varied Humidity: Filtration Efficiency, Breathability, and Quality Factor. *Appl. Sci.* **2022**, *12*, 9360. https://doi.org/10.3390/app12189360

Academic Editor: Rocco Furferi

Received: 2 August 2022
Accepted: 16 September 2022
Published: 18 September 2022

1. Introduction

Respiratory infections are the most common illnesses and are one of the leading causes of mortality worldwide [1,2]. Infection transmission comes from the fomite route and exposure to infectious aerosols. Airborne transmission can cause large outbreaks even when individuals have minimal contact with fomites [3–5]. The airborne transmission mechanism involves infectious aerosols in particle diameter (d_p) ranging from 0.01 to 100 μm [6]. Typically, larger particles ($d_p > 5$ μm) have been classified as droplets, whereas those with $d_p < 5$ μm are classified as aerosols [7]. Aerosols remain suspended in the air for hours, long enough to be inhaled, and can contain multiple viral copies [8–10].

Aerosol inhalation can be reduced by using a face mask as the material provides filtration of inhaled and exhaled air, protecting the users and those around them. Filtration occurs via five different mechanisms: interception, inertial impaction, diffusion, electrostatic attraction, and gravitational settling [11,12]. The effectiveness of these filtration mechanisms depends on the particle size and the type of filtration media. Gravitational settling is prevalent in removing droplets $d_p > 50$ μm. Inertial impaction is primarily seen on aerosols larger than 1 μm, whereas diffusion primarily affects the smallest particles [12]. As the flow rate increases, so does the effect of interception and inertial impaction, while the gravitational settling, diffusion, and electrostatic attraction become less impactful.

Face masks should provide the wearer with high filtration efficiency (FE), while remaining comfortable. Mask comfort can be attributed to the fit of the mask, a mask's ability to transfer heat and moisture away from the face, and its breathability [13]. Breathability is described by the permeability and pressure drop across the material [13,14]. Part of the certification testing for mask materials that public health organizations, such as the National Institute for Occupational Health (NIOSH), conduct is testing a materials pressure drop. Materials may not exceed the pressure drop values at the corresponding flow rate to ensure the breathability of the mask. Respirators can reach pressure differentials ranging from 210 to 350 Pa, whereas surgical masks should range between 40 Pa/cm^2 and 70 Pa/cm^2 [14,15]. At the start of the COVID-19 pandemic, high demand resulted in a scarcity of masks [16–19], leading to many using homemade masks made from various textiles. However, the protection from aerosolized transmission provided by these masks remains unclear, and the public has not received clear guidance from health officials.

Previous studies compared the performance of homemade masks and those certified by organizations such as NIOSH [20–44]. However, the accuracy of the results and the methodology used by some of these studies have come under scientific scrutiny [14,45–47]. Some studies did not provide sufficient details on the experimental methodology. For example, one study erroneously stated that N95 respirators did not provide 95% FE when tested with no leakage points [21,46,47]. Certified masks are subjected to a standardized testing procedure to ensure their ability to filter out 95% of aerosols. The NIOSH testing procedure uses a TSI Respirator Fit Tester to test twenty filters. With extensive testing procedures, masks that receive a certified N95 rating have a high-accuracy standard and excellent consistency between the samples. Real-time particle sizing instruments can significantly simplify and expedite the test procedures. An increasing number of studies use real-time particle sizers for calculating collection and transmission aerosols [48–51], for sensor calibration studies [52–56], and recently became a well-accepted method for testing the filtration efficiency, e.g., refs [57–63].

Environmental conditions and breathing rates affect local relative humidity (RH) in the fiber matrix. In a high RH environment, hygroscopic fibers absorb water and swell [64], affecting the material's porosity and pressure drop [65]. Few studies have examined the effects of humidity on the overall performance of textiles. The "quality factor" index can be used to parameterize this effect; it describes a mask's performance based on the ability to filter out particulates and the breathability [11].

This paper aims to complement previous studies describing the FE of common mask materials while adding new insight into the effects of RH on a material's filtration performance. In our study, materials were challenged with polydisperse sodium chloride (NaCl) aerosols in a size range of ~0.3 to 2.5 μm. The face velocity through the material was varied at 0.05 m/s, 0.10 m/s, and 0.19 m/s to simulate the variability in the breathing rate [15]. RH was varied in the 30% to 70% range. The FE and the pressure drop were measured to determine the quality factor. The certified materials by NIOSH outperform other textiles by a factor of three. Additionally, it is found that humidity has minimal effects on both the FE and the breathability of the materials in the tested RH range.

2. Materials and Methods

2.1. Materials

The materials tested in this study range from the masks certified by NIOSH, natural, synthetic, blended fabrics, and other non-traditional materials such as Thinsulate, heavyweight surgical wrap, and coffee filters. Certified masks include a 3M disposable N95 respirator filter, a 3M P100 filter, and a disposable 3-ply surgical mask (BYD care).

Fabrics tested in this study consisted of knitted cotton, muslin, and Kona cotton, along with silk and rayon. One natural-synthetic fabric blend was also tested, which was 82% rayon and 18% knitted cotton. Finally, more non-traditional materials were tested due to their commercial accessibility and their use as mask materials. These materials consisted of Thinsulate, heavyweight spunbond-meltblown-spunbond (SMS) polypropylene, and paper coffee filters. Optical microscopy images of these materials are shown in Figure 1, and additional details on each material are shown in Table S1 of the Supplemental Information.

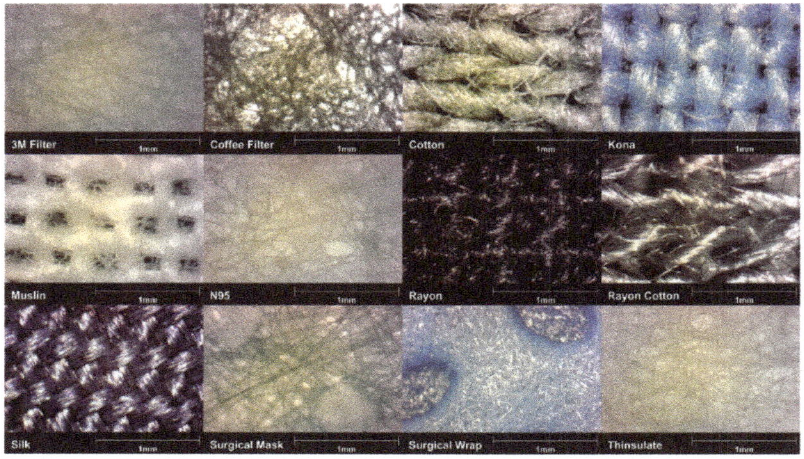

Figure 1. Microscopic images of the tested materials. Photos are taken at 2× magnification and are shown with a 1 mm scale.

The multiple material layers were evaluated, as double masking becomes a common practice for people to get more protection, affecting the FE, pressure drop and overall quality factor of the mask. The materials selected for testing with additional layers were the surgical mask, muslin, and coffee filter. The surgical mask was tested with two layers, whereas the muslin and coffee filter materials were tested with two and three layers.

2.2. Aerosol Chamber Setup

The experiments were conducted using a custom-built aerosol chamber (0.56 m × 0.52 m × 0.42 m) [66] with two 3D printed material holders attached. Two medical nebulizers (VixOne Small Volume Jet Nebulizer, Westmed, Tuscon, AZ, USA) were placed in the chamber. One of the nebulizers was used to generate polydisperse NaCl aerosols. The aerosolization was stopped when the particle concentration reached 900 to 1200 particles/cc, measured by the Aerodynamic Particle Sizer (APS 3321, TSI, Shoreview, MN, USA). The second nebulizer controlled the RH inside the chamber set to RH of 30%, 50%, or 70%. The two mixing fans inside provided homogeneous particle distribution in the chamber [2,66]. A 3D rendering of the aerosol chamber and the position of these components are shown in Figure 2(left).

Attached to the aerosol chamber are two sample holders: one holds the textile sample, and the other is empty, serving as a reference channel. Magnetic clips were used for alignment, and two binder clips were used to ensure an airtight seal, which was verified before each experiment. Particle-laden air from the aerosol chamber was aspirated through the material holder at a rate of 1 L per minute (LPM) from the APS. An additional make-up flow was provided by the building vacuum line at flow rates from 0.1 LPM to 3 LPM to reach the designed face velocity [50]. The make-up air was filtered by an in-line high-efficiency particulate air (HEPA) filter. A differential manometer (UEi EM201B) measured the pressure drop. The APS was connected to the holders through anti-static tubing to minimize deposition through electrostatic attraction.

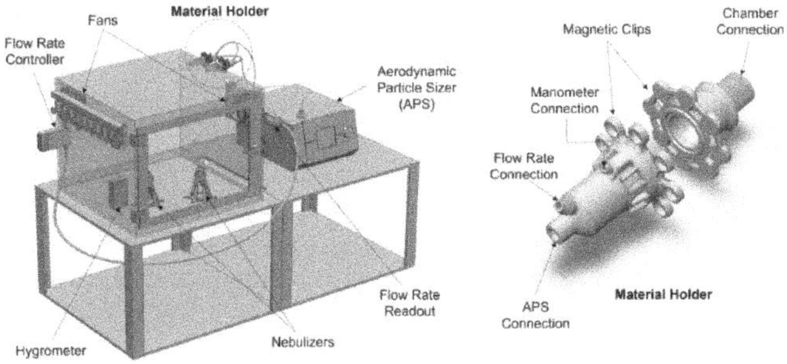

Figure 2. 3D render of the experimental setup. A custom-built aerosol chamber (**left**) contains two mixing fans, two nebulizers, and a hygrometer. Nebulizers generate polydisperse NaCl aerosols and control the humidity within the chamber. Attached to the chamber are two sample holders (**right**), one containing a material insert and one left empty to be used as a reference. The containers are connected to a mass flow rate controller and a manometer to measure the pressure drop. The flow passes through the material holders (one at a time) to the APS for data collection.

2.3. Pressure Drop and Filtration Efficiency Analysis

The breathability of the material was measured in terms of pressure drop. Pressure drop readings were taken over a range of face velocities between 0 m/s to 0.39 m/s and at each humidity level. Readings were taken using the differential manometer (UEI Test Instruments) connected to the material holder that measures the pressure difference across the sample material. Three pressure drop measurements for each face velocity were taken to determine a standard deviation.

To calculate FE, we took six measurements at each humidity and face velocities of 0.05 m/s, 0.10 m/s, and 0.19 m/s. Each reading was taken over ten seconds, alternating between filtered and reference streams. *FE* is calculated using Equation (1):

$$FE = \frac{n_{ref} - n_f}{n_{ref}} * 100\% \tag{1}$$

where the n_{ref} and n_f are the size-resolved aerosol concentrations measured by the APS for each size bin in the reference and in the filtered stream. The data collected from the APS is recorded in particle mass concentrations with an assumed density of 1.03 g/cm^3 for NaCl and in number density for the smaller particle bin (d_p = 0.3–0.52 µm). Though the APS records the particle in the range d_p = 0.3–20 µm, the data did not show a significant concentration of particles > 2.5 µm. To evaluate the FE as a function of particle size, we have binned the data into d_p = 0.3–0.52 µm, d_p < 0.97 µm (PM1), and d_p < 2.46 µm (PM2.5). Any calculations showing a negative FE due to the reference and filtered measurements variance are reported as "zero." The three replicates are averaged for each condition providing the standard deviation shown as error bars in the plots in the result section.

The filter quality factor Q combines the material's FE and pressure drop describing its overall performance [11]. By combining the FE and the pressure drop, the desired functions of a mask, such as comfort and high filtration abilities, can be presented by a single value. The best filter is the one that has the highest FE with the lowest pressure drop. The *Q* is calculated using the following equation:

$$Q = \frac{ln\left(\frac{1}{1-\frac{FE}{100}}\right)}{Pressure\ Drop} \tag{2}$$

3. Results and Discussion

3.1. Pressure Drop

Figure 3 shows pressure drop for three certified mask materials as a function of face velocity for three RH levels. The P100 3M filtration media had the highest pressure drop and the surgical mask had the lowest. Several materials (muslin, knitted cotton, and rayon cotton blend) had lower pressure drops than a surgical mask. A complete list of materials pressure drops can be found in the Supplemental Material Table S1. Pressure drop has a linear relationship to the face velocity through the material, as expected for the laminar airflow through the filter media. Not considering the dust load, the pressure drop (ΔP, Pa) is proportional to the air velocity at the face of a filter (U_0, m/s). $\Delta P = \beta U_0$, where β is the air resistance coefficient (Pa·s/m) [11].

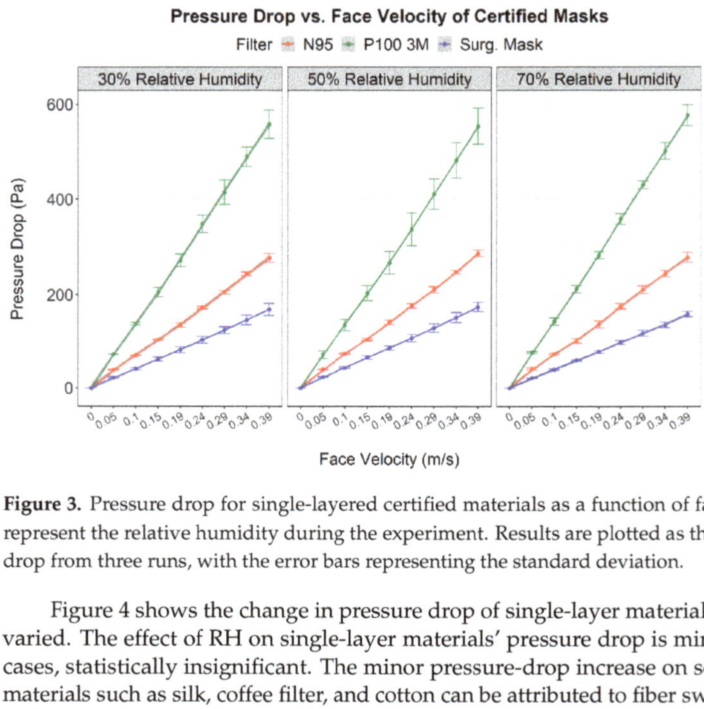

Figure 3. Pressure drop for single-layered certified materials as a function of face velocity. Facets represent the relative humidity during the experiment. Results are plotted as the average pressure drop from three runs, with the error bars representing the standard deviation.

Figure 4 shows the change in pressure drop of single-layer materials when the RH is varied. The effect of RH on single-layer materials' pressure drop is minimal or, in some cases, statistically insignificant. The minor pressure-drop increase on some hygroscopic materials such as silk, coffee filter, and cotton can be attributed to fiber swelling, a decrease in fabric porosity, and a possible increase in fabric thickness. A similar change in the resistance to convective flow at varying humidity levels was reported by Gibson, where the largest change was observed at higher RH above 0.8 [67]. Commercial surgical face masks typically have a three-layer structure. The middle layer is made of a melt-blown material that serves as the filter media, whereas the inner layer is for absorbing moisture, and the outer layer repels water. The moisture repelling by the outer layer might explain the small change in pressure drop and FE of the surgical mask at elevated RH levels. For multiple material layers, humidity has minimal to no effects on the breathability of the layered materials, as shown in Figure S2 of the Supplemental Information. As expected, the material's pressure drop increased with each additional textile layer.

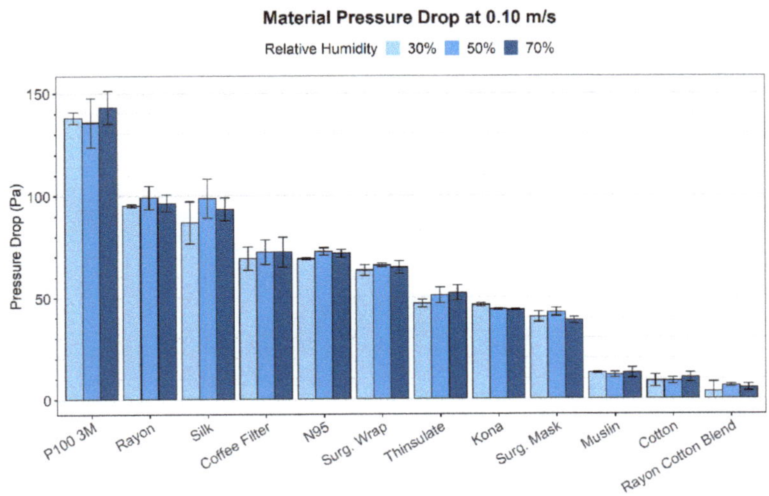

Figure 4. Pressure drop for single-layered materials as a function of RH. Bar colors represent the corresponding RH value, and results are plotted as the average pressure drop from three runs, with the error bars representing the standard deviation.

3.2. Filtration Efficiency

Figure 5 shows the filtration efficiency of materials at different face velocities for PM2.5, PM1, and particles in d_p = 0.3–0.52 μm range. The certified masks had the highest efficiency of all particle sizes; P100 and N95 had 100% flirtation for all particle channels and tested face velocities. Surgical mask filtration was lower, especially for particles smaller than 0.5 μm; however, the surgical mask outperformed all other tested materials for homemade masks. The high FE of the certified materials for the ultrafine particles is due to electrostatically charged fibers densely woven by the melt-blown extrusion process. The mechanical filtration mechanisms (e.g., inertial deposition, interception, and diffusion) are combined with electrostatic deposition to filter both large and small particles. Commonly used textiles, such as cotton and cotton blends, had filtration efficiency below 10%. More dense Kona cotton approached 25% FE; however, its pressure drop was 3 to 10 times greater than other textiles and about the same as the surgical mask. Synthetic materials showed increased FE at lower face velocities for smaller particles. This trend indicates that the primary filtration mechanisms for PM1 are diffusion and electrostatic attraction, as synthetic fibers are likely to carry permanent electrostatic charges or be charged by triboelectrification, which improves the deposition of small particles [27,68,69]. However, certified multilayer material such as surgical masks has clear advantages due to the removal of moisture in the outer layer and the strong attraction of electrostatically charged fibers. Natural fiber textiles did not show this trend, and in the case of the paper coffee filter, higher velocity resulted in higher filtration for all particle sizes.

Figure 6 shows the FE for multiple-layered materials. As the aerosol sizes increase, the effects from interception and inertial impaction filtration mechanisms begin to appear, increasing the FE with increasing face velocity. The trend between FE and face velocity seen by the single-layer materials continued as additional layers were added. Additional layers provided a higher FE, as seen by the surgical mask, which required two layers to reach a consistent FE above 95%. These results represent the FE when a mask tightly fits the wearer with no leakage. Masks that are not properly fitted would not be able to provide the same level of FE, making the fit of masks a vital component to ensure maximum protection for the wearer [27,40,70].

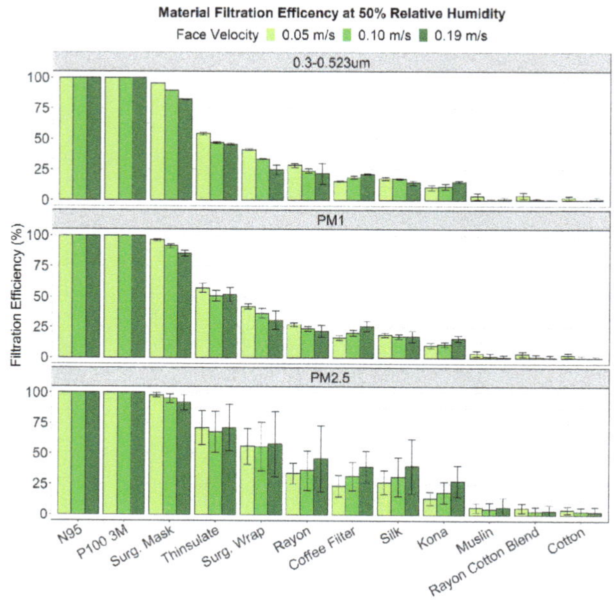

Figure 5. FE results for single-layered materials as a function of face velocity. Bar colors represent the corresponding face velocity values, with the results plotted at the average FE from three experimental runs with the error bars representing the standard deviation.

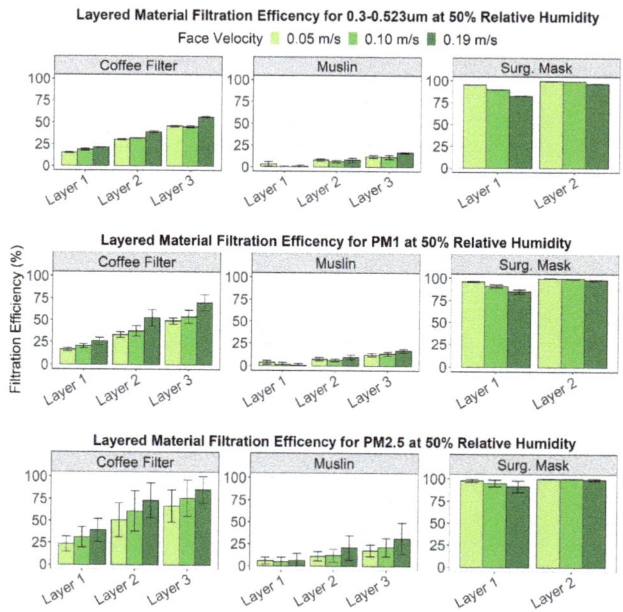

Figure 6. FE results for layered materials as a function of face velocity. Bar colors represent the corresponding face velocity values; the results are plotted at the average FE from three measurements, with the error bars representing the standard deviation.

3.3. Quality Factor

The overall performance of a material is quantified by using the quality factor to relate the FE and breathability a material provides. Figure 7 shows the quality factor of tested materials at three face velocities for PM2.5. The quality factors for other particle sizes can be found in the Supplemental Information. The quality factor was calculated using three measurements' average FE and pressure drop readings. In terms of the quality factor, the top three performing materials in this study were those commonly worn as PPE: the N95 mask, the surgical mask, and the P100 3M filter. Even though the N95 and P100 3M filters are considered more difficult to breathe through, their high FE makes them considerably better materials when compared to non-traditional mask materials. These certified materials were found to have, at a minimum, a quality factor three times higher than the top non-traditional mask material. The critical differences are the utilization of electrostatically charged fibers and the hydrophobic layers that are not considered in the homemade masks. Natural fabrics are all found to have the lowest quality factor ranging from 0.01 to 0.02. The effect of multiple material layers on the quality factor was also analyzed and included in the supplemental material. The quality factor tended to decrease as additional layers were added. While materials, such as the surgical mask, provided higher FE when double layers were used, the additional pressure from the second layer caused the overall mask performance to decline, as shown in Figure S5 of the Supplemental Information.

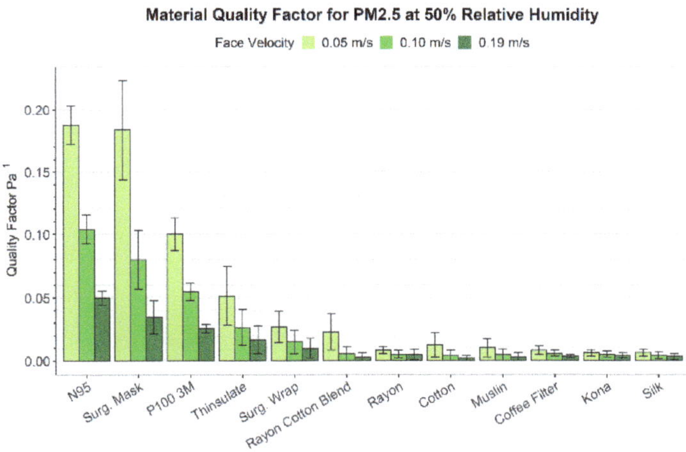

Figure 7. Quality factors for single-layered materials when face velocity is varied. Error bars in the figure represent the standard deviation across these three quality factor values, while the bar colors represent the corresponding face velocity values.

4. Conclusions

This study found that mask performance is dependent on the face velocity through the material rather than the relative humidity. Though varying humidity can inhibit particle growth and hygroscopic behavior in certain materials, the effects from humidity on mask FE and pressure drop are insignificant. While the impact of improper mask fittings is not analyzed in this study, ensuring the mask fits with no leakage will supply the wearer with filtered air within the filtration efficiencies shown in this study. In terms of overall performance, masks certified by health organizations such as NIOSH perform at least three times better than other non-certified materials. Non-certified materials provide the wearer with some protection and comfort; however, they do not provide as much protection as certified masks. While adding additional layers of material can increase a mask's FE, they can become increasingly difficult to breathe through, outweighing the increase in aerosol protection. These results suggest that single-layered surgical masks should be recommended, given their high FE, comfortability, and accessibility. Clips or additional accessories should be used to create a proper fit, reducing the risk of leakage. Higher-grade surgical masks should be used by those in areas where they are subjected to a high amount of air contaminants and users should refrain from long-term wear.

Supplementary Materials: The following supporting information can be downloaded at: https://www.mdpi.com/article/10.3390/app12189360/s1, Figure S1: Photos of some of the tested materials. Listing the materials starting from the top left and moving across are: Kona, Muslin, Cotton, Rayon, Rayon Cotton blend, and Surg. Wrap; Figure S2: Pressure drop results for layered materials as a function of RH. Bar colors represent the corresponding RH value and results are plotted as the average pressure drop from three runs, with the error bars representing the standard deviation; Figure S3: Filtration efficiency results for single layered materials as a function of relative humidity. Bar colors represent the corresponding RH values with the results plotted at the average filtration efficiency from three experimental runs with the error bars representing the standard deviation; Figure S4: Filtration efficiency results for layered materials as a function of relative humidity. Bar colors represent the corresponding RH values with the results plotted at the average filtration efficiency from three experimental runs with the error bars representing the standard deviation.; Figure S5: Quality factor results for layered materials when face velocity is varied. The quality factor results use the average filtration efficiency and pressure drop readings taken from three experiments to calculate three individual quality factor values. Error bars in the figure represent the standard deviation across these three quality factor values while the bar colors represent the corresponding face velocity values; Table S1: Material list, description, and average filtration efficiency, pressure drop, and quality factor values. Materials are listed from highest quality factor to lowest. Average and standard deviation values are calculated across three experiments.

Author Contributions: Conceptualization, J.H. and I.N.; methodology, J.H. and I.N.; software, J.M.S. and N.Y.; formal analysis, J.M.S. and C.-H.H.; data curation, N.Y. and J.M.S.; writing—original draft preparation, J.M.S. and M.M.; writing—review and editing, J.M.S. and J.H.; visualization, J.M.S.; supervision, I.N. and J.H.; project administration, J.H. and I.N.; funding acquisition, I.N. All authors have read and agreed to the published version of the manuscript.

Funding: This research received no external funding.

Conflicts of Interest: The authors declare no conflict of interest.

References

1. Monto, A.S. Epidemiology of viral respiratory infections. *Am. J. Med.* **2002**, *112*, 4–12. [CrossRef]
2. Kutter, J.S.; Spronken, M.I.; Fraaij, P.L.; Fouchier, R.A.M.; Herfst, S. Transmission routes of respiratory viruses among humans. *Curr. Opin. Virol.* **2018**, *28*, 142–151. [CrossRef]
3. Yu, I.T.; Li, Y.; Wong, T.W.; Tam, W.; Chan, A.T.; Lee, J.H.; Leung, D.Y.C.; Ho, T. Evidence of Airborne Transmission of the Severe Acute Respiratory Syndrome Virus. *N. Engl. J. Med.* **2004**, *350*, 1731–1739. [CrossRef]
4. Miller, S.L.; Nazaroff, W.W.; Jimenez, J.L.; Boerstra, A.; Buonanno, G.; Dancer, S.J.; Kurnitski, J.; Marr, L.C.; Morawska, L.; Noakes, C. Transmission of SARS-CoV-2 by inhalation of respiratory aerosol in the Skagit Valley Chorale superspreading event. *Indoor Air* **2021**, *31*, 314–323. [CrossRef]
5. Li, Y.; Qian, H.; Hang, J.; Chen, X.; Cheng, P.; Ling, H.; Wang, S.; Liang, P.; Li, J.; Kang, M.; et al. Probable airborne transmission of SARS-CoV-2 in a poorly ventilated restaurant. *Build. Environ.* **2021**, *196*, 107788. [CrossRef]
6. Klejnowski, K.; Krasa, A.; Rogula-Kozłowska, W.; Błaszczak, B. Number Size Distribution of Ambient Particles in a Typical Urban Site, The First Polish Assessment Based on Long-Term (9 Months) Measurements. *Sci. World J.* **2013**, *2013*, 539568. [CrossRef]
7. Morawska, L.; Milton, D.K. It Is Time to Address Airborne Transmission of Coronavirus Disease 2019 (COVID-19). *Clin. Infect. Dis.* **2020**, *71*, 2311–2313. [CrossRef]
8. Lin, K.; Marr, L.C. Humidity-Dependent Decay of Viruses, but Not Bacteria, in Aerosols and Droplets Follows Disinfection Kinetics. *Environ. Sci. Technol.* **2020**, *54*, 1024–1032. [CrossRef]
9. Milton, D.K.; Fabian, M.P.; Cowling, B.J.; Grantham, M.L.; McDevitt, J.J. Influenza Virus Aerosols in Human Exhaled Breath, Particle Size, Culturability, and Effect of Surgical Masks. *PLOS Pathog.* **2013**, *9*, e1003205. [CrossRef]
10. Ehsanifar, M. Airborne aerosols particles and COVID-19 transition. *Environ. Res.* **2021**, *200*, 111752. [CrossRef]
11. Hinds, W.C. *Aerosol Technology: Properties, Behavior, and Measurement of Airborne Particles*; Wiley: Hoboken, NJ, USA, 1999.
12. Lindsley, W. Filter Pore Size and Aerosol Sample Collection. In *NIOSH Manual of Analytical Methods*; NIOSH: Washington, DC, USA, 2016; FP1-14.
13. Lee, K.-P.; Yip, J.; Kan, C.-W.; Chiou, J.-C.; Yung, K.-F. Reusable Face Masks as Alternative for Disposable Medical Masks: Factors that Affect their Wear-Comfort. *Int. J. Environ. Res. Public Health* **2020**, *17*, 6623. [CrossRef] [PubMed]

14. Kwong, L.H.; Wilson, R.; Kumar, S.; Crider, Y.S.; Reyes Sanchez, Y.; Rempel, D.; Pillarisetti, A. Review of the Breathability and Filtration Efficiency of Common Household Materials for Face Masks. *ACS Nano* **2021**, *15*, 5904–5924. [CrossRef] [PubMed]
15. Jones, R.M.; Rempel, D. Standards for Surgical Respirators and Masks: Relevance for Protecting Healthcare Workers and the Public during Pandemics. *Ann. Work. Expo. Health* **2021**, *65*, 495–504. [CrossRef]
16. Ranney, M.L.; Griffeth, V.; Jha, A.K. Critical Supply Shortages—The Need for Ventilators and Personal Protective Equipment during the COVID-19 Pandemic. *N. Engl. J. Med.* **2020**, *382*, e41. [CrossRef]
17. Wu, H.-L.; Huang, J.; Zhang, C.J.P.; He, Z.; Ming, W.-K. Facemask shortage and the novel coronavirus disease (COVID-19) outbreak: Reflections on public health measures. *EClinicalMedicine* **2020**, *21*, 100329. [CrossRef] [PubMed]
18. Kim, H. Lesson Learned from the Power of Open Data: Resolving the Mask Shortage Problem Caused by COVID-19 in South Korea. *Sustainability* **2021**, *13*, 278. [CrossRef]
19. Kirubarajan, A.; Khan, S.; Got, T.; Yau, M.; Bryan, J.M.; Friedman, S.M. Mask shortage during epidemics and pandemics: A scoping review of interventions to overcome limited supply. *BMJ Open* **2020**, *10*, e040547. [CrossRef]
20. Arumuru, V.; Samantaray, S.S.; Pasa, J. Double masking protection vs. comfort—A quantitative assessment. *Phys. Fluids (1994)* **2021**, *33*, 077120. [CrossRef]
21. Konda, A.; Prakash, A.; Moss, G.A.; Schmoldt, M.; Grant, G.D.; Guha, S. Aerosol Filtration Efficiency of Common Fabrics Used in Respiratory Cloth Masks. *ACS Nano* **2020**, *14*, 6339–6347. [CrossRef]
22. Lustig, S.R.; Biswakarma, J.J.; Rana, D.; Tilford, S.H.; Hu, W.; Su, M.; Rosenblatt, M.S. Effectiveness of Common Fabrics to Block Aqueous Aerosols of Virus-like Nanoparticles. *ACS Nano* **2020**, *14*, 7651–7658. [CrossRef]
23. Radney, J.G.; Weaver, J.L.; Vicenzi, E.P.; Staymates, M.E.; Zangmeister, C.D. Filter Inserts Impact Cloth Mask Performance against Nano- to Micro-Sized Particles. *ACS Nano* **2021**, *15*, 12860–12868. [CrossRef] [PubMed]
24. Zangmeister, C.D.; Radney, J.G.; Vicenzi, E.P.; Weaver, J.L. Filtration Efficiencies of Nanoscale Aerosol by Cloth Mask Materials Used to Slow the Spread of SARS-CoV-2. *ACS Nano* **2020**, *14*, 9188–9200. [CrossRef] [PubMed]
25. Pei, C.; Ou, Q.; Kim, S.C.; Chen, S.-C.; Pui, D.Y.H. Alternative Face Masks Made of Common Materials for General Public: Fractional Filtration Efficiency and Breathability Perspective. *Aerosol Air Qual. Res.* **2020**, *20*, 2581–2591. [CrossRef]
26. Jung, H.; Kim, J.; Lee, S.; Lee, J.; Kim, J.; Tsai, P.; Yoon, C. Comparison of Filtration Efficiency and Pressure Drop in Anti-Yellow Sand Masks, Quarantine Masks, Medical Masks, General Masks, and Handkerchiefs. *Aerosol Air Qual. Res.* **2014**, *14*, 991–1002. [CrossRef]
27. Drewnick, F.; Pikmann, J.; Fachinger, F.; Moormann, L.; Sprang, F.; Borrmann, S. Aerosol filtration efficiency of household materials for homemade face masks: Influence of material properties, particle size, particle electrical charge, face velocity, and leaks. *Aerosol Sci. Technol.* **2021**, *55*, 63–79. [CrossRef]
28. Rengasamy, S.; Eimer, B.; Shaffer, R.E. Simple Respiratory Protection—Evaluation of the Filtration Performance of Cloth Masks and Common Fabric Materials against 20–1000 nm Size Particles. *Ann. Occup. Hyg.* **2010**, *54*, 789–798.
29. Hao, W.; Parasch, A.; Williams, S.; Li, J.; Ma, H.; Burken, J.; Wang, Y. Filtration performances of non-medical materials as candidates for manufacturing facemasks and respirators. *Int. J. Hyg. Environ. Health* **2020**, *229*, 113582. [CrossRef]
30. Teesing, G.R.; van Straten, B.; de Man, P.; Horeman-Franse, T. Is there an adequate alternative to commercially manufactured face masks? A comparison of various materials and forms. *J. Hosp. Infect.* **2020**, *106*, 246–253. [CrossRef]
31. Hao, W.; Xu, G.; Wang, Y. Factors influencing the filtration performance of homemade face masks. *J. Occup. Environ. Hyg.* **2021**, *18*, 128–138. [CrossRef]
32. Wang, D.; You, Y.; Zhou, X.; Zong, Z.; Huang, H.; Zhang, H.; Yong, X.; Cheng, Y.; Yang, L.; Du, L.; et al. Selection of homemade mask materials for preventing transmission of COVID-19, A laboratory study. *PLoS ONE* **2020**, *15*, e0240285. [CrossRef]
33. Davies, A.; Thompson, K.-A.; Giri, K.; Kafatos, G.; Walker, J.; Bennett, A. Testing the Efficacy of Homemade Masks: Would They Protect in an Influenza Pandemic? *Disaster Med. Public Health Prep.* **2013**, *7*, 413–418. [CrossRef] [PubMed]
34. Ma, Q.-X.; Shan, H.; Zhang, H.-L.; Li, G.-M.; Yang, R.-M.; Chen, J.-M. Potential utilities of mask-wearing and instant hand hygiene for fighting SARS-CoV-2. *J. Med. Virol.* **2020**, *92*, 1567–1571. [CrossRef] [PubMed]
35. Zangmeister, C.D.; Radney, J.G.; Staymates, M.E.; Vicenzi, E.P.; Weaver, J.L. Hydration of Hydrophilic Cloth Face Masks Enhances the Filtration of Nanoparticles. *ACS Appl. Nano Mater.* **2021**, *4*, 2694–2701. [CrossRef]
36. Sharma, S.K.; Mishra, M.; Mudgal, S.K. Efficacy of cloth face mask in prevention of novel coronavirus infection transmission: A systematic review and meta-analysis. *J. Educ. Health Promot.* **2020**, *9*, 192. [CrossRef]
37. Neupane, B.B.; Mainali, S.; Sharma, A.; Giri, B. Optical microscopic study of surface morphology and filtering efficiency of face masks. *PeerJ* **2019**, *7*, e7142. [CrossRef] [PubMed]
38. Reutman, S.R.; Reponen, T.; Yermakov, M.A.; Grinshpun, S. Homemade facemasks: Particle filtration, breathability, fit, and other performance characteristics. *J. Occup. Environ. Hyg.* **2021**, *18*, 334–344. [CrossRef] [PubMed]
39. Crilley, L.R.; Angelucci, A.A.; Malile, B.; Young, C.J.; VandenBoer, T.C.; Chen, J.I. Non-woven materials for cloth-based face masks inserts: Relationship between material properties and sub-micron aerosol filtration. *Environ. Sci. Nano* **2021**, *8*, 1603–1613. [CrossRef]
40. Hill, W.C.; Hull, M.S.; MacCuspie, R.I. Testing of commercial masks and respirators and cotton mask insert materials using SARS-CoV-2 virion-sized particulates: Comparison of ideal aerosol filtration efficiency versus fitted filtration efficiency. *Nano Lett.* **2020**, *20*, 7642–7647. [CrossRef]

41. Bhattacharjee, S.; Bahl, P.; De Silva, C.; Doolan, C.; Chughtai, A.A.; Heslop, D.; MacIntyre, C.R. Experimental Evidence for the Optimal Design of a High-Performing Cloth Mask. *ACS Biomater. Sci. Eng.* **2021**, *7*, 2791–2802. [CrossRef]
42. Asadi, S.; Cappa, C.D.; Barreda, S.; Wexler, A.S.; Bouvier, N.M.; Ristenpart, W.D. Efficacy of masks and face coverings in controlling outward aerosol particle emission from expiratory activities. *Sci. Rep.* **2020**, *10*, 15665. [CrossRef]
43. Kähler, C.J.; Hain, R. Fundamental protective mechanisms of face masks against droplet infections. *J. Aerosol Sci.* **2020**, *148*, 105617. [CrossRef] [PubMed]
44. Zhao, M.; Liao, L.; Xiao, W.; Yu, X.; Wang, H.; Wang, Q.; Lin, Y.L.; Price, A.; Chu, L.; Cui, Y.; et al. Household Materials Selection for Homemade Cloth Face Coverings and Their Filtration Efficiency Enhancement with Triboelectric Charging. *Nano Lett.* **2020**, *20*, 5544–5552. [CrossRef] [PubMed]
45. Rule, A.; Ramachandran, G.; Koehler, K. Comment on Aerosol Filtration Efficiency of Common Fabrics Used in Respiratory Cloth Masks: Questioning Their Findings. *ACS Nano* **2020**, *14*, 10756–10757. [CrossRef]
46. Konda, A.; Prakash, A.; Moss, G.; Schmoldt, M.; Grant, G.; Guha, S. Correction to Aerosol Filtration Efficiency of Common Fabrics Used in Respiratory Cloth Masks. *ACS Nano* **2020**, *14*, 10742–10743. [CrossRef] [PubMed]
47. Carr, I.A.; Hariharan, P.; Guha, S. Letter to the Editor Regarding Aerosol Filtration Efficiency of Common Fabrics Used in Respiratory Cloth Masks. *ACS Nano* **2020**, *14*, 10754–10755. [CrossRef] [PubMed]
48. Alderman, S.L.; Parsons, M.S.; Hogancamp, K.U.; Waggoner, C.A. Evaluation of the effect of media velocity on filter efficiency and most penetrating particle size of nuclear grade high-efficiency particulate air filters. *J. Occup. Environ. Hyg.* **2008**, *5*, 713–720. [CrossRef] [PubMed]
49. Novosselov, I.V.; Ariessohn, P.C. Rectangular slit atmospheric pressure aerodynamic lens aerosol concentrator. *Aerosol Sci. Technol.* **2014**, *48*, 163–172. [CrossRef]
50. Mahamuni, G.; Ockerman, B.; Novosselov, I. Electrostatic Capillary Collector for In-Situ Spectroscopic Analysis of Aerosols. *Aerosol Sci. Technol.* **2019**, *53*, 688–700. [CrossRef]
51. Vaddi, R.S.; Guan, Y.; Novosselov, I. Behavior of ultrafine particles in electro-hydrodynamic flow induced by corona discharge. *J. Aerosol Sci.* **2020**, *148*, 105587. [CrossRef]
52. Keady, P.B. *Getting Data you Need with Particle Measurements: TSI Incorporated*; TSI Inc.: Shoreview, MN, USA, 2012; Volume 8.
53. Chien, C.-H.; Theodore, A.; Wu, C.-Y.; Hsu, Y.-M.; Birky, B. Upon correlating diameters measured by optical particle counters and aerodynamic particle sizers. *J. Aerosol Sci.* **2016**, *101*, 77–85. [CrossRef]
54. Njalsson, T.; Novosselov, I. Design and optimization of a compact low-cost optical particle sizer. *J. Aerosol Sci.* **2018**, *119*, 1–12. [CrossRef] [PubMed]
55. Huang, C.-H.; He, J.; Austin, E.; Seto, E.; Novosselov, I. Assessing the value of complex refractive index and particle density for calibration of low-cost particle matter sensor for size-resolved particle count and PM2. 5 measurements. *PLoS ONE* **2021**, *16*, e0259745. [CrossRef] [PubMed]
56. Austin, E.; Novosselov, I.; Seto, E.; Yost, M.G. Laboratory Evaluation of the Shinyei PPD42NS Low-Cost Particulate Matter Sensor. *PLoS ONE* **2015**, *10*, e0137789.
57. Leith, D.; L'Orange, C.; Volckens, J. Quantitative protection factors for common masks and face coverings. *Environ. Sci. Technol.* **2021**, *55*, 3136–3143. [CrossRef] [PubMed]
58. Stan, A.; Steiner, S.; Majeed, S.; Weber, S.S.; Gosh, S.; Semren, T.Ž.; Guy, P.A.; Lebrun, S.; Steinhauser, J.; Tardy, Y.; et al. Aerosol filtration testing of fabrics for development of reusable face masks. *Aerosol Air Qual. Res.* **2021**, *21*, 210052. [CrossRef]
59. Duncan, S.; Bodurtha, P.; Naqvi, S. The protective performance of reusable cloth face masks, disposable procedure masks, KN95 masks and N95 respirators: Filtration and total inward leakage. *PLoS ONE* **2021**, *16*, e0258191. [CrossRef]
60. Pan, J.; Harb, C.; Leng, W.; Marr, L.C. Inward and outward effectiveness of cloth masks, a surgical mask, and a face shield. *Aerosol Sci. Technol.* **2021**, *55*, 718–733. [CrossRef]
61. Tong, Y.; Pan, J.; Kucukdeger, E.; Johnson, A.L.; Marr, L.C.; Johnson, B.N. 3D Printed Mask Frames Improve the Inward Protection Efficiency of a Cloth Mask. *ACS ES T Eng.* **2021**, *1*, 1000–1008. [CrossRef]
62. Cappa, C.D.; San Francisco Opera Costume Department; Ristenpart, W.D.; Barreda, S.; Bouvier, N.M.; Levintal, E.; Roman, S.A.; Wexler, A.S. A highly efficient cloth facemask design. *Aerosol Sci. Technol.* **2021**, *56*, 12–28. [CrossRef]
63. Whyte, H.E.; Montigaud, Y.; Audoux, E.; Verhoeven, P.; Prier, A.; Leclerc, L.; Sarry, G.; Laurent, C.; Le Coq, L.; Pourchez, J.; et al. Comparison of bacterial filtration efficiency vs. particle filtration efficiency to assess the performance of non-medical face masks. *Sci. Rep.* **2022**, *12*, 1188. [CrossRef]
64. Zhang, J.; Dichiara, A.B.; Novosselov, I.; Gao, D.; Chung, J.-H. Polyacrylic acid coated carbon nanotube–paper composites for humidity and moisture sensing. *J. Mater. Chem. C* **2019**, *7*, 5374–5380. [CrossRef]
65. Gibson, P.; Rivin, D.; Kendrick, C.; Schreuder-Gibson, H. Humidity-Dependent Air Permeability of Textile Materials1. *Text. Res. J.* **1999**, *69*, 311–317. [CrossRef]
66. He, J.Y.; Novosselov, I.V. Design and evaluation of an aerodynamic focusing micro-well aerosol collector. *Aerosol Sci. Technol.* **2017**, *51*, 1016–1026. [CrossRef] [PubMed]
67. Gibson, P.W.; Elsaiid, A.E.; Kendrick, C.E.; Rivin, D.; Charmchi, M. A test method to determine the relative humidity dependence of the air permeability of woven textile fabrics. *J. Test. Eval.* **1997**, *25*, 416–423.
68. Huang, S.-H.; Chen, C.-W.; Kuo, Y.-M.; Lai, C.-Y.; McKay, R.; Chen, C.-C. Factors affecting filter penetration and quality factor of particulate respirators. *Aerosol Air Qual. Res.* **2013**, *13*, 162–171. [CrossRef]

9. Yim, W.; Cheng, D.; Patel, S.H.; Kou, R.; Meng, Y.S.; Jokerst, J.V. KN95 and N95 respirators retain filtration efficiency despite a loss of dipole charge during decontamination. *ACS Appl. Mater. Interfaces* **2020**, *12*, 54473–54480. [CrossRef] [PubMed]
0. Rengasamy, S.; Eimer, B.C.; Szalajda, J. A Quantitative Assessment of the Total Inward Leakage of NaCl Aerosol Representing Submicron-Size Bioaerosol through N95 Filtering Facepiece Respirators and Surgical Masks. *J. Occup. Environ. Hyg.* **2014**, *11*, 388–396. [CrossRef]

materials

Article

Numerical and Field Investigations of Acoustic Emission Laws of Coal Fracture under Hydro-Mechanical Coupling Loading

Jie-Fang Song [1], Cai-Ping Lu [1,*], Zhao-Wei Zhan [2], Hai-Feng Cui [2], Yan-Min Wang [2] and Jian-Hua Wang [2]

1 Key Laboratory of Deep Coal Resource Mining, Ministry of Education, School of Mines, China University of Mining and Technology, Xuzhou 221116, China
2 Xiaoyun Coal Mine, Jining Energy Group, Jining 272000, China
* Correspondence: cplucumt@126.com

Highlights:

- The relationships among stress, acoustic emission (AE), and energy during coal fracture under hydro-mechanical coupling loading were analyzed.
- Moment tensor reveals event distribution, source type change, and b value of AE during crack initiation and propagation in coal.
- The relationships among stress, number and type of cracks, AE, number of contacts and K_E were revealed under different water and confining pressures.

Abstract: Taking coal under hydro-mechanical coupling as the research object, the discrete element software PFC3D (particle flow code) was used to analyze the relationships among the force, acoustic emission (AE), and energy during coal fracture. Based on the moment tensor (MT) inversion, we revealed the AE event distribution and source type during crack initiation and propagation until the final failure of coal. Meanwhile, we examined the relationships among the stress, number and type of cracks, magnitude, K_E, and b value of AE under different water and confining pressures. The results show that the numerical simulation can effectively determine the microscopic damage mechanism of coal under different conditions. Moreover, the rupture type of the numerical simulation is consistent with the field investigations, which verifies the rationality of the simulation. These research results can provide reference for safety production evaluation of water inrush mines.

Keywords: hydro-mechanical coupling; failure mechanism; confining pressure; acoustic emission; moment tensor inversion

Citation: Song, J.-F.; Lu, C.-P.; Zhan, Z.-W.; Cui, H.-F.; Wang, Y.-M.; Wang, J.-H. Numerical and Field Investigations of Acoustic Emission Laws of Coal Fracture under Hydro-Mechanical Coupling Loading. *Materials* **2022**, *15*, 6510. https://doi.org/10.3390/ma15196510

Academic Editors: Vasily Novozhilov and Cunlu Zhao

Received: 28 July 2022
Accepted: 26 August 2022
Published: 20 September 2022

1. Introduction

With the continuous increase in the mining depth of coal mines, geological conditions have become particularly complex [1]. Discontinuous joints such as microcracks in a rock stratum provide necessary locations for the storage and migration of groundwater [2,3]. In the seepage process, the generated water stress affects the stress field of a coal and rock mass [4]. Under the superposition of multiple stress fields, the frequency and intensity of catastrophic accidents such as rock bursts and water inrush remarkably increase [5,6]. When a water inrush accident occurs, the overall strength and plastic deformation of a coal pillar significantly change, which has a notable impact on the support capacity and resistance of the roadway confining pressure system, thereby affecting the width of the coal pillar [7].

Studies on the mechanical properties of coal and rock under hydro-mechanical coupling have mainly focused on the interaction between the seepage and stress fields, and the research methods have primarily included laboratory experiments and numerical simulations. For example, Hui et al. [8] conducted a series of laboratory acoustic emission (AE) tests to study the influence of fluid-filled rock joints on the propagation and attenuation of stress waves. Gardner et al. [9] investigated the interaction relationship between a rock mass and a fluid based on the three-dimensional discrete element and lattice Boltzmann methods. For the purpose of improving the calculation efficiency and accuracy, an extensive discussion on aspects that still need to be optimized was conducted. Chen et al. [10] studied the characteristics of the permeability evolution during sandstone failure under triaxial compression by experiments and simulations. Cai et al. [11] analyzed the evolution of the crack growth, coordination number, volumetric strain, and permeability under hydro-mechanical coupling in sandstone and established a synergistic mechanism among the microstructure, accumulated damage, and permeability evolution. Liu et al. [12] analyzed various parametric changes of a rock mass with hydro-mechanical coupling under unloading conditions. Wang et al. [13] investigated the mechanical characteristics and permeability evolution of red sandstone under hydro-mechanical coupling. Yao et al. [14] analyzed the mechanical properties and failure precursor characteristics of a coal and rock system caused by water content using AE and infrared radiation. Wang et al. [15] studied the deformation characteristics of granite under hydro-mechanical coupling and analyzed the changes in the permeability before and after expansion under different confining pressures. Zhang et al. [16] studied the influence of water and confining pressures on the permeability of fractured sandstone. Many scholars have studied the changes of rock mechanical parameters under the hydro-mechanical coupling loading in the laboratory, but there are few studies on the influence of seepage under different water and confining pressure [17,18]. In addition, few scholars have focused on the source rupture type in the process of rock seepage through AE.

In recent years, AE has been commonly used in the analysis of coal and rock failure. As an accompanying phenomenon in this process, it contains substantial information about the internal damage process of rocks, which helps to understand the mechanism of rock failure [19–21]. For instance, Shimizu et al. [22] discussed the effects of different fluid viscosity and particle size distributions on fracture propagation. Jia et al. [23] analyzed the relationship between AE and spatial fractal dimensions under different water soaking times. Ye et al. [24] obtained the relationships among AE events and the joint angle, water injection pressure, stress, and deformation amount in the process of a rock slip instability. Makhnenko et al. [25] observed that AE activity coincides with the onset of inelastic response in a fluid-saturated rock and explored the relationship between the clustering of AE events and inhomogeneous deformation. Li et al. [26] applied AE to dynamic disaster monitoring of coal and rock systems. Li et al. [27] concluded that there is a good positive correlation between the change in coal AE events and the water pressure curve under the action of triaxial hydraulic coupling. Chitrala et al. [28] reported the changes in the AE during hydraulic pressure damage of sandstone under different applied stresses. Chen et al. [29] used AE to analyze the evolution characteristics of granite damage during uniaxial compression and established a relationship between the permeability and the confining pressure. Makoto et al. [30] studied the relationship between the AE and fluid pressure during the failure of the Eagle Ford Shale, and analyzed the main types of AE. Lu et al. [31] successfully reproduced the stress redistribution, AE event evolution, permeability change, and formation of water inrush channels during a mining process. Jiang et al. [32] analyzed the AE event, energy, peak frequency, and crack type under the action of hydraulic fracturing in a layered rock. Most scholars widely use AE to locate the crack position of rock, and then study the force of the specimen [33]. However, there are few studies on the focal mechanism of AE, and analysis of the rupture form of rock/coal mass under different stress conditions is still lacking.

In summary, most studies mainly focused on the basic mechanical parameters of coal and rock masses under the action of hydro-mechanical coupling. However, they did not conduct detailed investigations on the focal mechanism of this action, and rarely used an energy method to evaluate the impact tendency of coal under hydro-mechanism coupling [34,35]. To better understand the characteristics of AE and its impact tendency under hydro-mechanism coupling, this study used the PFC3D software to analyze the relationships among the force, AE distribution (type), energy, and b value in the failure process of a coal sample under different water pressures. In addition, it is consistent with the results of our field investigation. These results are highly significant for guiding the safe production of coal mines.

2. Damage Mechanism of Coal under Hydro-Mechanism Coupling

To ensure no shear failure occurs during the loading of a rock, the limit value of the rock force can be obtained by the Mohr–Coulomb criterion [36] as follows:

$$|\tau| = C + \gamma_s \sigma \tag{1}$$

where $|\tau|$ is the shear stress of the rock, MPa, C is the internal cohesion, Pa, and γ_s is the internal friction coefficient, which can be also expressed as

$$\gamma_s = \tan\varphi \tag{2}$$

It is assumed that the minimum principal stress on the rock is σ_3. When the rock is in the critical limit state, the maximum principal stress, σ_1, can be calculated. Owing to the existence of water pressure in the rock pores, a part of the confining pressure and vertical stress is offset; therefore, the water pressure eliminates the confining pressure (reducing the servo stress). The triaxial stress state of the rock under water pressure is shown in Figure 1.

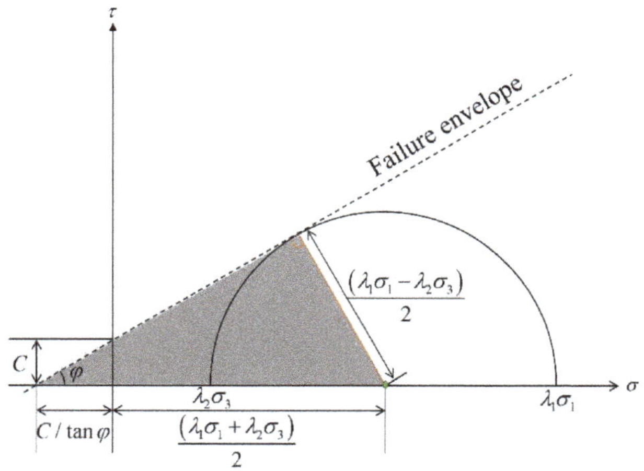

Figure 1. Schematic of critical stress state under influence of water pressure based on Mohr–Coulomb criterion.

When the Mohr–Coulomb circle is tangent to the envelope curve, the length of the right-angle side of the grey triangle area is $(\lambda_1\sigma_1 - \lambda_2\sigma_3)/2$, and the hypotenuse length is $C/\tan\varphi + (\lambda_1\sigma_1 + \lambda_2\sigma_3)/2$. Based on trigonometric functions,

$$(\lambda_1\sigma_1 - \lambda_2\sigma_3)/2 = [C/\tan\varphi + (\lambda_1\sigma_1 + \lambda_2\sigma_3)/2]\sin\varphi \tag{3}$$

Combined with Equation (2), the maximum principal stress under the critical stress state of the sample can be expressed as

$$\sigma_1 = \frac{2C + \lambda_2\left(\gamma_s + \sqrt{\gamma_s^2 + 1}\right)\sigma_3}{\lambda_1\left(\sqrt{\gamma_s^2 + 1} - \gamma_s\right)} \tag{4}$$

In the experiments, when the internal cohesion and the friction angle have certain values, because λ_1 and λ_2 are less than 1, the stress–strain curve of the sample under water pressure is above that of the sample without water pressure; however, the peak stress in the former is smaller. This is attributed to the effect of the pore water pressure reducing the effective average stress of the rock, making it easier for the rock to reach the ultimate strength when the deviatoric stress remains constant.

3. Modeling Methodology

3.1. Engineering Background

A water inrush accident occurred in the Xiaoyun coal mine in Shandong province on 11 September 2018. The coal body of the 1318 working face was severely affected by the water pressure of proximately 3 MPa, and the soaking time was 100 days. A sample was taken from the soaking water coal body of the above working face, which has a simple structure and is relatively stable. To make the sample moisture non-volatile, it was stored in plastic wrap before experiments and processed into a cylinder with a diameter of 50 mm and a length of 100 mm. An MTS Landmark 370.50 rock testing machine was used for loading.

3.2. Model Setup

In the discrete element model, the core of the fluid flow algorithm is the topological structure formed by pipe domains. As shown in Figure 2, many domains that can store the pore water pressure are formed between the particles, and these domains are represented by red squares. A fracture between adjacent particles is regarded as a seepage pipe to simulate the flow of water. A pipeline is represented by a yellow line. When there is a pressure difference between adjacent domains, the fluid flows through the pipe to them.

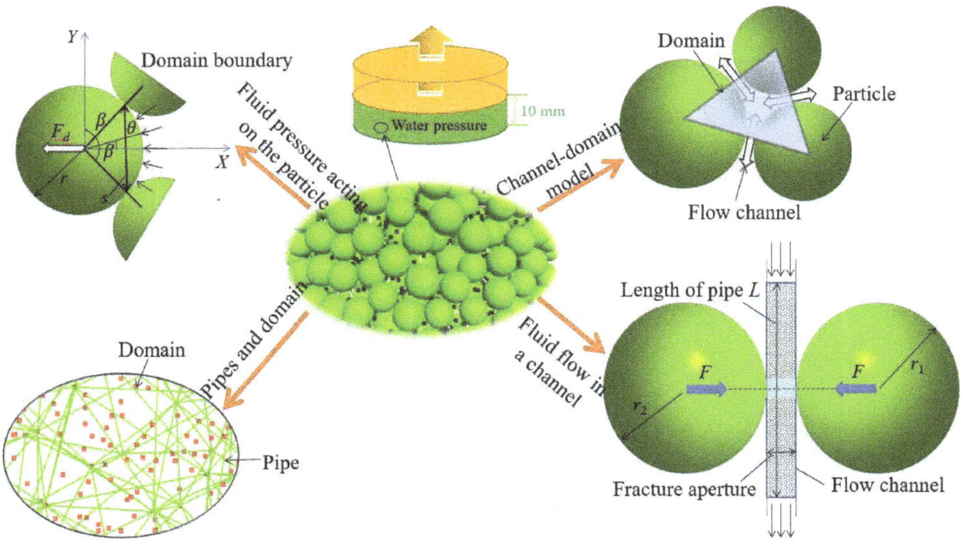

Figure 2. Schematic of incorporation of fluid coupling.

To study the mechanical characteristics and instability mechanism of coal under the action of seepage, a model was set up, which is shown in Figure 2. For the model, the standard size (Φ50 mm × 100 mm) was adopted, and a total of 3117 particles were generated. The radius was 2–2.5 mm. The linear contact bond model was used between particles, and the servo mechanism was used to apply different confining pressures around the model. When the confining pressure reached the target stress, the particles in contact with the wall were fixed. A high-pressure water area was applied within the range of

0–10 mm in the z-axis direction at the bottom of the model. Under the action of the pressure difference, a water body seeped from the bottom to the top of the model. At this time, the water pressure is close to linear distribution along the longitudinal direction, which can be regarded as the equilibrium state of hydro-mechanical coupling. After the balance was reached, the upper and lower walls were controlled to apply an axial pressure at a speed of 0.03 m/s, and the loading was stopped when the residual stress reached 70% peak stress.

The model parameter calibration was based on the uniaxial compressive strength, strain, and failure form of the specimen. Based on Figure 3, the peak stresses from the experiments and numerical simulations are 6.09 MPa and 6.05 MPa, respectively, and the corresponding strains are 1.93% and 1.95%. The difference between them is relatively small, and the failure modes are also similar. From the variation trends of the AE events, the microscopic properties inside the coal body are changed owing to the softening effect of water. When the strain is approximately 5%, the number of AE events shows remarkable fluctuations, which is followed by a quiet period. When the strain reaches 1.6%, the AE events abruptly increase, subsequently cracks gradually penetrate to form macroscopic cracks, and, finally, the number of AE reaches the maximum. Based on the above analysis, the rationality of the microparameter values can be verified, which are listed in Table 1.

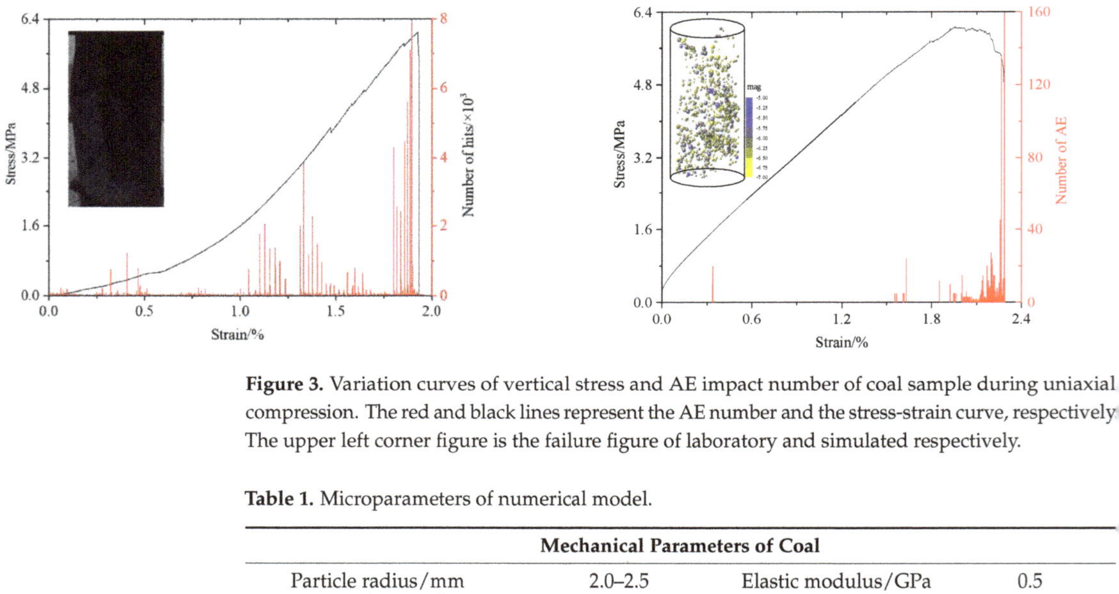

Figure 3. Variation curves of vertical stress and AE impact number of coal sample during uniaxial compression. The red and black lines represent the AE number and the stress-strain curve, respectively. The upper left corner figure is the failure figure of laboratory and simulated respectively.

Table 1. Microparameters of numerical model.

Mechanical Parameters of Coal			
Particle radius/mm	2.0–2.5	Elastic modulus/GPa	0.5
Density/kg·m^3	1250	Friction angle/°	45
Cohesive strength/MPa	4.5	Porosity	0.18
Parameters for Flow Model			
Ap_zero/mm	1.3×10^{-3}	Bulk_W/MPa	2.2×10^3
Flow_perm/mm/s	7.0×10^{-2}	Flow_dt/s	1.0×10^{-3}
P_give/MPa	2.5	Gap_mul	0

3.3. *MT Calculation in the PFC*

In seismology, MT is a mathematical representation of fault motion and an important tool for source characterization. We use M to represent the source MT, which is a 3×3 symmetric matrix with nine couples and six independent components, as shown in Figure 4.

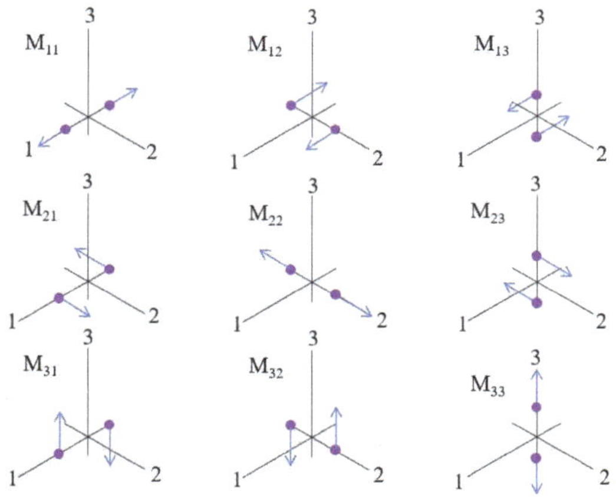

Figure 4. The nine force couples representing the components of MT.

Figure 4 shows the force couples in the Cartesian coordinate system, which is related to source strength and fault direction. We can further decompose the MT into elementary sources, e.g., compression, shear, and tensile. Then, we can intuitively obtain the rupture type of A.E. Hudson's source type plot [37,38], which is a way to vividly represent decomposition of an MT into isotropic, compensated linear vector dipole (CLVD) and double-couple (DC) components. The MT decomposition is shown in the following Equations (5) and (6):

$$\begin{aligned} M &= M_{DC} + M_{iso} + M_{v-clvd} \\ &= M_1(\phi_S) + M_2(\phi_D) + M_{iso} + M_{v-clvd} \end{aligned} \tag{5}$$

$$\begin{aligned} \begin{pmatrix} M_{11} & M_{12} & M_{13} \\ M_{21} & M_{22} & M_{23} \\ M_{31} & M_{32} & M_{33} \end{pmatrix} &= \begin{pmatrix} -\frac{1}{2}(M_{22}-M_{11}) & M_{12} & 0 \\ M_{21} & \frac{1}{2}(M_{22}-M_{11}) & 0 \\ 0 & 0 & 0 \end{pmatrix} + \begin{pmatrix} 0 & 0 & M_{13} \\ 0 & 0 & M_{23} \\ M_{31} & M_{32} & 0 \end{pmatrix} \\ &+ \frac{1}{3}(M_{11}+M_{22}+M_{33}) \begin{pmatrix} 1 & 0 & 0 \\ 0 & 1 & 0 \\ 0 & 0 & 1 \end{pmatrix} + \frac{1}{3}\left(\frac{1}{2}(M_{11}+M_{22}) - M_{33}\right) \begin{pmatrix} 1 & 0 & 0 \\ 0 & 1 & 0 \\ 0 & 0 & -2 \end{pmatrix} \end{aligned} \tag{6}$$

The PFC 5.0 software (Itasca Consulting Group, Inc., Minneapolis, MN, USA) can directly obtain the force magnitude, direction, and displacement of particles, and then obtain the MT according to the definition of AE. The calculation process of the AE in PFC is shown in Figure 5.

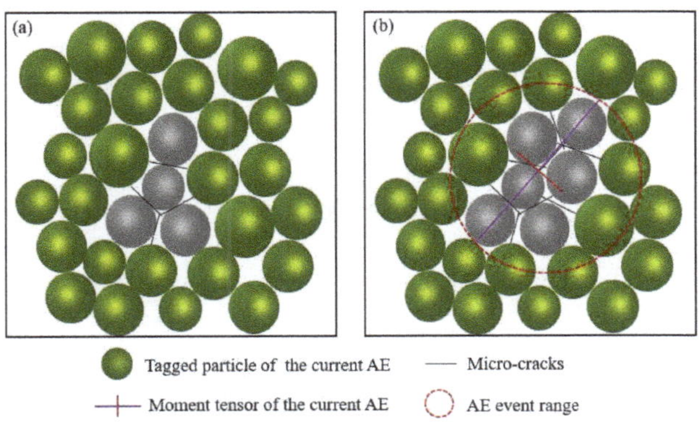

Figure 5. Generation process of an AE event: (**a**) is the start of an event and (**b**) is the finish state of this event.

An AE event is composed of particles around microcracks in PFC 5.0 software. When the contact between two particles is broken, a crack will occur. Two particles in contact with the crack and particles in contact with these two particles are marked as an AE event [39,40]. From Figure 5, (a) is the AE start stage, (b) is the AE end stage, and the range delineated by the red circle is the AE event. Through MT decomposition, tensile, shear, and compression failure mode can be obtained. The AE rupture mode corresponding to PFC 5.0 numerical simulation software is shown in Figure 6.

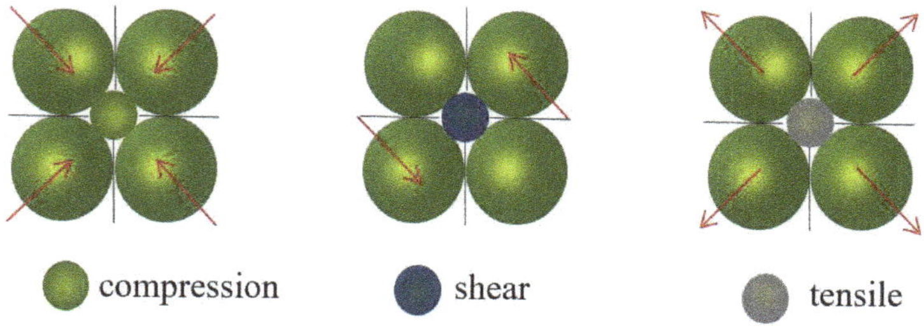

Figure 6. Schematic diagram of AE event failure mode in particle model.

Similar to earthquakes, the intensity of AE can also be expressed by magnitude. In PFC software, the AE intensity is calculated according to the strain energy (E_k) change before and after the event [41]. For each step of calculation, the E_k of all particles in the AE event can be written as Equation (7):

$$E_k = \sum_{i=1}^{n} \frac{1}{2} \left(\frac{\left(F_n^l\right)^2}{k_n} + \frac{\left\|F_s^l\right\|^2}{k_s} \right) \tag{7}$$

where F_n^l, F_n^s, k_n, and k_s are normal stress, shear stress, normal stiffness, and shear stiffness, respectively; n is the number of contacts within the AE range. For an AE event, its magnitude can be measured by the change of E_k at the beginning and end of AE [42]:

$$M_e = \frac{2}{3}(\log \Delta E_k - 6.0) \tag{8}$$

Based on the knowledge of seismology, the relationship between the magnitude and the frequency is

$$\log N = a - bM_e \tag{9}$$

where M_e is the magnitude, N is the frequency, and a and b are constants. The b value can be used to evaluate the damage degree of a rock mass and the size of a crack. A large b value implies a large proportion of small-sized cracks in the rock mass and a slow initiation and development of new cracks.

4. Mechanical Characteristics and Failure Mechanism of Water-Soaked Coal

4.1. AE and Stress Changes during Failure

In the simulations, the AE and the energy were monitored to investigate the characteristics of coal failure. To study the coal failure mechanism under hydro-mechanical coupling, the simulations described in Section 4.1 were taken as an example, and the confining and water pressures were set as 1 MPa and 2.5 MPa, respectively. The microscopic characteristics of the AE distribution (type) and energy change during the failure process were analyzed in detail.

4.1.1. Relationship between AE and Stress

To study the coal failure characteristics, the fish language and the history module were used to monitor the vertical stress, strain, AE, energy, etc. Simultaneously, the monitoring circle module was used to record the parameter, and the results are shown in Figures 7 and 8.

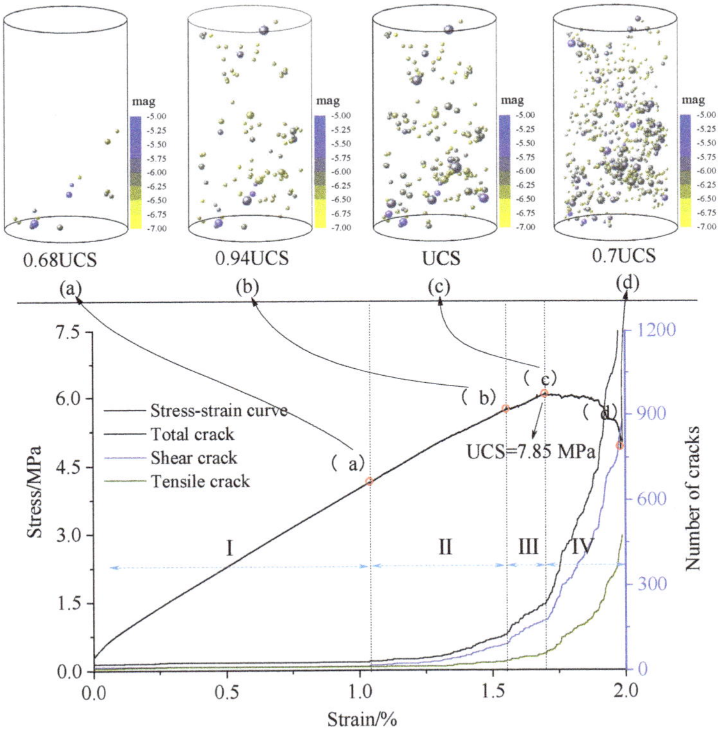

Figure 7. Stress–strain curve and number of cracks; upper snapshots show source distributions at different stages. I, II, III, and IV are the stages of crack generation, which are pore compaction stage, fracture expansion stage, peak stress stage and post peak stage, respectively.

The simulated stress–strain curve, cracks, and locations of the AE sources during uniaxial compression of water-soaked coal are shown in Figure 7. The coal failure process can be divided into four stages. The first stage is the compaction stage, in which the stress gradually increases without cracks. The second stage is the crack initiation stage, and the crack initiation stress is 5.33 MPa. The crack distribution is mainly observed in the lower part of the model, and the increase in the number of shear cracks is significantly greater than that in the number of tensile cracks. The third stage shows a notable increase in the cracks. In this stage, the increase rate of the shear cracks is much larger than that of the tensile cracks, and microcracks gradually develop upward. However, the damage degree of the lower part of the entire sample is much greater than that of the upper part, and the peak stress reaches 7.8 MPa. The fourth stage is the post-peak stage, in which the number of cracks sharply increases, the cracks gradually penetrate, and, finally, the entire instability occurs.

By depicting the distribution of the micro contact force in the particle system, the macro mechanical mechanism can be revealed. Figure 8 shows the contact force between particles in the model, which can better reflect the force in different directions of the model. From Figure 8, as the axial stress increases, the internal contact force shows a gradually increasing trend. The shape of the contact force gradually evolves from "spherical" to "elliptical", the stress levels at different positions gradually change, and at the final stage of failure, the local stress abruptly increases. The vertical stresses at monitoring points F1–F4 are consistent with the stress–strain curve. Except for at point F1, closeness to the water injection area implies a high vertical stress. Based on the change in the yy-stress, when the strain is 1.8%, the horizontal stress remarkably changes, and the model shows an instability at this stage.

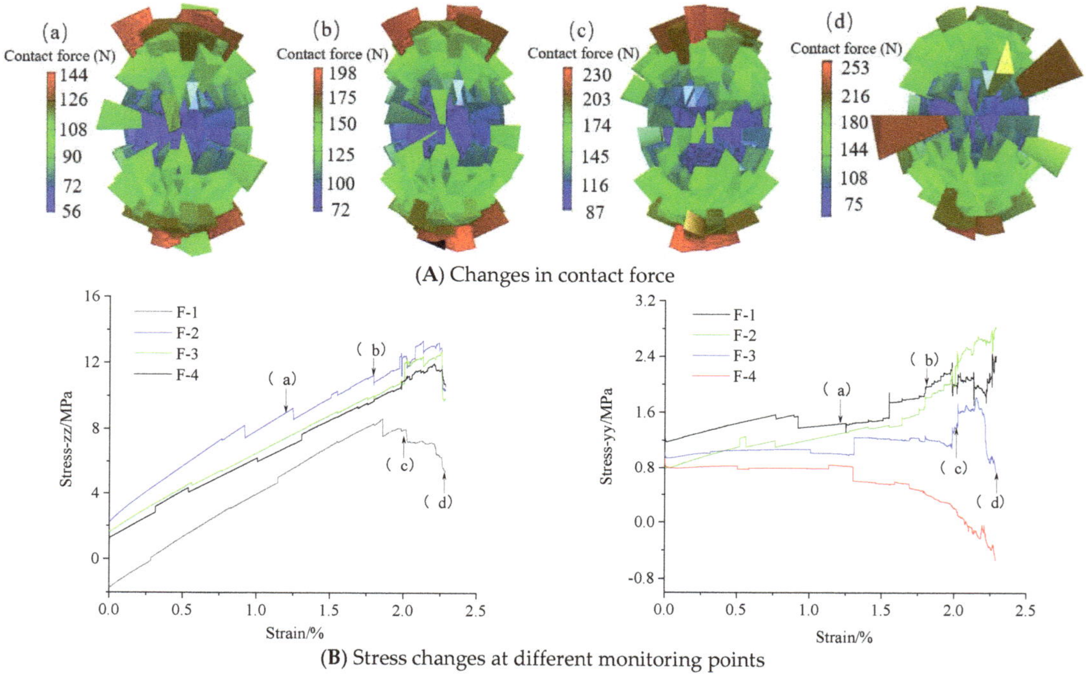

Figure 8. Changes in model internal stress and contact force. Panel (**A**) shows the internal contact force at different stages of the model. The warmer the color and the thicker the network, the greater the contact force. Panel (**B**) is the stress change at the monitoring point under the different stage corresponding to (**A**). (a–d) are the stages of crack generation, which are pore compaction stage, fracture expansion stage, peak stress stage and post peak stage, respectively.

4.1.2. AE Event Distribution and Field Verification

In the loading process, a total of 509 AE events were recorded. Figure 9 shows the T–K diagram of the MT obtained using the standard inversion method.

From Figure 9, at the four stages of loading, total of 15, 77, 65, and 417 AE events are recorded, respectively. The distribution of the AE events is relatively scattered, and most of them are distributed in the shear area. At stages (a) and (b), 15 and 77 events are generated, respectively, with magnitudes of between −6.68 and −5.3. The corresponding stresses at the end of these two stages are 5.32 MPa and 7.4 MPa, which are approximately 68% and 94% of peak stress, respectively. The AE events during this period are mainly dominated by the water pressure. At stages (c) and (d), the axial stress gradually reaches the peak strength of the coal. During this period, the effect of the axial stress is far greater than that of the water pressure, and the number of AE events sharply increases (mainly shear failure), which leads to instability of the sample.

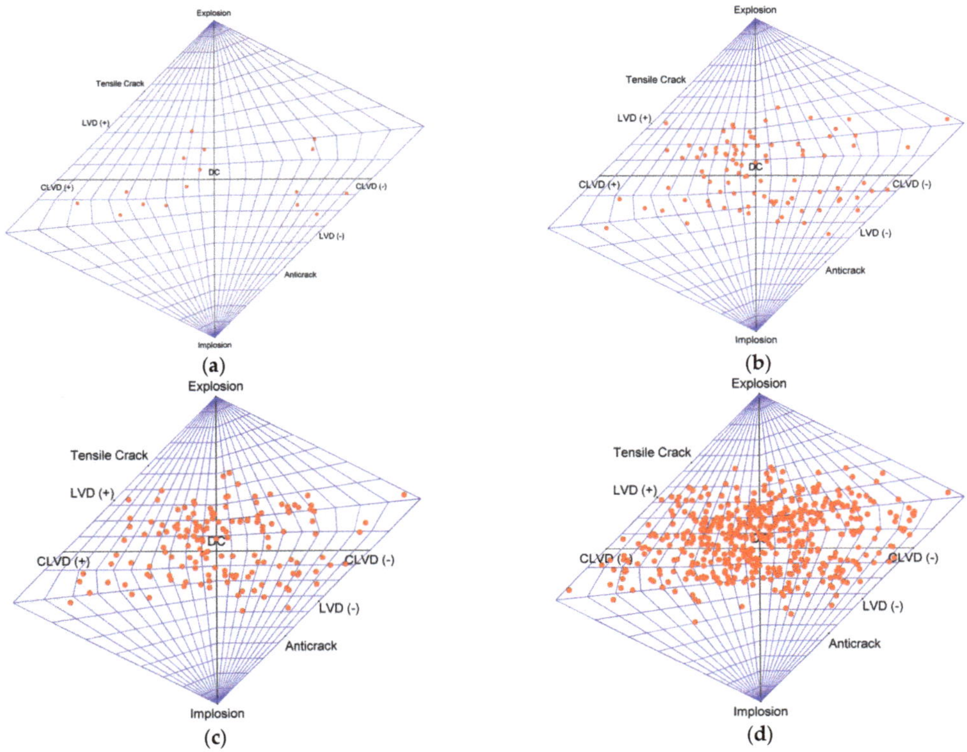

Figure 9. Hudson T–K locations of source models from (**a**–**d**).

4.1.3. Proportions of Source Type

To study the types of AE events in coal under the action of hydro-mechanical coupling, the number and proportion of tensile, shear, and implosion source types at different failure stages were quantitatively counted. The results are shown in Figure 10.

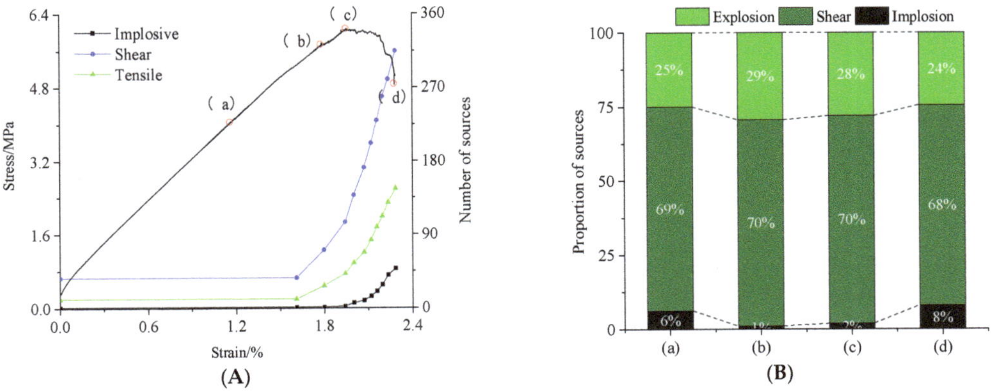

Figure 10. Nature of source versus time in uniaxial compression tests: (**A**) number and ratio of sources and (**B**) proportion of sources. (a–d) are the stages of crack generation, which are pore compaction stage, fracture expansion stage, peak stress stage and post peak stage, respectively.

From Figure 10, as the axial stress increases, the numbers of implosions and shears sharply increase; however, their proportions first increase and subsequently decrease. The entire damage is dominated by shear cracks. In stages (a) and (b), the proportions of shear and tension source types become larger, from 69% and 25% to 70% and 29%, respectively. Tensile failure increases in a slightly larger proportion than shear cracks. Observing the AE distribution shows that the damage is mainly concentrated in the lower part of the model. The particles in the water pressure area are subjected to tension. Simultaneously, tensile failure is prone to occur under axial stress. With the increase in the axial stress, the influence of the water pressure gradually weakens. Therefore, in stages (c) and (d), the proportion of tensile source type slightly decreases, whereas that of implosion source type increases to 8%. In summary, the water pressure mainly affects the early stage of loading and is primarily influenced by the axial stress in the later failure process.

4.1.4. Energy Change

Coal failure process is accompanied by generation, accumulation, dissipation, and release of energy, and the evolution of energy is closely consistent with the expansion of internal cracks in coal. Using the energy module of the PFC3D, the energy variation was recorded during coal failure process.

From Figure 11, the kinetic energy tends to increase in the initial stage of loading. The main cause is the force unbalance between some particles under the action of the axial pressure, which destroys the linear parallel bonds between them. There is no increase in the slip energy within a certain period. When the vertical stress of the coal body reaches approximately 50% peak stress, the kinetic and slip energies gradually increase. When the strain reaches 2%, the model slips overall. At this time, the slip and kinetic energies gradually reach maximum. The total strain energy is consistent with the overall change trend of the stress–strain curve.

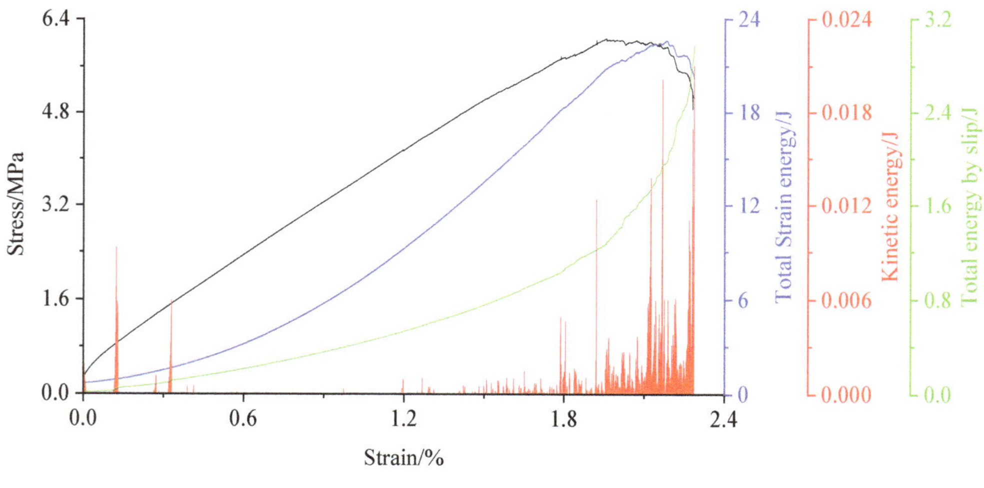

Figure 11. Various energy changes in coal failure process.

4.2. *Effects of Water and Confining Pressures on Coal*

The confining and water pressures were set as 0.5, 1.5, and 2.5 MPa, respectively. The specific settings are summarized in Table 2.

Table 2. Setting of confining and water pressures.

Model	P-1	P-2	P-3	P-4	P-5	P-6	P-7	P-8	P-9	P-10	P-11	P-12
Confining pressure/MPa	0.5	0.5	0.5	0.5	1	1	1	1	1.5	1.5	1.5	1.5
Water pressure /MPa	0	0.5	1.5	2.5	0	0.5	1.5	2.5	0	0.5	1.5	2.5

4.2.1. Failure Characteristics

A total of 12 models were established, and each model number and its corresponding parameters are listed in Table 2. To analyze the internal damage of coal under the effects of different water and confining pressures, the AE events distribution of the different models were examined, and the results are shown in Figure 12.

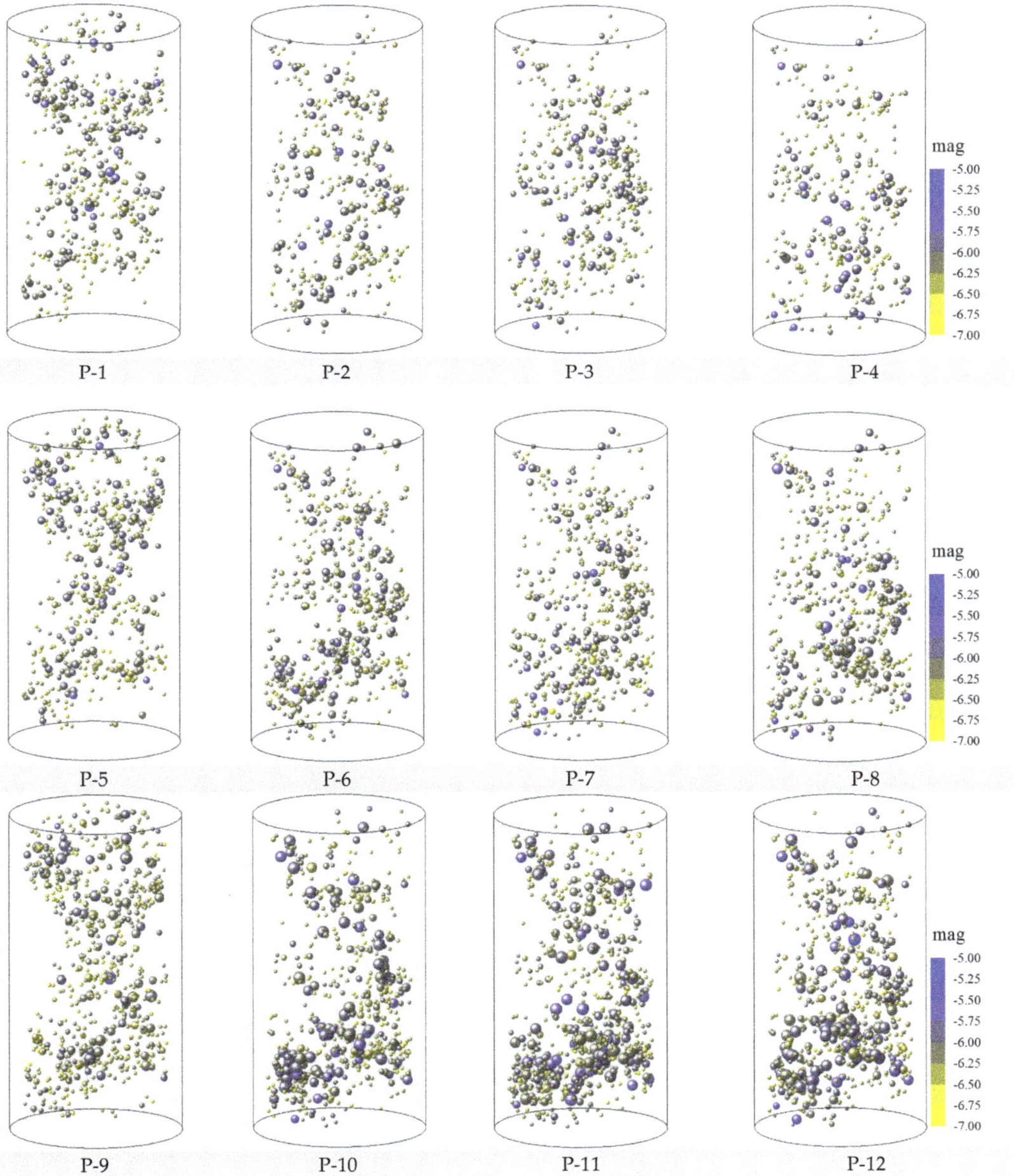

Figure 12. Source distributions of different models at different stages.

From Figure 12, it is seen that the confining and water pressures significantly affect the failure pattern of coal. When the model is damaged, the AE events are mainly concentrated in the water pressure area. Based on the comparative analysis of P-1–P-4, P-5–P-8, and P-9–

P-12, as the water pressure increases with the same confining pressure, the internal damage and high-energy events increase, and most of the AE events are mainly concentrated in the lower part of the model. When the water pressure is kept constant, the model accumulates more elastic energy before it reaches failure with increasing confining pressure. When the model is damaged, more high-energy events are generated.

From Figure 13, the force of the model has a large anisotropy under the effects of the confining, water, and axial pressures. As the water pressure increases, the number of contacts in the lower part is much smaller than that in the upper part, mainly owing to the greater damage in the lower part. As the confining pressure increases, the number of contacts increases significantly. The maximum number of contacts is increased from 28 to 29, and the influence of the water pressure is gradually weakened, indicating that the strengthening of a support can help protect the stability of the roadway.

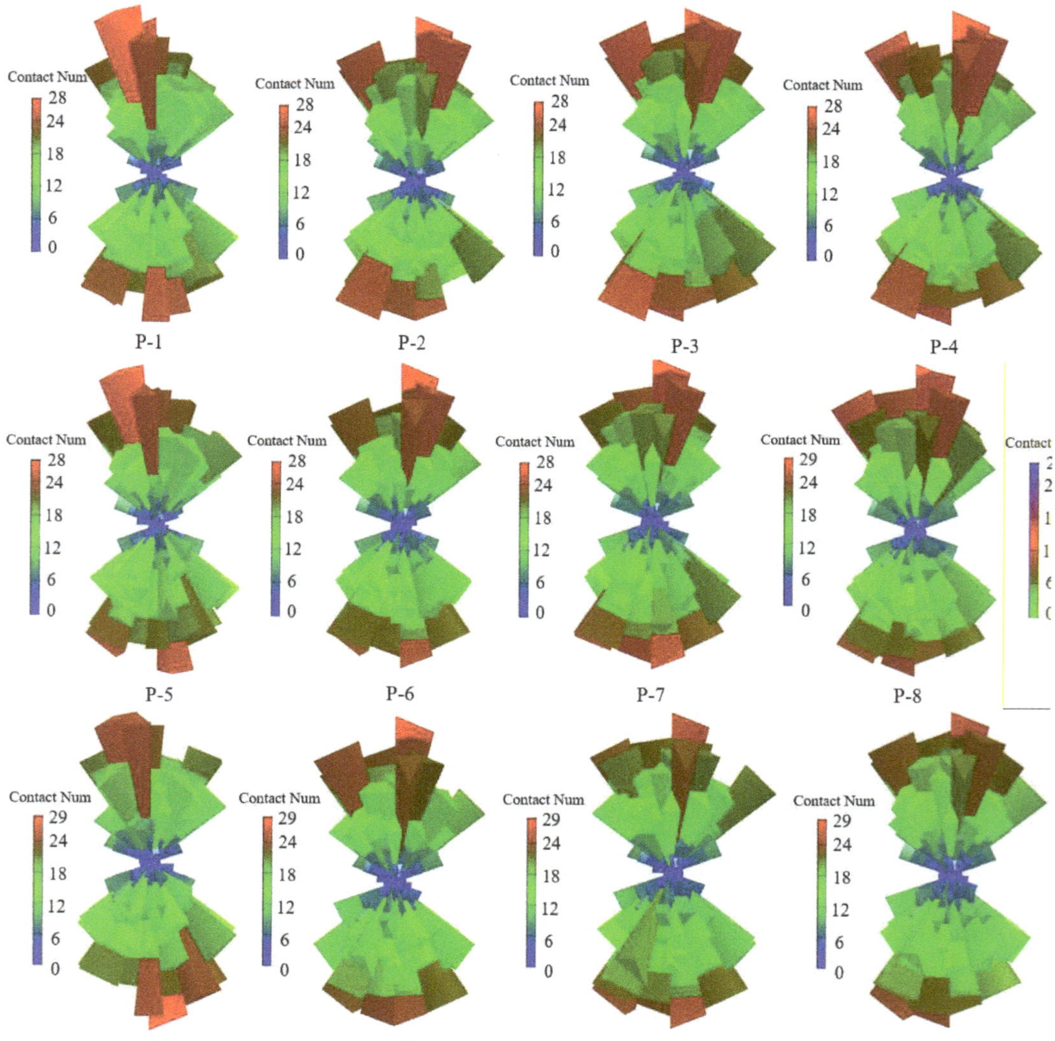

Figure 13. Number of contacts in different orientations.

4.2.2. Relationship between AE and Stress–Strain Curve

To analyze the effects of different water and confining pressures on the overall strength of the model, the stress–strain curves of the P-1–P-12 models were examined, and are shown in Figure 14.

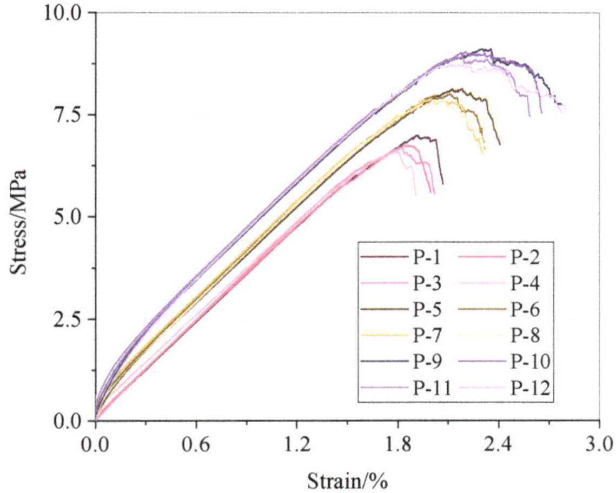

Figure 14. Stress–strain curves of different models.

Based on Figure 14, when the water pressure remains constant, a large confining pressure implies a high peak strength of the sample and a large deformation before failure. When the confining pressure is constant, a large water pressure implies a low peak strength and a small axial deformation before the peak; however, the slope increases, which is consistent with the analysis results of Equation (3).

Figure 15 shows the changes in the numbers of shear and tensile cracks during the failure of different models. The number of shear cracks is approximately twice that of tensile cracks. As the water pressure increases before uniaxial compression, more cracks are generated during the seepage process, and the number of shear cracks is greater than that of tensile cracks. In the uniaxial compression process, when the confining pressure is small, the water pressure has a significant effect on the number of AE events ($N_{P\text{-}2} \approx 1.23N_{P\text{-}3} \approx 2.01N_{P\text{-}4}$). When the confining pressure is high, the effect of the water pressure significantly weakens ($N_{P\text{-}10} \approx 1.05N_{P\text{-}11} \approx 1.4N_{P\text{-}12}$), and the degree of weakening significantly strengthens.

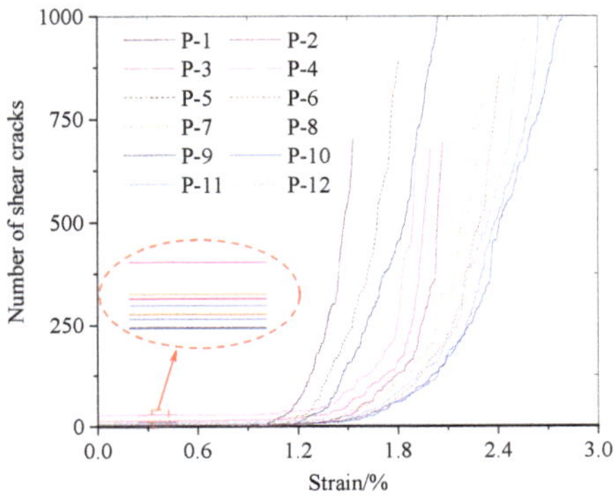

Figure 15. *Cont.*

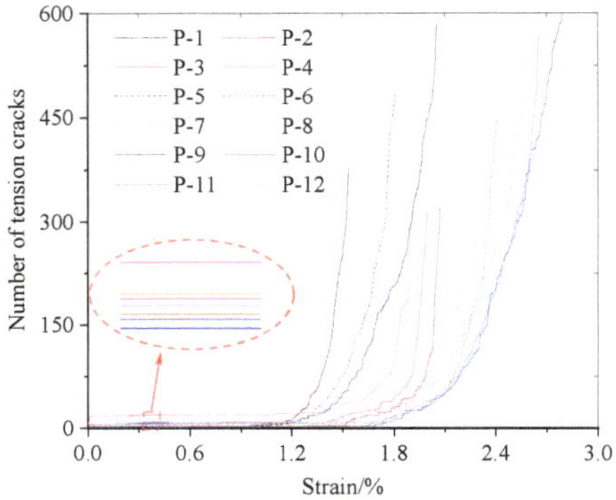

Figure 15. Development of cracks in coal body failure process in different models.

4.2.3. Frequency–Magnitude Curve and *b* Value

Smaller earthquakes occur more frequently than large earthquakes [43]. This trend can be expressed by the magnitude–frequency relationship and evaluated by *b* value.

Figure 16 shows that there is a good linear relationship between the water and confining pressures and the *b* value. If the confining and water pressures are high, the value of *b* is small, which proves that the large-scale cracks inside the model gradually increase and rapidly and unsteadily expand, and the AE distribution becomes more diffusing. Therefore, as the water and confining pressures increase, the internal stress concentration of the model rises. When cracks occur inside the model, microcracks are more likely to grow and penetrate large-scale cracks. Simultaneously, a high confining pressure implies a large number of AE events inside the rock mass. Thus, the water and confining pressures have remarkable impacts on the AE of the coal body.

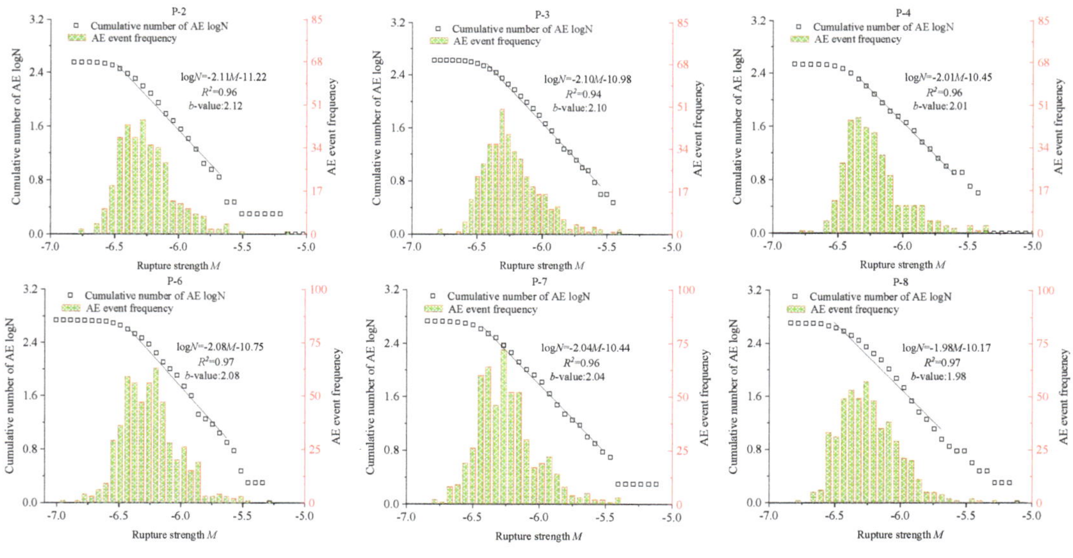

Figure 16. *Cont.*

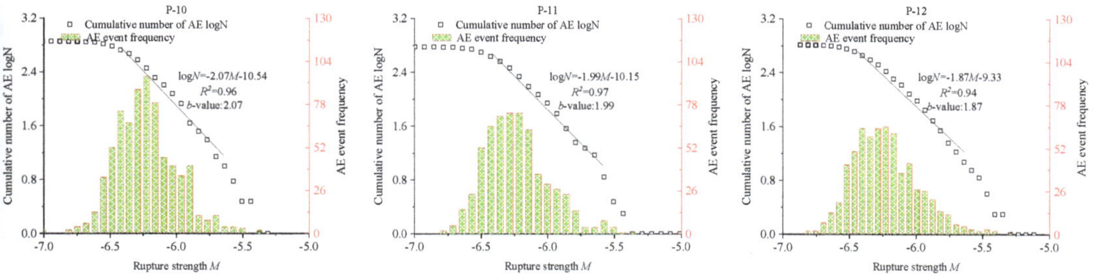

Figure 16. Frequency–amplitude curve and b value.

4.3. Burst Tendency of Coal under Hydro-Mechanism Coupling

There is a certain relationship between the dissipated and strain energies during the uniaxial compression of coal, which is shown in Figure 17.

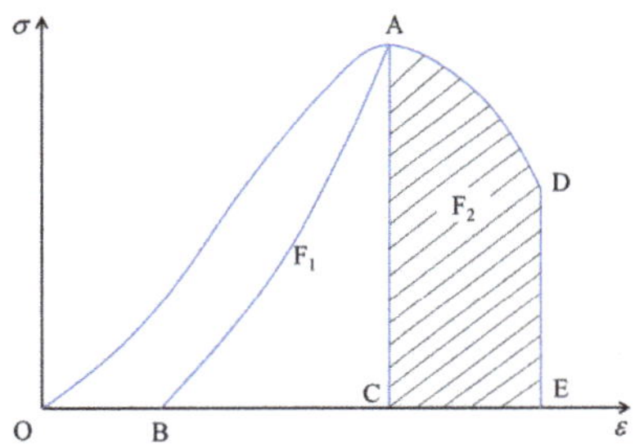

Figure 17. Relationship between dissipated and strain energies on stress–strain curve during uniaxial compression of coal. O, A and D correspond to the starting, peak stress and post peak value of the stress-strain curve respectively. B is the strain value of the model after unloading. C and D are the strain values corresponding to the peak and post peak stresses, respectively.

A schematic of the energy burst tendency is shown in Figure 17, where F1 represents the difference between the elastic energy accumulated before the peak and the dissipated energy consumed during the plastic deformation and crack development of the coal. F2 represents the energy consumed when the coal is damaged after the peak, i.e., F1/F2 is an energy burst tendency indicator of the coal. In the energy module part of the PFC3D 5.0 software (Itasca Consulting Group, Inc., Minneapolis, MN, USA), the total strain energy includes the strain energy (E_c) and the bond strain energy (E_{pb}), and the total dissipated energy includes the frictional energy (E_μ) and the kinetic energy (E_k). Among them, the slip energy remains equal to zero until the bond breaks. It can be seen that the particle flow can monitor the change in the energy in the deformation and failure process of the model in real time. Therefore, in the full stress–strain curve, the elastic and dissipation energies at any time can be expressed in PFC3D as

$$\begin{cases} U_i^e = E_{pb} + E_c \\ U_i^d = E_f + E_k \end{cases} \quad (10)$$

After deducting the energy consumed by the plastic deformation of the rock and crack development, the burst tendency of the coal body can be determined using the energy method as follows:

$$\begin{cases} F_1 = U_{peak}^{\mathrm{e}} + U_{peak}^{d} \\ F_2 = U_{peak}^{\mathrm{e}} + U_{residual}^{e} \\ K_E = F_1/F_2 \end{cases} \quad (11)$$

Figure 18 shows the calculation results of the burst tendency of coal under different confining and water pressures. When the water and confining pressures are 0 MPa and 1.5 MPa, respectively, the coal body has the lowest burst tendency, and K_E is 2.4. When the water and confining pressures are 1.5 MPa and 0 MPa, respectively, the coal body has the highest burst tendency, and K_E is 2.67. From the K_E distribution, when the water pressure remains constant, as the confining pressure increases, K_E becomes smaller. When the confining pressure remains constant, as the water pressure increases, K_E becomes larger. Therefore, the burst tendency under different conditions can be accurately determined.

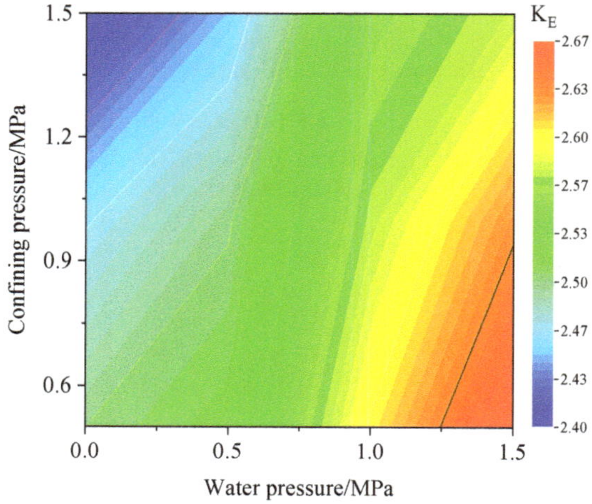

Figure 18. Change in K_E under different water and confining pressures.

5. Onsite Investigations

The 1314 working face is located at the level of −790 m; the mining seam is the #3 coal with the thickness of 1.5–3.7 m, and the buried depth is 648–700 m. The roadway is arranged along the roof of the #3 coal seam. The coal seam structure of the working face is simple, with an inclination angle of 13°–26°.

On 11 September 2018, a water inrush accident occurred in this working face, and the coal was immersed in a water pressure of approximately 3 MPa for approximately 100 days, which had a certain impact on the stability of the roadway under the influence of water. By counting MS events during soaking, the Hudson's source type is obtained according to focal mechanism inversion, as shown in Figure 19. There are a wide range of failure types in the coal, mainly shear and tensile cracks, and its failure mainly occurs in the type of shear cracks. The field observation results are consistent with the simulation conclusions.

In terms of forces, a shear crack can be expressed by two perpendicular force dipoles with zero angular momentum. Therefore, the shear failure source is usually called DC. In order to intuitively show the internal damage form of the coal, several MS events around the working face were selected for detailed analysis. Three types of beachballs, full, deviatoric, and DC, were drawn, as shown in Figure 20. It is found that the shear component accounts for a large proportion in the above MS events, and it is concluded that the force of the shear component is the main factor leading to the failure of the coal. In summary, the failure of coal under water pressure is mainly dominated by shear source, mixed with a small amount of other components, which is strictly consistent with the simulation conclusions. From Figure 21, several fault plane solutions were solved, including two normal fault, three reverse fault, and one normal oblique slip, most of which face the interior of the working face.

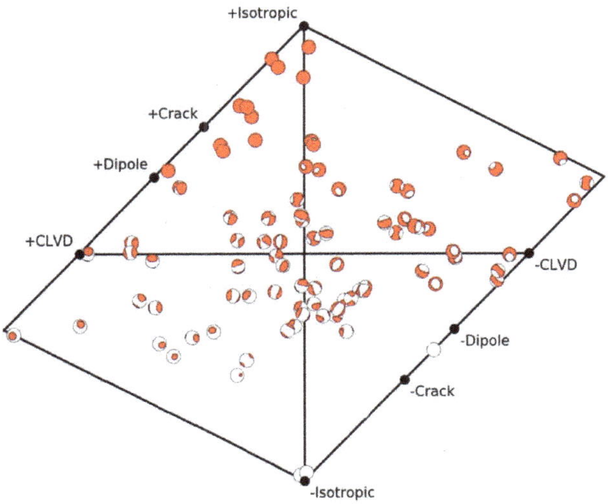

Figure 19. Hudson's source type of coal failure under water pressure.

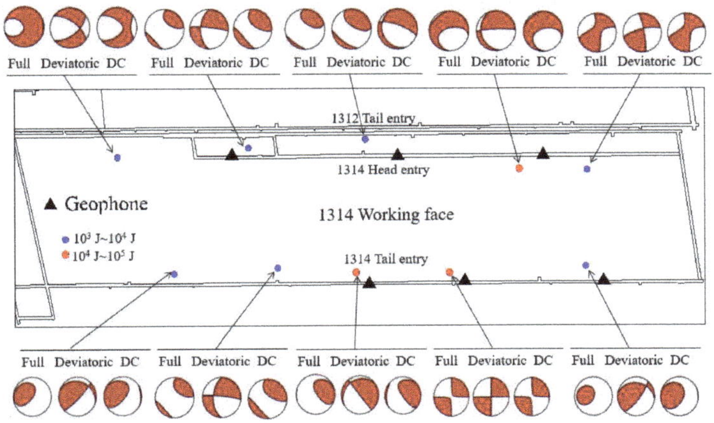

Figure 20. Beachballs in the 1314 working face.

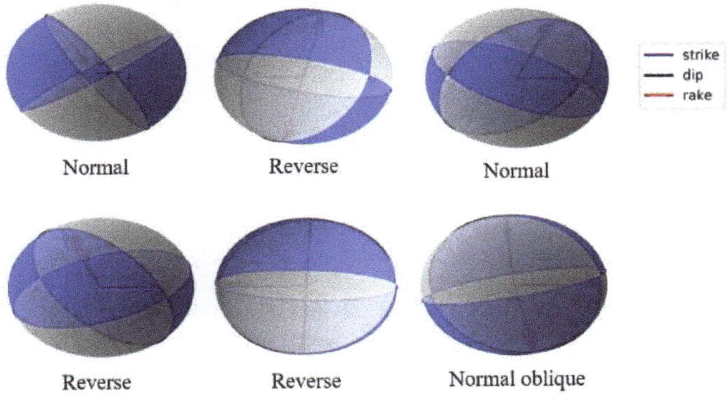

Figure 21. Fault plane solutions of the 1314 working face.

6. Conclusions

(1) Affected by hydro-mechanical coupling, the damage degree of coal in the flooded part is relatively large, and its damage mainly occurs as shear cracks; this is verified by field observation. The closeness of coal to the water injection area implies a high vertical stress.

(2) Monitoring of the energy changes shows that the kinetic and slip energies increase slightly in the initial stage of loading. When the peak stress is greater than 50%, a sharp energy increase occurs, and the increase rate becomes increasingly high.

(3) There is a good linear relationship between the water pressure, confining pressure, and *b* value. With increasing water and confining pressures, the damage degree and AE energy inside the model increase, the value of *b* reduces, and most of the AE events are mainly concentrated in the water injection area. When the confining pressure is low, the water pressure has a significant effect on the number of AE events; otherwise, the effect of water pressure gradually weakens.

(4) When the water pressure is constant, as the confining pressure increases, K_E decreases. When the confining pressure is kept constant, as the water pressure increases, K_E increases.

Author Contributions: Conceptualization, Y.-M.W.; Data curation, J.-H.W.; Investigation, Z.-W.Z. and H.-F.C.; Methodology, C.-P.L.; Software, J.-F.S. All authors have read and agreed to the published version of the manuscript.

Funding: This research was funded by the National Natural Science Foundation of China (51574225).

Informed Consent Statement: Written informed consent has been obtained from the patient(s) to publish this paper.

Data Availability Statement: The data in this manuscript are available from the authors.

Acknowledgments: We gratefully wish to acknowledge the collaborative funding support from the National Natural Science Foundation of China (51574225). In addition, the data in this manuscript are available from the authors.

Conflicts of Interest: The authors declare that there are no conflicts of interest regarding the publication of this paper.

References

1. Ma, D.; Miao, X.; Bai, H.; Huang, J.; Pu, H.; Wu, Y.; Zhang, G.; Li, J. Effect of mining on shear sidewall groundwater inrush hazard caused by seepage instability of the penetrated karst collapse pillar. *Nat. Hazards.* **2016**, *82*, 73–93. [CrossRef]
2. Yu, Y.; Cheng, W.; Zhou, G. Numerical simulation study on permeability properties of coal seam during high pressure water injection process. *Electron. J. Geotech. Eng.* **2014**, *19*, 10315–10326.
3. Kou, M.M.; Liu, X.R.; Wang, Z.Q.; Tang, S.D. Laboratory investigations on failure, energy and permeability evolution of fissured rock-like materials under seepage pressures. *Eng. Fract. Mech.* **2021**, *247*, 107694. [CrossRef]
4. Grgic, D.; Amitrano, D. Creep of a porous rock and associated acoustic emission under different hydrous conditions. *J. Geophys. Res. Solid Earth* **2009**, *114*, B10201. [CrossRef]
5. Li, H.; Li, H.; Gao, B.; Jiang, D.; Feng, J. Study of Acoustic Emission and Mechanical Characteristics of Coal Samples under Different Loading Rates. *Shock. Vib.* **2015**, *2015*, 1–11. [CrossRef]
6. Wen, Z.; Wang, X.; Chen, L.; Lin, G.; Zhang, H. Size Effect on Acoustic Emission Characteristics of Coal-Rock Damage Evolution. *Adv. Mater. Sci. Eng.* **2017**, *2017*, 1–8. [CrossRef]
7. Chen, X.; Yu, J.; Tang, C.A.; Li, H.; Wang, S. Experimental and Numerical Investigation of Permeability Evolution with Damage of Sandstone Under Triaxial Compression. *Rock Mech. Rock Eng.* **2017**, *50*, 1529–1549. [CrossRef]
8. Hui, Y.; Hfda, B.; Jz, C. Effects of filling fluid type and composition and joint orientation on acoustic wave propagation across individual fluid-filled rock joints. *Int. J. Rock Mech. Min. Sci.* **2020**, *128*, 104248.
9. Gardner, M.; Sitar, N. Modeling of Dynamic Rock–Fluid Interaction Using Coupled 3-D Discrete Element and Lattice Boltzmann Methods. *Rock Mech. Rock Eng.* **2019**, *52*, 5161–5180. [CrossRef]
10. Chen, Z.; Wang, M. Pore-scale modeling of hydromechanical coupled mechanics in hydrofracturing process. *J. Geophys. Res. Solid Earth* **2017**, *122*, 3410–3429. [CrossRef]
11. Cai, Y.; Chen, X.; Yu, J.; Zhou, J. Numerical Study on the Evolution of Mesoscopic Properties and Permeability in Sandstone Under Hydromechanical Coupling Conditions Involving Industrial Internet of Things. *IEEE Access* **2018**, *6*, 11804–11815. [CrossRef]

12. Liu, S.; Zhu, Q.; Shao, J. Deformation and mechanical properties of rock: Effect of hydromechanical coupling under unloading conditions. *Bull. Eng. Geol. Environ.* **2020**, *79*, 1–18. [CrossRef]
13. Wang, Q.; Hu, X.; Zheng, W.; Li, L.; Zhou, C.; Ying, C.; Xu, C. Mechanical Properties and Permeability Evolution of Red Sandstone Subjected to Hydro-mechanical Coupling: Experiment and Discrete Element Modelling. *Rock Mech. Rock Eng.* **2021**, *54*, 2403–2405. [CrossRef]
14. Yao, Q.; Wang, W.; Zhu, L.; Xia, Z.; Tang, C.; Wang, X. Effects of moisture conditions on mechanical properties and AE and IR characteristics in coal–rock combinations. *Arab. J. Geosci.* **2020**, *13*, 1–15. [CrossRef]
15. Wang, Z.H.; Ren, W.G.; Tan, Y.L.; Konietzky, H. Experimental and Numerical Study on Hydromechanical Coupled Deformation Behavior of Beishan Granite considering Permeability Evolution. *Geofluids* **2020**, *2020*, 1–14. [CrossRef]
16. Zhang, S.; Zhang, D.; Wang, Z.; Chen, M. Influence of Stress and Water Pressure on the Permeability of Fissured Sandstone Under Hydromechanical Coupling. *Mine Water Environ.* **2018**, *37*, 774–785. [CrossRef]
17. Martyushev, D.A.; Galkin, S.V.; Shelepov, V.V. The Influence of the Rock Stress State on Matrix and Fracture Permeability under Conditions of Various Lithofacial Zones of the Tournaisian–Fammenian Oil Fields in the Upper Kama Region. *Mosc. Univ. Geol. Bull.* **2019**, *74*, 573–581. [CrossRef]
18. Riabokon, E.; Turbakov, M.; Popov, N.; Kozhevnikov, E.; Poplygin, V.; Guzev, M. Study of the Influence of Nonlinear Dynamic Loads on Elastic Modulus of Carbonate Reservoir Rocks. *Energies* **2021**, *14*, 8559. [CrossRef]
19. Zhang, Q.; Wang, L. Research Status and Prospect of Rock Acoustic Emission Technology. *Appl. Mech. Mater.* **2012**, *170*, 461–464. [CrossRef]
20. Zhang, Q.; Wang, L. Research Status and Prospect of Rock Mass Acoustic Emission Instrument. *Adv. Mater. Res.* **2013**, *671*, 280–283. [CrossRef]
21. Zhang, Y.H.; Li, Y.; Zhang, Y. The Research of Shaoguo Tunnel Failure Characteristics of Triaxial and Acoustic Emission Characteristics. *Appl. Mech. Mater.* **2014**, *501*, 1757–1760. [CrossRef]
22. Shimizu, H.; Murata, S.; Ishida, T. The distinct element analysis for hydraulic fracturing in hard rock considering fluid viscosity and particle size distribution. *Int. J. Rock Mech. Min. Sci.* **2011**, *48*, 712–727. [CrossRef]
23. Jia, Z.; Ren, L.; Liu, Q.; Peng, Y.; Xu, D.; Zha, E. Influence of water-soaking time on the acoustic emission characteristics and spatial fractal dimensions of coal under uniaxial compression. *Therm. Sci.* **2017**, *21*, 327–334. [CrossRef]
24. Ye, Z.; Ghassemi, A. Investigation of Microseismicity and Permeability Evolution in Shale Fractures during Stimulation. *SPE Prod. Oper.* **2020**, *35*, 0797–0808. [CrossRef]
25. Makhnenko, R.Y.; Ge, C.; Labuz, J.F. Localization of deformation in fluid-saturated sandstone. *Int. J. Rock Mech. Min. Sci.* **2020**, *134*, 104455. [CrossRef]
26. Li, J.G.; Liu, H. Application Conditions on Acoustic Emission (AE) Technique Monitoring Coal and Rock Dynamic Disasters in Mines. *Adv. Mater. Res.* **2012**, *413*, 235–240. [CrossRef]
27. Li, N.; Sun, W.; Huang, B.; Chen, D.; Zhang, S.; Yan, M. Acoustic Emission Source Location Monitoring of Laboratory-Scale Hydraulic Fracturing of Coal Under True Triaxial Stress. *Nat. Resour. Res.* **2021**, *30*, 2297–2315. [CrossRef]
28. Chitrala, Y.; Moreno, C.; Sondergeld, C.; Rai, C. An experimental investigation into hydraulic fracture propagation under different applied stresses in tight sands using acoustic emissions. *J. Pet. Sci. Eng.* **2013**, *108*, 151–161. [CrossRef]
29. Chen, L.; Liu, J.F.; Wang, C.P. Characterization of damage evolution in granite under compressive stress condition and its effect on permeability. *Int. J. Rock Mech. Min. Sci.* **2014**, *71*, 340–349. [CrossRef]
30. Makoto, N.; Chen, Y.; Yamamoto, K.; Morishige, Y.; Imakita, K.; Tsutumi, N.; Kawakata, H.; Ishida, T.; Tanaka, H.; Arima, Y.; et al. Tensile-dominant fractures observed in hydraulic fracturing laboratory experiment using eagle ford shale. *Geophys. J. Int.* **2020**, *222*, 769–780.
31. Lu, Y.; Wang, L. Numerical simulation of mining-induced fracture evolution and water flow in coal seam floor above a confined aquifer. *Comput. Geotech.* **2015**, *67*, 157–171. [CrossRef]
32. Jiang, Z.; Li, Q.; Hu, Q.; Liang, Y.; Xu, Y.; Liu, L.; Wu, X.; Li, X.; Wang, X.; Hu, L.; et al. Acoustic emission characteristics in hydraulic fracturing of stratified rocks: A laboratory study—ScienceDirect. *Powder Technol.* **2020**, *371*, 267–276. [CrossRef]
33. Stanchits, S.; Surdi, A.; Gathogo, P.; Edelman, E.; Suarez-Rivera, R. Onset of Hydraulic Fracture Initiation Monitored by Acoustic Emission and Volumetric Deformation Measurements. *Rock Mech. Rock Eng.* **2014**, *47*, 1521–1532. [CrossRef]
34. Li, B.; Liu, J.; Bian, K.; Ai, F.; Hu, X.; Chen, M.; Liu, Z. Experimental study on the mechanical properties weakening mechanism of siltstone with different water content. *Arab. J. Geosci.* **2019**, *12*, 1–14. [CrossRef]
35. Chen, Y. Permeability Evolution in Granite Under Compressive Stress Condition. *Geotech. Geol. Eng.* **2017**, *36*, 641–647. [CrossRef]
36. He, M.; Li, Q.; Li, X.; Xu, L.; Kühn, M. Hydromechanical behaviors of andesite under different stress states during fluid injection. *J. Rock Mech. Geotech. Eng.* **2021**, *13*, 727–744. [CrossRef]
37. Aki, K.; Richards, P.G. *Quantitative Seismology*, 2nd ed.; University Science Books: Sausalito, CA, USA, 2002; ISBN 0-935702-96-2.
38. Hudson, J.A.; Pearce, R.G.; Rogers, R.M. Source type plot for inversion of the moment tensor. *J. Geophys. Res.* **1989**, *94*, 765–774. [CrossRef]
39. Zhou., Z.; Zhou, J.; Zhao, Y.; Chen, L.; Li, C. Microscopic Failure Mechanism Analysis of Rock Under Dynamic Brazilian Test Based on Acoustic Emission and Moment Tensor Simulation. *Front. Phys.* **2021**, *8*, 592483. [CrossRef]
40. Zhao, Y.; Zhao, G.; Zhou, J.; Ma, J.; Cai, X. Failure mechanism analysis of rock in particle discrete element method simulation based on moment tensors. *Comput. Geotech.* **2021**, *136*, 104215. [CrossRef]

41. Itasca Consulting Group Inc. *PFC2D (Particle Flow Code in 2 Dimensions)*; ICG: Minneapolis, MN, USA, 1999.
42. Hanks, T.C.; Kanamori, H. A moment magnitude scale. *J. Geophys. Res.* **1979**, *84*, 2348–2350. [CrossRef]
43. Shearer, P.M. *Introduction to Seismology*; Birkhauser: Basel, Switzerland, 1979.

mathematics

MDPI

Article

Instability of Viscoelastic Liquid Sheets in a Transverse Electric Field

Lu Niu [1] and Xiangdong Deng [2,*]

1 School of Mathematics and Statistics, Beijing Jiaotong University, Beijing 100044, China
2 School of Aerospace Engineering, Beijing Institute of Technology, Beijing 100181, China
* Correspondence: 3120205056@bit.edu.cn

Abstract: The temporal linear instability of a viscoelastic liquid sheet moving around an inviscid gas in a transverse electrical field is analyzed. The fluid is described by the leaky dielectric model, which is more complex than existing models and enables a characterization of the liquid electrical properties. In addition, the liquid is assumed to be viscoelastic, and the dimensionless dispersion relation of the sinuous and varicose modes between the wavenumber and the temporal growth rate can be derived as a 3×3 matrix. According to this relationship, the effects of the liquid properties on the sheet instability are determined. The results suggest that, as the electrical Euler number and the elasticity number increase and the time constant ratio decreases, the sheet becomes more unstable. Finally, an energy budget approach is adopted to investigate the instability mechanism for the sinuous mode.

Keywords: linear instability analysis; viscoelastic liquid sheet; leaky dielectric model; energy approach

MSC: 76E25; 76A05; 76E15

1. Introduction

Sheet instabilities have been extensively studied in recent years because of their vital importance in the fields of mathematical physics and industry [1,2]. In scientific fields, such instabilities are classic issues in solving differential equations that are especially worthy for studying the behavior of the liquid sheet in an electrical field. Moreover, the instabilities of a liquid sheet are regularly encountered in spray combustion processes, ultra-fine water mist fire suppression systems [3], gas turbines, inkjet printing, and even liquid rocket engines [4].

Squire [5] conducted pioneering work on sheet instabilities by exploring the instability of a thin inviscid sheet surrounded by still air. Hagerty and Shea [6] developed this theory through linear analysis and concluded that there existed two modes, namely, the sinuous mode and the varicose mode. On this foundation, numerous researchers became interested in sheet electrodynamic behavior, and the stability of the sheet in an electric field was extensively investigated. In earlier studies, the sheets were considered as perfect conductors or dielectrics. Melcher and Schwarz [7,8] described how the disturbed surface waves of the viscous liquid sheet, which was treated as a perfectly insulating fluid, propagated along the lines of electric field intensity. They showed that the dominant effect of the charge relaxation was to improve the instability. El-Sayed [9] extended the analysis of Melcher [8] to the case of a fluid moving in the same direction as an air stream. It was concluded that the aerodynamic force reduced the stability when the Weber number was less than some critical value, whereas the electric force increased the stability. Under the perfect conductor model, Yang et al. [10] conducted an electrified viscoelastic liquid sheet injected into a dielectric stationary ambient gas.

In 1969, Taylor and Melcher [11] published a breakthrough paper in which they proposed a more accurate electric–leaky dielectric model (the Taylor–Melcher model).

Citation: Niu, L.; Deng, X. Instability of Viscoelastic Liquid Sheets in a Transverse Electric Field. *Mathematics* **2022**, *10*, 3488. https://doi.org/10.3390/math10193488

Academic Editor: Vasily Novozhilov

Received: 11 August 2022
Accepted: 20 September 2022
Published: 24 September 2022

Saville [12] comprehensively summarized the basic equations of this model and, through the analysis of numerous experimental results, concluded that the theory was consistent with the practice. Research in this field then flourished. Cimpeanu [13] theoretically examined the classical Rayleigh–Taylor instability of thin films in a horizontal electric field, and Savettaseranee [14] examined the competition among surface tension, van der Waals, viscous, and electrically induced forces when a sheet was placed in an electric field parallel to its velocity. Moreover, Tilly et al. [15] concentrated on the nonlinear stability of an inviscid sheet between two electrodes. They found that the electric field led to a nonlinear stability for the sheet, delaying the formation of its singularity.

In fact, for many industrial and commercial applications, such as atomization and liquid rocket engines, the fluid of the sheet is non-Newtonian and can be characterized by viscoelastic properties. Therefore, it is necessary to understand the instability and breakup of viscoelastic sheets. Liu [16] performed a linearized stability analysis in which two-dimensional non-Newtonian liquid sheets moved in an inviscid gaseous environment. It was found that non-Newtonian liquid sheets had a higher growth rate than Newtonian liquid sheets for both symmetric and antisymmetric disturbances. Brenn et al. [17] extended this analysis to three-dimensional disturbances. Jia et al. [18] manipulated the linear temporal instability of viscoelastic planar liquid sheets in the presence of gas velocity oscillations. They found that the absence of shear viscosity, the stress relaxation time, and the deformation retardation time all affected the unstable regions.

To date, there have been few studies considering the instability of a viscoelastic sheet in an electric field, especially those that simulated the liquid sheet using the leaky dielectric model. The present study focused on the linear temporal instability of viscoelastic liquid sheets subjected to a transverse electric field and explained the physical mechanism using an energy approach. The influence of various parameters related to the non-Newtonian characteristics (the elasticity number and the time constant ratio) of the fluid on the stability of the jet is considered in detail. This research can better understand the mechanism of liquid film breaking. It has important academic value for science and engineering applications. In practical engineering applications, such as liquid rocket engines, most propellants are non-Newtonian fluids. Our contributions were important for two reasons. Firstly, we considered the effect of electrical properties that were similar to real-world scenarios alongside the viscoelasticity of the liquid sheet, which has not previously been explored. Secondly, this study used energy analysis to quantitatively describe the contribution of various forces, including the electric force and the elastic force, to the sheet stability.

The remainder of this paper is organized as follows. Section 2 describes the theoretical model and derives the dispersion relation by solving the governing equations and boundary conditions. Section 3 presents an energy analysis to explain the mechanism of the instability. Section 4 analyzes the influence of physical parameters. Finally, the conclusions to this study are presented in Section 5.

2. Theoretical Model

In the case of this paper, the physical model is extracted from the atomization experiment. The liquid film is sprayed into the static air ($U_g = 0$) from the nozzle, and the electric field is applied to the external field to study the physical mechanism at the initial stage of the atomization process. As shown in Figure 1, a viscoelastic sheet moving through an inviscid gas in a transverse electric field is considered. The coordinates are chosen such that the x-axis is parallel to the direction of the liquid sheet flow and the y-axis is normal to the liquid sheet. The leaky dielectric model is employed to describe the electrical properties of the liquid with finite conductivity and permittivity, while the gas is assumed to be a perfect dielectric with the associated permittivity in a vacuum. The liquid sheet is characterized by a viscoelastic model that includes three main parameters: zero shear viscosity μ_0, the stress relaxation time λ_1, and the deformation retardation time λ_2. Gravity is ignored because the liquid film is very thin, and gravity is negligible compared to other forces for this model.

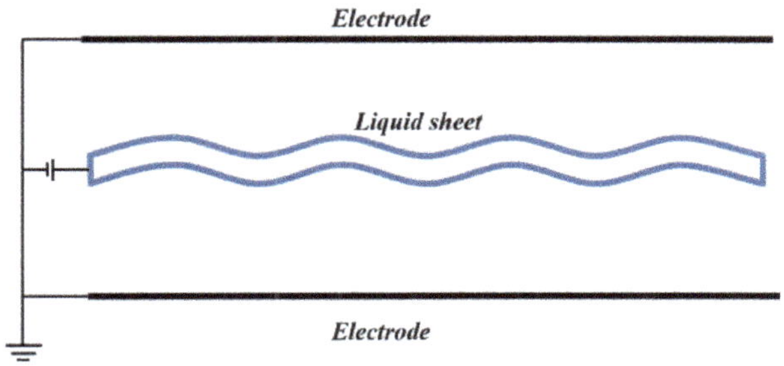

Figure 1. Schematic diagram of moving liquid sheet in an electric field.

In the present case, the relationship between the stress tensor and the velocity field can be described by the Oldroyd-B constant equation [19,20], which can be written as

$$\begin{aligned} &\boldsymbol{\tau} + \lambda_1 \left(\frac{\partial \boldsymbol{\tau}}{\partial t} + (\mathbf{v} \cdot \nabla)\boldsymbol{\tau} - (\nabla \mathbf{v}) \cdot \boldsymbol{\tau} - \boldsymbol{\tau} \cdot (\nabla \mathbf{v})^T \right) \\ &= \mu_0 \left(\dot{\boldsymbol{\gamma}} + \lambda_2 \left[\frac{\partial \dot{\boldsymbol{\gamma}}}{\partial t} + (\mathbf{v} \cdot \nabla)\dot{\boldsymbol{\gamma}} - (\nabla \mathbf{v}) \cdot \dot{\boldsymbol{\gamma}} - \dot{\boldsymbol{\gamma}} \cdot (\nabla \mathbf{v})^T \right] \right), \end{aligned} \tag{1}$$

where $\boldsymbol{\tau}$ is the extra stress tensor and $\dot{\boldsymbol{\gamma}} = \nabla \mathbf{v} + (\nabla \mathbf{v})^{\mathrm{T}}$ is the strain rate tensor.

Physical quantities are defined with appropriate values, and the corresponding scales are presented in Table 1. Both the liquid and the gas are assumed to be incompressible and the force of gravity is ignored.

Table 1. Scales of dimensionless parameters.

Dimensionless Parameter	Meaning	Scale
x	Distance parallel to the basic flow	a
y	Distance normal to the basic flow	a
t	Time	a/U
u	x-direction velocity	U
v	y-direction velocity	V
ϕ_g	Gas phase velocity	Ua
P_0	Basic flow pressure	$\rho_l U^2$
P	Liquid phase pressure	$\rho_l U^2$
$\boldsymbol{\tau}$	Liquid stress tensor	$\rho_l U^2$
η_0	Initial disturbance amplitude	a
d	Distance from electrode to sheet surface	a
λ_1	Stress relaxation time	a/U
λ_2	Deformation retardation	a/U
η_0	Initial disturbance amplitude	a

2.1. Governing Equations

The governing equations for the liquid and gas phases are given as follows. The mass conservation equations are

$$\nabla \cdot \mathbf{v} = 0, -a + \eta < y < a + \eta, \tag{2}$$

$$\nabla^2 \phi_g = 0, y < -a + \eta \text{ or} y > a + \eta, \tag{3}$$

where the liquid velocity vector is $\mathbf{v} = \begin{pmatrix} u & v & 0 \end{pmatrix}$.

The equations of momentum for the liquid phase are expressed as

$$\frac{\partial \mathbf{v}}{\partial t} + (\mathbf{v} \cdot \nabla)\mathbf{v} = -\nabla p + \nabla \cdot \boldsymbol{\tau}, -a + \eta < y < a + \eta, \tag{4}$$

$$p_g = -\rho \left[\frac{\partial \phi_g}{\partial t} + \frac{1}{2} (\nabla \phi_g)^2 \right], y > a + \eta \text{ or } y < -a + \eta. \tag{5}$$

2.2. Solutions for Electrical Field

An electrical potential function V is first introduced to satisfy the Laplace equations for the gas phase and the liquid phase, which are written as follows:

$$\nabla^2 V_g = 0, \tag{6}$$

$$\nabla^2 V_l = 0. \tag{7}$$

The strength of the electrical field can be expressed as

$$\mathbf{E} = -\nabla V. \tag{8}$$

The electric boundary conditions for the gas are simple:

$$\begin{aligned} V_g &= V_0, y = a + \eta, \\ V_g &= 0, y = a + d. \end{aligned} \tag{9}$$

The electric stress in the liquid phase $\mathbf{T}_l{}^{\mathbf{e}}$ and in the gas $\mathbf{T}_{\mathbf{g}}^{\mathbf{e}}$ can be obtained as follows:

$$\mathbf{T}_l{}^{\mathbf{e}} = \varepsilon \mathbf{E}_l \mathbf{E}_l - \frac{1}{2} \delta \varepsilon \mathbf{E}_l \cdot \mathbf{E}_l \mathbf{I}, \mathbf{T}_{\mathbf{g}}^{\mathbf{e}} = \mathbf{E}_{\mathbf{g}} \mathbf{E}_{\mathbf{g}} - \frac{1}{2} \delta \mathbf{E}_{\mathbf{g}} \cdot \mathbf{E}_{\mathbf{g}} \mathbf{I}, \tag{10}$$

where δ is the Kronecker delta and $\mathbf{I}$ is the identity matrix.

The boundary conditions for the liquid sheet are much more complicated. The model requires the continuity of the tangential electrical field strength, Gauss' law, and the conservation law of interface charge on the gas-to-liquid surface, which are expressed, respectively, as

$$\mathbf{n} \times (\mathbf{E}_l - \mathbf{E}_g) = 0, y = \pm a + \eta, \tag{11}$$

$$(\varepsilon_1 \mathbf{E}_l - \varepsilon_2 \mathbf{E}_g) \cdot \mathbf{n} = q_s, y = \pm a + \eta, \tag{12}$$

$$\frac{\partial q_s}{\partial t} + \mathbf{v} \cdot \nabla q_s - q_s \mathbf{n} \cdot (\mathbf{n} \cdot \nabla) \cdot \mathbf{v} - (\sigma \mathbf{E}_l) \cdot \mathbf{n} = q_s, y = \pm a + \eta. \tag{13}$$

2.3. Boundary Conditions

The kinematic boundary conditions for the liquid and gas phases are

$$\frac{\partial H}{\partial t} + v \cdot \nabla H = 0, y = \pm a + \eta, \tag{14}$$

$$\frac{\partial H}{\partial t} + \nabla \phi_g \cdot \nabla H = 0, y = \pm a + \eta, \tag{15}$$

where $H = y - \eta(x,t)$ and $H = 0$ denote the two gas-to-liquid interfaces.

Because the gas is inviscid and there exists a tangential electrical stress $\mathbf{T}_t^e$, the dynamic boundary condition parallel to the interfaces should be

$$(\mathbf{n} \cdot \boldsymbol{\tau}) \times \mathbf{n} = \mathbf{T}_{\mathbf{t}}^{\mathbf{e}}, y = \pm a + \eta, \tag{16}$$

where $\mathbf{n} = \nabla H / |\nabla H|$ is the interface unit normal vector.

The dynamic boundary condition normal to the interface is written as

$$-p+(\mathbf{n}\cdot\boldsymbol{\tau})\cdot\mathbf{n}+\frac{1}{We}(\nabla\cdot\mathbf{n})-\rho\left[\frac{\partial\phi_{gj}}{\partial t}+\frac{1}{2}\left(\frac{\partial\phi_{gj}}{\partial x}\right)^2+\frac{1}{2}\left(\frac{\partial\phi_{gj}}{\partial y}\right)^2\right]=\mathbf{T}_{\mathbf{t}}^{\mathbf{e}},\ y=\pm a+\eta. \tag{17}$$

2.4. Linear Stability Analysis

To solve Equations (1)–(17), we employ linear instability analysis [12,13] using the form of the normal mode. For a sufficiently small disturbance, the following equation is obtained:

$$\begin{aligned}&(\eta,u,v,p,\boldsymbol{\tau},\dot{\boldsymbol{\gamma}},\phi_g,p_g,V_g,V_l)=\\&[\eta(y),u(y),v(y),p(y),\boldsymbol{\tau}(y),\dot{\boldsymbol{\gamma}}(y),\phi_g(y),p_g(y),V_g(y),V_l(y)]\cdot\exp(ikx+\omega t),\end{aligned} \tag{18}$$

where k is the wavenumber and ω is the complex frequency $\omega=\omega_r+i\omega_i$, where ω_r represents the temporal growth rate of the disturbance and ω_i represents the disturbance frequency). Substituting Equation (18) into Equation (1) yields

$$\boldsymbol{\tau}+\lambda_1\left(\frac{\partial\boldsymbol{\tau}}{\partial t}+\frac{\partial\boldsymbol{\tau}}{\partial x}\right)=\frac{1}{Re}\left[\dot{\boldsymbol{\gamma}}_1+\lambda_2\left(\frac{\partial\dot{\boldsymbol{\gamma}}}{\partial t}+\frac{\partial\dot{\boldsymbol{\gamma}}}{\partial x}\right)\right], \tag{19}$$

$$Re_1=\frac{1+\lambda_1(ik+\omega)}{1+\lambda_2(ik+\omega)}Re, \tag{20}$$

where Re_1 is the effective Reynolds number for the viscoelastic fluid.

The two modes of liquid film are shown in Figure 2 below. The left figure describes the pattern in which the upper and lower boundaries are disturbed in the opposite direction, which is called varicose or symmetric, while the right figure is the pattern in which the upper and lower boundaries are disturbed in the same direction, which is called sinusoidal or antisymmetric.

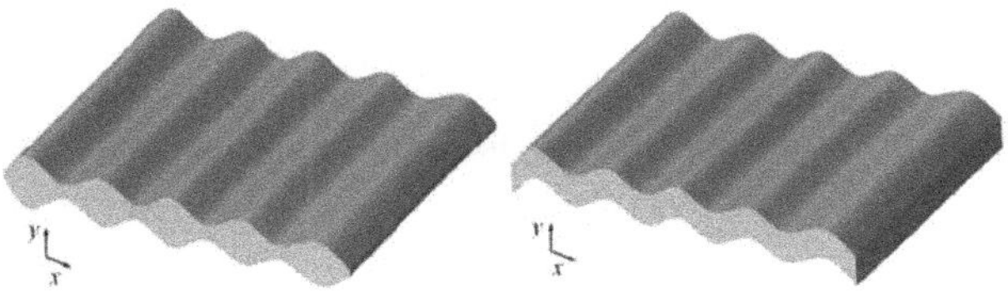

Figure 2. The figure of the liquid sheet "varicose" and "sinuous" modes.

Substituting Equation (18) into Equations (2)–(5) and the constitutive relation in Equation (19), the solution of the disturbance flow field can be obtained with a set of integration constants, which can be determined from the boundary conditions in Equations (14)–(17). The requirement of a nontrivial solution to the linear homogeneous equations leads to the dispersion relation.

$$D_{sin}(\omega_1,k,Re_1)=0,\quad\text{or}\quad D_{var}(\omega_1,k,Re_1)=0, \tag{21}$$

where the subscripts "sin" and "var" represent the sinuous (antisymmetric) mode and the varicose (symmetric) mode, respectively. The two modes exist at the same time for the sheet instability, and their final expressions are as follows.

For the sinuous mode, the dispersion relation should be

$$\begin{aligned} \det(A) &= (\tau - \tfrac{L^2-K^2}{\mathrm{Re}}\tfrac{1+\lambda\cdot El(\mathrm{i}K+\Omega)}{1+El(\mathrm{i}K+\Omega)})(-2\cosh(K))D_{33} \\ &+\varepsilon\tfrac{L^2-K^2}{\mathrm{Re}}\tfrac{1+\lambda\cdot El(\mathrm{i}K+\Omega)}{1+El(\mathrm{i}K+\Omega)}\tfrac{\cosh(K)}{KD\sinh(K)}\tfrac{V_0}{d}D_{32} \\ &+(\tau - \tfrac{L^2-K^2}{\mathrm{Re}}\tfrac{1+\lambda\cdot El(\mathrm{i}K+\Omega)}{1+El(\mathrm{i}K+\Omega)})\tfrac{V_0}{d}D_{32} \\ &-\varepsilon\tfrac{L^2-K^2}{\mathrm{Re}}\tfrac{1+\lambda\cdot El(\mathrm{i}K+\Omega)}{1+El(\mathrm{i}K+\Omega)}2\cosh(K)D_{33}= 0, \end{aligned} \tag{22}$$

where

$$\begin{aligned} \tfrac{V_0}{d}D_{32} &= -\tfrac{K^2+L^2}{K^2-L^2}2\coth(K)\cosh(K)\tfrac{KEu}{D^2} + \tfrac{4K^2L}{(K^2-L^2)}\tfrac{\cosh(K)}{\tanh(L)}\tfrac{Eu}{D^2} \\ &-2\sinh(K)\tfrac{KEu}{D^2}, \\ D_{33} &= -\tfrac{(K^2+L^2)^2}{K}\tfrac{1}{\mathrm{Re}^2}\tanh(K) + \tfrac{4K^2L}{\mathrm{Re}^2}\tanh(L) - \tfrac{K^2}{We} - \bar{\rho}\tfrac{\Omega^2}{K}. \end{aligned} \tag{23}$$

Similarly, the varicose mode can be given as

$$\begin{aligned} \det(A) &= (\tau - \tfrac{L^2-K^2}{\mathrm{Re}}\tfrac{1+\lambda\cdot El(\mathrm{i}K+\Omega)}{1+El(\mathrm{i}K+\Omega)})(-2\sinh(K))D'_{33} \\ &+\bar{\varepsilon}\tfrac{L^2-K^2}{\mathrm{Re}}\tfrac{1+\lambda\cdot El(\mathrm{i}K+\Omega)}{1+El(\mathrm{i}K+\Omega)}\tfrac{\sinh(K)}{KD\cosh(K)}\tfrac{V_0}{d}D'_{32} \\ &+(\tau - \tfrac{L^2-K^2}{\mathrm{Re}}\tfrac{1+\lambda\cdot El(\mathrm{i}K+\Omega)}{1+El(\mathrm{i}K+\Omega)})\tfrac{V_0}{d}D'_{32} \\ &-\varepsilon\tfrac{L^2-K^2}{\mathrm{Re}}\tfrac{1+\lambda\cdot El(\mathrm{i}K+\Omega)}{1+El(\mathrm{i}K+\Omega)}2\sinh(K)D'_{33}= 0. \end{aligned} \tag{24}$$

where

$$\begin{aligned} \tfrac{V_0}{d}D'_{32} &= -\tfrac{K^2+L^2}{K^2-L^2}2\tanh(K)\sinh(K)\tfrac{KEu}{D^2} + \tfrac{4K^2L}{(K^2-L^2)}\tfrac{\sinh(K)}{\coth(L)}\tfrac{Eu}{D^2} \\ &-2\sinh(K)\tfrac{KEu}{D^2}, \\ D'_{33} &= -\tfrac{(K^2+L^2)^2}{K}\tfrac{1}{\mathrm{Re}^2}\coth(K) + \tfrac{4K^2L}{\mathrm{Re}^2}\coth(L) - \tfrac{K^2}{We} - \bar{\rho}\tfrac{\Omega^2}{K} \\ &-\tfrac{(K^2+L^2)}{K^2-L^2}\tfrac{KEu}{D^2}\tanh(K) + \tfrac{2K^2L}{(K^2-L^2)}\tfrac{Eu}{D^2}\tanh(L) + \tfrac{Eu}{D^3}. \end{aligned} \tag{25}$$

3. Energy Budget

To investigate the mechanism of the sheet instability, we conduct an energy budget analysis [21,22], which can trace the perturbation kinetic energy and the effects of various forces, especially the elastic force and the electric field force.

Firstly, we linearize the momentum equation and the constitutive relations of the liquid are as follows:

$$\frac{\partial \mathbf{v}}{\partial t} + \frac{\partial \mathbf{v}}{\partial x} = -\nabla p + \nabla\cdot\boldsymbol{\tau}, \tag{26}$$

$$\boldsymbol{\tau} = \frac{1}{Re}\dot{\boldsymbol{\gamma}} - \lambda_1\left(\frac{\partial \boldsymbol{\tau}}{\partial t} + \frac{\partial \boldsymbol{\tau}}{\partial x}\right) + \frac{\lambda_2}{Re}\left(\frac{\partial \dot{\boldsymbol{\gamma}}}{\partial t} + \frac{\partial \dot{\boldsymbol{\gamma}}}{\partial x}\right). \tag{27}$$

Combining Equations (24) and (25), the following equation is obtained

$$\frac{\partial \mathbf{v}}{\partial t} + \frac{\partial \mathbf{v}}{\partial x} = -\nabla p + \frac{1}{Re}\nabla^2\mathbf{v} - \lambda_1\nabla\cdot\left(\frac{\partial \boldsymbol{\tau}}{\partial t} + \frac{\partial \boldsymbol{\tau}}{\partial x}\right) + \frac{\lambda_2}{Re}\nabla\cdot\left(\frac{\partial \dot{\boldsymbol{\gamma}}}{\partial t} + \frac{\partial \dot{\boldsymbol{\gamma}}}{\partial x}\right). \tag{28}$$

Secondly, we multiply both sides of Equation (26) by the perturbation velocity $\mathbf{v}$ and integrate over one wavelength $\lambda = 2\pi/k$ to get the energy equation. The results can be written as

$$\dot{e}_k = \dot{w}_p + \dot{w}_{vis} + \dot{w}_{El} + \dot{w}_{\overline{\lambda}}, \tag{29}$$

where the left-hand side $\dot{e}_k$ represents the rate of change in the kinetic energy of the disturbance; the terms on the right-hand side represent different mechanisms: $\dot{w}_p$ is the total contribution of liquid pressure; $\dot{w}_{vis}$ is the viscous dissipation; $\dot{w}_{El}$ is the effect of elasticity; and $\dot{w}_{\overline{\lambda}}$ is the retardation of the deformation. The specific expressions for these terms are

$$\dot{e}_k = \frac{k}{2\pi}\int_0^{\frac{2\pi}{k}}\int_{-1}^{1} \mathbf{v}\cdot\left(\frac{\partial \mathbf{v}}{\partial t}+\frac{\partial \mathbf{v}}{\partial x}\right)\mathrm{d}y\mathrm{d}x, \tag{30}$$

$$\dot{w}_p = \frac{k}{2\pi}\int_0^{\frac{2\pi}{k}}\int_{-1}^{1} -\mathbf{v}\cdot\nabla p\mathrm{d}y\mathrm{d}x, \tag{31}$$

$$\dot{w}_{vis} = \frac{k}{2\pi}\int_0^{\frac{2\pi}{k}}\int_{-1}^{1} \mathbf{v}\cdot\frac{1}{Re}\nabla^2\mathbf{v}\mathrm{d}y\mathrm{d}x, \tag{32}$$

$$\dot{w}_{El} = \frac{k}{2\pi}\int_0^{\frac{2\pi}{k}}\int_{-1}^{1} -\mathbf{v}\cdot\lambda_1\nabla\cdot\left(\frac{\partial \boldsymbol{\tau}}{\partial t}+\frac{\partial \boldsymbol{\tau}}{\partial x}\right)\mathrm{d}y\mathrm{d}x, \tag{33}$$

$$\dot{w}_{\overline{\lambda}} = \frac{k}{2\pi}\int_0^{\frac{2\pi}{k}}\int_{-1}^{1} \mathbf{v}\cdot\frac{\lambda_2}{Re}\nabla\cdot\left(\frac{\partial \dot{\gamma}}{\partial t}+\frac{\partial \dot{\gamma}}{\partial x}\right)\mathrm{d}y\mathrm{d}x. \tag{34}$$

Finally, we substitute Equations (2) and (17) into Equation (29) and the results should be

$$\dot{w}_p = \dot{w}_g + \dot{w}_\sigma + \dot{w}_{\tau yy} + \dot{w}_{ew} + \dot{w}_{ez}, \tag{35}$$

$$\dot{w}_g = \frac{k}{2\pi}\int_0^{\frac{2\pi}{k}} \rho\left[\left(v\frac{\partial \phi_g}{\partial t}\right)_{y=a} - \left(v\frac{\partial \phi_g}{\partial t}\right)_{y=-a}\right]\mathrm{d}x, \tag{36}$$

$$\dot{w}_\sigma = \frac{k}{2\pi}\int_0^{\frac{2\pi}{k}} \frac{1}{We}\left[\left(v\frac{\partial^2 \eta}{\partial x^2}\right)_{y=a} + \left(v\frac{\partial^2 \eta}{\partial x^2}\right)_{y=-a}\right]\mathrm{d}x, \tag{37}$$

$$\dot{w}_{\tau yy} = \frac{k}{2\pi}\int_0^{\frac{2\pi}{k}} \left[-(v\tau_{yy})_{y=a} + (v\tau_{yy})_{y=-a}\right]\mathrm{d}x, \tag{38}$$

$$\dot{w}_{ew} = \frac{k}{2\pi}\int_0^{\frac{2\pi}{k}} \left[-\left(v\frac{2E_u}{D^3}\right)_{y=a} + \left(v\frac{2E_u}{D^3}\right)_{y=-a}\right]\mathrm{d}x, \tag{39}$$

$$\dot{w}_{ez} = \frac{k}{2\pi}\int_0^{\frac{2\pi}{k}} \left[-\left(v\frac{2KE_u}{D^2}sinh(k)\right)_{y=a} + \left(v\frac{2KE_u}{D^2}sinh(k)\right)_{y=-a}\right]\mathrm{d}x, \tag{40}$$

where $\dot{w}_g$ is the result of gas pressure; $\dot{w}_\sigma$ is the effect of surface tension; $\dot{w}_{\tau yy}$ is the work of additional surface stress; $\dot{w}_{ew}$ is the product of the normal electrical force; and $\dot{w}_{ez}$ is the comprehensive effect of the tangential and normal electrical forces.

In summary, we have

$$\dot{e}_k = \dot{w}_{vis} + \dot{w}_{El} + \dot{w}_{\overline{\lambda}} + \dot{w}_g + \dot{w}_\sigma + \dot{w}_{\tau yy} + \dot{w}_{ew} + \dot{w}_{ez}. \tag{41}$$

In the Equation (41), the rate of change in the kinetic energy of the disturbance $\dot{e}_k$ represents the degree of instability of the viscoelastic sheet according to the energy analysis. The more rapidly the kinetic energy grows, the more unstable the sheet becomes. Additionally, the amount of work performed by any force component can be used to measure the contribution to the sheet instability. The positivity or negativity of this value determines the stability or instability of the liquid film, and the magnitude reflects the contribution to the instability [23].

4. Results and Discussion

4.1. Basic Case

In view of reality, the magnitude range of physical property parameters of viscoelastic fluid was selected by referring to relevant parameters in previous literature [24], and we adopted one set of real fluid parameters (PIB Boger 4000 ppm [24] and air) for analysis. According to the physical properties of the fluid, the following dimensionless values were chosen to discuss the basic case see (Table 2), which are displayed as

$$[El, \lambda, \mathrm{Re}, We, \rho, Eu, \varepsilon, \tau, D] = [1, 0.5, 100, 400, 0.0012, 5, 0.0125, 5, 40]. \quad (42)$$

Table 2. Definitions of dimensionless numbers and their appropriate values.

Dimensionless Number	Definition	Appropriate Values
Weber number	$We = \rho_l U^2 a/\gamma$	$100 \sim 1000$
Reynolds number	$Re = \rho_l U a/\mu$	$1 \sim 1000$
Gas-to-liquid density ratio	$\overline{\rho} = \rho_g/\rho_l$	$0.001 \sim 0.1$
Electrical Euler number	$Eu = \varepsilon_1 V_0^2/\rho_l U^2 a^2$	$0 \sim 10$
Electrical relaxation time	$\tau = \sigma U/\varepsilon_1 a$	$0.1 \sim 10^7$
Dielectric constant ratio	$\varepsilon = \varepsilon_2/\varepsilon_1$	$0.01 \sim 1$
Dimensionless distance	$D = d/a$	$10 \sim 50$
Elasticity number	$El = \lambda_1 \mu_0/\rho_l a^2 = \lambda_1/Re$	$1 \sim 5$
Time constant ratio	$\overline{\lambda} = \lambda_2/\lambda_1$	$0 \sim 1$

In addition, Table 2 introduces some dimensionless numbers and their appropriate values.

Figure 3 illustrates the dispersion relations for Oldroyd-B and Newtonian fluid sheets in the electric field under the sinuous and varicose modes. In both modes, the growth rates are greater when the sheet is a non-Newtonian fluid. The sheet becomes more unstable when non-Newtonian rheological properties are considered. In this sense, research on the instability of a viscoelastic liquid sheet could identify potential methods for breaking liquid sheets into droplets. Additionally, in the varicose mode, there is greater discrepancy between the two kinds of fluids. Overall, the sinuous mode plays a leading role in the sheet instability of the two fluids. However, the unstable areas are almost the same for both modes. The following sections analyze the effects of the physical properties of the sheet on its instability.

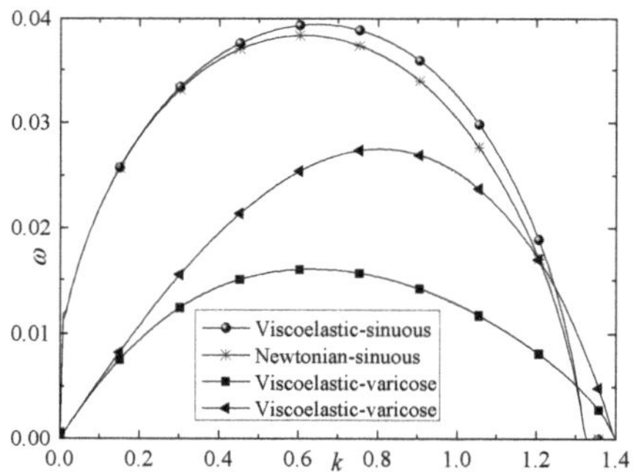

Figure 3. Temporal instability of an Oldroyd-B and Newtonian fluid sheet in the electric field for sinuous and varicose modes.

4.2. Electrical Properties

This section explores the influence of the electric field strength (Eu, D), the relaxation time of the surface charge (τ), and electric permittivity (ε) on the sheet instability.

The effect of the electric strength Eu on the sheet instability is depicted in Figure 4. It can be observed obviously that the electric strength causes the liquid sheet to become dramatically more unstable in both modes. Furthermore, the unstable range of the wavenumber widens as the electric strength increases. It can be seen that the maximum unstable growth rate for the sinuous mode is greater than that for the varicose mode, indicating that the sinuous mode plays a dominant role in the sheet instability. A liquid sheet with a larger maximum unstable growth rate can behave with greater instability. The dominant wavenumber is slightly larger in the varicose mode. According to the studies of Yin [21] et al., increasing the *Eu* can promote the breakup process and obtain a smaller main drop size.

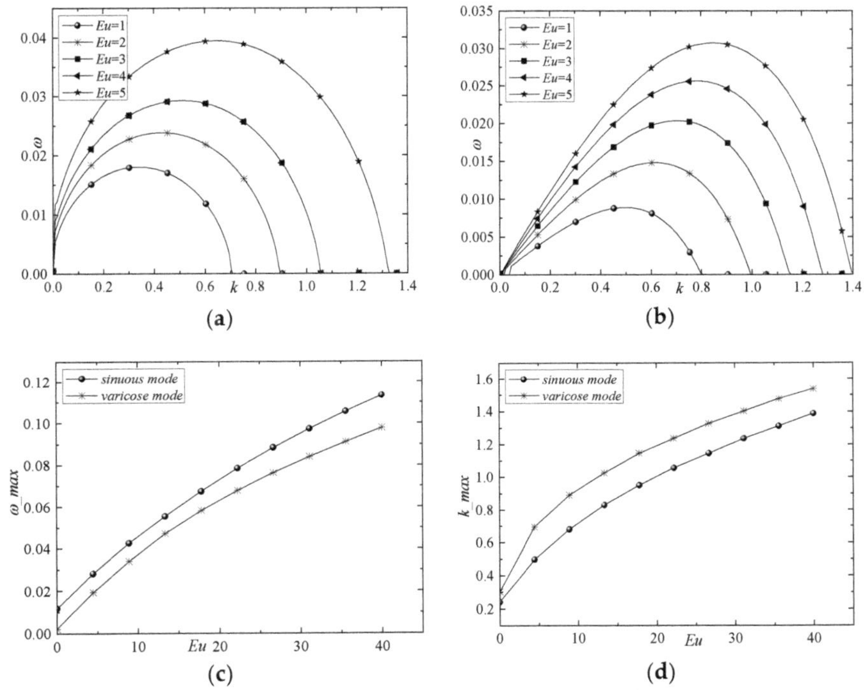

Figure 4. Effects of *Eu* on (**a**) sheet instability for sinuous mode, (**b**) sheet instability for varicose mode, (**c**) maximum growth rate, and (**d**) dominant wavenumber.

The effect of the dimensionless distance D on the instability of the viscoelastic liquid sheet is further studied in the present study, as shown in Figure 5. According to Equation (8), the electric strength increases sharply as D decreases. In this situation, it is easy to conclude that the sheet could become much more unstable in both modes if the distance between the electrodes was decreased, as the unstable area would widen considerably (Figure 4). Note that the varicose mode is more sensitive to variations in D. In brief, the electrical force accelerates the sheet's breakup. The mechanism is complicated because the sheet has been formulated using the leaky dielectric model. According to recent studies, decreasing the distance is beneficial to obtain a smaller main drop size after the breakup process. This is discussed in detail in Section 4.3.

The fluid conductivity is represented by the electrical relaxation time τ and the relative electrical permittivity ε. In fact, the effects of τ and ε are limited, which has been reported in previous studies [23], because the two parameters only influence the conductive term of the surface charge conservation in Equation (13). Compared with the electrical relaxation time τ, the relative electrical permittivity ε has a slightly greater effect because it is included in the electric field strength. As Figure 6 shows, enhancing ε causes the maximum growth rate $\omega_{\max}$ to increase. In this sense, the relative electrical permittivity is an unstable factor for a sheet.

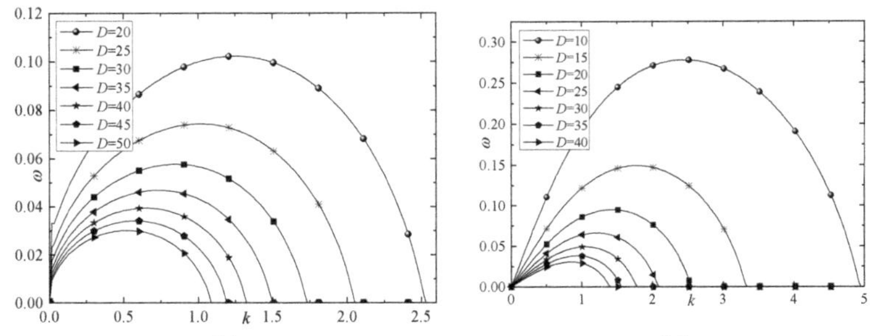

(a) (b)

Figure 5. Effects of D on sheet instability for (**a**) sinuous mode and (**b**) varicose mode.

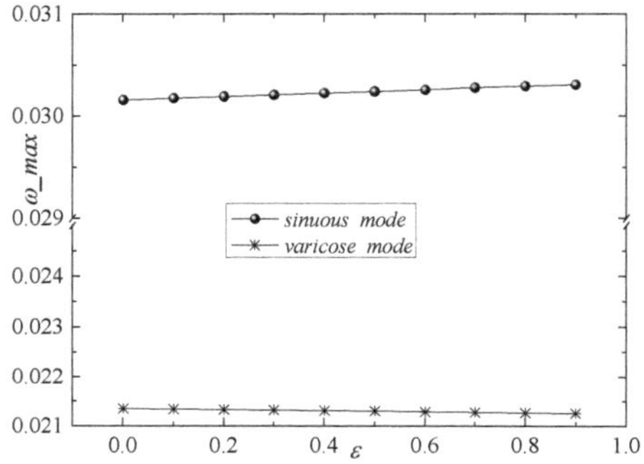

Figure 6. Effects of ε on sheet instability for sinuous and varicose modes.

4.3. Rheological Properties

This section explores the effect of the sheet rheological properties, including the elasticity and time constant ratio, on sheet instability.

Liquid elasticity is an important parameter in the stability of the sheet. Figure 7 displays curves of the temporal growth rate varies with the wavenumber. It can be seen clearly that the temporal growth rate increases as the elasticity number increases. The results suggest that the elasticity has a destabilizing effect on the viscoelastic sheet. The effect of the elasticity number in the linear analysis can be explained by the equation in Table 2: increasing El results in an increase in λ_1, leading to a larger effective Reynolds number. Figure 7 also shows that the temporal growth rate varies only slightly for different elasticity numbers, so the destabilizing effect of viscoelasticity is weak according to linear analysis. Further aspects of the mechanism of the elasticity number El on the linear temporal growth rate ω are examined in Section 4.4.

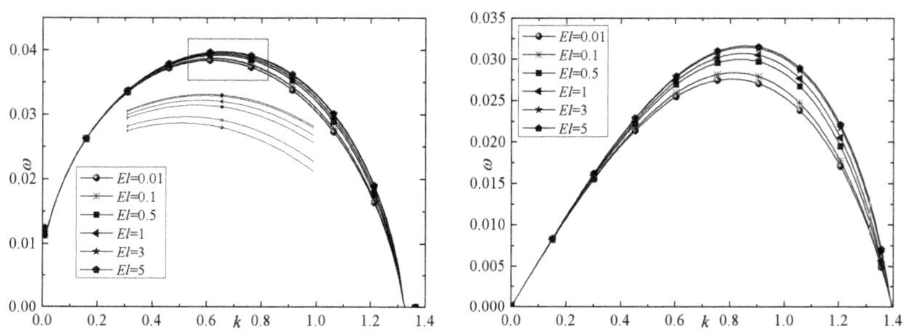

Figure 7. Effects of El on sheet instability for sinuous and varicose modes.

In the present study, the time constant ratio is defined as the ratio of the deformation retardation time λ_2 to the stress relaxation time λ_1. Figure 8 illustrates the effects of the time constant ratio on sheet instability under the sinuous and varicose modes, respectively. It can be seen obviously that the temporal growth rate decreases with an increasing time constant ratio. It means that the viscoelastic planar liquid sheet behaves with greater stability as the time constant ratio increases, i.e., decreasing the time constant ratio promotes the breakup process of the liquid sheet. Figure 8 also shows that the time constant ratio has a relatively large effect. As the stress tensor of a viscoelastic liquid sheet can be increased by enlarging the deformation retardation time λ_2, according to Equation (1), the liquid sheet will become more unstable. The effects of other properties, such as viscosity, surface tension, and density, are well-known [9,10,16], so a detailed description is omitted here.

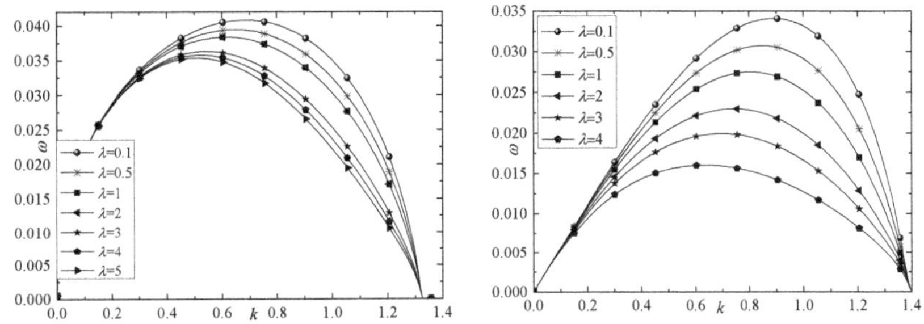

Figure 8. Effects of λ on sheet instability for sinuous and varicose modes.

4.4. Mechanism Analysis

In this section, an energy budget is derived to explain the mechanism of the onset of instability. For the sinuous mode, the rates of change in the kinetic energy and various forces are shown as a function of the wavenumber at the onset of instability $(t = 0)$ in Figure 9, according to Equation (39) in Section 3. Firstly, $\dot{w}_{ez}, \dot{w}_{ew}, \dot{w}_g, \dot{w}_{El}$ are positive, so the electrical strength, ambient gas, and elasticity have a destabilizing effect, which explains the trend in the temporal growth rate identified in Sections 4.2 and 4.3. In contrast, $\dot{w}_{\overline{\lambda}}, \dot{w}_\sigma, \dot{w}_{vis}$ are negative, so deformation retardation, surface tension, and viscosity have a stabilizing effect. This explains the phenomenon observed in Section 4.3, whereby an increase in the time constant ratio $\overline{\lambda}$ reduces the temporal growth rate ω. Additionally, $\dot{w}_{\tau yy}$ is very small, but positive, so it slightly destabilizes the sheet. Secondly, the absolute values of $\dot{w}_\sigma, \dot{w}_{ez}$ are much greater than those of the other forces, so the electric force and aerodynamic force dominate the sheet instability. The magnitudes of $\dot{w}_{\overline{\lambda}}, \dot{w}_g, \dot{w}_{El}$ are also considerable, so they are secondary factors in the instability of the viscoelastic sheet.

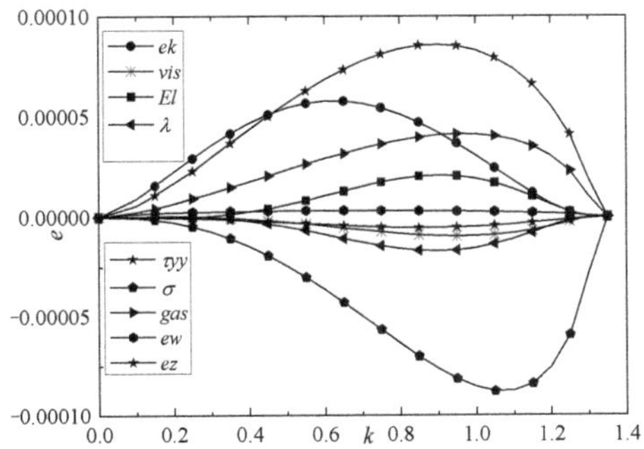

Figure 9. Rates of change in kinetic energy and various forces for sinuous mode.

The present study focuses on the influence of viscoelasticity and electric strength on instability, so the effects of the elasticity number El, time constant ratio $\overline{\lambda}$, and electrical Euler number Eu are further examined in Figures 10–12. Generally, the rates of all forces exhibit minor variations with respect to the elasticity number, as shown in Figure 10. As the elasticity number increases, $\dot{w}_{ez}$, $\dot{w}_{ew}$, $\dot{w}_{El}$ increase, whereas $\dot{w}_{\sigma}$ decreases slightly, but the absolute value of the rate of change in kinetic energy increases, which means that the force created by elasticity does more work to enhance the instability as the elasticity increases. However, this provides a limited explanation for the effect of elasticity on the temporal growth rate shown in Figure 6, whereby ω increases slightly as El increases.

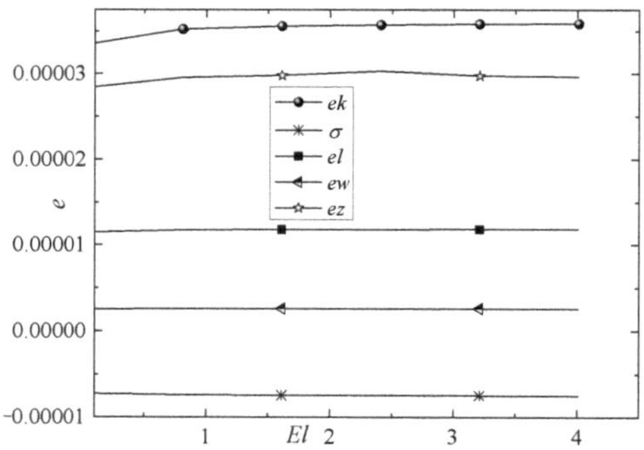

Figure 10. Effect of elasticity number El on the rate of change in the kinetic energy and other forces.

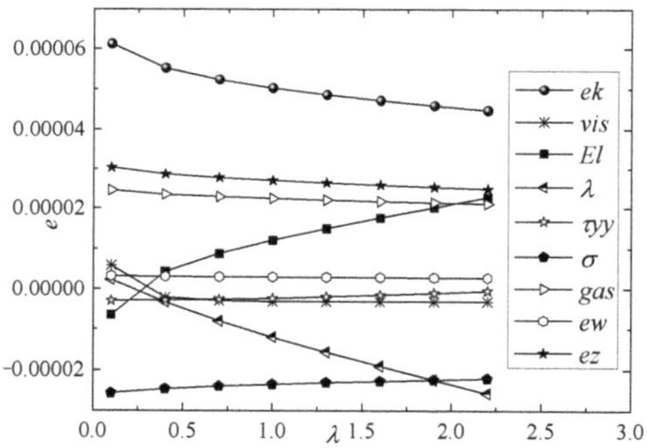

Figure 11. Effect of time constant ratio on rate of change in the kinetic energy and other forces.

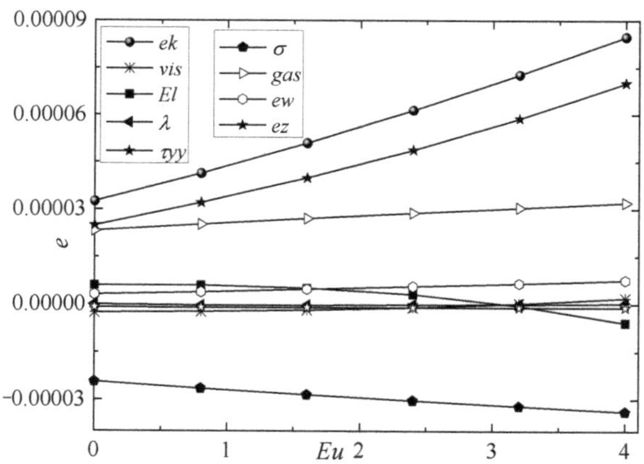

Figure 12. Effect of electrical Euler number Eu on rate of change in kinetic energy and other forces.

The effect of the time constant ratio λ is shown in Figure 11. The kinetic energy and other forces vary considerably as λ changes. Firstly, as λ increases, $\dot{w}_{El}, \dot{w}_{\sigma}$ increase while $\dot{w}_{ez}, \dot{w}_{g}, \dot{w}_{\overline{\lambda}}$ decrease. The other forces remain basically unchanged. Note the significant change in the kinetic energy, which implies that the time constant ratio effectively enhances the sheet instability, as shown in Figure 7. Secondly, increasing the time constant ratio weakens the effect of the electric force. Finally, as λ increases, $\dot{w}_{El}$ increases markedly, whereas $\dot{w}_{\overline{\lambda}}$ decreases rapidly. The elasticity enhances the instability, but is not sufficient to compensate for most of the energy dissipated by deformation retardation. As a result, the time constant ratio stabilizes the sheet.

Figure 12 displays the effect of the electrical Euler number Eu on the rate of change in the kinetic energy and other forces. When Eu increases, the $\dot{w}_{ez}, \dot{w}_{g}$ increase, while $\dot{w}_{El}, \dot{w}_{\sigma}$ decrease. Note that $\dot{w}_{ez}$ grows rapidly as Eu changes, resulting in the kinetic energy increasing dramatically. Thus, the electric force is a leading factor in the instability of the sheet, as illustrated by Figure 4. It is worth noting that $\dot{w}_{El}$ decreases markedly as Eu increases, which reduces the destabilizing effect of the liquid elasticity on the sheet. In this sense, the effect of surface tension has been enhanced.

5. Conclusions

The present study made two main contributions. Firstly, using temporal linear analysis, the instability of a non-Newtonian liquid sheet moving through inviscid gas in a transverse electrical field was studied with the leaky dielectric model. Secondly, an energy budget was used to investigate the mechanism of sheet breakup. The results presented herein explained the interaction between the electric field and the rheological properties.

The electric field strength and liquid elasticity caused the liquid sheet to be dramatically unstable in both sinuous and varicose modes. Furthermore, the unstable range of the wavenumber widened as the electric strength increased. A viscoelastic planar liquid sheet behaved with greater stability as the time constant ratio increased.

According to the energy budget, the aerodynamic forces, electrical forces, and liquid elasticity enhanced the sheet instability by doing positive work, whereas the surface tension, viscous stresses, and deformation retardation performed negative work, indicating their stabilizing effects. The electrical force was the most significant destabilizer of the sheet under the leaky dielectric model. Additionally, increasing the time constant ratio weakened the effect of the electric force, while the electric force could reduce the destabilizing effect of liquid elasticity on the sheet.

Author Contributions: Conceptualization, X.D.; Methodology, X.D.; Software, X.D.; Validation, X.D. and L.N.; Formal Analysis, X.D. and L.N.; Investigation, L.N.; Resources, L.N.; Data Curation, X.D.; Writing—Original Draft Preparation, X.D.; Writing—Review and Editing, L.N.; Visualization, X.D.; Supervision, L.N.; Project Administration, L.N.; Funding Acquisition, L.N. All authors have read and agreed to the published version of the manuscript.

Funding: This research was funded by [China Postdoctoral Science Foundation], grant number [No. 2020M680326]. The APC was funded by [China Postdoctoral Science Foundation].

Institutional Review Board Statement: Not applicable.

Informed Consent Statement: Not applicable.

Acknowledgments: We would particularly like to acknowledge the editors for theirs patient supports. We also appreciate the anonymous reviewers for their professional and helpful comments and suggestions.

Conflicts of Interest: The authors declare no conflict of interest. The funders had no role in the design of the study; in the collection, analyses, or interpretation of data; in the writing of the manuscript; or in the decision to publish the results.

References

1. Lefebvre, A.H.; Wang, X.F.; Martin, C.A. Spray characteristics of aerated-liquid pressure atomizers. *J. Propuls. Power* **1988**, *4*, 293–298. [CrossRef]
2. Lefebvre, A.H. *Atomization and Sprays*; CRC Press: New York, NY, USA, 1989. [CrossRef]
3. Ahmadi, M.; Sellens, R.W. A simplified maximum-entropy-based drop size distribution. *At. Sprays* **1993**, *3*, 291–310. [CrossRef]
4. Bayvel, L.; Orzechowski, Z. *Liquid Atomization*; Taylor and Francis: Washington, DC, USA, 1993; ISBN 9780891169598.
5. Squire, H.B. Investigation of the Instability of a Moving Liquid Film. *Br. J. Appl. Phys.* **1953**, *4*, 167–169. [CrossRef]
6. Hagerty, W.W.; Shea, J.F. A study of the stability of plane fluid sheets. *J. Appl. Mech.* **1955**, *22*, 509–514. [CrossRef]
7. Melcher, J.R. *Field-Coupled Surface Waves: A Comparative Study of Surface-Coupled Electrohydrodynamic and Magnetohydrodynamic Systems*; MIT Press: Cambridge, MA, USA, 1963.
8. Melcher, J.R.; Schwarz, W.J., Jr. Interfacial Relaxation Overstability in a Tangential Electric Field. *Phys. Fluids* **1968**, *11*, 2604. [CrossRef]
9. El-Sayed, M.F. Electro-Aerodynamic Instability of a Thin Dielectric Liquid Sheet Sprayed with an Air Stream. *Phys. Rev. E* **1999**, *60*, 7588–7591. [CrossRef]
10. Yang, L.J.; Liu, Y.X.; Fu, Q.F.; Wang, C. Linear Stability Analysis of Electrified Viscoelastic Liquid Sheets. *At. Sprays* **2012**, *22*, 951–982. [CrossRef]
11. Melcher, J.R.; Taylor, G.I. Electrohydrodynamics: A review of the role of interfacial shear stresses. *Annu. Rev. Fluid Mech.* **1969**, *1*, 111–146. [CrossRef]
12. Saville, D.A. Electrohytrodynamics: The Taylor–Melcher Leaky Dielectric Model. *Annu. Rev. Fluid Mech.* **2003**, *29*, 27–64. [CrossRef]
13. Cimpeanu, R.; Papageorgiou, D.T.; Petropoulos, P.G. On the Control and Suppression of the Rayleigh–Taylor Instability Using Electric Fields. *Phys. Fluids* **2014**, *26*, 022105. [CrossRef]
14. Savettaseranee, K.; Papageorgiou, D.T.; Petropoulos, P.G.; Tille, B.S. The effect of electric fields on the rupture of thin viscous films by van der Waals forces. *Phys. Fluids* **2003**, *15*, 641–652. [CrossRef]
15. Tilley, B.S.; Petropoulos, P.G.; Papageorgiou, D.T. Dynamics and rupture of planar electrified liquid sheets. *Phys. Fluids* **2001**, *13*, 3547. [CrossRef]
16. Liu, Z.; Günter, B.; Durst, F. Linear analysis of the instability of two-dimensional non-Newtonian liquid sheets. *J. Non-Newton. Fluid Mech.* **1998**, *78*, 133–166. [CrossRef]
17. Brenn, G.; Liu, Z.; Durst, F. Three-dimensional temporal instability of non-Newtonian liquid sheets. *At. Sprays* **2001**, *11*, 49–84. [CrossRef]
18. Jia, B.Q.; Xie, L.; Yang, L.J.; Fu, Q.F.; Cui, X. Linear instability of viscoelastic planar liquid sheets in the presence of gas velocity oscillations. *J. Non-Newton. Fluid Mech.* **2019**, *273*, 104169. [CrossRef]
19. Oldroyd, J.G. On the Formulation of Rheological Equations of State. *Proc. R. Soc. Lond. Ser. A* **1950**, *200*, 523–541. [CrossRef]
20. Oldroyd, J.G. Non-Newtonian Effects in Steady Motion of Some Idealized Elastico-Viscous Liquids. *Proc. R. Soc. Lond. A* **1958**, *245*, 278–297. [CrossRef]
21. Li, F.; Yin, X.Y.; Yin, X.Z. Axisymmetric and non-axisymmetric instability of an electrically charged viscoelastic liquid jet. *J. Non-Newton. Fluid Mech.* **2011**, *166*, 1024–1032. [CrossRef]
22. Weder, M.; Gloor, M.; Kleiser, L. Decomposition of the temporal growth rate in linear instability of compressible gas flows. *J. Fluid Mech.* **2015**, *778*, 120–132. [CrossRef]
23. Ruo, A.C.; Chen, K.H.; Chang, M.H. Instability of a charged non-Newtonian liquid jet. *Phys. Rev. E* **2012**, *85*, 016306. [CrossRef]
24. Carroll, C.P.; Joo, Y.L. Electrospinning of viscoelastic Boger fluid: Modeling and experiments. *Phys. Fluids* **2006**, *18*, 053102. [CrossRef]

energies

Article

Effect of Multistage Circulation Control on Blade Aerodynamic Performance

Hai Du [1,2,*], Lejie Yang [3], Shuo Chen [3], Wenxiao Zhang [1] and Shengchun Han [3]

1 School of Aeronautics and Astronautics, Xihua University, Chengdu 610039, China
2 National Key Laboratory of Science and Technology on Aerodynamic Design and Research, Northwest Polytechnical University, Xi'an 710072, China
3 Key Laboratory of Fluid and Power Machinery, Ministry of Education, Xihua University, Chengdu 610039, China
* Correspondence: duhai2017@163.com; Tel.: +86-151-9668-6983

Abstract: To improve the low aerodynamic efficiency and reduce the high energy consumption of a single-stage circulation control wing, a multistage circulation control wing was designed. By combining force measurement and particle image velocimetry (PIV), the aerodynamic and flow-field characteristics of an aerofoil were investigated with respect to the increase in the number of blowing slots, changes in the blowing coefficient, and different blowing ratios for three slots. The force measurement results revealed that the maximum lift-to-drag ratio resulting from simultaneous blowing into the three slots increased by 95.3% compared with that in the absence of circulation control. With an increase in the blowing coefficient, two stages were observed: separation control and supercirculation control. In the separation control stage, the lift and drag coefficients significantly increased and decreased, respectively. In the supercirculation control stage, the lift coefficient gradually increased with the blowing coefficient, whereas the drag coefficient remained unchanged. When the blowing ratio (blowing flow ratio of three slots) in the three slots was 3:1:2, the maximum lift-to-drag ratio of the wing could reach 143.48%. The effects of different slot positions on the aerodynamic control were found to vary. The effects of Slot.1 and Slot.3 in terms of the drag reduction and lift, respectively, were evident, and the influence of Slot.2 on blowing between these two slots played a role in jet relay. The PIV results revealed that multistage blowing circulation increased the curvature of the trailing-edge streamline, thus increasing the equivalent aerofoil camber and improving the wing lift. At a high angle of attack, this circulation demonstrated a flow separation control effect.

Keywords: wind turbine; circulation control; lift enhancement; wind tunnel experiments; PIV

Citation: Du, H.; Yang, L.; Chen, S.; Zhang, W.; Han, S. Effect of Multistage Circulation Control on Blade Aerodynamic Performance. *Energies* **2022**, *15*, 7395. https://doi.org/10.3390/en15197395

Academic Editors: Vasily Novozhilov and Cunlu Zhao

Received: 15 September 2022
Accepted: 29 September 2022
Published: 9 October 2022

1. Introduction

Environmental pollution and other problems have become more severe owing to the global energy crisis. Consequently, environmental pollution, energy conservation, and emission reduction have gained more attention. Wind power, as one of the clean energy sources, has attracted considerable interest. Wind power generation is a type of power production method that converts the kinetic energy of wind into electrical energy through a wind turbine. Wind turbine blades, which are important in capturing wind energy, directly affect the utilisation rate of energy [1]. Study results indicate that the best approach to improve the utilisation rate of wind energy is to increase the aerodynamic efficiency of wind turbine blades.

With the development of flow control technology, an increasing number of flow control methods have been used to improve the aerodynamic efficiency of wind turbine blades [2]. Currently, the main flow control methods include blowing/suction, synthetic jets, vortex generators, slotted aerofoils, bionic nodes, grooves, and tip winglets [3,4].

Zhang et al. [5] adopted synthetic jet technology to replace the traditional blowing control mode to improve the aerodynamic performance of wind turbine blades; however, the lift increment obtained was evidently insufficient. Stalnov et al. [6] applied zero-mass jet technology to wind turbine blades to increase the lift. Experiments reveal that the zero-mass jet can double the maximum lift coefficient of the aerofoil; however, the lift force barely increases in the linear segment of the lift. Krentel et al. [7] constructed the trailing edge of an aerofoil into a stepped and zigzag shape and relieved the periodic vortex shedding of the trailing edge of the aerofoil using a passive flow control method to achieve drag reduction. Remarkably, this control method reduced the drag by 29%; however, it also lowered the lift. Kotsonis et al. [8] placed the plasma controller on the round trailing edge of the aerofoil. Through force measurement experiments, they found that the lift could be increased by a maximum of 20% only when the magnitude of the angle of attack approached the angle corresponding to the occurrence of stall.

Circulation control is currently the most widely used active flow control technology and has great potential for improving the aerodynamic performance of aerofoils. It is primarily implemented through the trailing-edge open-air blowing slot using tangential flow to produce the Coanda effect [9] (i.e., the delay in boundary layer separation [10] and the increase in circulation along an aerofoil [11]), consequently increasing the lift. The circulation control device is composed of Coanda trailing edge, air blowing slot and air compression chamber as the main control components [12]. Currently, the circulation control technology can be used not only to improve the aerodynamic performance of aerofoils but also to reduce noise [13,14].

Compared with other flow control technologies, the boundary layer entrained by the Coanda face jet was found to prevent the separation of the tail airflow [15,16], significantly contributing to the control of boundary layer separation [17]. Moreover, the jet adheres to the circular surface without separation because of the equilibrium between the centrifugal force of jet gas molecules and the local pressure drop [18]. As the blowing speed increases, the two stagnation points continue to move, resulting in considerable circulation [19]. As the blowing air separates from the arc surface, it mixes with the mainstream, resulting in a downward deflection of the streamline; this is similar to the lift system of a conventional rudder surface [20] shown in Figure 1. At the same time, the parameters such as the blowing coefficient, the size of the blowing slot and the location of the blowing slot also affect the lifting efficiency of the circulation control.

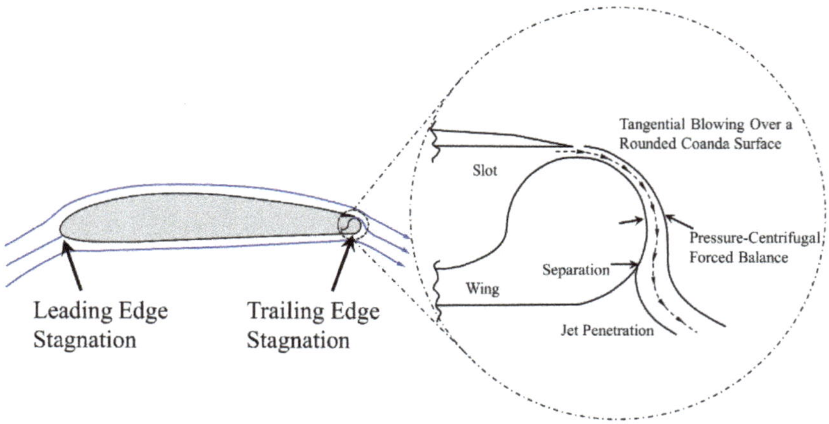

Figure 1. The influence of Coanda effect on streamlines and the concept of circulation control.

The studies on the efficiency of circulation control technology show that circulation control aerofoils have a significant lifting effect compared with traditional aerofoils. Englar et al. [21–24] systematically studied the circulation control technology. The results show that under certain blowing conditions, the lift gain of the aerofoil using circulation control technology is two to three times higher than that of the traditional aerofoil. Xu et al. [25,26]

studied the effects of a circulation control device on the aerofoil of a wind turbine. They found that the use of circulation control technology increased the lift and reduced the drag. Although increasing the blowing coefficient or decreasing the height of the blowing gap can further enhance the aerodynamic efficiency, more external energy is consumed.

Numerous studies have been conducted on circulation control technology in terms of theory and application. However, most of the current single-stage circulation control technologies must satisfy the conditions of high-speed jet flow at the blowing slot; this, in turn, requires significant energy consumption. To resolve this, a multistage circulation control aerofoil was developed in this study. Through pneumatic measurement and a particle image velocimetry (PIV) experiment, the advantages of the proposed multistage circulation control were compared. The aerofoil aerodynamic characteristics, which vary with the blowing coefficient and ratio of multistage circulation control, were obtained. Moreover, the effectiveness of the multistage cycle control and changing blowing conditions in improving the aerodynamic efficiency of the wind turbine aerofoil was verified.

The structure of this paper is as follows. The first part introduces the advantages and disadvantages of most flow control technologies used to improve the aerodynamic performance of wind turbine blades. The advantages of circulation control technology and the problems encountered in the application of single-stage circulation control technology are introduced. The second section introduces the experimental model and conditions. The analysis of force measurement results and their discussion in terms of three aspects (varying the number of blowing slots, blowing coefficient, and blowing proportion) are presented in the third section. The fourth part elaborates on the study of the flow control mechanism of the trailing edge when the number of blowing slots, blowing coefficient, and blowing ratio are modified. The fifth part summarises the study.

2. Experimental Equipment and Programme

2.1. Wind Tunnel Experimental System

The experiment was conducted in the low-turbulence direct-flow wind tunnel of the Key Laboratory of Fluid and Power Machinery of the Ministry of Education, Xihua University. The dimensions of the experimental section were 2.0 m × 0.3 m × 0.5 m (length × width × height), the turbulence of the wind tunnel was less than 0.5%, the airflow deviation angle was less than 0.5°, and the adjustable steady wind speed was 5–35 m/s. The wind tunnel experimental system is shown in Figure 2.

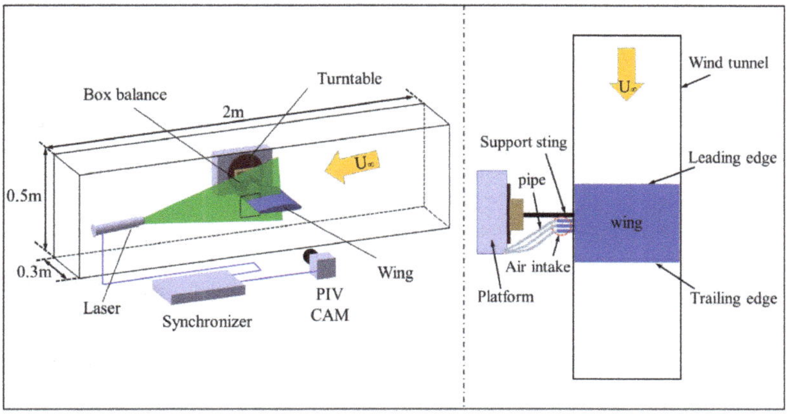

Figure 2. Wind tunnel experiment system and experiment arrangement.

To avoid the influence of gravity and other impact forces on the force measurement data when the blowing pipe was connected with the air inlet, in the experiment, the blowing pipe was fixed to the mobile platform and then connected to the air source. The aerodynamic measurement results recorded when the pipe was connected and not connected to the wing are shown in Figure 3. In the two cases, the lift and drag coefficients of the wing were observed to be fundamentally the same, indicating that the installation of the blowing pipe has no influence on the aerodynamic measurement.

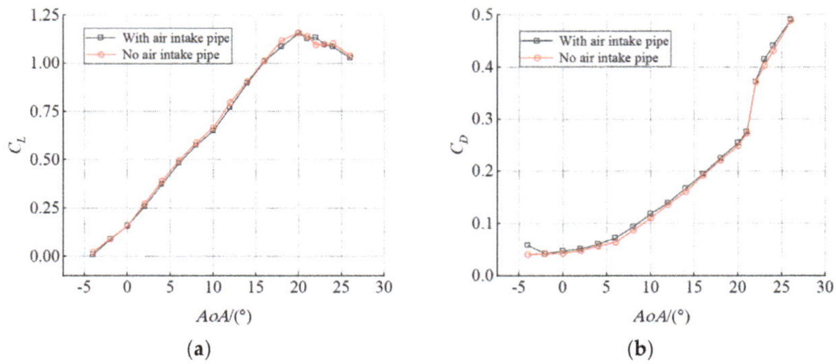

Figure 3. Influence of air pipe on aerodynamic force. (**a**) Lift coefficient. (**b**) Drag coefficient.

Similarly, to avoid the influence of air intake on the flow field, the wingspan was 296 mm, which also avoided collision between the wing and the wall of the wind tunnel and eliminated the influence on the force measurement results. Moreover, the air intake was extended outside the wind tunnel and then connected to the blowing pipe. Thus, the influence of air inlet on the flow field was effectively avoided.

As shown in Figure 4a, the experiment was to modify the pointed trailing edge of the original Clark-Y airfoil into a circular trailing edge with a radius of 4.5 mm, and the modified airfoil chord length was 228 mm. The modified Clark-Y airfoil was used to design a two-dimensional equivalent wing with a length of 296 mm. The Clark-Y airfoil was selected in the experiment mainly because of its large lift-to-drag ratio and good stall performance. The multistage circulation control device is shown in Figure 4b. The design of the air inlet and the three layers of the compressed air chamber could realise the independent blowing control of the three slots. The trailing-edge shape, blow slot position, and PIV experimental shooting area are shown in Figure 4c. The trailing-edge radius was 4.5 mm, and three slots were open from top to bottom. The height (h) of the slots was 0.4 mm (h/c = 0.0018). Slot.1 was located at the beginning of the rear edge of the arc; Slot.2 was located 31.3% below the arc; and Slot.3 was located 69.9% below the position of the arc. In this experiment, single-slot blowing means that the trailing edge opens only one blowing slot (Slot.1). Double-slot blowing indicates that the trailing edge opens two blowing slots (Slot.1 and Slot.2). Three-slot blowing denotes that the trailing edge opens three blowing slots (Slot.1, Slot.2, and Slot.3). The wings and the multistage circulation control device are all 3D printed in one piece.

A high-pressure centrifugal fan was used to supply air to the blowing device. The maximum flow was 330 m^3/h, and the dynamic pressure was 36 kPa. The fan outlet was divided into three outputs. The high precision flow meter was connected to the fan outlet and model inlet. The different flow sizes were controlled by adjusting the flow meter.

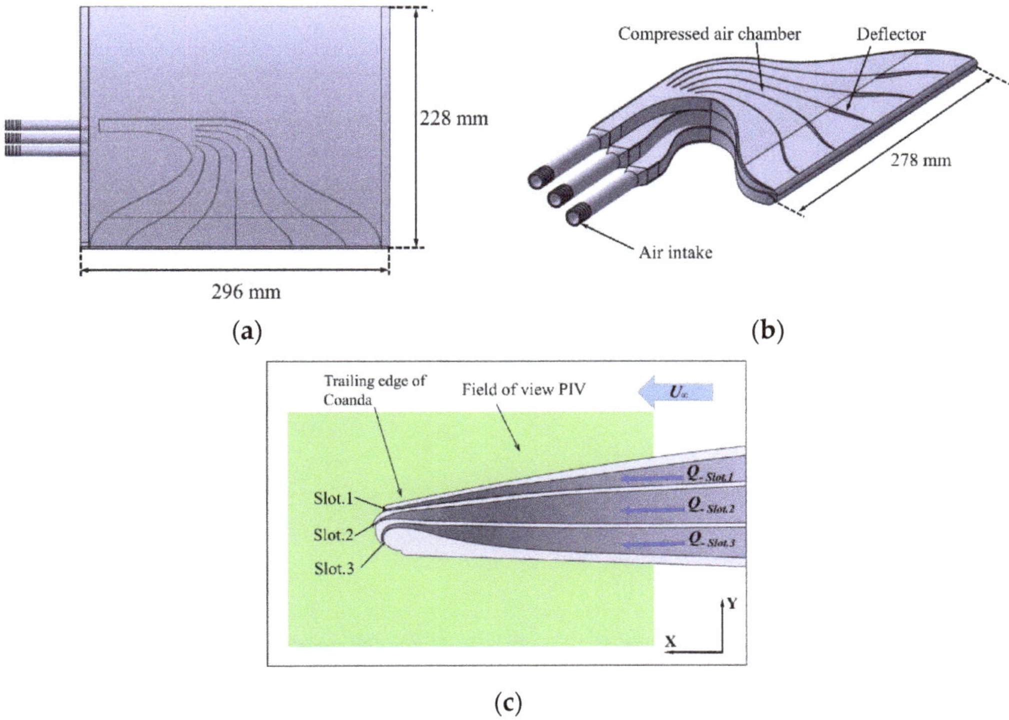

Figure 4. Multistage circulation control model. (**a**) Drawing of experimental model. (**b**) Multistage circulation control device. (**c**) Internal section view of wing trailing-edge and PIV shooting area.

To describe the blowing volume and facilitate comparison, the blowing coefficient, C_μ, is calculated as follows:

$$C_\mu = \frac{\dot{m} U_J}{0.5 \rho_\infty U_\infty^2 S} \tag{1}$$

where $\dot{m}$ is the jet mass flow rate; U_J is the jet velocity, which is calculated by flow meter reading and jet outlet area; ρ_∞ is the free-stream density; U_∞ is the free-stream velocity; and S is the reference area of the wing. The blowing coefficient, C_μ, is the ratio of the momentum flux of the jet to that of the free-stream.

2.2. Aerodynamic Force and Flow-Field Measurement System

The force measurement system mainly includes a DH8300N dynamic signal acquisition system, balance, connecting piece, and rotating mechanism. The measurement range, calibration accuracy and calibration accuracy of the six-component box strain balance used in this experiment are shown in Table 1.

Table 1. Measurement range, calibrated precision, and accuracy of the balance.

	X (kg)	Y (kg)	Z (kg)	Mx (kg·m)	My (kg·m)	Mz (kg·m)
Force	5	20	5	1	1	3
Precision (%)	0.48	0.31	0.5	0.27	0.3	0.42
Accuracy (%)	0.18	0.2	0.2	0.1	0.085	0.19

To verify the stability of the force measurement system, five experiments were repeatedly conducted when the experimental wind speed was 8 m/s, and the trailing edge was not blown. The experimental results are shown in Figure 5. The calculated repeatability errors of the lift and drag coefficients were 0.971% and 0.616%, respectively, indicating that the experimental system was stable and reliable.

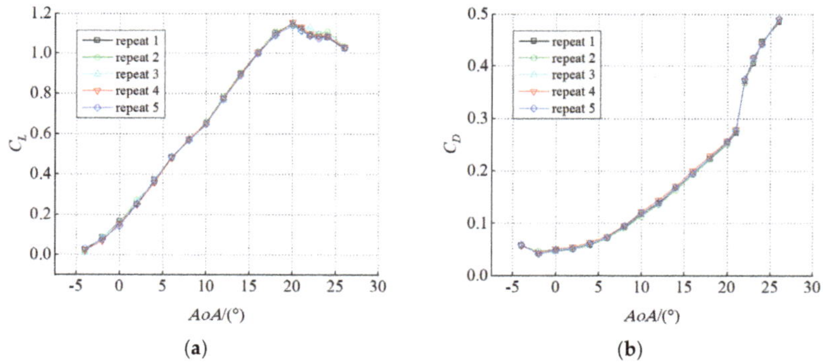

Figure 5. Results of repeated experiments. (**a**) Lift coefficient. (**b**) Drag coefficient.

When measuring the aerodynamic force under the circulation control, due to the proportion of the jet reverse thrust F_{Jet} to the aerodynamic force $F_{Aerodynamic}$ of the model body under typical working conditions $F_{Jet}/F_{Aerodynamic} \approx O(10^{-3})$, this paper ignores the influence of the jet reverse thrust on the aerodynamic force in the analysis.

High-time resolution PIV measures the velocity distribution on the wing surface and wake. The speed measurement system includes a Fastcam AX-100 high-speed camera (1024 pixel × 1024 pixel (12 bits)) (Motion Engineering Company, Westfield, IN, USA), Nikon 85 mm lens, Vlite-Hi-527 double-pulse laser (energy: 20 mJ; output wavelength: 532 nm) (Beamtech Company, Beijing, China), tracer particle generator (Dongfang Fluid Measurement Technology Co., Ltd., Beijing, China), computer and an ILA synchroniser (pulse interval: 30 µs) (ILA 5150 Company, Rotter Bruch 26a 52068 Aachen, Germany).

The ILA synchroniser controls the digital synchronisation between the high-speed cameras and lasers. The smoke particle generator produces approximately 1 µm oil droplets in the flow field. The laser generates an approximately 1 mm thick slice of light. After calibration, the physical space size of the camera shooting area was 121 mm × 121 mm, so the pixel ratio was 0.118 mm/pixels. Due to the limitation of the laser range, it also captured the region where $x/c = 1$ as the reference at the trailing edge, and the flow direction was from $x/c = 0.8$ to $x/c = 1.2$, and the vertical direction was from $y/c = -0.11$ to $y/c = 0.125$ area (length 91.2 mm, width 53.8 mm). The camera shooting frequency was 2000 Hz, the recording time was 4 s, and 8000 consecutive images were recorded. The PIV-view software was used to process the original images, and a cross-correlation algorithm was adopted. The interrogation window was 64 × 64, and the overlap rate was 50%.

3. Analysis and Discussion of Experimental Results of Force

3.1. *Influence of Number of Slots on Aerodynamic Characteristics*

To study the influence of different numbers of slots on the wing aerodynamic force, a force measurement experiment with a variable number of slots was conducted. In the experiment, the single slot blowing generated a jet at Slot.1, and Slot.2 and Slot.3 were closed. Two slot blowing produced jets at Slot.1 and Slot.2, and closed Slot.3. Table 2 lists the experimental parameters for the variable number of slots. The total input flow was consistently maintained, and the input flow of each slot was evenly distributed. The experimental wind speed was 8 m/s, and the Reynolds number was 1.21×10^5.

Table 2. Experimental parameters of variable number of slots.

	Single-Slot Blowing Test			Double-Slot Blowing Test			Three-Slot Blowing Test		
Slot	**Slot.1**	**Slot.2**	**Slot.3**	**Slot.1**	**Slot.2**	**Slot.3**	**Slot.1**	**Slot.2**	**Slot.3**
Total flow, Q (m^3/h)		41.91			41.91			41.91	
Flow rate, Q, of each slot (m^3/h)	41.91	/	/	20.95	20.95	/	13.97	13.97	13.97

The aerodynamic results of the wing when the total input flow was $Q = 41.91\ m^3/h$ and the number of blowing slots of the trailing edge changes are shown in Figure 6. The lift coefficient and lift-to-drag ratio increased with the number of slots, as shown in Figure 6a,c, respectively. The maximum lift-to-drag ratios corresponding to single-slot blowing, double-slot blowing, and three-slot blowing increased by 13.2, 21.5, and 95.3%, respectively, compared with those in the absence of blowing. The drag coefficient significantly decreased when three slots were simultaneously used, as shown in Figure 6b.

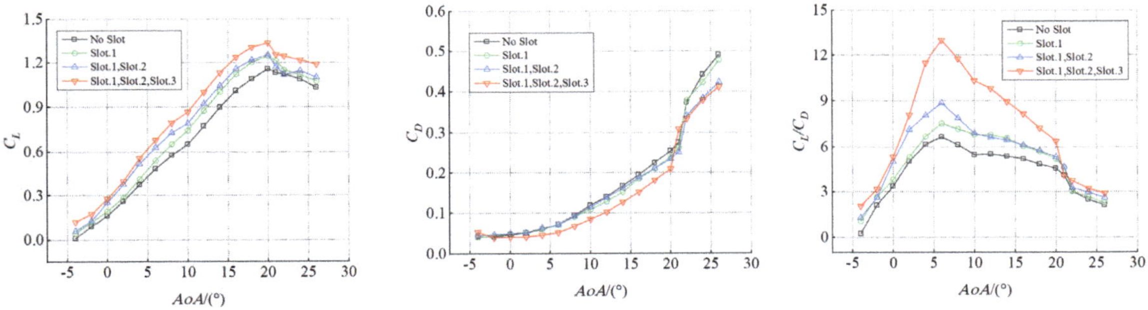

(a) (b) (c)

Figure 6. Comparison of wing aerodynamic characteristics with different numbers of slots ($U_\infty = 8\ m/s$ and $Q = 41.91\ m^3/h$). (**a**) Lift coefficient. (**b**) Drag coefficient. (**c**) Lift-to-drag ratio.

The multistage circulation control (double-slot and triple-slot blowing) can further enhance the aerodynamic force of the wing compared with the single-stage circulation control (single-slot blowing) under the premise that the amount of external energy consumed is the same.

3.2. Influence of Blowing Coefficient on Aerodynamics of Wing with Circulation Control

According to the force measurement experiment with a variable number of slots, under the same input flow rate, the best control effect is achieved when the three slots are simultaneously blown. Accordingly, the effect of varying the blowing coefficient on the aerodynamic performance of the wing was studied when the three slots were concurrently blown.

Air was independently supplied to the three slots during the experiment, and the same flow rate was applied. The experimental conditions are summarised in Table 3. The experimental wind speed was 8 m/s, and the Reynolds number was 1.21×10^5.

Table 3. Experimental parameters of variable blowing coefficient.

Total Flow Q (m^3/h)	$Q_{\text{-Slot.1}}$	$C_{\mu\text{-Slot.1}}$	$Q_{\text{-Slot.2}}$	$C_{\mu\text{-slot.2}}$	$Q_{\text{-Slot.3}}$	$C_{\mu\text{-slot.3}}$
0	0	0	0	0	0	0
41.91	13.97	0.01	13.97	0.01	13.97	0.01
59.28	19.76	0.02	19.76	0.02	19.76	0.02
72.57	24.19	0.03	24.19	0.03	24.19	0.03
83.82	27.94	0.04	27.94	0.04	27.94	0.04
93.72	31.24	0.05	31.24	0.05	31.24	0.05
118.53	39.51	0.08	39.51	0.08	39.51	0.08
132.54	44.18	0.1	44.18	0.1	44.18	0.1

The aerodynamic characteristic curves of the wing with different blowing coefficients recorded, when the three slots were simultaneously blown into, are shown in Figure 7. The lift coefficient curve of the wing gradually shifted upward when the blowing coefficient increased, and the maximum lift coefficient increased from 1.15 (without blowing) to 1.81 ($C_{\mu\text{-Slot.1}} = C_{\mu\text{-Slot.2}} = C_{\mu\text{-Slot.3}} = 0.1$), as shown in Figure 7a. The drag coefficients in the presence of blowing were smaller than those in the absence of blowing, as shown in Figure 7b. As shown in Figure 7c, the lift-to-drag ratio also increased significantly with the increase in the blowing coefficient. The results show that an increase in the blowing coefficient had a significant effect on improving the wing aerodynamic performance under the action of simultaneous blowing of the three slots.

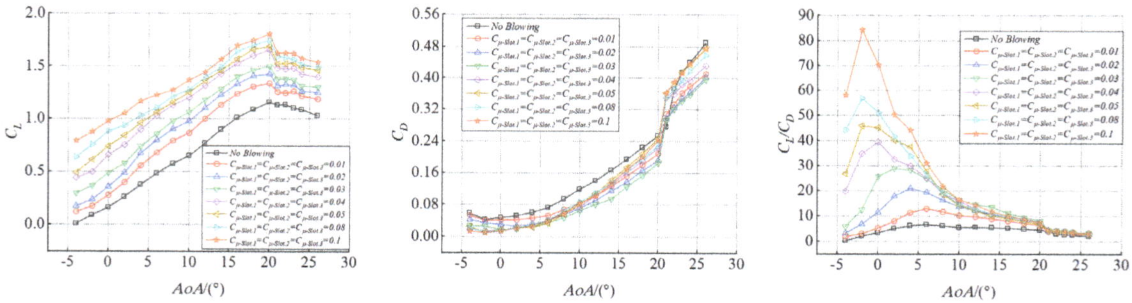

(a) (b) (c)

Figure 7. Aerodynamic curves of wing characteristics with different blowing coefficients ($U_\infty = 8$ m/s). (**a**) Lift coefficient. (**b**) Drag coefficient. (**c**) Lift-to-drag ratio.

As shown by the C_L–C_μ curve in Figure 8a, the wing lift coefficient gradually increased with the blowing coefficient; however, the slope of the lift coefficient curve decreased after $C_{\mu\text{-Slot.1}} = C_{\mu\text{-Slot.2}} = C_{\mu\text{-Slot.3}} = 0.04$. As shown by the C_D–C_μ curve in Figure 8b, the drag coefficient did not monotonically decrease; instead, it increased after the blowing coefficient reached $C_{\mu\text{-Slot.1}} = C_{\mu\text{-Slot.2}} = C_{\mu\text{-Slot.3}} = 0.03$ and then finally flattened out. The C_L/C_D–C_μ curve in Figure 8c shows that the slope of the lift-to-drag ratio curve reached the maximum at a small angle of attack and tended to be gentle as the angle of attack gradually increased.

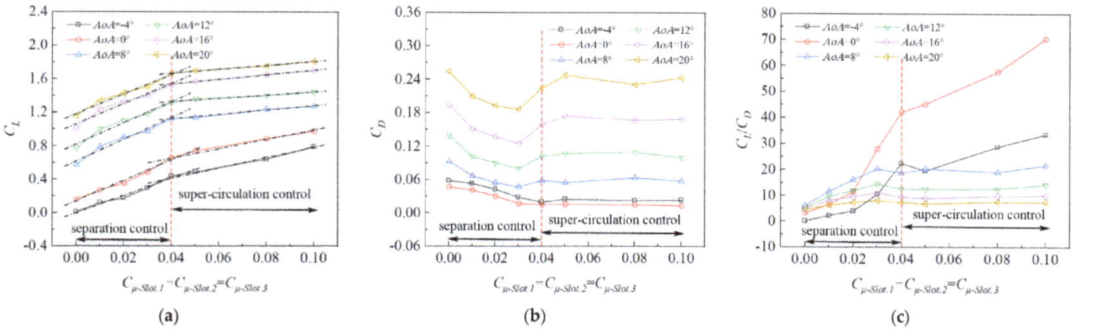

Figure 8. Wing aerodynamic characteristic curve with increasing blowing coefficient (U_∞ = 8 m/s). (**a**) Lift coefficient. (**b**) Drag coefficient. (**c**) Lift-to-drag ratio.

This is a typical phenomenon in which the lift and drag coefficients of a circulation control aerofoil appear to be critical points as the blowing coefficients increase [27]. At critical points, the blowing coefficients are also critical values. The critical point is preceded by the separation control stage; that is, when the blowing coefficient gradually increases, the jet separation position moves downward along the Coanda surface under the action of a high-speed jet and gradually moves away from the slot. After attaining the critical point, the supercirculation control stage follows (i.e., the separation position of the jet does not vary with the increase in the blowing coefficient). Compared with the supercirculation control stage, the effect of lift increase and drag reduction on the separation control stage was better.

3.3. *Effect of Different Proportions of Air Blowing on Wing Aerodynamics*

The experiment on the blowing coefficient showed that increasing this coefficient could improve the lift and reduce the drag. However, as the blowing coefficient increased, the external input flow gradually increased, also increasing the energy consumption and weakening the lifting effect in the supercirculation control. Therefore, increasing the blowing coefficient has a specific limit in improving the aerodynamic performance of the wing. Accordingly, the aerodynamic efficiency of the wing under multistage circulation control must be enhanced without increasing the energy consumption, and the efficiency of each slot must be maximised. To accomplish this, an experiment in which the flow distribution of the three slots was varied when the total input flow was constant was conducted.

To evaluate the contribution of the different slots to the aerodynamic force of the wing, three slots were independently blown into. Then, a combined blowing control experiment on the three slots was performed.

3.3.1. Individual Blowing Experiments for Each Slot

In the experiment, air was independently supplied to the three slots at the same flow rate. The experimental conditions are summarised in Table 4. The experimental wind speed was 8 m/s, and the Reynolds number was 1.21 × 10^5.

Table 4. Experimental parameters of independent blowing of three slots.

	Slot.1 Independently Blown			Slot.2 Independently Blown			Slot.3 Independently Blown		
Slot	**Slot.1**	**Slot.2**	**Slot.3**	**Slot.1**	**Slot.2**	**Slot.3**	**Slot.1**	**Slot.2**	**Slot.3**
Total flow, Q (m^3/h)	27.94	/	/	/	27.94	/	/	/	27.94
C_μ	0.04	/	/	/	0.04	/	/	/	0.04

The aerodynamic characteristic curves of the wing without blowing and with each slot blown separately are shown in Figure 9. The lift coefficient curves for each slot blown separately are shown in Figure 9a. The minimum and maximum lift coefficients were attained at Slot.1 and Slot.3, respectively. The increase in lift in Slot.1 was small because its position was close to the suction surface, the entrainment effect of the slot jet on the mainstream was weak, and the jet was prematurely separated from the Coanda surface. In contrast, the lift increased in Slot.3 because its position was close to the pressure surface, and the jet flow increased the circulation around the aerofoil.

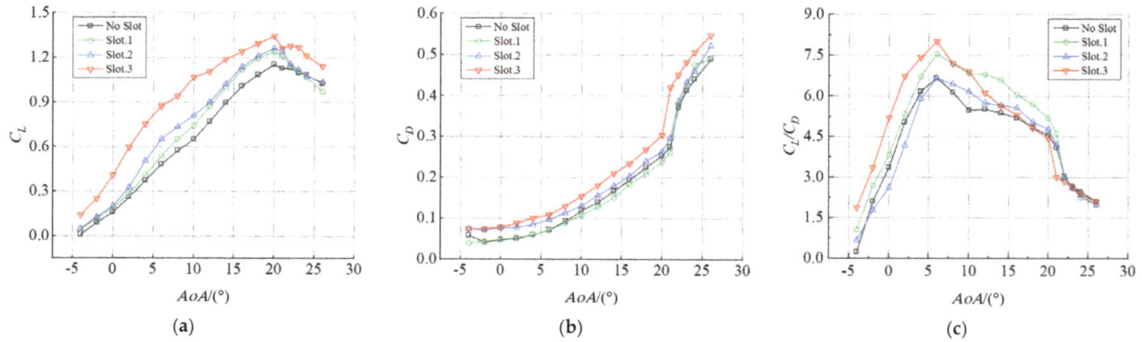

Figure 9. Aerodynamic curve of wing with each slot independently blown (U_∞ = 8 m/s, Q = 27.94 m^3/h). (**a**) Lift coefficient. (**b**) Drag coefficient. (**c**) Lift-to-drag ratio.

The drag coefficients of the three independently blown slots are compared in Figure 9b. The drag coefficients were minimum and maximum when Slot.1 and Slot.3 were subjected to blowing. Because Slot.1 is close to the suction surface, more energy is added to the boundary layer during blowing. This delays flow separation and reduces the drag on the wing. In the independent blowing experiment on the three slots, the drag reduction effect was excellent when Slot.1 was blown, whereas the increase in lift was evident when Slot.3 was blown.

3.3.2. Results of Experiment on Different Blowing Ratios for Three Slots

In the experiment with different blowing ratios, air was independently supplied to the three slots. In other words, the total input flow was fixed, and the input flow of each slot was allocated according to different proportions. The experimental conditions are summarised in Table 5. The experimental wind speed was 8 m/s, and the Reynolds number was 1.21 × 10^5.

Table 5. Experimental parameters of different blowing proportions.

Total Input Flow, Q (m^3/h)	**Flow Rate, Q, through Each Slot (m^3/h)**			**Blowing Ratio**
	Slot.1	**Slot.2**	**Slot.3**	
41.91	13.97	13.97	13.97	1:1:1
41.91	10.47	10.47	20.95	1:1:2
41.91	6.98	13.97	20.94	1:2:3
41.91	20.94	13.97	6.98	3:2:1
41.91	20.94	6.98	13.97	3:1:2

The characteristic aerodynamic curves of the wing when the three slots were blown at different proportions when the total input flow, Q = 41.91 m^3/h, are shown in Figure 10. When the five blowing ratios shown in the figure were compared, the best control effect was observed when the 3:1:2 ratio was adopted. When the blowing ratios were 1:1:1 and

3:1:2, the maximum lift-to-drag ratios increased by 97.1 and 143.48%, respectively. The lift coefficient was highest when the blowing ratio was 1:2:3; however, the lift-to-drag ratio was low. This is because Slot.3 is close to the pressure surface, and the blowing flow distribution is considerable in this slot, thus increasing the lift. However, Slot.1 has a low flow distribution and no substantial drag-reduction effect; consequently, the lift–drag ratio does not significantly increase. These results are consistent with the conclusions of the slot-independent blowing experiment.

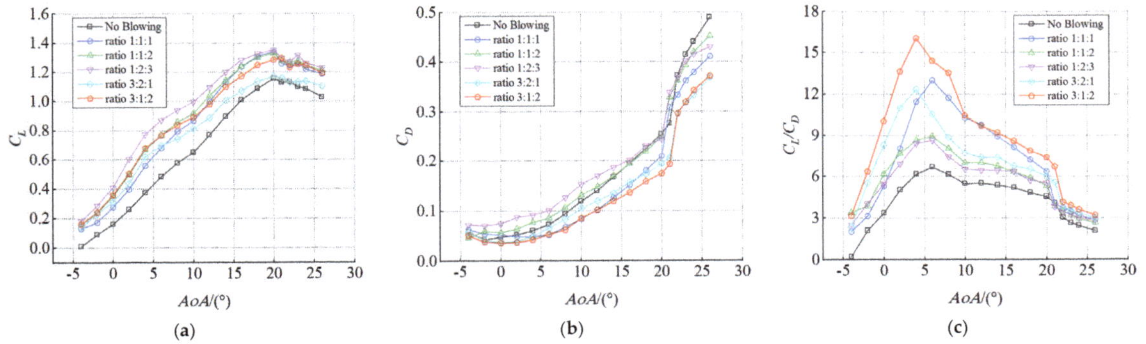

Figure 10. Wing aerodynamic characteristic curves at different blowing ratios (U_∞ = 8 m/s, Q = 41.91 m^3/h). (**a**) Lift coefficient. (**b**) Drag coefficient. (**c**) Lift-to-drag ratio.

The slots have the following advantages. Slot.1 reduces the wing drag, Slot.2 has a transfer jet effect that inhibits the premature separation of the jet from the arc surface, and Slot.3 moves the stagnation point at the trailing edge further down, thus increasing the circulation and lift.

4. Mechanism of Multistage Circulation Control

This section mainly discusses the flow-field structure of the trailing edge when the experimental wind speed is 8 m/s and the Reynolds number is 1.21×10^5. The flow mechanism was analysed under the conditions of increasing the number of slots, increasing blowing coefficient, and varying blowing ratio.

4.1. PIV Flow-Field Effect on Wing Aerodynamics at Different Number of Slots

In the flow experiment, the trailing edge was captured with no blowing and with single-slot blowing, double-slot blowing, and three-slot blowing under the condition that the total input flow was Q = 41.91 m^3/h.

4.1.1. Control Mechanism of Variation in Number of Slots at a Low Angle of Attack

The time-averaged flow-field diagram of blowing air resulting from the increase in the number of trailing slots and when the angle of attack was 0° is shown in Figure 11. The flow-field diagram of blowing air with the angle of attack at 0° and increasing number of trailing slots is shown in Figure 11. According to the velocity cloud and streamline distribution, the trailing edge exhibited a distinct flow separation phenomenon without air blowing. This occurs because the trailing edge is modified compared with a normal aerofoil (blunt leading edge and pointed trailing edge). As the number of slots increased, the flow separation zone and range decreased. The deflection curvature of the trailing-edge streamlines was also found to increase with the number of slots.

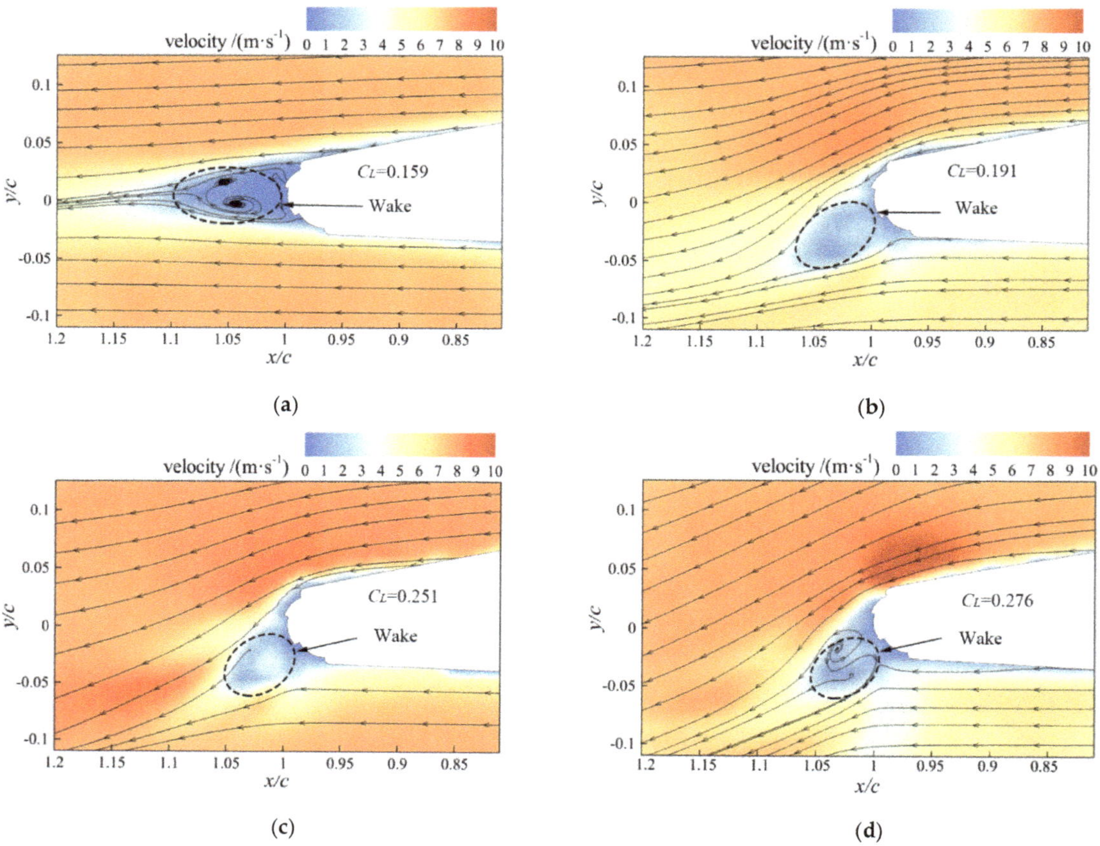

Figure 11. Comparison of velocity cloud and streamline diagrams at different numbers of slots (U_∞ = 8 m/s, AoA = 0°). (**a**) No slot. (**b**) Single slot. (**c**) Two slots. (**d**) Three slots.

The flow separation moved downward along the surface under the influence of the Coanda effect with the increasing number of blowing slots. First, multiple-slot blowing had a relay effect. This influence intensifies the shear action of blowing, accelerates the flow near the jet, and increases the downward bending of the streamlines near the trailing edge. Consequently, the equivalent aerofoil camber changes, increasing the circulation around the aerofoil, which in turn increases lift.

4.1.2. Control Mechanism of Variation in Number of Slots at High Angle of Attack

The time-averaged flow-field diagram of blowing air resulting from the increase in the number of trailing slots and when the angle of attack was 20° is shown in Figure 12. When blowing was not applied (Figure 12a), flow separation occurred on the suction surface, and a backflow zone appeared near the trailing edge. With one slot open (Figure 12b), the streamlines near the pressure surface exhibited a distinct deflection; however, a large reflux area remained on the suction surface. With two slots open (Figure 12c), flow separation was suppressed, and some areas attached to the suction surface until fluid started to flow. When all the three slots were open (Figure 12d), the flow separation was significantly inhibited, and the fluid attachment area further increased.

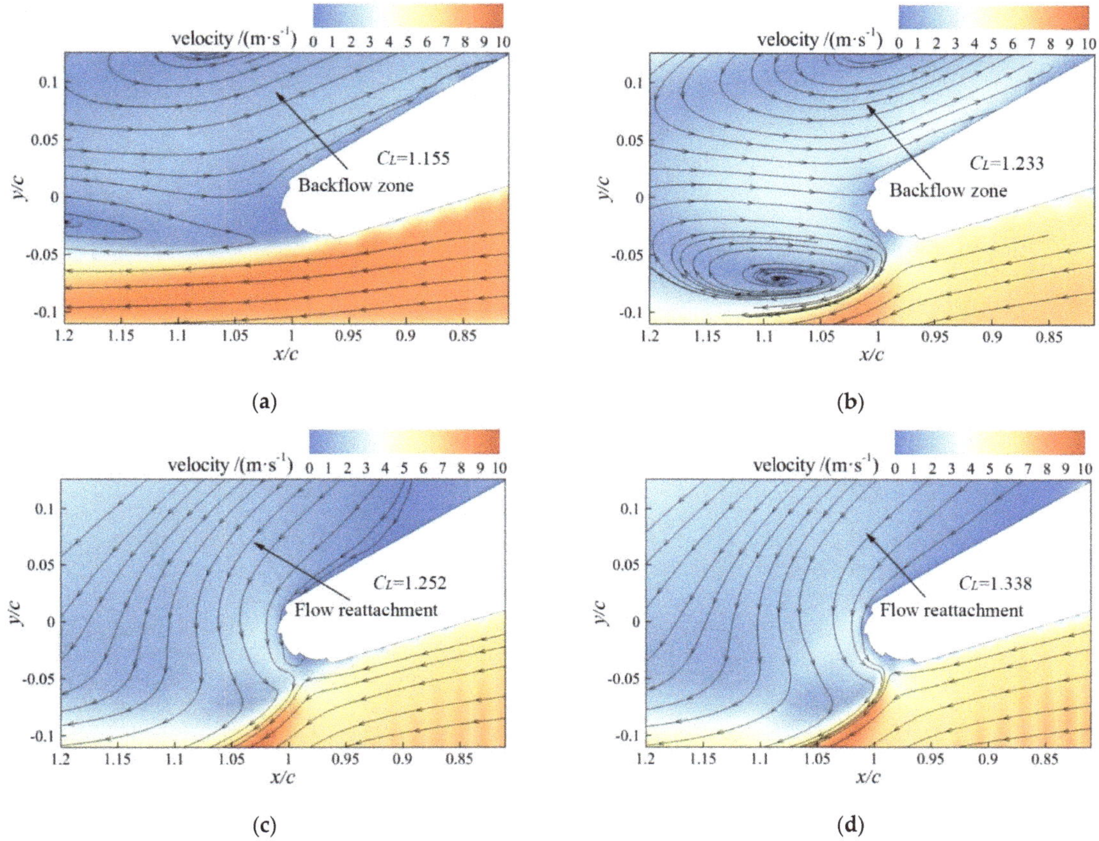

Figure 12. Comparison of velocity cloud and streamline diagrams with different number of slots (U_∞ = 8 m/s, AOA = 20°). (**a**) No slot. (**b**) Single slot. (**c**) Two slots. (**d**) Three slots.

The foregoing results indicate that under a high angle of attack, the control effect of single-slot blowing on the separation zone of the suction surface was weak. However, with double-slot and triple-slot blowing, the separation zone gradually decreased. The phenomenon of flow reattachment on the suction surface also increasingly became evident, improving the wing lift and reducing the pressure differential drag.

The extraction of the time-averaged composite velocity distribution curves at the trailing edge at x/c = 1 in Figures 11 and 12 is shown in Figure 13. As shown in Figure 13a, when the trailing edge was not blown, the range of the wake area was $-0.05 < y/c < 0.05$ (shaded area in the figure), and the velocity on both suction and pressure surfaces was approximately 0.98 U_∞. When the number of blowing slots was increased to three, the range of the wake area was decreased to $-0.06 < y/c < 0.03$. Moreover, the velocity in regions $y/c > 0.05$ (suction surface) and $y/c < -0.05$ (pressure surface) increased to 1.3 U_∞ and decreased to 0.5 U_∞, respectively. Therefore, multiple-slot blowing can cause the wake area to move down. This increases and decreases the velocities of the suction and pressure surfaces, respectively.

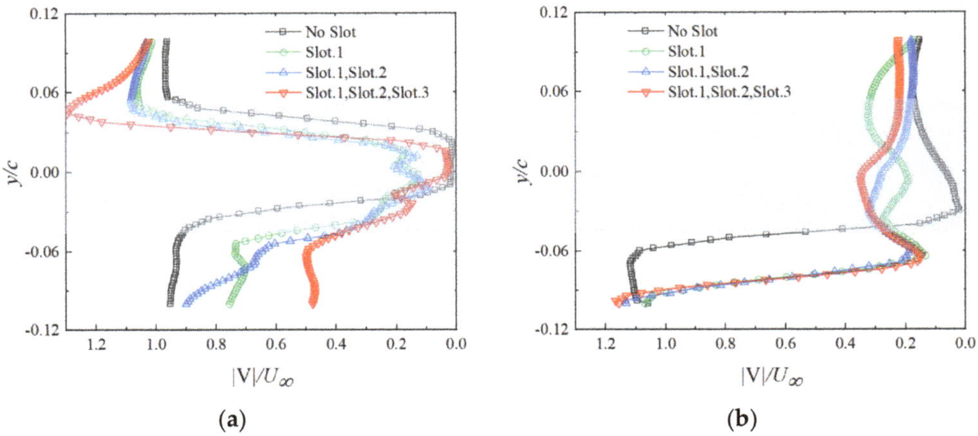

Figure 13. Near-field wake profiles with different number of slots (U_∞ = 8 m/s, x/c = 1). (**a**) AoA = 0°. (**b**) AoA = 20°.

As shown in Figure 13b, when the trailing edge was not subjected to blowing, the lowest speed in the trailing-edge wake zone (shaded area in Figure 13b) was 0.02 U_∞. This figure also shows that the speed in the $y/c > 0.05$ area was approximately 0.15 U_∞. When the number of blowing slots was increased to three, the velocity in the $y/c > 0.05$ (suction surface) area also increased to 0.3 U_∞. The wake area moved downward, and the velocity increased to 0.3 U_∞. Therefore, with the increase in the number of slots at a high angle of attack, the velocity loss in the wake area decreased. Moreover, the velocity in the separation area significantly increased, reducing the drag on the wing. This observation is consistent with the experimental results of force measurement shown in Figure 6.

4.2. PIV Flow-Field Effect on Wing Aerodynamics Due to Blowing Coefficient

The force measurement experiment indicated the successive appearance of separate control stage and supercirculation control stages with an increase in the blowing coefficient. In the experiments, the flow field no blowing control and the flow fields with blowing coefficients of 0.03, 0.05 and 0.1 were captured.

4.2.1. Control Mechanism of Variation in Blowing Coefficient at Low Angle of Attack

The time-averaged flow-field diagrams at different blowing coefficients when the three slots were used for simultaneous blowing are shown in Figure 14. The velocity cloud and streamline distribution indicate that the trailing edge exhibits a distinct flow separation phenomenon when no air is blown. With an increase in the blowing coefficient, the velocity on the suction surface and the curvature of the downward deflection of the streamline increased. By comparing the velocities at the coordinate points (x/c = 1; y/c = 0.05) shown in Figure 14, the velocity was found to increase to 9.8, 15.2, and 23.1% when the blowing coefficients were 0.03, 0.05, and 0.1, respectively.

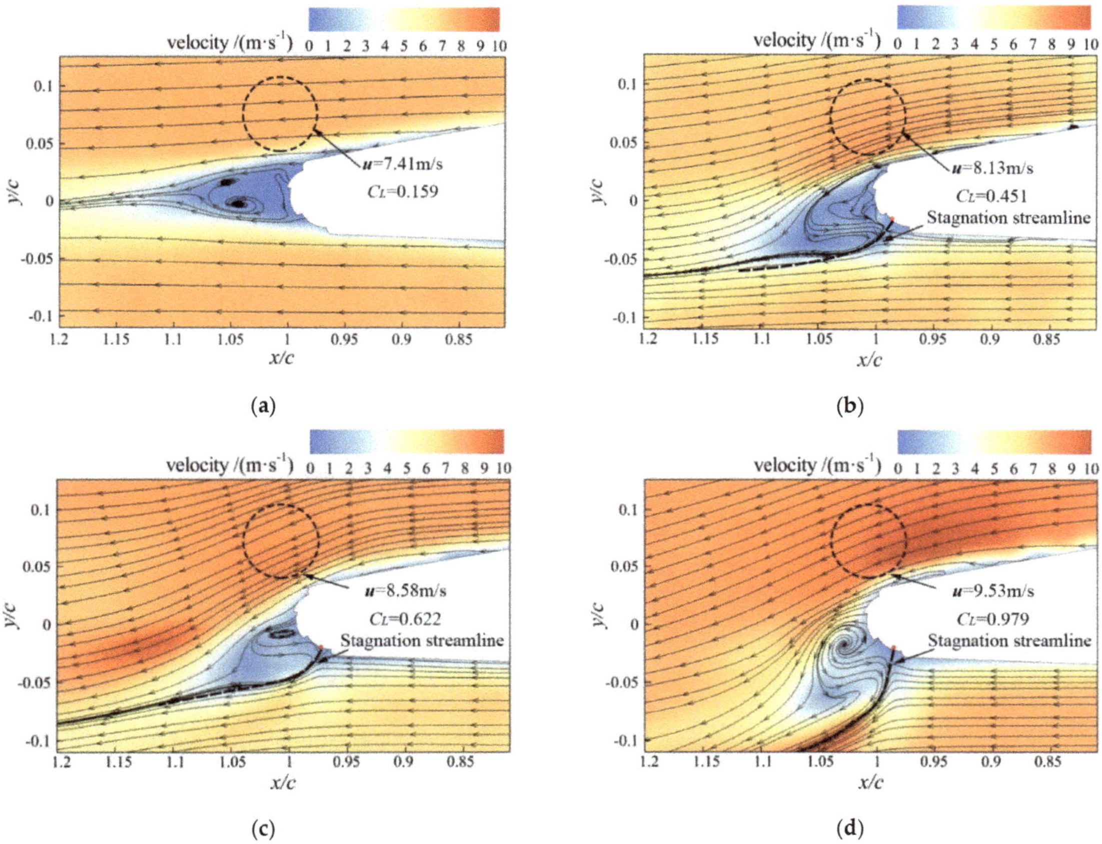

Figure 14. Comparison of velocity cloud and streamline diagrams at different blowing coefficients (U_∞ = 8 m/s, AOA = 0°). (**a**) No Blowing. (**b**) $C_{\mu\text{-}Slot.1} = C_{\mu\text{-}Slot.2} = C_{\mu\text{-}Slot.3} = 0.03$. (**c**) $C_{\mu\text{-}Slot.1} = C_{\mu\text{-}Slot.2} = C_{\mu\text{-}Slot.3} = 0.05$. (**d**) $C_{\mu\text{-}Slot.1} = C_{\mu\text{-}Slot.2} = C_{\mu\text{-}Slot.3} = 0.1$.

By comparing the positions of the stagnation points (red dot in Figure 14) of the trailing edge, the stagnation points of the trailing edge moved down significantly when the blowing coefficients were 0.03 and 0.05. However, when the blowing coefficient was 0.1, the position of the stagnation points of the trailing edge was virtually the same as that when the blowing coefficient was 0.05. This occurs because when the blowing coefficients are 0.05 and 0.1, the trailing edge is under supercirculation control. With an increase in the blowing coefficient, the stagnation point of the trailing edge ceases to move down; this is also consistent with the force measurement experiment result shown in Figure 8.

At a low angle of attack (AoA = 0°), the strong shear action of the jet has a significant acceleration effect on the suction surface with an increase in the blowing coefficient. Moreover, the curvature of the downward deflection of the streamline gradually increases. In the separation control stage, the trailing-edge stagnation point moves downward along the arc surface by increasing the blowing coefficient. In the supercirculation control stage, the trailing-edge stagnation point never varies.

4.2.2. Control Mechanism of Variation in Blowing Coefficient at High Angle of Attack

The time-averaged flow-field diagram, when the three slots were blown simultaneously at different blowing coefficients and the angle of attack was 20°, is shown in Figure 15. The velocity cloud and streamline distribution indicate that flow separation occurred on the suction surface without air blowing, and a backflow zone appeared. When the blowing coefficient was increased to 0.03, the backflow area near the slot on the suction surface tended to move downward because of the entrainment of jet flow. The streamlines began to deflect downward near the pressure surface on the back edge. When the blowing coefficient was increased to 0.05, the backflow area decreased, and air attached to most of the suction surface. When the blowing coefficient was increased to 0.1, the flow attachment area on the suction surface increased, and the deflection curvature of the trailing-edge streamline also significantly increased. At this time, the flow separation was significantly inhibited, and the backflow area disappeared. With an increase in the blowing coefficient, the variation laws of the stagnation streamline and stagnation point (red dot in Figure 15) of the trailing edge are consistent with the law when the angle of attack is 0°.

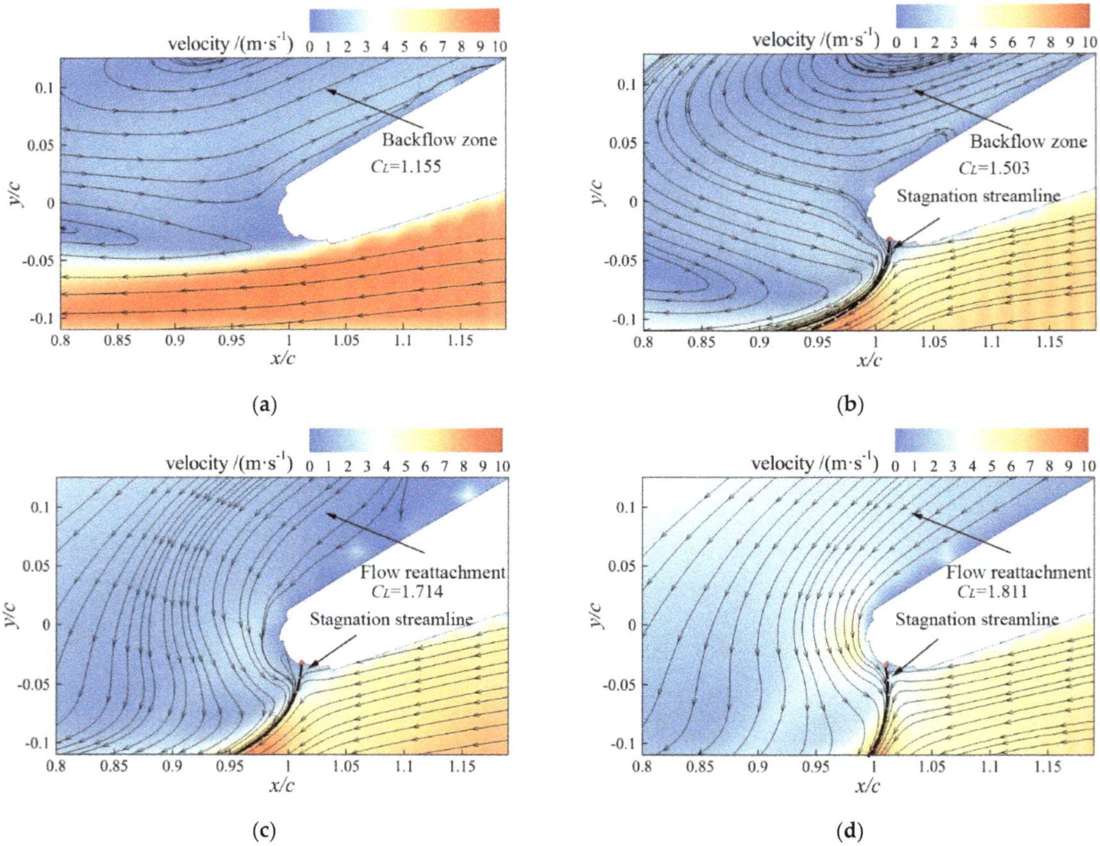

Figure 15. Comparison of velocity cloud and streamline diagrams at different blowing coefficients (U_∞ = 8 m/s, AoA = 20°). (**a**) No Blowing. (**b**) $C_{\mu\text{-}Slot.1} = C_{\mu\text{-}Slot.2} = C_{\mu\text{-}Slot.3} = 0.03$. (**c**) $C_{\mu\text{-}Slot.1} = C_{\mu\text{-}Slot.2} = C_{\mu\text{-}Slot.3} = 0.05$. (**d**) $C_{\mu\text{-}Slot.1} = C_{\mu\text{-}Slot.2} = C_{\mu\text{-}Slot.3} = 0.1$.

At a high angle of attack (AoA = 20°), when the blowing coefficient was small, the jet energy was weak. This could not overcome the large inverse pressure gradient on the suction surface, and the control effect on the backflow zone was weak. When the blowing coefficient gradually increased, the high-speed jet drove and accelerated the flow in the backflow zone on the suction surface. It had the effect of controlling flow separation.

The time-averaged composite velocity distribution curve derived from the trailing edge at x/c = 1 (Figures 14 and 15) is shown in Figure 16. As shown in Figure 16a, without blowing, the trailing-edge wake velocity decreased to 0.02 U_∞. When the blowing coefficient increased to 0.1, the wake area speed increased to 0.3 U_∞. Therefore, increasing the blowing coefficient can reduce the trailing-edge wake speed loss. Based on the distribution on the wake area in y/c, without blowing, the wake area was in the range $-0.05 < y/c < 0.05$. When the blowing coefficient was increased to 0.03, the wake area range was decreases to $-0.06 < y/c < 0.03$. At this time in the separation control phase, the wake area was reduced so that the drag was reduced. When the blowing coefficient was increased to 0.1, the wake area range was increased to $-0.09 < y/c < 0.05$. At this point, the wake area was in the supercirculation control stage, and the lift and resistance increased.

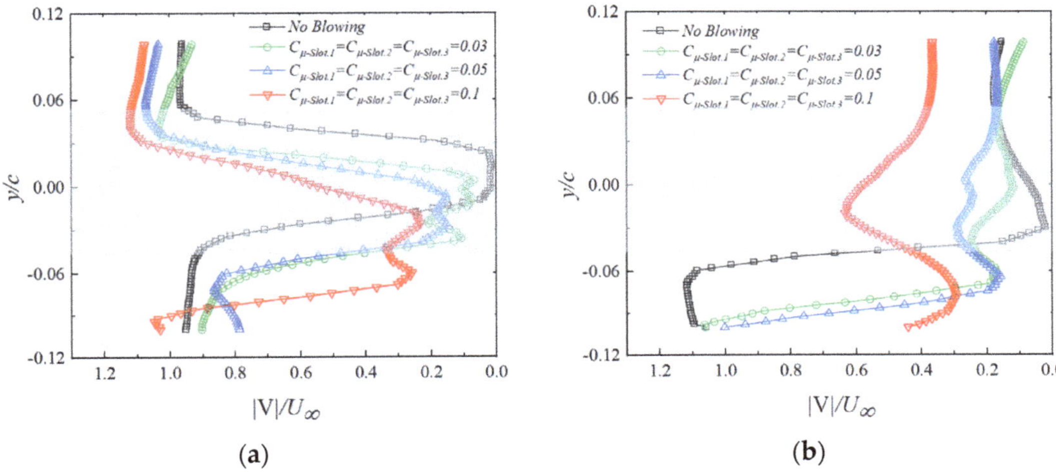

Figure 16. Near-field wake profiles with different blowing coefficients (U_∞ = 8 m/s, x/c = 1). (**a**) $AoA = 0°$. (**b**) $AoA = 20°$.

As shown in Figure 16b, the velocity was approximately 0.2 U_∞ in the region $y/c > 0.05$ (suction surface) under the condition of no air blowing and a low air-blowing coefficient. When the blowing coefficient was increased to 0.1, the velocity increased to 0.4 U_∞. Therefore, the acceleration effect of the small blowing coefficient on the suction surface separation zone was weak. However, as the blowing coefficient increased, the high-speed jet significantly enhanced the suction surface velocity through entrainment.

4.3. PIV Flow-Field Effect on Wing Aerodynamics at Different Air Blowing Proportions

Under the condition of the same external input flow, the control mode with a blowing ratio of 3:1:2 could further improve the wing lift and reduce the drag according to the force measurement experiment. The flow display experiment captured the flow field at the trailing edge when the blowing ratios of the three slots were 1:1:1 and 3:1:2 and the total input flow was $Q = 93.72\ m^3/h$.

4.3.1. Control Mechanism under Different Blowing Proportions at Low Angle of Attack

The time-averaged flow-field diagrams for different blowing ratios when the angle of attack was 0° are shown in Figure 17. The velocity cloud image indicates that when the blowing ratio was 3:1:2, the streamline curvature of the trailing edge was larger compared with that when the ratio was average. By comparing the speed at the coordinate points (x/c = 1; y/c = 0.05), the control result of the relative average blowing ratio could be determined. When the blowing ratio was 3:1:2, the speed at this position increased by 4.6%. Moreover, the low-speed area of the trailing edge (shown by the red dotted line) significantly decreased, and the velocity on the suction surface increased.

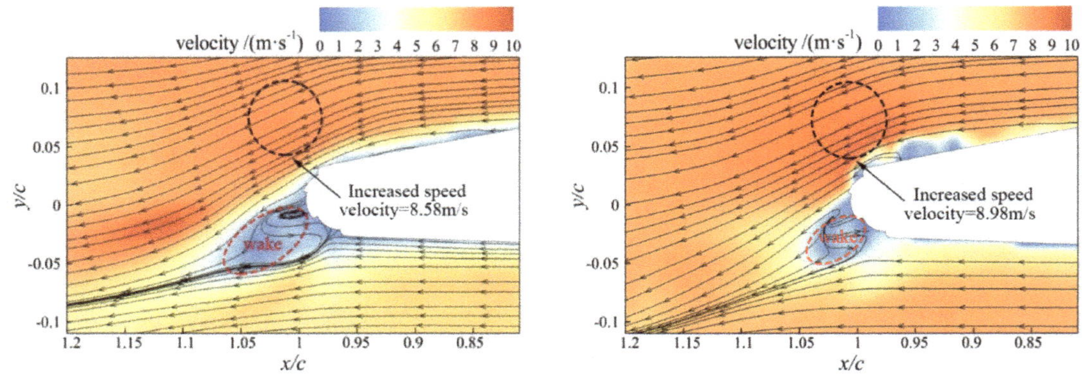

(a) (b)

Figure 17. Comparison of velocity cloud and streamline diagrams at different blowing ratios (U_∞ = 8 m/s, AoA = 0°). (**a**) 1:1:1 Blowing ratio. (**b**) 3:1:2 Blowing ratio.

With the same energy consumption, a blowing ratio of 3:1:2 can further increase the flow rate on the suction surface and increase the curvature of the downward deflection of the trailing-edge streamline. Moreover, it promotes the stagnation point of the trailing edge to move downward, thus further improving the wing lift. In addition, the trailing-edge low-speed zone significantly decreases, thus reducing the wing drag.

4.3.2. Control Mechanism under Different Blowing Proportions at High Angle of Attack

The time-averaged flow-field diagrams at different blowing ratios when the angle of attack was 20° are shown in Figure 18. According to the velocity cloud and streamline distribution, the wake velocity was higher at the slot position on the trailing edge of the wing when the blowing ratio was 3:1:2 than when it was 1:1:1.

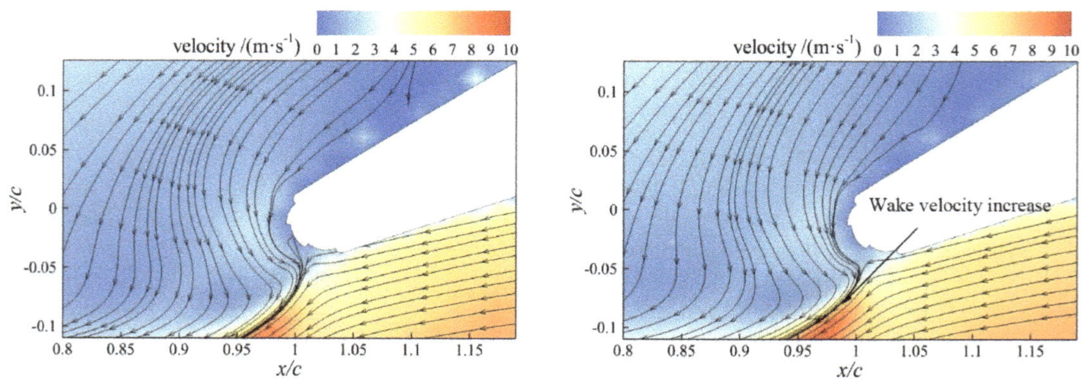

(a) (b)

Figure 18. Comparison of velocity cloud and streamline diagrams at different blowing ratios (U_∞ = 8 m/s, AoA = 20°). (**a**) 1:1:1 Blowing ratio. (**b**) 3:1:2 Blowing ratio.

The foregoing results show that the blowing ratio of 3:1:2 at a high angle of attack had a weak control effect on flow separation, but it could improve the wake velocity. This can greatly reduce the drag of the wing.

The time-averaged composite velocity distribution curve derived from the trailing edge of x/c = 1 (Figures 17 and 18) is shown in Figure 19. At a low angle of attack, the minimum velocity in the wake area (shaded area in Figure 18a) was virtually the same as that when the average proportion blowing control was 3:1:2. However, the wake area evidently decreased from $-0.05 < y/c < 0.03$ to $-0.06 < y/c < 0.01$ when the blowing ratio was 3:1:2. Furthermore, the size range of the y/c distribution decreased. At a high angle of attack, the 3:1:2 blowing ratio of the control mode increased the velocity on the suction and pressure surfaces.

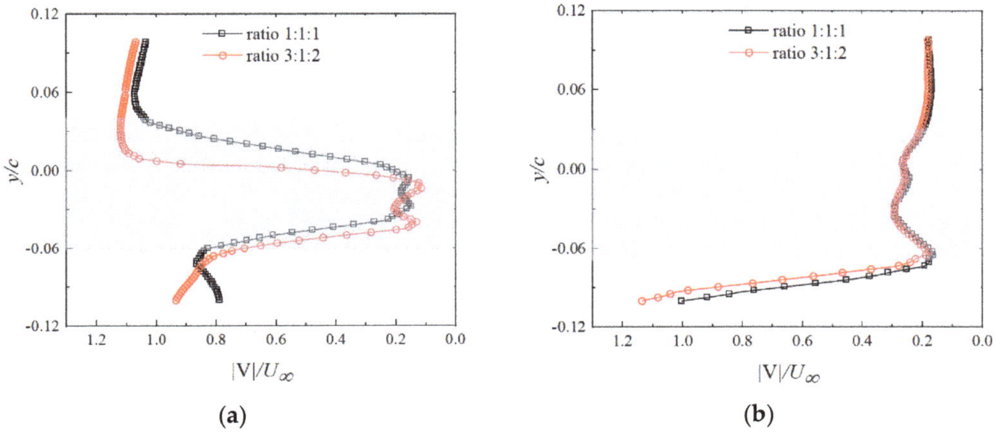

Figure 19. Near-field wake profiles with different blowing ratios (U_∞ = 8 m/s, x/c = 1). (**a**) AoA = 0°. (**b**) AoA = 20°.

The above results indicate that the 3:1:2 ratio of blowing control increases the velocity on the suction surface and reduces the velocity deficit in the wake area regardless of the angle of attack. In addition, the wake area moves downward, and the distribution area of y/c significantly decreases (AoA = 0°), thus reducing the drag.

5. Conclusions

In this study, the aerodynamic control effect of a multistage circulation control wing was investigated through force measurements and PIV experiments. Then, based on a three-slot blowing experiment, the aerodynamic force and flow-field characteristics of the wing were studied when different blowing coefficients and blowing ratios were applied. The conclusions are as follows:

(1) The lift and lift-to-drag ratio increased, and the drag decreased as the number of blowing slots in the trailing edge increased. When the experimental wind speed was 8 m/s, the maximum lift-to-drag ratios corresponding to single-slot blowing, double-slot blowing, and triple-slot blowing (compared with no blowing) increased by 13.2, 21.5, and 95.3%, respectively.

(2) When the three slots were subjected to blowing simultaneously, the aerodynamic characteristics of the wing could be significantly improved by increasing the blowing coefficient. In the separation control stage, the lift coefficient significantly increased, and the drag coefficient gradually decreased with an increase in the blowing coefficient. In the supercirculation control, the lift coefficient gradually increased, and the drag coefficient first increased and then flattened. The lift increase and aerodynamic efficiency were better in the separation control than in the supercirculation control.

(3) When different blowing ratios were adopted for the three slots, the wing control effect could be further improved with the same total flow. The maximum lift-to-drag ratios increased by 97.1 and 143.48% at 1:1:1 and 3:1:2, respectively (compared with those in the absence of blowing). Remarkably, blowing through Slot.1 reduced the wing drag, whereas blowing through Slot.3 increased the wing lift.

(4) In terms of the flow mechanism, at a low angle of attack, increasing the number of slots, had a relay effect on the trailing edge. This effect increased the curvature of the streamline near the Coanda surface, thus increasing the equivalent aerofoil camber and enhancing the wing lift. When air was blown simultaneously through the three slots, the increase in the blowing coefficient had a distinct acceleration effect on the suction surface. Moreover, velocity circulation on the wing increased, and the lift significantly improved. Furthermore, in the separation control, the entrainment effect of the blowing jet of the trailing edge through the upper surface enhanced the wake velocity. This reduced speed loss and drag. The control mode with a blowing ratio of 3:1:2 further reduced the velocity loss in the wake area of the trailing edge.

At a high angle of attack, increasing the number of slots and blowing coefficient could inhibit the formation of a backflow zone and enable the reattachment of flow to the suction surface. The control method with a blowing ratio of 3:1:2 could cause flow reattachment in the separation area of the suction surface and reduce velocity loss at the trailing edge.

Author Contributions: H.D. wrote the paper. L.Y. participated in the wind tunnel experiment and analysed the calculated results. S.C. provided constructive guidance in the process of preparing this paper. S.H. and W.Z. participated in the analysis and collation of literature materials. All authors have read and agreed to the published version of the manuscript.

Funding: This work was supported by the National Natural Science Foundation of China (Grant No. 51806181) and Foundation of the National Key Laboratory of Science and Technology on Aerodynamic Design and Research (No. 614220121030205).

Institutional Review Board Statement: Not applicable.

Informed Consent Statement: Not applicable.

Data Availability Statement: The data supporting the findings of this study are available from the corresponding author upon reasonable request.

Conflicts of Interest: The authors declare no conflict of interest.

References

1. Hassanzadeh, A.; Hassanabad, A.H.; Dadvand, A. Aerodynamic shape optimization and analysis of small wind turbine blades employing the Viterna approach for post-stall region. *Alex. Eng. J.* **2016**, *55*, 2035–2043. [CrossRef]
2. Mm, A.; Mjm, B. Improvement of wind turbine aerodynamic performance by vanquishing stall with active multi air jet blowing. *Energy* **2021**, *224*, 120–176.
3. Gen, M.S.; Akel, H.H.; Koca, K. Effect of partial flexibility over both upper and lower surfaces to flow over wind turbine airfoil. *Energy Convers. Manag.* **2020**, *219*, 113042.
4. Wang, H.; Jiang, X.; Chao, Y.; Li, Q.; Li, M.; Zheng, W.; Chen, T. Effects of leading edge slat on flow separation and aerodynamic performance of wind turbine. *Energy* **2019**, *182*, 988–998. [CrossRef]
5. Zhang, P.; Yan, B.; Dai, C. Lift enhancement method by synthetic jet circulation control. *Sci. China Technol. Sci.* **2012**, *55*, 2585–2592. [CrossRef]
6. Stalnov, O.; Kribus, A.; Seifert, A. Evaluation of active flow control applied to wind turbine blade section. *J. Renew. Sustain. Energy* **2010**, *2*, 12080. [CrossRef]
7. Krentel, D.; Nitsche, W. Investigation of the near and far wake of a bluff airfoil model with trailing edge modifications using time-resolved particle image velocimetry. *Exp. Fluids* **2013**, *54*, 1551. [CrossRef]
8. Kotsonis, M.; Pul, R.; Veldhuis, L. Influence of circulation on a rounded-trailing-edge airfoil using plasma actuators. *Exp. Fluids* **2014**, *55*, 1772. [CrossRef]
9. Harris, M.J. Investigation of the Circulation Control Wing/Upper Surface Blowing High-Lift System on a Low Aspect Ratio Semispan Model. In *Report DTNSRDC/ASED-81/10*; David Taylor Naval Ship R&D Center, Aviation and Surface Effects Department: Bethesda, MD, USA, 19 May 1981.
10. Kind, R.J.; Maull, D.J. An Experimental Investigation of a Low-Speed Circulation-Controlled Aerofoil. *Aeronaut. Q.* **2016**, *19*, 170–182. [CrossRef]
11. Zhu, H.; Hao, W.; Li, C.; Ding, Q.; Wu, B. Application of flow control strategy of blowing, synthetic and plasma jet actuators in vertical axis wind turbines. *Aerosp. Sci. Technol.* **2019**, *88*, 468–480. [CrossRef]
12. Xiao, T.; Zhu, Z.; Deng, S.; Gui, F.; Li, Z.; Zhou, Z. Effects of nozzle geometry and active blowing on lift enhancement for upper surface blowing configuration. *Aerosp. Sci. Technol.* **2021**, *111*, 106536. [CrossRef]

13. Sommerwerk, K.; Krukow, I.; Haupt, M.C.; Dinkler, D. Investigation of Aeroelastic Effects of a Circulation Controlled Wing. *J. Aircr.* **2016**, *53*, 1746–1756. [CrossRef]
14. Li, Y.; Wang, X.; Zhang, D. Control strategies for aircraft airframe noise reduction. *Chin. J. Aeronaut.* **2013**, *26*, 249–260. [CrossRef]
15. Luo, Z.; Zhao, Z.; Liu, J.; Deng, X.; Zheng, M.; Yang, H.; Chen, Q.; Li, S. Novel roll effector based on zero-mass-flux dual synthetic jets and its flight test. *Chin. J. Aeronaut.* **2022**, *35*, 1–6. [CrossRef]
16. Cao, S.; Li, Y.; Zhang, J.; Deguchi, Y. Lagrangian analysis of mass transport and its influence on the lift enhancement in a flow over the airfoil with a synthetic jet. *Aerosp. Sci. Technol.* **2019**, *86*, 11–20. [CrossRef]
17. Miklosovic, D.; Imber, R.; Britt-Crane, M. Measurements of Midspan Flow Interactions of a Low-Aspect-Ratio Circulation Control Wing. *J. Aircr.* **2016**, *53*, 1969–1974. [CrossRef]
18. Wetzel, D.A.; Griffin, J.; Cattafesta, L.N. Experiments on an elliptic circulation control aerofoil. *J. Fluid Mech.* **2013**, *730*, 99–144. [CrossRef]
19. Jonathan, K.; Panthe, C.C.; Smit, J.E. Applications of Circulation Control, Yesterday and Today. *Int. J. Eng.* **2010**, *4*, 411.
20. Jones, G.; Viken, S.; Washburn, A.; Jenkins, L.; Cagle, C. An Active Flow Circulation Controlled Flap Concept for General Aviation Aircraft Applications, AIAA-2002-3157. In Proceedings of the 1st Flow Control Conference, St. Louis, MO, USA, 24–26 June 2002.
21. Englar, R.J.; Huson, G.G. Development of advanced circulation control wing high-lift airfoils. *J. Aircr.* **1984**, *21*, 476–483. [CrossRef]
22. Englar, R.J.; Smith, M.J.; Kelley, S.M.; Rover, R.C., III. Application of Circulation Control to Advanced Subsonic Transport Aircraft, Part I—Airfoil Development. *J. Aircr.* **1994**, *31*, 1160–1168. [CrossRef]
23. Englar, R. Circulation Control Pneumatic Aerodynamics: Blown Force and Moment Augmentation and Modification; Past, Present and Future. *AIAA J.* **2013**, *12*. [CrossRef]
24. Englar, R.J. Circulation Control for High Lift and Drag Generation on STOL Aircraft. *J. Aircr.* **1975**, *12*, 457–463. [CrossRef]
25. Xu, H.Y.; Qiao, C.L.; Yang, H.Q.; Ye, Z.Y. Active Circulation Control on the Blunt Trailing Edge Wind Turbine Airfoil. *AIAA J.* **2018**, *56*, 554–570. [CrossRef]
26. Xu, H.Y.; Dong, Q.L.; Qiao, C.L.; Ye, Z.Y. Flow Control over the Blunt Trailing Edge of Wind Turbine Airfoils Using Circulation Control. *Energies* **2018**, *11*, 619. [CrossRef]
27. Jones, G.S.; Lin, J.C.; Allan, B.G.; Milholen, W.E.; Rumsey, C.L.; Swanson, R.C. Overview of CFD Validation Experiments for Circulation Control Applications at NASA, AIAA Paper 2008-030041. In *2008 International Powered Lift Conference*; NASA Langley Research Center: Hampton, VA, USA, 2008; p. 23681.

applied sciences

MDPI

Article

Accelerated Parallel Numerical Simulation of Large-Scale Nuclear Reactor Thermal Hydraulic Models by Renumbering Methods

Huajian Zhang †, Xiao-Wei Guo †, Chao Li, Qiao Liu, Hanwen Xu and Jie Liu *

Institute for Quantum Information & State Key Laboratory of High Performance Computing, College of Computer Science and Technology, National University of Defense Technology, Changsha 410073, China
* Correspondence: liujie@nudt.edu.cn
† These authors contributed equally to this work.

Abstract: Numerical simulation of thermal hydraulics of nuclear reactors is widely concerned, but large-scale fluid simulation is still prohibited due to the complexity of components and huge computational effort. Some applications of open source CFD programs still have a large gap in terms of comprehensiveness of physical models, computational accuracy and computational efficiency compared with commercial CFD programs. Therefore, it is necessary to improve the computational performance of in-house CFD software (YHACT, the parallel analysis code of thermohydraulices) to obtain the processing capability of large-scale mesh data and better parallel efficiency. In this paper, we will form a unified framework of meshing and mesh renumbering for solving fluid dynamics problems with unstructured meshes. Meanwhile, the effective Greedy, RCM (reverse Cuthill-Mckee), and CQ (cell quotient) grid renumbering algorithms are integrated into YHACT software. An important judgment metric, named median point average distance (MDMP), is applied as the discriminant of sparse matrix quality to select the renumbering methods with better effect for different physical models. Finally, a parallel test of the turbulence model with 39.5 million grid volumes is performed using a pressurized water reactor engineering case component with 3*3 rod bundles. The computational results before and after renumbering are also compared to verify the robustness of the program. Experiments show that the CFD framework integrated in this paper can correctly perform simulations of the thermal engineering hydraulics of large nuclear reactors. The parallel size of the program reaches a maximum of 3072 processes. The renumbering acceleration effect reaches its maximum at a parallel scale of 1536 processes, 56.72%. It provides a basis for our future implementation of open-source CFD software that supports efficient large-scale parallel simulations.

Keywords: mesh renumbering; thermal engineering hydraulics; CFD; YH-ACT; RCM; parallel calculation

Citation: Zhang, H.; Guo, X.-W.; Li, C.; Liu, Q.; Xu, H.; Liu, J. Accelerated Parallel Numerical Simulation of Large-Scale Nuclear Reactor Thermal Hydraulic Models by Renumbering Methods. *Appl. Sci.* **2022**, *12*, 10193. https://doi.org/10.3390/app122010193

Academic Editors: Vasily Novozhilov and Cunlu Zhao

Received: 19 September 2022
Accepted: 4 October 2022
Published: 11 October 2022

1. Introduction

Because of the importance of nuclear reactions and the complexity of their physical models, research exploration using large-scale numerical simulations is essential. Some early studies of nuclear reactor thermodynamics were based on several parameters at large scale or one-dimensional system analysis methods, which mainly studied the average characteristics of physical quantities in nuclear power systems. With the boom in high performance computers, nuclear power thermohydraulic studies based on more refined computational fluid dynamics (CFD) methods and thermohydraulic software have been widely studied and applied [1,2].

The internal structure of nuclear reactors is complex, and the dominant reactor type today is the pressurized water reactor. Core fuel rod bundles are an important component of a pressurized water reactor. The fuel rods are surrounded by a flowing liquid coolant.

There is a violent flow of liquid between the bar bundles and between the bar bundles and the walls, and they are responsible for the transfer of momentum and heat. It is widely believed that the flow pulsations in the gap are closely related to the large-scale vortices. It has been studied that vortexes will form in pairs on both sides of the slit [3,4]. The flow and heat transfer characteristics of coolant flowing through the core of a nuclear reactor are extremely complex. To analyze the turbulent motion and thermodynamics of large-scale fuel rod bundles [5–10], large-scale CFD numerical simulations based on FVM (finite volume method) are a widely accepted solution. However, large-scale CFD numerical simulations require an extremely large amount of mesh data. The usual solution is to encrypt the mesh of the physical model. The larger the amount of data, the larger the solution size and the accuracy of the solution are naturally improved. However, this will lead to more and more complex and difficult to solve calculations. Today, the use of large-scale CFD techniques in thermal hydraulic analysis is still very difficult due to high technical requirements and shortage of computational resources [11].

In the CFD, simulation of fine flow field structures in real-world engineering cases is still forbidden [12]. As a result, there is widespread interest in improving computational efficiency by well-developed parallel technology. The CFD solving process involves three parts: pre-processing, numerical solving, and post-processing. The use of appropriate meshing methods and mesh numbering in the preprocessing process will greatly improve the computational performance of fluid dynamics simulations. Taking FVM [13,14] as an example, after meshing the physical model, assembly of linear systems in parallel from the number of grid cells of each divided component is necessary. Then, solving them iteratively,. In addition, the numbering of the grid determines the order of access to the individual processes for parallel computation and affects the cache hit rate. Since modern computer architecture is characterized by multiple storage levels, reasonable grid numbering has a significant impact on the efficient utilization of computational performance.

One of the key technologies in CFD preprocessing is the meshing technique. Its theoretical basis is becoming more and more mature. In recent years, the research field of finite element meshing has shifted from two-dimensional plane problems to three-dimensional entities, and the research focus has changed from triangular (tetrahedral) meshes to quadrilateral (hexahedral) meshes, focusing on the fully automatic generation of meshes, mesh adaption and other research. Mesh partitioning can be abstracted as graph partitioning. The relationship of cells and neighbors can be abstracted as the relationship of vertices and edges. The large-scale physical model is divided into multiple blocks by coarse-grained and fine-grained lattice partitioning, which facilitates parallel computation [15]. It is a challenge to achieve high quality meshing as quickly as possible and under the condition that the load balance is satisfied. The efficiency and cell quality of meshing need to be further improved.

Another key technique in preprocessing is the grid numbering technique. The effect of grid numbering techniques on the computational efficiency of solving sparse linear systems is a classical topic with many research results, examples of which can be found in the work of Gibbs et al. [16], Cuthill and McKee [17], Sloan [18], Akhras and Dhatt [19], and George et al. are reflected in the work of Reverse Cuthill–Mckee [20]. Because the heuristic renumbering algorithm consumes a lot of time during preprocessing. Therefore, heuristic algorithms may be more appropriate when computational performance is not a critical issue [21,22]. In addition, this paper studies how to accelerate the computational performance of general-purpose CFD software, so heuristic algorithms are not considered for the time being.

In general, meshing and grid numbering techniques are a traditional approach to performance optimization, and there are many existing methods to choose from. However, these techniques still lack quantitative analysis for large-scale thermal-hydraulic cases. In particular, the optimization effectiveness for general-purpose CFD software executed on supercomputers is still lacking. Therefore, this paper focuses on the effectiveness of the FVM-based computational framework for use in unstructured grids, integrating typical meshing methods and mesh numbering techniques. Ultimately, a suitable mesh renumbering method is selected to enhance the parallel scale and computational speed of numerical simulation.

The main innovative points of this paper are as follows: (1) A general framework for further integration of parallel meshing methods and renumbering schemes in general-purpose CFD software for unstructured meshes; and (2) improved runtime speed and parallel scale for real-world numerical simulations of large-scale complex cases. The case is a thermal hydraulic analysis of a 3*3 fuel rod bundle with 39.5 million grid volumes.

The structure of this paper is as follows: Section 2 describes the generic CFD software used in this paper and aims to demonstrate the advantages of the software between data structure and parallel communication. Section 3 shows the meshing strategy and the integrated RCM renumbering algorithm. Section 4 demonstrates the correctness of the massively parallel numerical simulations before and after renumbering and the huge improvement of the renumbering algorithm on the numerical simulation efficiency. Conclusions are included in Section 5.

2. General CFD Software Based on Finite Volume Method

The general-purpose CFD software used in this paper, YHACT, was initially used for thermal-hydraulic analysis. YHACT uses a modular development architecture based on scalability. It is based on a framework blueprint of CFD program modules consisting of key steps in CFD solving such as data loading, physical pre-processing, iterative solving, and result output. YHACT mainly addresses the thermal-hydraulic CFD program requirements for nuclear reactors, and more details about it can be found in [23].

2.1. *Scalability of YHACT Software Architecture*

Thanks to the modern features of object-oriented technology, YHACT can be developed modularly and has flexible scalability. YHACT is based on common parallel algorithms and parallel support, combined with existing advanced software engineering methods. It uses physical model module, numerical discretization module, linear equation system module, and parallel communication module as the main components to form the general framework of CFD program based on supercomputing system [24].

The pre-processing toolbox has functions such as mesh control and physical pre-processing. The physical model and numerical solution constitute the CFD solver, encapsulating the physical mathematical model and efficient mathematical solution methods with parametric configuration capabilities. The common algorithm implements parallel computation such as sparse iterative methods and preconditioners. Through close coupling with the computer architecture, YHACT can obtain support for massively parallel computational optimization and provide fundamental computational power for the solver. Parallel support can encapsulate the structural features of high-performance computer systems and provide fundamental computing, communication, storage, and debugging functions, thereby supporting the efficient operation of large-scale CFD computing applications. Therefore, the good scalability of YHACT makes it easier to integrate meshing techniques and mesh rearrangement schemes. As shown in Figure 1, it does not require much change to the software architecture, but only requires the implementation of the required functions by calling the interfaces of each module. The overall framework of the grid partitioning technique and grid rearrangement scheme will be presented in Section 3.

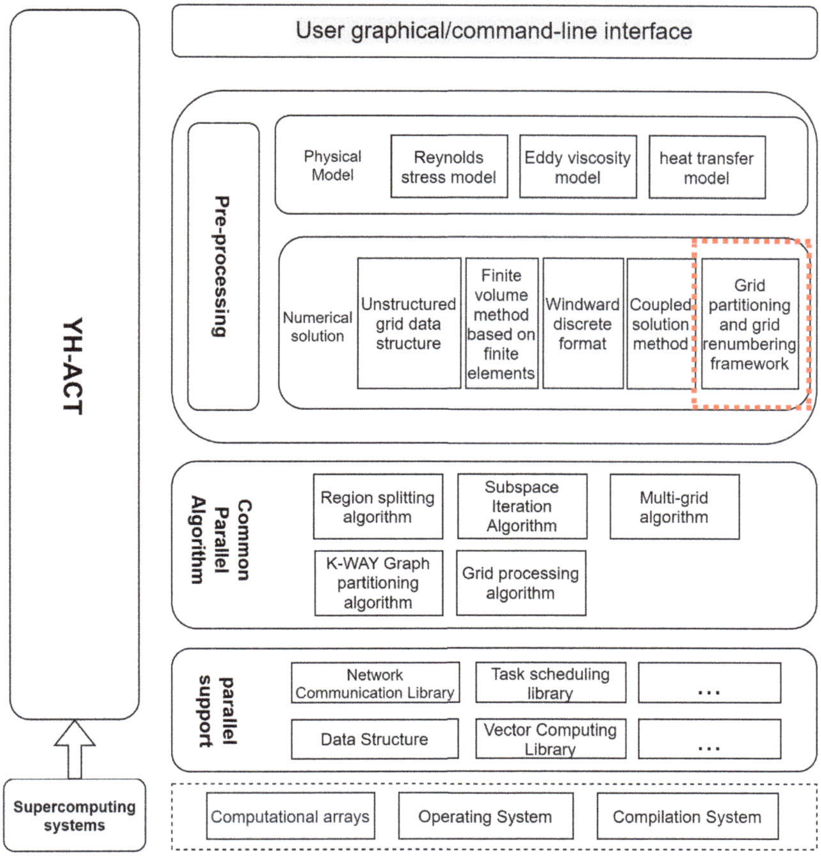

Figure 1. Overall framework of YHACT, framework after integration of meshing and mesh renumbering functions. Red boxes indicate new modules for integration.

2.2. Parallel Solving Algorithm

2.2.1. Parallel Decomposition of Grid Data

A simple and intuitive way of parallelism is to decompose the grid data. Similar to Cartesian partitioning, the grid data are divided into as many copies as required and then distributed to different processes for computation. The processes rely on communication to exchange data between them.

The data decomposition is shown in Figure 2. The grid to be solved is divided into non-overlapping blocks of grid sub-cells, and each process reads only one piece of grid data. After data decomposition, dummy cells are generated on the physical boundaries adjacent to each sub-grid and other grids for data communication only. Dummy cells are only used for data exchange and are not part of the physical model, which makes each cell consider only the data of its neighbors. The data on the dummy cell can be used as boundary data. This simple data decomposition does not share the grid data. In addition, a discrete format of up to 2nd order is used, and only the values of adjacent grid cells are used in the grid calculation. Therefore, the cross-process data to be acquired can be stored on the boundary line (dummy cell in Figure 2). The solution can be divided into the following three steps: (1) boundary initialization (fetching data on adjacent processors); (2) process local computation; and (3) boundary update (updating data on adjacent processors).

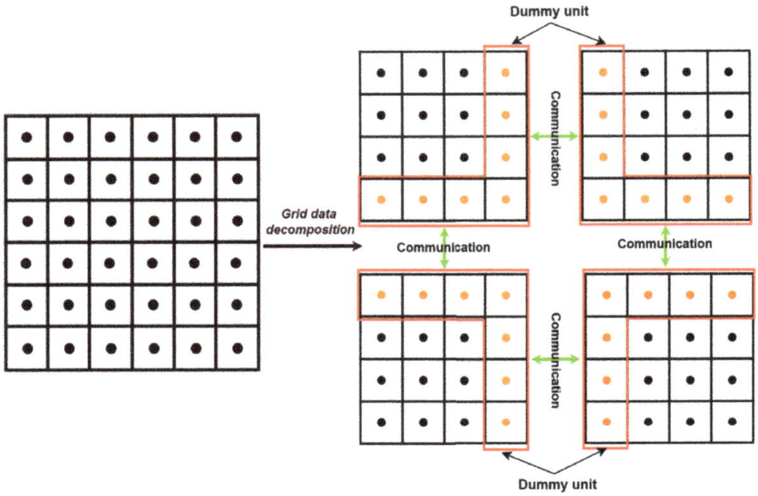

Figure 2. Grid data decomposition. The left figure shows the original grid model. The right side indicates the four sub-grids after data decomposition. Dummy cells are generated at the boundary of each subgrid. It is only used for data communication during grid computation of different processes.

2.2.2. Communication Patterns after Parallel Decomposition

Since each grid block only needs to communicate with its neighbors. The first step in establishing the communication relationship is to find all the neighboring grid blocks and store them in the local array. Each grid block then only has information about itself and its neighbors, which are "localized" grid blocks. When the parallel scale increases, the current grid only needs to traverse the "localized" grid blocks when communicating, which greatly improves the efficiency of communication. MPI asynchronous communication is also used to further improve communication efficiency. An example of a communication model is shown in Figure 3.

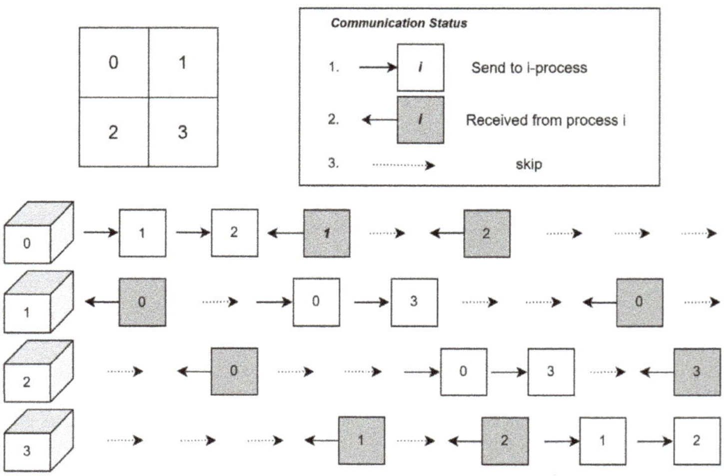

Figure 3. Example of communication between four processes. The top left indicates the subgrid to be calculated. The upper right indicates the three different process states during communication. The bottom indicates the state of each process at the same time step. Note that each process performs a time step of computation after receiving data, hence skip.

2.2.3. Parallel Architecture Based on MPI and OpenMP

Based on the grid model decomposition described above, two or more levels of data partitioning can be carried out. Depending on the computational resources or data structure, a certain number of "sub-grids" can be used as input to a process to perform computation. A hybrid MPI/OpenMP parallelism mode can be used. MPI parallelism is used on the first-level partition, and then the first-level partition is divided into several second-level sub-partitions. The number of second-level sub-partitions is generally equivalent to the cores inside the compute node. Coarse-grained OpenMP parallelism is performed on the second-level partition.

This multi-tier parallel architecture is adapted to the current architecture of supercomputer clusters, where primary partitions are assigned to individual compute nodes, while secondary partitions are on top of the same compute node. The above is also applicable to heterogeneous systems. The first layer still uses MPI communication based on the grid's first-level partition interfacing information.The second layer is OpenMP parallelism, i.e., coarse-grained parallelism between compute cores, based on secondary subpartitions of the grid. Data transfer between CPUs and coprocessors is performed using the corresponding channels. When there is no coprocessor, the above heterogeneous parallel mode automatically degenerates to homogeneous parallel mode. In order to provide better control over the different partitioning layers, the whole computational domain is divided into two layers: Layer 1 is the **Region** partition and Layer 2 is the **Zone** sub-partition. In parallel computation, the **Region** partition is assigned to a compute node, corresponding to an MPI process, and the **Zone** subpartition is assigned to a compute core thread, corresponding to an OpenMP thread.

A typical CFD solver execution process mainly includes residual initialization, time step calculation, flux calculation, time advancement, flow field update, and intersection communication. The solver is applied to the computational domain **Zone**. For different physical models, multiple solvers may be loaded on a **Zone**. During the computation, **Region** traverses each of these **Zones** in turn, calling the corresponding solver to perform the computational task. When OpenMP parallelism is used, the computational tasks can be performed concurrently for each **Zone**.

2.2.4. Parallel Communication for Data Packaging

In the multi-level parallel architecture described above, there may be multiple subgrid blocks on each MPI process. If we directly follow the above model of cyclic traversal of grid blocks for MPI/OpenMP hybrid parallel communication, there will be the problem of multiple communications between processes. Take Figure 4 as an example, Region0 and Region1 have 2 sets of adjacent subgrid blocks, and each exchange of data requires two send and receive processes. Therefore, the grid partition data between processes can be packaged as shown in Figure 5 to reduce the number of communications.

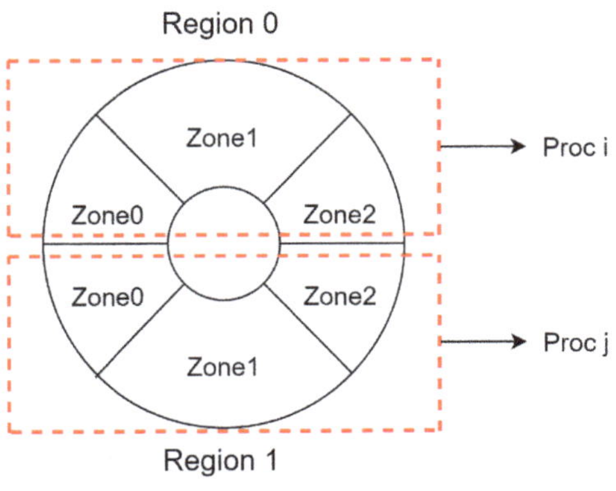

Figure 4. Partition level of two processes.

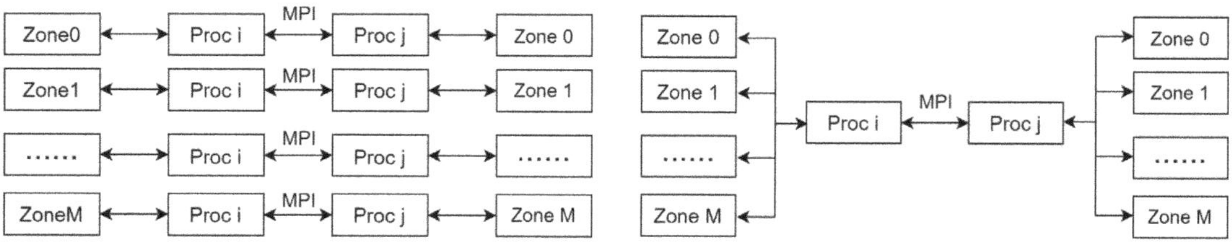

Figure 5. Example of packaging of process data. (**a**) unpacked transfer of data; (**b**) packet transmission of data.

3. A General Framework for Grid Partitioning Techniques and Grid Renumbering Methods

3.1. *Parmetis*-Based Grid Partitioning Strategy

Metis and **ParMetis** are tools that work on physical model partitioning. Usually, using the **Metis** library for mesh slicing of physical models using a serial strategy is an ideal solution. However, it is slow under large-scale mesh partitioning. **ParMetis** is a parallel version of **Metis**, an **MPI** parallel library. It has many built-in algorithms for mesh partitioning and repartioning of unstructured graphs.

ParMetis started out as a tool for large-scale graph partitioning, but mesh models in CFD can also be abstracted to graphs. Therefore, **ParMetis** is particularly well suited to deal with unstructured meshing problems in large scale numerical simulations. Simulations of fluid problems often require thousands of cluster nodes to be computed simultaneously. Before the computation, a regional decomposition of the computational grid has to be performed.Communication is required between different regions in the computation. Therefore, grid region decomposition has a greater impact on load balancing and interprocess communication. **PartMetis** enables the partitioned area grid to be approximately the same and the number of intersections to be minimized. Therefore, the amount of data exchanged is minimized, which in turn significantly reduces the communication time between processes.

ParMetis provides the **ParMETIS_V3_PartGeomKway** routine for computing partitions of graphs derived from finite element meshes where vertices have coordinates associated with them. Given a graph distributed between processor and vertex coordinates, **ParMETIS_V3_PartGeomKway** uses the space-filling curve method to quickly compute an initial partition. The graph is redistributed according to this partition and **ParMETIS_V3_PartKway** is then called to compute the final high quality partition. Information about the partitioning process of **Parmetis** is available in [25].

3.2. Grid Renumbering Framework

The neighbor relationship between adjacent cells reflects the memory access order of the numerical simulation. The system of discrete equations is obtained by discretizing on each interval of the control volume of the entire domain to be solved. Discrete systems of equations are presented as sparse matrices in a computer [13], where the non-zero elements of each row denote the coefficients of the unknown quantities of the current set of equations. The matrix usually contains a few nonzero elements , and the number of non-zero elements is exactly equal to the number of elements associated with this row.

For different grid types, the coefficient matrix of a structured grid is banded, while the coefficient matrix of an unstructured grid is relatively irregular and only has the characteristics of a diagonal matrix. Therefore, based on the above characteristics, many solutions have been proposed. The methods for solving large-scale equations can be divided into direct and iteration approaches. Since the real case grid size of thermal hydraulics is usually large, the solution process of direct method requires a large amount of computational resources and the cost of solving the inverse is very high. Therefore, iteration approaches are usually used to addressing massive thermal-hydraulic problems. The iterative method means that the algorithm needs to be used several times to reach a convergence state. For example, ILU decomposition method, and the conjugate gradient method can be used as representatives of the iterative method for solving.

The key to improving the speed of solving is that, by writing part of the memory data to the cache, the solver can access the required solved data with higher efficiency. In other words, by rearranging the mesh without altering the program structure [26], more valid data can be written to the cache to prevent frequent memory reads for a higher cache hit rate. Therefore, this section integrates part of the grid renumbering framework in YHACT, including but not limited to the Greedy algorithm and CQ algorithm implementation, which can be found in [27], and the RCM algorithm is elaborated below.

3.2.1. RCM Algorithm

After careful algorithm design, the RCM algorithm was shown to be able to adapt to large-scale distributed memory parallelism [28] and has been widely used in software such as OpenFoam and MatLab. When the numerical stage of sparse matrix computation is reached, the matrix is usually distributed. At this point, the RCM algorithm can improve the sparse structure of the sparse matrix, making the non-zero elements more concentrated, which can theoretically improve the storage efficiency of the computer.

The RCM algorithm uses BFS to traverse the nodes in the graph. At each layer of the graph, nodes are sorted according to their degree labels for the renumbering of that layer. In this paper, due to the modular nature of the numerical simulation framework and the complexity of the case, the algorithm's feature of sorting nodes according to their degrees is weakened to reduce the time cost of partially traversing the nodes. Algorithm 1 shows the overall program structure.

Algorithm 1 RCM numbering algorithm

input: $mesh_0$
output: $mesh_n$
1: Select the initial unit to be numbered, set to $mesh_k$. Set its number to the value of the maximum number of units
2: Arrange the unnumbered neighbors of $mesh_k$ in the container L according to the descending order of degree
3: Set the first element of L to $mesh_k$ and renumber it. The value is the maximum number of cells minus one
4: Implementation Step 2
5: Implementation of step 3, and the numbered values in decreasing order
6: Repeat until the numbering is complete

Note that the initial numbered cells of the original RCM algorithm are selected by traversing the entire grid of the computational solution domain. The degree of meshes is used to achieve the desired optimization. Here, we require that the initial elements be selected as boundary cells or close to the boundary. This method prevents excessive time overhead due to the large and complex size of the case. It makes the algorithm more adaptive in increasingly complex cases. The key step is the numbering. The RCM algorithm sorts the meshes in reverse according to the sequence of degrees, from "nElement -1" to "0".

3.2.2. Selection of Grid Renumbering Methods

Usually the sparse structure of the generated sparse matrices will vary for different renumbering methods. Bandwidth and profile are two common methods to discriminate the quality of sparse matrices. In the computer's storage structure, the sparse matrix eventually participates in the CFD solver computation, and its sparse structure has a significant impact on the cache miss rate during the computation, which in turn affects the computational efficiency of the numerical simulation. Nevertheless, the above only represents the sparse structure of the matrix as a whole and ignores the degree of aggregation around the diagonal elements of the sparse matrix. The more elements close to the diagonal, the smaller the cache miss rate will be. Therefore, this paper adopts the MDMP discriminant method, and more details can be found in [27]. This ensures that more data are hit in the same cache:

The MDMP discriminant formula is:

$$\bar{d} = \sum_{k=1}^{\frac{n}{2}} \frac{d_k}{k} \tag{1}$$

It has been shown that the MDMP metric can describe the dispersion of this sparse matrix very well.

4. Parallel Experiments for Numerical Simulation of Large Scale Nuclear Reactor Thermodynamics

This section runs a typical configuration on a high performance machine in modern times. The computer consists of about 120 computing nodes. Details of the HPC are shown in Table 1. The hardware configuration CPU is an Intel(R) Xeon(R) Gold 6240R processor with 48 cores and 192G RAM.

Table 1. The platform configuration used in the study.

Parameter	Configuration
Processor model	Intel(R) Xeon(R) Gold 6240R @2.40GHz
Number of cores	48 cores
Memory	192G
OS	Red Hat 8.3.1-5
Compiler	GNU Compiler Collection (GCC 8.3.1)

4.1. CFD Physical Modeling of Nuclear Reactor Thermal Hydrology

In the process of nuclear reactor thermodynamic analysis, the flow of coolant heat transfer process needs to be modeled. Figure 6 illustrates the thermal hydraulic model of a nuclear reactor that is the main focus of this paper. The theoretical model needed to completely describe the flow in the fluid as well as the heat transfer process contains three types of conservation equations for mass, momentum and energy [29].

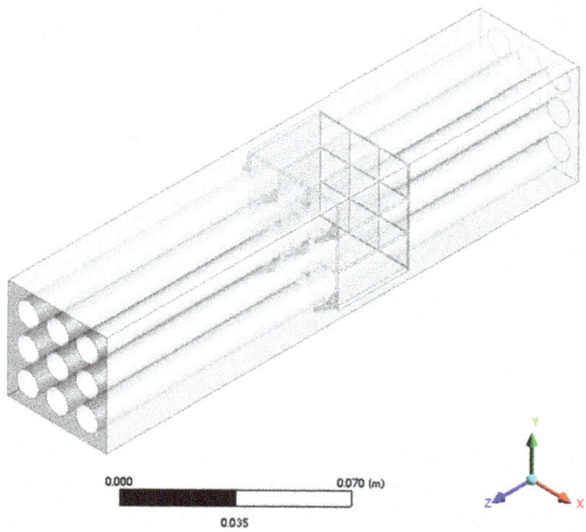

Figure 6. Fluid computing domain.

The structure inside the pressure vessel is complex, the range of flow scales is large, and the practical application of engineering is more concerned with the average amount of physical quantities. Therefore, Reynolds averaging is generally performed for the three major equations to calculate the average physical quantities of fluid flow. In addition, then the turbulence model needs to be added in order to make the averaged mass, momentum and energy equation set closed. This is used to perform simulations of core cooling fluid flow and heat transmission. Therefore, for the thermodynamic model, a smaller grid size will lead to insufficient accuracy of its simulation. Therefore, in this paper, a 3*3 assembly with about 39.5 million grid size is simulated numerically for a fuel rod bundle. The **SIMPLE** iterative algorithm and **PETSc** solver are used to perform the calculations. The relevant parameters of the calculation, such as the relaxation factor, are adjusted. Pressure relaxation factor and velocity relaxation factor are set to 0.3 and 0.5, respectively, and a turbulence physical quantity relaxation factor of 0.5 are used to calculate the turbulent flow between fuel rod bundles with stirred wings.

It is worth noting that YHACT currently mainly considers the implementation of the Reynolds time-averaged turbulence model. As for the fluid–solid coupled heat transfer model, the heat transfer model between fluid–solid domains is generally established through the intersection of fluid and solid domains. When calculating the energy equation in the solid domain, the relevant physical quantities at the intersection of the fluid domain are used as boundary conditions. In addition, when calculating the energy equation in the fluid domain, the heat (temperature) distribution at the solid intersection is used as the source term of the fluid energy equation to finally realize the coupled fluid-solid heat transfer. Since this paper mainly concentrates on the turbulent heat transfer model, the solid models of fuel rod and control rod are neglected. The boundary condition of the solid interface on the pure fluid is a fixed heat flux. This heat flux is roughly calculated based on the solid bar size and the volumetric heat source at the time of fluid–solid coupling.

4.2. Select the Grid Renumbering Algorithm

This subsection focuses on how to select the appropriate grid renumbering algorithm. Based on the integrated renumbering strategy, the sparse structure of the sparse matrices generated by each type of grid renumbering is discriminated by MDMP metrics for the above fuel rod bundle model. Finally, the most suitable algorithm is selected for numerical simulation.

Due to the large grid size, it is more difficult to obtain the grid numbering geometry information of the physical model by serial means. Therefore, this subsection collects the numbering geometry information of the subgrids in process 0 for the two parallel scales of 48 and 96 processes to determine. Table 2 shows the sparse matrices generated by four different numbering methods for each of the two parallel scales. Below, each row of the sparse matrix corresponds to its MDMP index.

A simple comparison can be made to conclude that the original numbering of the grid drawn by the **ICEM** library is confusing. After renumbering, the properties of the sparse matrix are better improved. Meanwhile, the RCM algorithm will be better compared to several other numbering methods. Therefore, in the following parallel tests, the RCM renumbering algorithm is chosen.

Table 2. The sparse matrix of the grid after being processed by different renumbering algorithms. The blue dots indicate the non-zero elements in the matrix.

Numbering Strategy	48 Process	96 Process
original	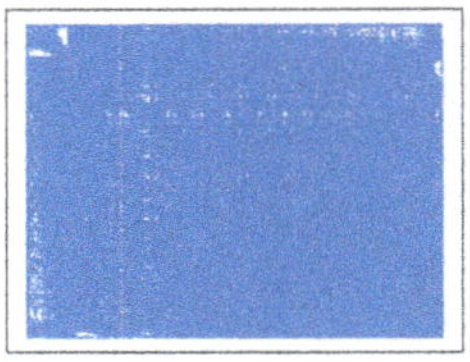	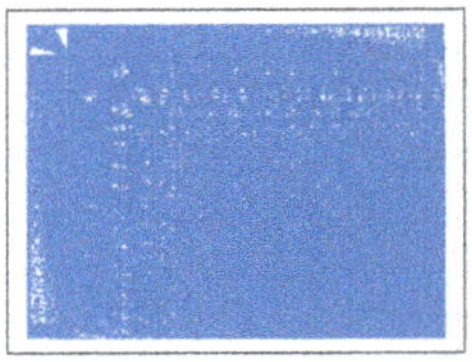
MDMP	32,387.88	16,588.94
RCM	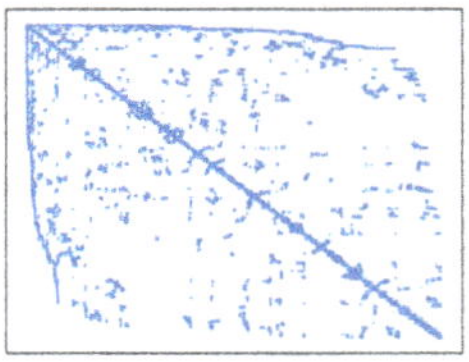	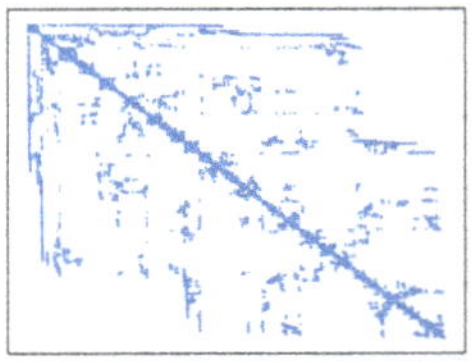
MDMP	44.61	44.35
greedy	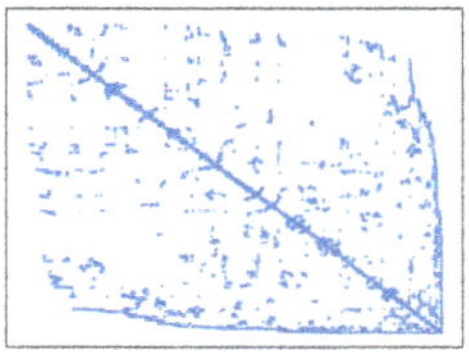	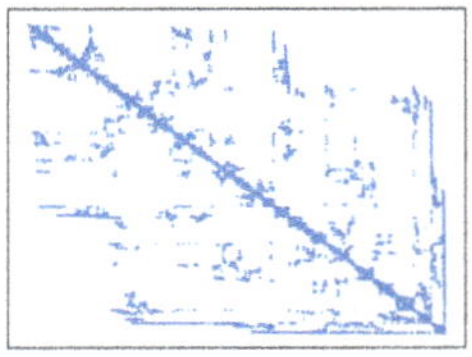
MDMP	55.24	54.16
CQ	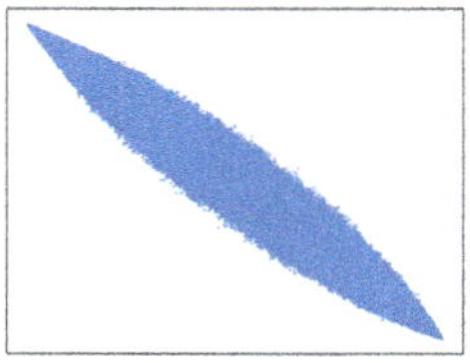	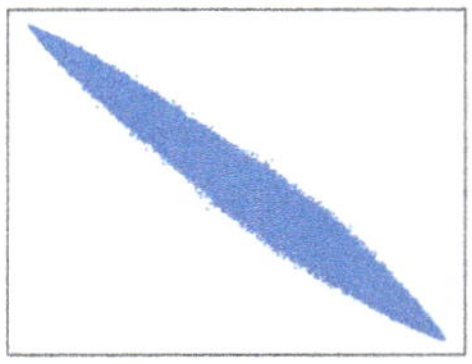
MDMP	838.29	378.98

4.3. Massively Parallel Performance Testing

4.3.1. Proof of Correctness

Parallel performance tests must ensure the correctness of the numerical simulations. After optimization by renumbering the modules, proof of program robustness is essential. Figure 7 illustrates the two slice models tested in this subsection. (a) shows the slice with Z = 0.01, (b) shows the slice with Z = −0.01. In addition, (b) belongs to the shelf section, which has more complex physical properties.

Tables 3 and 4 show the profile cloud plots for the slice at Z = 0.01 and the profile cloud plots for the slice at Z = −0.01, respectively, for comparing the correctness of the numerical simulations before and after renumbering. Each table shows the contour plots of temperature, pressure, and velocity. The first column represents the different physical quantities. The second column represents the range of physical quantities in the cloud plot. The third and fourth columns show the sliced cloud plots at the same locations before and after the renumbering algorithm. The white numbers in each cloud represent the values of physical quantities at the current position. When the same physical quantities are compared, both numbering methods take the same position. Therefore, the corresponding positions of the white numbers before and after renumbering are the same. The white numbers in the cloud plot are the values of physical quantities at the same locations, which are used to determine whether the accuracy of the renumbering optimization meets the requirements and whether the robustness of the program is guaranteed. It can be seen that the renumbering does not change the correctness of the numerical simulation either at the fuel bar (Z = 0.01) or at the shelf (Z = −0.01). The robustness of the procedure is guaranteed.

Table 3. Contour clouds for slices with Z = −0.01. The first column indicates the different physical quantities. The second column indicates the range of physical quantities in the cloud map. The third and fourth columns show the sliced clouds at the same locations before and after the renumbering algorithm. The white numbers in each cloud represent the values of physical quantities at the current position. When the same physical quantities are compared, both numbering methods take the same position.

Physical Properties	Range	Old	RCM
Temperature	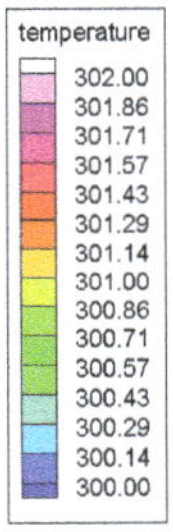	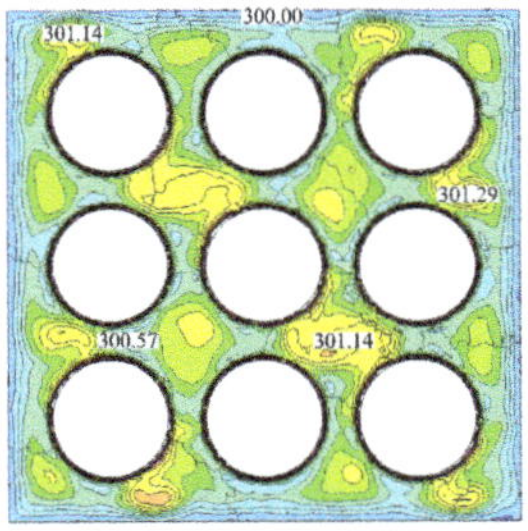	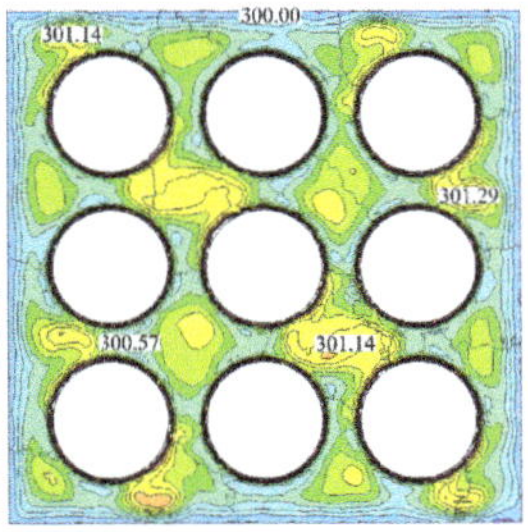
Pressure	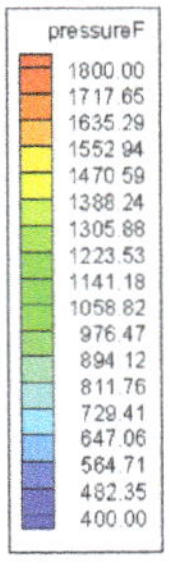	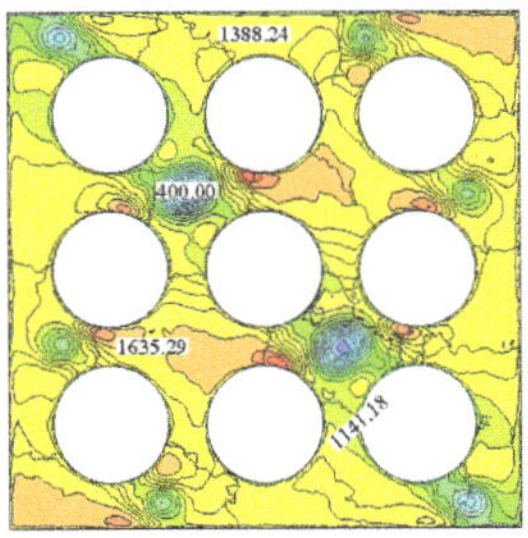	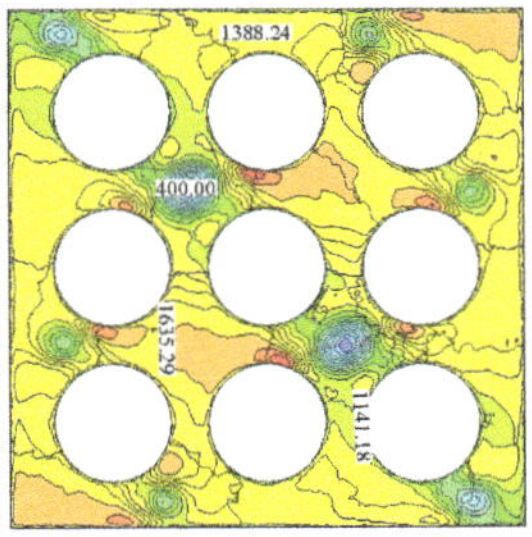
Velocity	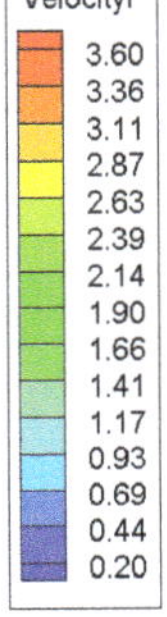	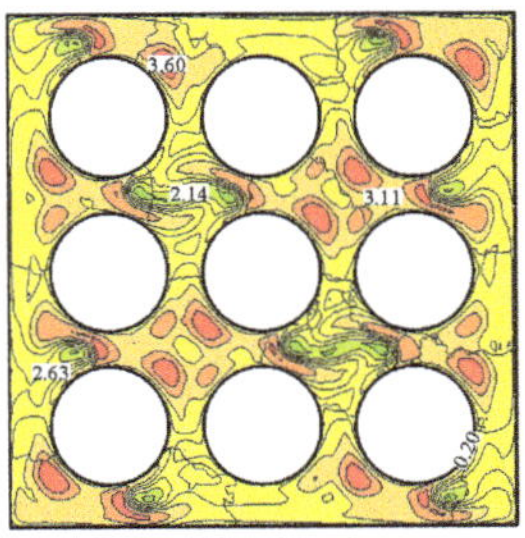	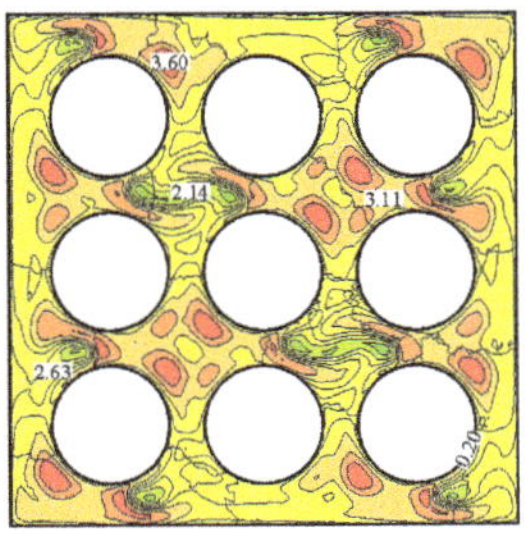

Table 4. Contour clouds for slices with Z = −0.01. The specific information is the same as the legend in Table 3.

Physical Properties	Range	Old	RCM
Temperature	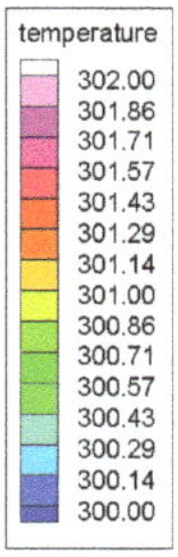	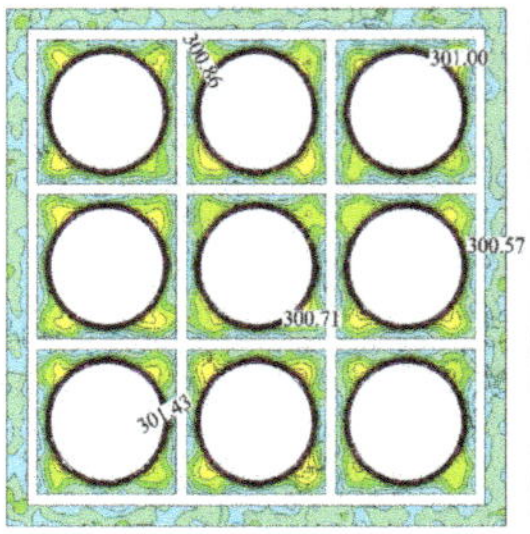	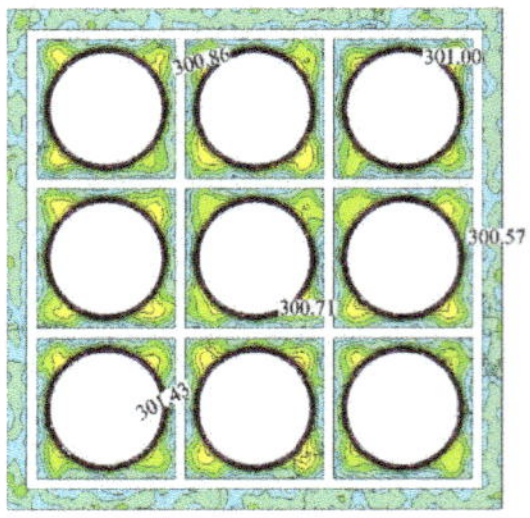
Pressure	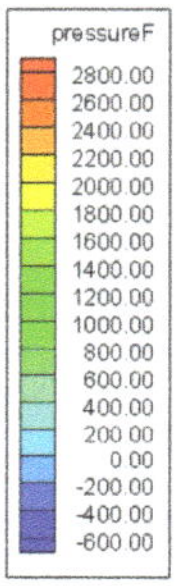	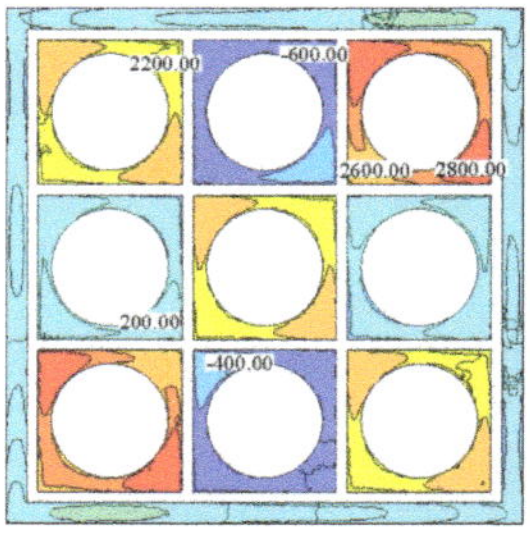	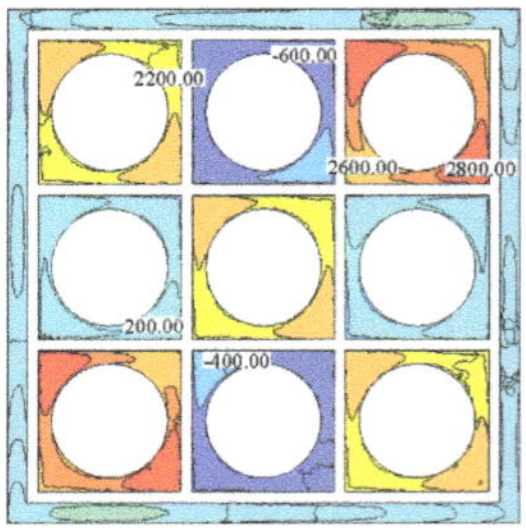
Velocity	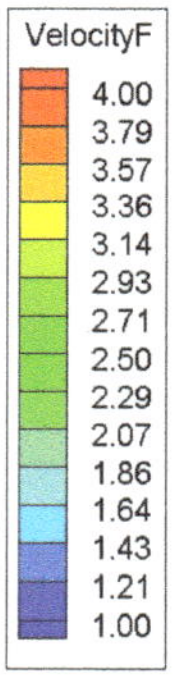	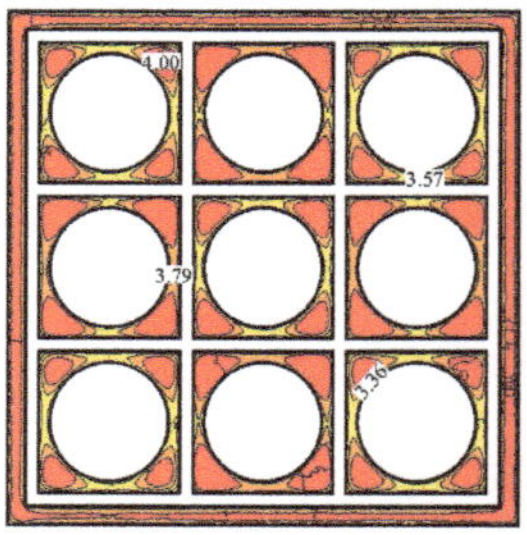	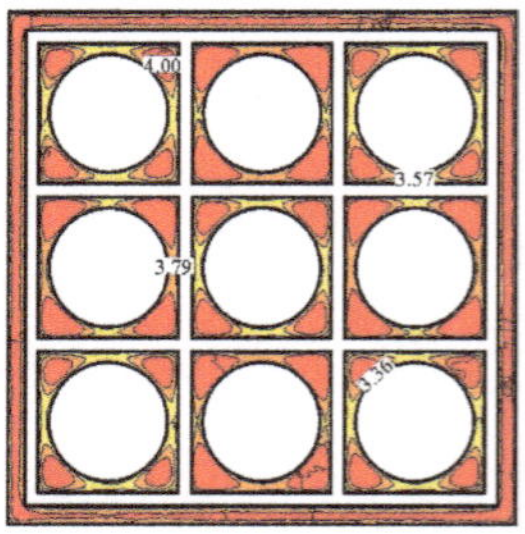

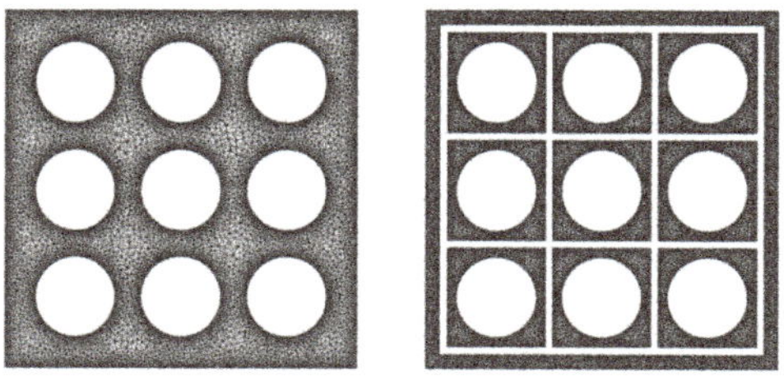

(a) (b)

Figure 7. Slicing model. (**a**) slice with Z = 0.01; (**b**) slice with Z = −0.01. In addition, slice (**b**) is located in the shelf.

4.3.2. Parallel Testing of Large-Scale Numerical Simulations

Finally, in this paper, parallel numerical simulations are performed for a fuel rod bundle of 39.5 million grid size with a different number of processes. Table 5 shows the time to iterate 100 steps on different processes for the original grid and the grid renumbered by RCM. The first column represents the amount of nodes used. The second column indicates the number of processes used, trying to ensure that each node is used 100%. The third and fourth columns indicate the total simulation time for the original mesh and the RCM renumbered mesh. The fifth column indicates the speedup of the numerical simulation of the renumbered grid compared to the original grid. It can be obtained that the renumbering module has a good effect on the computational effectiveness of the mathematical simulation of the fuel rod technique. In addition, the parallel scale reaches a maximum of 3072 processes. At the parallel scale of 1536 processes, the renumbering module has the maximum speedup effect of 56.72%. In addition, at the parallel scale of 3072 processes, the time difference is not much. This may be due to the fact that, when the grid size is certain, the larger the parallel scale, the matrix computation time within the processes decreases dramatically, while the percentage of communication time increases dramatically.

Table 5. The degree of enhancement of the RCM algorithm for numerical simulation of fuel rod bundles at different parallel scales. The first column represents the number of nodes. The second column represents the amount of processes. The third and fourth columns indicate the total simulation time for the original grid and the grid renumbered by RCM. The fifth column indicates the numerical simulation speedup of the renumbered grid compared to the original grid.

No. of Nodes	No. of Processes	Time		Speedup Rate
		Original(s)	RCM(s)	
2	96	19,731.4	13,682.5	30.66%
4	192	8635.3	6727.38	22.09%
8	384	3761.11	3342.41	11.13%
16	768	1794.29	1568.07	12.6%
32	1536	1618.58	700.528	56.72%
64	3072	748.43	716.175	4.31%

5. Conclusions

In this paper, we focus on how to apply suitable renumbering algorithms to enhance the performance of large-scale nuclear reactor fuel rod technology in a CFD numerical simulation framework. We do this by further integrating the meshing and mesh renumbering framework of YHACT, a general-purpose CFD software. The RCM algorithm is easily implemented by relying on the modular nature of the software. In this paper, we first perform parallel meshing of the model using **Parmetis** to ensure that each process obtains a similar size of mesh blocks and thus try to ensure that the computation time of each process is not significantly different. In addition, then, through rigorous experiments, we use MDMP as the discriminative index of sparse matrix quality to select a more suitable renumbering algorithm for the experimental case of this paper. Through the MDMP metric, it is easy to select the RCM algorithm is what we need. In addition, by plotting the contour clouds at the bar bundle and at the shelf, it is concluded in detail that our renumbering algorithm has a strong scalability in YHACT. This is also due to the modular development of YHACT, and the easy setup of the API. By observing the three contour plots of temperature, pressure, and velocity, it can be seen that the renumbered grid does not have any effect on the final solution results. The solution accuracy is well within the acceptable range. Therefore, we boldly and confidently conducted a large-scale parallel test. In the end, the number of processes for the 3*3 fuel rod bundle problem reached 3072, and the most significant speedup of 56.72% was achieved when the number of processes reached 1536. Meanwhile, the overall speedup increases to 56.72% when the speedup efficiency is gradually increased from 96 to 1536 threads. The acceleration efficiency does not decrease with the increase of parallelism size. Thus, the goodness of the strong scalability of parallelism was tested and proved.

As industry develops and advances, the complexity of CFD numerical simulations will gradually increase. The gradual development of supercomputers will inevitably lead to an increasing parallel scale. It is necessary to improve the efficiency of numerical simulation through renumbering strategy. It is also necessary to select a suitable renumbering algorithm without affecting the scalability of the numerical simulation. It can be seen that the final test results of this paper have good scalability and obvious acceleration effect.The difficulty lies in the easy combination of different acceleration methods. In addition, the best way to do this is to continuously enrich the functionality of the general CFD software through modular development. The significance of this is that it does have a strong acceleration effect on real engineering cases. Continuously increasing the complexity and parallel scale of numerical simulations is the only way to better identify problems in the development process. Continuous problem solving leads to faster operation and more accurate numerical simulations. The next step is naturally to investigate which renumbering algorithm is suitable for different types of cases, such as turbulence, laminar flow, steady state, transient, etc., in order to further improve the software solution accuracy, parallel scale, and running speed. Another direction is to dive into the Solver and explore how the renumbering algorithm affects the assembly and solution of the matrix.

Author Contributions: Conceptualization, J.L. and X.-W.G.; methodology, H.Z.; software, X.-W.G., C.L., H.Z.; investigation, X.-W.G.; data curation, H.X. and Q.L.; writing—original draft preparation, H.Z.; writing—review and editing, H.Z.; funding acquisition, J.L., X.-W.G. and C.L. All authors have read and agreed to the published version of the manuscript.

Funding: This work was partially supported by the National Natural Science Foundation of China (Grant Nos. 61902413, 12102468, and 12002380) and the National University of Defense Technology Foundation (No. ZK21-02, No. ZK20-52).

Conflicts of Interest: The authors declare no conflict of interest.

References

1. Chen, R.; Tian, M.; Chen, S.; Tian, W.; Su, G.; Qiu, S. Three dimensional thermal hydraulic characteristic analysis of reactor core based on porous media method. *Ann. Nucl. Energy* **2017**, *104*, 178–190. [CrossRef]
2. Wang, M.; Zuo, Q.; Yu, H.; Tian, W.; Su, G.; Qiu, S. Multiscale thermal hydraulic study under the inadvertent safety injection system operation scenario of typical pressurized water reactor. *Sci. Technol. Nucl. Install.* **2017**, *2017*, 2960412. [CrossRef]
3. Krauss, T.; Meyer, L. Experimental investigation of turbulent transport of momentum and energy in a heated rod bundle. *Nucl. Eng. Des.* **1998**, *180*, 185–206. [CrossRef]
4. Guellouz, M.; Tavoularis, S. The structure of turbulent flow in a rectangular channel containing a cylindrical rod–Part 1: Reynolds-averaged measurements. *Exp. Therm. Fluid Sci.* **2000**, *23*, 59–73. [CrossRef]
5. Kaiser, H.; Zeggel, W. Turbulent flows in complex rod bundle geometries numerically predicted by the use of FEM and a basic turbulence model. In Proceedings of the Third International Topical Meeting on Reactor Thermal Hydraulics, Newport, RI, USA, 15–18 October 1985.
6. Lee, K.B.; Jang, H.C. A numerical prediction on the turbulent flow in closely spaced bare rod arrays by a nonlinear k–ε model. *Nucl. Eng. Des.* **1997**, *172*, 351–357. [CrossRef]
7. Baglietto, E.; Ninokata, H. A turbulence model study for simulating flow inside tight lattice rod bundles. *Nucl. Eng. Des.* **2005**, *235*, 773–784. [CrossRef]
8. Baglietto, E.; Ninokata, H.; Misawa, T. CFD and DNS methodologies development for fuel bundle simulations. *Nucl. Eng. Des.* **2006**, *236*, 1503–1510. [CrossRef]
9. Cheng, X.; Tak, N. CFD analysis of thermal–hydraulic behavior of heavy liquid metals in sub-channels. *Nucl. Eng. Des.* **2006**, *236*, 1874–1885. [CrossRef]
10. Guo, R.; Oka, Y. CFD analysis of coolant channel geometries for a tightly packed fuel rods assembly of Super FBR. *Nucl. Eng. Des.* **2015**, *288*, 119–129. [CrossRef]

11. Hu, R.; Fanning, T. *Progress Report on Development of Intermediate Fidelity Full Assembly Analysis Methods*; Technical report; Argonne National Lab. (ANL): Argonne, IL, USA, 2011.
12. Slotnick, J.P.; Khodadoust, A.; Alonso, J.; Darmofal, D.; Gropp, W.; Lurie, E.; Mavriplis, D.J. *CFD Vision 2030 Study: A Path to Revolutionary Computational Aerosciences*; Technical report; NASA: Washington, DC, USA, 2014.
13. Darwish, M.; Moukalled, F. *The Finite Volume Method in Computational Fluid Dynamics: An Advanced Introduction with OpenFOAM® and Matlab®*; Springer: Berlin/Heidelberg, Germany, 2021.
14. Zou, S.; Yuan, X.F.; Yang, X.; Yi, W.; Xu, X. An integrated lattice Boltzmann and finite volume method for the simulation of viscoelastic fluid flows. *J. Non-Newton. Fluid Mech.* **2014**, *211*, 99–113. [CrossRef]
15. Karypis, G.; Kumar, V. A fast and high quality multilevel scheme for partitioning irregular graphs. *SIAM J. Sci. Comput.* **1998**, *20*, 359–392. [CrossRef]
16. Gibbs, N.E.; Poole, W.G., Jr.; Stockmeyer, P.K. An algorithm for reducing the bandwidth and profile of a sparse matrix. *SIAM J. Numer. Anal.* **1976**, *13*, 236–250. [CrossRef]
17. Cuthill, E.; McKee, J. Reducing the bandwidth of sparse symmetric matrices. In Proceedings of the 1969 24th National Conference, New York, NY, USA, 26–28 August 1969; pp. 157–172.
18. Sloan, S. An algorithm for profile and wavefront reduction of sparse matrices. *Int. J. Numer. Methods Eng.* **1986**, *23*, 239–251. [CrossRef]
19. Akhras, G.; Dhatt, G. An automatic node relabelling scheme for minimizing a matrix or network bandwidth. *Int. J. Numer. Methods Eng.* **1976**, *10*, 787–797. [CrossRef]
20. George, A.; Liu, J.W. *Computer Solution of Large Sparse Positive Definite*; Prentice Hall Professional Technical Reference: Hoboken, NJ, USA, 1981.
21. de Oliveira, S.G.; Silva, L.M. An ant colony hyperheuristic approach for matrix bandwidth reduction. *Appl. Soft Comput.* **2020**, *94*, 106434. [CrossRef]
22. Pop, P.; Matei, O.; Comes, C.A. Reducing the bandwidth of a sparse matrix with a genetic algorithm. *Optimization* **2014**, *63*, 1851–1876. [CrossRef]
23. Xiao-wei, G.; Chao, L.; Jie, L.; Chuan-fu, X.; Chun-ye, G.; Li-juan, C. A highly scalable general purpose CFD software architecture and its prototype implementation. *Comput. Eng. Sci.* **2020**, *42*, 2117–2124.
24. Xu, C.; Deng, X.; Zhang, L.; Fang, J.; Wang, G.; Jiang, Y.; Cao, W.; Che, Y.; Wang, Y.; Wang, Z.; et al. Collaborating CPU and GPU for large-scale high-order CFD simulations with complex grids on the TianHe-1A supercomputer. *J. Comput. Phys.* **2014**, *278*, 275–297. [CrossRef]
25. Karypis, G.; Schloegel, K.; Kumar, V. *Parmetis: Parallel Graph Partitioning and Sparse Matrix Ordering Library*; University of Minnesota: Minneapolis, MN, USA, 1997.
26. Burgess, D.; Giles, M.B. Renumbering unstructured grids to improve the performance of codes on hierarchical memory machines. *Adv. Eng. Softw.* **1997**, *28*, 189–201. [CrossRef]
27. Zhang, H.; Guo, X.W.; Li, C.; Liu, Q.; Xu, H.; Liu, J. Accelerating FVM-Based Parallel Fluid Simulations with Better Grid Renumbering Methods. *Appl. Sci.* **2022**, *12*, 7603. doi: 10.3390/app12157603. [CrossRef]
28. Azad, A.; Jacquelin, M.; Buluç, A.; Ng, E.G. The reverse Cuthill-McKee algorithm in distributed-memory. In Proceedings of the 2017 IEEE International Parallel and Distributed Processing Symposium (IPDPS), Buena Vista, FL, USA, 29 May–2 June 2017; pp. 22–31.
29. Versteeg, H.K.; Malalasekera, W. *An Introduction to Computational Fluid Dynamics: The Finite Volume Method*; Pearson Education: London, UK, 2007.

mathematics

MDPI

Article

The Modified Local Boundary Knots Method for Solution of the Two-Dimensional Advection–Diffusion Equation

Karel Kovářík * and Juraj Mužík

Department of Geotechnics, Faculty of Civil Engineering, University of Žilina, 01026 Žilina, Slovakia
* Correspondence: karel.kovarik@uniza.sk

Abstract: This paper deals with a new modification of the local boundary knots method (LBKM) which will allow the irregular node distribution and the arbitrary shape of the solution domain. Unlike previous localizations, it has no requirements on the number of nodes in the support or on the number of virtual points. Owing to the limited number of virtual points, the condition number of boundary knots matrix remains relatively low. The article contains the derivation of the relations of the method for steady and unsteady states and shows its effectiveness in three control examples.

Keywords: boundary knots method; particular solution; finite collocation; advection–diffusion

MSC: 65M80; 65M99

Citation: Kovářík, K.; Mužík, J. The Modified Local Boundary Knots Method for Solution of the Two-Dimensional Advection–Diffusion Equation. *Mathematics* **2022**, *10*, 3855. https://doi.org/10.3390/math10203855

Academic Editors: Vasily Novozhilov and Cunlu Zhao

Received: 15 September 2022
Accepted: 14 October 2022
Published: 18 October 2022

1. Introduction

In recent decades, the significant evolution of meshless methods for solving partial differential equations is evident. The first sign of this trend can be considered the boundary element method (BEM) [1,2], which is not yet an utterly network-free method. However, it has significantly reduced the necessary network of elements. On the other hand, this method required solving the integrals of the fundamental solution, which was sometimes very complicated. The removal of integration is the main advantage of the method of fundamental solution (MFS) [3–5], which uses fundamental solutions as basis functions to approximate the solution without needing integration. This property has contributed to the significant expansion of this method and its considerable popularity. However, this method also has its problems, especially related to using a network of fictitious virtual points. The singular boundary method (SBM) [6–8] tries to eliminate this disadvantage using real points at the boundary of an area to be identified with fictitious points. At the same time, however, this leads to the need to solve the problems of the singularity of the fundamental solution in case the two points are identical. SBM solves this by introducing the so-called origin intensity factors (OIF) [6,9], which are calculated in a more or less complicated way and are essentially the main weakness of this method. Another way to solve the singularity uses the boundary knot method (BKM) [10–12], which uses the general solution of the governing differential equation as the basis function instead of the fundamental solution.

Unfortunately, the condition number of the BKM interpolation matrix is very high, and its inversion is overburdensome. Recently, a local BKM solution [13,14] can help keep the interpolation matrix conditional on a reasonable level, thus enabling even more extensive tasks. Nevertheless, even so, with more local support, problems can arise. These should be removed by the presented modification of the local BKM. It is based partly on the LBIEM principle [15,16] and separates the virtual area around each point from its support. It allows keeping the condition number of the matrix independent of the number of points in the support.

Our article focuses on the solution of the steady and non-steady advection–diffusion problem, which is of great practical importance, e.g., in modeling the transport of substances in a flowing liquid. The first two sections describe the connection between BKM and FC and its application to the advection–diffusion problem in the two-dimensional domain. The following sections present the control examples and compare the results with the exact solution.

2. Governing Equations

The governing equation of the unsteady hydrodynamic dispersion in the domain $\Omega \in \mathbb{R}_2$ with boundary Γ is

$$R_d \frac{\partial C(\mathbf{x},t)}{\partial t} = D\Delta C(\mathbf{x},t) - \mathbf{v} \cdot \nabla C(\mathbf{x},t) - \lambda C(\mathbf{x},t) \qquad \mathbf{x} \in \Omega, \tag{1}$$

where C is the concentration of the tracer, D is the coefficient of dispersion, R_d is the retardation factor, λ is the decay coefficient, $\mathbf{x}$ are spatial coordinates, $\mathbf{v}$ is the vector of velocity, and t is the time.

The usual boundary conditions of Equation (1) are as follows:

- The Dirichlet boundary conditions, where the value of the concentration C on the part of boundary Γ_1 is prescribed, i.e., $C = C_0(\mathbf{x},t)$ $\mathbf{x} \in \Gamma_1$;
- The Neumann boundary conditions, where the flux q_0 with concentration C_0 perpendicular to the boundary Γ_2 is given, i.e.,

$$D\frac{\partial C}{\partial x_i} n_i = (C - C_0) q_0(\mathbf{x},t) \quad \mathbf{x} \in \Gamma_2, \tag{2}$$

where n_i is the i component of the outer normal vector, perpendicular to the boundary Γ_2.

These are in addition to the whole boundary $\Gamma = \Gamma_1 \cup \Gamma_2$. The initial condition is defined by the prescribed value of the concentration at time $t_0 = 0$.

All boundary conditions can be simply expressed as

$$\mathcal{B}(u) = b_0(\mathbf{x},t) \quad \mathbf{x} \in \Gamma, \tag{3}$$

where $\mathcal{B}(u)$ is the boundary operator.

3. Numerical Solution

Time-dependent tasks are solved in two main ways when using meshless methods. We can use a time-dependent fundamental or general solution of a differential equation, or approximate the time term using a finite difference (FD) scheme. Since the time-dependent general solution of the advection–diffusion equation is difficult to find, we replaced the time derivative on the left side of (1) with a finite difference scheme.

The backward (Euler) scheme (4) is the simplest and can be defined as

$$\frac{\partial C^{n+1}}{\partial t} = \frac{1}{\Delta t}\left(C^{n+1} - C^{n}\right). \tag{4}$$

When we applied the scheme (4) to the presented method, it performed poorly for long time series. Therefore, we are looking for a more suitable and accurate scheme.

The Houbolt method [8,17] is an implicit and unconditionally stable FD scheme that can be obtained by the cubic-Lagrange interpolation of the concentration C from time $(n-2)\Delta t$ through to time $(n+1)\Delta t$. This scheme can be written as

$$\frac{\partial C^{n+1}}{\partial t} = \frac{1}{6\Delta t}\left(11C^{n+1} - 18C^{n} + 9C^{n-1} - 2C^{n-2}\right), \tag{5}$$

where Δt is the time step and the superscripts $n-2$, $n-1$, n, and $n+1$ of u represent the time level. The differential Equation (1) is now changed to

$$D\Delta C^{n+1} - \mathbf{v}\cdot\nabla C^{n+1} - \lambda C^{n+1} = \frac{R_d}{6\Delta t}\left(11C^{n+1} - 18C^{n} + 9C^{n-1} - 2C^{n-2}\right). \tag{6}$$

The simple Euler formula is used in the first two steps to obtain the needed data C^n and C^{n-1} to start the Houboldt scheme (C^{n-2} is a given initial solution).

To solve the unsteady diffusion Equation (1), we represent the solution as a sum of homogeneous and particular solutions. The solution of (6) can now be defined as the sum

$$C^{n+1} = C_H^{n+1} + C_P^{n+1}, \tag{7}$$

where C_H^{n+1} is a solution of homogeneous differential equation in time level $n+1$ that satisfies boundary conditions and C_P^{n+1} is a particular solution of the non-homogeneous Equation (6). The homogeneous problem has been solved using the modified local boundary knots method (LBKM). The particular solution could be solved using the local method of approximating particular solutions (LMAPS) [9,18].

3.1. Homogeneous Solution

As is usual with most local methods, we assume that the domain Ω is covered by individual points. We find a group of the nearest points for each point $i \in \Omega$ that form the support. In this modified version, in addition to support, we need to define a circular virtual area around each point and regularly spaced virtual points at its boundary (Figure 1). In this area, we now approximate the value of the concentration at a given point i and time interval $n+1$ using the general solution of a homogeneous differential equation in the form

$$C(x_i)_H^{n+1} = \sum_{j=1}^{n} \alpha_j G^*(r_{ij}) \qquad x_i \in \Omega, \tag{8}$$

where n is the number of virtual points, $r_{ij} = \|\mathbf{x}_i - \mathbf{x}_j\|$ is the distance between point i and virtual point j, and G^* is the general solution. For the 2D advection–diffusion differential Equation (6), the non-singular general solution is given as

$$G^*(r_{ij}) = \frac{1}{2\pi}\exp\left(\frac{\mathbf{v}\cdot\mathbf{r}}{2D}\right) I_0(\mu r_{ij}), \tag{9}$$

where $r_{ij} = \|x_i - x_j\|$, I_0 is the modified Bessel function of the first kind and

$$\mu = \sqrt{\left(\frac{\|\mathbf{v}|}{2D}\right)^2 + \frac{\lambda}{D}}. \tag{10}$$

The coefficients α_j are unknown and we determine them by applying (8) to all virtual points and we obtain

$$\sum_{j=1}^{n} A_{kj}\alpha_j = C_k \qquad k = 1\ldots n, \tag{11}$$

where C_k are the values of concentrations in virtual points for homogeneous solution.

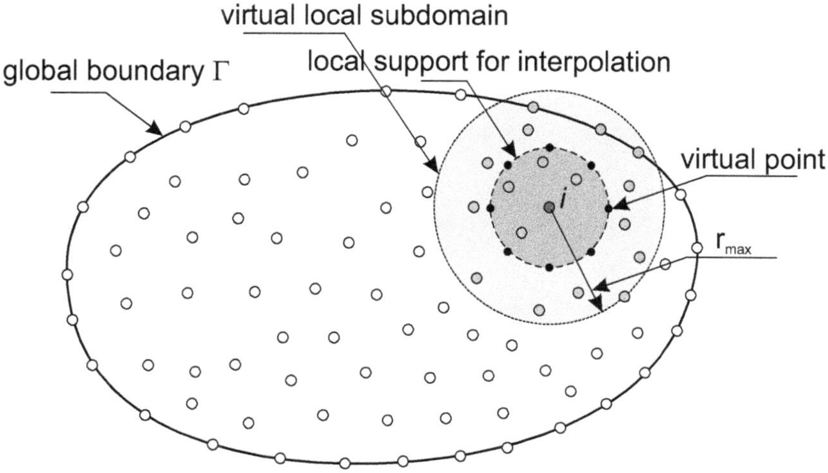

Figure 1. Virtual and supporting nodes in the domain Ω.

3.2. Particular Solution

The particular solution C_P^{n+1} is approximated by radial basis (RBF) functions as

$$C_P^{n+1} = \sum_{j=1}^{p} \beta_j^{n+1} \Theta(r_j), \tag{12}$$

where $\Theta(r_j)$ are radial basis functions, β_j^{n+1} are unknown coefficients, and p is the number of internal virtual points in the virtual subdomain of the point i. These points can be regularly placed inside the subdomain (see Figure 2). The function Θ is defined as a solution of the following equation [2,8]

$$D\Delta\Theta(r_j) - \mathbf{v} \cdot \nabla\Theta(r_j) - \lambda\Theta(r_j) = \varphi(r_j), \tag{13}$$

where $\varphi(r_j)$ are also the radial basis functions. There are various possibilities for how to choose these functions [2,19]. Instead of choosing the simple form of the function $\varphi(r_j)$ on the right-hand side of (13), we choose the simple expression for the basis functions $\Theta(r_j)$. By substituting into (13), we can obtain the corresponding formula for the function $\varphi(r_j)$. In our paper, we chose $\Theta(r_j)$ functions as multiquadrics (MQ) and we obtain

$$\begin{aligned} \Theta(r) &= \sqrt{r^2 + R_P^2} \\ \varphi(r) &= D\frac{r^2+2R_P^2}{\left(r^2+R_P^2\right)^{3/2}} - \frac{\mathbf{r}\cdot\mathbf{v}}{\sqrt{r^2+R_P^2}} - \lambda\sqrt{r^2+R_P^2}, \end{aligned} \tag{14}$$

where R_P is the shape factor of the particular solution. This factor can be different from the factor R used in (21). According to (6) and (13), we can write

$$\sum_{j=1}^{p} \beta_j^{n+1}\varphi(r_j) = R_d \frac{11\left(C_P^{n+1} + C_H^{n+1}\right) - 18C^n + 9C^{n-1} - 2C^{n-2}}{6\Delta t}. \tag{15}$$

Matrix $\mathbf{A}$ from Equation (11) is now extended to $(n+p) \times (n+p)$ dimension and it has the following structure

$$\mathbf{A} = \begin{bmatrix} \mathbf{K} & \mathbf{L} \\ \mathbf{M} & \mathbf{N} \end{bmatrix}, \tag{16}$$

where

$$\begin{array}{ll} K_{kj} = G^*(r_{kj}) & k,j = 1\ldots n \\ L_{kl} = \Theta(r_{kl}) & l = 1\ldots p \end{array} \quad (17)$$

in the boundary virtual points (see Figure 2) and

$$\begin{array}{ll} M_{lj} = -\frac{11R_d}{6\Delta t}G^*(r_{lj}) & l = 1\ldots p, j = 1\ldots n \\ N_{lq} = \varphi(r_{lq}) - \frac{11R_d}{6\Delta t}\Theta(r_{lq}) & l,q = 1\ldots p \end{array} \quad (18)$$

in the internal virtual points (see Figure 2). For the first two time intervals, when we use the Euler scheme, Formula (18) has the form

$$\begin{array}{ll} M_{lj} = -\frac{R_d}{\Delta t}G^*(r_{lj}) & l = 1\ldots p, j = 1\ldots n \\ N_{lq} = \varphi(r_{lq}) - \frac{R_d}{\Delta t}\Theta(r_{lq}) & l,q = 1\ldots p \end{array} \quad (19)$$

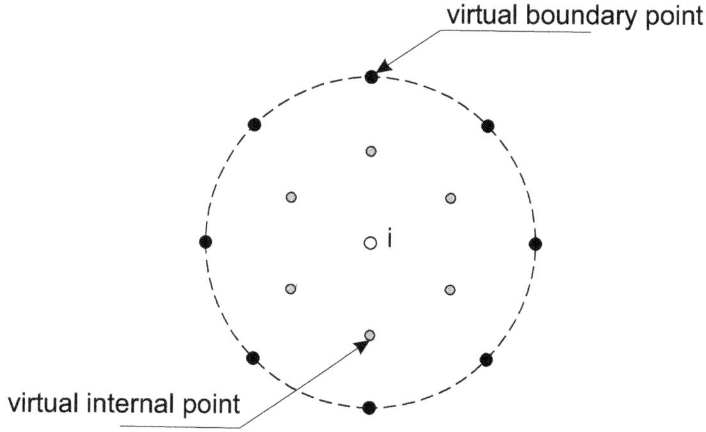

Figure 2. Virtual subdomain around the node *i*.

Since the virtual points do not correspond to the nodes in the support, we must express C_k as a function of the concentrations in the support using some interpolation method. In our article, we have chosen the combination of the weighted radial basis functions and polynomials.

The unknown values C_k in virtual source points are approximated in a support of the point *i* as

$$C_k = \sum_{l=1}^{m} \beta_l^k \mathcal{R}(r_{kl}) + \sum_{l=1}^{M} \chi_l^k p_l(x_k, y_k), \quad k = 1\ldots n, \quad (20)$$

where β_l^k and χ_l^k are the weights, $\mathcal{R}(r_{kl})$ are the radial basis functions, and p_l are polynomials with degree *M*-1. *M* is the order of $\mathcal{R}$, and *m* is the number of nodes in the support of a reference point. In our paper, the multiquadrics functions [19] have been used

$$\mathcal{R}(r_{kl}) = \sqrt{R^2 + r_{kl}^2}, \quad (21)$$

where *R* is a so-called shape factor of the multiquadric function. We can determine the weighting coefficients β_j^k and χ_j^k in Equation (20) by requiring that this equation is fulfilled at all *m* support points. Then, by the procedure described, e.g., in [20], we obtain a set of RBF shape functions Φ_l^k and we can write

$$C_k = \sum_{l=1}^{m} \Phi_l^k C_l. \quad (22)$$

Now, we can substitute (22) to the right side of (11) and obtain

$$\sum_{j=1}^{n} A_{kj}\alpha_j = \sum_{l=1}^{m} \Phi_l^k C_l \quad k = 1 \dots n \tag{23}$$

in matrix notation

$$\mathbf{A}\boldsymbol{\alpha} = \boldsymbol{\Phi}\mathbf{C} \tag{24}$$

and we can obtain the unknown coefficients α_j as

$$\boldsymbol{\alpha} = \mathbf{A}^{-1}\boldsymbol{\Phi}\mathbf{C}. \tag{25}$$

For a homogeneous (steady) solution without internal virtual points, matrix $\mathbf{A}$ has dimensions $(n \times n)$ and matrix $\boldsymbol{\Phi}(n \times m)$. The concentration in point i can now be expressed as

$$C_i = \mathbf{G}_i^T \boldsymbol{\alpha} = \mathbf{G}_i^T \mathbf{A}^{-1}\boldsymbol{\Phi}\mathbf{C} = \mathbf{W}_i^T \mathbf{C}, \tag{26}$$

where $\mathbf{W}^i$ is a weight vector of the point i [21]. The vector $\mathbf{W}$ can be used to assemble the global system of equations to solve homogeneous problems. This system is sparse and can be defined as

$$\begin{aligned} C_i &= C_{0i} \quad \forall i \in \Gamma_1 \\ C_i - \sum_{j=1}^{m} W_j^i u_j &= 0 \quad \forall i \in \Omega \\ \sum_{j=1}^{m} \frac{\partial W_j^i}{\partial \mathbf{n}} u_j &= q_{0i} C_{0i} \quad \forall i \in \Gamma_2, \end{aligned} \tag{27}$$

where m is the number of points in the i-th support.

For a non-stationary problem, it is necessary to add a particular solution (see Section 3.2), and we must extend the matrices $\mathbf{A}$ and $\boldsymbol{\Phi}$ to the dimensions $(n+p) \times (n+p)$ and $(n+p) \times m$, respectively. Then, (26) remains formally the same but the weight vector $\mathbf{W}$ also consists of two parts

$$\mathbf{W}^i = \{\ \mathbf{W}_1^i \quad \mathbf{W}_2^i\ \}. \tag{28}$$

The resulting system of sparse linear equations in the time t^{n+1} can be written as

$$\begin{array}{ll} C_i^{n+1} = C_{0i} & \forall i \in \Gamma_1 \\ C_i^{n+1} - \sum_{j=1}^{m} W_{1j}^i C_j^{n+1} = -\frac{R_d}{6\Delta t}\sum_{j=1}^{p} W_{2j}^i \left(18C_j^n - 9C^{n-1} + 2C^{n-2}\right) & \forall i \in \Omega \\ \sum_{j=1}^{m} \frac{\partial W_{1j}^i}{\partial \mathbf{n}} C_j^{n+1} = q_{0i}(C_i^n - C_{0i}) & \forall i \in \Gamma_2. \end{array} \tag{29}$$

We can solve these N equations to obtain values of concentration at all nodes in $n + 1$ time step.

In the case of the steady problem, the algorithm of the method can be clearly described by the following steps:

1. We define the support of each point in the area.
2. We generate virtual points around each point.
3. We prepare RBF shape functions $\boldsymbol{\Phi}$ according to (22).
4. We calculate the matrix $\mathbf{A}$ according to (11) for each point, except for the points where the Dirichlet boundary condition is prescribed.
5. At these points, we solve the system of linear Equation (26) and obtain the weight vector $\mathbf{W}$.
6. We use this weight vector to construct a sparse global matrix of linear equations according to (27).

7. We multiply the prescribed values of the Dirichlet boundary condition by the corresponding values of the weight vector **W** and, thus, create the right side.
8. By solving the equations, we obtain the concentration values at the points of the area.

The following style will modify the algorithm, describing the unsteady state: The first three steps will be the same as in the steady problem, and we start with step No. 4.

4. We generate internal virtual points.
5. We calculate the matrix **A** according to (19) for each point, except for the points where the Dirichlet boundary condition is prescribed.
6. At these points, we solve the system of linear Equation (26) and obtain the weight vector **W**.
7. We use this weight vector to construct a sparse global matrix of linear equations according to (29).
8. We use the initial conditions and create the right side of the global system (29).
9. We multiply the prescribed values of the Dirichlet boundary condition by the corresponding values of the weight vector **W** and add the results to the right side.
10. By solving the global equations, we obtain the concentration values at the points of the area in the next time step.
11. We can use the results and change the right side of the global system.
12. We repeat steps No. 9 to 11 in the first two time steps.
13. We calculate the new matrix **A** according to (18) and reassemble the global system of equations.
14. We prepare the system's right side using all previous values of concentrations.
15. By solving the global equations, we obtain the concentration values at the points of the area in the next time step.
16. We repeat steps No. 14 to 15 in all remaining time steps.

4. Results

To test the possibilities of the proposed method, we present the results of several test examples in this chapter. In all these cases, the exact analytical solution is known; therefore, it is possible to compare the error of the numerical method. The root mean squared error (RMSE) and R_∞ are employed to evaluate accuracy. These errors are defined as

$$RMSE = \sqrt{\frac{\sum_{i=1}^{N}(\bar{C}_i - C_i)^2}{N}} \qquad R_\infty = \max_{i=1\ldots N}(|\bar{C}_i - C_i|), \tag{30}$$

where $\bar{C}_i$ is the exact value of concentration in point i.

When solving advection–diffusion problems, the Peclet number Pe is often used to assess the effect of advection

$$Pe = \frac{\|\mathbf{v}\|L}{D}. \tag{31}$$

All examples have been computed on a PC computer with an Intel(R) Core(M) i7-8550U processor (1.8 GHz CPU), a 64-bit Windows 11 operating system, and a 16 GB internal memory. The programming language has been Visual C++ and Eigen library for sparse matrix operations.

4.1. Example No. 1, Steady Case

In the first example, we consider the steady advection–diffusion problem with the Dirichlet boundary conditions [22]. The domain is a unit square with prescribed concentrations

$$C(x,0) = \frac{1 - e^{(x-1)v_x/D}}{1 - e^{-v_x/D}} \tag{32}$$

$$C(0,y) = \frac{1 - e^{(y-1)v_y/D}}{1 - e^{-v_y/D}}. \tag{33}$$

The dispersion coefficient is $D = 1$ and the velocity vector is $\mathbf{v} = (v_x, v_y)$ and $v_x = v_y$. The exact solution to this problem is [22]

$$C(x,y) = \frac{\left(1 - e^{(x-1)v_x/D}\right)\left(1 - e^{(y-1)v_y/D}\right)}{\left(1 - e^{-v_x/D}\right)\left(1 - e^{-v_y/D}\right)}. \quad (34)$$

For the numerical solution of this example, three meshes were used. Two meshes were regular with 21 × 21 and 51 × 51 points. The third mesh was irregular with 5426 points. This example was solved with three different Peclet numbers, namely, Pe = 10, 30, and 50. Table 1 shows the solution RMSEs for all previously mentioned meshes and three different Peclet numbers.

Table 1. Example No. 1—comparison of RMSE and R_∞ for different meshes and Pe values.

Pe	Mesh 21 × 21		Mesh 51 × 51		Irregular Mesh	
	RMSE	R_∞	RMSE	R_∞	RMSE	R_∞
10	9.1395×10^{-4}	2.7817×10^{-3}	1.4861×10^{-4}	4.4874×10^{-4}	9.5202×10^{-5}	4.0198×10^{-4}
30	6.0220×10^{-3}	3.5671×10^{-2}	9.3171×10^{-4}	4.6437×10^{-3}	7.3789×10^{-4}	5.1634×10^{-3}
50	1.3147×10^{-2}	9.4959×10^{-2}	2.1142×10^{-3}	1.3600×10^{-2}	1.8039×10^{-3}	2.0027×10^{-2}

The course of the absolute error in the profile x = y is also interesting (Figure 3). It is clear that for higher Peclet numbers, the most significant error is concentrated in the largest concentration gradient at the upper right corner of the area. In Figure 4, we can see the contours of the absolute error of the solution for a regular network of 51 × 51 points and a Peclet number of 10 and 30.

In this example, we also tested the effect of the number of virtual points n and support points m on the method's accuracy. It has been shown that increasing the number of virtual points does not lead linearly to reducing errors (Table 2).

Table 2. Example No. 1—comparison of RMSE and R_∞ for different number of virtual points.

n	Regular Mesh 51 × 51		Irregular Mesh	
	RMSE	R_∞	RMSE	R_∞
6	1.4862×10^{-4}	4.4850×10^{-4}	9.6585×10^{-5}	4.0193×10^{-4}
8	1.4861×10^{-4}	4.4789×10^{-4}	9.6592×10^{-5}	4.0197×10^{-4}
12	1.4783×10^{-4}	4.4629×10^{-4}	9.6779×10^{-5}	4.0168×10^{-4}
16	1.4884×10^{-4}	4.4984×10^{-4}	9.6639×10^{-5}	4.0198×10^{-4}

As for supporting points, point i itself has been also included in the support. The shape of the support for a regular network has been a square with sides formed by an odd number of points. For an irregular network, the algorithm described in [23] has been used. The principle is to divide the vicinity of point i into identical segments, and the point in the segment closest to point i is taken into support.

As seen from Table 3, there is a certain optimal number of points in the support, and further increasing the number of points will not cause an increase in the accuracy of the method.

We also tested the effect of the virtual area's radius on the solution's accuracy. The results are shown in Table 4. We express the radius size as the ratio of the distance from the nearest network point r/d_{min}.

Table 3. Example No. 1—comparison of RMSE and R_∞ for different numbers of supporting points.

m	Regular Mesh 51 × 51 RMSE	R_∞	m	Irregular Mesh RMSE	R_∞
9	1.4860×10^{-4}	4.4792×10^{-4}	5	1.5962×10^{-2}	5.4667×10^{-2}
25	8.1076×10^{-5}	2.5550×10^{-4}	7	9.5201×10^{-5}	4.0198×10^{-4}
49	1.0941×10^{-4}	3.7766×10^{-4}	13	2.0092×10^{-5}	5.3685×10^{-5}
81	3.4093×10^{-4}	1.5674×10^{-3}	19	1.6715×10^{-5}	2.0910×10^{-4}

Table 4. Example No. 1—comparison of RMSE and R_∞ for different radius of the virtual area.

r/d_{min}	Regular Mesh 51 × 51 RMSE	R_∞	Irregular Mesh RMSE	R_∞
0.6	4.3285×10^{-4}	1.3045×10^{-3}	1.6194×10^{-4}	4.9429×10^{-4}
0.8	3.0831×10^{-4}	9.2960×10^{-4}	1.3312×10^{-4}	4.4993×10^{-4}
1.0	1.4861×10^{-4}	4.4789×10^{-4}	9.6592×10^{-5}	4.0197×10^{-4}
1.2	4.6538×10^{-5}	1.4011×10^{-4}	5.6234×10^{-5}	3.4336×10^{-4}
1.4	2.7695×10^{-4}	8.3469×10^{-4}	4.0517×10^{-5}	2.7423×10^{-4}
1.6	5.4251×10^{-4}	1.6342×10^{-3}	8.7447×10^{-5}	3.1388×10^{-4}

With this dependence, it is interesting that there is an optimal radius of the virtual area, which is slightly larger than the minimum distance ($r \approx 1.4d_{min}$).

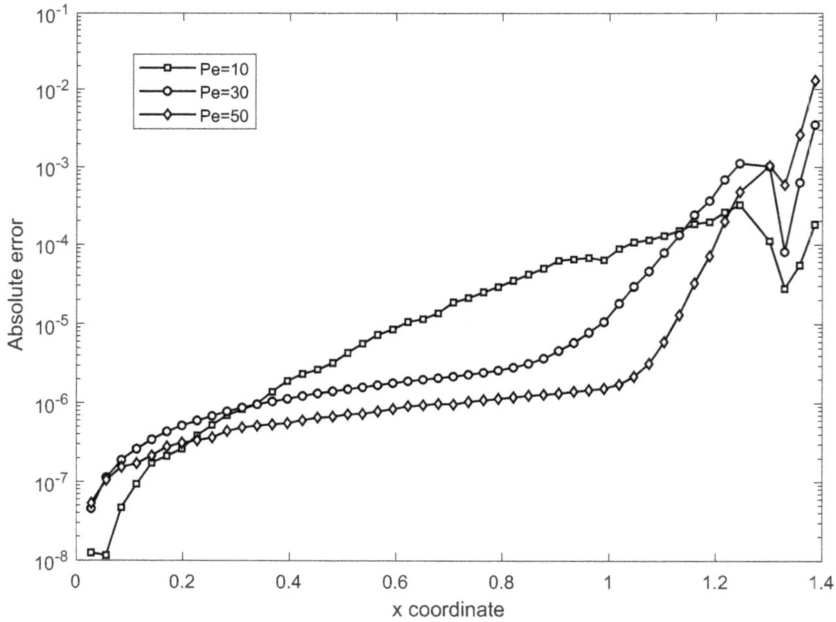

Figure 3. Example No. 1—irregular mesh, absolute errors in the profile x = y.

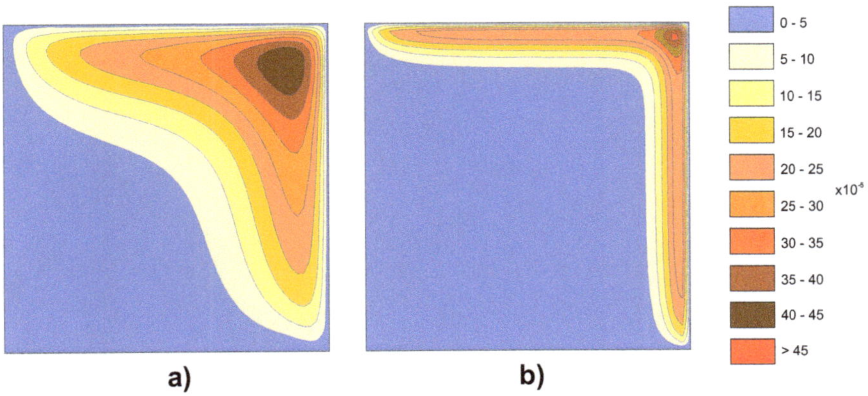

Figure 4. Example No. 1—contours of absolute errors, (**a**) Pe = 10, (**b**) Pe = 30.

Since the accuracy of interpolation using multiquadric functions depends on the shape factor R, we also performed tests for the optimal value of this factor. The result is presented in Figure 5, where the minimum error at the value of $R \approx 0.47$ is obvious.

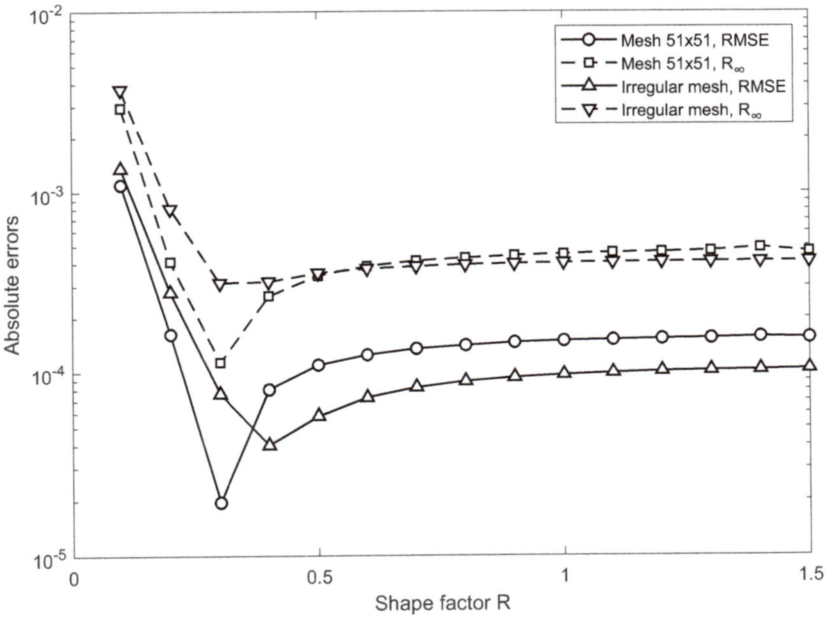

Figure 5. Example No. 1—RMSE and R_∞ as functions of the shape factor R.

4.2. Example No. 2, Steady Case with Decay

The second example tests steady advection along with tracer decay. For the test, we used an irregular area with two different networks of points—the sparser one has 3216 nodes and the denser one has 6530 points (Figure 6). The coordinates of the boundary points have been computed according to the following formula [8,24]

$$r(\theta) = (37 - 12\cos(5\theta))/25, \;\; 0 \leq \theta \leq 2\pi \qquad (35)$$
$$\{x, y\} = \{r\cos(\theta), r\sin(\theta)\}.$$

The internal points in both networks have been generated using the Poisson disc algorithm [25,26].

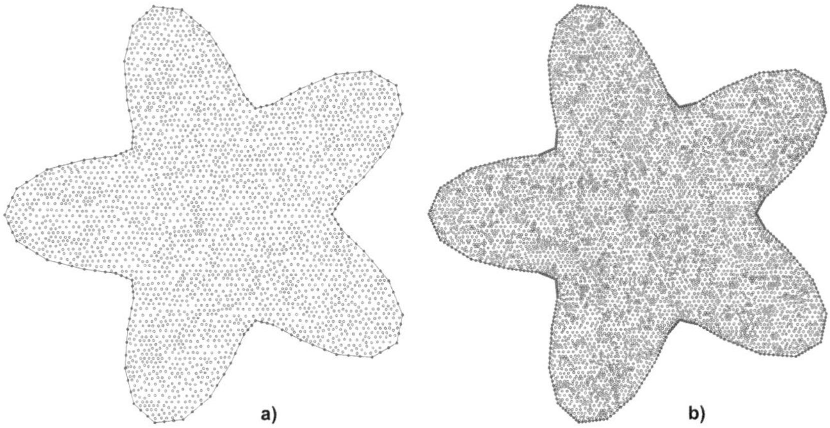

Figure 6. Example No. 2—irregular meshes: (**a**) 3216 points, (**b**) 6530 points.

The dispersion and decay coefficients are $D = 5$ and $\lambda = 4$, respectively. The vector of velocity is constant, $\mathbf{v} = (v_x, v_y) = (1, 1)$. Dirichlet boundary conditions are prescribed at the boundary Γ as

$$C_0(\mathbf{x}) = e^x + e^y \quad \mathbf{x} \in \Gamma. \tag{36}$$

The exact solution is

$$C(\mathbf{x}) = e^x + e^y \quad \mathbf{x} \in \Omega. \tag{37}$$

Similar to the first example, we present a comparison of the accuracy of our modified method for different numbers of virtual points n. We can see from Table 5 that this influence of the number of virtual points on accuracy is negligible.

Table 6 shows the dependence of the accuracy of the method on the number of points in the support. The situation is now different; the number of supporting points m affects the accuracy significantly. The results for both networks used are different.

Table 5. Example No. 2—comparison of RMSE and R_∞ for a different number of virtual points.

n	3216 Points		6530 Points	
	RMSE	R_∞	RMSE	R_∞
6	1.0489×10^{-4}	1.8703×10^{-3}	3.3324×10^{-5}	6.8178×10^{-5}
8	8.0688×10^{-5}	3.6588×10^{-4}	3.6236×10^{-5}	7.3594×10^{-5}
10	8.5035×10^{-5}	8.2199×10^{-4}	3.4279×10^{-5}	7.0187×10^{-5}
16	8.2809×10^{-5}	7.0545×10^{-4}	3.5056×10^{-5}	7.1266×10^{-5}

The error decreases with the increasing number of supporting points in the first sparser network. In the second denser one, the initial decrease is followed by an increase in the error; thus, we can find the optimal number of points in the support (see also Figure 7 and Table 6). The slight deterioration in accuracy when increasing the number of supporting points is probably since the criterion according to [23] at higher numbers leads to an unsatisfactory selection. In these cases, it would probably be better to return to a simple choice based on the distance from point i.

Table 6. Example No. 2—comparison of RMSE and R_∞ for different numbers of supporting points.

m	3216 Points		6530 Points	
	RMSE	R_∞	RMSE	R_∞
7	3.57×10^{-4}	2.56×10^{-3}	1.87×10^{-4}	3.79×10^{-4}
9	6.13×10^{-4}	2.68×10^{-3}	2.55×10^{-4}	1.54×10^{-3}
13	8.07×10^{-5}	3.66×10^{-4}	3.62×10^{-5}	7.36×10^{-5}
19	1.01×10^{-5}	3.47×10^{-5}	3.85×10^{-6}	7.96×10^{-6}
25	3.25×10^{-6}	2.29×10^{-5}	8.01×10^{-6}	6.65×10^{-5}
31	7.63×10^{-7}	1.61×10^{-5}	3.23×10^{-6}	6.09×10^{-5}

The distribution of absolute errors in the area for both solved networks is presented in Figure 8.

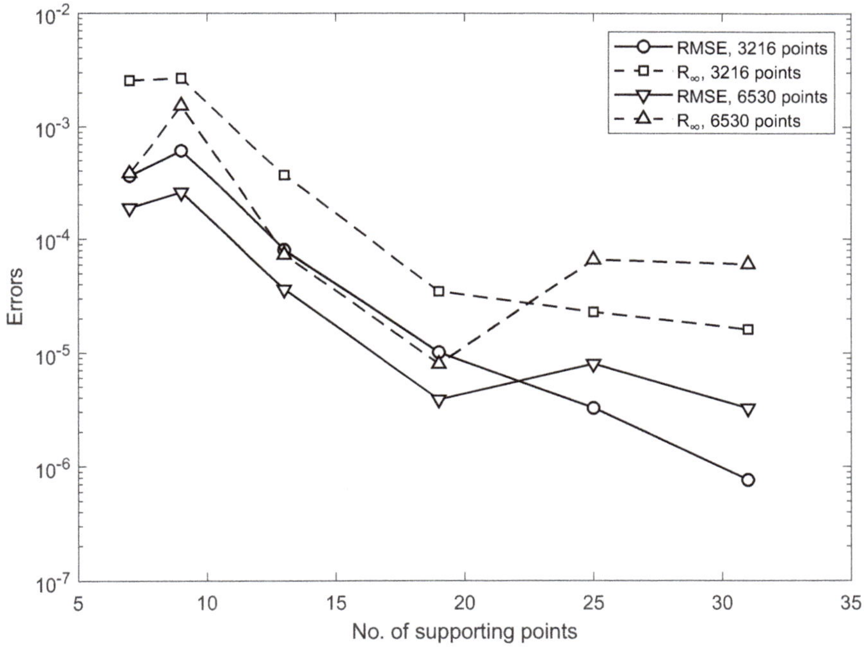

Figure 7. Example No. 2—course of RMSE and R_∞ for different numbers of supporting points.

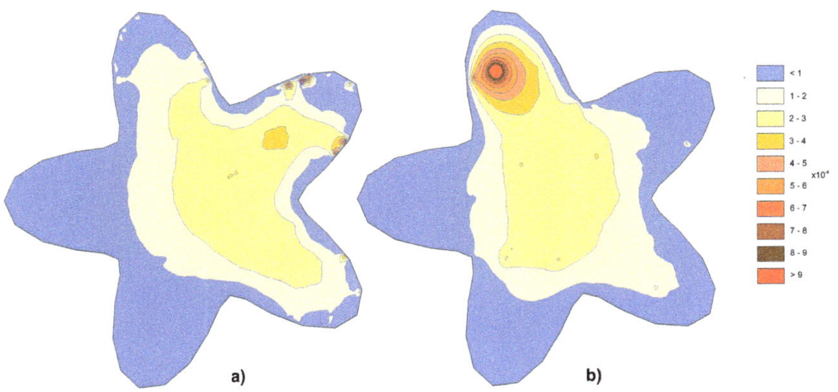

Figure 8. Example No. 2—contours of absolute errors, (**a**) 3216 points, (**b**) 6530 points.

As in the previous example, we can also see in Table 7 that it is possible to increase the accuracy of the solution by approximately one order of magnitude by slightly increasing the radius of the virtual area to $r \approx 1.4d_{min}$.

Table 7. Example No. 2—comparison of RMSE and R_∞ for different radii of virtual area.

r/d_{min}	3216 Points		6530 Points	
	RMSE	R_∞	RMSE	R_∞
0.6	1.8708×10^{-5}	3.7636×10^{-5}	5.4936×10^{-6}	1.1633×10^{-5}
0.8	1.5363×10^{-5}	5.6583×10^{-5}	4.9468×10^{-6}	1.0110×10^{-5}
1.0	1.0094×10^{-5}	3.4681×10^{-5}	3.8808×10^{-6}	8.0037×10^{-6}
1.2	4.1740×10^{-6}	3.5152×10^{-5}	2.6612×10^{-6}	5.7076×10^{-6}
1.4	2.7209×10^{-6}	5.8848×10^{-5}	1.2998×10^{-6}	4.9287×10^{-6}
1.6	5.8256×10^{-6}	8.4728×10^{-5}	2.0681×10^{-6}	2.1252×10^{-5}

Further, in this example, we tested the shape factor's influence on the method's accuracy. It turns out that the accuracy increases slightly with the increasing value of R (Figure 9).

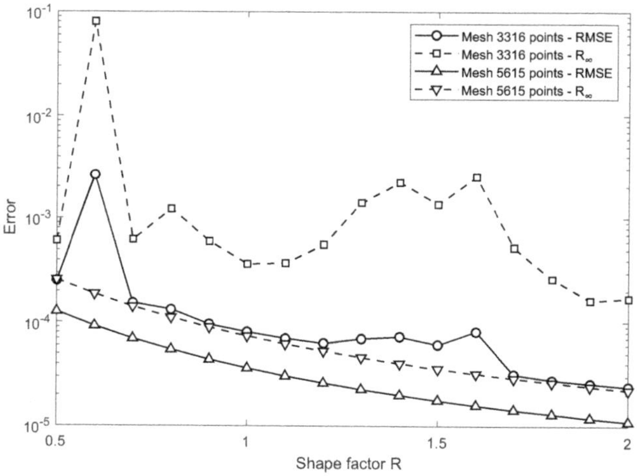

Figure 9. Example No. 2—RMSE and R_∞ as functions of the shape factor R.

4.3. *Example No. 3, Unsteady Case*

The third example is the usual case used for testing of the unsteady problem [27–29]. The rectangular domain $[0, 20] \times [0, 2]$ has the initial concentration $C_0 = 0$ and the Dirichlet boundary conditions $C(0, y) = 1$ and $C(20, y) = 0$. The Neumann boundary conditions are prescribed as $q(0, x) = q(2, x) = 0$. The exact solution is (see e.g., [30])

$$C(x, t) = \frac{C_0}{2}\left[erfc(z_1) + \exp\left(\frac{v_x x}{D}\right) erfc(z_2)\right], \tag{38}$$

where

$$z_1 = \frac{x - v_x t}{\sqrt{4Dt}} \qquad z_2 = \frac{x + v_x t}{\sqrt{4Dt}}. \tag{39}$$

The horizontal velocity $v_x = 1$ and three diffusion coefficients $D = 0.1$, $D = 0.05$, and $D = 0.02$ have been used. Then, the Peclet numbers (31) are Pe = 200, Pe = 400, and Pe = 1000, respectively. The RBFs with shape functions $\varphi(r)$ according to (14) are used in this example. Two regular meshes of 101×11 and 161×17 points have been used. The total simulation time is $t = 10$.

Figure 10 presents the concentration for both networks in the profile $y = 1$. Figure 11 then represents the course of the absolute error in this profile. All these results are plotted for times t = 2, 4, 6, 8, and 10.

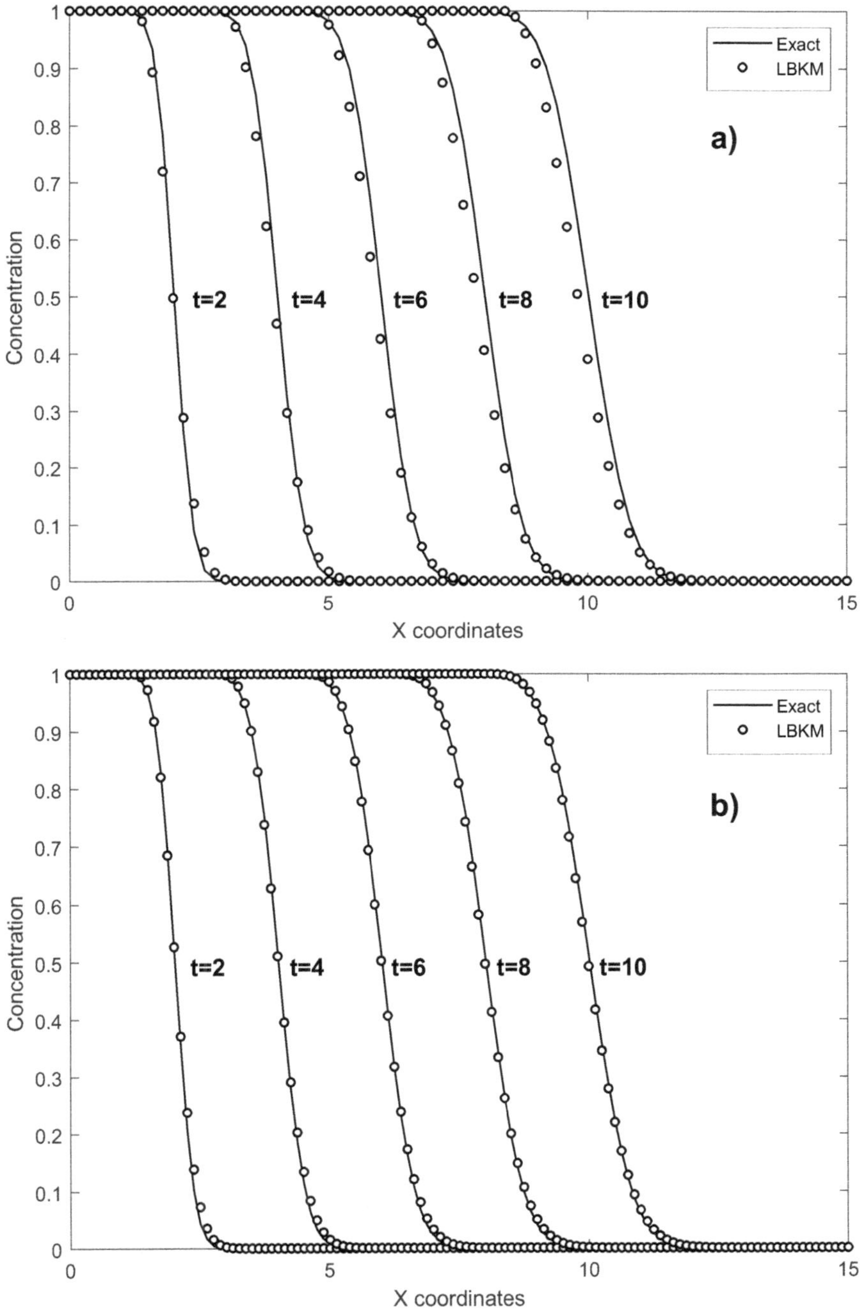

Figure 10. Example No. 3—concentration profiles for Pe = 1000: (**a**) Grid 101 × 11, (**b**) Grid 161 × 17.

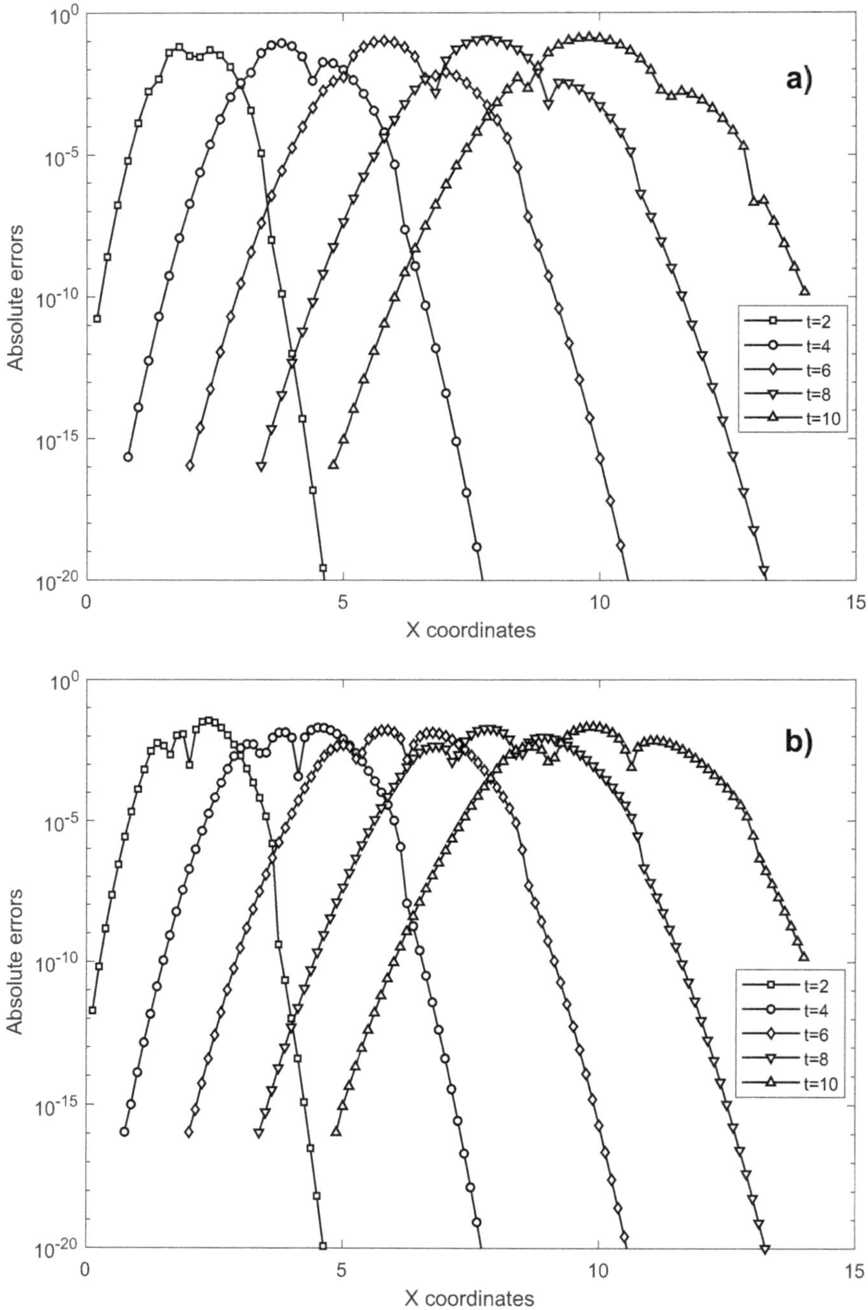

Figure 11. Example No. 3—absolute errors for Pe = 1000, (**a**) Grid 101 × 11, (**b**) Grid 161 × 17.

Table 8 clearly shows that the values of RMSE and R_∞ decrease when increasing the density of the grid.

Table 8. Example No. 3—RMSE and R_∞ for different Peclet numbers.

Pe	Mesh 101 × 11		Mesh 161 × 17	
	RMSE	R_∞	RMSE	R_∞
200	4.6244×10^{-3}	1.4834×10^{-2}	8.3319×10^{-4}	3.3870×10^{-3}
400	1.0440×10^{-2}	3.9621×10^{-2}	1.2844×10^{-3}	5.6586×10^{-3}
1000	2.2662×10^{-2}	1.0775×10^{-1}	4.1826×10^{-3}	2.1395×10^{-2}

In this example, because it is an unsteady problem, we focused primarily on testing the influence of the time step size and the values of the two shape factors R and R_P on the accuracy of the solution.

Figure 12 shows a comparison of the RMSE and R_∞ errors using the various sizes of time steps and Pe = 200. It is clear from Figure 12 that there is an optimal time step for every mesh. Its further refinement only reduces the accuracy of the solution and increases the CPU time.

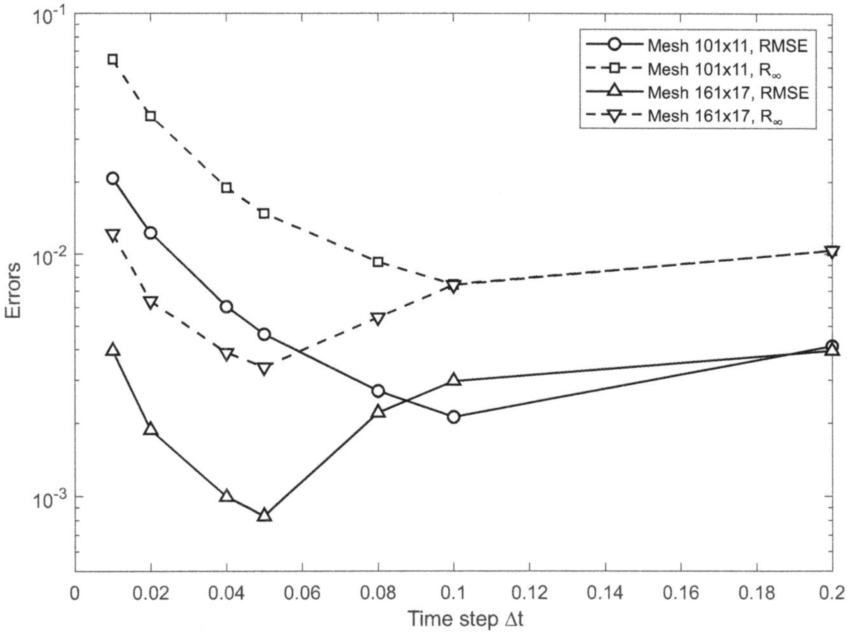

Figure 12. Example No. 3—RMSE and R_∞ as functions of the time step Δt.

Similar to the previous examples, we monitored the dependence of the accuracy of the solution on the shape factor R of the RBF interpolation for the third example. We also tested the influence of the R_P factor, which is used for the particular solution approximation. These dependencies are plotted in Figures 13 and 14. It can be seen that initially, the error of the method decreases to an insignificant minimum and then the values of RMSE and R_∞ stabilize.

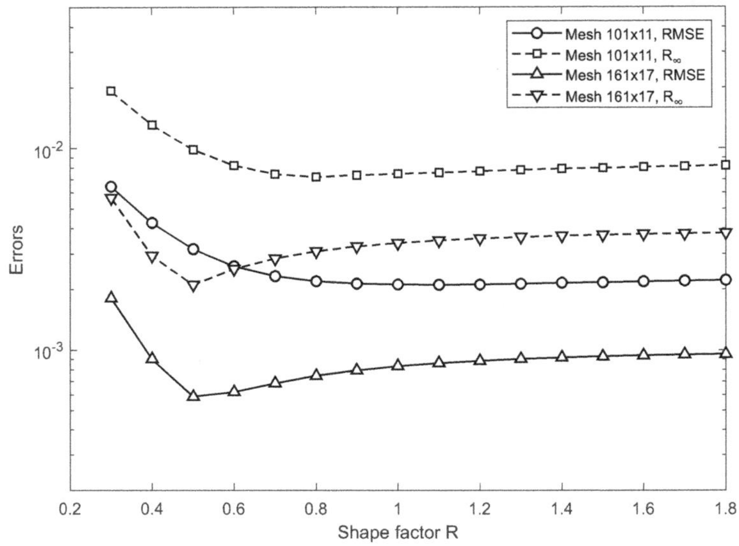

Figure 13. Example No. 3—RMSE and R_∞ as functions of the shape factor R.

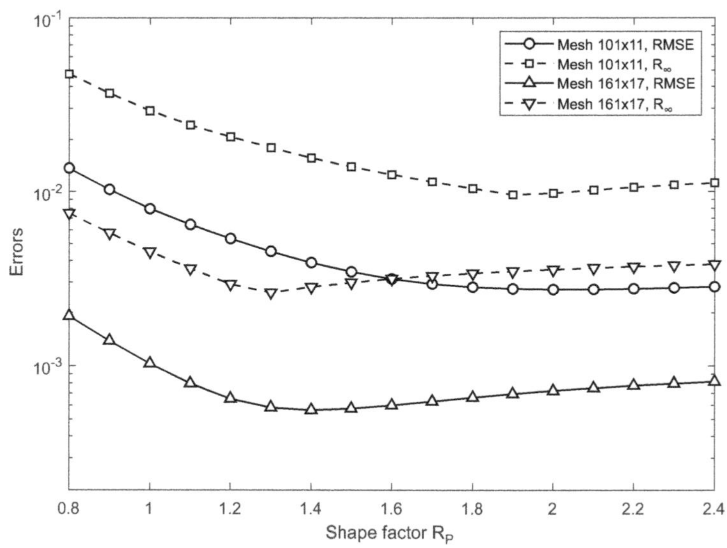

Figure 14. Example No. 3—RMSE and R_∞ as functions of the shape factor R_P.

The same rectangular domain with the same two different point meshes as well as boundary conditions is used to test the effect of tracer decay; the decay coefficient values $\lambda = 0.1$ and $\lambda = 0.3$ are entered. The exact solution is then given as [30]

$$C(x,t) = \frac{C_0}{2}\exp(\frac{v_x x}{2D})[\exp(-x\beta)erfc(z_1) + \exp(x\beta)erfc(z_2)], \tag{40}$$

where

$$\beta = \sqrt{\frac{v_x^2}{4D^2} + \frac{\lambda}{D}} \tag{41}$$

and z_1 and z_2 are now

$$z_1 = \frac{x - t\sqrt{v_x^2 + 4\lambda D}}{\sqrt{4Dt}} \qquad z_2 = \frac{x + t\sqrt{v_x^2 + 4\lambda D}}{\sqrt{4Dt}}. \tag{42}$$

The course of exact concentration and LBKM results in the profile $y = 2$ at time t = 2, 4, 6, 8, and 10 can be seen in Figure 15 for the two different values of the decay coefficient. Figure 16 then shows the course of the absolute errors at the same time intervals.

Table 9 shows the RMSE values for λ = 0.1 and 0.3 for both used point networks.

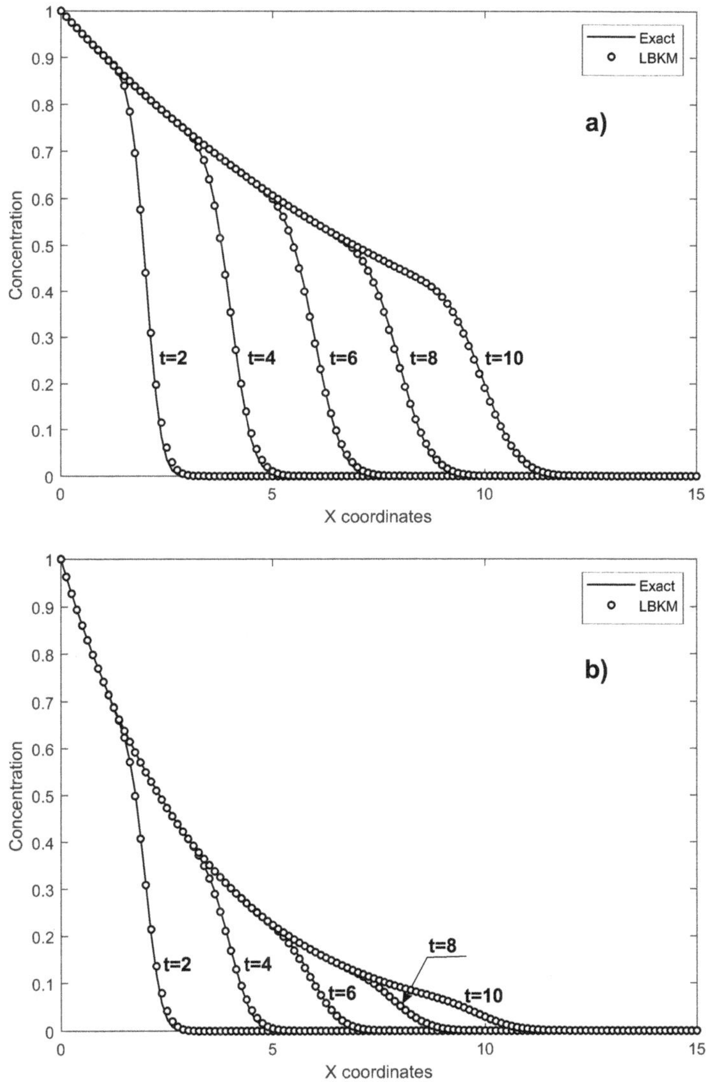

Figure 15. Example No. 3—concentration profiles for Pe = 1000, mesh 161 × 33: (**a**) $\lambda = 0.1$, (**b**) $\lambda = 0.1$.

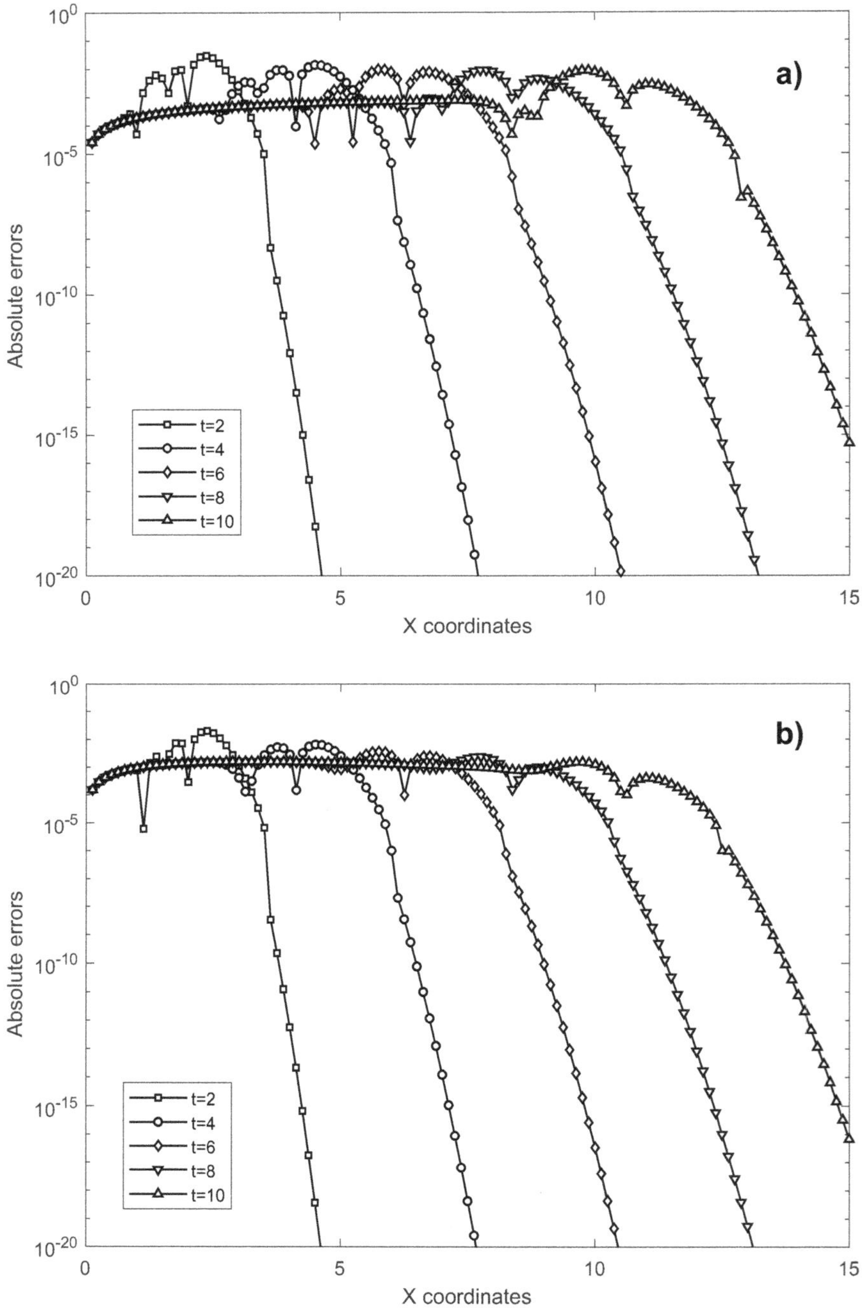

Figure 16. Example No. 3—absolute errors for Pe = 1000, mesh 161 × 33: (**a**) $\lambda = 0.1$, (**b**) $\lambda = 0.5$.

Table 9. Example No. 3—comparison of RMSE for Pe = 1000, $\lambda = 0.1$, $\lambda = 0.3$, and different meshes.

Time	$\lambda = 0.1$		$\lambda = 0.3$	
	Mesh 101 × 11	Mesh 161 × 17	Mesh 101 × 11	Mesh 161 × 17
2	8.0725×10^{-3}	4.0627×10^{-3}	5.9374×10^{-3}	2.7769×10^{-3}
4	8.0044×10^{-3}	2.3824×10^{-3}	4.5891×10^{-3}	1.2482×10^{-3}
6	8.8779×10^{-3}	1.8665×10^{-3}	4.0667×10^{-3}	9.8041×10^{-4}
8	9.2876×10^{-3}	1.7438×10^{-3}	3.6041×10^{-3}	9.4922×10^{-4}
10	9.2755×10^{-3}	1.7132×10^{-3}	3.2915×10^{-3}	9.4261×10^{-4}

5. Discussion and Conclusions

In the article, we presented a modification of the local knots method applied to the solution of the advection–diffusion equation. Unlike the previous localizations of the node method, this method uses a regular circular virtual region with evenly spaced virtual points. The boundary knots method is applied to this area. In the next step, this area is connected to the support of the resolved node. Although this procedure is a bit more complicated than the previous methods, it has some significant advantages.

5.1. Condition Numbers

Probably the most significant advantage concerns the reduction of the order of the boundary knots matrix and, thus, also the decrease of the condition number of this matrix. It is possible owing to the fact that the virtual points number is small. It also remains constant for all nodes. Therefore, it was possible to work with this matrix in the presented method using only simple algorithms for solving linear equations or matrix inversion. In addition, it is possible (especially for regular networks of nodes) to design this virtual region equal for all nodes and, thus, to calculate the inverse matrix of the method only once.

In our method, we can distinguish three different condition numbers (CN): local, global, and RBF. The local CN is the condition number of the local matrix **A** (16). The global CN is the condition number of the global system of equations and is significantly lower than the local one. The RBF condition number refers to the interpolation matrix of radial basic functions (see Table 10). Table 11 contains condition numbers of unsteady case (Example No. 3).

The local CN depends substantially on the number of virtual boundary points (see Figure 17). As the number of these points does not influence the precision of our method, we recommend using a maximum of 8 points.

Table 10. Example No. 1—values of condition numbers, eight virtual boundary points.

Pe	Local	Global	RBF
10	4.9691×10^{7}	5.0797×10^{2}	7.5114×10^{5}
30	2.6072×10^{5}	3.0437×10^{2}	7.5114×10^{5}
50	4.6362×10^{5}	2.4988×10^{2}	7.5114×10^{5}

Table 11. Example No. 3—values of condition numbers, eight virtual boundary and six internal points.

Mesh	Local	Global	RBF
101 × 11	6.9144×10^{5}	1.4806×10^{3}	1.8830×10^{7}
167 × 16	7.9764×10^{6}	3.8825×10^{3}	2.8587×10^{7}

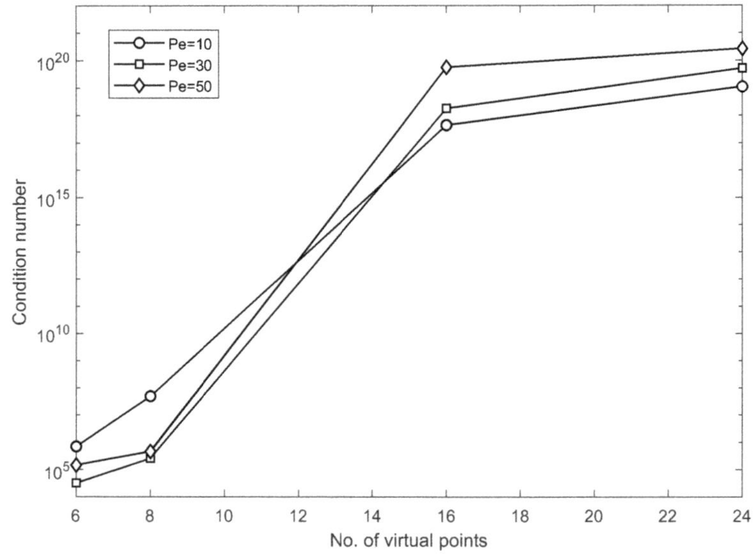

Figure 17. Example No. 1—connection of the local condition number and number of virtual points.

5.2. Convergence Rate

For all three examples, we performed tests of the speed of convergence of the method. For the purposes of these tests, we have additionally added one more sparse network for each example. For the first example, the grid had 16 × 16 points; in the second example it was an irregular grid with 4305 points; and in the third example, we used a grid of 81 × 9 points. Figure 18 shows the dependency of RMSE on the number of points. To demonstrate the convergence rate (CR) of the present method, the following formula is introduced

$$CR = -\frac{\log(RMSE_1) - \log(RMSE_2)}{\log(N_1) - \log(N_2)}. \tag{43}$$

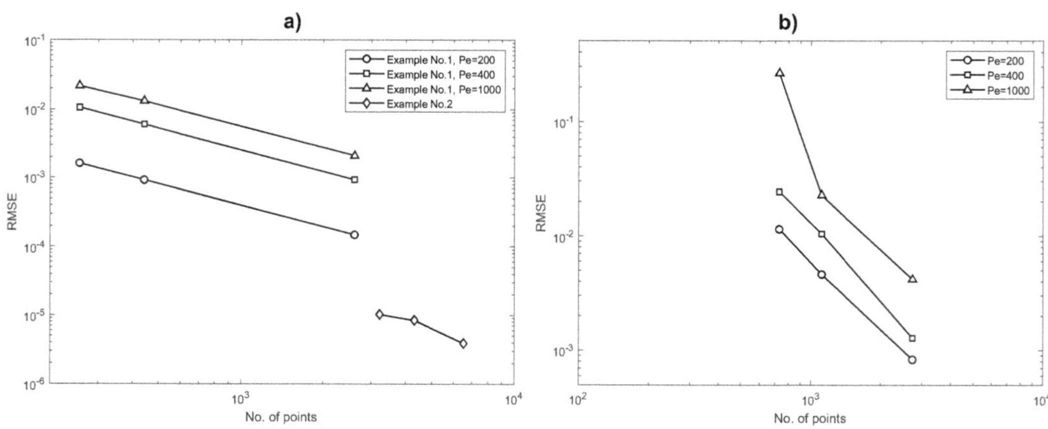

Figure 18. Connection of the RMSE and number of points: (**a**) steady solution, (**b**) unsteady solution.

Table 12 contains values of RMSE and convergence rates for examples of the steady case and Table 13 for those of the unsteady transport.

Table 12. Example Nos. 1 and 2—values of RMSE and the convergence rates (CR).

No. of Points	RMSE, Example No. 1			No. of Points	RMSE, Example No. 2
	Pe = 10	Pe = 30	Pe = 50		
256	1.619×10^{-3}	1.060×10^{-2}	2.162×10^{-2}	3216	1.023×10^{-5}
441	9.140×10^{-4}	6.022×10^{-3}	1.315×10^{-2}	4305	8.390×10^{-6}
2601	1.486×10^{-4}	9.317×10^{-4}	2.114×10^{-3}	6530	3.881×10^{-6}
CR	1.024	1.052	1.030	CR	1.368

Table 13. Example No. 3—values of RMSE and the convergence rates (CR).

No. of Points	Pe = 20	Pe = 400	Pe = 1000
729	1.150×10^{-2}	2.425×10^{-2}	2.630×10^{-1}
1111	4.624×10^{-3}	1.044×10^{-2}	2.266×10^{-2}
2737	8.332×10^{-4}	1.284×10^{-3}	4.183×10^{-3}
CR	1.984	2.221	3.130

From the values in Table 12, we can conclude that with a regular network of points, the order of the method is slightly above the value of one, and with an irregular network, it is about 30% higher, which may be caused by the different geometric configuration of the irregular networks. In the unsteady state, we see that Houbolt's method confirms its effectiveness in this case as well, and the values of the rate of convergence are above two.

In the further development of the method, a logical step will be to extend it to 3D tasks or non-linear problems.

Author Contributions: Conceptualization and methodology, K.K. and J.M.; software, J.M.; validation, K.K. and J.M.; project administration, J.M. All authors have read and agreed to the published version of the manuscript.

Funding: This research was funded by Vedecká Grantová Agentúra MŠVVaŠ SR a SAV (VEGA) grant number 1-0879-21.

Institutional Review Board Statement: Not applicable.

Informed Consent Statement: Not applicable.

Data Availability Statement: Not applicable.

Conflicts of Interest: The authors declare no conflict of interest.

Abbreviations

The following abbreviations are used in this manuscript:

CN	Condition Number
CR	Convergence Rate
MDPI	Multidisciplinary Digital Publishing Institute
LBKM	Local Boundary Knots Method
LMAPS	Local Method of Approximating Particular Solutions
Pe	Peclet number
RBF	Radial Basis Functions
RMSE	Root Mean Square Error
R_∞	Maximum Absolute Error

References

1. Brebbia, C.A.; Telles, J.C.F.; Wrobel, L.C. *Boundary Element Techniques*; Springer: Berlin, Germany; New York, NY, USA, 1984.
2. Partridge, P.W.; Brebbia, C.A.; Wrobel, L.C. *The Dual Reciprocity Boundary Element Method*; CM Publications: Southampton, UK, 1992.
3. Golberg, M. The method of fundamental solutions for Poisson's equations. *Eng. Anal. Bound. Elem.* **1995**, *16*, 205–213. [CrossRef]

4. Golberg, M.A.; Chen, C.S. The method of fundamental solutions for potential, Helmholtz and diffusion problems. In *Boundary Integral Methods-Numerical and Mathematical Aspects*; Golberg, M.A., Ed.; CM Publications: Southampton, UK, 1998; pp. 103–176.
5. Chen, C.S.; Karageorghis, A. On choosing the location of the sources in the MFS. *Numer. Algorithms* **2016**, *72*, 107–130. [CrossRef]
6. Chen, W.; Fu, Z.; Wei, X. Potential problems by singular boundary method satisfying moment condition. *CMES-Comput. Model. Eng. Sci.* **2009**, *54*, 65–85.
7. Chen, W.; Gu, Y. Recent Advances on Singular Boundary Method. In Proceedings of the Joint International Workshop for Trefftz Method, Kaohsiung, Taiwan, 15–18 March 2011; Volume 4, pp. 543–558.
8. Kovářík, K.; Mužík, J.; Bulko, R.; Sitányiová, D. Singular boundary method using dual reciprocity for two-dimensional transient diffusion. *Eng. Anal. Bound. Elem.* **2017**, *83*, 256–264. [CrossRef]
9. Kovářík, K.; Mužík, J.; Masarovičová, S.; Sitányiová, D. Regularized singular boundary method for 3D potential flow. *Eng. Anal. Bound. Elem.* **2018**, *95*, 85–92. [CrossRef]
10. Hon, Y.C.; Chen, W. Boundary knot method for 2D and 3D Helmholtz and convection–diffusion problems under complicated geometry. *Int. J. Numer. Method Eng.* **2003**, *56*, 1931–1948. [CrossRef]
11. Chen, W.; Shen, L.J.; Shen, Z.J.; Yuan, G.W. Boundary knot method for Poisson equations. *Eng. Anal. Bound. Elem.* **2005**, *29*, 756–760. [CrossRef]
12. Mužík, J. Boundary Knot Method for Convection-diffusion Problems. *Procedia Eng.* **2015**, *111*, 582–588. [CrossRef]
13. Wang, F.; Wang, C.; Chen, Z. Local knot method for 2D and 3D convection–diffusion–reaction equations in arbitrary domains. *Appl. Math. Lett.* **2020**, *105*, 106308. [CrossRef]
14. Yue, X.; Wang, F.; Li, P.W.; Fan, C.M. Local non-singular knot method for large-scale computation of acoustic problems in complicated geometries. *Comput. Math. Appl.* **2021**, *84*, 128–143. [CrossRef]
15. Zhu, T.; Zhang, J.D.; Atluri, S. A local boundary integral equation (LBIE) method in computational mechanics, and a meshless discretization approach. *Comput. Mech.* **1998**, *21*, 223–235. [CrossRef]
16. Sellountos, E.J.; Sequeira, A. An advanced meshless LBIE/RBF method for solving two-dimensional incompressible fluid flows. *Comput. Mech.* **2008**, *41*, 617–631. [CrossRef]
17. Young, D.; Gu, M.; Fan, C. The time-marching method of fundamental solutions for wave equations. *Eng. Anal. Bound. Elem.* **2009**, *33*, 1411–1425. [CrossRef]
18. Kovářík, K.; Mužík, J.; Bulko, R.; Sitányiová, D. Local singular boundary method for two-dimensional steady and unsteady potential flow. *Eng. Anal. Bound. Elem.* **2019**, *108*, 168–178. [CrossRef]
19. Golberg, M.; Chen, C.; Bowman, H. Some recent results and proposals for the use of radial basis functions in the BEM. *Eng. Anal. Bound. Elem.* **1999**, *23*, 285–296. [CrossRef]
20. Kovářík, K.; Mužík, J.; Mahmood, M.S. A meshless solution of two dimensional unsteady flow. *Eng. Anal. Bound. Elem.* **2012**, *36*, 738–743. [CrossRef]
21. Stevens, D.; Power, H.; Meng, C.Y.; Howard, D.; Cliffe, K.A. An alternative local collocation strategy for high-convergence meshless PDE solutions, using radial basis functions. *J. Comput. Phys.* **2013**, *294*, 52–75. [CrossRef]
22. Reddy, J.; Martinez, M. A dual mesh finite domain method for steady-state convection–diffusion problems. *Comput. Fluids* **2021**, *214*, 104760. [CrossRef]
23. Kovářík, K.; Mužík, J. A meshless solution for two dimensional density-driven groundwater flow. *Eng. Anal. Bound. Elem.* **2013**, *37*, 187–196. [CrossRef]
24. Wang, F.; Chen, W. Accurate empirical formulas for the evaluation of origin intensity factor in singular boundary method using time-dependent diffusion fundamental solution. *Int. J. Heat Mass Transf.* **2016**, *103*, 360–369. [CrossRef]
25. Dunbar, D.; Humphreys, G. A spatial data structure for fast Poisson-disk sample generation. *ACM Trans. Graph.* **2006**, *25*, 503–508. [CrossRef]
26. Wei, L.Y. Parallel Poisson Disk Sampling. *ACM Trans. Graph.* **2008**, *27*, 1–9.
27. Singh, K.M.; Tanaka, M. Dual reciprocity boundary element analysis of transient advection-diffusion. *Int. J. Numer. Method Heat Fluid Flow* **2003**, *13*, 633–646. [CrossRef]
28. Kovářík, K. Numerical simulation of groundwater flow and pollution transport using the dual reciprocity and RBF method. *Communications* **2010**, *12*, 5–10. [CrossRef]
29. Wang, F.; Chen, W.; Tadeu, A.; Correia, C.G. Singular boundary method for transient convection–diffusion problems with time-dependent fundamental solution. *Int. J. Heat Mass Transf.* **2017**, *114*, 1126–1134. [CrossRef]
30. Bear, J. *Dynamics of Fluids in Porous Media*; American Elsevier Publishing Co.: New York, NY, USA, 1972.

applied sciences

MDPI

Article

Numerical Study on Single-Bubble Contraction–Rebound Characteristics in Cryogenic Fluids

Shaohang Yan [1], Tianwei Lai [1,*], Qi Zhao [1], Mingchen Qiang [1], Mingzhe Liu [1], Wenjing Ding [2], Yutao Liu [3] and Yu Hou [1]

1 State Key Laboratory of Multiphase Flow in Power Engineering, Xi'an Jiaotong University, Xi'an 710049, China
2 Beijing Institute of Satellite Environmental Engineering, Beijing 100029, China
3 Beijing Institute of Aerospace Test Technology, Beijing 100074, China
* Correspondence: laitianwei@mail.xjtu.edu.cn; Tel.: +86-152-0295-2214

Abstract: In cryogenic fluid storage and delivery, the rapid contraction and rebound of bubbles are prone to occur during bubble collapse due to the pressure saltation. With the contraction and rebound of bubbles, the pressure and temperature in the bubbles fluctuate greatly, which affects the service life of fluid machinery. During bubble contraction and rebound, there is an accompanied complex heat and mass transfer process. According to the thermal properties of cryogenic fluids, a single-bubble collapse model is proposed considering the temperature variations inside the bubble. In order to study the variation in temperature and pressure during bubble collapse in cryogenic fluids, the contraction and rebound of a single bubble in liquid hydrogen are investigated numerically under various operating pressures and supercooling degrees. The numerical results of the model indicate that there are periodic contraction and rebound of the bubble when the pressure rises suddenly. Furthermore, the periods and attenuation rates of bubbles in different media are studied and compared. For the most concerned pressure and temperature characteristics, the relationship between the peak pressure, the attenuation rate of the temperature and the dimensionless number is proposed.

Keywords: single bubble; oscillation; attenuation rate; periodicity; supercooling degree; operating pressure

Citation: Yan, S.; Lai, T.; Zhao, Q.; Qiang, M.; Liu, M.; Ding, W.; Liu, Y.; Hou, Y. Numerical Study on Single-Bubble Contraction–Rebound Characteristics in Cryogenic Fluids. *Appl. Sci.* **2022**, *12*, 10839. https://doi.org/10.3390/app122110839

Academic Editor: Vasily Novozhilov

Received: 24 September 2022
Accepted: 24 October 2022
Published: 26 October 2022

Publisher's Note: MDPI stays neutral with regard to jurisdictional claims in published maps and institutional affiliations.

1. Introduction

For long-distance delivery and large-scale storage of cryogenic fluids, liquid storage and delivery are the most widely used methods in favor of high security and cost effectiveness [1–4]. Among the cryogenic fluids, liquid hydrogen (normal boil point is 20.37 K) and liquid oxygen (normal boil point is 90.19 K) have been widely used in rocket engines [5,6]. Taking liquid hydrogen as an example, its delivery cost is only 0.181 EUR/kg [7] and energy consumption is only 0.37 kWh/kg [8], which is much lower than 0.63~0.76 EUR/kg and 2.43 kWh/kg for high-pressure hydrogen delivery.

In the transportation of cryogenic fluids, substantial sloshing cannot be avoided easily due to the jounce of transport tankers and LNG carrier ships. This results in a large variation in local pressure for the delivered cryogenic fluids. Under this condition, cavitation bubbles grow when the local pressure drops below the saturated pressure of liquids [9,10]. When the operating pressure increases suddenly, the bubbles contract and rebound rapidly. During this contraction–rebound process, there is noise, vibration, pressure pulsation and damage to the fluid machinery due to the high pressure and temperature of the bubbles [11].

Single-bubble growth in ordinary fluids has been preliminarily studied. The single bubbles growth in water and freon has been focused on in many experiments [12–14]. Compared with bubble experiments in ordinary fluids, cavitation experiments in cryogenic fluids are relatively rare. Among cryogenic research, experiments of liquid nitrogen and hydrogen flow around a hydrofoil and an ogive were carried out by Hord in the 1970s [15–18]. The temperature and pressure distributions along the cavitation region of the hydrofoil surface were measured experimentally. Moreover, Ball et al. [19] experimentally investigated liquid hydrogen cavitation characteristics in various inducers. The experimental results provided a reference for the design of an inducer for cryogenic fluids. Recently, Ito et al. [20] studied the difference in cavitation characteristics between liquid nitrogen and water in an inducer using a visualization experiment. The experimental results indicated that the bubble size in liquid nitrogen was much smaller than that in water. In addition to cavitation flow experiments in fluid machinery, single-bubble growth and collapse in cryogenic fluids are the basis for the study of the cavitation mechanism in cryogenic fluids. However, there is relatively little experimental research in the open literature due to its extremely rapid growth and collapse process. Hewitt et al. [21] carried out the bubble growth experiment in liquid nitrogen under depressurization and studied the bubble radius growth process. Due to the limitations of harsh low-temperature conditions, the pressure and temperature inside the nitrogen bubble were not investigated in the experiments.

In addition to experimental studies, there have also been theoretical and numerical results focused on the heat and mass transfer process at the bubble interface. The isothermal hypothesis is often adopted in cavitation models of ordinary fluids. Transport cavitation models based on the Rayleigh–Plesset equation are used widely. Among these models, the Zwart–Gerber–Belamri model [22], Schnerr–Sauer model [23] and Singhal model [24] are applied in commercial CFD software. The growth and heat and mass transfer between the bubble and liquid are predicted using the single-bubble model [25,26]. The bubble radius growth results calculated with the model are verified with data from experiments in water and freon. For ordinary fluids, the thermodynamic temperature drop is small and the liquid vapor density ratio is large. Additionally, the transition time of bubble growth from the dynamic growth stage to the thermal growth stage is equivalent to the bubble existence time. Therefore, the isothermal assumption is prone to be suitable for the cavitation in ordinary fluids.

Different from ordinary fluids, the temperature drop of the bubble growth process cannot be ignored in cryogenic liquids. The existing isothermal hypothesis is not viable. Therefore, for cryogenic fluids, thermodynamic effects must be considered on cavitation development and bubble growth. Zhang et al. [27] modified the cavitation model for ordinary fluids considering thermal effects and used it to study cavitation in liquid nitrogen. The application of the cavitation model for cryogenic fluids was improved in prediction. Comparing to the experimental results, the numerical results were within the error limits. Recently, a large number of numerical studies popped up focusing on cryogenic fluid cavitation. Xu et al. [28] used the modified cavitation model with thermodynamic effects to compare the cavitation process of water and liquid nitrogen. The numerical results indicated that thermal effects could inhibit the cavitation in liquid nitrogen but had almost no suppression effect for water. Moreover, the particularity of liquid hydrogen cavitation, which is different from other cryogenic fluids, was further investigated by Le et al. [29]. The empirical evaporation and condensation constant in liquid hydrogen was obviously higher than that in other fluids. In addition to the modification of the cavitation model, Li et al. [30] also used the modified model to study the cavitation model of liquid oxygen in turbopumps. Cavitation was mainly distributed in the leading edge of the inlet of the inducer and the central blade head of the impeller.

In most of the existing cavitation models, modifications have focused on the influence of the thermal effect. There are relatively few models which have been proposed for single bubble growth and collapse. Ito [31] considered the equation of heat conduction outside the bubble and used it to calculate the phase transition in the thermal boundary layer. In addition, this model was also used to study the bubble development process in liquid hydrogen, liquid nitrogen and liquid oxygen. The growth process of bubbles in the inducer was studied numerically based on this model [32]. However, the temperature difference between the center and boundary of the bubble was not considered.

During the growth and collapse of a single bubble, the temperature and pressure of the vapor inside the bubble vary greatly [33]. Nevertheless, most of the existing bubble growth and collapse models assume that the temperature inside the bubble is the same as that of the fluid at the bubble boundary. For the more accurate prediction of the growth–collapse process of single bubbles in cryogenic fluids, existing bubble models should be modified. In this paper, based on the existing bubble model, a single-bubble collapse model was proposed for cryogenic fluids. The modified Rayleigh–Plesset equation, heat balance equation, heat diffusion equation and state equation were solved simultaneously in various cryogenic fluids. By comparing the tendency of bubble bursting and rebound in different fluids, the influence of the physical properties of cryogenic fluids on bubbles was analyzed. In addition, the influence of the bubble internal temperature on the bubble collapse and rebound process was considered and calculated. The bubble contraction–rebound process in liquid hydrogen was selected as the main research object. The oscillations of the bubble radius, radius growth rate, temperature at the bubble boundary and temperature and pressure inside the bubble were studied using the model. The amplitude, attenuation rate and period of bubble collapse were analyzed under different supercooling degrees and operating pressures.

2. Derivation and Verification of the Single-Bubble Collapse Model

During bubble collapse, the temperature and pressure inside the bubble are affected by the operating pressure, supercooling degree and thermophysical property of different fluids. In order to accurately predict and compare the bubble collapse process in various fluids, the temperature and pressure inside the bubble were considered in the model. The bubble collapse model is derived and the results of this model are verified in this section.

2.1. Derivation of the Single-Bubble Collapse Model

The equation for bubble growth was first derived and used by Rayleigh [34] in the absence of surface tension and viscosity. Subsequently, this equation was improved by Plesset [35] and was applied to the problem of traveling cavitation bubbles.

$$R\frac{d^2R}{dt^2}+\frac{3}{2}\left(\frac{dR}{dt}\right)=\frac{P_v(T_v)-P_{\text{inf}}}{\rho_l}-\frac{4\nu_l}{R}\frac{dR}{dt}-\frac{2S}{\rho_l R} \tag{1}$$

where R is the bubble radius, T_v is the temperature inside the bubble, P_v is the pressure inside the bubble, ρ_l is the liquid density, P_{inf} is the operating pressure, ν_l is the kinematic viscosity of the liquid and S is the surface tension of the fluid.

In addition to the Rayleigh–Plesset equation, the noncondensable gas in the bubble was considered. In general, it was assumed that there is no mass transfer between the noncondensable gas and the liquid. When the initial partial pressure of the noncondensable gas is P_{G0}, the partial pressure, P_G, varies with the bubble radius, as shown in Equation (2).

$$P_G=P_{G0}\left(\frac{T_v}{T_{\text{inf}}}\right)\left(\frac{R_0}{R}\right)^{3k} \tag{2}$$

where R_0 is the initial bubble radius, T_{inf} is the operating temperature and k is the polytropic exponent; k = 1 for bubble growth and k = 1.4 for bubble collapse.

In the process of bubble collapse, the contraction speed of the bubble interface is greater than the bubble growth rate. When the Mach number ($|dR/dt|/c$) is higher than 0.3, the influence of liquid compressibility should be considered [10]. For this point, the near-acoustic solutions of Herring [36] and Trilling [37] modified the Rayleigh–Plesset equation, and the results are shown in Equation (3):

$$R\frac{d^2R}{dt^2}\left(1-\frac{2}{c_{l0}}\frac{dR}{dt}\right)+\frac{3}{2}\left(\frac{dR}{dt}\right)^2\left(1-\frac{4}{3c_{l0}}\frac{dR}{dt}\right)=\frac{R}{\rho_{l0}c_{l0}}\frac{dP}{dt}+\frac{P_v-P_{\text{inf}}}{\rho_{l0}} \tag{3}$$

where c_{l0} and ρ_{l0} are the constant sound speed and the density of the liquid, respectively.

The Rayleigh–Plesset equation adopted by the model could be modified by substituting Equations (1) and (2) into Equation (3):

$$R\frac{d^2R}{dt^2}\left(1-\frac{2}{c_{l0}}\frac{dR}{dt}\right)+\frac{3}{2}\left(\frac{dR}{dt}\right)^2\left(1-\frac{4}{3c_{l0}}\frac{dR}{dt}\right)=\frac{R}{\rho_{l0}c_{l0}}\frac{d}{dt}\left(P_v(T_v)+P_{G_0}\left(\frac{T_v}{T_{\text{inf}}}\right)\left(\frac{R_0}{R}\right)^{3k}-\frac{2S}{R}-\frac{4\mu_l}{R}\frac{dR}{dt}\right)+\frac{P_v(T_v)+P_{G_0}\left(\frac{T_v}{T_{\text{inf}}}\right)\left(\frac{R_0}{R}\right)^{3k}-\frac{2S}{R}-\frac{4\mu_l}{R}\frac{dR}{dt}-P_{\text{inf}}}{\rho_{l0}} \tag{4}$$

where μ_l is the dynamic viscosity of the liquid.

In order to further consider the thermal effect in the bubble model, the thermal equilibrium equation was introduced. The vapor density ρ_v in the bubble varies greatly with the temperature. Additionally, the variation in temperature inside the bubble could not be ignored during the bubble growth–collapse process. The transient variation in vapor density in the bubble was also considered. In order to obtain the temperature inside the bubble, T_B, and the temperature at the bubble boundary, T_l, it was assumed that only heat conduction exists in the thermal boundary layer around the bubble. According to the thermal balance equation, Equation (5) was introduced:

$$4\pi R^2 k_l\left(\frac{dT_l}{dr}\right)_{r=R}=L\frac{d}{dt}\left(\frac{4}{3}\pi R^3\rho_v\right)=4\pi R^2 L\rho_v\frac{dR}{dt}+\frac{4}{3}\pi R^3 L\frac{d\rho_v}{dT_l}\left(\frac{dT_l}{dt}\right)_{r=R} \tag{5}$$

where L is the latent heat of the fluid and k_l is the thermal conductivity of the liquid. The derivative of the temperature at the bubble boundary, T_l, with respect to time could be derived from Equation (5):

$$\frac{dT_l}{dt}=\left(k_l\frac{dT_l}{dr}-L\rho_v\frac{dR}{dt}\right)/\left(\frac{1}{3}RL\frac{d\rho_v}{dT}\right) \tag{6}$$

In Equation (6), the temperature gradient at the bubble boundary, *dTl/dr*, was unknown. The temperature gradient at the bubble boundary could be obtained by solving the energy equation in spherical coordinates for the moving boundary. The one-dimensional energy equation is as follows.

$$\frac{\partial T}{\partial t}+u\frac{\partial T}{\partial r}=\alpha_l\left(\frac{\partial^2 T}{\partial r^2}+\frac{2}{r}\frac{\partial T}{\partial r}\right) \tag{7}$$

where α_l is the thermal diffusivity of the liquid. In Equation (7), u is the radial velocity which is a function of the bubble growth rate and bubble radius. The radial velocity could be solved:

$$u(R,t)=\frac{dR}{dt}\left(\frac{R}{r}\right)^2 \tag{8}$$

By substituting Equation (8) into Equation (7), the thermal diffusion equation of the bubble boundary could be obtained:

$$\frac{\partial T}{\partial t}+\frac{dR}{dt}\left(\frac{R}{r}\right)^2\frac{\partial T_l}{\partial r}=\frac{\alpha_l}{r^2}\frac{\partial}{\partial r}\left(r^2\frac{\partial T_l}{\partial r}\right) \tag{9}$$

The boundary conditions and initial condition are shown in Equation (10). The temperature of the entire flow field is constant and uniform. Moreover, the temperature at the bubble boundary could be derived from the assumption of discontinuity at the bubble interface. The liquid temperature is constant at an infinite distance.

$$\begin{cases} T(r,0) = T_{\text{inf}} \\ T(R,t) = T_l \\ T(R_{\text{inf}},t) = T_{\text{inf}} \end{cases} \tag{10}$$

By the external discretization of bubbles, Equations (4) and (9) could be solved using the fourth-order Runge–Kutta method. Photos and a schematic diagram of the bubble collapse–rebound process are shown in Figure 1. It was assumed that the liquid beyond the 5 mm bubble boundary is not affected by the temperature and pressure change inside the bubble [1], where the temperature and pressure are equal to T_{inf} and P_{inf}, respectively. Based on the bubble radius and the temperature at the bubble boundary, the temperature and pressure inside the bubble could be obtained.

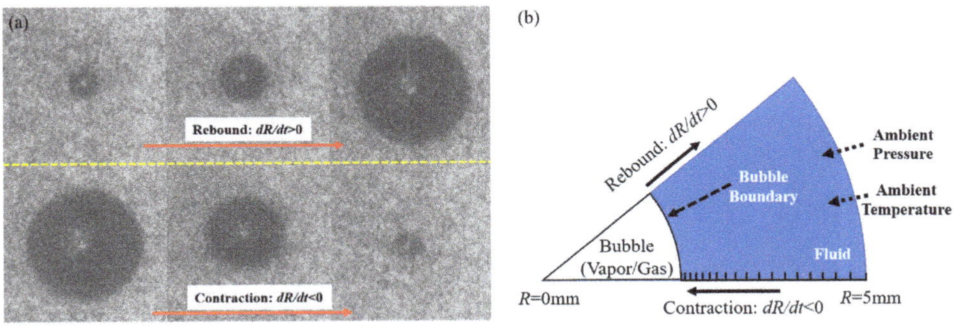

Figure 1. Single-bubble experimental photos [38] (**a**) and collapse model schematic diagram (**b**).

At first, it was assumed that there is no noncondensable gas inside the bubble. During bubble collapse, the bubble boundary contracts inward rapidly. Inside the bubble, there exists vapor compression and phase transformation simultaneously. Taking the boundary layer at the bubble interface as the control volume, the heat flowing into the boundary layer is positive and the heat flowing out is negative. The latent heat from phase transformation flows into the boundary layer, and the conducted heat flows out of the boundary layer. Assuming that the mass in the boundary layer is constant and equals m, the energy balance equation of the thermal boundary layer is shown in Equation (11).

$$T_l^{(i+1)} - T_l^{(i)} = \frac{4\pi\left(R^{(i)}\right)^2\left[\Delta n M\left(R^{(i)} - R^{(i+1)}\right)L - k_l\frac{dT_l}{dr}\right]}{m} \tag{11}$$

where Δn is the variation in the amount of vapor substances and the superscript (i) indicates the i-th time step. The total amount of vapor in the bubble could be obtained from Equation (12)

$$n^{(i+1)} = n^{(i)} + \Delta n \tag{12}$$

If only the phase transition process was being considered, the bubble radius R' could be obtained using the Clapeyron Equation (13).

$$\frac{4}{3}\pi(R\prime)^3 P_v^{(i)} = n^{(i+1)}\mathbb{R}T_v^{(i)} \tag{13}$$

where $\mathbb{R}$ is the molar gas constant which has a value of 8.3145 J/(mol·K).

At last, the pressure and temperature inside the bubble could be solved with the polytropic compression/expansion Equation (14). In the equation, k is the polytropic exponent. Generally, the growth and rebound of bubbles are deemed to be isothermal expansion processes with $k = 1$. The collapse process is an adiabatic compression wit[h] $k = 1.4$.

$$\begin{cases} P_v^{(i+1)} = P_v^{(i)}\left(\frac{R\prime}{R^{(i+1)}}\right)^{3k} \\ T_v^{(i+1)} = T_v^{(i)}\left(\frac{R\prime}{R^{(i+1)}}\right)^{3(k-1)} \end{cases} \quad (14)$$

For bubbles containing noncondensable gas, the solution process could be simplifie[d]. The variation in the partial pressure of the noncondensable gas could be solved directl[y] according to the polytropic process Equation (15). Then, the temperature of the noncon-densable gas was obtained. According to the assumption of uniform temperature insid[e] the bubble, the temperature inside the bubble was represented by the temperature of th[e] noncondensable gas, $T_v^{(i+1)}$.

$$\begin{cases} P_G^{(i+1)} = P_G^{(i)}\left(\frac{R^{(i+1)}}{R^{(i)}}\right)^{3k} \\ T_v^{(i+1)} = T_v^{(i)}\left(\frac{P_G^{(i+1)}}{P_G^{(i)}}\right)^{(k-1)/k} \end{cases} \quad (15)$$

The model solving process is summarized as a flow chart shown in Figure 2. Th[e] operating of the flow field, bubble radius and bubble growth time "t_{end}" were initialize[d] at first. The temperature inside the bubble (T_v), temperature at the bubble boundary (T_l), pressure inside the bubble (P_v), bubble radius (R) and bubble growth rate (dR/dt) wer[e] investigated.

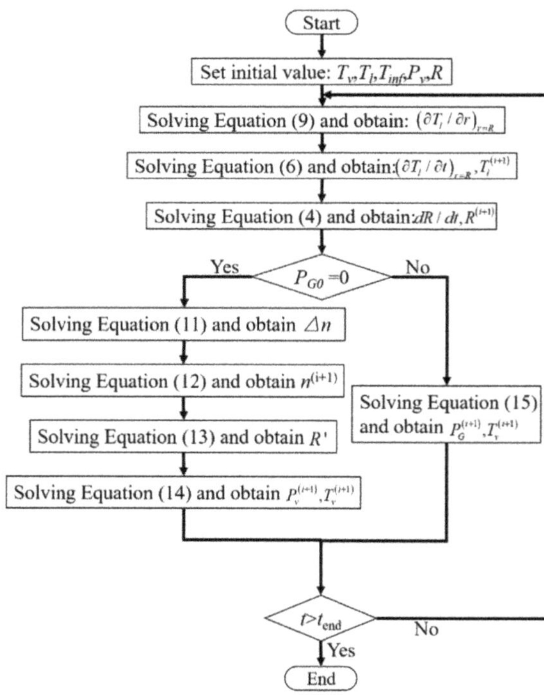

Figure 2. The flow chart of solving the bubble collapse model.

2.2. Verification of the Single-Bubble Collapse Model

Due to the rapid contraction and rebound of the bubble, there are relatively few experiments on the collapse of a single bubble, especially in cryogenic fluids. In order to verify the model, the experimental results [38] of water bubbles under the condition of an alternating pressure field and the numerical results of Yukio et al. [39] were selected in this paper. The verification of the bubble collapse and rebound process is shown in Figure 3. Under the operating pressure of 1 atm, the initial radius of the bubbles in the saturated water was 6.18 μm. With the condition of a sinusoidal pressure field with an applied amplitude of 1.29×10^5 Pa and frequency of 25 kHz, the experimental and numerical results of the bubble radius are shown in Figure 3a. The numerical results using the model presented in this paper were within 10% of the experimental measurements. The numerical results of this model were compared with the calculation results of the Yukio model, shown in Figure 3b. The difference between the numerical results of the two models were within 10%.

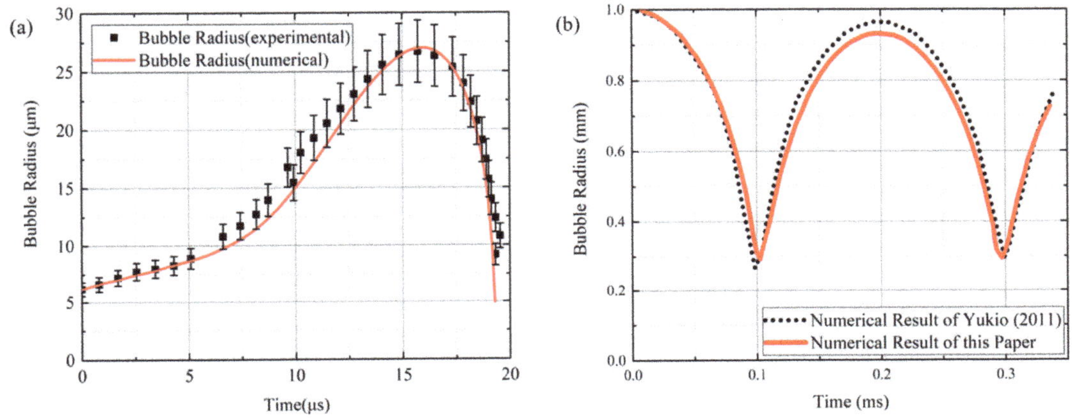

Figure 3. Comparison of the calculation results of the growth–collapse model with the experimental results (**a**) and the calculation results of the Yukio [39] model (**b**).

3. Comparison of Bubble Collapse in Different Fluids

According to the open literature [40], the pressure of 0.1 MPa~0.51 MPa and the temperature of 20~27.4 K are often used in liquified hydrogen storage and delivery. Therefore, in this section, the operating pressure of 1~3 atm and the supercooling degree of 1~5 K were selected for calculation. The effects of operating pressure and supercooling degree on the radius of a single bubble, bubble growth rate, temperature and pressure inside a bubble were investigated. In order to study the particularity of bubble behavior in liquid hydrogen, the bubble collapse process of liquid nitrogen, liquified methane and water were introduced and compared.

In the process of bubble collapse, there is a bubble period with rapid contraction and rebound. In the collapse process, the variation in the bubble radius with time under different operating pressures and supercooling degrees is shown in Figure 4a,b, respectively. Under the condition of low operating pressure, the liquid hydrogen force acting on the bubble was small, so the bubble growth rate was slow. Therefore, the bubble contraction–rebound period increased with higher operating pressure. Moreover, the amplitude of the bubble radius increased with lower operating pressure because of the small kinetic energy attenuation. When the operating pressure of the bubble remained the same, the velocity of bubble contraction increased greatly with the supercooling degree. Therefore, the period of the bubble radius decreased with the supercooling degree. Meanwhile, due to the fast contraction of the bubble, the bubble radius became larger with the same kinetic energy attenuation.

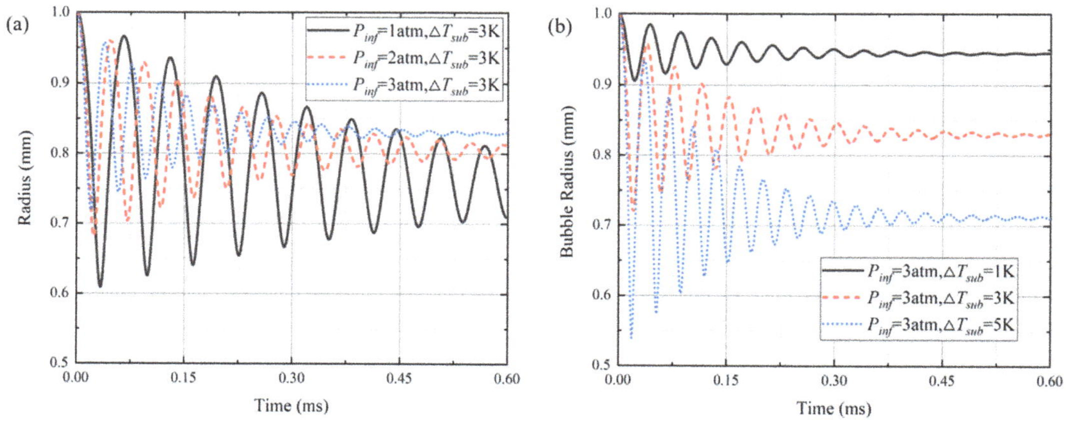

Figure 4. Radius of liquid hydrogen bubbles under different operating pressures (**a**) and different supercooling degrees (**b**) during collapse.

In addition to the bubble radius, the bubble growth rate over time was also investigated. The bubble growth rate under different operating pressures and supercooling degrees is shown in Figure 5. As analyzed above, the maximum bubble growth rate was not significantly correlated with operating pressure. Initially, the peak and valley values of the bubble growth rate were almost the same under the same supercooling degree. However, the bubble growth rate was affected by the high operating pressure and the attenuation rate was rapidly accelerated. In contrast, the amplitude of the bubble growth rate increased greatly with the supercooling degree of liquid hydrogen. The peak value and attenuation rate of the bubble growth rate were affected by the operating pressure and the supercooling degree. The period of the bubble growth rate decreased with the operating pressure and the supercooling degree.

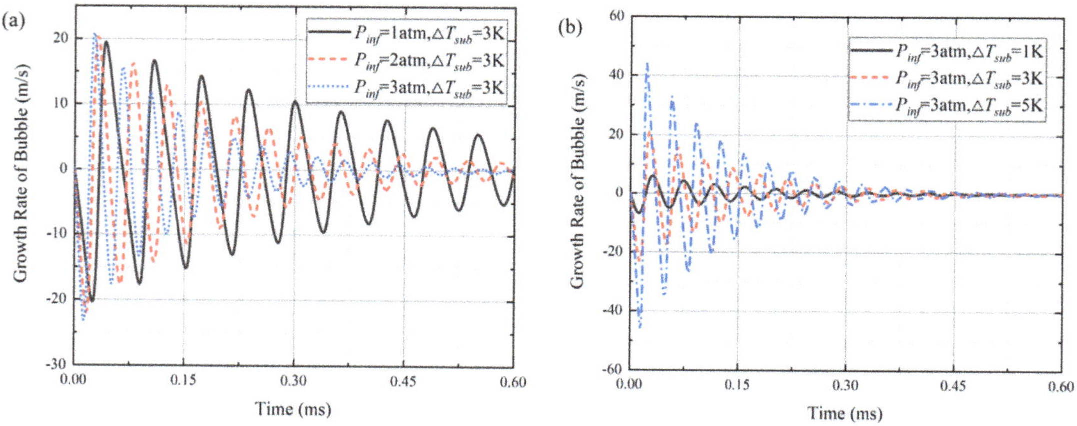

Figure 5. Radius growth rate of liquid hydrogen bubbles under different operating pressures (**a**) and different supercooling degrees (**b**) during collapse.

During contraction–rebound, the pressure and temperature of the bubble also varied greatly. The pressure inside the bubble during oscillation is shown in Figure 6. The peak and equilibrium of the pressure inside the bubble was promoted under larger operating pressure. Similarly, the peak in pressure inside the bubble was also enlarged sharply with supercooling degree. Accompanied with the variation in bubble radius, rapid contraction led to the fast compression of hydrogen inside the bubble. The period of the bubble pressure was almost the same as the bubble radius period. The fluctuation of pressure inside the bubble was possibly caused by the variation in the bubble radius.

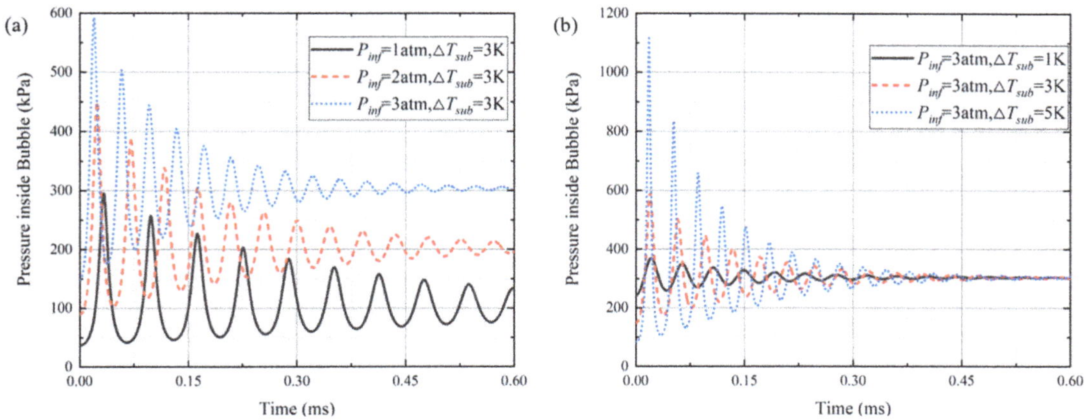

Figure 6. Pressure inside liquid hydrogen bubbles under different operating pressures (**a**) and different supercooling degrees (**b**) during collapse.

Due to the different operating pressures, the absolute value of the peak pressure inside the bubble did not reflect the impact intensity during bubble collapse clearly. Therefore, the pressure inside the bubble was nondimensionalized using the operating pressure of infinity. The comparison results are shown in Figure 7. The dimensionless pressure inside the bubble was reduced. This result indicated that the impact caused by bubble collapse was strong under low operating pressure.

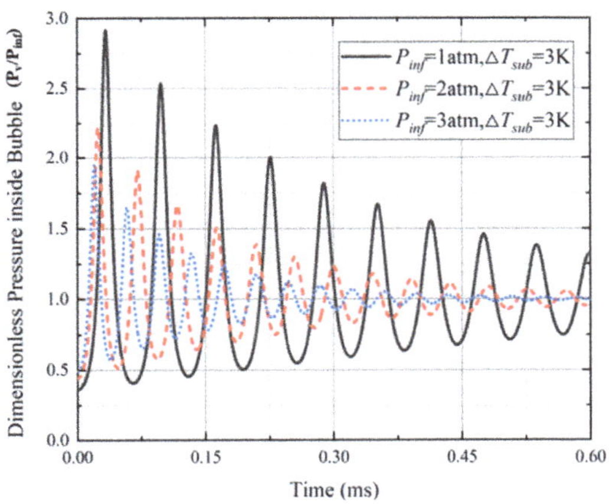

Figure 7. Dimensionless pressure inside liquid hydrogen bubbles under different operating pressures.

The temperature variation at the bubble boundary with time is shown in Figure 8. The initial and final temperature at the bubble boundary were upraised at higher operating pressure, but the amplitude decayed. With increment of supercooling degree, the lower initial temperature at the bubble boundary led to larger temperature undulation. However, the influence of supercooling degree on the final bubble boundary temperature was not significant.

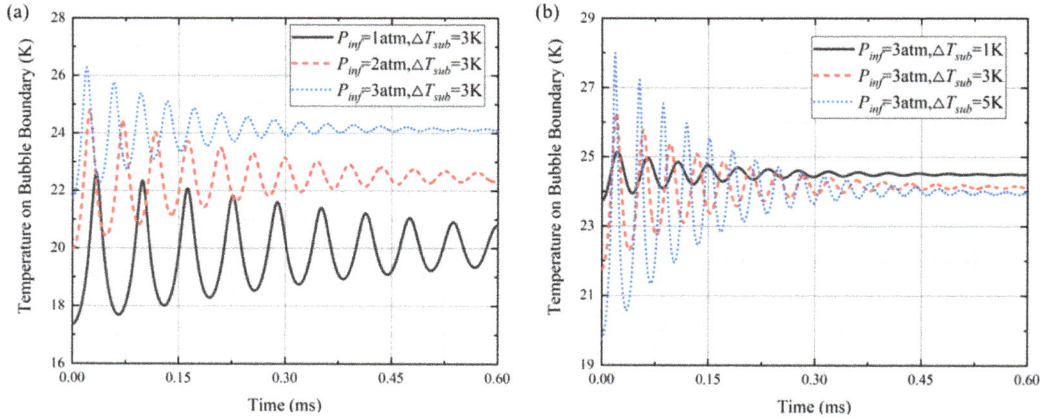

Figure 8. Temperature at liquid hydrogen bubble boundaries under different operating pressures (**a**) and different supercooling degrees (**b**) during collapse.

Different from the liquid temperature at the bubble boundary, there was a larger undulation in vapor temperature due to its small heat capacity. Moreover, the oscillation of temperature inside the bubble was enhanced due to the vapor compression process. As shown in Figure 9a, the initial temperature inside the bubble increased under higher operating pressure. Under different operating pressures, the peak value of the temperature remained almost the same during the first few periods. The influence of the supercooling degree on the bubble temperature is shown in Figure 9b. The equilibrium temperature inside the bubble was almost unaffected by the supercooling degree. The temperature peak increased with the supercooling degree.

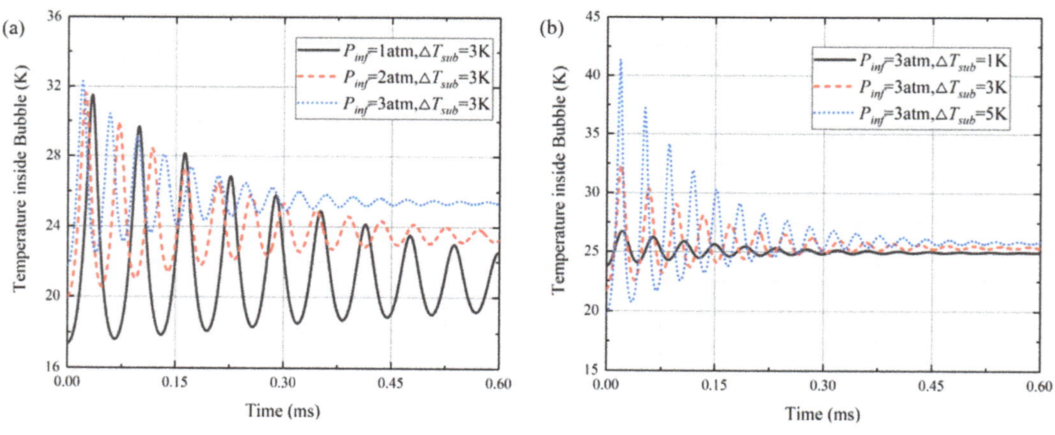

Figure 9. Temperature inside liquid hydrogen bubbles under different operating pressures (**a**) and different supercooling degrees (**b**) during collapse.

In order to compare the temperature inside the bubble and that at the bubble boundary, the condition was selected with the operating pressure equal to 3 atm and supercooling degree equal to 1 K. The two temperatures are shown in Figure 10. The temperature peak inside the bubble was much larger than that at the bubble boundary. As the thermal conductivity of the vapor phase was much lower than that of the liquid phase, the temperature inside the bubble accumulated gradually, and its valley value gradually separated from the temperature at the bubble boundary.

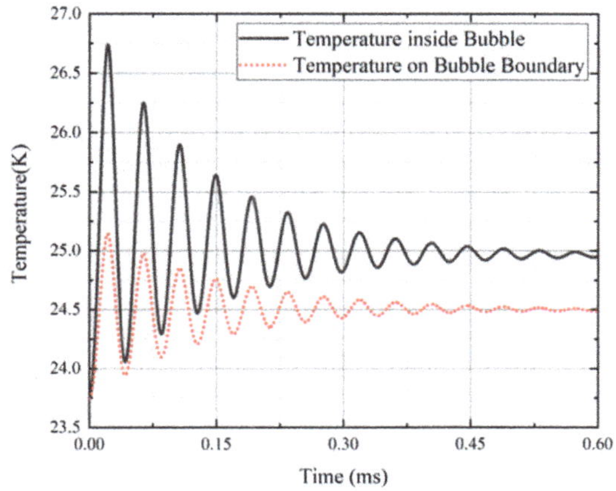

Figure 10. Comparison of boundary temperature and internal temperature of liquid hydrogen bubble during collapse.

Based on the investigation of bubble collapse in liquid hydrogen, the effect of fluids' properties on bubble collapse was studied. The comparison of the bubble radius and bubble growth rate in different fluids is shown in Figure 11a,b, respectively. The oscillation frequency and amplitude of the liquid hydrogen bubble were much larger than those in the other three fluids. The radius and growth rate of the bubble in liquid nitrogen and liquid methane had similar peak values. The bubble radius in liquid nitrogen had a relatively lower valley value.

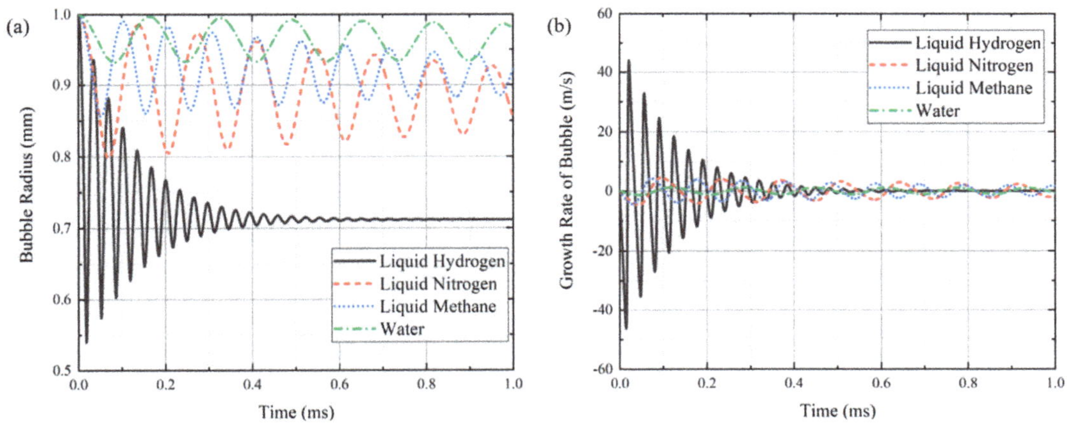

Figure 11. Bubble radius (**a**) and radius growth rate (**b**) in different fluids under 3 atm operating pressure and 5 K supercooling degree during collapse.

Due to the great differences in standard boiling points among different fluids, it was meaningless to compare temperature inside the bubble and that at the bubble boundary directly. Therefore, the bubble temperature was nondimensionalized with the fluid critical temperature. As shown in Figure 12, the amplitude and frequency of the dimensionless temperature and the equilibrium temperature of the bubble in liquid hydrogen were larger than those in the other fluids. Moreover, it is worth noting that only the peak temperature inside the liquid hydrogen bubble was higher than the critical temperature during temperature oscillation. Additionally, the bubble oscillation time of liquid hydrogen was much smaller than that in the other fluids.

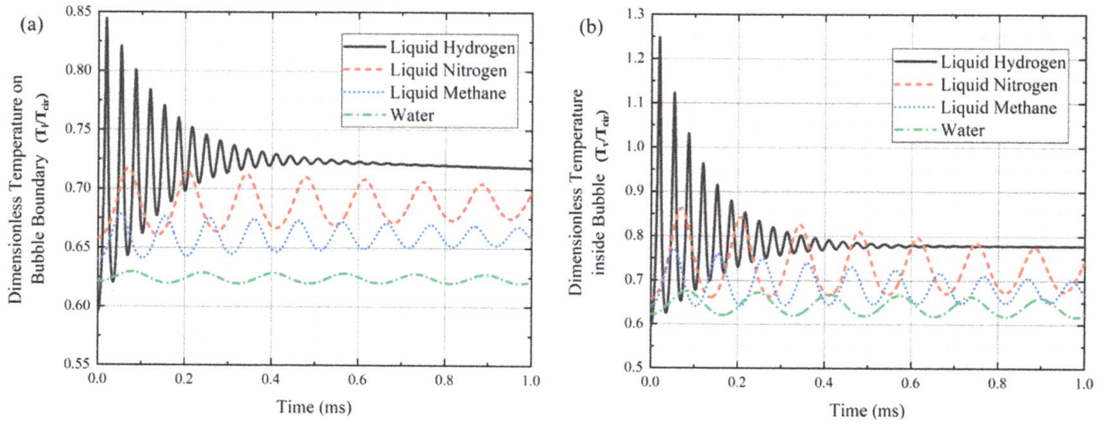

Figure 12. Dimensionless temperature at bubble boundaries (**a**) and inside bubbles (**b**) in different fluids under 3 atm operating pressure and 5 K supercooling degree during collapse.

The variation in bubble pressure in the different fluids is shown in Figure 13. The pressure amplitude and frequency in the different fluids were in the same order as the other parameters. The peak pressure during bubble collapse in liquid hydrogen was approximately four times the operating pressure, which proved that the impact intensity on the nearby surface was the largest. In other words, the damage of liquid hydrogen cavitation to hydraulic machinery could be much greater than that of the other fluids.

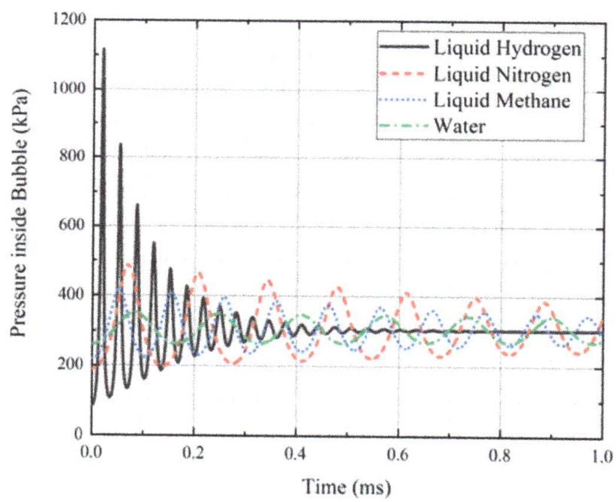

Figure 13. Pressure inside bubbles in different fluids under 3 atm operating pressure and 5 K supercooling degree during collapse.

It is generally believed that the high pressure during bubble collapse is the main cause of cavitation damage [33]. In order to investigate the potential damage, it was necessary to compare the maximum pressure peaks in the different fluids, as well as different conditions. The maximum peak pressure and dimensionless number (Weber number divided by the Reynolds number: Wb/Re) during bubble collapse are shown in Figure 14. The maximum peak pressure monotonously increased with the Wb/Re roughly. Therefore, the intensity of cavitation damage could be roughly predicted using the magnitude of Wb/Re.

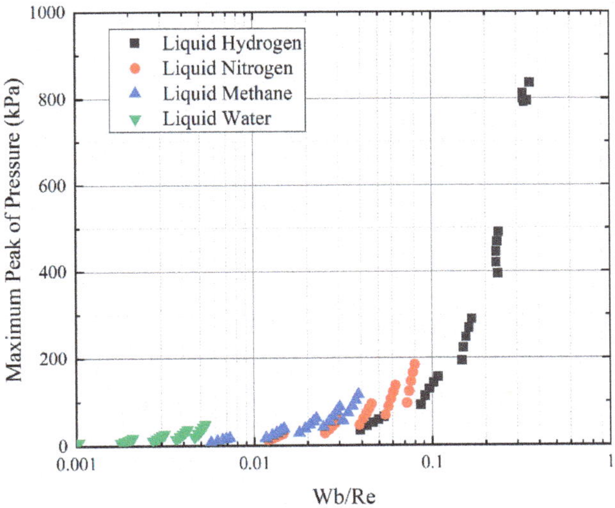

Figure 14. The relationship between the maximum peak pressure and Wb/Re.

4. Analysis of Periodicity and Attenuation Rate of Bubble Collapse

From Section 3, all the physical parameters of the bubbles featured oscillation characteristics. The oscillation period and attenuation rate were influenced by the operating pressure and supercooling degree. In this section, the attenuation rate and periodicity of the liquid hydrogen bubble collapse process were studied and compared with the other fluids. Moreover, the relationship between the temperature attenuation rate inside the bubble and dimensionless number was drawn.

The period length of the bubble oscillation was defined as the length of time between peak values in terms of radius, radius growth rate, temperature and pressure, respectively. The period of the different parameters was almost same. Therefore, the first period of the bubble radius oscillation was studied and is shown in Figure 15. The statistical results indicated that the bubble oscillation period was the shortest in liquid hydrogen and the longest in water. The period in liquid nitrogen was slightly longer than that in liquid methane. The period length of the bubble oscillation decreased with the operating pressure. Moreover, the increase in the supercooling degree also reduced the bubble oscillation period. In addition, the influence of supercooling degree on period was weaker than that of the operating pressure in the different fluids.

Besides the period, the attenuation rate was also an important parameter. The attenuation rate was the dimensionless result of dividing the difference between the amplitude of the current period and that of the next period with the amplitude of the current period. The attenuation rate in the first thirty oscillation periods is shown in Figure 16. The attenuation rate of the bubble radius was suppressed with operating pressure and lessened with supercooling degree. The attenuation rate was almost not affected by operating pressure. The influence of the supercooling degree on the radius attenuation rate was not as great as that of the operating pressure. However, under the condition of smaller supercooling degree, the attenuation rate of the bubble radius almost did not vary with period. With larger supercooling degree, the slope of the radius attenuation rate increased obviously in both the ascending and descending parts.

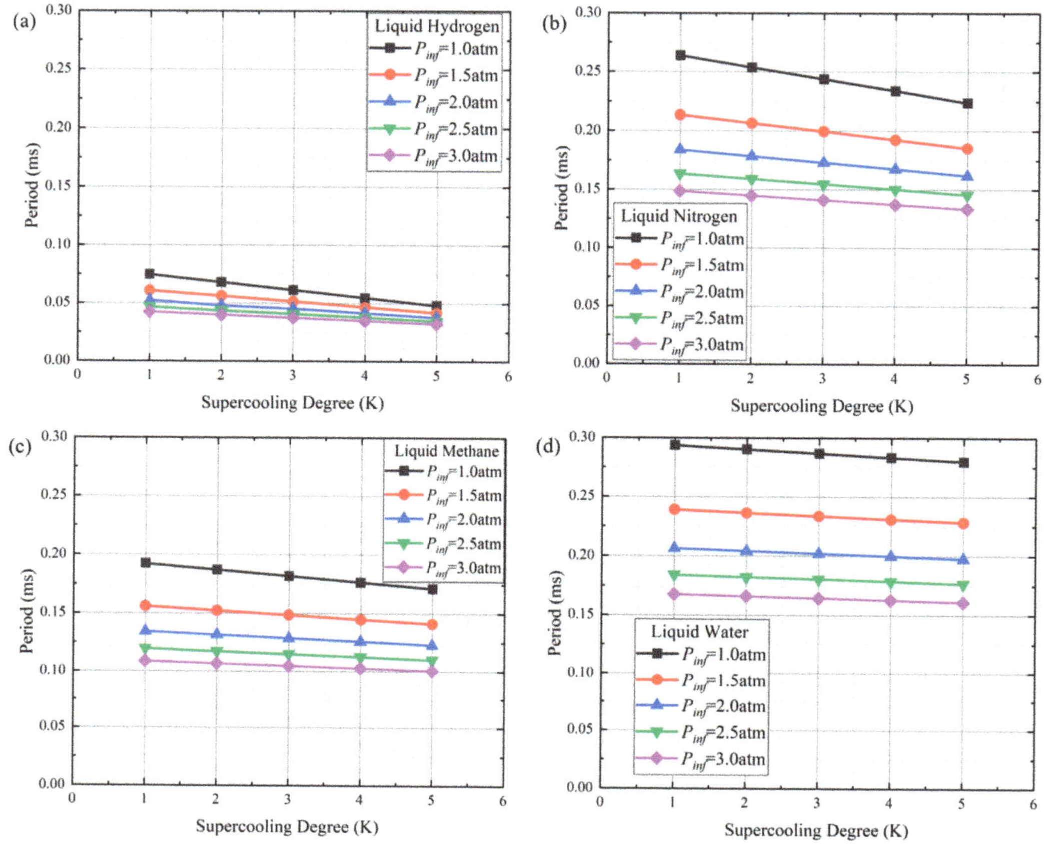

Figure 15. Bubble period of liquid hydrogen (**a**), liquid nitrogen (**b**), liquid methane (**c**) and liquid water (**d**) during collapse.

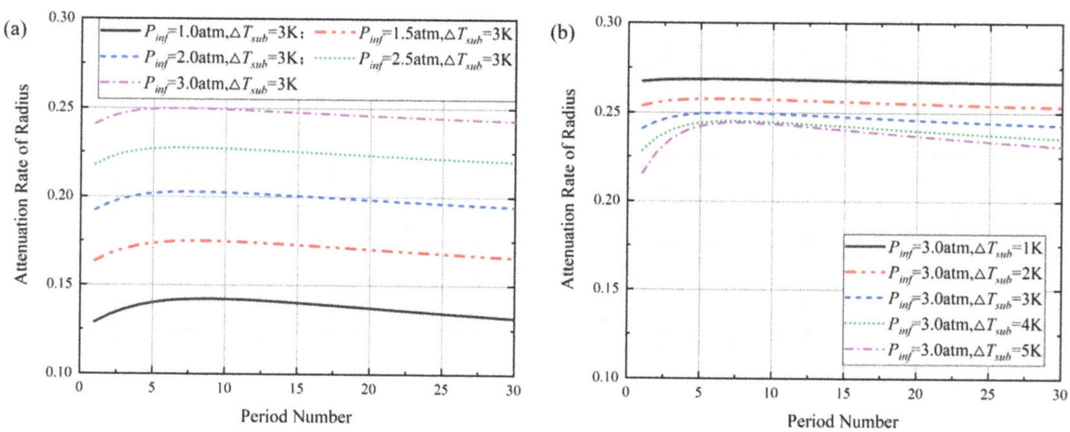

Figure 16. Attenuation rate of liquid hydrogen bubble radii under different operating pressures (**a**) and different supercooling degrees (**b**) during collapse.

Furthermore, the attenuation rate of the bubble growth rate (dR/dt) was investigated in the first thirty periods. The attenuation rate with different operating pressures and supercooling degrees is shown in Figure 17. Similar to the influence of operating pressure and supercooling degree on the attenuation rate of bubble radius, operating pressure greatly enlarged the attenuation rate *dR/dt*, while it was reduced slightly with higher supercooling degree. The difference was that the higher supercooling degree only raised the slope of the attenuation rate *dR/dt* but did not change its monotonicity.

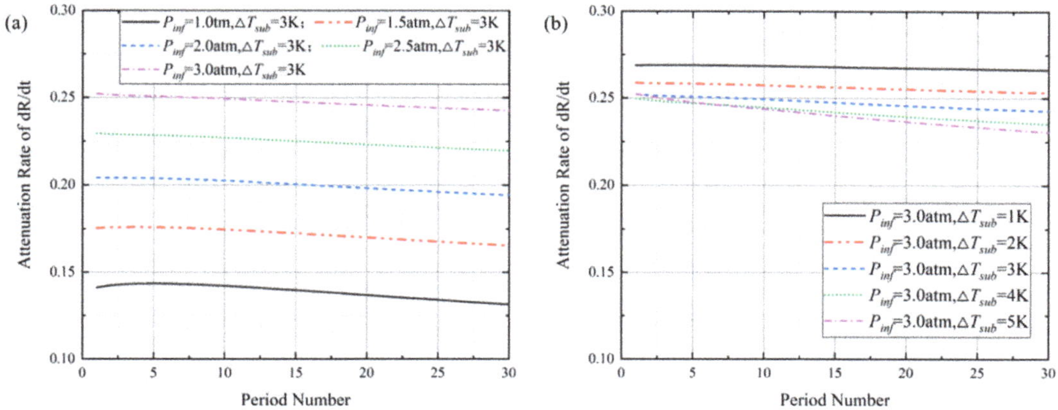

Figure 17. Attenuation rate of liquid hydrogen bubble growth rate under different operating pressures (**a**) and different supercooling degrees (**b**) during collapse.

Different from the attenuation rate mentioned above, the attenuation rate of the temperature inside the bubble (T_v) dropped significantly in the first thirty periods, as shown in Figure 18. Therefore, the period investigated was extended. The attenuation rate of T_v was like an "S" curve. Its slope reached a maximum in the middle 20 periods, while its decline was very slow at the beginning and ending. In addition to the T_v attenuation rate, its maximum slope was enlarged with the operating pressure. The influence of the supercooling degree on the maximum value, minimum value and slope of the T_v attenuation rate was slight.

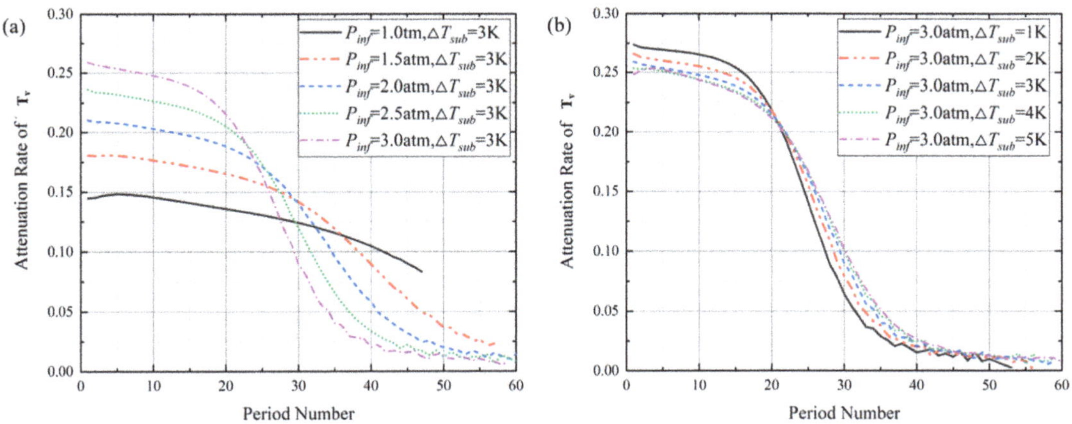

Figure 18. Attenuation rate of temperature inside liquid hydrogen bubbles under different operating pressures (**a**) and different supercooling degree (**b**) during collapse.

The slope of the attenuation rate of the bubble pressure (P_v) was large in the first few periods and gradually decreased in the subsequent periods, as shown in Figure 19. The influence of the operating pressure on the attenuation rate of P_v was similar to the attenuation rate of the other parameters. It increased with higher operating pressure, but its maximum slope varied slightly.

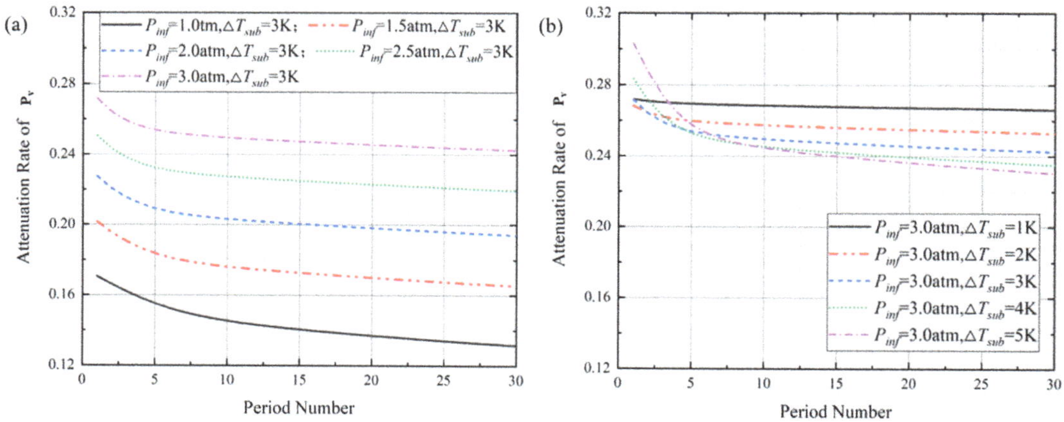

Figure 19. Attenuation rate of pressure inside liquid hydrogen bubbles under different operating pressures (**a**) and different supercooling degrees (**b**) during collapse.

In order to study the difference in the bubble collapse process among the different fluids, the comparison results of bubble attenuation rates are shown in Figure 20. The attenuation rates of all physical parameters in liquid hydrogen were the highest, and they were the lowest in water. The attenuation rates of bubbles in liquid nitrogen and liquid methane were very similar. However, the attenuation rate of T_v in the liquid hydrogen bubbles suddenly declined in the 20th period and fell below that of water bubbles.

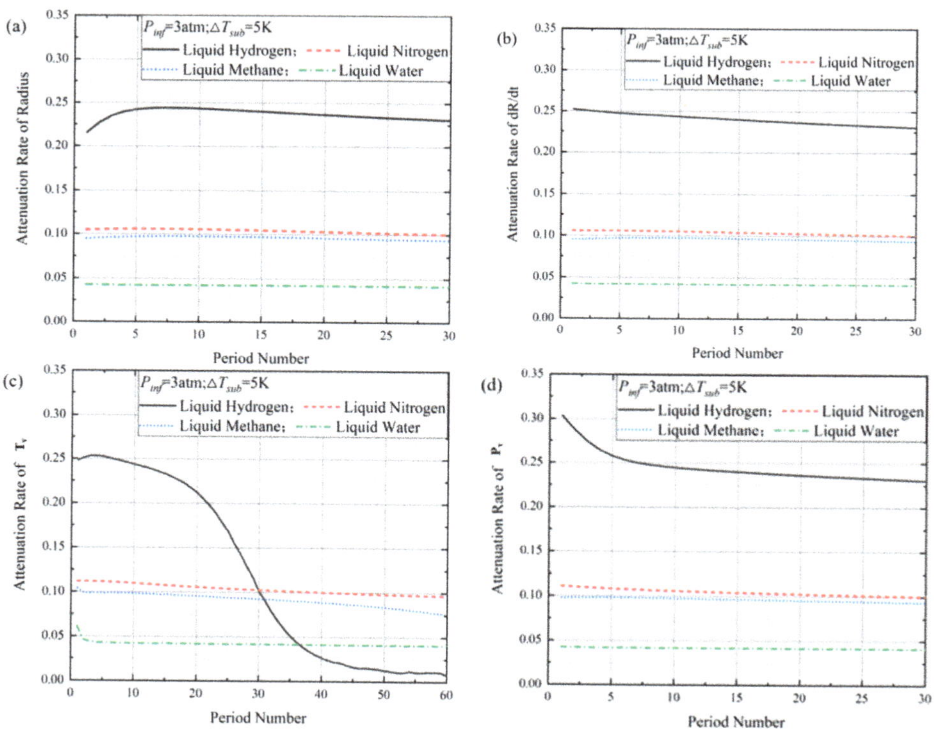

Figure 20. Attenuation rates of bubble radius (**a**), radius growth rate (**b**), temperature inside bubble (**c**) and pressure inside bubble (**d**) in different fluids under 3 atm operating pressure and 5 K supercooling degrees during collapse.

In order to understand the difference in bubble attenuation rates in the different fluids, the standard temperature drop and thermodynamic effect parameter of the different fluids were investigated. The standard temperature drop (ΔT) and thermodynamic effect parameter ($\sum(T_{inf})$) are defined as follows:

$$\Delta T = \frac{\rho_v L}{\rho_l C_{Pl}} \tag{16}$$

$$\sum(T_{\text{inf}}) = \frac{L^2 \rho_v^2}{\rho_l^2 C_{pl} T_{\text{inf}} \alpha_l^{0.5}} \tag{17}$$

where ρ_v and ρ_l are the density of vapor and liquid, L is the latent heat, C_{pl} is the specific heat capacity of the liquid at constant pressure and αl is the thermal diffusivity of the liquid.

Under the operating pressure of 3 atm, the trends in $\sum(T_{inf})$ and ΔT with the supercooling degree are shown in Figure 21. $\sum(T_{inf})$ and ΔT in liquid hydrogen were much larger than in the other fluids. The two parameters in liquid nitrogen and methane were very close. Moreover, the magnitude of $\sum(T_{inf})$ and ΔT in the four fluids were almost the same as the magnitude of the attenuation rate. Therefore, the magnitude of the bubble attenuation rate could be qualitatively determined according to $\sum(T_{inf})$ and ΔT in the different fluids.

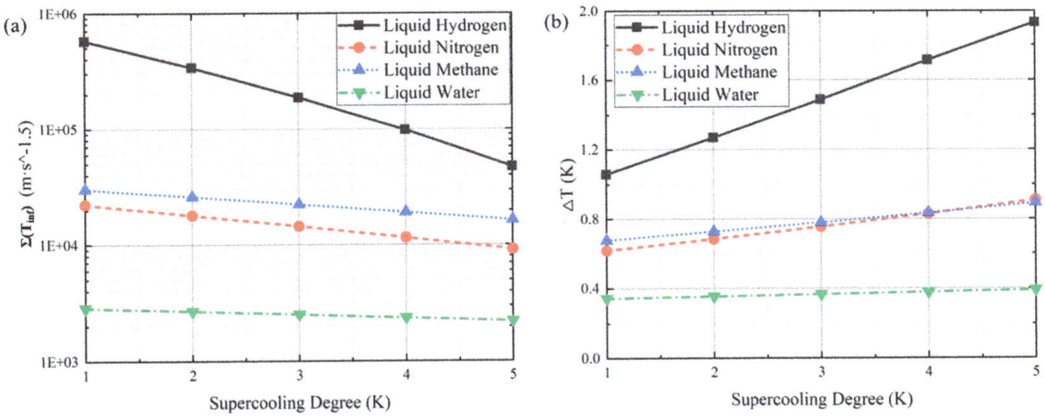

Figure 21. Thermodynamic effect parameter $\sum(T_{inf})$ (**a**) and standard temperature drop ΔT (**b**) in different fluids at 3 atm.

The relationship between $\Delta T/T_{inf}$ and T_v maximum attenuation rates in the different fluids was investigated using the statistical method and is shown in Figure 22. The maximum T_v attenuation rate was enhanced with $\Delta T/T_{inf}$ monotonously and presented a roughly logarithmic relationship. The magnitude could be qualitatively determined according to the relationship between $\Delta T/T_{inf}$ and the maximum attenuation rate of T_v.

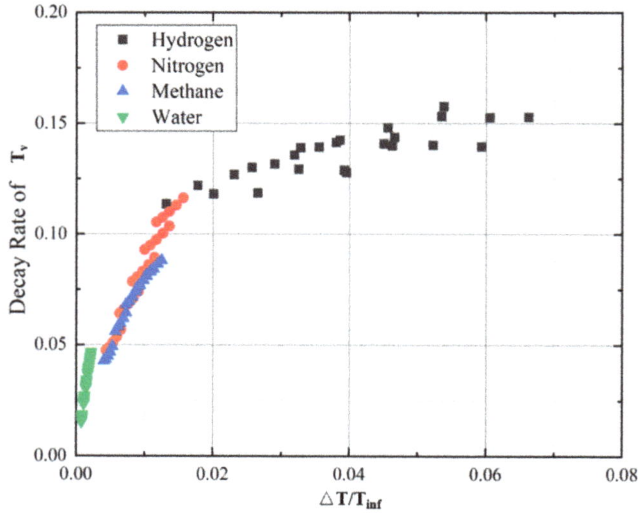

Figure 22. Relationship between attenuation rate of temperature inside the bubble and $\Delta T/T_{inf}$.

5. Conclusions and Discussions

In this paper, based on the existing bubble growth model, a mathematical model suitable for the collapse process of a single bubble was proposed, considering the vapor compression and expansion process inside the bubble. The collapse–rebound process of bubbles in liquid hydrogen was studied. The variation in the period length and the attenuation rates of different parameters with operating pressure and supercooling degree were investigated using statistical methods and comparisons. The following conclusions can be drawn.

1. The amplitude of physical parameters increased with the supercooling degree and decreased with the operating pressure. The influence of the supercooling degree on the bubble collapse process was larger than that of the operating pressure. However, the supercooling degree and the operating pressure had different effects on the attenuation rates of the physical parameters. The attenuation rate of every parameter was suppressed with higher operating pressure obviously, and slightly decreased with the supercooling degree. For bubbles in various fluids, the oscillation period was shortened under larger operating pressure and supercooling degree.
2. The amplitudes and frequencies of various parameters of bubbles in cryogenic media were higher than those in normal media. Specifically, the amplitude and frequency of every bubble parameter in liquid hydrogen were much larger than that in the other fluids. Moreover, it took the liquid hydrogen bubble the shortest time to reach equilibrium. For different liquids, the magnitude of the attenuation rates of parameters was roughly consistent with that of the standard temperature drop (ΔT) and thermal effect parameters ($\Sigma(T_{inf})$). Moreover, all attenuation rates of bubbles increased with the ambient pressure and decreased with the supercooling degree of the fluid.
3. The maximum peak pressure and maximum attenuation rate of the temperature could be roughly predicted using dimensionless numbers. The maximum pressure peak increased monotonically with the dimensionless parameter (Wb/Re) and presented a roughly exponential relationship. Moreover, the maximum attenuation rate of temperature inside the bubble increased with the dimensionless parameter ($\Delta T/T_{inf}$) monotonously and presented a roughly logarithmic relationship.

Author Contributions: Data curation, S.Y.; formal analysis, S.Y., M.Q., M.L., Y.L. and W.D.; resources, Q.Z.; supervision, T.L., Y.H.; validation, S.Y.; writing—original draft, S.Y.; writing—review and editing, S.Y. and T.L. All authors have read and agreed to the published version of the manuscript.

Funding: This project is supported by the National Natural Science Foundation of China (51976150), the China Postdoctoral Science Foundation (2021M692533) and the Youth Innovation Team of Shaanxi Universities.

Institutional Review Board Statement: This paper does not involve humans or animals.

Informed Consent Statement: This paper does not involve humans or animals.

Conflicts of Interest: The authors declare no conflict of interest.

Nomenclature

L	latent heat of the fluid
P	pressure
R	bubble radius
S	surface tension
T	temperature
c	speed of sound
k	polytropic exponent
k_l	thermal conductivity of liquid
m	mass in boundary layer
n	species of substances
t	time
u	radial velocity
Greek symbols	
α	thermal diffusivity
μ	dynamic viscosity
ν	kinematic viscosity
ρ	density
$\mathbb{R}$	molar gas constant
Subscripts	
G	noncondensable gases
v	vapor inside bubble
l	liquid near bubble boundary
inf	operating parameters at infinity
0	initial constant
Subscripts	
i	time step

References

1. Frank, D.; Nast, T.; Mix, J. Concept for on orbit liquid hydrogen test bed. *Cryogenics* **2012**, *52*, 226–230. [CrossRef]
2. Zou, D.; Hao, Y.; Wu, H.; Sun, J.; Xu, L.; Li, J. Safety assessment of large-scale all steel LNG storage tanks under wind-borne missile impact. *Thin-Walled Struct.* **2022**, *174*, 109078. [CrossRef]
3. Li, C.; Zheng, S.; Chen, Y.; Zeng, Z. Proposal and parametric analysis of an innovative natural gas pressure reduction and liquefaction system for efficient exergy recovery and LNG storage. *Energy* **2021**, *223*, 120022. [CrossRef]
4. Nitin, B.; Sandilya, P.; Chakraborty, G. Revisiting the dewar design for liquid oxygen storage in fuel cell energy systems. *Int. Commun. Heat Mass Transf.* **2022**, *134*, 105975. [CrossRef]
5. Yang, Q.; Shi, W.; Chang, J.; Bao, W. Maximum thrust for the rocket-ejector mode of the hydrogen fueled rocket-based combined cycle engine. *Int. J. Hydrogen Energy* **2015**, *40*, 3771–3776. [CrossRef]
6. Sun, X.-W.; Guo, Z.-Y.; Huang, W. Passive zero-boil-off storage of liquid hydrogen for long-time space missions. *Int. J. Hydrogen Energy* **2015**, *40*, 9347–9351. [CrossRef]
7. Mayer, T.; Semmel, M.; Morales, M.A.G.; Schmidt, K.M.; Bauer, A.; Wind, J. Techno-economic evaluation of hydrogen refueling stations with liquid or gaseous stored hydrogen. *Int. J. Hydrogen Energy* **2019**, *44*, 25809–25833. [CrossRef]
8. Bauer, A.; Mayer, T.; Semmel, M.; Morales, M.A.G.; Wind, J. Energetic evaluation of hydrogen refueling stations with liquid or gaseous stored hydrogen. *Int. J. Hydrogen Energy* **2019**, *44*, 6795–6812. [CrossRef]
9. Chen, T.; Wang, G.; Huang, B.; Wang, K. Numerical study of thermodynamic effects on liquid nitrogen cavitating flows. *Cryogenics* **2015**, *70*, 21–27. [CrossRef]
10. Brennen, C.E. *Cavitation and Bubble Dynamics*; Oxford University Press: Oxford, UK, 1995.
11. Li, G.; Caldwell, S.; Clark, J.A.; Gulick, S.; Hecht, A.; Lascar, D.D.; Levand, T.; Morgan, G.; Orford, R.; Savard, G.; et al. A compact cryogenic pump. *Cryogenics* **2016**, *75*, 35–37. [CrossRef]
12. Lien, Y.C. *Bubble Growth Rates at Reduced Pressure*; Massachusetts Institute of Technology: Cambridge, MA, USA, 1969.
13. Board, S.; Duffey, R. Spherical vapour bubble growth in superheated liquids. *Chem. Eng. Sci.* **1971**, *26*, 263–274. [CrossRef]
14. Abdelmessih, A.H. *Cocurrent Gas-Liquid Flow*; Plenum Press: New York, NY, USA, 1969.
15. Anderson, L.M.; Hord, J.; Hall, W.J. *Cavitation in Liquid Cryogens. 1: Venturi*; NASA Center for Aerospace Information (CASI): Cleveland, OH, USA, 1972. Available online: http://ntrs.nasa.gov/search.jsp?R=19720016713 (accessed on 23 September 2022).
16. Hord, J. *Cavitation in Liquid Cryogens. 2: Hydrofoil*; NASA Center for Aerospace Information (CASI): Cleveland, OH, USA, 1973. Available online: http://ntrs.nasa.gov/search.jsp?R=19730007528 (accessed on 23 September 2022).

17. Hord, J. *Cavitation in Liquid Cryogens. 3: Ogives*; NASA Center for Aerospace Information (CASI): Cleveland, OH, USA, 1973. Available online: http://ntrs.nasa.gov/search.jsp?R=19730019421 (accessed on 23 September 2022).
18. Hord, J. *Cavitation in Liquid Cryogens. 4: Combined Correlations for Venturi, Hydrofoil, Ogives, and Pumps*; NASA Center for Aerospace Information (CASI): Cleveland, OH, USA, 1974. Available online: http://ntrs.nasa.gov/search.jsp?R=19740026591 (accessed on 23 September 2022).
19. Ball, C.L.; Meng, P.R.; Reid, L. *Cavitation Performance of 84deg Helical Inducer Operated in 37degR and 42degR Liquid Hydrogen*; National Aeronautics and Space Administration: Washington, DC, USA, 1969; Report No. TM X-1360.
20. Ito, Y. The World's First Test Facility That Enables the Experimental Visualization of Cavitation on a Rotating Inducer in Both Cryogenic and Ordinary Fluids. *J. Fluids Eng.* **2021**, *143*, 121105. [CrossRef]
21. Hewitt, H.C.; Parker, J.D. Bubble Growth and Collapse in Liquid Nitrogen. *J. Heat Transf.* **1968**, *90*, 22–26. [CrossRef]
22. Zwart, P.J.; Gerber, A.G.; Belamri, T. A two-phase flow model for predicting cavitation dynamics. In Proceedings of the Fifth International Conference on Multiphase Flow, Yokohama, Japan, 30 May–3 June 2004.
23. Habil, S.I. Physical and Numerical Modeling of Unsteady Cavitation Dynamics. In Proceedings of the ICMF-2001, 4th International Conference on Multiphase Flow, New Orleans, LA, USA, 27 May–1 June 2001.
24. Singhal, A.K.; Athavale, M.M.; Li, H.; Jiang, Y. Mathematical Basis and Validation of the Full Cavitation Model. *J. Fluids Eng.* **2002**, *124*, 617–624. [CrossRef]
25. Robinson, A.; Judd, R. Bubble growth in a uniform and spatially distributed temperature field. *Int. J. Heat Mass Transf.* **2001**, *44*, 2699–2710. [CrossRef]
26. Robinson, A.; Judd, R. The dynamics of spherical bubble growth. *Int. J. Heat Mass Transf.* **2004**, *47*, 5101–5113. [CrossRef]
27. Zhang, X.; Qiu, L.; Gao, Y. Computational fluid dynamic study on cavitation in liquid nitrogen. *Cryogenics* **2008**, *48*, 432–438. [CrossRef]
28. Xu, B.; Feng, J.; Wan, F.; Zhang, D.; Shen, X.; Zhang, W. Numerical investigation of modified cavitation model with thermodynamic effect in water and liquid nitrogen. *Cryogenics* **2020**, *106*, 103049. [CrossRef]
29. Le, A.D.; Okajima, J.; Iga, Y. Numerical simulation study of cavitation in liquefied hydrogen. *Cryogenics* **2019**, *101*, 29–35. [CrossRef]
30. Li, D.; Ren, Z.; Li, Y.; Gong, R.; Wang, H. Thermodynamic effects on the cavitation flow of a liquid oxygen turbopump. *Cryogenics* **2021**, *116*, 103302. [CrossRef]
31. Ito, Y. Numerical Model and Validation for Cryogenic High-Speed Cavitating Flow Based on Bubble Size Distribution Model in Consideration of Rigorous Heat Transfer around Bubble and Bubble Oscillation. *J. Jpn. Soc. Aeronaut. Space Sci.* **2008**, *56*, 456–463. [CrossRef]
32. Ito, Y.; Zheng, X.; Nagasaki, T. One-way Coupling Numerical Simulation of Cryogenic Cavitation Around an Inducer. *Int. J. Fluid Mach. Syst.* **2019**, *12*, 235–243. [CrossRef]
33. Kimoto, H. *Experimental Evaluation of the Effects of a Water Microjet and a Shock Wave by a Local Pressure Sensor*; American Society of Mechanical Engineers, Fluids Engineering Division (Publication) FED: New York, NY, USA, 1987; Volume 57, pp. 217–224. Available online: https://search.ebscohost.com/login.aspx?direct=true&db=edselc&AN=edselc.2-52.0-0023566640&lang=zh-cn&site=eds-live. (accessed on 23 September 2022).
34. Rayleigh, L. VIII. On the pressure developed in a liquid during the collapse of a spherical cavity. *Lond. Edinb. Dublin Philos. Mag. J. Sci.* **1917**, *34*, 94–98. [CrossRef]
35. Plesset, M.S. The Dynamics of Cavitation Bubbles. *J. Appl. Mech.* **1949**, *16*, 277–282. [CrossRef]
36. Herring, G. Theory of the pulsations of the gas bubble produced by an underwater explosion. *OSRD Rpt.* **1950**, *236*, 1941.
37. Trilling, L. The Collapse and Rebound of a Gas Bubble. *J. Appl. Phys.* **1952**, *23*, 14–17. [CrossRef]
38. Liu, Y.; Chen, W.; Huang, W.; Gao, X.; Jiang, L.; Xu, J.; Zhu, Y. High Precision Measurement Technique of Steady State Acoustic Cavitation Bubble. *Chin. Sci. Bull.* **2005**, *50*, 2458–2462. (In Chinese) [CrossRef]
39. Akira, S.; Yukio, T. On the Behavior of a Spherical Bubble and the Impulse Pressure in a Viscous Compressible Liquid. *Bull. JSME* **2011**, *20*, 1453–1460.
40. Moradi, R.; Groth, K.M. Hydrogen storage and delivery: Review of the state of the art technologies and risk and reliability analysis. *Int. J. Hydrogen Energy* **2019**, *44*, 12254–12269. [CrossRef]

applied sciences

MDPI

Article

Modeling and Experimental Study of the Dual Cylinder Fluid Inerter

Fu Du [1,2], Chao Wang [1,2,*] and Wei Nie [1,2]

1 School of Mechanical Engineering, Beijing Institute of Technology, Beijing 100081, China
2 Suspension System Technology Department, China North Vehicle Research Institute, Beijing 100072, China
* Correspondence: 13488684909@163.com

Abstract: The fluid inerter is a new mechanical element which has received great attention in the field of vibration reduction. However, due to the influence of secondary flow in the curved channel, the damping force is too large and the inertia force is relatively small, which limits the engineering applications of the single-cylinder fluid inerter. To eliminate the influence of secondary flow in the single-cylinder fluid inerter, this paper proposes a dual-cylinder fluid inerter that has a straight tube instead of the spiral pipe or spiral groove. We Analyze the working principle, derive conditions of free movement, establish the damping force and inertia force model, and prove the validity of the model through bench testing. Contrastingly, it is found that the maximum parasitic damping force is only 40.32% of the single-cylinder structure, but the inertia force increases to 180.96% of the single-cylinder structure. The proposed inerter greatly increases the proportion of inertia force, and provides a new scheme for engineering applications.

Keywords: dual cylinder fluid inerter; damping force; inertia force; bench test; mechanical property

Citation: Du, F.; Wang, C.; Nie, W. Modeling and Experimental Study of the Dual Cylinder Fluid Inerter. *Appl. Sci.* **2022**, *12*, 10849. https://doi.org/10.3390/app122110849

Academic Editors: Vasily Novozhilov and Cunlu Zhao

Received: 12 September 2022
Accepted: 29 September 2022
Published: 26 October 2022

1. Introduction

In 2002, Smith of Cambridge University proposed the concept of an inerter [1] and constructed a new vibration-reduction theoretical system of "inerter–spring–damping". Inerters can help solve common problems in the field of vibration reduction, improving the performance of vehicle suspensions [2], enhancing the cross-country mobility of multi-axle vehicles [3], improving the safety of high-rise buildings, and greatly attenuating vibrations from earthquakes [4] and typhoons [5]; inerters also extend to other technical fields, such as resonance-suppression of ships [6], long-span cables [7] and bridges [8], high-speed rail suspensions [9], and aerospace [10].

Reference [11] first proposed the concept of a fluid inerter. The main source of inertia force is the fluid in the slender spiral tube or spiral groove, and the generation device of inertia force is a dual-rod single cylinder. It has the advantages of a large transmission ratio, small back gap, long lifespan, simple structure, and so on.

Swift and Smith [12] established the first mechanical model of single-cylinder fluid inerters. Through the quantitative analysis of inertia force and parasitic damping force, it was proven that when fluid reciprocates through a slender spiral tube or spiral slot at high speed, a huge parasitic damping force will be generated, and the proportion of inertia force is relatively small. To control the parasitic damping force, researchers from the University of Sheffield [13,14] and the University of Bristol [15] used magnetorheological fluid as a fluid medium to adjust the size of the parasitic damping force through an external magnetic field. Wang Le and Mao Ming [16] adopted the method of fluid mechanics analysis to build a mechanical model of single-cylinder fluid inerters, proving that the parasitic damping force mainly comes from the secondary turbulence of the helical pipeline section. The magnitude and source of parasitic damping force have been fully studied, and reducing the resistance of fluid media has become a new hotspot in this field, but with little effect.

Because of this situation, this paper draws on the dual-cylinder transmission mechanism of a hydraulic piston mass inerter [17] and proposes a new structural scheme, a dual-cylinder fluid inerter [18]. Where a single-cylinder fluid inerter has a spiral tube or groove, we instead it with a straight tube in order to eliminate the secondary flow of the spiral tube section, and the use of two double-acting single-rod cylinders instead of the original double-rod cylinder greatly reduces the parasitic damping force, while increasing the inertia force.

2. Structure and Principle of the Dual-Cylinder Fluid Inerter

2.1. Structure and Working Principle

The dual-cylinder fluid inerter consists of two cylinders, as shown in Figure 1. The inertia force output cylinder is composed of cylinder 13 and piston 12. The flow-regulating cylinder is composed of cylinder 7 and piston 8. Terminal 1 is on the left side of cylinder 13, and terminal 16 on the right side of piston rod 18. Terminal 1 and terminal 16 are the two inertia force output terminals. Pipeline 4 is used to connect oil port 3 and oil port 5, so that the main chamber 2 and 6 are connected, and pipeline 11 is connected to oil port 14 and oil port 10, so that the annular chamber 9 and 15 are connected.

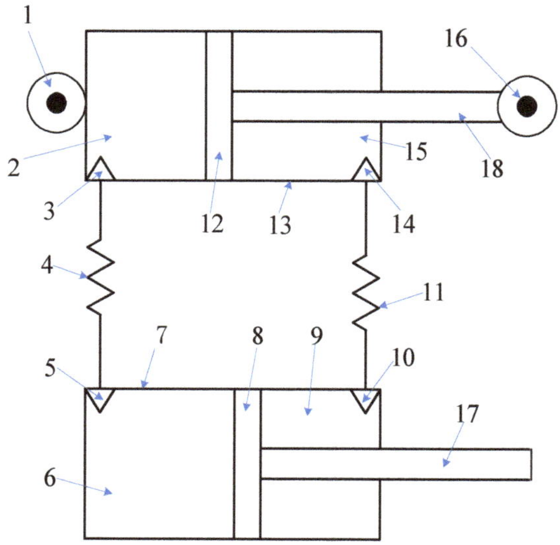

Figure 1. Schematic drawing of dual-cylinder fluid inerter. 1—terminal; 2—main chamber; 3—oil port; 4—pipeline; 5—oil port; 6—main chamber; 7—cylinder; 8—piston; 9—annular chamber; 10—oil port; 11—pipeline; 12—piston; 13—cylinder; 14—oil port; 15—annular chamber; 16—terminal; 17—piston rod; 18—piston rod.

A closed liquid-flow loop is formed by main chamber 2, oil port 3, pipeline 4, oil port 5 and main chamber 6. Another closed liquid flow loop is constructed by annular chamber 9, oil port 10, pipeline 11, oil port 14 and annular chamber 15. When terminal 1 and 16 move relative to each other, they force the oil to flow back-and-forth in the two loops, creating inertia forces.

2.2. Condition of Free Motion

Suppose that the effective area of main chamber 2 is A_2, that of main chamber 6 is A_6, that of annular chamber 15 is A_{15}, and that of annular chamber 9 is A_9. Suppose that the relative movement velocity between terminal 1 and terminal 16 is $\dot{x}$, and the volume flow discharged from main chamber 2 is $A_2\dot{x}$. According to the conservation rule of oil volume, the volume flow Q_9 out of annular chamber 9 is equal to the flow Q_{15} into annular chamber 15:

$$Q_{15} = Q_9 = \frac{A_2 \dot{x}}{A_6} \cdot A_9 \tag{1}$$

To make sure there is no motion interference in the reciprocating process of the piston, the discharge (or inflow) Q_2 of main chamber 2 and the inflow (or outflow) Q_{15} of annular chamber 15 should obey the following relationship:

$$\frac{Q_2}{Q_{15}} = \frac{A_2}{A_{15}} \tag{2}$$

Substitute Equation (1) into Equation (2), and obtain:

$$\frac{A_6}{A_9} = \frac{A_2}{A_{15}} \tag{3}$$

A_2/A_{15} is the ratio of the effective area of the main chamber of the inertia force output cylinder to the annular chamber. A_6/A_9 is the ratio of the effective area of the main chamber and the annular chamber of the flow-regulating cylinder. As long as they are equal, they can move back-and-forth without motion interference.

3. Modeling of Mechanical Properties of a Dual-Cylinder Fluid Inerter

3.1. Modeling of Inertia Forces

The inertia force at both terminals of the inerter is proportional to its relative acceleration. The ratio is called the inertance, and the unit is kg. Assuming that the inertance of the fluid inerter is B, the stored kinetic energy E is:

$$E = \frac{1}{2} B \dot{x}^2 \tag{4}$$

The kinetic energy of the system is provided by the high-speed reciprocating flow of the fluid in two closed loops, so:

$$E = \frac{1}{2} \rho_4 L_4 A_4 v_4^2 + \frac{1}{2} \rho_{11} L_{11} A_{11} v_{11}^2 \tag{5}$$

where ρ_4 is the density of the fluid in pipeline 4, L_4 is the length of pipeline 4, v_4 is the linear velocity of the fluid in pipeline 4, ρ_{11} is the density of the fluid in pipeline 11, L_{11} is the length of pipeline 11, and v_{11} is the linear velocity of the fluid in pipeline 11. Since the volume of oil is conserved, it can be obtained:

$$\begin{aligned} v_4 &= \tfrac{A_2}{A_4} \dot{x} \\ v_{11} &= \tfrac{A_{15}}{A_{11}} \dot{x} \end{aligned} \tag{6}$$

In combination with (4)–(6), the inertance B is:

$$B = \rho_4 L_4 A_4 \left(\frac{A_2}{A_4} \right)^2 + \rho_{11} L_{11} A_{11} \left(\frac{A_{15}}{A_{11}} \right)^2 \tag{7}$$

where $\rho_4 L_4 A_4$ and $\rho_{11} L_{11} A_{11}$ are the masses of the fluid in pipelines 4 and 11, respectively. Therefore, the size of B has nothing to do with the structural size of the flow-regulating cylinder, but only with the inertia force output cylinder and the pipeline.

3.2. Damping Force Modeling

Part of the damping force is caused by the friction between cylinder and piston, and the other part is caused by the viscous resistance of oil flowing back-and-forth. When the relative velocities of the two terminals of the dual-cylinder fluid inerter are low, damping and inertia forces are small, and friction dominates. Therefore, in this paper, a 0.01 Hz quasi-static sinusoidal displacement signal is used as the input, and the mean value $\overline{|F|}$ of the absolute values of the output forces at both terminals is used to approximate the friction force F_f; that is to say:

$$F_f = \overline{|F|}\mathrm{sign}\left(\dot{S}\right) \tag{8}$$

where sign is the sign function, indicating that the direction of friction force F_f is opposite to the relative velocity S. The pressure-loss along the path when the fluid flows through pipelines 4 and 11 is the source of damping force, and the pressure-loss coefficient λ [19] is:

$$\lambda = \begin{cases} \frac{75}{R_e}, R_e \leqslant 2320 \\ 0.0025R_e^{\frac{1}{3}}, 2320 < R_e \leqslant 4000 \\ \frac{0.3164}{R_e^{0.25}}, 4000 < R_e \leqslant 10^5 \\ 0.0032 + \frac{0.221}{R_e^{0.237}}, 10^5 < R_e \leqslant 10^6 \end{cases} \tag{9}$$

where R_e is the Reynolds number; the calculation formula is:

$$R_e = \frac{\rho v d}{\mu} \tag{10}$$

where ρ is fluid density, v is fluid velocity, d is flow-channel diameter, and μ is fluid dynamic viscosity. The pressure-drop Δp [20] generated by liquid flow damping is:

$$\Delta p = \frac{1}{2}\rho v^2 \lambda \frac{L}{d} \tag{11}$$

where L is the length of a straight pipe. The damping-loss Δp_4 along pipeline 4 is:

$$\Delta p_4 = p_2 - p_6 = \frac{1}{2}\rho_4 v_4^2 \lambda_4 \frac{L_4}{d_4} \tag{12}$$

where p_2 is the pressure in main chamber 2 and p_6 is the pressure in main chamber 6. Similarly, the damping-loss Δp_{11} along pipeline 11 is:

$$\Delta p_{11} = p_9 - p_{15} = \frac{1}{2}\rho_{11} v_{11}^2 \lambda_{11} \frac{L_{11}}{d_{11}} \tag{13}$$

where p_9 is the pressure in annular chamber 9 and p_{15} is the pressure in annular chamber 15. The damping force F_c is:

$$F_c = A_2 p_2 - A_{15} p_{15} + F_f \tag{14}$$

In combination with Formulas (3), (12)–(14), the following can be obtained:

$$F_c = \Delta p_4 A_2 + \Delta p_{11} A_{15} + F_f \tag{15}$$

3.3. Simulation of Mechanical Properties

The force generated by the dual-cylinder fluid inerter is the resultant force of inertia force and damping; i.e., F is:

$$F = B\ddot{x} + F_c \tag{16}$$

MATLAB was used to write programs to establish the inertial force F_b, damping force F_c and resultant force F by combining Equations (7), (15), and (16). The structural parameters are shown in Table 1, and the results are shown in Table 2.

As can be seen from Figures 3 and 4, the mechanical properties of a dual-cylinder fluid inerter are equivalent to nonlinear damping and linear inerter in parallel.

Table 1. Table of structure and fluid parameters of the single-cylinder fluid inerter.

Structure Size	
Diameter of main chamber cylinder 2 D_2 (mm)	80
Diameter of main chamber cylinder 6 D_6 (mm)	80
Diameter of piston rod 18 D_{18} (mm)	50
Diameter of piston rod 17 D_{17} (mm)	50
Diameter of pipeline 4 D_4 (mm)	12
Diameter of pipeline 11 D_{11} (mm)	12
Length of pipeline 4 L_2 (mm)	5000
Length of pipeline 11 L_{11} (mm)	5000
Parameters of the Fluid Medium	
Fluid type	water
Density ρ (kg/m^3)	1000
Dynamic viscosity μ (cSt)	1
Input Signal Parameters	
Input type	sinusoidal velocity signal
Amplitude (mm)	60
Frequency ω (Hz)	1

Table 2. Table of result.

Inertance	Result
Inertance generated by pipeline 4 B_4 (kg)	1115
Inertance generated by pipeline 11 B_{11} (kg)	412
The total inertance B (kg)	1527

The inertial and damping forces are shown in Figure 2.

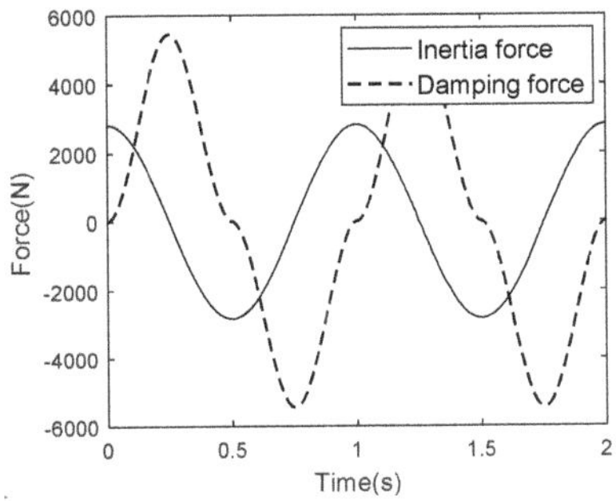

Figure 2. The curve of damping and inertia force of a dual-cylinder fluid inerter.

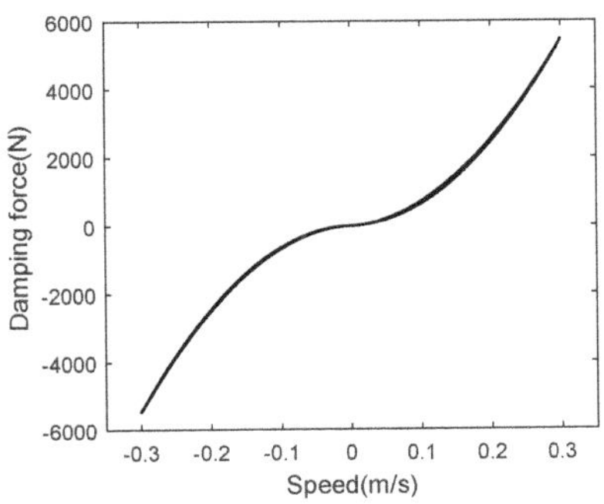

Figure 3. The curve of damping force and velocity.

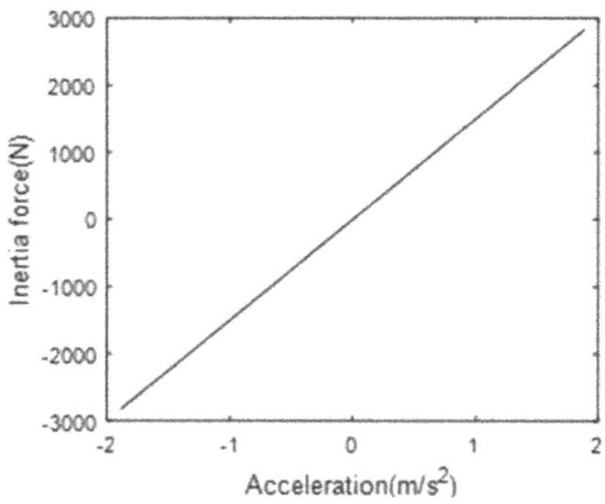

Figure 4. The curve of inertia force and acceleration.

In Figure 5, the solid line is the resultant force calculated by Formula (16), and the dotted line is the result calculated by Simulink/Hydraulic Segmented Pipe LP. The Segmented Pipe LP module could consider the inertia force and damping force of the fluid simultaneously. "The Segmented Pipe LP" module in Simulink was based on the finite-element idea. Except in the initial stage of loading, there was shock in the Segmented Pipe LP module, and after stabilizing, the results of the above two methods were highly consistent, which proved the correctness of the established dual-cylinder fluid inerter mechanical model before the test comparison.

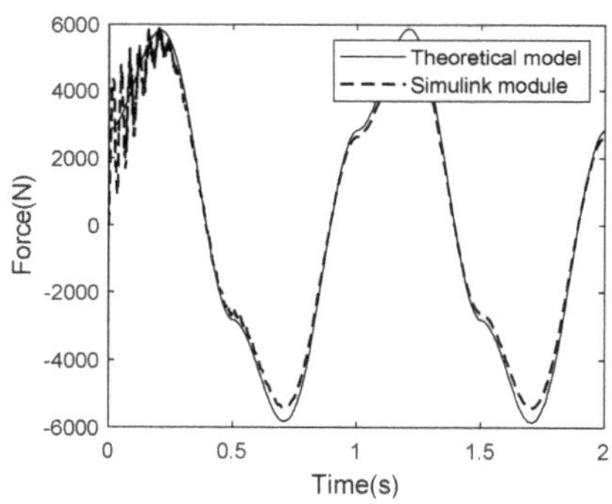

Figure 5. The comparison curve between theoretical model and pipe module in Simulink.

The oscillations in the calculation results of Simulink are a normal physical phenomenon. At the moment of loading, the internal pressure of the fluid inerter transiently fluctuates, and gradually converges to a steady state after a period of oscillation. The mechanical model established by Formula (16) is based on the empirical formula of fluid mechanics and is a description of a steady state, so there is no oscillation in Figure 5.

4. Comparative Study with Single-Cylinder Fluid Inerter

According to the literature [16], the mechanical model of a single-cylinder fluid inerter is constructed. The diameter of the cylinder, the diameter of the piston rod, and the length and diameter of the pipeline are the same as those of the dual-cylinder fluid inerter. The specific parameters are shown in Table 3.

Table 3. Table of structure and fluid parameters of the dual cylinder fluid inerter.

Structure Size	
Diameter of cylinder (mm)	**80**
Diameter of piston rod (mm)	**50**
Diameter of pipe (mm)	**12**
Length of pipeline (m)	**10**
Parameters of the Fluid Medium	
Fluid type	**Water**
Density ρ (kg/m^3)	**1000**
Dynamic viscosity μ (cst)	**1**
Result	
Inertance (kg)	**828**

Inertance of the single-cylinder fluid inerter is 828 kg, and inertance of the dual-cylinder fluid inerter is 1529 kg. When $L_4 + L_{11} = 10$, that is to say, the total length of the pipeline is 10 m, inertance is in a linear relationship with L_4, as shown in Figure 6. When $L_4 = 0$, inertance is the same as in a single-cylinder fluid inerter, and inertance increases linearly with an increase in L_4. Under the condition that the total pipe length is the same, different inertance values can be obtained by changing the lengths of L_4 and L_{11}.

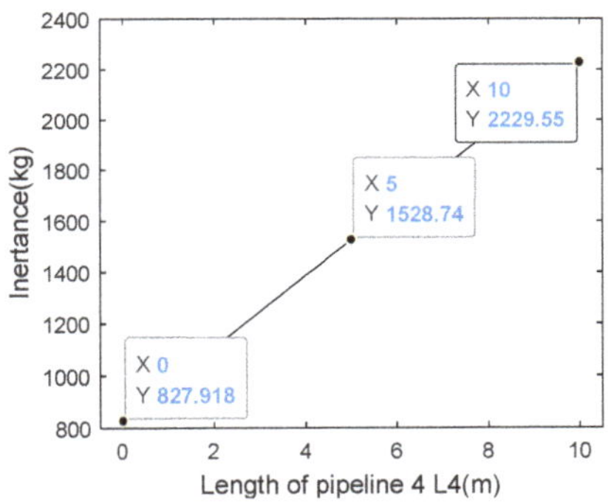

Figure 6. The relationship between inertance and L_4.

In the single-cylinder fluid inerter, as the secondary flow on the section of the pipe increases the damping force when the fluid flows in the spiral pipe, the inertia force is relatively small [16]. When sinusoidal speed signals shown in Table 2 are used, the change curves of inertia and damping forces over time are shown in Figure 7.

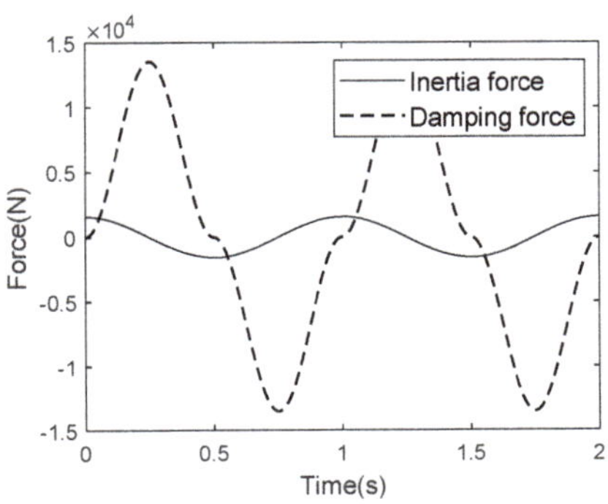

Figure 7. The curve of damping and inertia force over time of a single-cylinder fluid inerter.

As can be seen from Figures 8 and 9, the maximum damping force of a dual-cylinder fluid inerter is 40.32% (5451/13,519 = 0.4032) of that in a single-cylinder fluid inerter, and the inertia force is 180.96% (2823/1560 = 1.8096).

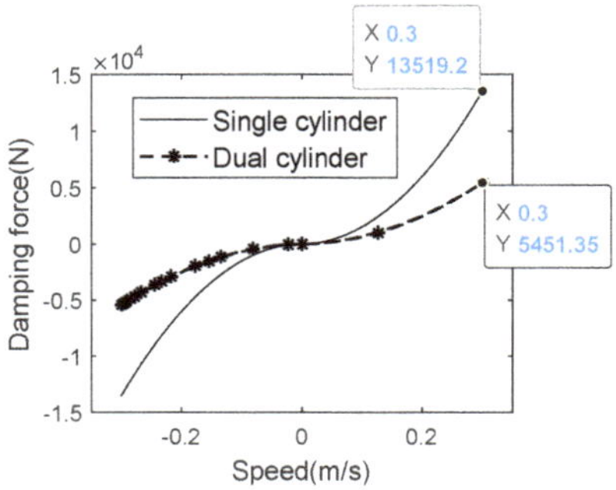

Figure 8. The comparison curve of damping force and velocity.

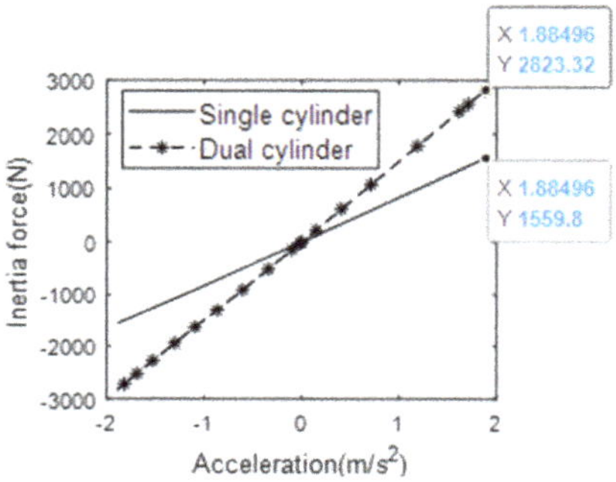

Figure 9. The comparison curve of inertia force and acceleration.

5. Bench Test

According to the structural parameters shown in Table 1, a dual-cylinder fluid inerter prototype is produced. We adopt two hydraulic cylinders of the same size; one is the inertia force output cylinder, the other is the flow-regulating cylinder. We use hydraulic pipeline to connect the main chambers and annular chambers of the two cylinders, respectively, as shown in Figure 10.

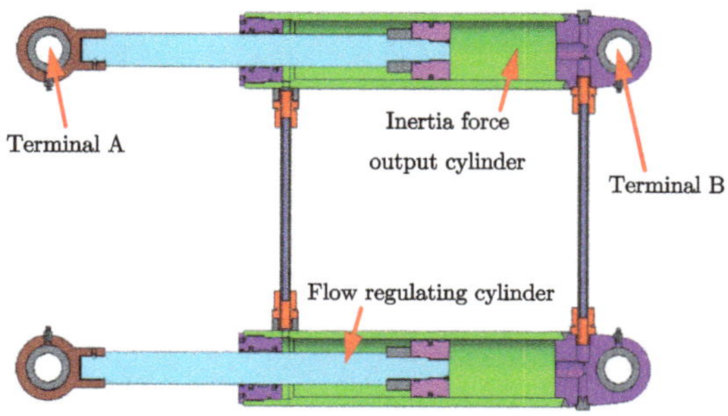

Figure 10. Drawing of the dual-cylinder fluid inerter.

Sinusoidal speed signals of different frequencies are loaded into the fluid inerter, and signals such as time, displacement, and load at both terminals are collected. The specific operation is as follows: Connect hinge A (terminal 16) and hinge B (terminal 1) of the inertia force output cylinder with the base and exciter of the test bench, respectively, as shown in Figure 11. Connect the pipe and inject water, as shown in Figure 12.

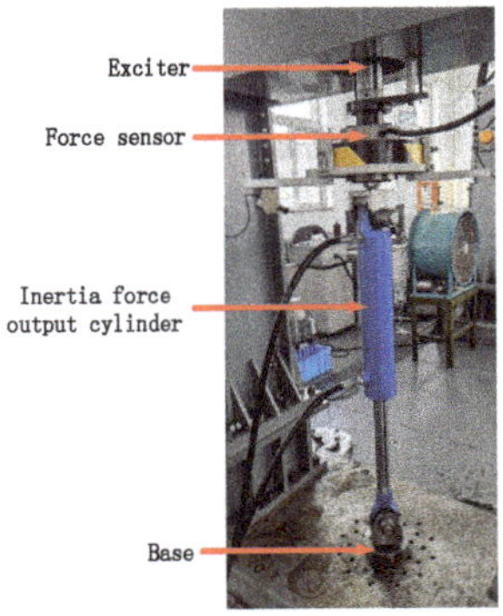

Figure 11. Prototype of dual-cylinder fluid inerter.

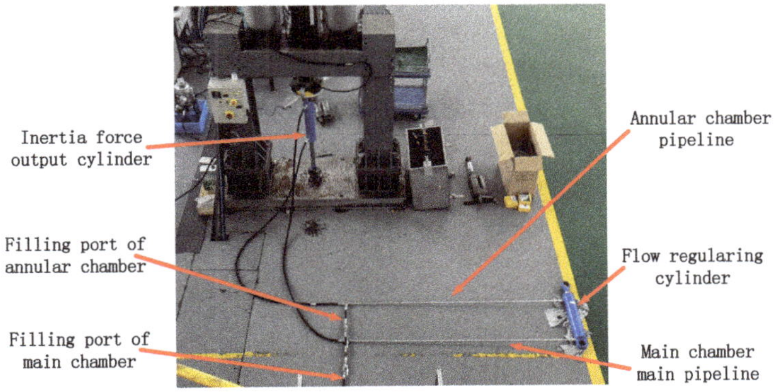

Figure 12. Test layout of dual-cylinder fluid inerter.

Figure 13 shows the output force curves at both terminals of the dual-cylinder inerter when the quasi-static sinusoidal speed signal of 0.01 Hz is used as the input. According to the statistics, the average of the absolute value of F: $\overline{|F|} = 178\text{N}$, and the approximate value of friction force F_f can be obtained by combining Formula (8).

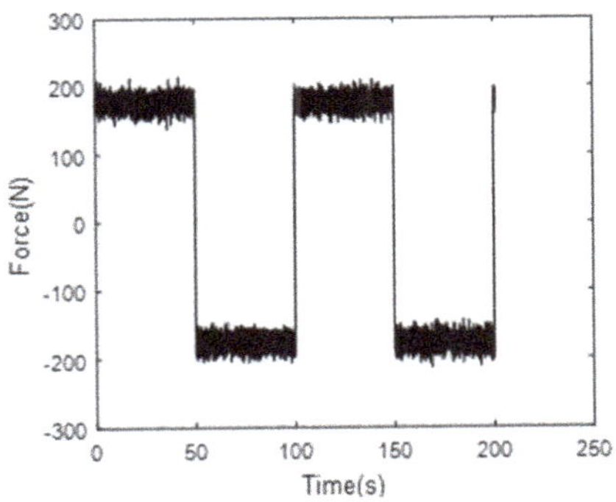

Figure 13. The force value of 60 mm and 0.01 Hz.

As can be seen from Figure 14 and Table 4, as the frequency increases, the deviation between the experimental value and the theoretical value becomes smaller and smaller, and the variation trend and value of the experimental value and the theoretical value are highly consistent. The mechanical model established in this paper can guide the structural design and parameter-optimization of dual-cylinder fluid inerters.

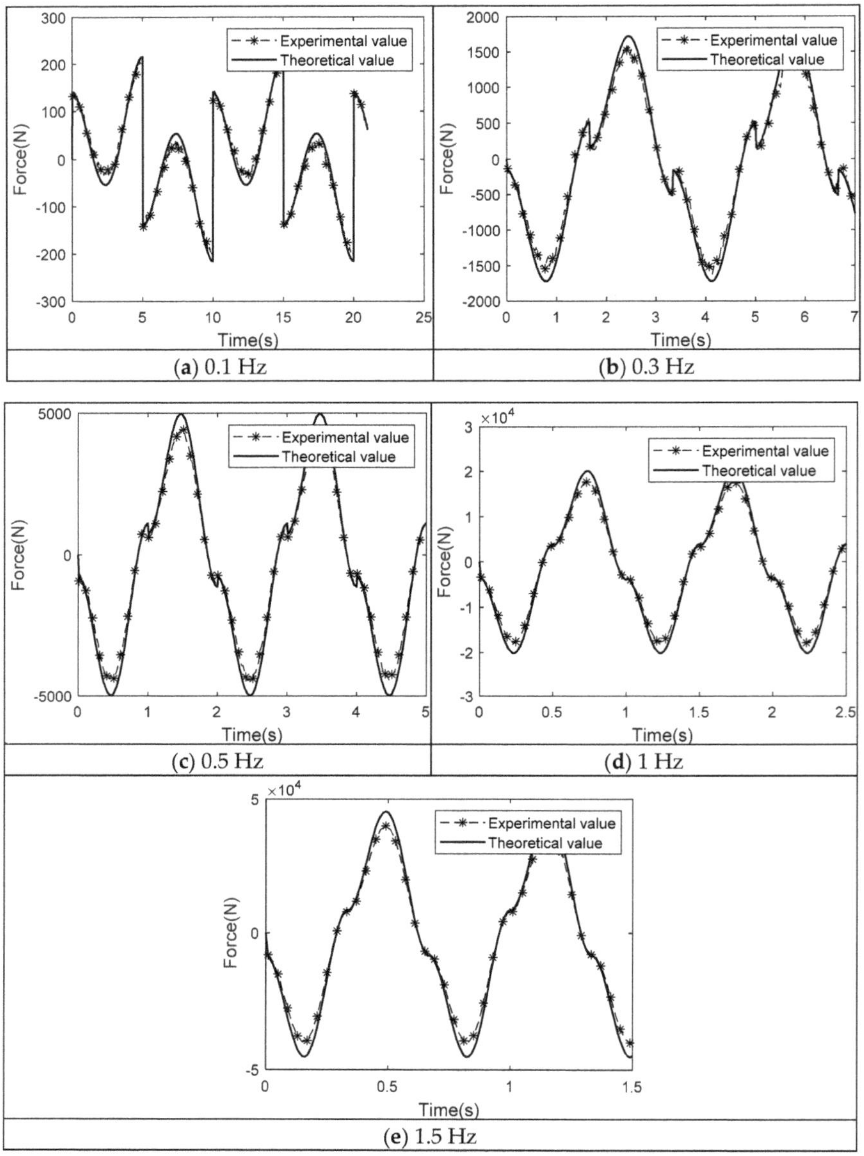

Figure 14. The force value of 60 mm at different frequencies.

Table 4. Table of theoretical and experimental force values.

Frequency (Hz)	Max of Experimental Value (kN)	Max of Simulation Value (kN)	Size Error (%)
0.1	0.24	0.213	11.25
0.3	1.68	1.83	8.93
0.5	4.55	4.88	7.25
1	19.32	20.16	4.35
1.5	45.43	47.1	3.68

Note: Size error $= \frac{|\text{max of experimental value} - \text{max of simulation value}|}{\text{max of experimental value}} \times 100\%$.

6. Conclusions

Through the principal analysis, mechanical-characteristics modeling, and bench test of a dual-cylinder fluid inerter, the following conclusions are drawn:

(1) Dual-cylinder fluid inerters exists in theory, are feasible in principle, and can be achieved with engineering.

(2) Compared with single-cylinder fluid inerters, the damping force of dual-cylinder fluid inerters is significantly reduced, with 40.32% of the single-cylinder fluid inerter, and the inertia force is significantly increased, with 180.96% of the single-cylinder fluid inerter.

(3) The mechanical model established in this paper can effectively simulate the mechanical properties of a dual-cylinder fluid inerter, and can guide its structural design and parameter-optimization.

(4) Dual-cylinder fluid inerters can provide greater inertial force, and have great potential in vehicle suspensions, building vibration reduction, and other applications.

Author Contributions: Conceptualization, F.D.; methodology, F.D.; Software, C.W.; Validation, W.N.; Formal analysis, F.D.; Investigation, C.W.; Resources, C.W.; Data curation, W.N.; Writing—original draft preparation, F.D.; Writing—review and editing, F.D.; Supervision, F.D.; Project administration, F.D. All authors have read and agreed to the published version of the manuscript.

Funding: National Natural Science Foundation of China (51875536) and Fengtai District science and technology Nova program (Kjxx202005).

Institutional Review Board Statement: Not Applicable.

Informed Consent Statement: Not Applicable.

Data Availability Statement: Not Applicable.

Conflicts of Interest: The authors declare no conflict of interest.

References

1. Smith, M.C. Synthesis of mechanical networks: The inerter. *IEEE Trans. Autom. Control* **2002**, *47*, 1648–1662. [CrossRef]
2. Shen, Y.; Chen, L.; Liu, Y. Analysis of vibration isolation performance of vehicle suspension with nonlinear fluid inerter. *Automot. Eng.* **2017**, *39*, 789–795.
3. Du, F.; Mao, M.; Chen, Y. Structure design and performance analysis of inerter-spring-damper suspension based on dynamic model and parameter optimization. *J. Vib. Shock* **2014**, *33*, 59–65.
4. Lazar, I.F.; Neild, S.A.; Wagg, D.J. Using an inerter-based device for structural vibration suppression. *Earthq. Eng. Struct. Dyn.* **2014**, *43*, 1129–1147. [CrossRef]
5. Giaralis, A.; Petrini, F. Wind-induced vibration mitigation in tall buildings using the tuned mass-damper-inerter. *J. Struct. Eng.* **2017**, *143*, 04017127. [CrossRef]
6. Wen, H.; Guo, J. *Performance Analysis of Habitual Capacity Vibration Isolation System and Exploration of Its Application in Ship-Sea Field*; National Defence Industry Press: Beijing, China, 2018; pp. 1–20.
7. Lazar, I.F.; Neild, S.A.; Wagg, D.J. Vibration suppression of cables using tuned inerter dampers. *Eng. Struct.* **2016**, *122*, 62–71. [CrossRef]
8. Xu, K.; Bi, K.; Han, Q. Using tuned mass damper inerter to mitigate vortex-induced vibration of long-span bridges: Analytical study. *Eng. Struct.* **2019**, *182*, 101–111. [CrossRef]
9. Jiang, J.Z.; Matamoros-Sanchez, A.Z.; Goodall, R.M.; Smith, M.C. Passive suspensions incorporating inerters for railway vehicles. *Veh. Syst. Dyn.* **2012**, *50* (Suppl. S1), 263–276. [CrossRef]
10. Li, Y.; Jiang, J.Z.; Neild, S. Inerter -based configurations for main-landing-gear shimmy suppression. *J. Aircr.* **2016**, *54*, 684–693. [CrossRef]
11. Robin, T. Fluid Inerter: EP. EP2528757A1, 5 December 2012.
12. Swift, S.J.; Smith, M.C.; Glover, A.R. Design and modelling of a fluid inerter. *Int. J. Control* **2013**, *86*, 2035–2051. [CrossRef]
13. Deastra, P.; Wagg, D.J.; Sims, N.D. The realisation of an inerter-based system using fluid inerter. In *Dynamics of Civil Structures*; Springer: Cham, Switzerland, 2019; Volume 2, pp. 127–134.
14. Tipuric, M.; Deastra, P.; Wagg, D. Semi-active inerters using magnetorheological fluid: A feasibility study. In *Active & Passive Smart Structures & Integrated Systems XII, Proceedings of the SPIE Smart Structures and Materials + Nondestructive Evaluation and Health Monitoring, Denver, CO, USA, 4–8 March 2018*; SPIE: Bellingham, WA, USA, 2018.
15. Liu, X. Fluid Inerter Based Vibration Suppression: Modelling Methodology. Ph.D. Thesis, University of Bristol, Bristol, UK, 2019.
16. Wang, L.; Mao, M.; Lei, Q.; Chen, Y.; Zhang, X. Modeling and testing for a hydraulic inerter. *J. Vib. Shock* **2018**, *37*, 146–152.

17. Shen, Y.; Liu, Y.; Chen, L. Optimal design and experimental research of vehicle suspension based on a hydraulic electric inerter. *Mechatronics* **2019**, *61*, 12–19. [CrossRef]
18. Du, F.; Yang, Y.; Wang, Y. Dual Cylinder Fluid Inerter. CN113719572A, 30 November 2021.
19. Wu, W. *Fluid Mechanics*; Peking University Press: Beijing, China, 2018.
20. Dixon, J.C. *The Shock Absorbe Handbook*; Professional Engineering Publishing Ltd.: London, UK; John Wiley & Sons Ltd.: Chichester, UK, 2007.

applied sciences

MDPI

Article

Software for Monitoring the In-Service Efficiency of Hydraulic Pumps

Alin-Adrian Anton [1,*], Adrian Cococeanu [2] and Sebastian Muntean [3,*]

[1] Computer and Information Technology Deparment, Faculty of Automation and Computing, Politehnica University Timişoara, 2nd Vasile Pârvan Ave., 300223 Timişoara, Romania
[2] AQUATIM S.A. Timişoara, 11/A Gheorghe Lazăr Str., 300081 Timişoara, Romania
[3] Center for Fundamental and Advanced Technical Research, Romanian Academy Timişoara Branch, 24th Mihai Viteazu Ave., 300222 Timişoara, Romania
* Correspondence: alin.anton@upt.ro (A.-A.A.); sebastian.muntean@academiatm.ro (S.M.)

Abstract: The present paper introduces the creation of an algorithm and the software used to determine the energetic performance and monitor the efficiency of hydraulic pumps working in various industrial applications, such as water supply systems, water treatment processes, and irrigation systems, particularly in the cases where there is no permanent monitoring. Our field investigations and the surveyed literature show that the only parameter that is neither monitored nor computed is the efficiency of the pumps. The software implementation allows for determining the in-service efficiency of the pumps and comparing it to the value associated with the best efficiency point (BEP). The solution is user-friendly and can be easily installed on any computer or smartphone. The software has been applied and tested in the Hydraulic Machines Laboratory at the "Politehnica" University Timişoara and at the AQUATIM S.A. regional water supply company. The software module monitors the operating regimes of the pumps and supports the deployment of predictive maintenance and servicing.

Keywords: cellular phones; energy consumption; hydraulic pumps; software algorithm; water resources

Citation: Anton, A.-A.; Cococeanu, A.; Muntean, S. Software for Monitoring the In-Service Efficiency of Hydraulic Pumps. *Appl. Sci.* **2022**, *12*, 11450. https://doi.org/10.3390/app122211450

Academic Editors: Vasily Novozhilov and Cunlu Zhao

Received: 13 October 2022
Accepted: 8 November 2022
Published: 11 November 2022

1. Introduction

Water and energy are two vital sources of any urban community's daily life. In the United States, 19% of all the energy produced is consumed for the supply of drinking water, including treatment, transportation, storage, distribution, collecting of sewage water, treatment and discharge [1].

The situation is almost the same in the European Union, where 22% of the total electricity consumption of the industrial systems fitted with electric engines runs the pumps; the annual electricity consumption was 109 TWh in 2005 for water pumping, and in 2020, the increase in energy consumption was 136 TWh, a growth of 25% [2].

The life cycle cost of centrifugal pumps installed in industrial applications consists mainly of maintenance and energy costs. The energy cost is the most significant contribution in the total life cycle cost of a pump [3,4]. Depending on the industry in which the centrifugal pumps are installed, the total life-cycle cost comes from the consumption of electric power (e.g., chemical industry 26%, pulp and paper 31%, petroleum 59%, water treatment and supply systems from 55% to 90%) [5].

Proper selection of the pump in correlation with the system requirements can reduce energy costs by an average of 20%. This means that the pump operates mostly with maximum efficiency in the vicinity of the best efficiency point (BEP). In contrast, the hydraulic pumps that operate at off-design conditions lead to premature damage, wasted costs and costly repairs or replacements [6]. All these issues can be avoided by selecting the appropriate pump, monitoring it and providing well-planned maintenance [7]. This applies to all situations involving the operation of pumps and particularly in cases where an adequate monitoring system is missing.

Once the pump is installed, its efficiency is determined by the process conditions. The major factors affecting performance include the efficiency of the pump and system components, overall system design, efficient pump control and appropriate maintenance cycles. To achieve the efficiencies available from the mechanical design, pump manufacturers must work closely with end users and design engineers to consider all of these factors when specifying pumps [8].

The only parameters that are neither determined nor examined are the pump and pumping station efficiencies, although all the other variables (absorbed power, pumping head) are measured to determine and calculate these parameters.

Energy savings represent lower operating costs and higher performance. The most obvious energy savings are those associated with improvements in pump efficiency [9]. Although worth pursuing, they are small compared to the efficiency gains that can be achieved through a proper analysis of the pumping system to obtain the best fit of the pump to the system and the operating requirements. It may be comforting to select the nominal flow rate (at the highest value of the flow rate). However, the energy penalty can be significant when the pump is operating at partial flow rates.

A correctly selected maximum flow rate is a key value where energy optimization is targeted. Most pumps run far from their BEP. For reasons ranging from short-sighted or overly conservative design, specification and procurement to decades of incremental changes in operating conditions, pipes and control valves are too large or too small. In anticipation of future load growth, the end user, supplier and design engineers routinely add 10–50% safety margins to ensure the pump and motor can accommodate anticipated capacity increases. Important energy savings can be made if the safety margins imposed to obtain the rated service condition are not excessive [3].

The average pumping accounted for 80% of total electricity use in public water systems, as reported by EPRI [5]. The use of electricity for water and wastewater treatment has grown during the last 20 years and will continue to grow. The efficiency of the pump affects the pumping performance significantly. The efficiency of pumping units is often relatively low because pumps are typically oversized [10].

The relative significance of different energy-using systems will vary depending on the system, yet a "typical" treatment system can be developed and presented. The distribution of energy within the water treatment conveyance, treatment, and distribution cycle of a surface water system shows one approach based on certain key assumptions. The data do not apply to all water treatment systems but instead provide context as to the energy issues within water treatment facilities. In this case, pumping sewage accounts for 67%, water treatment for 14%, raw water pumping for 11%, and in-plant water pumping for 8% [1] of total energy use.

In addition to the energy costs, inefficient operation of the pump units could impact system reliability because the mechanical reliability of a pumping unit is linked to the pump efficiency, and any damage to a pump unit may entail substantial additional costs. Consequently, efficient pump operation over the lifespan of a system is a key element of any cost-reduction program.

In general, adjusting the flow rate using the variable speed drive (VSD) of centrifugal pumps is efficient. When a VSD does not exist, the pump's duty point can be tuned using the valve installed on the pump discharge line or by modifying the number of operating pumps. These pump-tuning methods are frequently selected but are inefficient, leading to the adjustment of the flow rate through hydraulic losses. The surveyed literature underlines that studies of and patents for the monitoring of pump operating parameters exist, mostly oriented towards the determination of the pumped flow rate and the pumping head [11] by measuring the parameters of the electric motor that drives the pump. There are also concerns noted in studies and research about the reduction in the pumps' efficiency in their duty point [12–18]. The only parameter that is not measured, computed or informatively shown is the pump efficiency and the pumping station efficiency. Researchers state that the pump efficiency can be determined, but they do not provide a methodology or a solution to do so.

At the same time, it is rather simple to measure the pressure, the flow rate and various electric parameters, regardless of the operating mode of the pumping stations (manual or automated), to particularize the operation of the pumps and the pumping unit and optimize the system in service.

The paper presents the software solution developed for monitoring hydraulic pumps installed in pumping stations. The software identifies the pump's operating regime by determining the efficiency at the duty point and fitting this value within the boundaries set by the operator.

To begin, one must know the efficiency behavior of the pump from the catalog or laboratory investigations. Efficiency can also be determined in situ using various indirect methods [15–18] and subsequent calculations or direct methods, such as the thermodynamic method, which can be used in the case of large-capacity pumps [19,20].

A pump's energetic behavior (efficiency curves) can be determined based on the manufacturer's catalog data or data obtained from previous tests [21]. The main geometric parameters can be determined in situ. All the geometric parameters and the coefficients of the longitudinal and local hydraulic losses are inserted in a visible barcode for each pump. After scanning the barcode with a smartphone, the operator inputs two parameters, p_s—suction pressure and p_d—discharge pressure, available on the two manometers already installed on the pumps. The software identifies the duty point of the pump and establishes the operating regime, comparing the value of the efficiency at the duty point to the maximum efficiency of the pump. Finally, the software issues a recommendation to the operator about its normal/abnormal operation and suggests to the operator how to act [22].

The pumps installed in different systems need to be operated safely at a low cost. Monitoring and preventive maintenance of centrifugal pumps are crucial issues to increase their reliability and diminish the costs [23,24]. Industry 4.0 principles must be implemented in drinking water systems to provide an efficient link between the pumping units and monitoring systems so the operator can make the best decisions [25]. The software module presented in this manuscript is a basic component of this system.

Our field investigations and the surveyed literature show that the only parameter that is neither monitored nor computed is the efficiency of the pumps. Even though the power factor and the efficiency of the electric motors are provided by their manufacturers, Ferreira et al. [26] conducted extensive tests on 435 three-phase induction electric motors from 38 different manufacturers between 2015 and 2016, showing that 58% of the values measured for the power factor of the motors were lower than those reported by the manufacturer [26], and also that 55% of the measured values for the performance were lower than the values reported by the manufacturer [26]. The pump efficiency value **is ignored in practice** and many industry studies such as [7] prove that improper design and poor pump performance may affect the plant operation such as maintenance cost, downtime and loss of production. Consequently, the software implementation allows for determining the in-service efficiency of the pumps and comparing it to the value associated with the best efficiency point (BEP).

The development of a robust algorithm to assess hydraulic pump efficiency is detailed in Section 2. Free software packages have been selected to implement the algorithm, ensuring its portability on any operating system. The implementation of the algorithm and the selected software packages is detailed in Section 3. Within Section 4 the application of the software module in the operating of the PCN 65/200 centrifugal pump installed in the laboratory is discussed, which allows the determination of percentage deviations of the results supplied for the full range of operation. Section 4.1 contains the available data for the PCN 65/200 centrifugal pump that have been used for the sensitivity analysis of the algorithm regarding the input values and the parameters associated with the investigated hydraulic configuration for various operating points, covering the full operating range of the pump. The sensitivity analysis allows determining the influence of the input values and the parameters associated with the hydraulic configuration over the output values of the algorithm and their ranking according to the level of percentage deviation. The results obtained with the software module implemented in the regional water supply company, AQUATIM S.A., for the monitoring of the in situ operation of the Worthington 500 LNN-775A double flux pumps of 1 MW, ensuring the water supply of Timişoara city, are shown in Section 5. The conclusions regarding the development and testing of the software module to monitor hydraulic pump efficiency are summarized in Section 6.

2. Algorithm for Assessing Pump Efficiency

A robust algorithm is developed to monitor the efficiency of in-service pumps. The η_P pump efficiency equation is:

$$\eta_P = \frac{P_h}{P_m} = \frac{g\rho QH}{P_m} \tag{1}$$

where $P_h = g\rho QH$ [W] is the hydraulic power, representing the power transferred to the liquid, Q [m^3/s] is the pump discharge (volumetric flow rate), H [m] represents the pumping head, ρ [kg/m^3] is the density of the working liquid (water), g [m/s^2] is the gravity acceleration, and P_m [W] represents the mechanical power at the pump shaft.

The pumping head H is presented in Equation (2).

$$H = \frac{(p_d - p_s)}{g\rho} + \frac{8Q^2}{g\pi^2}[\frac{1}{D_d^4} - \frac{1}{D_s^4}] + (z_d - z_s) \tag{2}$$

This equation is valid if the instruments are installed on the pump flanges. In practical terms, the instruments cannot be installed on the pump flanges in most cases. That is why the algorithm will be considered with a generalized form of the equation of the pumping head H given in Equation (3). This equation includes the distributed and local hydraulic losses due to the positioning of the instruments in relationship to the pump flanges and the static pressures measured on the suction and discharge lines.

$$H = \frac{(p_d - p_s)}{g\rho} + \frac{8Q^2}{g\pi^2}[\frac{1}{D_d^4}(\lambda_d\frac{l_d}{D_d} + \zeta_d + 1) + \frac{1}{D_s^4}(\lambda_s\frac{l_s}{D_s} + \zeta_s - 1)] + (z_d - z_s) \tag{3}$$

where p_s and p_d are the static pressures measured at the pump's suction and discharge, l_s, D_s, λ_s, ζ_s are the length, diameter, coefficient of distributed hydraulic loss, or Darcy's coefficient, and the local hydraulic loss coefficient, in connection with the position of the instrument installed on the suction line compared to the pump flange, l_d, D_d, λ_d, ζ_d are the length, diameter, coefficient of distributed hydraulic loss, or the Darcy's coefficient and the local hydraulic loss coefficient, in connection with the position of the instrument installed on the discharge line compared to the pump flange, and $z_d - z_s$ is the quota difference between the position of the instrument installed on the discharge line and the suction line. The generalized equation of the pumping head (3) is reduced to (2) when the instruments are installed on the pump flanges, because the distributed hydraulic loss coefficients $\lambda_s = \lambda_d = 0$ and the local hydraulic loss coefficients $\zeta_s = \zeta_d = 0$.

It is important to mention that for the present algorithm, we need only the readings of the static suction pressure p_s and discharge pressure p_d to determine the efficiency η_P of the pump. The remaining values needed for the calculus are **determined** (e.g., the volumetric flow rate that has passed through the pump Q—a value difficult to determine in industrial applications), or **assessed** (the mechanical power by the pump P_m).

The pumping head is defined as the specific energy difference of the fluid between the inlet section and the outlet section of the pump. The pumping head is given by the $H(Q)$ curve from the pump supplier's catalog or by the measurements performed in situ. As a result, the $H(Q)$ curve can be expressed in the form of a 2nd-degree polynomial for a radial centrifugal pump. The coefficients h_2, h_1 and h_0 are determined by a fitting procedure using the available data.

$$H(Q) = h_2 Q^2 + h_1 Q + h_0 \tag{4}$$

The algorithm for assessing the efficiency of hydraulic pumps is based on knowing the pump's catalog features (or the pump's curves, determined experimentally) and the determination of geometric data in situ, which will be introduced as parameters of the algorithm.

The value of the volumetric flow rate Q [m^3/s] is obtained from the equality of the relationships (3) and (4) by solving the 2nd-degree equation below.

$$Q^2\{h_2 - \frac{8}{g\pi^2}[\frac{1}{D_d^4}(\lambda_d \frac{l_d}{D_d} + \zeta_d + 1) + \frac{1}{D_s^4}(\lambda_s \frac{l_s}{D_s} + \zeta_s - 1)]\} + h_1 Q + [h_0 - (z_d - z_s - \frac{(p_d - p_s)}{g\rho})] = 0 \tag{5}$$

The 2nd-degree equation for the volumetric flow rate in Equation (5) is written in the following manner:

$$t_2 Q^2 + t_1 Q + t_0 = 0 \tag{6}$$

for which the following terms have been used:

$$t_0 = h_0 - \Delta z - \frac{(p_d - p_s)}{g\rho}$$
$$t_1 = h_1 \tag{7}$$
$$t_2 = h_2 - \frac{8}{g\pi^2}[\frac{1}{D_d^4}(\lambda_d \frac{l_d}{D_d} + \zeta_d + 1) + \frac{1}{D_s^4}(\lambda_s \frac{l_s}{D_s} + \zeta_s - 1)]$$

where Δz is defined by Equation (8):

$$\Delta z = z_d - z_s \tag{8}$$

The solutions of the 2nd-degree polynomial equation presented in Equation (6) for flow rate are:

$$Q_{1,2} = \frac{-t_1 \pm \sqrt{{t_1}^2 - 4t_2 t_0}}{2t_2} \tag{9}$$

Of the two solutions of Equation (6), the one with the positive value shall be retained ($Q_1 > 0$).

Once the Q_1 value is known, the coefficients h_2, h_1 and h_0 for the $H(Q)$ curve in Equation (4), the p_3, p_2, p_1 and p_0 coefficients for the $P_m(Q)$ curve in Equation (10), respectively, e_2, e_1 and e_0 coefficients for $\eta_P(Q)$ curve in Equation (11) are used for calculating the pumping head value $H(Q_1)$, the mechanical power value $P_m(Q_1)$ and the pump efficiency $\eta_P(Q_1)$. These coefficients are determined by a fitting procedure, applied to the available data.

$$P_m(Q) = p_3 Q^3 + p_2 Q^2 + p_1 Q + p_0 \tag{10}$$

$$\eta_P(Q) = e_2 Q^2 + e_1 Q + e_0 \tag{11}$$

The value of the volumetric flow rate corresponding to the BEP is obtained from Equation (12):

$$Q(\eta_{P_{BEP}}) = -\frac{e_1}{2e_2} \tag{12}$$

The value is positive for all the cases $Q(\eta_{P_{BEP}}) > 0$ because the coefficient $e_2 < 0$ is associated with the efficiency curve of η_P pump, which is a vertex-up parabola. The $\eta_{P_{BEP}}$ value is obtained by introducing the value $Q(\eta_{P_{BEP}})$ in Equation (11). This value of the maximum efficiency $\eta_{P_{BEP}}$ corresponding to each constant speed pump is used as a reference value to define the pump's operating regime. We notice that the maximum efficiency value $\eta_{P_{BEP}}$ of the BEP is determined from the input values and is different for each pump.

The pump's operating regime is determined by comparing the efficiency value η_P of the pump's duty point with the reference value corresponding to the pump's maximum efficiency $\eta_{P_{BEP}}$. Three operating regimes are defined for a hydraulic pump using color vision deficiency codes, containing both text and color flag information as follows:

- Normal operation (NO) from an efficiency point of view (GREEN):

$$0.9\eta_{P_{BEP}} < \eta_P < 1.05\eta_{P_{BEP}}$$

  Normal operation (NO) means the pump is operating at the proper efficiency. That is, the costs of energy consumption for pumping are justified. The operation of the pump in this range will be marked with GREEN ;
- Operation at the limit (LO), from an efficiency point of view: needs scheduled maintenance (YELLOW):

$$0.8\eta_{P_{BEP}} < \eta_P < 0.9\eta_{P_{BEP}} \text{ or}$$
$$1.05\eta_{P_{BEP}} < \eta_P < 1.1\eta_{P_{BEP}}$$

  A pump operating at the limit (LO) suggests the pump is operating with an acceptable efficiency but requires more power than is ideal. A pump operating under these conditions requires scheduled maintenance to identify and address any issues that may have occurred. The operation of the pump under these conditions will be signaled using the color YELLOW ;
- Abnormal operation (AO) from an efficiency point of view: needs urgent maintenance (RED):

$$\eta_P < 0.8\eta_{P_{BEP}} \text{or} \eta_P > 1.1\eta_{P_{BEP}}.$$

  A pump operating under abnormal operation (AO) is pumping with low efficiency and requires additional power to pump. A pump operating under these conditions requires urgent maintenance to identify and address any issues that may have occurred. The operation of the pump under these conditions will be signaled using the color RED .

The green stripe corresponds to Preferred Operating Region (POR) in our selection, while the extreme limits of the yellow area define the Allowable Operating Region (AOR) [4]. To be more explicit, the allowable operating region (AOR) is the operating zone provided by the manufacturer and it includes the yellow zones together with the green zone in our selection.

The reference values predefined by $\eta_{P_{BEP}}$ to circumscribe the fields of operation can be selected by each user based on the pump's operating conditions and on the recommendations issued by its manufacturer. The algorithm allows us to define these reference values for each pump by introducing them into their set of input values. Thus the predefined reference levels can be customized from one pump to another, depending on the required operating conditions and the specific in situ conditions [27] ch. 3.

The boundaries of the operating regions (e.g., POR, AOR) in our algorithm are selected based on the pump efficiency boundaries. The limits of the operating regions defined in other references are related to the volumetric flow rate limits (e.g., range from −30% Q_{BEP} to +15% Q_{BEP} [28,29] range from −10% Q_{BEP} to +10% Q_{BEP} [4] and range from −20% Q_{BEP} to +20% Q_{BEP} reference [30]. We consider that this concept of defining operating region boundaries using pump efficiency is better suited when the pump operation strategy is targeted, rather than defining operating region boundaries by selecting volumetric flow rate.

3. Implementation of an Algorithm to Assess Pump Efficiency

Free software packages that provide portability on any operating system have been selected to implement the software solution. The GNU GSL scientific software library provides robust interpolation, extrapolation and curve-fitting instruments and assessment of the approximation errors [31].

The free package Qt [32] was used to develop the graphic interface. The algorithm's implementation is structured on independent components, materialized by individual processes and tools in the command line, separated from the user interface.

Figure 1 shows the data flow diagram of the software solution.

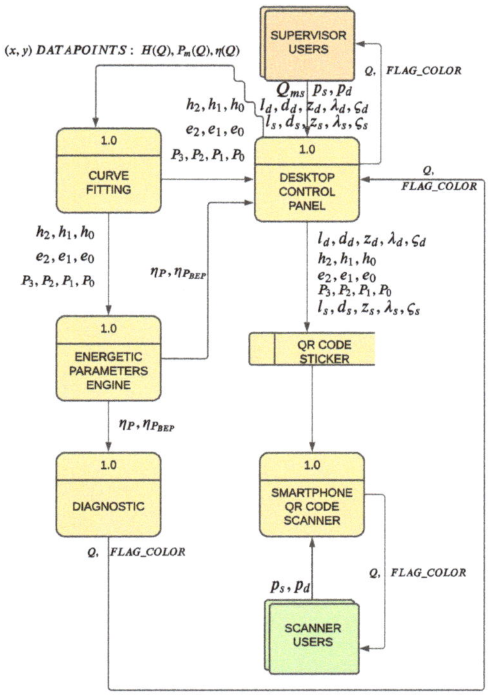

Figure 1. Diagram of the data flow in the software system.

A series of data are read from the additions $H(Q)$, $\eta_P(Q)$ and $P_m(Q)$, made available from the manufacturer's catalog or experimental data. Based on these, the fitting coefficients h_2, h_1, and h_0 are determined for the approximation of the $H(Q)$ curve, which takes its shape from Equation (4). The fitting coefficients e_2, e_1, and e_0 are determined for the approximation of the $\eta(Q)$ curve, which takes its shape from Equation (11), and the fitting coefficients p_3, p_2, p_1 and p_0 are determined for the approximation of the $P_m(Q)$ curve, which takes its shape from Equation (10).

Using these coefficients, the current efficiency value η_P and the maximum efficiency value of the BEP ($\eta_{P_{BEP}}$) are determined, located on the vertex of the efficiency parabola.

The desktop application generates and prints the QR code with the parameters of the hydraulic pathway, l_s, d_s, λ_s, ζ_s l_d, d_d, λ_d, ζ_d. The static pressures measured during suction p_s and discharge p_d lines are input by the user in both smartphone and desktop applications, optionally accompanied by the volumetric flow rate running through the pump Q_{ms}.

The mobile phone application scans the coefficients of the hydraulic pathway from the printed QR code and, based on the measured values of the static suction pressure p_s and those of the discharge p_d computes the efficiency values η_P and $\eta_{P_{BEP}}$ based on which they can diagnose the operating regime of the pump and return the red, yellow or green code using *FLAG_COLOR*.

Figure 2 shows the diagram of the efficiency assessment algorithm for constant speed hydraulic pumps—Algorithm 1.

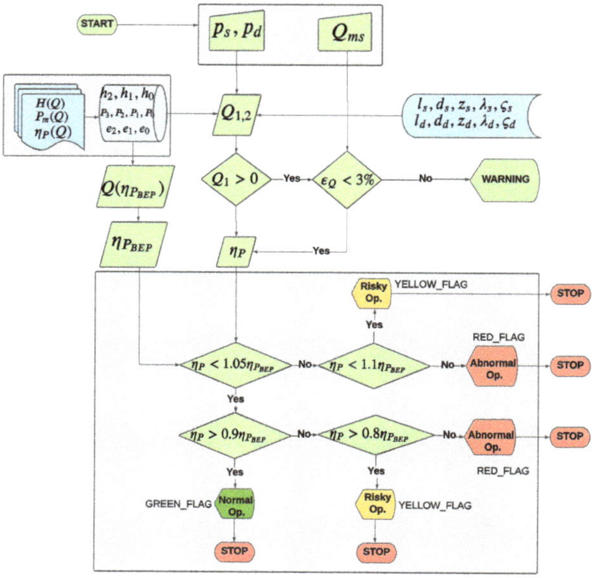

Figure 2. Diagram of the assessment algorithm for hydraulic pump efficiency while operating at constant speed.

Algorithm 1 The assessment algorithm for hydraulic pump efficiency while operating at constant speed

$HQ[] \leftarrow read(HQ)$ ▷ $H(Q)$ data points
$EQ[] \leftarrow read(EQ)$ ▷ $\eta(Q)$ data points
$PQ[] \leftarrow read(PQ)$ ▷ $P_m(Q)$ data points
$h[] \leftarrow fit(HQ)$ ▷ get 2nd order polynomial coefficients
$e[] \leftarrow fit(EQ)$ ▷ get 2nd order polynomial coefficients
$P[] \leftarrow fit(PQ)$ ▷ get 3rd order polynomial coefficients
procedure DIAGNOSTIC(p_s, p_d, Q_{ms})
    $\eta_P, \eta_{P_{BEP}} \leftarrow \eta(h, e, P)$
    if ($\eta_P < 1.05\eta_{P_{BEP}}$) **then**
        if ($\eta_P > 0.9\eta_{P_{BEP}}$) **then**
            $FLAG\_COLOR \leftarrow GREEN$
        else if ($\eta_P > 0.8\eta_{P_{BEP}}$) **then**
            $FLAG\_COLOR \leftarrow YELLOW$
        else
            $FLAG\_COLOR \leftarrow RED$
        end if
    else if ($\eta_P < 1.1\eta_{P_{BEP}}$) **then**
        $FLAG\_COLOR \leftarrow YELLOW$
    else
        $FLAG\_COLOR \leftarrow RED$
    end if
end procedure

Using the reading and data fitting component in Figure 2 for a finite set of data received as input for $H(Q)$, $P_m(Q)$, respectively, $\eta_P(Q)$, the best 2nd- and 3rd-order polynomial functions are obtained and, automatically, the values for these curves' coefficients, with a minimal square error:

$$\chi^2 = (\sum_i^{w_i} y_i - \sum_j^{x_{ij}} c_j)^2 = ||y - X_c||_{w^2} \quad (13)$$

where y is the input data vector of order n, X is a $n\,x\,p$ matrix having predictor variables, and c is the vector with those p fitting coefficients that must be estimated. The matrix $w = diag(w_1 w_2 \ldots w_n)$ contains the weights of the observed vector. The χ^2 square error also takes the form in Equation (17) where $w_i = (y_i - f(x_i))^{-2}$.

The determination of the values of the polynomial curve coefficients has been performed using the GSL Scientific Software Library [31] available as free software and tested in numerous scientific and engineering applications on the market.

The measured volumetric flow rate passing through the pump, Q_{ms}, is used to verify the percentage error ε_Q compared to the value estimated by the algorithm. The user receives a warning if the error exceeds the 3% threshold. The flowmeters used by the water supply company have an accuracy limit of $\pm 3\%$.

Figure 3 shows a diagram of the application's use cases. The operator uses a mobile phone to enter the suction and discharge pressure values, p_s and p_d, and then scans the QR barcode with the parameters configured and printed in the desktop application. Optionally, this software application can also be used to assess the percentage error ε_Q if the measured volumetric flow rate value Q_{ms} is available as input data; see Figure 2.

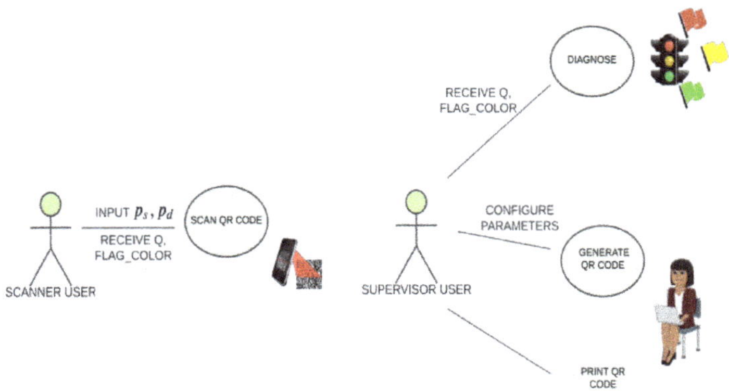

Figure 3. Use case diagram for the software solution.

After reading the barcode that identifies each pump, containing its hydraulic configuration, and after the input of the two read data, p_s and p_d, the duty point of the hydraulic pump is determined. Then, the pump efficiency of the duty point is compared with the BEP value. The hydraulic pump operates normally if the duty point falls in the green region. The pump operation is defective if the duty point falls into the other regions, and the operator must look for the causes and determine the appropriate intervention.

The R^2 parameter will be used to approximate the set of points with the selected polynomial function, defined as follows:

$$R^2 = 1 - \frac{\chi^2}{TSS} \quad (14)$$

where

$$TSS = \sum_i^n (x_i - \bar{x})^2 \quad (15)$$

is the total sum of the squares of the variations compared to the average value

$$\bar{x} = \frac{1}{n}\sum_{i}^{n} ix_i \tag{16}$$

and

$$\chi^2 = \sum_{i}^{n}(y_i - f(x_i))^2 \tag{17}$$

is determined with the approximation function for those n experimental read data.

The approximation of the set of points n with the selected polynomial function is more precise when the R^2 parameter is closer to 1. Moreover, $R^2 < 1$ as long as $n + 1 < N$ where n is the approximation polynomial's order, and N is the total number of read data values in the input set.

4. The Laboratory Validation of the Software Module for the Pump's Full Operating Range

The algorithm provides the values of the computed variables Q, H, P_m and η_P of the pump's operating point. Moreover, the following values are found as output data: $Q(\eta_{P_{BEP}})$, $H(\eta_{P_{BEP}})$ and $P(\eta_{P_{BEP}})$, and the 10 coefficients of the polynomial functions that approximate the input data for the pump: three coefficients (h_2, h_1 and h_0) for the $H(Q)$ curve, four coefficients (p_3, p_2, p_1 and p_0) for the $P_m(Q)$ curve, three coefficients (e_2, e_1 and e_0) for the $\eta_P(Q)$ curve, and, finally, those three R^2 coefficients that allow for accurate estimation of the read data. In the end, the user is also provided with the regime and the color of the regime where the pump's operating point is found. Additionally, the relative percentage error of the Q flow rate is presented, computed in relation to the measured flow rate value Q_{ms} if this variable has been included in the input data set. If the value of Q_{ms} was not mentioned in the input data set, then the relative percentage error of the flow rate will be missing.

We have used experimental data obtained for centrifugal pump PCN 65/200 at speed of $n = 2900$ rpm to validate the results supplied by the software. Preliminary validation of the software was performed at a speed of $n = 2500$ rpm. The data measured on the test rig available at the Hydraulic Machinery Laboratory of the "Politehnica" University Timişoara, presented in Figure 4, have been obtained using the IEC60193 methodology [33].

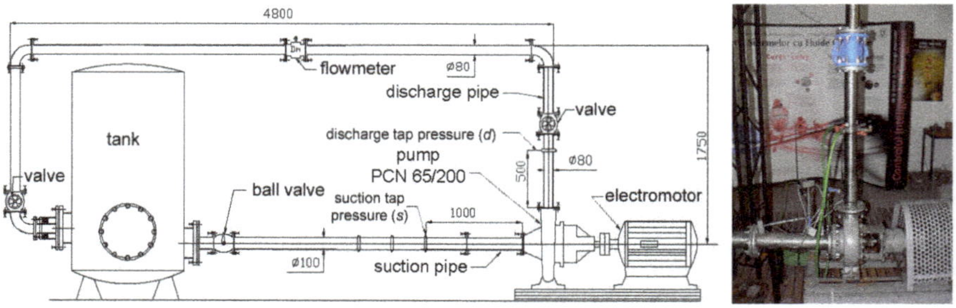

Figure 4. The test rig available at the Hydraulic Machinery Laboratory of the "Politehnica" University Timişoara. Schematic view of the test rig with actual dimensions in mm and photo.

A 22 kW asynchronous electrical motor is installed on the test rig to actuate the PCN 65/200 centrifugal pump. An ACS 850 45 kW Direct Torque Control (DTC) [34] inverter is used to vary the speed of the electrical motor from 500 rpm up to 3000 rpm [35]. A software platform completely controls the test rig. Firstly, an acquisition system was implemented to acquire sensor data for overall pressure, temperature, discharge, torque, speed and electrical power. The acquisition system with 32 input channels (voltage/current differential inputs) and a maximum 100 kb/s acquisition frequency was developed. The data is transferred to a computer using an RS232 interface. A remote control system was implemented, increasing the operability of the test rig [36]. Next, a SCADA platform was implemented to acquire the needed variables (suction and discharge pressures, temperature, discharge, speed and torque) and store them in a log file.

The total uncertainty (f_t) is obtained by combining the uncertainties due to systematic (f_s) and random (f_r) errors. The systematic and random errors are evaluated taking into account the measuring system and the operating conditions of the pump (18).

$$f_t = \sqrt{(f_s)^2 + (f_r)^2} \tag{18}$$

The first step in the estimation of the pump efficiency uncertainty is to identify each component that can influence its value. As a result, the total uncertainty of the pump efficiency ($f_{t_{\eta_P}}$) of the discharge (f_{tQ}), suction pressure ($f_{t_{p_s}}$), discharge pressure ($f_{t_{p_d}}$), speed (f_{t_n}) and torque (f_{tT}) (19).

$$f_{t_{\eta_P}} = \sqrt{(f_{tQ})^2 + (f_{t_{p_s}})^2 + (f_{t_{p_d}})^2 + (f_{t_n})^2 + (f_{tT})^2} \tag{19}$$

The systematic errors are provided by the measuring devices. An electromagnetic flowmeter is used to measure discharge values up to 45×10^{-3} m^3/s with ±0.4% accuracy. This device is installed on the rig's top pipe. The suction pressure sensor range is –1 ÷ +2.5 bars with an accuracy of ±0.25%. The discharge pressure sensor range is 0 ÷ 6 bar with accuracy reported by the manufacturer of ±0.25%. A T22 torque transducer manufactured by HBM is installed on the test rig between the electrical motor and the centrifugal pump. The range from 0 to 100 Nm is covered by the torque transducer with an accuracy of ±0.5%. ROP520 incremental encoder is linked to the electrical motor shaft to measure the speed with an accuracy of ±0.01%. A systematic error of ±0.732% is obtained for the pump efficiency ($f_{s_{\eta_P}}$).

The random error is determined for the quantity acquired by each measuring device installed on the test rig using a set of ten values for ten operating points. As a result, the following maximum random errors are obtained for the measured quantities: the suction pressure of ±0.616%, the discharge pressure of ±0.658%, the discharge of ±0.18%, the speed of ±0.471% and the torque of ±0.387%. As a result, a random error value of ±1.103% is determined for the pump efficiency ($f_{r_{\eta_P}}$). Then, the total uncertainty of the pump efficiency ($f_{t_{\eta_P}}$) by ±1.323% is obtained on the test rig.

The ACS850 45 kW (DTC) inverter [34] is also used to acquire the electrical power, the mechanical power and the speed on the test rig. In this case, the total uncertainty for the mechanical power is determined using the accuracy of ±4% with nominal torque in an open loop and the speed control in a closed loop with an accuracy of 0.01% from the nominal speed. Then, the systematic error ($f_{s_{\eta_P}}$) of ±4.035% is obtained and the total uncertainty of ±4.183% for the pump efficiency ($f_{t_{\eta_P}}$). In conclusion, the total uncertainty of the pump efficiency is 3 times smaller if the torque transducer is installed on our test rig.

The software module was verified for the operation of the pump across its full operating range, covering all three operating regimes shown in Figure 5.

The curves $H = f(Q)$, $P_m = f(Q)$, and $\eta_P = f(Q)$ for the centrifugal pump PCN 65/200 at speed of $n = 2900$ rpm are shown in Figure 6. The experimental data are marked with dots, whilst the polynomial functions approximating them are plotted as continuous lines. These figures show a proper correlation between the experimental data and the approximation curves. The value of the R^2 parameter is shown for each curve, quantifying the accuracy of the approximation degree of the experimental data with the polynomial function selected in the software module.

Table 1 contains experimental data for centrifugal pump PCN 65/200. Tables 2–4 show the fitting coefficients for Equations (4), (10) and (11).

Table 1. The experimental data determined for centrifugal pump PCN 65/200 at speed of 2900 rpm.

OPno.	$Q_{ms} \times 10^3$ [m³/s]	H_{ms} [m]	$P_{m_{ms}}$ [kW]	$\eta_{P_{ms}}$ [%]
OP7	6.727	49.678	11.815	27.737
OP8	12.289	48.861	13.497	43.629
OP9	18.640	47.295	15.816	54.662
OP10	24.232	45.104	17.109	62.645
OP11	29.604	42.387	18.145	67.818
OP12	33.668	38.439	18.827	67.411
OP13	37.213	36.266	18.888	70.070
OP14	40.037	33.721	19.110	69.282
OP15	42.560	31.468	18.939	69.348
OP16	44.617	28.774	18.781	67.036

Table 2. The coefficients of the 2nd-degree approximation polynomial function for the pumping head $H = f(Q)$ at speed of 2900 rpm.

n [rpm]	h_0	h_1	h_2	R^2
2900	49.859	105.330	−12,759.798	0.99791

Table 3. The coefficients of the 3rd-degree approximation polynomial function for the pump mechanical power $P_m = f(Q)$ at speed of 2900 rpm.

n [rpm]	p_0	p_1	p_2	p_3	R^2
2900	3.554	881.109	−13,978.015	40,315.701	0.96804

Table 4. The coefficients of the 2nd-degree approximation polynomial function for the pump efficiency $\eta_P = f(Q)$ at speed of 2900 rpm.

n [rpm]	e_0	e_1	e_2	R^2
2900	1.911	3834.803	−53,651.835	0.99599

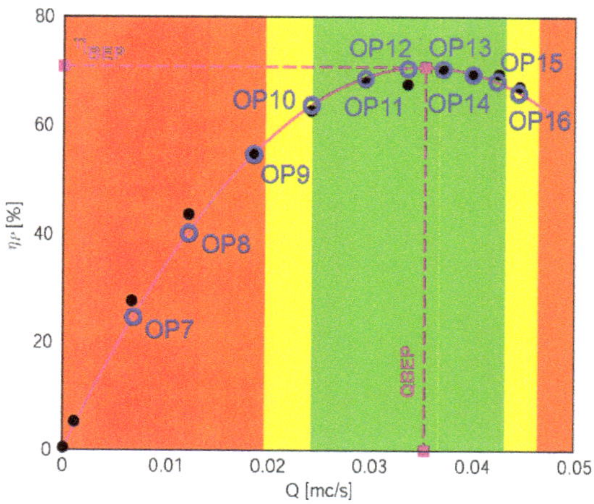

Figure 5. The validation of the software module for the pump's full operating range.

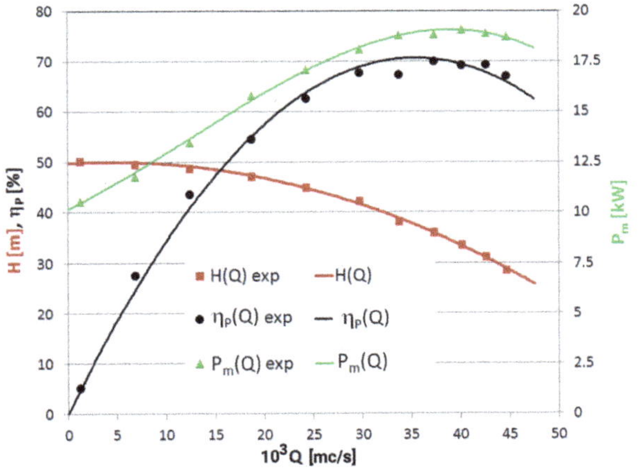

Figure 6. The experimental data for pumping head $H = f(Q)$ (■), mechanical power at the pump shaft $P_m = f(Q)$ (▲) and pump efficiency $\eta_P = f(Q)$ (•) at speed of 2900 rpm, together with polynomial curve fittings (solid lines).

The input parameters corresponding to the configuration of the test rig available at the "Politehnica" University Timişoara are as follows: the diameter of the suction line $D_s = 0.11$ m, the diameter of the discharge line $D_d = 0.08$ m, the position of the instrument on the suction line compared to the pump flange $l_s = 1.0$ m, the position of the instrument on the discharge line compared to the pump flange $l_d = 0.5$ m, the distributed hydraulic loss coefficient for the suction line $\lambda_s = 0.0158835$, the distributed hydraulic loss coefficient for the discharge line $\lambda_d = 0.0166889$, the local hydraulic loss coefficients for the suction and discharge lines $\zeta_s = \zeta_d = 0$ because no elbow was installed in the pump suction or discharge, the level difference between the position of the instrument on the discharge line and on the suction line $\Delta_z = 0.85$ m.

The last two columns in Table 5 ε_Q and ε_{η_P} are defined as $\varepsilon_\otimes$ in Equation (20):

$$\varepsilon_\otimes = \frac{\otimes_{ms} - \otimes}{\otimes_{ms}} 100[\%] \tag{20}$$

(corresponding to $y_i - f(x_i)$ in Equation (17)). $\eta_{P_{ms}}$ is the experimental data from Table 1, and η_P is the value calculated with curve fitting in Equation (11) with coefficients from Table 4. Q_{ms} is experimental read data in Table 1 and Q is calculated based on Equation (9).

The $\eta_{P_{ms}}$ experimental data in Table 1 are shown in Figures 5 and 7 with a black dot (•) and the calculated η_P values are shown with a white circle (○).

The input variables of each operating point are the static pressure measured at suction p_s [Pa], static pressure measured at discharge p_d [Pa] and optionally, the value of the measured volumetric flow rate Q_{ms} [m^3/s]. The following data are obtained for the pump speed of $n = 2900$ rpm: $H(\eta_{P_{BEP}}) = 37.33$ m, $P_m(\eta_{P_{BEP}}) = 17.19$ kW, $Q(\eta_{P_{BEP}}) = 35.7 \times 10^{-3}$ m^3/s.

See Figure 7a for operating point number 7 ($OP7$) set on partial flow rate during abnormal regime (AO). Our input values were: static pressure measured at suction $p_s = -9933.191$ Pa, static pressure measured at discharge p_d =47,0631.463 Pa and the value of the volumetric flow rate measured by the flowmeter $Q_{ms} = 6.727 \times 10^{-3}$ m^3/s. The results provided by the software module for OP7 are: $Q = 6.35 \times 10^{-3}$ m^3/s, with a percentage error of -5.6% compared to $Q_{ms} = 6.727 \times 10^{-3}$ m^3/s, $\eta_P = 24.10\,\%$, $H = 50.01$ m, $P_m = 8.6$ kW. The abnormal operation diagnostic RED supplied is correctly determined.

Operating point number 9 (OP9) set on partial flow rate during abnormal regime (AO) (see Figure 7b) was checked with the following input values: static pressure measured at suction $p_s = -17{,}270.447$ Pa, static pressure measured at discharge $p_d = 435{,}108.521$ Pa and the value of the volumetric flow rate measured by the flowmeter $Q_{ms} = 18.64 \times 10^{-3}\,\mathrm{m}^3/\mathrm{s}$. The results provided by the software module for OP9 are: $Q = 17.968 \times 10^{-3}\,\mathrm{m}^3/\mathrm{s}$, with a percentage error of -3.61% compared to $Q_{ms} = 18.64 \times 10^{-3}\,\mathrm{m}^3/\mathrm{s}$, $\eta_P = 53.49\,\%$, $H = 47.63\,\mathrm{m}$, $P_m = 15.1\,\mathrm{kW}$. The negative value of the percentage error shows that the flow rate value, determined by the software module, is lower than the value measured by the flowmeter. The abnormal operation diagnostic (AO) RED supplied is correctly determined.

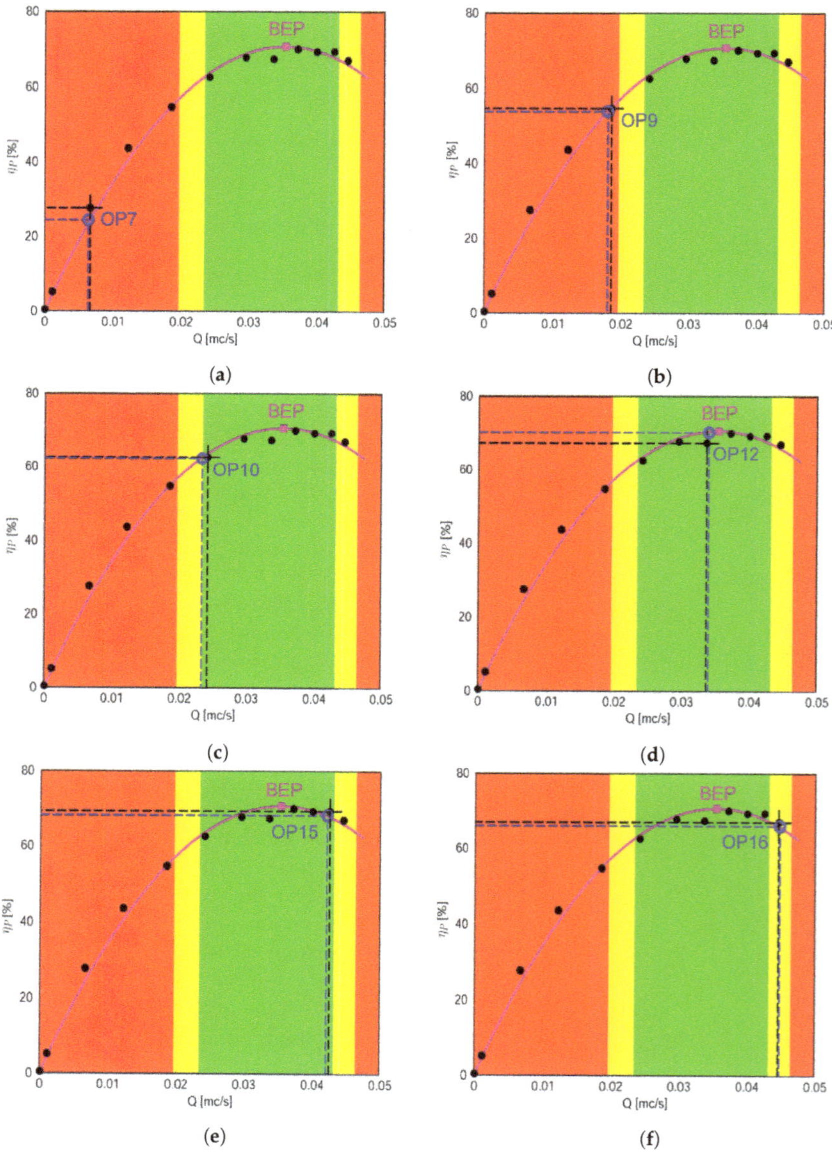

Figure 7. The validation of the algorithm for six operating points: (**a**) *OP7*: Q = $6.35\times10^{-3}\,\mathrm{m}^3/\mathrm{s}$, $\eta_P = 24.10\,\%$ (AO) RED; (**b**) *OP9*: Q = $17.968\times10^{-3}\,\mathrm{m}^3/\mathrm{s}$, $\eta_P = 53.49\,\%$ (AO) RED; (**c**) *OP10*: Q = $23.21\times10^{-3}\,\mathrm{m}^3/\mathrm{s}$, $\eta_P = 62.02\,\%$ (NO) GREEN (**d**) *OP12*: Q = $37.33\times10^{-3}\,\mathrm{m}^3/\mathrm{s}$, $\eta_P = 70.21\,\%$ (NO) GREEN (**e**) *OP15*: Q = $41.86\times10^{-3}\,\mathrm{m}^3/\mathrm{s}$, $\eta_P = 68.43\,\%$ (NO) GREEN (**f**) *OP16*: Q = $44.463\times10^{-3}\,\mathrm{m}^3/\mathrm{s}$, $\eta_P = 66.35\,\%$; (LO) YELLOW.

Table 5. The data computed by the software module for centrifugal pump PCN 65/200 at speed of 2900 rpm.

OPno.	$Q_{ms} \times 10^3$ [m^3/s]	H [m]	P_m [kW]	η_P [%]	ε_Q [%]	ε_{η_P} [%]
OP7	6.350	50.013	8.595	24.099	−5.602	13.114
OP8	12.321	49.220	12.363	41.014	0.257	5.993
OP9	17.968	47.632	15.106	53.494	−3.605	2.137
OP10	23.216	45.427	16.980	62.022	−4.191	0.994
OP11	29.691	41.738	18.447	68.473	0.295	−0.966
OP12	33.707	38.913	18.916	70.213	0.115	−4.157
OP13	37.792	35.616	19.065	70.208	1.555	−0.197
OP14	40.651	33.055	18.981	69.139	−1.647	0.206
OP15	41.859	31.911	18.901	68.425	−0.346	1.331
OP16	44.463	29.316	18.640	66.350	−0.346	1.023
BEP	35.738	37.327	17.190	70.435	—	—

The input variables for operating point number 10 (OP10) set on partial flow rate during normal regime (NO) (see Figure 7c) are: static pressure measured at suction $p_s = -18{,}530.6$ Pa, static pressure measured at discharge $p_d = 408{,}535.72$ Pa and the value of the volumetric flow rate measured by the flowmeter $Q_{ms} = 24.23 \times 10^{-3}\,\text{m}^3/\text{s}$. The results provided by the software module for OP10 are: $Q = 23.21 \times 10^{-3}\,\text{m}^3/\text{s}$, with a percentage error of −4.192% compared to $Q_{ms} = 24.23 \times 10^{-3}\,\text{m}^3/\text{s}$, $\eta_P = 62.02\,\%$, $H = 45.427\,\text{m}$, $P_m = 16.98\,\text{kW}$. The normal operation diagnostic (NO) GREEN supplied is correctly determined.

The validation for the operating point number 12 (OP12) set in the vicinity of the maximum efficiency value during normal operation (NO) (see Figure 7d) shows the following input variables: static pressure measured at suction $p_s = -17{,}665.65$ Pa, static pressure measured at discharge $p_d = 335{,}325.2$ Pa and the value of the volumetric flow rate measured by the flowmeter $Q_{ms} = 33.67 \times 10^{-3}\,\text{m}^3/\text{s}$. The results provided by the software module for OP12 are: $Q = 37.33 \times 10^{-3}\,\text{m}^3/\text{s}$, with a percentage error of +0.113% compared to $Q_{ms} = 33.67 \times 10^{-3}\,\text{m}^3/\text{s}$, $\eta_P = 70.21\,\%$, $H = 38.913\,\text{m}$, $P_m = 18.96\,\text{kW}$. The normal operation diagnostic (NO) GREEN supplied was predicted correctly.

The input variables for the operating point number 15 (OP15) set on overflow rate during normal regime (NO) (see Figure 7e) are: static pressure measured at suction $p_s = -27{,}807.4$ Pa, static pressure measured at discharge $p_d = 246{,}015.5$ Pa and the value of the volumetric flow rate measured by the flowmeter $10^3 Q_{ms} = 42.56\,\text{m}^3/\text{s}$. The results provided by the software module for OP15 are: $Q = 41.86 \times 10^{-3}\,\text{m}^3/\text{s}$, with a percentage error of −1.647% compared to $Q_{ms} = 42.56 \times 10^{-3}\,\text{m}^3/\text{s}$, $\eta_P = 68.43\,\%$, $H = 31.911\,\text{m}$, $P_m = 18.9\,\text{kW}$. The normal operation diagnostic (NO) GREEN supplied is correctly identified.

The input variables for the operating point number 16 (OP16) set on overflow rate during normal regime (NO) (see Figure 7f) are: static pressure measured at suction $p_s = -29{,}996.15$ Pa, static pressure measured at discharge $p_d = 214{,}548.83$ Pa and the value of the volumetric flow rate measured by the flowmeter $Q_{ms} = 44.617 \times 10^{-3}\,\text{m}^3/\text{s}$. The results provided by the software module for OP16 are: $Q = 44.463 \times 10^{-3}\,\text{m}^3/\text{s}$, with a percentage error of −0.345% compared to $Q_{ms} = 44.617 \times 10^{-3}\,\text{m}^3/\text{s}$, $\eta_P = 66.35\,\%$, $H = 29.316$ m, $P_m = 18.64$ kW. The operation at limit diagnostic (LO) YELLOW supplied is predicted correctly.

The analysis of the results supplied by the software module for the six operating points set on the full operating range of the pump shows the capacity of the software module to correctly identify the pump's operating regimes. The flow rate is a fundamental hydraulic quantity that must be determined while the pump is in service. Determining the flow rate in situ is a challenge and a requirement for the water treatment technological process. The relative error of the flow rate estimated by the software module compared to the value measured by the flowmeter fits within the limit of −5.6% for the operating points with a flow rate below 25% of the value of the operating point with maximum efficiency. In exchange, the limit of the relative error is between −4.192% and +0.113% for all the operating points of the pump, except those with a flow rate below 25% of the value associated with the operating point with maximum efficiency. Based on the validated results, we can conclude that the software module allows estimating the flow rate of the pump's operating point when there is no flowmeter installed (or where there is no possibility to install one). The estimation of the flow rate using the software module is an additional gain to the assessment of the operating regime of the pump.

4.1. Sensitivity Analysis of the Algorithm for Various Operating Points

In some situations, particularly in situ, the input parameters in the software module cannot be precisely determined (e.g., the inner diameter of the line or the coefficient of longitudinal losses). This is why we have performed the sensitivity analysis of the output value (pump efficiency) in relation to the input values using the algorithm parameters. The variables that can influence the pump efficiency are: (1) input values read by the operator p_s [Pa] and p_d [Pa] and (2) the parameters of the algorithm corresponding to the in situ configuration, which are scanned from the unique barcode generated for each pump: (i) diameter of the line at suction/discharge $D_{s/d}$ [m]; (ii) the discharge transducer's quota compared to the suction transducer's quota $\Delta z_{d/s}$ [m]; (iii) the length of the line from suction/discharge up to the location of the pressure manometers (transducers) compared to the pump flanges $l_{s/d}$ [m]; and (iv) the coefficient of longitudinal losses along the suction/discharge line $\lambda_{s/d}$ [-].

To gain a more relevant view of how these parameters influence the pump's efficiency, the percentage variation of the pump efficiency has been represented in relation to the percentage variation of each input value. Moreover, to assess the sensitivity of the algorithm parameters for the pump's full range of operation, we have considered five operating points (OP8, OP9, OP10, OP12, OP14) from the range of flow rates corresponding to the defined areas (NO) GREEN, (LO) YELLOW, and (AO) RED on the efficiency curve.

In addition to the parameters used in the algorithm, for each configuration of a pump unit, the algorithm needs the input of the static pressure p_s, p_d read/recorded at the pump suction/discharge for each operating point. These values of the static pressure at the pump suction/discharge boast a great degree of erroneous readings. The error can come from the reading of analogical measurement tools, which are not all that accurate, or the reading of the values in the first part of the instrument scale, where there is a lower degree of accuracy. Errors can also come from the oscillation of these measurement instruments used in the operation of the pumps during transient or unsteady operating conditions.

The sensitivity analysis of the algorithm regarding the determination of the percentage deviation in the pump efficiency ϵ_{η_P} for the percentage static pressure variation ϵ_{p_d} for five operating points: OP8, OP9, OP10, OP12 and OP14 is shown in Figure 8. The percentage deviation of the pump efficiency ϵ_{η_P} is determined using Equation (21) where $\eta_P(\epsilon_x)$ is the value of the efficiency, determined for the percentage deviation of the input quantities (e.g., $p_d, p_s, D_d, D_s, l_d, l_s, \lambda_d, \lambda_s, \Delta z$).

$$\epsilon_{\eta_P} = \frac{\eta_P(\epsilon_x) - \eta_P}{\eta_P} 100 \ [\%] \tag{21}$$

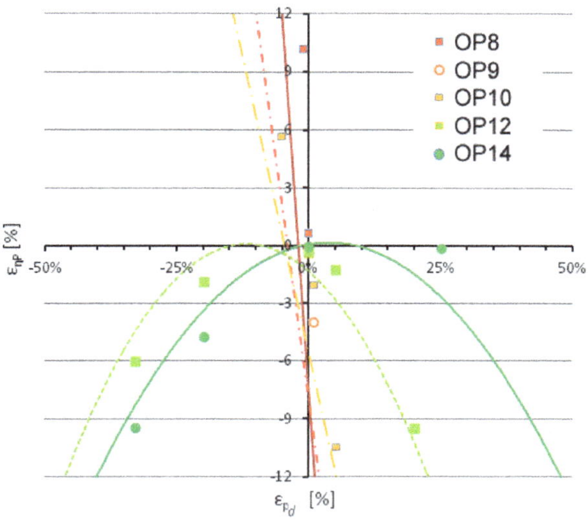

Figure 8. Sensitivity analysis of the algorithm regarding the determination of the pump efficiency $\epsilon_{\eta P}$ for deviations in the reading of the discharge pressure ϵ_{p_d} for five operating points: OP8, OP9, OP10, OP12 and OP14.

One can notice the significant influence of the static pressure variation read at the pump discharge ϵ_{p_d} on the pump efficiency. The operating points laid out in the vicinity of the maximum efficiency influence the pump efficiency within the limit $\epsilon_{\eta P} < \pm 12\%$ when the deviation of the static pressure determined at the pump discharge is read with a deviation within the limit $\epsilon_{p_d} < \pm 25\%$.

For the operating points laid out in the yellow and red zones, a slight deviation in the reading of the static pressure at the pump discharge $\epsilon_{p_d} > \pm 5\%$ leads to deviations in the values of the pump efficiency greater than $\epsilon_{\eta P} > \pm 12\%$. For greater positive deviations at the reading of the static pressure at the pump discharge, an error message is returned because the input of the pressure difference between the pump suction and discharge is greater than the pumping head prescribed from the pump's catalog curve.

Figure 9 shows the algorithm sensitivity analysis for the determination of pump efficiency for the percentage variation of static pressure ϵ_{p_s} measured at the pump suction for the five operating points selected across the full range. One can notice a linear distribution of the pump efficiency variation, with the percentage variation of the static pressure ϵ_{p_s} measured at the pump suction. As expected, the variation of the static pressure read at the pump suction ϵ_{p_s} has a significant influence over the pump efficiency variation. The operating points located in the green area are an exception; the influence on the pump efficiency is less $\epsilon_{\eta P} < \pm 1\%$ when the variation of the static pressure determined at pump suction is read with a deviation within the limit of $\epsilon_{p_s} < \pm 50\%$. Otherwise, for the operating points located in the yellow area, the pump variation can reach $\epsilon_{\eta P} < \pm 5\%$ when the variation of the static pressure taken at pump suction is read with a deviation within the limit of $\epsilon_{p_s} < \pm 50\%$.

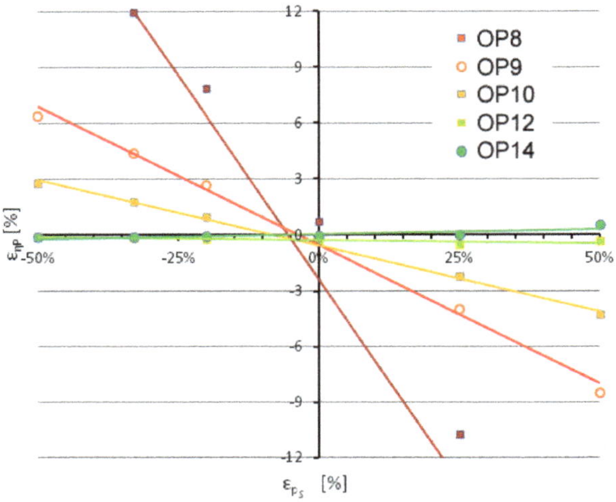

Figure 9. The algorithm sensitivity analysis regarding the determination of the pump efficiency $\epsilon_{\eta P}$ for deviations in the reading of the suction pressure ϵ_{p_s} for five operating points: OP8, OP9, OP10, OP12 and OP14.

The algorithm sensitivity analysis for determining the pump efficiency for the variation of the line inner diameter at discharge ϵ_{D_d} for the five operating points across the operating range is shown in Figure 10. One can observe a significant variation of the pump efficiency ($\epsilon_{\eta P} > \pm 3\%$) when we consider a deviation of the line inner diameter at discharge greater with $\epsilon_{D_d} > 25\%$ compared to its real value. The assessments for the two operating points located in the green area are an exception; the variation of the pump efficiency is below 3% ($\epsilon_{\eta P} < -3\%$) even for a deviation of the line inner diameter at discharge greater up to $\epsilon_{D_d} < 50\%$ compared to its real value. For the situations where the exact value of the discharge line inner diameter D_d cannot be determined, it is preferable to consider a value higher than the real one.

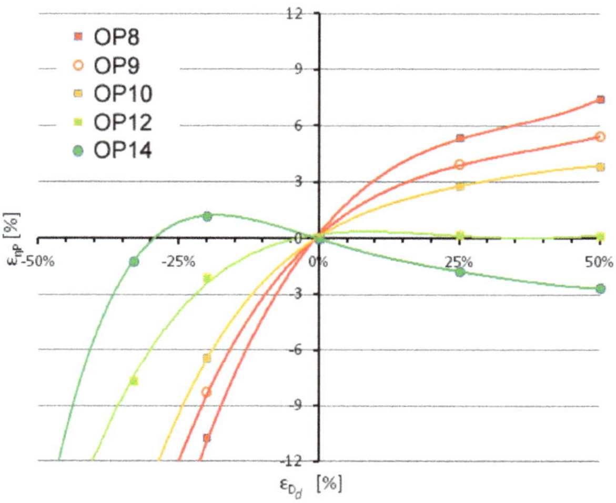

Figure 10. Algorithm sensitivity analysis regarding the determination of the pump efficiency $\epsilon_{\eta P}$ or deviations in the input of the discharge line diameter ϵ_{D_d} for five operating points: OP8, OP9, OP10, OP12 and OP14.

Figure 11 shows the algorithm sensitivity analysis regarding the determination of pump efficiency for the variation of the suction line inner diameter ϵ_{D_s} for the five operating points across the full range. In Figure 11, one can notice a variation of the pump efficiency of up to 3% lower for the five operating points when the deviation of the suction pipe inner diameter is greater by up to $\epsilon_{D_s} < 50\%$ compared to its real value. In exchange, one can notice a significant variation of the pump efficiency ($\epsilon_{\eta_P} > \pm 5\%$) when it is considered a deviation of the suction line inner diameter below 33%, compared to its real value. An exception regarding the deviation of the suction line inner diameter is the situation where the assessment of the output is performed for the BEP. For the situations where the exact value of the suction line inner diameter D_s cannot be determined, it is preferable to consider a value higher than the real one. A reasonable approximation would be the outer diameter. In most cases, the approximation falls within the efficiency deviation $\epsilon_{\eta_P} < -3\%$.

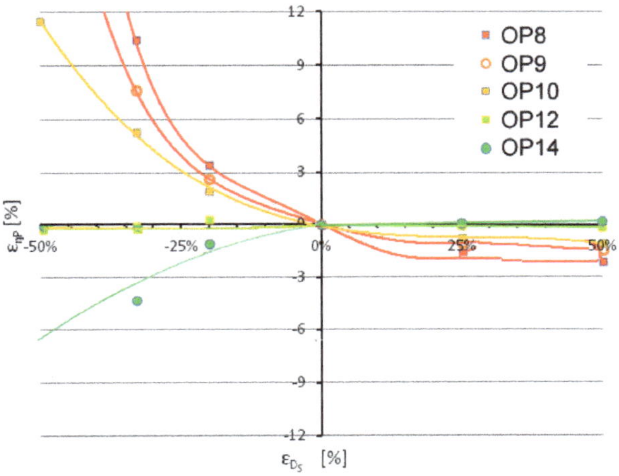

Figure 11. The algorithm sensitivity analysis for the determination of the pump efficiency ϵ_{η_P} for deviations in the input of the suction line diameter ϵ_{D_s} for the five operating points: OP8, OP9, OP10, OP12 and OP14.

If the suction and discharge diameters can be correctly measured on the outside, the inner values can be found in tables showing the thicknesses of the pipe wall. When establishing the inner diameters, it is preferable to consider a value that is greater than the real one. Another input parameter with an important influence on the efficiency algorithm is the variation of discharge pressure, Figure 8. The greatest deviations in the efficiency values are obtained at off-design conditions far away from the BEP corresponding to the maximum efficiency $\eta_{P_{BEP}}$.

The algorithm sensitivity analysis regarding the determination of the pump efficiency for the level variation between the two instruments that measure the suction/discharge pressure for the five operating points across the full range is shown in Figure 12.

One can notice a linear distribution of the pump efficiency variation with the deviation of the level $\epsilon_{\Delta z}$ defined in Equation (8) between the taps of the two instruments measuring the suction/discharge pressure compared to the reference value for the five operating points. As a result, a greater/smaller value of the pump efficiency is obtained when the deviation of the level $\epsilon_{\Delta z}$ is smaller or greater than the reference value. The pump efficiency value determined with the aid of the algorithm is within the limit of $\epsilon_{\eta_P} < \pm 1.5\%$ when the deviation of the level $\epsilon_{\Delta z}$ between the taps of the two instruments measuring the suction/discharge pressure deviate by $\epsilon_{\Delta z} < \pm 50\%$ compared to the reference value. The situations in which the operating points are laid out within the red area are an exception; the value of the pump efficiency determined with the algorithm is greater than $\epsilon_{\eta_P} > \pm 1.5\%$. The variation of the pump efficiency is more sensitive to the deviation of the inner diameters of the suction/discharge lines $\epsilon_{D_s}/\epsilon_{D_d}$ than to the level deviation $\epsilon_{\Delta z}$ between the taps of the two instruments measuring the suction/discharge pressure.

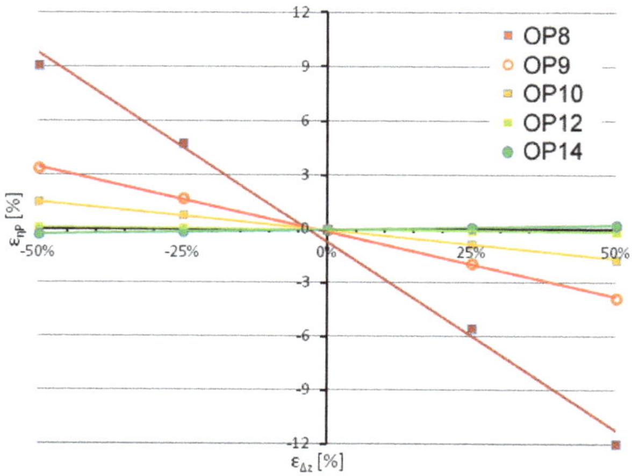

Figure 12. Algorithm sensitivity analysis regarding the determination of the pump efficiency ϵ_{η_P} for deviations in the input of the level difference $\epsilon_{\Delta z}$ between the locations of the transducers at suction and at discharge for the five operating points: OP8, OP9, OP10, OP12 and OP14.

Figure 13 shows the algorithm sensitivity analysis regarding the determination of pump efficiency for the variation of the layout position of the pressure tap length ϵ_{l_d} on the discharge line, compared to the pump's discharge flange for the five operating points across the full range. One would remark the fact that the variation of the layout position of the pressure tap length ϵ_{l_d} on the discharge line compared to the pump discharge flange has an insignificant influence on the pump efficiency variation $\epsilon_{\eta_P} < 1\%$ for all the operating points investigated.

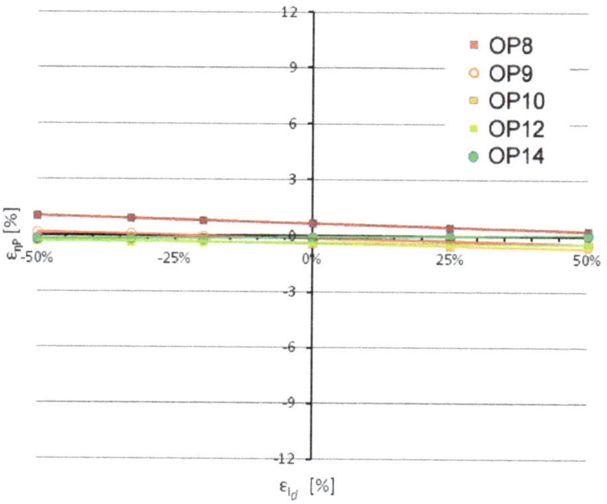

Figure 13. The algorithm sensitivity analysis regarding the determination of the pump efficiency ϵ_{η_P} for deviation in the length of the discharge line ϵ_{l_d} between the pressure transducer and the pump flange for the five operating points: OP8, OP9, OP10, OP12 and OP14.

Figure 14 shows the algorithm sensitivity analysis regarding the determination of the pump efficiency for the layout position of the pressure tap length ϵ_{l_s} on the suction line compared to the pump's suction flange for the five operating points across the entire range. Note the same behavior of the algorithm regarding the variation of the pump efficiency $\epsilon_{\eta P} < 1\%$ corresponding to the variation of the length ϵ_{l_s} for all the operating points assessed and for that of the length variation ϵ_{l_d}.

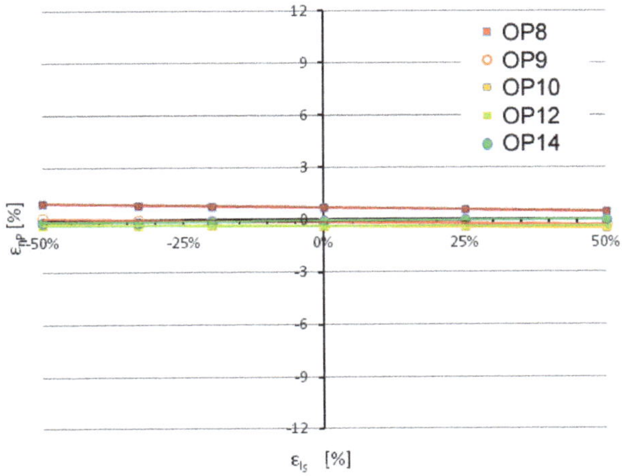

Figure 14. The algorithm sensitivity analysis regarding the determination of the pump efficiency $\epsilon_{\eta P}$ or deviations in the length of the suction line ϵ_{l_s} between the pressure transducer and the pump flange for the five operating points: OP8, OP9, OP10, OP12 and OP14.

We can conclude that the influence of the layout positions of the pressure tap lengths ϵ_{l_s} and ϵ_{l_d} on the suction and discharge lines compared to the pump's suction and discharge flanges over the pump efficiency variation $\epsilon_{\eta P} < 1\%$ is insignificant for the two operating points while functioning on a wider range.

Figures 15 and 16 show the algorithm sensitivity analysis regarding the determination of the pump efficiency for the variation of the hydraulic losses coefficient $\epsilon_{\lambda_d}/\epsilon_{\lambda_s}$ on the discharge/suction line, up to the layout position of the pressure tap, towards the pump's discharge/suction flange, for the five operating points across the entire operating range. The sensitivity of the algorithm regarding the determination of the pump efficiency for the variation of the hydraulic losses coefficient $\epsilon_{\lambda_d}/\epsilon_{\lambda_s}$ is identical to that produced by the length variation $\epsilon_{l_d}/\epsilon_{l_s}$. This situation is explained by the fact that the product of the two parameters $\epsilon_{\lambda_d}\epsilon_{l_d}/\epsilon_{\lambda_s}\epsilon_{l_s}$ are found in the equation of the pumping head. The influence of the hydraulic losses coefficient variation $\epsilon_{\lambda_d}/\epsilon_{\lambda_s}$ on the discharge/suction line up to the location of the pressure tap as opposed to the pump's discharge/suction flange over the pump efficiency variation $\epsilon_{\eta P} < 1\%$ is insignificant for the five operating points across the entire range.

The results of the sensitivity analysis performed on the parameters used by the algorithm for the determination of the pump efficiency, implemented in the software module presented in this paper are summarized below. The parameters used in the algorithm are listed in the order of their importance on the pump efficiency:

- p_d [Pa]—static pressure at pump discharge;
- p_s [Pa]—static pressure at pump suction;
- D_d [m]—line diameter at discharge;
- D_s [m]—line diameter at suction;
- Δ_z [m]—transducer's level from discharge compared to that from suction;
- $l_{s/d}$ [m]—length of the suction/discharge line up until the location of the pressure manometers (transducers) compared to the pump flanges;
- $\lambda_{s/d}$ [-]—longitudinal hydraulic losses coefficient along the suction/discharge line;

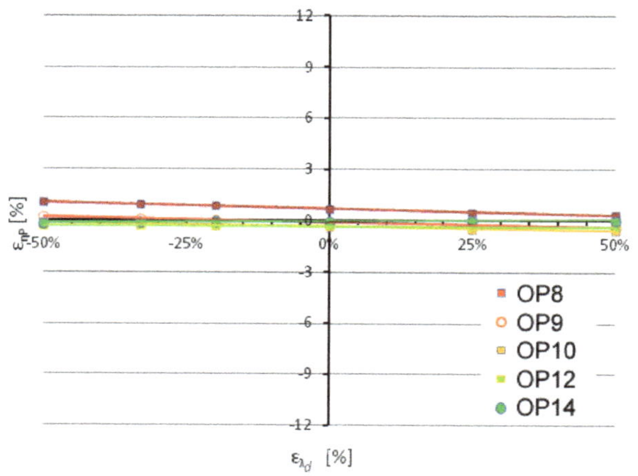

Figure 15. Algorithm sensitivity analysis regarding the determination of the pump efficiency $\epsilon_{\eta P}$ for deviations in the input of the longitudinal losses coefficient for the discharge line ϵ_{λ_d} for the five operating points: OP8, OP9, OP10, OP12 and OP14.

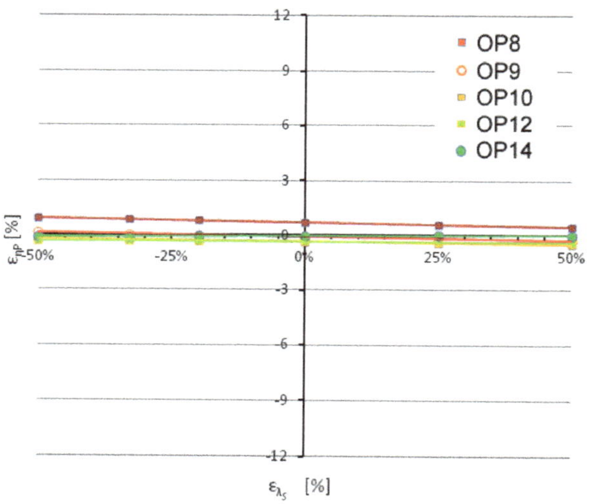

Figure 16. Algorithm sensitivity analysis regarding the determination of the pump efficiency $\epsilon_{\eta P}$ or deviations in the input of the longitudinal losses coefficient for the suction line ϵ_{λ_s} for the five operating points: OP8, OP9, OP10, OP12 and OP14.

In conclusion, the determination of the input data that are characteristic to the pump, and the input of the read data, respectively, p_s and p_d, must be performed as accurately as possible to have the lowest influence on the pump efficiency η_P. It is recommended that the data acquired by the pressure transducers are taken directly into the software module to avoid any reading and data input errors.

The line diameter at discharge D_d has a greater influence on the pump's efficiency than the line diameter at suction D_s. This situation is because the centrifugal pumps tend to reach a higher pumping head, which is observed in a greater static pressure at discharge.

5. Assessment of a Pump's In Situ Operating Conditions Using the Software Module

The software module has been implemented at AQUATIM S.A. to monitor the operation of the double suction Worthington 500 LNN-775A double suction pumps of 1 MW, that supply Timişoara city with drinkable water.

The experimental data determined at the speed of 993 rpm by the manufacturer of the 1 MW double suction pump, installed at AQUATIM S.A., is listed in Table 6 and is marked with black dots (•) in Figure 17. Application of the algorithm presented in Section 2 has allowed the determination of the approximation polynomials' coefficients, which are shown in Table 7 for $H = f(Q)$, Table 8 for $P_m = f(Q)$ and Table 9 for $\eta_P = f(Q)$ and marked with continuous lines in Figure 17 of the polynomial functions approximating these experimental data. For the speed of 993 rpm, the $R^2 = 0.996$ parameter has had a value close to 1, indicating a good degree of approximation of the set of points with the selected polynomial function.

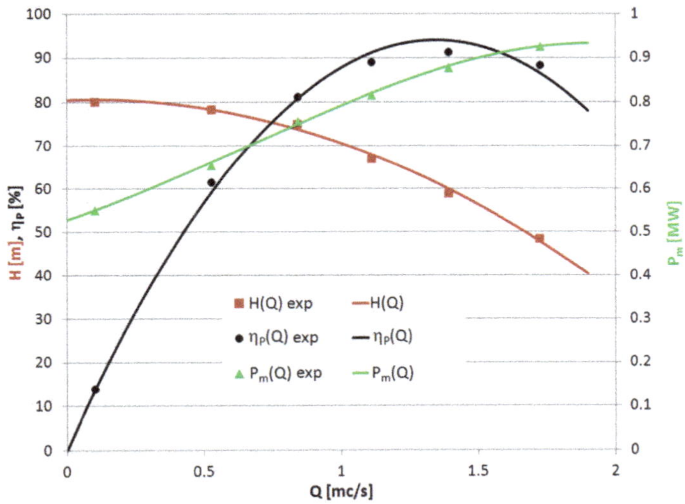

Figure 17. The experimental data provided by the manufacturer (■, ●, ▲) and the polynomial functions $H(Q)$, $P_m(Q)$ and $\eta_P(Q)$ determined for Worthington 500 LNN-775A double suction pump at speed of 993 rpm.

Table 6. The data for Worthington 500 LNN-775A double suction pump at speed of 993 rpm.

P no.	Q [m^3/s]	H [m]	P_m [MW]	η_P [%]
P1	0.0991	80.3	0.552	14.1
P2	0.5238	78.5	0.655	61.5
P3	0.8379	75.1	0.759	81.4
P4	1.1091	67.1	0.818	89.2
P5	1.3887	59.0	0.879	91.4
P6	1.7221	48.6	0.927	88.5
BEP	1.3498	61.189	0.874	93.804
DP	1.7191	48.071	0.926	86.860

Table 7. The coefficients of the 2nd-degree polynomial approximation function for the pumping head $H = f(Q)$ of the 1 MW Worthington pump at speed of 993 rpm.

n [rpm]	h_0	h_1	h_2	R^2
993	80.499	2.347	−12.338	0.995

Table 8. The coefficients of the 3rd-degree polynomial approximation function for the mechanical power $P_m = f(Q)$ of the 1 MW Worthington pump at speed of 993 rpm.

n [rpm]	p_0	p_1	p_2	p_3	R^2
993	0.528	0.209	0.113	−0.058	0.998

Table 9. The coefficients of the 2nd-degree polynomial approximation function for the efficiency $\eta_P = f(Q)$ of the 1 MW Worthington pump at speed of 993 rpm.

n [rpm]	e_0	e_1	e_2	R^2
993	1.056	137.427	−50.908	0.997

Figure 18 shows the curve of the pumping head ($H(Q)$) of the Worthington 500LNN-775A double suction pump of 1 MW, installed at AQUATIM S.A., at the speed of 993 rpm (continuous red line), determined based on the experimental data provided by the manufacturer. The curve for hydraulic losses ($H_r(Q)$) of the Timişoara city supply network (continuous blue line) is determined based on experimental data (♦ in Figure 18) obtained in situ. The duty point (DP, marked with a red circle (○) of this pump was determined to be at the intersection of the two curves $H(Q)$ and $H_r(Q)$.

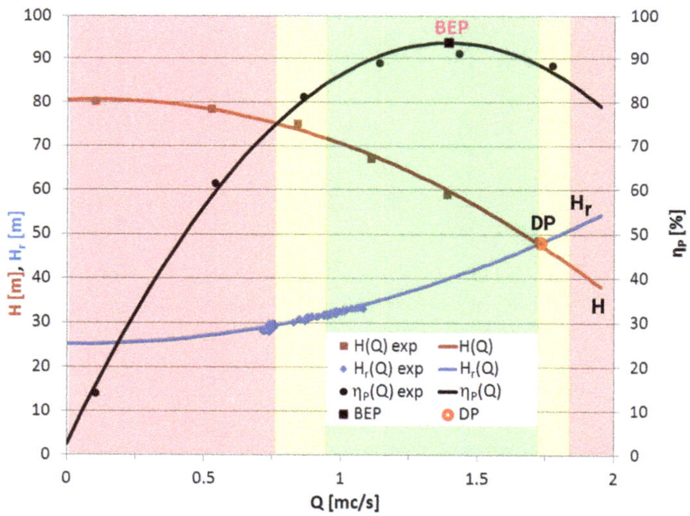

Figure 18. The duty point (DP) of a Worthington 500LNN-775A double suction pump of 1 MW at speed of 993 rpm operating in Timişoara drinking water network.

The input parameters for the Worthington 500 LNN-775A double suction pump of 1 MW, installed at AQUATIM S.A., read by the software module are: the suction line diameter $D_s = 0.6\,m$, and the discharge line diameter $D_d = 0.5$ m. The instruments are installed on the pump flanges, meaning that $l_s = l_d = 0$ m and the level difference between the location of the instruments installed on the discharge and suction flanges $\Delta z = 0.6$ m. The hydraulic loss coefficients, distributed for the suction lines λ_s and discharge lines λ_d, are not relevant in this case because the instruments are installed on the flanges, and the local hydraulic loss coefficients for the suction and discharge lines $\zeta_s = \zeta_d = 0$ because no elbow was installed in the pump suction or discharge. In this case, the pumping head is computed using the simplified formula presented in Equation (2) instead of the generalized formula presented in Equation (3).

The variable speed drive for Worthington 500 LNN-775A double suction pump includes an ACS 800-7 1000 kW DTC closed-loop speed inverter [37], but no torque transducer is installed in situ. This DTC inverter is manufactured by the same company as the one installed in the "Politehnica" University laboratory [34]. Both DTC inverters have the same torque and speed control performances [34,37]. In these conditions, the total uncertainty for the mechanical power is determined using the accuracy of ±4% with nominal torque in an open loop and the speed control in a closed loop with an accuracy of 0.01% from the nominal speed. As a result, a total uncertainty of over ±4% is predicted for the pump efficiency under in situ conditions.

The input data associated with pump performance over time have to be updated by the user. Certainly, the input data associated with pump performance have to be updated if repair work is applied to the geometry of the impeller blades. Assessing degraded pump performance in-service should be taken into account. Degraded pump performance methodology based on in-service test data is introduced by Gaiewski [38]. The methodology is based on the prediction of a small difference in the head at near pump shut-off head when compared to a new pump. An extensive degraded pump performance study on 150 pumps ranged from 22 kW to 3 MW installed in the municipal water supply and distribution system was conducted by Papa et al. [39]. The results obtained in this study revealed that the pump efficiency degraded by 9.3% on average when the pump operated at the BEP. Moreover, the efficiency degraded on average by 12.7% when the actual operating point and the original best efficiency point were compared. Papa et al. [39] noticed that while the ages of the pumps ranged from 1 year to 61 years with an average of 25 years, there was no discernible correlation between pump age and efficiency degradation. The above statement was not entirely surprising given that the pumps in service undergo various forms of routine maintenance, repair, refurbishment and modification throughout their working life [39].

The hydraulic performance along with the vibration response of an industrial scale centrifugal pump of 7.5 kW was experimentally investigated by [40,41] to assess the degradation in pump performance. The results delivered by all these investigations can be used to identify the degradation level in pump performance. As a result, the time when input data associated with pump performance has to be updated in the algorithm due to degradation is determined based on these investigations.

The marking of the $\eta_P(Q)$ curve for this pump allows the marking of the maximum efficiency value of $\eta_{P_{BEP}} = 93.6\,\%$ (marked with ■ in Figure 18) with a $+2.4\,\%$ deviation compared to the value indicated by the manufacturer, that of 91.4 %. The pump's operating regime (normal operation—NO, operation at limit—LO and abnormal operation—AO) have been determined according to the criteria defined in Section 2 and are marked by green, yellow and red.

The pump's duty point at a speed of 993 rpm can be found at the boundary between the green and the yellow areas (at a greater flow than that corresponding to the BEP flow rate). The selection of the pump's duty point for the maximum speed of 993 rpm is justified by the maximum flow rate of drinkable water supplied by this water treatment plant for Timişoara city. The operation of the pump in situ with VSD is considered to provide the variable flow rate needed by the consumers in Timişoara city.

6. Conclusions and Perspectives

This paper presents the development, implementation and validation of software developed for monitoring hydraulic pumps in service. The only parameter which is neither determined nor examined is the pump efficiency, although all the other variables (absorbed power, pumping head) are measured to determine and calculate it. The goal of this software is to identify the hydraulic pumps' operating regimes in situ and to alert the user when they operate outside their preset efficiency range. The algorithm developed for assessing the efficiency of the hydraulic pump is based on knowing its features (available in the manufacturer's catalog or determined based on experimental investigations) and in situ geometric data. The input data of the algorithm are the geometric and hydraulic values related to the location of instruments compared to the position of suction/discharge flanges. The input data of each hydraulic pump installed in a station is unique, even though a station may use several identical pumps. The implemented algorithm provides the value of the pump efficiency η_P corresponding to the duty point. As a result, the operating regime is identified by collecting only the values of the suction p_s and discharge p_d static pressures. In addition to efficiency η_P, the algorithm delivers the following values for the duty point: pumping head H, the mechanical power at the pump shaft P_m, and the volumetric flow rate passed through the pump Q, which is a useful quantity in industrial applications but is difficult to determine in situ.

The algorithm is robust, and easy to use and implement, regardless of the user's training. The major advantage of the software, compared to other means of monitoring, is the easy in situ implementation because it does not require additional expenses. The pressure gauges are normally available at the suction and discharge of the hydraulic pump.

Free software packages have been selected for the algorithm to ensure its portability on any operating system. For example, the scientific software library GNU GSL provides robust interpolation, extrapolation and curve-fitting tools for the available data and for the assessment of the deviations. The free Qt package was selected for the graphic interface. The implementation of the algorithm is structured on independent components, materialized by individual procedures and tools in the command line, disconnected from the user interface. The data flow from the implemented software solution shown in the diagram in Section 3 provides a synoptic view of the processing and data transfer flow.

The software has been first used to analyze the operation of the centrifugal pump PCN 65/200, installed in the Hydraulic Pump Laboratory of the "Politehnica" University Timişoara. The research conducted on this pump was used to validate the output results delivered by the software module against experimental data over the full operating range. This software validation over the full operating range cannot be performed on pumps installed in situ due to the particular installing conditions and the limited operating range imposed on each pump. Moreover, the investigations carried out in the laboratory allowed the determination of the deviations of the results delivered by the software depending on the selected operating point over the full range of the pump. The analysis of the results delivered by the software for the six operating points laid out over the full pump range highlighted the capability to correctly identify the operating regime. The software module estimated the pump efficiency η_P within limits that do not affect the prediction of the operating regime.

We have previously stated that the volumetric flow rate is a fundamental hydraulic quantity that has to be determined while the pump is in service. This output quantity provided by the pump unit is directly related to the requirements of the process or system in which it operates. Therefore, the investigations conducted on the hydraulic pump installed in the laboratory have shown that the relative error of the volumetric flow rate estimated by the software module against the value measured by the flowmeter is up to -5.6% for the operating points with a flow rate below 25% of the value of the BEP. In exchange, the limit of the relative error is between -4.192% and $+0.113\%$ for all operating points, except those with a volumetric flow rate below 25% Q_{BEP}. The experimental data supports the conclusion that the software module allows estimating the flow rate of the operating point when there is no (or no possibility of installing) flowmeter installed. Estimating the volumetric flow rate using the software module is an additional gain on top of the assessment of the operating regime of the pump.

The sensitivity analysis of the input variables and parameters used in the algorithm highlighted their impact on the deviation of the output quantity (pump efficiency) and the operating point. The most significant impact on the efficiency deviation is provided by the pressure ($p_{d/s}$) values acquired at the discharge and the suction of the pump.

That is why the input data must be accurate enough to diminish the influence of the output value (e.g., the pump efficiency η_P). It is recommended that the data recorded by the pressure transducers are taken over directly to the software module to avoid any reading and data input errors. The impact of the discharge and suction pipe diameters ($D_{d/s}$) on the deviation efficiency is all the more significant when the selected values are smaller than the actual ones. Therefore, it is recommended that the outer diameters of the discharge and suction pipes be selected in the software module if it is not possible to measure the inner ones. In this case, the influence of the length ($l_{s/d}$) of the suction/discharge line between the pressure manometers (transducers) as well as the pump flanges and the longitudinal hydraulic losses coefficient ($\lambda_{s/d}$) along the suction/discharge line on the efficiency deviation are negligible. These input data can have a greater impact on the pump efficiency deviation if the measuring instruments are located at large distances from the pump and the pipelines have been in operation for a long time. The impact of the operating point on the pump efficiency deviation is all the greater the further we move away from the BEP.

The implementation of the software within the regional water supply company AQUATIM S.A. for monitoring in situ operations of pumps highlighted the demand for software products that can monitor high-efficiency pump operating regimes, justifying electricity costs according to European Union requirements, and supporting the predictive maintenance.

Note that the regional water supply company manages a fleet of hundreds of different hydraulic pumps, ranging from a few kilowatts to megawatts, that are distributed over a geographical area of hundreds of square kilometers. The development and implementation of software solutions to monitor the operation of the pump fleets are crucial in the management of the available resources in critical infrastructures.

The last section of the paper contains the results obtained with the software module for the Worthington 500 LLN-775A double-suction pump of 1 MW, which supplies drinkable water to the Timişoara city. The Worthington pumps are the most powerful pumps in the AQUATIM fleet and have been in service for 23 years. The pump's duty point at speed of 993 rpm is identified at the boundary between the green and the yellow region (at a greater flow rate than that corresponding to the BEP flow rate value). The choice of the pump's duty point for the maximum speed of 993 rpm is justified by the maximum flow rate of drinkable water supplied from this water treatment plant to Timişoara city.

In situ operation of the pumps with variable speed drive (VSD) is used to supply the variable flow rate needed by the consumers. The next stage in the development of the software consists of expanding its capabilities to monitor the efficiency of the VSD pumps. The work of this next stage is based on the obtained results and on the experience gathered to date.

Author Contributions: Conceptualization, A.-A.A. and S.M.; Data curation, A.-A.A., A.C. and S.M.; Funding acquisition, A.-A.A.; Investigation, A.-A.A. and S.M.; Methodology, A.-A.A. and S.M.; Project administration, A.-A.A.; Resources, A.C.; Software, A.-A.A.; Supervision, A.-A.A. and S.M.; Validation, A.-A.A., A.C. and S.M.; Visualization, A.-A.A. and S.M.; Writing-original draft, A.-A.A.; Writing-review & editing, A.-A.A. and S.M. All authors have read and agreed to the published version of the manuscript.

Funding: This research received no external funding

Institutional Review Board Statement: Not applicable.

Informed Consent Statement: Not applicable.

Data Availability Statement: Not applicable.

Acknowledgments: The authors acknowledge Eng. Daniel Moş for conducting experimental measurements on the PCN 65/200 pump. The author affiliated with "Politehnica" University Timişoara has been supported by the research contract BC27/27.03.2018 (https://staff.cs.upt.ro/~alin.anton/BC27/27.03.2018) (accessed on 7 November 2022). The author affiliated with Romanian Academy—Timişoara Branch has been supported by the research program of the Hydrodynamic and Cavitation Laboratory from the Center for Fundamental and Advanced Technical Research.

Conflicts of Interest: The authors declare no conflict of interest.

References

1. Pabi, S.; Amarnath, A.; Goldstein, R.; Reekie, L. *Electricity Use and Management in the Municipal Water Supply and Wastewater Industries*; Technical Report 3002001433; Water Research Foundation and Electric Power Research Institute: Palo Alto, CA, USA, 2013.
2. European Union Commission Regulation No 547/2012. Available online: https://data.europa.eu/eli/reg/2012/547/oj (accessed on 1 October 2022).
3. Sahoo, T.; Guharoy, A. Energy cost savings with centrifugal pumps. *World Pumps* **2009**, *2009*, 35–37. [CrossRef]
4. Bloch, H.P.; Budris, A.R. *Pump User'S Handbook: Life Extension*; River Publishers: New York, NY, USA, 2021.
5. EPRI. *Water and Wastewater Industries: Characteristics and Energy Management Opportunities*; Technical Report CR-106941; Electric Power Research Institute: Palo Alto, CA, USA, 1996.
6. Merkle, T. *Damages on Pumps and Systems: The Handbook for the Operation of Centrifugal Pumps*, 1st ed.; Elsevier: Oxford, UK, 2014.
7. Matlakala, M.; Kallon, D.; Simelane, S.; Mashinini, P. Impact of Design Parameters on the Performance of Centrifugal Pumps. *Procedia Manuf.* **2019**, *35*, 197–206. [CrossRef]
8. Guelich, J.F. *Centrifugal Pumps*, 1st ed.; Springer, Schmidt and Voeckler GbR: Leipzig, Germany, 2008.
9. Shiels, S. Centrifugal Pump Academy: Locating the greatest centrifugal pump energy savings. *World Pumps* **1998**, *1998*, 56–59. [CrossRef]
10. U.S. Environmental Protection Agency. *Fiscal year 2011: Drinking Water and Ground Water Statistics*; Technical Report EPA 816-R-13-003; U.S. Environmental Protection Agency, Office of Water: Washington, DC, USA, 2013.
11. KSB Guard Operating Manual. 2021. Available online: https://products.ksb.com/ (accessed on 1 October 2022).
12. de Almeida, A.T.; Fonseca, P.; Falkner, H.; Bertoldi, P. Market transformation of energy-efficient motor technologies in the EU. *Energy Policy* **2003**, *31*, 563–575. [CrossRef]
13. Ahonen, T.; Kortelainen, J.T.; Tamminen, J.K.; Ahola, J. Centrifugal pump operation monitoring with motor phase current measurement. *Int. J. Electr. Power Energy Syst.* **2012**, *42*, 188–195. [CrossRef]
14. Jussi Tamminen, T.A.; Ahola, J. Method and Arrangement for Estimating Flow Rate of Pump. U.S. Patent US9416787B2, 16 August 2016.
15. Ahonen, T. Monitoring of Centrifugal Pump Operation by a Frequency Converter. Ph.D. Thesis, LUT University, Lappeenranta, Finland, 2011.
16. Bertil Ohlsson, U.W.; Zahrai, S. Device, System and Method for On-Line Monitoring of Flow Quantities. U.S. Patent US6918307B2, 19 July 2005.
17. Gopalakrishnan, S.; Hanson, L. Pump with Integral Flow Monitoring. U.S. Patent US20030047008A1, 25 March 2003.
18. Sabini, E.P.; Lorenc, J.A. Centrifugal Pump Performance Degradation Detection. U.S. Patent US7112037B2, 26 September 2006.
19. Stuparu, A.; Baya, A.; Bosioc, A.; Anton, L.; Mos, D. Modelling the operation curves of two similar high power centrifugal pumps. *Am. Inst. Phys. Conf. Ser.* **2018**, *1978*, 030005. [CrossRef]
20. Stuparu, A.; Baya, A.; Bosioc, A.; Anton, L.; Mos, D. Experimental investigation of a pumping station from CET power plant Timisoara. *IOP Conf. Ser. Earth Environ. Sci.* **2019**, *240*, 032018. [CrossRef]
21. Brekke, H. A proposal for improving the thermodynamic method. In Proceedings of the 9th International Conference on Hydraulic Efficiency Measurements, Trondheim, Norway, 27–30 June 2012.
22. Carrasco, H.; Kengskool, K. Solutions to Industrial Engineering problems using integrated software environments. *Comput. Ind. Eng.* **1988**, *15*, 204–209. [CrossRef]
23. Wang, J.; Zhang, L.; Zheng, Y.; Wang, K. Adaptive prognosis of centrifugal pump under variable operating conditions. *Mech. Syst. Signal Process.* **2019**, *131*, 576–591. [CrossRef]
24. Johnson, H.A.; Simon, K.P.; Slocum, A.H. Data analytics and pump control in a wastewater treatment plant. *Appl. Energy* **2021**, *299*, 117289. [CrossRef]
25. Roozbeh Nia, A.; Awasthi, A.; Bhuiyan, N. Industry 4.0 and demand forecasting of the energy supply chain: A literature review. *Comput. Ind. Eng.* **2021**, *154*, 107128. [CrossRef]
26. Ferreira de Souza, D.; Antonio Marino Salotti, F.; Luis Sauer, I.; Tatizawa, H.; Gakiya Kanashiro, A. A Comparison Between Reported Values and Measured Values of Power Factor and Efficiency for Electric Induction Motors. *IEEE Lat. Am. Trans.* **2021**, *19*, 173–181. [CrossRef]
27. Takacs, G. *Electrical Submersible Pumps Manual: Design, Operations, and Maintenance*, 1st ed.; Gulf Professional Publishing: Oxford, UK, 2009.

28. Barringer, H. How to use reliability engineering principles for business issues. In Proceedings of the Proceedings, YPF Reliability Symposium, La Platta, Argentina, 30 November 1998.
29. Barringer, H.P.; Barringer & Associates, Inc. A life cycle cost summary. In Proceedings of the International Conference of Maintenance Societies, Mesa Perth, Australia, 20–23 May 2003 ; pp. 20–23.
30. Lai, Z.; Li, Q.; Zhao, A.; Zhou, W.; Xu, H.; Wu, D. Improving Reliability of Pumps in Parallel Pump Systems Using Particle Swam Optimization Approach. *IEEE Access* **2020**, *8*, 58427–58434. [CrossRef]
31. Galassi, M.; Davies, J.; Theiler, J.; Gough, B.; Jungman, G.; Alken, P.; Booth, M.; Rossi, F.; Ulerich, R. GNU Scientific Library Manual. Release 2.7, 2021. Available online: https://www.gnu.org/software/gsl/doc/latex/gsl-ref.pdf (accessed on 1 October 2022).
32. Qt Documentation. 2022. Available online: https://doc.qt.io/ (accessed on 1 October 2022).
33. *IEC 60193*; Hydraulic Turbines, Storage Pumps and Pump-Turbines—Model Acceptance Tests. International Electrotechnical Commission: Geneva, Switzerland, 2019.
34. *Low Voltage AC Drives ABB Machinery Drives ACS850 0.37 to 560 kW Catalog*; ABB Asea Brown Boveri Ltd.: Zurich, Switzerland, 2014.
35. Stanciu, R.; Ginga, G.; Muntean, S.; Anton, L. Low-speed-small-load direct torque control ripples filtering. *Proc. Rom. Acad. Ser. A Math. Phys. Tech. Sci. Inf. Sci.* **2012**, *13*, 125–132.
36. Stanciu, R.; Turcin, I.; Muntean, S.; Anton, L. Cellular wind-power integration using remotely controlled pump hydro energy storage. *Proc. Rom. Acad. Ser. A Math. Phys. Tech. Sci. Inf. Sci.* **2013**, *14*, 242–249.
37. *AC Drives ABB Machinery Drives ACS800*; Hardware Manual; ABB Asea Brown Boveri Ltd: Zurich, Switzerland, 2013.
38. Gaiewski, D. A Methodology for Determining Degraded Pump Performance Based on In-Service Test Criteria or Data. In Proceedings of the ASME 2011 Power Conference Collocated with JSME ICOPE 2011, Denver, CO, USA, 12–14 July 2011; Volume 2, pp. 255–259. [CrossRef]
39. Papa, F.; Radulj, D.; Karney, B.; Robertson, M. Pump energy efficiency field testing and benchmarking in Canada. *J. Water Supply Res. Technol.-Aqua* **2014**, *63*, 570–577. [CrossRef]
40. Eaton, A.; D'Alessandro, F.; Ahmed, W.; Hassan, H. On the Performance Degradation of Centrifugal Pumps. In Proceedings of the 5th International Conference of Fluid Flow, Heat and Mass Transfer (FFHMT'18), Niagara Falls, ON, Canada, 7–9 June 2018. [CrossRef]
41. Eaton, A.; Ahmed, W.H.; Hassan, M. Evaluating the Performance Degradation of Centrifugal Pumps Using the Principal Component Analysis. *J. Press. Vessel Technol.* **2021**, *144*, 021405. [CrossRef]

applied sciences

MDPI

Article

Flood Prediction with Two-Dimensional Shallow Water Equations: A Case Study of Tongo-Bassa Watershed in Cameroon

Alain Joel Elong [1], Ling Zhou [2,*], Bryan Karney [3], Haoyu Fang [1], Yun Cao [1] and Steve L. Zeh Assam [4]

1 College of Water Conservancy and Hydropower Engineering, Hohai University, Nanjing 210098, China
2 College of Water Conservancy and Hydropower Engineering, Hohai University and Yangtze Institute for Conservation and Development, 1 Xikang Road, Nanjing 210098, China
3 Department of Civil Engineering, University of Toronto, 35 St. George St., Toronto, ON M5S 1A4, Canada
4 College of Harbour Coastal and Offshore Engineering, Hohai University, Nanjing 210098, China
* Correspondence: zlhhu@163.com

Abstract: As a result of urbanization, combined with the anthropogenic effects of climate change, natural events such as floods are showing increasingly adverse impacts on human existence. This study proposes a new model, based on shallow water equations, that is able to predict these floods and minimize their impacts. The first-order finite volume method (FVM), the Harten Lax and van Leer (HLL) scheme, and the monotone upwind scheme for conservation laws (MUSCL) are applied in the model. In addition, a virtual boundary cell approach is adopted to achieve a monotonic solution for both interior and boundary cells and flux computations at the boundary cells. The model integrates the infiltration parameters recorded in the area, as well as the Manning coefficient specific to each land-cover type of the catchment region. The results provided were mapped to highlight the potential flood zones and the distribution of water heights throughout the catchment region at any given time, as well as that at the outlet. It has been observed that when standard infiltration and the Manning parameters were selected, the floodable surface increased, as expected, with the increasing rainfall intensity and duration of the simulation. With sufficient infiltration, only a portion of the water tends to stagnate and flow off on the surface toward the outlet. A sensitivity analysis of certain parameters, such as rainfall data and the final infiltration coefficient in the lower watershed of the littoral region, was conducted; the results show that the model simulates well the general character of water flow in the watershed. Finally, the model's validation using field-collected parameters during the flood of 25 July 2017 and 18 to 22 July 2016 in the Grand Ouaga basin in Burkina reveals Nash–Sutcliffe values of 0.7 and 0.73, respectively.

Keywords: flood; Godunov-type scheme; HLL scheme; rainfall intensity; infiltration parameters; Manning

Citation: Elong, A.J.; Zhou, L.; Karney, B.; Fang, H.; Cao, Y.; Assam, S.L.Z. Flood Prediction with Two-Dimensional Shallow Water Equations: A Case Study of Tongo-Bassa Watershed in Cameroon. *Appl. Sci.* **2022**, *12*, 11622. https://doi.org/10.3390/app122211622

Academic Editor: Vasily Novozhilov

Received: 11 October 2022
Accepted: 7 November 2022
Published: 16 November 2022

1. Introduction

The global increase in urban populations has led to a corresponding increase in the demand for infrastructure, including buildings, roads, and drainage networks [1]. The implementation of these infrastructures has led to a reduction in rainwater absorption capacities, exposing urban vulnerabilities during flooding in the presence of heavy rain events [2,3]. Furthermore, the climatic changes that have become accentuated over the last few decades have tended to cause an increase in the frequency, intensity, and duration of rain events in many countries [1,4]. Natural disasters resulting from high-water events, such as glacial melting, droughts, storms, and floods are being exacerbated by frequent extreme weather occurrences that are driven by climate change [5,6]. These "high-water events" could be described as "water overflowing its natural or manmade banks into ordinarily dry terrain, such as a river inundating its floodplain" [7].

Flooding is perhaps the most catastrophic natural disaster, wreaking havoc on human lives and the economy while damaging infrastructure (bridges, buildings, etc.) [8]. For example, significant damage was recorded in Australia in the Brisbane region at the beginning of 2011, as a result of flooding [9].

Furthermore, massive floods in the Mississippi River basin in the United States have been caused by rainfall, coupled with spring snowmelt, resulting in damage estimated to cost billions of dollars. The 2007 floods in the United Kingdom (UK) caused 42,000 individuals to lose electricity for more than 24 h, while 350,000 others lost their water supply for up to 17 days, and over 10,000 of the latter group became stranded on roads and trains [10]. The most expensive flood damage in history occurred in Thailand's Chao Praya River Basin, as a result of persistent floods. The cost of the damage was estimated to be over USD 45 billion [11,12] reported that in the USA alone, flood management infrastructure interventions have helped to save 10 million acres of land and almost one million acres from flooding, averting more than USD 110 billion in flood damage costs. The lack of flood management infrastructure has long been a concern for researchers. More recently, the Ministry of Humanitarian Affairs, Disaster Management, and Social Development of Nigeria published a report of the damage caused by the floods earlier in October 2022. The balance sheet shows that more than 1.4 million people have been affected by the floods; an estimated 500 people were killed, with 1546 others injured and 790,254 persons displaced. One month later, the National Emergency Management Agency (NEMA) of Nigeria said that at least 300 people had died and more than 100,000 others had been displaced since the start of the rainy season and with the release of excess water from a dam in neighboring Cameroon. This assessment was recorded as the most catastrophic event compared to the last major floods of 2012 (https://edition.cnn.com/2022/10/13/africa/hundreds-killed-nigeria-floods-intl, accessed on 13 October 2022). Ref. [13] studied the history of flash-flood occurrence in a southern Italy drainage basin, to identify the real cause of the increment in flood occurrence. The study compared rainfall to anthropogenic modifications and concluded that the rainfall in the study area had a decreasing trend while flood damage still increased, showing that the real cause of flash floods in the area was mismanagement and land-use modification. Floods have become more common in various nations around the world in recent decades, prompting governments and experts to collaborate on flood management and prevention techniques [14,15].

It is, therefore, important to pay attention to the behavior of water and flood systems. Numerical and experimental approaches have been applied to investigate the flooding issue and water behavior [16] using the shallow water equations developed by Saint-Venant in 1871 [17]. These equations constitute the fundamental basis for the development of numerical models for open channel flow. Two types of numerical models were developed to investigate flood-wave propagation in one dimension [18–22] and two dimensions [23–25]. Following the Nari typhoon in Taiwan, ref. [26] reproduced the flood occurrence by comparing the numerical model output with the pumped water volume obtained from the subway system, along with the failure time of several pump stations. They compared the results of their numerical model to the data gathered in the field. The massive flooding event in Sumacarcel, Spain, was caused by the breakdown of the Tous Dam in October 1982 and was also investigated numerically by [27]. They collected water-depth elevation data by conducting interviews and reviewing the Centre for Studies and Experimentation of the Ministry of Public Works (CEDEX, Spain) reports. Ref. [28] simulated the Nîmes flood event and the numerical results were compared to the historical data. Ref. [29] represented the roughness effects of structures in flood-prone urban areas, eventually formulating a numerical simulation model. The creation of dedicated computer tools has heavily contributed to the development and implementation of software capable of efficiently simulating floods. Using the Medjerda River in Tunisia as an example, a comparison of 1Dimension and 2Dimension hydraulic models for floods has been performed by [30].

The Hydrological Engineering Center's river analysis system (HEC RAS) and MIKE 11 are used in the one-dimensional models, whereas TELEMAC 2D is used in two-dimensional models. In addition, [31] developed an operational forecasting system (OFS) for better flash flood prediction and warnings in mountainous regions. The developed system uses high-density and high-accuracy airborne LiDAR DEM data to simulate rapid water-level rises and flooding as the result of intense rainfall within relatively small watersheds. The results revealed that the water levels and flood extent derived from the OFS produced agreements between the measured and surveyed data. More recently, [32] developed a high-performance two-dimensional hydrodynamic model based on the finite element method and unstructured grids. They combined the weather research and forecasting (WRF) model, the storm water management model (SWMM), a two-dimensional hydrodynamic model, and a map-oriented visualization tool to operational high performance. They concluded that the extent of flooding during historical inundation events, as derived from the forecasting systems, agrees well with the surveyed data for plain areas in southwestern Taiwan. Despite the development of several numerical flood simulation models, such as the finite difference method (FDM) [33], the finite volume method (FVM) [34], and dedicated software [35], the issue of prediction and prevention is still relevant. Some of these previous models failed to address the issue of flow simulation, not only due to the difficulty in access (due to the high cost) and handling for the majority of the population but also due to the rigidity of certain parameters used in the software, which prevented them from being adjusted to the case study and for sensitivity to the basin's topography. Another drawback was their inability to forecast and emphasize those regions that would flood and how the water levels in the study area were distributed. In order to considerably reduce the damage and disasters linked to the sudden occurrence of floods, therefore, it appears necessary to develop a model capable of predicting the flood zones as well as the water depths for a given watershed. This would not only identify those areas at risk where populations should be prohibited from settling but also provide estimates of the time required for a rapid evacuation operation in the event of an emergency.

The new flood simulation method which is a combination of previous numerical models and field-estimated parameters, developed in this paper gives a more detailed representation of the surface flow and makes a reasonable approximation of the hydrology. Moreover, the model can predict the distribution of water height in a catchment area, as well as the height and flow discharge at the outlet. To build the model, the finite volume method (FVM)–Godunov type is adopted to solve the 2D shallow-water equation, and the monotonic upstream-centered scheme for conservation laws (MUSCL)–Hancock scheme has been associated with the model to provide high-accuracy numerical solutions to the partial differential equations, which can involve solutions that exhibit shocks or discontinuities. The main goal of using the MUSCL scheme in the model is to replace the piecewise constant approximation of Godunov's scheme with reconstructed states, derived from cell-averaged states obtained from the previous time step. The structured grid was chosen for its ease of implementation, and the fluxes were computed using the Harten Lax and van Leer (HLL) approximation Riemann solution. The model is sensitive to the topography of the free surface of water flow, which results in the integration of roughness or the Manning coefficient. A virtual boundary cell is also used to compute the flux in different cells by coupling the Riemann invariant with a head-flow boundary at time M and N respectively in the x and y direction [36]. Furthermore, the model also considers the daily rainfall or rainfall data recorded in a given space, along with the infiltration parameters specific to each type of soil. Estimates of surface evaporation of the basin and the flow rate will then be integrated into the numerical equations to add further realism. The Courant number used in the stability of unstable numerical methods and the simulation time have also been incorporated into the model.

To demonstrate the model's power, this work models the response of the Tongo-Bassa catchment in Douala-Cameroon by using a new model integrating the parameters from the field and validating the model by means of another catchment called Grand-Ouaga in Burkina Faso.

The following sections briefly discuss the 2D governing equation of the surface flow problem, before providing general information on the research area, the numerical strategy for solving it, and an examination of how well the different numerical approaches perform in the field.

2. Numerical Model

2.1. Mathematical Equations

The conservative form of the 2D shallow water equations, derived from the Navier–Stokes equations by considering the viscous, source, and friction terms are presented as follows:

$$\frac{\partial U}{\partial t}+\frac{\partial F(U)}{\partial x}+\frac{\partial G(U)}{\partial y}=S_s+S_f+S_p \quad (1)$$

where x and y represent the Cartesian coordinates; t is the time; U is the conservative variables; F and G are the flux vectors in the *x* and *y* directions, respectively. The bottom slope, the friction terms, and the source term are expressed by S_s, S_f, and S_p, respectively. The bottom slope, friction terms, source term, and flux vectors are defined as:

- **The flux vectors F and G**:

$$F(U)=\begin{bmatrix} hu \\ hu^2+\frac{1}{2}gh^2 \\ huv \end{bmatrix};\ G(U)=\begin{bmatrix} hv \\ huv \\ hv^2+\frac{1}{2}gh^2 \end{bmatrix} \quad (2)$$

where h is the water height; g is the gravitational acceleration; u and v are the depth-averaged velocity components in the *x* and *y* directions.

- **The bottom slope S_s, friction S_f, and source S_p**:

$$S_s=\begin{bmatrix} 0 \\ -gh\frac{\partial z}{\partial x} \\ -gh\frac{\partial z}{\partial y} \end{bmatrix};\ S_f=\begin{bmatrix} 0 \\ -nm^2\frac{\sqrt{u^2+v^2}}{h^{\frac{4}{3}}}u \\ -nm^2\frac{\sqrt{u^2+v^2}}{h^{\frac{4}{3}}}v \end{bmatrix};\ S_p=\begin{bmatrix} P-(I-E-q) \\ 0 \\ 0 \end{bmatrix} \quad (3)$$

where $-\frac{\partial z}{\partial x}$ and $-\frac{\partial z}{\partial y}$ are the bed slopes in the *x* and *y* directions, respectively; z is the bed elevation; nm is the Manning–Strickler coefficient; P and I are the rainfall intensity (mm/min) and infiltration rate; E is the surface evaporation (mm/min) and q is the flow rate. Horton's infiltration model, developed in the 1930s, has been used to calculate the infiltration rate [37]. The model indicates that infiltration tends to decrease exponentially if the rainfall exceeds the infiltration capacity. The formula is expressed as:

$$I(t_s)=i_f+(i_0-i_f)\exp(-rt_s) \quad (4)$$

where i_0, i_f, r, and t_s are, respectively, the initial infiltration rate (mm/min), the final infiltration rate (mm/min), a constant depending on soil properties (min^{-1}), and the duration of the storm (min).

2.1.1. Discretization of the Shallow-Water Equation

The FVM-Godunov approach solves the two-dimensional shallow-water equations by integrating the source term.

$$U_t+F(U)_x+G(U)_y=S(U) \quad (5)$$

By integrating Equation (6) into each cell and time step, the formulation becomes:

$$\int_{x_{i-1/2}}^{x_{i+1/2}}\int_{y_{j-1/2}}^{y_{j+1/2}}\int_{t^n}^{t^{n+1}}\left(\frac{\partial U}{\partial t}+\frac{\partial(F(U))}{\partial x}+\frac{\partial(G(U))}{\partial y}\right)dxdydt=\int_{x_{i-1/2}}^{x_{i+1/2}}\int_{y_{j-1/2}}^{y_{j+1/2}}\int_{t^n}^{t^{n+1}}(S(U))dxdydt\,. \quad (6)$$

For a better resolution of Equation (7) and to ensure its stability, system decomposition is recommended. This will have to be performed in the x-direction on the one hand and the y-direction on the other hand.

The resolution is split into a sequence of one-dimensional equations.

In the x-direction:

$$\int_{x_{i-1/2}}^{x_{i+1/2}}\int_{t^n}^{t^{n+1}}\left(\frac{\partial U}{\partial t}+\frac{\partial(F(U))}{\partial x}\right)dxdt=0\,. \tag{7}$$

The numerical solution at time $t_{n+1/4}$ is obtained from the resolution of Equation (8), as follows:

$$\frac{U_{i,j}^{n+1/4}-U_{i,j}^{n}}{\Delta t}+\frac{F_{i+1/2,j}^{n}-F_{i-1/2,j}^{n}}{\Delta x}=0 \tag{8}$$

where $U_{i,j}^{n}$ is the first solution along each row of cells, with j fixed; $F_{i+1/2,j}^{n}$ is the numerical flux between two consecutive cells $(i-1, j)$ and (i, j).

Then, we solve the equation again, this time along the y-direction.

$$\int_{y_{i-1/2}}^{y_{i+1/2}}\int_{t^n}^{t^{n+1}}\left(\frac{\partial U}{\partial t}+\frac{\partial(G(U))}{\partial y}\right)dydt=0 \tag{9}$$

The numerical solution at time $t_{n+2/4}$ is obtained from the resolution of Equation (10), as follows:

$$\frac{U_{i,j}^{n+2/4}-U_{i,j}^{n+1/4}}{\Delta t}+\frac{G_{i+1/2,j}^{n+1/4}-G_{i-1/2,j}^{n+1/4}}{\Delta y}=0 \tag{10}$$

where $U_{i,j}^{n+1/4}$ are present as data along each row of cells, with i fixed; $G_{i-1/2,j}^{n+1/4}$ is the numerical flux between two consecutive cells $(i,j-1)$ and (i,j).

The integration of the friction and source terms (rainfall, infiltration, and bottom slope) are established as follows:

$$\frac{U_{i,j}^{n+\frac{3}{4}}-U_{i,j}^{n+\frac{2}{4}}}{\Delta t}=S_f\left(U_{i,j}^{n+\frac{2}{4}}\right)+S_S\left(U_{i,j}^{n+\frac{2}{4}}\right); \tag{11}$$

$$\frac{U_{i,j}^{n+1}-U_{i,j}^{n+\frac{3}{4}}}{\Delta t}=S_p\left(U_{i,j}^{n+\frac{3}{4}}\right). \tag{12}$$

2.1.2. Integration of the Topography

The method of solving the equations by decomposition, which considers the system as homogeneous, as seen above, could not be used in the context of topography because such an approach yields poor results when U corresponds to a stationary state. It can be expressed as:

$$\begin{cases} u=0 \\ \frac{\partial\left(\frac{1}{2}gh^2\right)}{\partial x}=-gh\frac{\partial Z}{\partial x} \end{cases}. \tag{13}$$

The following Equation (11) can be resumed as a constant, namely:

$$H=h+Z \tag{14}$$

where H is the water level.

H is a constant; this state physically corresponds to the state of equilibrium of the watercourse or steady state. The hydrostatic reconstruction proposed by [38], coupled with a positive numerical flux, allows us to verify important mathematical and physical properties, such as the positivity of the water height, and, thus, to avoid instabilities when dealing with dry zones. The equations are as follows:

$$\begin{cases} h_{i+\frac{1}{2}L} = \max(h_i + Z_i - \max(Z_i, Z_{i+1}), 0) \\ h_{i+\frac{1}{2}R} = \max(h_{i+1} + Z_{i+1} - \max(Z_i, Z_{i+1}), 0) \\ U_{i+\frac{1}{2}L} = \left(h_{i+\frac{1}{2}L}, h_{i+\frac{1}{2}L} * u_i\right) \\ U_{i+\frac{1}{2}R} = \left(h_{i-\frac{1}{2}R}, h_{i-\frac{1}{2}R} * u_i\right) \end{cases} \tag{15}$$

The water height computed on the left and right sides of the cell, respectively, are expressed as $h_{i+\frac{1}{2}L}$ and $h_{i+\frac{1}{2}R}$; the general solutions found at the left and right sides of each cell are represented respectively as $U_{i+\frac{1}{2}L}$ and $U_{i+\frac{1}{2}R}$. The numerical schema can now be written as:

$$U_i^{n+1} - U_i^n + \frac{\Delta t}{\Delta x}\left(F^n_{i+\frac{1}{2}L} - F^n_{i-\frac{1}{2}R}\right) = 0 \tag{16}$$

where:

$$F^n_{i-\frac{1}{2}R} = F^n_{i-\frac{1}{2}} + S_{i-\frac{1}{2}R} \text{ and } F^n_{i+\frac{1}{2}L} = F^n_{i+\frac{1}{2}} + S_{i+\frac{1}{2}L} \tag{17}$$

$$S_{i-\frac{1}{2}R} = \begin{pmatrix} 0 \\ P(h_i^n) - P\left(h^n_{i+\frac{1}{2}L}\right) \end{pmatrix} \text{ and } S_{i+\frac{1}{2}L} = \begin{pmatrix} 0 \\ P(h_i^n) - P\left(h^n_{i+\frac{1}{2}L}\right) \end{pmatrix}, \tag{18}$$

where the hydrostatic pressure is given by $P(h) = \frac{1}{2}gh^2$.

2.1.3. Integration of Rainfall Intensity

The rainfall intensity can be computed through integration, as follows:

$$\int_{x_{i-\frac{1}{2}}}^{x_{i+\frac{1}{2}}} \int_{t^n}^{t^{n+1}} \left(\frac{\partial h}{\partial t}\right) dxdt = \int_{x_{i-\frac{1}{2}}}^{x_{i+\frac{1}{2}}} \int_{t^n}^{t^{n+1}} (P(h) - i(t))dxdt \tag{19}$$

The resolution of Equation (17) yields:

$$h_i^{n+\frac{1}{2}} = h_i^n + \frac{dt}{2}P_i^n - \left(I\left(t^{n+\frac{1}{2}}\right) - I(t^n)\right) \tag{20}$$

The Horton model appears to be the best for the management of infiltration but that model can face certain problems, due to the heterogeneity of the infiltration and rainfall parameters. Furthermore, the model's runoff is generated not only by rain that supplies water to a cell but also by flowing from another cell. The amount of water available for infiltration on a cell at time t^n is given as:

$$w_i^n = \frac{2}{dt}h_i^n + P_i^n \tag{21}$$

A comparison should then be made between the infiltration and the water available in each cell. Now we have:

$$\begin{cases} I_i^{n+\frac{1}{2}} = I_i^n + \frac{dt}{2}P_i^n + h_i^n & \text{if} \quad W \le i(t) \\ I_i^{n+\frac{1}{2}} - I_i^n = i_f t^n + \frac{(i_o - i_f)\left(1 - e^{-rt^n}\right)}{r} & \text{if} \quad W \ge i(t) \end{cases} \tag{22}$$

In the case where $W \le i(t)$, then water height is assumed to be constant, namely, $h_i^{n+\frac{1}{2}} = 0$, but if $W \ge i(t)$, the water height should be considered as:

$$h_i^{n+\frac{1}{2}} = h_i^n + \frac{dt}{2}P_i^n - \left(i_f t^n + \frac{(i_o - i_f)\left(1 - e^{-rt^n}\right)}{r}\right). \tag{23}$$

2.1.4. Computation of the Flow Rate

The surface of the basin where the runoff is supposed to take place can be assumed to be a rectangular channel, with l as the width, D-Ds as the height, and a slope of S. The flow can be computed using the Manning equation, as follows:

$$Q = \frac{1.49}{nm} S^{0.5} R_H^{2/3} A_S \tag{24}$$

where:

nm is the roughness coefficient;
A_s is the surface of the basin crossed by the runoff ($A_s = l * (D - Ds)$;
R_H is the hydraulic radius ($R_H = D - Ds$);
A is the surface of the basin or catchment;
D is the height from the impervious area in the soil until the surface runoff;
Ds is the thickness between the soil surface and the impervious area of the soil.
At this point, the flow rate can be determined using the following equation.
If (D > Ds):

$$q = \frac{1 * 49}{A.nm} l * S^{0.5} * (D - Ds)^{5/3}. \tag{25}$$

In the case where (D ≤ Ds), then $q = 0$.

2.2. Numerical Flux Computation

Among the different numeric fluxes developed in the past, [39] proposed the HLL scheme (Harten–Lax–Leer) [40]) as the most stable method and the one best able to compute fluxes F and G for this case. They suggested the following HLL fluxes at the cell interfaces.

- The computation of the F flux:

$$\mathrm{F}(\mathrm{U_L}, \mathrm{U_R}) = \begin{cases} \mathrm{F}(\mathrm{U_L}) & \text{if } 0 \le \mathrm{C_1^x} \\ \frac{\mathrm{C_2^x F(U_L) - C_1^x F(U_R) + C_1^x C_2^x (U_R - U_L)}}{\mathrm{C_2^x - C_1^x}} & \text{if } \mathrm{C_1^x} < 0 < \mathrm{C_2^x} \\ \mathrm{F}(\mathrm{U_R}) & \text{if } \mathrm{C_2^x} \le 0 \end{cases} \tag{26}$$

where $\mathrm{U_L}$ and $\mathrm{U_R}$ are, respectively, the left and right approximations of the solution in the cell. The coefficients $\mathrm{C_1^x}$ et $\mathrm{C_2^x}$ are calculated as:

$$\mathrm{C_1^x} = \min_{\mathrm{U=U_L,U_R}} \left(\min_{\mathrm{j=1,2,3}} \beta_j^x(\mathrm{U}) \right); \ \mathrm{C_2^x} = \max_{\mathrm{U=U_L,U_R}} \left(\max_{\mathrm{j=1,2,3}} \beta_j^x(\mathrm{U}) \right) \tag{27}$$

where the term β_j^x, with j = 1, 2, 3, represents the eigenvalues of the Jacobian of F and can be expressed as:

$$\beta_1^x = \mathrm{u} - \sqrt{\mathrm{gh}},\ \beta_2^x = \mathrm{u},\ \beta_3^x = \mathrm{u} + \sqrt{\mathrm{gh}}. \tag{28}$$

- The computation of the G flux:

$$\mathrm{G}(\mathrm{U_D}, \mathrm{U_U}) = \begin{cases} \mathrm{G}(\mathrm{U_D}) & \text{if } 0 \le \mathrm{C_1^y} \\ \frac{\mathrm{C_2^y G(U_D) - C_1^y G(U_U) + C_1^y C_2^y (U_U - U_D)}}{\mathrm{C_2^x - C_1^x}} & \text{if } \mathrm{C_1^y} < 0 < \mathrm{C_2^y} \\ \mathrm{G}(\mathrm{U_U}) & \text{if } \mathrm{C_2^y} \le 0 \end{cases} \tag{29}$$

where $\mathrm{U_D}$ and $\mathrm{U_U}$ are, respectively, the top and bottom approximations of the solution in the cell. The coefficients $\mathrm{C_1^y}$ et $\mathrm{C_2^y}$ are calculated as:

$$\mathrm{C_1^y} = \min_{\mathrm{U=U_D,U_U}} \left(\min_{\mathrm{j=1,2,3}} \beta_j^y(\mathrm{U}) \right); \ \mathrm{C_2^y} = \max_{\mathrm{U=U_D,U_U}} \left(\max_{\mathrm{j=1,2,3}} \beta_j^y(\mathrm{U}) \right) \tag{30}$$

where the term β_j^y with j = 1, 2, 3 is the eigenvalues of the Jacobian of G and can be expressed as:

$$\beta_1^y = \mathrm{v} - \sqrt{\mathrm{gh}},\ \beta_2^y = \mathrm{v},\ \beta_3^y = \mathrm{v} + \sqrt{\mathrm{gh}}. \tag{31}$$

Here, h is the water height; g is the gravitational acceleration; u and v are the depth-averaged velocity components in the *x* and *y* directions.

2.3. Computational Grid

The computational grid developed by [36] can be extended, in the case of the 2D dam-break problem, by labeling the interior cells with i = 1, 2, . . . , M and j = 1, 2, . . . N (Figure 1). The grid considers two ghost cells at all sides: for i = −1, 0 and i = M + 1 and j = −1, 0 and j = N + 1; the computation made in the ghost cell should also be based on data from the interior cells, namely:

$$U^{n+1}_{-1,j} = U^{n+1}_{0,j} = U_{1/2,j} \text{ and } U^{n+1}_{M+1,j} = U^{n+1}_{M+2,j} = U_{M+1/2,j},$$
$$U^{n+1}_{i,-1} = U^{n+1}_{i,0} = U_{i,1/2} \text{ and } U^{n+1}_{i,N+1} = U^{n+1}_{i,M+2} = U_{i,M+1/2}.$$

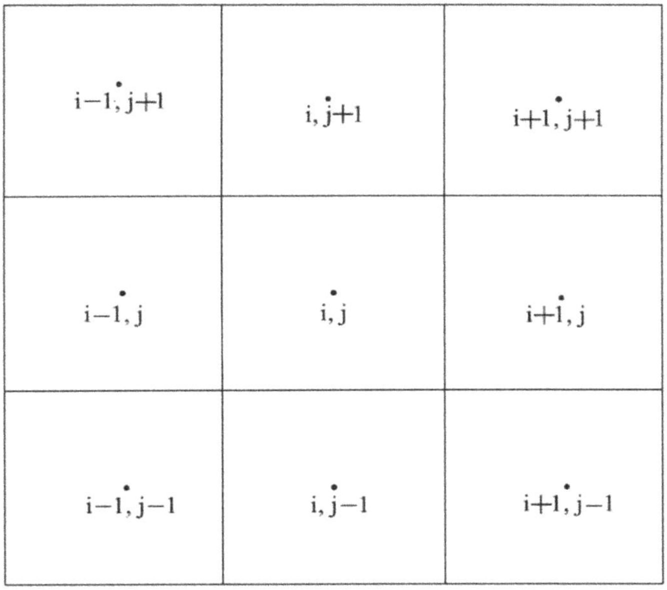

Figure 1. The structured mesh.

2.4. Courant Number

The Courant–Friedrichs–Lewy or CFL condition expresses the finding that the distance that any information travels during the time-step length within the mesh must be lower than the distance between the mesh elements. In other words, information from a given cell or mesh element must propagate only to its immediate neighbors (https://www.simscale.com/knowledge-base/what-is-a-courant-number/, 10 February 2020). The derivation of the CFL condition leads to the formula for the Courant number and is given by:

$$CFL = \frac{\Delta t}{\min\{\frac{\Delta x, \Delta y}{\left(C+\sqrt{u^2+v^2}\right)_{i,j}}\}} \quad (32)$$

where $C = \sqrt{gh}$ is the celerity and the CFL number is chosen to be equal to 0.9 for the stability criterion of the model used in this case study.

2.5. The Nash–Sutcliffe Efficiency (NSE)

The NSE was developed by Nash and Sutcliffe in 1970 as a way to normalize the statistic that determines the relative magnitude of the residual variance, compared to the measured data variance [41]. They found out that the NSE was able to indicate how well the plot of the observed versus the simulated data fitted the 1:1 line. The equation can be expressed as follows:

$$\mathrm{NSE} = 1 - \left[\frac{\sum_{i=1}^{n} \left(H_i^{obs} - H_i^{sim} \right)^2}{\sum_{i=1}^{n} \left(H_i^{obs} - \overline{H^{obs}} \right)^2} \right] \quad (33)$$

where H_i^{obs} is the observed water height obtained at time i; H_i^{sim} is the simulated water height at time i, and the average observed water height is represented by $\overline{H^{obs}}$.

3. Description of the Study Area

3.1. Description of the Watershed

Douala is the most populous city in Cameroon, with about 3 million people distributed across the 12 watersheds. Among the watersheds, the Tongo-Bassa watershed has been chosen for this study because it has been recorded as the place where enormous human and infrastructural damage has been caused by flooding each year. The Tongo Bassa watershed is located between longitude 9°46′00″ E and latitude 4°4′00″ N (WGS 84 UTM Zone 32N) and its area is estimated at 4161.6 ha (Figure 2). [42] estimated in 2009 that the watershed constituted 27 districts, with a population density ranging from 25 to more than 350 inhabitants per hectare. The watershed has a river called the "Tongo Bassa River", which flows into the major Wouri River in the region and then joins the Atlantic Ocean [43]. The soil of the area is essentially made up of sedimentary rocks, with colors that vary from brown to black and soils that change from sandy ferritic to deep sandy clayey soils [44].

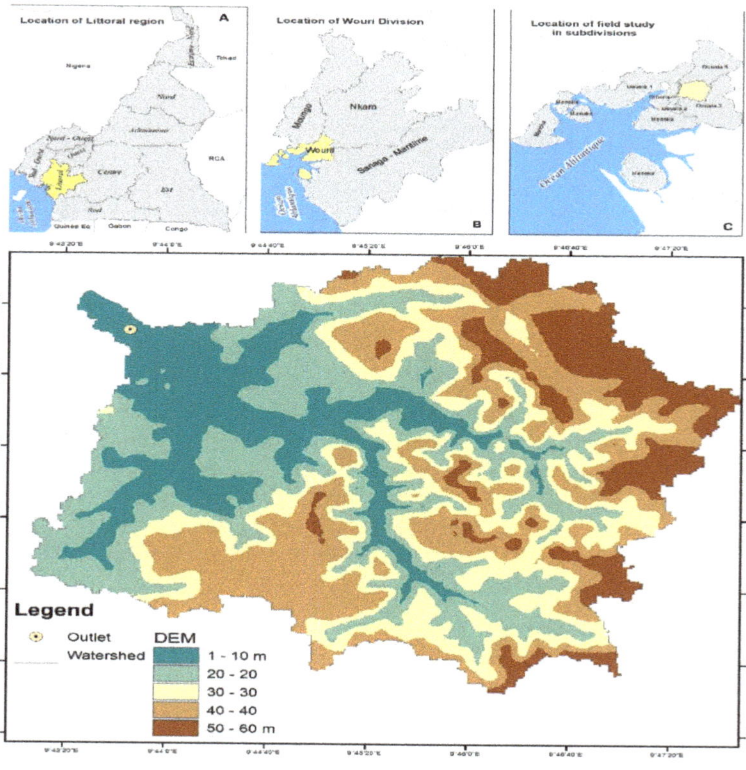

Figure 2. Location of the Tongo Bassa watershed in Cameroon: (**A**) (Location of Littoral Region on the Cameroon map), (**B**) (Location of Wouri Division on the Littoral Region), (**C**) (Location of Field study in subdivisons of Wouri) and Tongo Bassa Watershed.

The watershed is composed of valleys where the altitudes vary from about 5 to 60 m above sea level, on average. This area is dominated by two main seasons: dry (December to February) and wet (March to November). However, it should be noted that these seasons do not affect the monthly temperatures. Particularly in August, the temperature is 26 °C and reaches 28.6 °C in February, with an annual temperature average estimated at 27 °C [45]. They found that the lowest amount of precipitation is recorded in December (the monthly average is 35 mm) and the maximum is recorded in July and August (an average of 754 mm). The land cover of the watershed is occupied by vegetated soil at 6.43% (267.61 ha), bare soil at 5.03% (209.17 ha), farming at 26.23% (1091.46 ha), urban areas at 62.25% (2590.44 ha) and water at 0.07% (2.92 ha) (Figure 3).

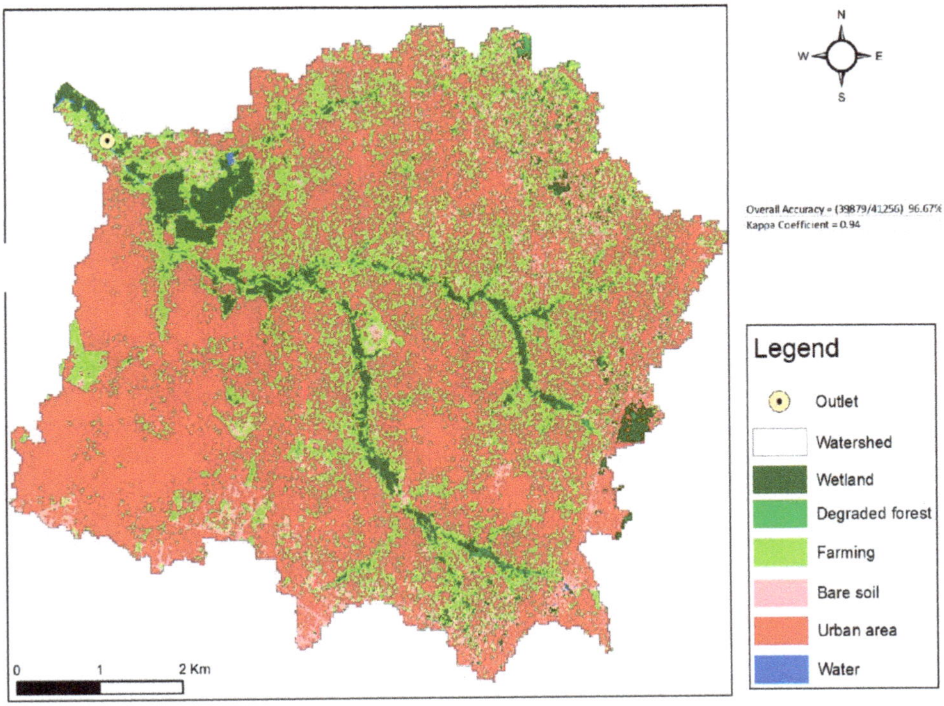

Figure 3. Tongo Bassa watershed land-cover map.

3.1.1. Map Processing

In this study, a digital elevation model (DEM) mapping 30 m from the study area was uploaded from the website (https://search.earthdata.nasa.gov/search accessed 21 June 2022) and processed using the ArcGIS software. This procedure highlighted the watershed, in addition to identifying the outlet and its coordinates; next, we imported "the whole catchment" which was previously delimited in the ArcGIS software, and the DEM uploaded using "global mapper software" for the next processing stage. The output file was created in the "xyz format", following this final processing of the map using a global mapper with a mesh of 20 m along the x and y directions. The coordinates were output in the UTM projection and the surface was projected in a Cartesian reference frame, with a 20-meter gap in the x and y directions. The results are shown in the Figure 4, as follows.

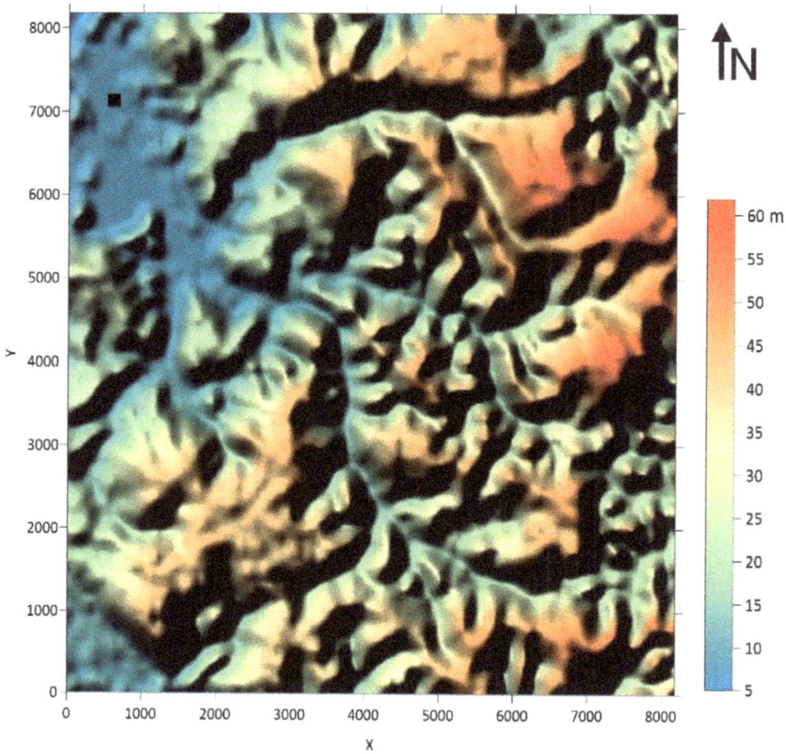

Figure 4. A visualization of the watershed in an orthonormal frame.

3.1.2. Manning's Roughness and Infiltration Parameters

We relied on the literature for sourcing data on Manning's roughness and infiltration parameters. This is owing to the lack of relevant data in our study area. In 1989, Arcement proposed a guide for selecting Manning's roughness coefficients for natural channels and flood plains [46]. Table 1 presents the infiltration parameters proposed by [47,48], with Manning's roughness coefficients [46].

Table 1. Infiltration parameters and Manning coefficients.

Land Cover	i_f	i_o	r (min^{-1})	n ($s/m^{1/3}$)
Vegetated	0.029	0.097	1.383	0.065
Bare	0.005	0.017	1.383	0.020
Cultivated	0.021	0.069	1.383	0.050
Urban	0.003	0.008	1.383	0.015

- ***Sensitivity to rainfall intensity in the model***

In this first case study, we used three values of rainfall intensity, as mentioned in Table 2.

Table 2. The parameters applicable for the study of sensitivity to rainfall intensity.

Land Cover	i_f	i_o	r (min^{-1})	n (s/m$^{1/3}$)	Rainfall Intensity (mm/h)
Vegetated	0.029	0.097	1.383	0.065	Case 1: 15 mm/h
Bare	0.005	0.017	1.383	0.020	Case 2: 50 mm/h
Cultivated	0.021	0.069	1.383	0.050	Case 2: 85 mm/h
Urban	0.003	0.008	1.383	0.015	

- ***Sensitivity to the infiltration parameters in the model***

The initial infiltration (i_o), the parameters depending on the type of soil (r), and the Manning coefficient (n) were maintained as mentioned above, but only the final infiltration (i_f) was subjected to several tests; i.e., $1/4i_f$, $1/2i_f$, $3/4i_f$, and i_f. The simulation was conducted with a rain intensity of 15 mm/h for 360 min (6 h).

4. Initial Conditions and Numerical Results

4.1. Initial Conditions

At time t = 0 s, the surface of the watershed is dry and is characterized by its topography. The velocities in the x and y directions are zero, which means u = v = 0. The watershed is traversed by a watercourse where the water height in the left and right positions is assumed to be equal to 0 m. Moreover, E and q are assumed to be 0 m^3/s during the whole simulation. It is also assumed that for each part of the land used in the whole watershed, the infiltration rate and Manning coefficient are uniform and constant.

4.2. Numerical Results

4.2.1. Sensitivity to Rainfall Intensity of the Model

Figure 5 shows the water height propagation in the Tongo-Bassa watershed for rainfall intensity of 15 mm/h. After 15 h, a surge in the water converges toward the watercourse located in low-altitude or valley-bottom areas, which eventually comes out of its minor bed; the water height reaches 1.2 m in the deeper section.

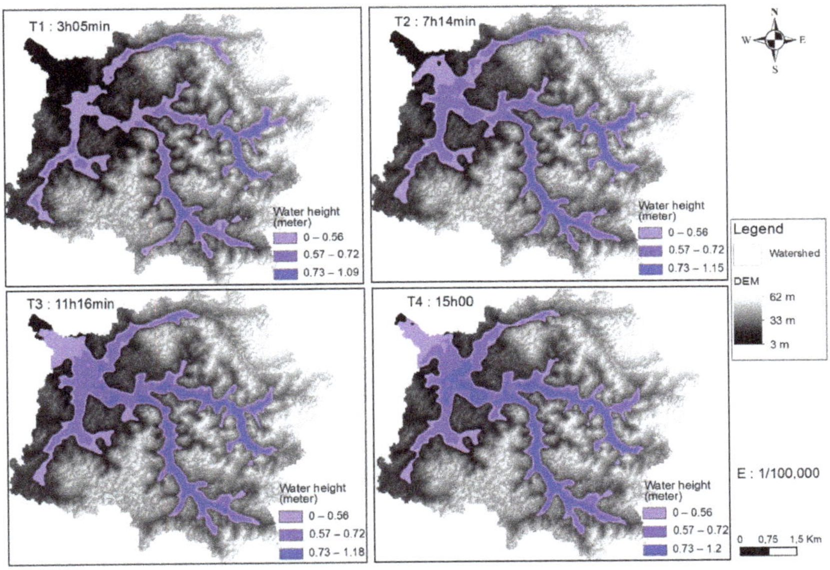

Figure 5. Water height propagation at different time steps for a constant rainfall intensity (15 mm/h).

The same simulation has been repeated, taking 85 mm/h as the rainfall intensity, and the results are illustrated in Figure 6. The distribution of water over the watershed appears greater than in case study 1 (at 15 mm/h). The floodable surface is more significant and covers the urge space of the watershed. The water heights in the lowest points (watercourse) reach a maximum value of 4 m after 11 h of simulation.

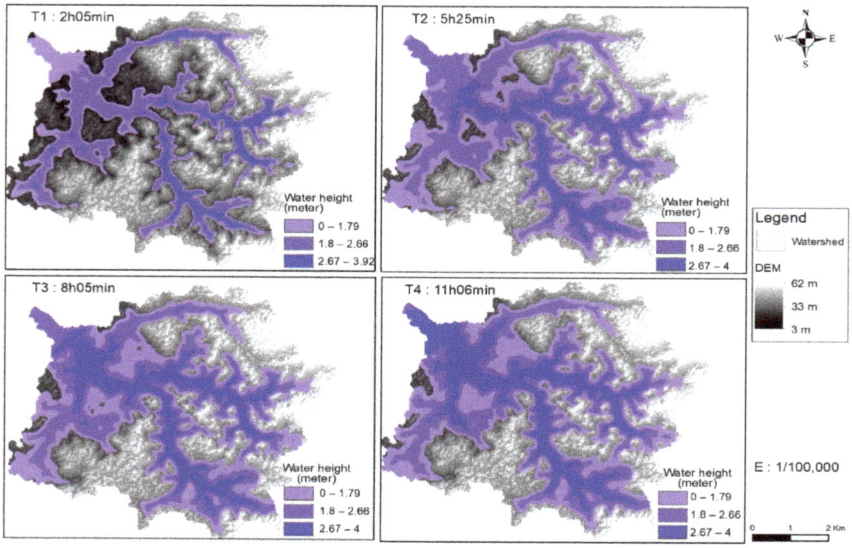

Figure 6. Water-height propagation at different time steps for a rainfall intensity of 85 mm/h.

Overall, the results show that the greater the rainfall intensity, the shorter the time needed for the water in the watershed to reach a significant height. The floodable surface is well-represented in the watershed, with an emphasis on flooded areas with a wide distribution of water height.

In summary, the results indicate that water depth in the watershed is sensitive to both rainfall intensity and the simulation duration.

The curves presented in Figure 7a,b show the heights of the water and flow discharge obtained at the outlet for each value of rainfall intensity (15, 50, and 85 mm/h). It transpires that the height at the outlet increases with rainfall intensity and simulation duration. This can be explained by the fact that the study area is located in a sedimentary zone; the water saturation rate of the soil is very important, resulting in the low infiltration of water. The runoffs on the surface are more important and all converge toward the outlet. This explains how the greatest water depth (4.04 m) was obtained with high rainfall intensity (85 mm/h) and the lowest value (0.87 m) was obtained with low rainfall intensity (15 mm/h).

All the curves start at point 0 because it was assumed that the point located at the outlet was dry at the start of the simulation. The flow rate curves (Figure 7b) for different rainfall intensities show an increase in flow rates that is proportional to the water height. The maximum value of the flow (35 m^3/s) is reached at 85 mm/h. When the rain intensity is 50 mm/h, the flow rate reaches 13.5 m^3/s while at 15 mm/h, 2.5 m^3/s is obtained as the maximum value of the flow at 12.14 m of water height.

4.2.2. Sensitivity of the Infiltration Parameters in the Model

The final infiltration parameters for each case were set, as follows: $1/4i_f$, $1/2i_f$, $3/4i_f$, and i_f; the results were obtained after 360 min (6 h). The results show that water propagation in the catchment is more significant when the final infiltration value (i_f) is high (Figure 8A). A significant quantity of the water is also flowing on the surface and toward the outlet. When the final infiltration decreases (Figure 8B–D), a large volume of water stagnates on the surface, forming small lakes where the maximum height can reach 1 m, while a very small percentage runs off to the outlet.

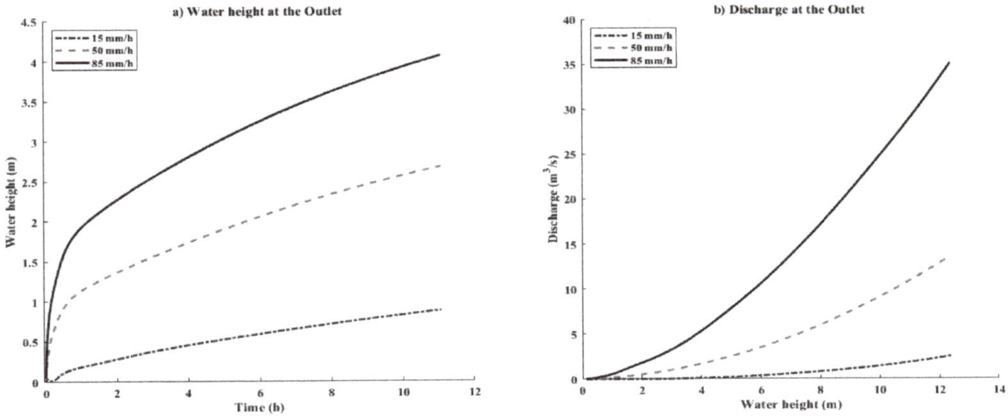

Figure 7. Water height (**a**) and discharge (**b**) at the outlet.

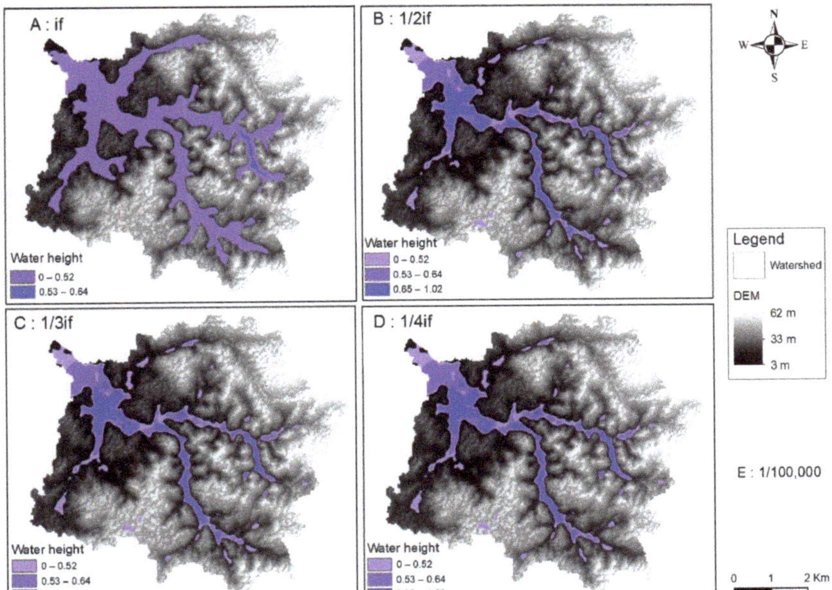

Figure 8. Water height propagation map after 6 h, with 15 mm/h as the rainfall intensity.

The water height and flow rates at the outlet corresponding to each final infiltration parameter are shown in Figure 9a,b, respectively. From Figure 9a, it can be seen that there is a significant difference between the curve obtained with i_f and the other curves. This shows that the maximum value of the water level is 1 m after a simulation of 11 h for (i_f), while the other curves tend toward 4.28×10^{-3} m. It should be noted that at the beginning of the simulation, i.e., at T = 1 min, the other curves ($1/4i_f$, $1/2i_f$, and $3/4i_f$) reach their maximum values (0.2 m); then, at the same time, the i_f curve is 0.13 m. Over a simulation time of 14 min, i.e., ranging from 0.0 to 17 min, the other curves ($1/4i_f$, $1/2i_f$, and $3/4i_f$) remain above the curve (i_f) before reversing proportionally from 17 min until the 11-hour point. This scenario supports the theory that the majority of the water stagnates on the surface and very little moves toward the outlet, which will lower the height of the water at the outlet. This same explanation is also reflected in the calculation of the rating curves in

Figure 9b; the maximum flow (2.59 m^3/s) is obtained with (i_f) while the rest of the curves converge toward 4.8×10^{-2} m^3/s, over a total water height at the outlet of 2 m.

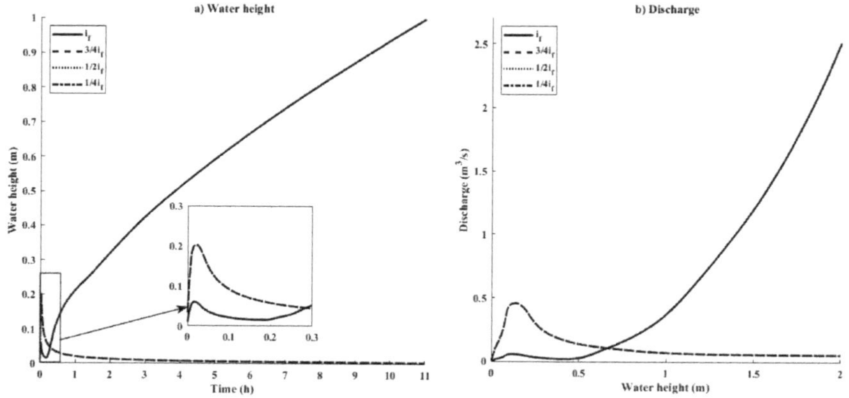

Figure 9. Water height (**a**) and flow rates (**b**) at the outlet of the catchment for different values of final infiltration, with 15 mm/h set as the rainfall intensity.

4.2.3. Model Validation

As part of the validation of the model, the Grand-Ouaga basin has been used, which is located in Burkina Faso, West Africa (Figure 10). The watershed is located between parallels 12.26° and 12.35° North in latitude and between meridians 1.43° and 1.54° West in longitude. This study is based on data from the floods that occurred in the area from 18 to 22 July 2016 (case study 1) and 25 July 2017 (case study 2). Prior to the flood, a water level gauge that allows the recording and transmission of the measurement of water height at the given point in a watercourse had been installed at the outlet. This work involved simulating the water height at the outlet, located at longitude: −1°26′ E and latitude: 12°18′ N, and comparing it to the measured data collected in the field. The mapping of the area was made according to the steps mentioned in Section 3.1 (Description of the watershed) in the current paper.

Table 3. Infiltration parameters from the Burkina Faso National Institute of Geography.

Land Cover	i_f	i_o	r (min^{-1})
Vegetated	0.22	2.95	0.503
Bare	0.54	3.46	0.116
Cultivated	0.66	4.07	0.097
Urban	0.25	3.15	0.238

Table 3 presents the infiltration parameters provided by the Burkina Faso National Institute of Geography (http://www.igb.bf/ accessed on 18 January 2018).

The rainfall data used in the case study were collected from project activities (https://www.amma2050.org/ 18 to 22 July 2016; 25 July 2017).

The results are presented in Figures 11a and 12a for case study 1 and case study 2, respectively, and the Manning coefficients were taken [49]. At the beginning of the long rains from 18 to 22 July 2016, the water height at the outlet was estimated to be 0.32 m, while on 25 July 2017, the water height was 0 m. The comparison made between the simulated water height and the measured height is illustrated in Figures 11b and 12b, for case 1 and case 2, respectively. The first comparison conducted for case study 1 (Figure 11a) over 24 h suggests a large difference in the maximum value between the observed data (2.21 m) and the simulated data (1.9 m). The Nash–Sutcliffe value is 0.7, which means a good correlation was observed between the two curves at the beginning and the end of the simulation.

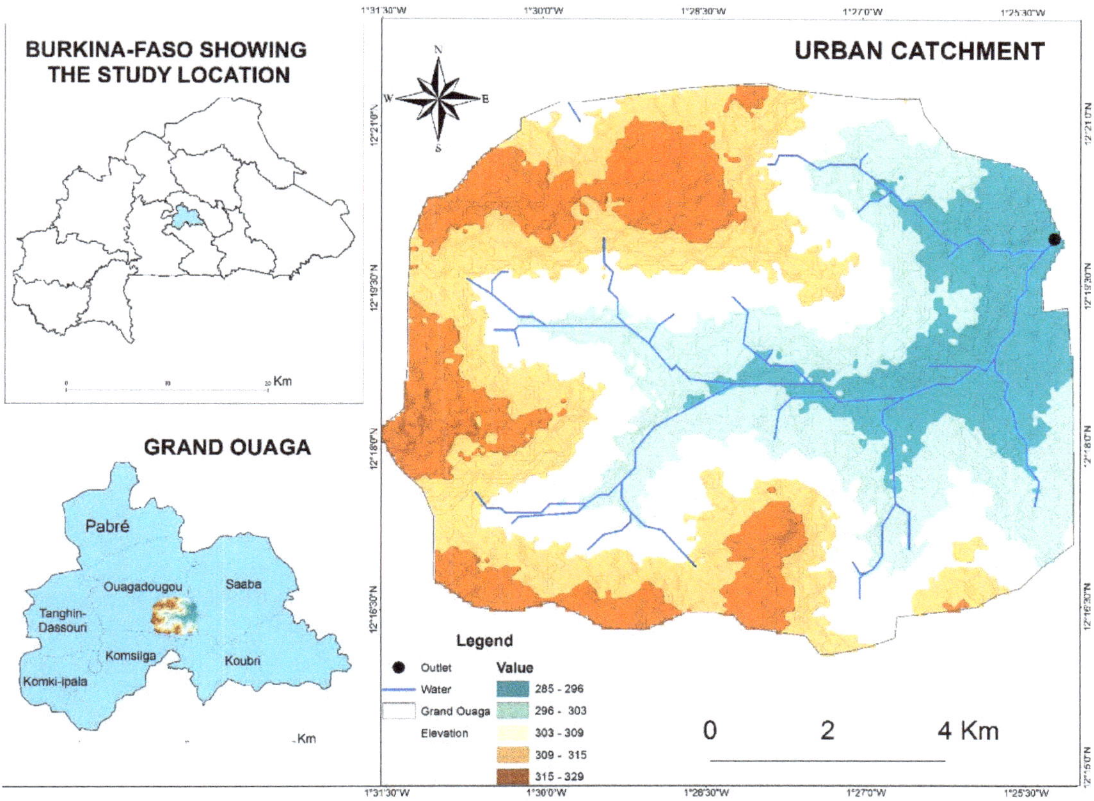

Figure 10. Location map of the Grand-Ouaga area in Burkina Faso (West Africa).

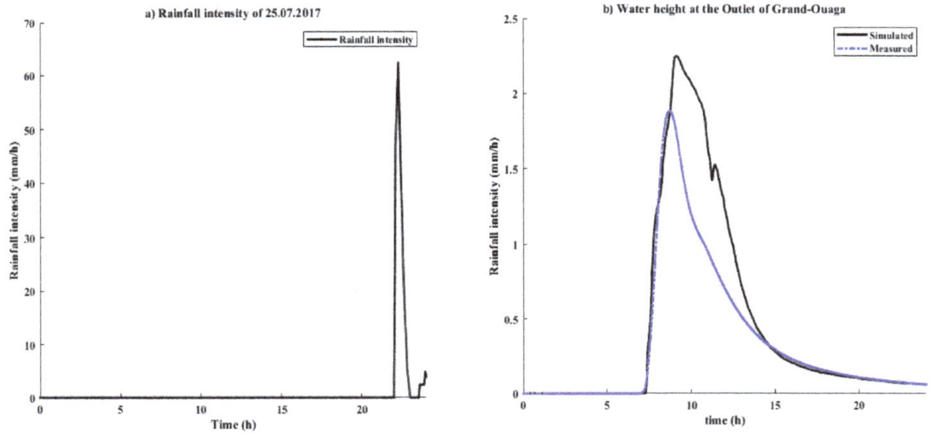

Figure 11. Rainfall intensity (a) and a comparative study of the water height (b), during the flood of 25 July 2017.

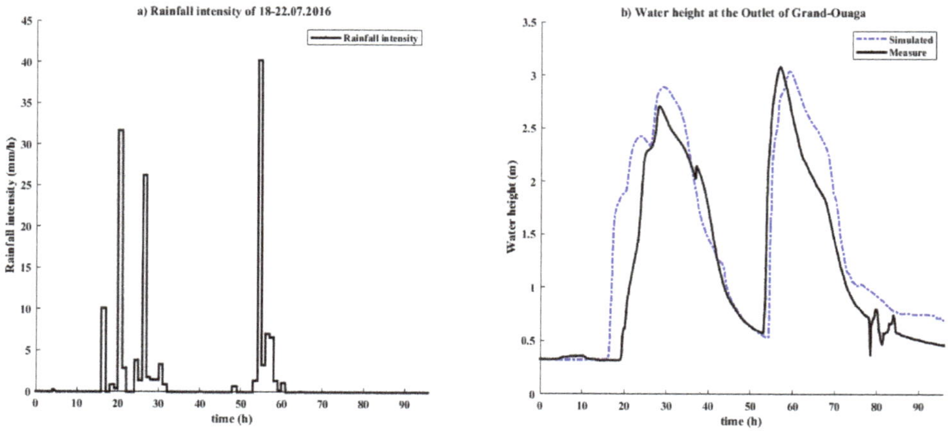

Figure 12. Rainfall intensity (**a**) and comparative study of the water height (**b**) during the flood of 18 to 22 July 2016.

According to the simulation results for case study 2 (Figure 12a), the water reaches its maximum height on the third day (of 4 days) of 3.02 m for the observed values and 3 m for the simulated values. Although the values differ somewhat, the two curves exhibit a nearly identical trend, and the Nash–Sutcliffe value is 0.73. Nevertheless, the findings show a strong link between the two trends.

The results from this validation show that the model can simulate the water height at the outlet when the coordinate points are known. The differences observed on the curves of the simulated and measured results are due to certain assumptions, such as the consideration of infiltration parameters and Manning coefficient as constant values over all the land cover. Furthermore, the digital elevation models available for free are long-dated and are not frequently updated according to the evolution of land use. In addition, these maps are downloaded with an elevation of 30 m (DEM30), which does not allow the exact land cover estimation to be established with precision.

5. Conclusions

This study applied the FVM-Godunov model to solve 2D shallow-water equations, in order to simulate the surface flow. The ghost-cell approach was integrated into the model, in order to maintain a uniform solution in both the interior and boundary cells. The model is demonstrated by simulating the surface runoff on a given catchment, by considering parameters such as infiltration, the roughness coefficient, and the rainfall intensity. The computed results were validated with the observed data on the catchment of Grand-Ouaga in Burkina Faso. The main conclusions are as follows:

1. The model adapts to both the topography and the field data. The model is sensitive to several parameters, including infiltration rate, rainfall, and surface slopes. Furthermore, the model can predict the flow propagation on a given catchment and provide the water height/discharge at the outlet. As expected, runoff intensity increases with rainfall intensity, but this is not in a linear fashion; more intense rainfall produces proportionally more severe flooding.
2. When the estimated final value infiltration is moderated, there is an increase in the rate of water infiltration in the catchment, which reduces the quantity of water that converges toward the outlet. This leads to a decrease in surface runoff; water tends to accumulate in the already saturated zone.
3. The results proposed in the validation section show good agreement between the measured data and the simulated model. These results reveal that the parameters collected on the field and the rainfall intensity registered with the rain gauge

present the results close to reality. However, it was acknowledged that the study needs improvement, which can help the model to simulate values closer to the observed values.

4. The consideration of a constant and uniform value of the infiltration coefficient and the Manning coefficients in each land-cover type of the watershed also increases the risk of overestimating the results, since these different coefficients should not be considered constant. Hence, there is a need to take samples in the field with the appropriate equipment.
5. Further elaboration of the model is required. For example, parameters such as evaporation, evapotranspiration, and water flow rate that are present in the study area will be considered in a subsequent study in order to improve the reach and accuracy of the final results. Although these parameters have been integrated into the numerical equations in this paper, their effects have yet to be analyzed.

Author Contributions: Conceptualization, A.J.E.; data curation, A.J.E.; formal analysis, S.L.Z.A.; funding acquisition, L.Z.; investigation, A.J.E., H.F. and Y.C.; methodology, H.F. and S.L.Z.A.; project administration, B.K.; resources, Y.C.; software, A.J.E.; supervision, L.Z.; validation, Y.C.; writing—original draft, L.Z. and B.K.; writing—review and editing, B.K. All authors have read and agreed to the published version of the manuscript.

Funding: The authors gratefully acknowledge the financial support given for this research by the National Natural Science Foundation of China (Grant Nos. 51839008 and 51679066), Fok Ying Tong Education Foundation (Grant No. 161068), and the China Scholar Council (File No. 201806715024).

Institutional Review Board Statement: Ling Zhou, Ph.D Professor, College of Water Conservancy and Hydropower Engineering, Hohai University; 1 Xikang Road, Nanjing, China; 210098. His research interests include hydraulic transients in pipe system, hydraulics in Hydropower/Pump/Pumped Storage Station, Transient Flow in Urban Storm Sewers. As principal investigator, Ling Zhou was awarded 25 research grants and 2 travel grants, including: 3 funds (general, young, international exchange) from National Natural Science Foundation of China, 1 fund from Fok Ying Tong Education Foundation, and Specialized Research Fund for the Doctoral Program of Higher Education of China, etc.

Informed Consent Statement: Informed consent was obtained from all subjects involved in this study.

Data Availability Statement: Not applicable.

Acknowledgments: The authors gratefully acknowledge the financial support given for this research by the National Natural Science Foundation of China (Grant Nos. 51839008 and 51679066), Fok Ying Tong Education Foundation (Grant No. 161068), and the China Scholar Council (File No. 201806715024).

Conflicts of Interest: The authors declare no conflict of interest.

Glossary

The following symbols are used in this paper:

C	Celerity
C_1^x; C_2^x	The coefficients in the x-direction
C_1^y; C_2^y	The coefficients in the y-direction
F	The flux vectors in the x-direction
$\mathrm{F}^n_{i+\frac{1}{2}L}$; $\mathrm{F}^n_{i-\frac{1}{2}R}$	The average cell value of flux to the left and right of the interface, at time $n\Delta t$
g	Gravitational acceleration
G	The flux vectors in the y-direction
h	Water height
H	Water level
$\mathrm{h}_{i+\frac{1}{2}R}$	The water height, computed in the right side of the cell
$\mathrm{h}_{i+\frac{1}{2}L}$	The water height, computed in the left side of the cell

i_0	Initial infiltration rate
i_f	The final infiltration rate
I	Infiltration rate
nm	The Manning Strickler coefficient
P	Rainfall intensity
$P(h)$	Hydrostatic pressure
r	A constant, depending on soil properties
S_f	The friction terms
S_p	The source terms
S_s	The bottom slope
t	Time
t_s	The duration of the storm
u	The depth-averaged velocity components in the x-direction
U	The conservative variables
$\mathrm{U}_{\mathrm{i}+\frac{1}{2}\mathrm{L}}$	The general solution at the left side of the cell
$\mathrm{U}_{\mathrm{i}+\frac{1}{2}\mathrm{R}}$	The general solution at the right side of the cell
U_D; U_U	The down and up approximations of the solution at the cell
U_L; U_R	The left and right approximations of the solution at the cell, respectively
v	The depth-averaged velocity components in the y-direction
w_i^n	The amount of water available for infiltration on a cell at a time t^n
x, y	Cartesian coordinates
z	The bed elevation
Δt	Time increment
Δx	The reach length in the x-direction
Δy	The reach length in the y-direction
β_1^y; β_2^y; β_3^y	The eigenvalues of the Jacobian of G
β_1^x; β_2^x; β_3^x	The eigenvalues of the Jacobian of F

Abbreviations

CEDEX	Centre for Studies and Experimentation of the Ministry of Public Works
CFL	The Courant–Friedrichs–Lewis condition
CLOCA	Central Lake Ontario Conservation Authority
CEDEX	Courrier d'Entreprise à Distribution Exceptionnelle (Centre for Decision Research and Experimental Economics)
Dfb	A warm-summer humid continental climate
DEM	The digital elevation model
FDM	The finite difference method
FVM	The finite volume method
HEC RAS	The Hydrological Engineering Centre river analysis system
HLL	Harten Lax Leer
HLLC	Harten-Lax–van Leer-Contact
NITI	The National Institution for Transforming India
IGPCC	The Intergovernmental Panel on Climate Change
LiDAR	Light detection and ranging
MUPH	The Ministry of Urban Planning and Housing
MUSCL	The monotone upwind scheme for conservation laws
NAD	The North American Datum
OFS	Operational forecasting system
SWMM	Stormwater management model
UNDP	The United Nations development program
UK	United Kingdom
UTM	The Universal Transverse Mercator coordinate system
WRF	Weather research and forecasting

References

1. Berggren, K.; Olofsson, M.; Viklander, M.; Svensson, G.; Gustafsson, A.-M. Hydraulic impacts on urban drainage systems due to changes in rainfall caused by climatic change. *J. Hydrol. Eng.* **2011**, *17*, 92–98. [CrossRef]
2. Jacobson, C.R. Identification and quantification of the hydrological impacts of imperviousness in urban catchments: A review. *J. Environ. Manag.* **2011**, *92*, 1438–1448. [CrossRef] [PubMed]
3. Fletcher, T.; Andrieu, H.; Hamel, P. Understanding, management and modelling of urban hydrology and its consequences for receiving waters: A state of the art. *Adv. Water Resour.* **2013**, *51*, 261–279. [CrossRef]
4. Semadeni-Davies, A.; Hernebring, C.; Svensson, G.; Gustafsson, L.-G. The impacts of climate change and urbanization on drainage in Helsingborg, Sweden: Combined sewer system. *J. Hydrol.* **2008**, *350*, 100–113. [CrossRef]
5. United Nations Development program (UNDP). *Adapting to the Inevitable: National Action and International Cooperation*; Human Development Report: New York, NY, USA, 2007; p. 36. Available online: http://hdr.undp.org/sites/default/files/hdr_20072008_fr.pdf (accessed on 10 October 2022).
6. Descroix, L.; Guichard, F.; Grippa, M.; Lambert, L.; Panthou, G.; Mahé, G.; Gal, L.; Dardel, C.; Quantin, G.; Kergoat, L.; et al. Evolution of surface hydrology in the Sahelo-Sudanian strip: An updated review. *Water* **2018**, *10*, 748. [CrossRef]
7. National Institution for Transforming India (NITI) Aayog. *Report of the Committee Constituted for Formulation of Strategy for Flood Management Works in Entire Country and River Management Activities and Works Related to Border Areas (2021–26)*; Rajiv Kumar: New Delhi, India, 2021. Available online: https://www.niti.gov.in/sites/default/files/2021-03/Flood-Report.pdf (accessed on 10 October 2022).
8. IGPCC. *Climate Change 2007: Synthesis Report*; IPCC: Geneva, Switzerland, 2007; p. 114. Available online: https://www.ipcc.ch/site/assets/uploads/2018/02/ar4_syr_fr.pdf (accessed on 10 October 2022).
9. van den Honert, R.C.; McAneney, J. The 2011 Brisbane Floods: Causes, Impacts and Implications. *Water* **2012**, *3*, 1149–1173. [CrossRef]
10. Pitt, M. *The Pitt Review: Learning Lessons from the 2007 Floods*; The Cabinet Office: London, UK, 2008.
11. Benfield, A. *Thailand Floods Event Recap Report 2011*; Impact Forecasting, Report; Wiley: Chicago, IL, USA, 2012; pp. 4–40. Available online: http://thoughtleadership.aonbenfield.com/Documents/20120314_impact_forecasting_thailand_flood_event_recap.pdf (accessed on 1 March 2012).
12. Mississippi River Commission. *MR & T Flood Report*; Mississippi River Commission: Vicksburg, MS, USA, 2011.
13. Petrucci, O.; Pasqua, A.A.; Polemio, M. Flash flood occurrences since the 17th century in steep drainage basins in southern Italy. *Environ. Manag.* **2012**, *50*, 807–818. [CrossRef] [PubMed]
14. Erpicum, S.; Dewals, B.; Archambeau, P.; Detrembleur, S.; Pirotton, M. Detailed 2D numerical modeling for flood extension forecasting. In Proceedings of the 4th International Conference on Fluvial Hydraulics, Izmir, Turkey, 3–5 September 2008.
15. Okyere, C.Y.; Yacouba, Y.; Gilgenbach, D. The problem of annual occurrences of floods in Accra: An integration of hydrological, economic and political perspectives. *Theor. Empir. Res. Urban Manag.* **2013**, *8*, 45–79.
16. Garcıa-Navarro, P.; Brufau, P.; Burguete, J.; Murillo, J. The shallow water equations: An example of hyperbolic system. *Monogr. Real Acad. Cienc. Zaragoza.* **2008**, *31*, 89–119.
17. Saint-Venant, A.B. *Theory and General Equations of the Non-Permanent Movement of Running Waters*; Reports of the Sessions of the Academy of Sciences: Paris, France, 1871; Volume 17, pp. 147–154.
18. Li, L.; Cargnelutti, M.; Mosca, C. Dam-break flood forecasting in Piemonte region, northwest Italy. *Water Resour. Manag.* **1991**, *5*, 261–270. [CrossRef]
19. Yanmaz, A.M.; Seckiner, G.; OZaydın, V. A method for optimum layout design of concrete gravity dams. *Water Eng. Res. Int. J. Korea Water Res.* **2001**, *2*, 199–207.
20. Petaccia, G.; Natale, L.; Savi, F. Simulation of the Sella Zerbino catastrophic Dam Break. In *River Flow; Proceedings of the International Conference on Fluvial Hydraulics, Izmir, Turkey, 3–5 September 2008*; Kubaba Congress Department and Travel Services: Ankara, Turkey, 2008; Volume 1, pp. 601–608.
21. Bozkus, Z. Failure Analysis for Cinarcik Dam. In Proceedings of the IV National Water Engineering Symposium, Istanbul, Turkey, 6–10 July 2009; pp. 89–98. (In Turkish)
22. Tsakiris, G.; Spilliotis, M. Dam-Breach Hydrograph Modelling: An Innovative Semi-Analytical Approach. *Water Resour. Manag.* **2013**, *27*, 1751–1762. [CrossRef]
23. Brufau, P.; Vazquez-Cendon, M.E.; Garcia-Navarro, P. A numerical model for flooding and drying of irregular domains. *Int. J. Numer. Methods Fluids* **2002**, *39*, 247–275. [CrossRef]
24. Singh, J.; Altinakar, M.S.; Ding, Y. Two-Dimensional Numerical Modeling of Dam-Break Flows over Natural Terrain Using a Central Explicit Scheme. *Adv. Water Resour.* **2011**, *34*, 1366–1375. [CrossRef]
25. Mahdizadeh, H.; Stansby, P.K.; Rogers, B.D. Flood wave modeling based on a two-dimensional modified wave propagation algorithm coupled to a full-pipe network solver. *J. Hydraul. Eng.* **2012**, *138*, 247–259. [CrossRef]
26. Chen, A.S.; Hsu, M.H.; Chen, T.S.; Chang, T.I. An integrated evacuation model for highly developed urban areas. *Water Sci. Technol.* **2005**, *51*, 221–229. [CrossRef]
27. Alcrudo, F.; Mulet, J. Description of the Tous Dam break case study (Spain). *J. Hydraul. Res.* **2007**, *45*, 45–57. [CrossRef]
28. Mignot, E.; Paquier, A.; Haider, S. Modelling floods in a dense urban area using 2D shallow water equations. *J. Hydrol.* **2006**, *321*, 186–199. [CrossRef]

29. Bellos, V.; Tsakiris, G. Comparing various methods of building representation for 2D flood modeling in built-up areas. *Water Resour. Manag.* **2015**, *29*, 379–397. [CrossRef]
30. Gharbi, M.; Soualmia, A.; Dartus, D.; Masbernat, L. Comparison of 1D and 2D hydraulic Models for Floods Simulation on the Medjerda Riverin Tunisia. *J. Mater. Environ. Sci.* **2016**, *7*, 3017–3026.
31. Yung, C.Y.; Che, L.H.; Hung, S.J.; Chih, C.H.; Wei, C.B.; Yi-Chiang, Y.; Ray, S.W.; Lee-Yaw, L. An Operational Forecasting System for Flash Floods in Mountainous Areas in Taiwan. *Water* **2019**, *11*, 2100.
32. Wei-Bo, C.; Tzu-Yin, C.; Hongey, C.; Huei-Shuin, F.; Yi-Chiang, Y.; Wen-Ray, S.; Lee-Yaw, L. An Operational High-Performance Forecasting System for City-Scale Pluvial Flash Floods in the Southwestern Plain Areas of Taiwan. *Water* **2021**, *13*, 405.
33. Benedini, M.; Tsakiris, G. *Water Quality Modelling for Rivers and Streams*; Springer: New York, NY, USA, 2013.
34. Bahtiar, S.; Chuai-Aree, S.; Busaman, A. A numerical algorithm and Visualization Software for Flood simulation in Urban Area: A Case Study of West Jakarta, Indonesia. *Int. J. Circuits Syst. Signal Proc.* **2018**, *12*, 147–153. Available online: https://www.naun.org/main/NAUN/circuitssystemssignal/2018/a422005-aej.pdf (accessed on 10 October 2022).
35. Yamaguchi, S.T.; Ikeda, I. Rapid Flood Simulation Software for Personal Computer with Dynamic Domain Defining Method. In Proceedings of the 4th International Symposium of Flood Defense: Managing Flood Risk, Reliability and Vulnerability, Toronto, ON, Canada, 6–8 May 2008.
36. Zhou, L.; Wang, H.; Liu, D.; Ma, J.; Wang, P.; Xia, L. A second-order Finite Volume Method for pipe flow with column. *J. Hydro-environ. Res.* **2017**, *17*, 47–55. [CrossRef]
37. Maugeri, A. *Capabilities of a Coupled 1D/2D Model for Flood Inundation Simulation*; Columbia Water Center: New York, NY, USA, 2012.
38. Audusse, E. Hyperbolic Models and Numerical Analysis for Shallow Water Flows. Doctoral Thesis, Pierre and Marie Curie University, Paris, France, 2004.
39. Gnenakantanhan, C.; Babacar, L.; Fowe, T.; Adjadi, M.L.; Harouna, K. Urban Flood Modeling Using 2D Shallow-Water Equations in Ouagadougou, Burkina Faso. *Water* **2020**, *12*, 2120. [CrossRef]
40. Toro, E.F. The HLL and HLLC Riemann Solvers. In *Riemann Solvers and Numerical Methods for Fluid Dynamics*; Springer: Berlin/Heidelberg, Germany, 2009; pp. 315–344.
41. Nash, J.A.; Sutcliffe, J.V. River flow forecasting through conceptual models, I: A discussion of principles. *J. Hydrol.* **1970**, *10*, 398–409. [CrossRef]
42. Kannan, N.; Sabu, J. Quality of Groundwater in shallow Aquifers of a Paddy Dominated Agricultural River Basin, Kerala, India. *World Acad. Sci. Eng. Technol.* **2009**, *3*, 1137–1155.
43. Ministry of Urban Planning and Housing of Cameroon (MUPHC). Urban Development and Poverty reduction Strategy: City of Douala and its Greater Urban Area. Final Report, December 2009, 0828 Rapport CDS final-141209-15.12.09-JG. Available online: https://knowledge-uclga.org/IMG/pdf/development_strategy_of_the_city_of_douala_and_its_metropolitan_area.pdf (accessed on 10 October 2022).
44. Regnoult, M. *Geological Synthesis of Cameroon*; DMG: Yaounde, Cameroon, 1986.
45. Nkot, S.; Ketchemen-Tandia, B.; Ndje, Y.; Emvouttou, H.; Ebonji, C.R.; Frederic, H. Origin of Mineralization of Groundwater in the Tongo Bassa Watershed (Douala-Cameroon). *Res. J. Environ. Sci.* **2015**, *7*, 29–41. Available online: https://www.researchgate.net/publication/332778338_Origin_of_Mineralization_of_Groundwater_in_the_Tongo_Bassa_Watershed_Douala-Cameroon (accessed on 10 October 2022). [CrossRef]
46. Arcement, G.J.; Schneider, V.R. *Guide for Selecting Manning's Roughness Coefficients for Natural Channels and Flood Plains*; US Geological Survey: Reston, VA, USA, 1989.
47. Pit, R.; Lantrip, J.; Harrison, R.; Henry, C.L.; Xue, D. *Infiltration through Disturbed Urban Soils and Compost-Amended Soil Effects on Runoff Quality and Quantity*; National Risk Management Research Laboratory: Duluth, GA, USA, 1999.
48. Hingray, B.; Picouet, C.; Musy, A. *Hydrologie. A Science for the Engineer*, 2nd ed.; PPUR Polytechnics and Universities Press: Lausanne, Switzerland, 2009; Volume 21.
49. Vidal, J.P. Assistance with the calibration of numerical models in river hydraulics—Contributions of artificial intelligence. Ph.D. Thesis, The National Polytechnic Institute of Toulouse-France, Toulouse, France, 2005.

mathematics MDPI

Article

Unsteady Water-Based Ternary Hybrid Nanofluids on Wedges by Bioconvection and Wall Stretching Velocity: Thermal Analysis and Scrutinization of Small and Larger Magnitudes of the Thermal Conductivity of Nanoparticles

Isaac Lare Animasaun [1,2,*], Qasem M. Al-Mdallal [2], Umair Khan [3,4] and Ali Saleh Alshomrani [5]

1 Fluid Dynamics and Survey Research Group, Department of Mathematical Sciences, Federal University of Technology, Akure PMB 704, Nigeria
2 Department of Mathematical Sciences, United Arab Emirates University, Al Ain, Abu Dhabi PMB 15551, United Arab Emirates
3 Department of Mathematical Sciences, Faculty of Science and Technology, Universiti Kebangsaan Malaysia, UKM, Bangi 43600, Selangor, Malaysia
4 Department of Mathematics and Social Sciences, Sukkur IBA University, Sukkur 65200, Sindh, Pakistan
5 Mathematical Modelling and Applied Computation (MMAC) Research Group, Department of Mathematics, Faculty of Science, King Abdulaziz University, P.O. Box 80203, Jeddah 21589, Saudi Arabia
* Correspondence: anizakph2007@gmail.com

Citation: Animasaun, I.L.; Al-Mdallal, Q.M.; Khan, U.; Alshomrani, A.S. Unsteady Water-Based Ternary Hybrid Nanofluids on Wedges by Bioconvection and Wall Stretching Velocity: Thermal Analysis and Scrutinization of Small and Larger Magnitudes of the Thermal Conductivity of Nanoparticles. *Mathematics* **2022**, *10*, 4309. https://doi.org/10.3390/math10224309

Academic Editors: Vasily Novozhilov and Cunlu Zhao

Received: 1 November 2022
Accepted: 16 November 2022
Published: 17 November 2022

Abstract: The uniqueness of nanofluids in the field of thermal analysis and engineering is associated with their thermal conductivity and thermodynamics. The dynamics of water made up of (i) single-walled carbon nanotubes with larger magnitudes of thermal conductivity of different shapes (i.e., platelet, cylindrical, and spherical) and (ii) moderately small magnitudes of thermal conductivity (i.e., platelet magnesium oxide, cylindrical aluminum oxide, spherical silicon dioxide) were explored in order to address some scientific questions. In continuation of the exploration and usefulness of ternary hybrid nanofluid in hydrodynamics and geothermal systems, nothing is known on the comparative analysis between the two dynamics outlined above due to the bioconvection of static wedges and wedges with stretching at the wall. Reliable and valid numerical solutions of the governing equation that models the transport phenomena mentioned above are presented in this report. The heat transfer through the wall increased with the wall stretching velocity at a smaller rate of 0.52 and a higher rate of 0.59 when the larger and smaller thermal conductivity of nanoparticles were used, respectively. Larger or smaller magnitudes of the thermal conductivity of nanoparticles were used; the wall stretching velocity had no significant effects on the mass transfer rate but the distribution of the gyrotactic microorganism was strongly affected. Increasing the stretching at the wedge's wall in the same direction as the transport phenomenon is suitable for decreasing the distribution of temperature owing to the higher velocity of ternary hybrid nanofluids either parallel or perpendicular to the wedge.

Keywords: ternary hybrid nanofluids; SWCNT nanoparticles; platelet magnesium oxide; cylindrical aluminum oxide nanoparticles; spherical silicon dioxide nanoparticles

MSC: 76R10; 76-10

1. Introduction

Experts dealing with microelectromechanical systems and nanoelectromechanical systems have expressed interest in single-walled carbon nanotubes because of their excellent electrical, structural, and mechanical capabilities, as proved by Baughman et al. [1]. Carbon nanotubes are excellent heat and cold conductors because of their outstanding thermal conductivity. Although silver and other metals are quite effective heat conductors, carbon nanotubes have thermal conductivity that is ten times greater than silver. According to Kim et al. [2], Prasek et al. [3], Dai [4], Li et al. [5], and Zhang et al. [6], the length of the nanotube, magnitude of temperature, reliance on phonons instead of electrons, axial strain, interaction between the nanotube with the substrate, chirality, and radius of the tube are likely factors that could influence the thermal conductivity of SWCNTs. Yu-Hua et al. [7] showed that beyond 30 degrees Celsius, single-walled carbon nanotube-based nanofluids change in thermal conductivity significantly quicker than ordinary water. It was also suggested in the same report that the Brownian motion of the particles drives the rise in thermal conductivity with temperature.

Nanda et al. [8] contributed to the study on the colloidal mixture of (a) ethylene glycol and single-wall carbon nanotubes and (b) (poly)-alpha-olefins and single-wall carbon nanotubes with the primary purpose of emphasizing the relevance of the interfacial thermal boundary layer and its corresponding resistance. The experimental findings demonstrated that the suspension of a single-wall carbon nanotube increased the thermal conductivity of ethylene glycol and (poly)-alpha olefins. The exceptional chemical and thermal stabilities, good tensile strength, and light weight of water-conveying single-walled carbon nanotubes were validated by Sabiha et al. [9]. The stability and characteristics of ordinary water and water-conveying single-walled carbon nanotubes indicate that from 0.05 vol% and 20 degrees Celsius to 0.25 vol% and 60 degrees Celsius, the thermal conductivity of water increases from 2.84% to 36.39%. The stability and thermo–physicality of ordinary water and water-conveying single-walled carbon nanotubes indicate that from 0.05 vol% and 20 degrees Celsius to 0.25 vol% and 60 degrees Celsius, the thermal conductivity of water increases from 2.84% to 36.39%. Additionally, the findings demonstrate that the nanofluid's viscosity dynamic decreases while temperature increases. As a result, the specific heat of the fluids rose from 6.73 to 28.96% at 20 °C to 60 °C temperature and volume concentrations of 0.05–0.25 vol%. However, compared to water, nanofluids possess smaller specific heat capacities. The results mentioned above were also corroborated by the findings of Said [10] on the thermal conductivity of water-carrying single-walled carbon nanotubes at temperatures between 20 and 60 degrees Celsius and volume percentages between 0.10 and 0.50. Namarvari et al. [11] performed a molecular dynamics simulation of a colloidal water mixture with single-walled carbon nanotubes at 298 and 313 K to investigate the impact of volume fraction on the viscosity of the nanofluid. The findings show that viscosity increases due to a rise in volume fraction. Additionally, the results indicate that as temperature ascended, the viscosity of the same nanofluid reduced.

The efficiency of hydrodynamics, geothermal systems, nuclear waste storage, magnetohydrodynamics, thermal insulation, heat exchangers, and crude oil exploration and extraction are significantly influenced by thermal conductivity; see Alqahtani et al. [12], Sarkar and Endalew [13]. Experts researching these areas within the scope of science are all interested in understanding and exploring the dynamics of fluid substances colloidally mixed with tiny particles over wedge-shaped surfaces. The wedge angle is important for the exploration of transonic flows over airfoils and wings because it is difficult to eliminate the corner owing to a finite wedge angle at the trailing edge; Jameson [14]. The dynamics of viscous fluid starting from the sharp edge of the wedge were reviewed by Stewartson [15] and Riccardi and Iafrati [16]; horizontal surfaces of the paraboloid were reviewed by Makinde and Animasaun [17,18]. The velocity of the laminar flow in the vicinity of the apex was elucidated by Schlichting [19] as a variable that varied with the wedge angle and distance away from the apex. Chemiski et al. [20] defined new healthcare accelerators as pieces of equipment that included a dynamic wedge option (i.e., dynamic movement pairs of collimator jaws), which suits dose–rate modulation.

Improvement of dose uniformity for the actualization of target volume within the scope of radiation oncology led Saminathan et al. [21] to recommend the introduction of wedge filters and their dynamics to alter the structures of isodose curves. In a study of fluids with significant viscosity around two symmetric wedges, when the tangential velocity at the wedge's vertex is zero, Miksis and Vanden-Broeck [22] observed that fluids of low viscosity contact back faster but fluids of high viscosity contact back at a longer time. The outcome of an analysis of bulk hydrodynamics on dynamics and statics wedge by Ruiz-Gutierrez et al. [23] revealed that the equilibrium contact angle of the fluid with the solid wedge and the opening angle of the wedge is needed for the characterization of confinement geometry and wetting properties of the fluid, respectively. The outcomes of a study on the importance of wedge angles within the interval $\pi/3 \leq \beta\pi \leq 5\pi/6$ and variation of acceleration rates $0 \leq p \leq 1$ suggest that fluid substance on the leeward side ends up closer to the spiral core. Moreover, the creation of the envelope that enfolds all other streaklines completely is possible due to fluid substances on the windward side of the wedge; see Xu [24]. Without losing generality, it is worth noting that there is either little information or no report on the dynamics of unsteady bioconvective ternary hybrid nanofluids on wedges focusing on a comparative analysis of the suspended three types of nanoparticles with small and moderately large thermal conductivity. The following are the main study areas that will aid in the discovery of transport phenomena:

1. What effect does rising stretching of wall velocity on the wedge have on the velocity of water conveying small and large thermal conductivity?
2. How does the wedge flow of ternary hybrid nanofluids carrying smaller and bigger thermal conductivity of nanoparticles affect heat transfer, friction at the wall, mass transfer, and the distribution rate of motile gyrotactic organisms?
3. What effect does increasing the Péclet number and lengthening the wall at the wedge have on the dispersion of motile microorganisms within the domain of ternary hybrid nanofluids transmitting smaller and bigger thermal conductivity of nanoparticles?

2. Research Methodology

We appraised a two-dimensional motion of a water-based ternary hybrid nanofluid made up of three types of nanoparticles with larger and smaller magnitudes of thermal conductivity on a fixed wedge and a wedge with a stretchable wall, as depicted in Figure 1. The $x-$axis and $y-$axis are the distances along the surface and perpendicular to the surface. The velocities are $v(t,x,y)$ and $u(t,x,y)$ in both directions. The changes in the potential flow velocity u_e with time and distance parallel to the wall greatly generate the transport phenomena in the dynamics of ternary hybrid nanofluids along the wedges. The pressure changes that induce the flow parallel to the $x-$axis for the unsteady flow of water-based ternary hybrid nanofluids on wedges are

$$-\frac{1}{\rho_{tf}}\frac{dp}{dx} = \frac{\partial u_e}{\partial t} + u_e\frac{\partial u_e}{\partial x} \tag{1}$$

where the external flow velocity far from the wall, wall velocity, and the Falkner–Skan power-law constant m are defined as

$$u_e(x,t) = \frac{ax^m}{1-ct}, \quad u_w(x,t) = \frac{bx^m}{1-ct}, \quad m = \frac{\beta}{2-\beta} = \frac{\Omega}{2\pi-\Omega}, \quad \beta = \frac{\Omega}{\pi} = \frac{2m}{1+m} \tag{2}$$

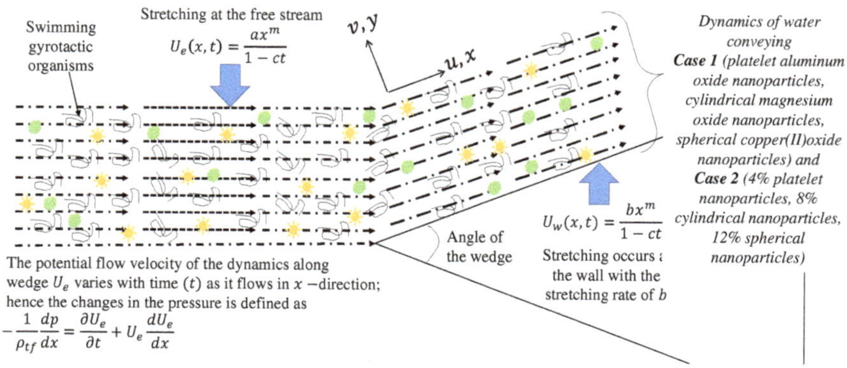

Figure 1. Physical configuration of the transport phenomena.

The ratio of the entire wedge angle $\Omega\pi$ is represented by the Hartree pressure gradient β. When $\Omega = 0$ implies $m = 0$, the dynamics on a converging channel are attainable for $\pi < \Omega < 2\pi$ and the well-known Blasius flow. In agreement with Raju et al. [25], Miksis and Vanden-Broeck [22], and Xu [24], the ideal set of equations to understand the comparative analysis is

$$\frac{tialu}{tialx} + \frac{tialv}{tialy} = 0, \tag{3}$$

$$\frac{\partial u}{\partial t} + v\frac{\partial u}{\partial y} + u\frac{\partial u}{\partial x} = u_e\frac{\partial u_e}{\partial x} + \frac{\partial u_e}{\partial t} + \frac{\mu_{tf}}{\rho_{tf}}\frac{\partial^2 u}{\partial y^2} + g\beta(T - T_\infty) + g\beta^*(C - C_\infty) + g\gamma(\rho_m - \rho_f)(N - N_\infty), \tag{4}$$

$$\frac{\partial T}{\partial t} + u\frac{\partial T}{\partial x} + v\frac{\partial T}{\partial y} = \frac{\kappa_{tf}}{(\rho c_p)_{tf}}\frac{\partial^2 T}{\partial y^2}, \tag{5}$$

$$\frac{\partial C}{\partial t} + u\frac{\partial C}{\partial x} + v\frac{\partial C}{\partial y} = D_B\frac{\partial^2 C}{\partial y^2}, \tag{6}$$

$$\frac{\partial N}{\partial t} + u\frac{\partial N}{\partial x} + v\frac{\partial N}{\partial y} + \frac{bw_c}{(C - C_\infty)}\frac{\partial}{\partial y}\left(N\frac{\partial C}{\partial y}\right) = D_m\frac{\partial^2 N}{\partial y^2} \tag{7}$$

For the static wedge, the wall velocity of the transport phenomenon is zero. The conditions worthy to be satisfied by the dynamics at the boundary are

$$u(t,x,0) = 0, \quad v(t,x,0) = 0, \quad T(t,x,0) = T_w, \quad C(t,x,0) = C_w, \quad N(t,x,0) = N_w \tag{8}$$

$$u(t,x,\infty) \to u_e(t,x), \quad T(t,x,\infty) \to T_\infty, \quad C(t,x,\infty) \to C_\infty, \quad N(t,x,\infty) \to C_\infty. \tag{9}$$

For a moving wedge, the motion of the wedge is equivalent to the wall velocity $u_w(t,x)$. The boundary conditions for the dynamics are

$$u(t,x,0) = u_w(t,x), \quad v(t,x,0) = 0, \quad T(t,x,0) = T_w, \quad C(t,x,0) = C_w, \quad N(t,x,0) = N_w \tag{10}$$

$$u(t,x,\infty) \to u_e(t,x), \quad T(t,x,\infty) \to T_\infty, \quad C(t,x,\infty) \to C_\infty, \quad N(t,x,\infty) \to C_\infty. \tag{11}$$

The dynamic viscosity, thermal conductivity, volume fraction, thermal volumetric expansion, density, and heat capacitance of the ternary hybrid nanofluid, defined by Wenhao Cao et al. [26], Song et al. [27], Timofeeva et al. [28], Elnaqeeb et al. [29], Ahammed et al. [30], Sahoo [31,32], Hamilton-Crosser [33], and Sahu and Sarkar [34], are defined as

$$\kappa_{hnf} = \frac{\kappa_{nf1}\phi_3 + \kappa_{nf2}\phi_2 + \kappa_{nf3}\phi_1}{\phi}, \quad \mu_{hnf} = \frac{\mu_{nf1}\phi_3 + \mu_{nf2}\phi_1 + \mu_{nf3}\phi_3}{\phi}, \quad \phi = \phi_3 + \phi_2 + \phi_1 \tag{12}$$

$$\rho_{hnf} = (1 - \phi_3 - \phi_2 - \phi_1)\rho_{bf} + \phi_3\rho_{sp3} + \phi_2\rho_{sp2} + \phi_1\rho_{sp1}. \tag{13}$$

$$\beta_{hnf} = (1-\phi_3-\phi_2-\phi_1)\beta_{bf} + \phi_3\beta_{sp3} + \phi_2\beta_{sp2} + \phi_1\beta_{sp1}. \tag{14}$$

$$(\rho c_p)_{hnf} = (1-\phi_3-\phi_2-\phi_1)(\rho c_p)_{bf} + \phi_3(\rho c_p)_{sp3} + \phi_2(\rho c_p)_{sp2} + \phi_1(\rho c_p)_{sp1} \tag{15}$$

Table 1 presents the details of the adopted platelet, spherical, and cylinder tiny particles for viscosity and thermal conductivity models

$$\mu_{nf1} = (6.2\phi^2 + 2.5\phi + 1)\mu_{bf}, \quad \mu_{nf2} = (904.4\phi^2 + 13.5\phi + 1)\mu_{bf},$$

$$\mu_{nf3} = (612.6\phi^2 + 37.1\phi + 1)\mu_{bf}, \quad \kappa_{nf3} = \kappa_{bf}\left[\frac{-4.7\phi(\kappa_{bf} - k_{sp3}) + 4.7\kappa_{bf} + \kappa_{sp3}}{\phi(\kappa_{bf} - \kappa_{sp3}) + 4.7\kappa_{bf} + \kappa_{3sp3}}\right],$$

$$\kappa_{nf1} = \kappa_{bf}\left[\frac{-2\phi(\kappa_{bf} - \kappa_{sp1}) + 2\kappa_{bf} + \kappa_{sp1}}{\phi(\kappa_{bf} - \kappa_{sp1}) + 2\kappa_{bf} + \kappa_{sp1}}\right] \quad \kappa_{nf2} = \kappa_{bf}\left[\frac{-3.9\phi(\kappa_{bf} - \kappa_{sp2}) + 3.9\kappa_{bf} + \kappa_{sp2}}{\phi(\kappa_{bf} - \kappa_{sp2}) + 3.9\kappa_{bf} + \kappa_{2sp2}}\right] \tag{16}$$

At the wall of the static and dynamic wedge, the distribution of the gyrotactic microorganism Nm_x, quantifier for heat transfer Nu_x, quantifier for friction at the wall C_{fx}, and quantifier for mass transfer Sh_x are defined as

$$Nm_x = \frac{-x}{(N_w - N_\infty)}\frac{\partial N}{\partial y}\bigg|_{y=0}, \quad Nu_x = \frac{-x\kappa_{tf}}{\kappa_{bf}(T_w - T_\infty)}\frac{\partial T}{\partial y}\bigg|_{y=0} \quad C_{fx} = \frac{\mu_{tf}}{\rho_{bf}U_w^2(x)}\frac{\partial u}{\partial y}\bigg|_{y=0},$$

$$Sh_x = \frac{-x}{(C_w - C_\infty)}\frac{\partial C}{\partial y}\bigg|_{y=0}. \tag{17}$$

The transport phenomenon does not occur at several crucial points in fluid flow systems under complex flow conditions despite the emergence of bioconvection described above, and the velocity of the gyrotactic microorganisms is not distinct from the velocity of the ternary hybrid nanofluid. Furthermore, the transport phenomena under discussion do not involve the interaction flow of two or more separate phases with common interfaces. Due to the lack of several states or phases and various chemical properties, the single-phase model was taken into consideration in light of the aforementioned facts.

Table 1. Details of platelet magnesium oxide nanoparticles from Animasaun et al. [35], cylindrical aluminum oxide nanoparticles from Ref. [36], spherical silicon dioxide nanoparticles from Refs. [36,37], single-wall carbon nanoparticles by Kandasamy et al. [38], and water.

	ρ (kgm^{-3})	κ (W/mK)	c_p (J/kgK)	P_r
Water H_2O	997.1	0.613	4180	6.1723
Platelet SWCNT	2600	6600	425	
Cylindrical SWCNT	2600	6600	425	
Spherical SWCNT	2600	6600	425	
Platelet MgO	3580	48.4	960	
Cylindrical Al_2O_3	3970	40	765	
Spherical SiO_2	3970	36	765	

Similarity Variables

The following variables are used to transform and non-dimensionalize the dimensional governing equation

$$\frac{\eta}{y} = \sqrt{\frac{(m+1)u_o}{2\vartheta_{bf}x}}, \quad \theta(\eta) = \frac{T - T_\infty}{T_w - T_\infty}, \quad \frac{\psi(x,y)}{f(\eta)} = \sqrt{\frac{2\vartheta_{bf}xu_o}{m+1}}, \quad \aleph(\eta) = \frac{-C_\infty + C}{C_w - C_\infty}, \quad \lambda = \frac{b}{a}$$

$$\chi(\eta) = \frac{-N_\infty + N}{N_w - N_\infty}, \quad v = -\frac{\partial\psi}{\partial x}, \quad u = \frac{\partial\psi}{\partial y}, \quad P_r = \frac{\mu_{bf}Cp_{bf}}{\kappa_{bf}}, \quad S_c = \frac{\vartheta_{bf}}{D_B}, \quad S_{mm} = \frac{\nu_{bf}}{D_m},$$

$$Gt_{bf} = \frac{gx\beta(T_w - T_\infty)}{u_e^2}, \quad Gc_{bf} = \frac{gx\beta^*(C_w - C_\infty)}{u_e^2}, \quad R_b = \frac{gx\gamma(\rho_m - \rho_f)(N_w - N_\infty)}{u_e^2},$$

$$A = \frac{c}{x^{m-1}a}, \quad P_e = \frac{bW_c}{D_m}, \quad E_8 = \frac{-4.7\phi(\kappa_{bf} - k_{sp3}) + 4.7\kappa_{bf} + \kappa_{sp3}}{\phi(\kappa_{bf} - \kappa_{sp3}) + 4.7\kappa_{bf} + \kappa_{3sp3}}, \quad E_{10} = E_6\phi_1 + E_7\phi_2 + E_8\phi_3,$$

$$E_1 = 6.2\phi^2 + 2.5\phi + 1, \quad E_2 = 904.4\phi^2 + 13.5\phi + 1, \quad E_3 = 612.6\phi^2 + 37.1\phi + 1,$$

$$E_4 = E_1\phi_1 + E_2\phi_2 + E_3\phi_3, \quad E_5 = 1 - \phi_1 - \phi_2 - \phi_3 + \phi_1\frac{\rho_{sp_1}}{\rho_{bf}} + \phi_2\frac{\rho_{sp_2}}{\rho_{bf}} + \phi_3\frac{\rho_{sp_3}}{\rho_{bf}},$$

$$E_6 = \frac{-2\phi(\kappa_{bf} - \kappa_{sp1}) + 2\kappa_{bf} + \kappa_{sp1}}{\phi(\kappa_{bf} - \kappa_{sp1}) + 2\kappa_{bf} + \kappa_{sp1}}, \quad E_7 = \frac{-3.9\phi(\kappa_{bf} - \kappa_{sp2}) + 3.9\kappa_{bf} + \kappa_{sp2}}{\phi(\kappa_{bf} - \kappa_{sp2}) + 3.9\kappa_{bf} + \kappa_{2sp2}},$$

$$E_9 = 1 - \phi_1 - \phi_2 - \phi_3 + \phi_1\frac{(\rho c_p)_{sp1}}{(\rho c_p)_{bf}} + \phi_2\frac{(\rho c_p)_{sp2}}{(\rho c_p)_{bf}} + \phi_3\frac{(\rho c_p)_{sp3}}{(\rho c_p)_{bf}}. \tag{18}$$

The dimensionless governing equations are

$$(m+1)\frac{E_4}{\phi E_5}\frac{d^3f}{d\eta^3} + 2A + 2m - A\eta\frac{d^2f}{d\eta^2} - 2A\frac{df}{d\eta} + f\frac{d^2f}{d\eta^2} + 2G_t\theta + 2\aleph G_c + 2R_b\chi = 0, \tag{19}$$

$$(m+1)\frac{E_{10}}{\phi E_9}\frac{d^2\theta}{d\eta^2} - P_rA\eta\frac{d\theta}{d\eta} + P_rf\frac{d\theta}{d\eta} = 0, \tag{20}$$

$$(m+1)\frac{d^2\aleph}{d\eta^2} - S_cA\eta\frac{d\aleph}{d\eta} + S_cf\frac{d\aleph}{d\eta} = 0, \tag{21}$$

$$(m+1)\frac{d^2\chi}{d\eta^2} - S_{mm}A\eta\frac{d\chi}{d\eta} + S_{mm}f\frac{d\chi}{d\eta} - P_e(m+1)\left(\chi\frac{d^2\aleph}{d\eta^2} + \frac{d\aleph}{d\eta}\frac{d\chi}{d\eta}\right) = 0. \tag{22}$$

For static wedge $\lambda = 0$, the fluid dynamics is bounded as

$$\theta = 1, \quad f = 0, \quad \frac{df}{d\eta} = \lambda, \quad \aleph = 1, \quad \chi = 1 \quad at \quad \eta = 0 \tag{23}$$

$$\theta \to 0, \quad \frac{df}{d\eta} \to 1, \quad \aleph \to 0, \quad \chi \to 0 \quad as \quad \eta \to \infty \tag{24}$$

For a moving wedge ($\lambda \neq 0$) at the wall ($\eta = 0$). As the boundary layer flow is induced, λ quantifies the movement of the wedge and the free stream flow. The dimensionless physical quantities are

$$\frac{Nm_x}{\sqrt{Re_x}}\sqrt{\frac{2}{m+1}} = -\chi'(0), \quad \frac{\phi Nu_x}{E_{10}\sqrt{Re_x}}\sqrt{\frac{2}{m+1}} = -\theta'(0), \quad \frac{\phi C_{fx}\sqrt{Re_x}}{E_4}\sqrt{\frac{2}{m+1}} = f''(0)$$

$$\frac{Sh_x}{\sqrt{Re_x}}\sqrt{\frac{2}{m+1}} = -\aleph'(0). \tag{25}$$

3. Numerical Integration and Validation

To obtain the numerical solution of Equations (19)–(24), the Runge–Kutta integration scheme (`rk4sh`) and collocation-s-hooting technique `bvp4c` package of MATLAB were employed. For the sake of conciseness, the descriptions of the numerical approaches given above by Kierzenka and Shampine [39], Al-Mdallal et al. [40], de Hoog and Weiss [41], and Animasaun [42] were rigorously adhered to. The dependent variables are $f(\eta)$, $f'(\eta)$, $f''(\eta)$, $\theta(\eta)$, $\theta'(\eta)$, $\aleph(\eta)$, $\aleph'(\eta)$, $\chi(\eta)$, $\chi'(\eta)$, $f''(0)$, $-\theta'(0)$, $-\aleph'(0)$, and $-\chi'(0)$. The nomenclature of the dependent variables listed in Section 3 that are needed in this study are diffusion of motile gyrotactic organisms away from the heated wall and across the ternary hybrid nanofluid, the distribution of motile gyrotactic organisms in the ternary hybrid nanofluid, the mass movement via the hot wedge and the immediately surrounding layer, the concentration of the species that formed the ternary hybrid nanofluid, the shear stress function, the horizontal velocity function, the vertical velocity function, the temperature gradient function, the temperature distribution function, the concentration gradient function, the gradient of distribution of motile gyrotactic organisms, the heat transfer through the heated wedge and immediate adjacent layer, and the friction between the upper surface of the wedge and the last layer of the flowing fluid.

Results Validation

By juxtaposing the results of the same problem by utilizing the various approaches to the solution (`bvp4c` and `rksh`), the authors established the credibility of the integration technique. The limiting case for the dynamics on a wedge involved a colloidal mixture of water with platelet magnesium oxide nanoparticles, cylindrical aluminum oxide nanoparticles, and spherical titanium dioxide nanoparticles. The data used for the validation were $m = 0.1$, $G_t = G_c = R_b = 1$, $P_r = 6.1723$, $S_c = S_{mm} = 0.62$, $P_e = 0.5$, $\phi_1 = \phi_2 = \phi_3 = 0.1$, $\rho_{bf} = 997.1$, $\kappa_{bf} = 0.613$, and $Cp_{bf} = 4,180$ when (i.e., $A = 0.1, 0.2, 0.3$). Tables 2 and 3 indicate that the numerical approach for both numerical methods is valid and that the findings are dependable to an acceptable degree for investigating transport phenomena.

Table 2. Reliance on numerical integration for static wedge $\lambda = 0$ and $\eta_\infty = 5$.

A	`rk4sh` $f''(0)$	`bvp4c` $f''(0)$	`rk4sh` $-\theta'(0)$	`bvp4c` $-\theta'(0)$
0.1	0.451206374602597	0.451206301040792	0.453340211815564	0.453340295041978
0.2	0.475722148811551	0.475722187415029	0.371908418391181	0.371908400147596
0.3	0.502553781403640	0.502553744719328	0.289615627410633	0.289615696031472

Table 3. Reliance on numerical integration for moving wedge $\lambda = 1$ and $\eta_\infty = 5$.

A	`rk4sh` $f''(0)$	`bvp4c` $f''(0)$	`rk4sh` $-\theta'(0)$	`bvp4c` $-\theta'(0)$
0.1	0.149083116414503	0.149083185741905	1.060073047726859	1.060073096024179
0.2	0.154286959072002	0.154286985741906	1.003808597826008	1.0038085997400126
0.3	0.160265225109621	0.160265254127903	0.944589109510400	0.9445891784519038

4. Analysis of Results and Discourse

This section presents an examination and discourse of the findings from the governing equation that describes cases 1 and 2 of the transport phenomena. This section discusses the first case and the dynamics of water composed of SWNCT nanoparticles with thermal conductivity of a larger magnitude. The second scenario is detailed, which is the transport phenomena composed of water conveying smaller magnitudes of thermal conductivity (i.e., platelet magnesium oxide, cylindrical aluminum oxide, and spherical silicon dioxide nanoparticles). It is worth noting that the nanoparticles created in scenario 2 of the transport phenomena are incredibly dense; see Table 1.

4.1. Analysis of Results

When $A = 0.5$, $m = 0.1$, $G_t = G_c = R_b = 1$, $P_r = 6.1723$, $S_c = S_{mm} = 0.62$, $P_e = 0.5$, $\phi_1 = \phi_2 = \phi_3 = 0.1$, $\rho_{bf} = 997.1$, $\kappa_{bf} = 0.613$, and $Cp_{bf} = 4,180$, the parameter that quantifies the stretching velocity at the wall of the wedge increases within the interval $0 \leq \lambda \leq 2$. It can be seen in Figures 2 and 3 that the velocities for the two transport phenomena increase due to a higher stretching rate b of the wall velocity $u_w(x,t)$. The differences between the two transport phenomena for the two cases are only significant when the wedge is stagnant and moves slowly (i.e., $0 \leq \lambda \leq 0.5$). As shown in Figure 4, the temperature distribution across the transport phenomenon declines with higher levels of stretching wall velocity (λ). This study observed optimal shear stress proportional to friction within the domain $0 \leq \eta \leq 5$. Moreover, higher temperature distributions across both ternary hybrid nanofluids and the maximum species (i.e., concentration) of the ternary hybrid nanofluids are not far-fetched. A wider distribution of motile gyrotactic organisms within the domain $0 \leq \eta \leq 5$ was found when the wedge was static (i.e., $\lambda = 0$). These results are not presented for brevity but are worthy to be mentioned. In the case of the static wedge (i.e., $\lambda = 0$), there exists an optimal significant difference in the velocity, temperature distribution, concentration, and distribution of motile gyrotactic organisms across the domain between case 1 and case 2.

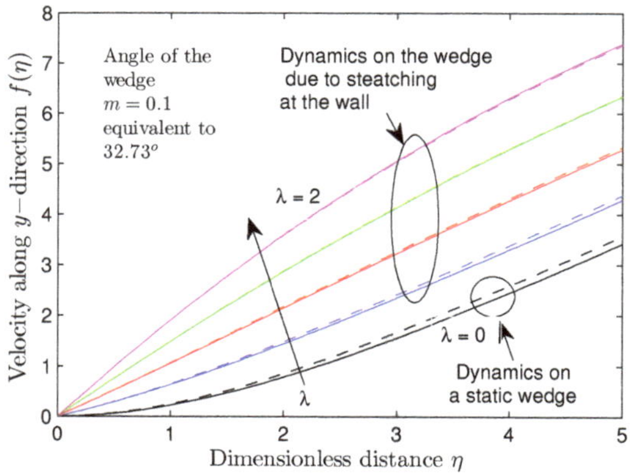

Figure 2. Effect of wall stretching velocity λ on the velocity of both fluids along the $y-$direction when $m = 0.1$

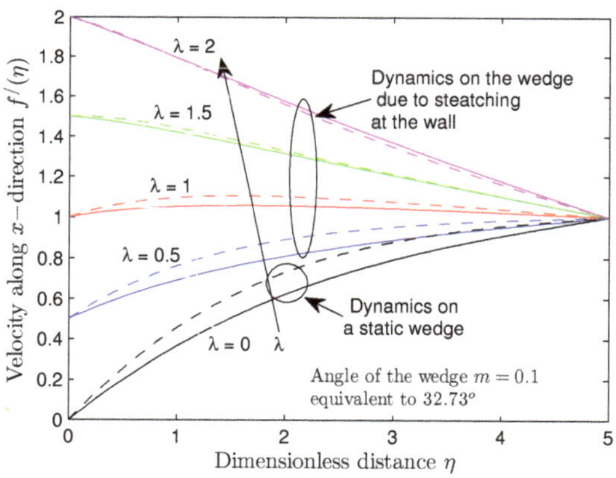

Figure 3. Effect of wall stretching velocity λ on the dynamics of both ternary hybrid nanofluids along $x-$direction when $m = 0.1$

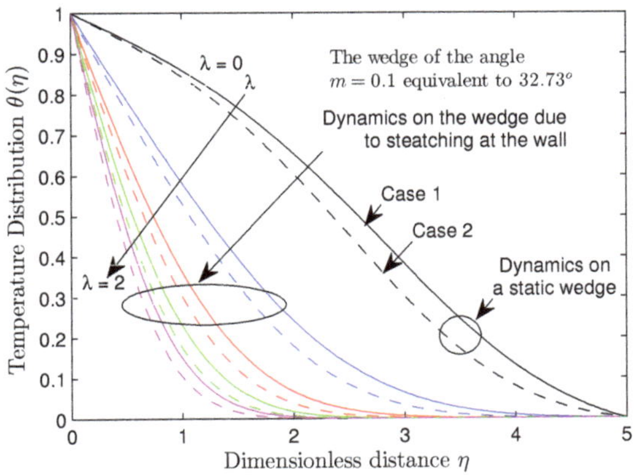

Figure 4. Effect of wall stretching velocity λ on the temperature distribution within the dynamics of both nanofluids when $m = 0.1$.

After generating the appropriate data using the integration technique mentioned above, data were moved to Surfer Version 11.1.719. The Kriging method of data gridding was used to generate the contour results presented in this section. Figures 5 and 6 show that $f''(\eta)$ declines as λ rises. Tables 4 and 5 show that the friction at the wall reduces due to a higher level of wall stretching velocity. These figures present the variations in the function that represents the shear stress at different levels of wall stretching on the wedge (λ) for the transport phenomena of cases 1 and 2. It is evident that the most minimum shear stresses, denoted as $\frac{\phi C_{fx}\sqrt{Re_x}}{E_4}\sqrt{\frac{2}{m+1}}$, occur at larger values of λ near the wall. At $\eta = 0$, $f''(\eta)$ decreases significantly with λ. At the other end, $\eta = 0$, $f''(\eta)$ decreases moderately with λ. In case 1 transport phenomenon presented as Table 4, $\frac{\phi C_{fx}\sqrt{Re_x}}{E_4}\sqrt{\frac{2}{m+1}}$ decreases with λ at the rate of -0.329789963. However, for the case 2 transport phenomenon, $\frac{\phi C_{fx}\sqrt{Re_x}}{E_4}\sqrt{\frac{2}{m+1}}$ decreases with λ at the rate of -0.356473325; see Table 5. As shown in Figures 7–10, the transfer of heat energy and transfer of species that form the nanofluids increase significantly with λ near the wall. In other words, increasing wall stretching on the wedge leads to a higher heat transfer commensurate with the Nusselt number and broader mass transfer commensurate with the Sherwood number in Equation (25).

Table 4. Variation of physical quantities for case 1—larger thermal conductivity of nanoparticles when $\eta_\infty = 5$.

λ	$\frac{\phi C_{fx}\sqrt{Re_x}}{E_4}\sqrt{\frac{2}{m+1}}$	$\frac{\phi Nu_x}{E_{10}\sqrt{Re_x}}\sqrt{\frac{2}{m+1}}$	$\frac{Sh_x}{\sqrt{Re_x}}\sqrt{\frac{2}{m+1}}$	$\frac{Nm_x}{\sqrt{Re_x}}\sqrt{\frac{2}{m+1}}$
0	0.501391652	0.126832395	0.164413180	0.198769191
0.5	0.313117110	0.425081500	0.322440050	0.449063565
1.0	0.145577661	0.721081217	0.492851011	0.704626364
1.5	−0.009437541	0.961784227	0.644943911	0.926617648
2.0	−0.161805929	1.163072510	0.776441110	1.116848885
S_{lp}	−0.329789963	0.521836591	0.309311944	0.462742694

$A = 0.5$, $m = 0.1$, $G_t = G_c = R_b = 1$, $P_r = 6.1723$, $S_c = S_{mm} = 0.62$, $P_e = 0.5$, $\phi_1 = \phi_2 = \phi_3 = 0.1$, $\rho_{bf} = 997.1$, $\kappa_{bf} = 0.613$, and $Cp_{bf} = 4180$.

Table 5. Variation of physical quantities for case 2—smaller thermal conductivity of nanoparticles when $\eta_\infty = 5$.

λ	$\frac{\phi C_{fx}\sqrt{Re_x}}{E_4}\sqrt{\frac{2}{m+1}}$	$\frac{\phi Nu_x}{E_{10}\sqrt{Re_x}}\sqrt{\frac{2}{m+1}}$	$\frac{Sh_x}{\sqrt{Re_x}}\sqrt{\frac{2}{m+1}}$	$\frac{Nm_x}{\sqrt{Re_x}}\sqrt{\frac{2}{m+1}}$
0	0.562250645	0.137735549	0.179214207	0.221766375
0.5	0.356021584	0.480527624	0.331789925	0.462646012
1.0	0.17532672	0.815315534	0.497225412	0.710701282
1.5	0.008681269	1.086212765	0.646341186	0.928566151
2.0	−0.15526251	1.312552075	0.77611485	1.116477
S_{lp}	−0.356473325	0.591063639	0.301670509	0.451068278

$A = 0.5$, $m = 0.1$, $G_t = G_c = R_b = 1$, $P_r = 6.1723$, $S_c = S_{mm} = 0.62$, $P_e = 0.5$, $\phi_1 = \phi_2 = \phi_3 = 0.1$, $\rho_{bf} = 997.1$, $\kappa_{bf} = 0.613$, and $Cp_{bf} = 4180$.

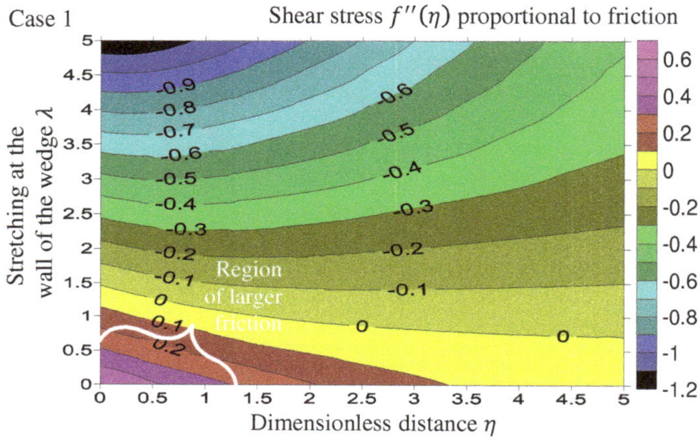

Figure 5. Changes in the shear stress function $f''(\eta)$ with the wall stretching velocity λ—case 1.

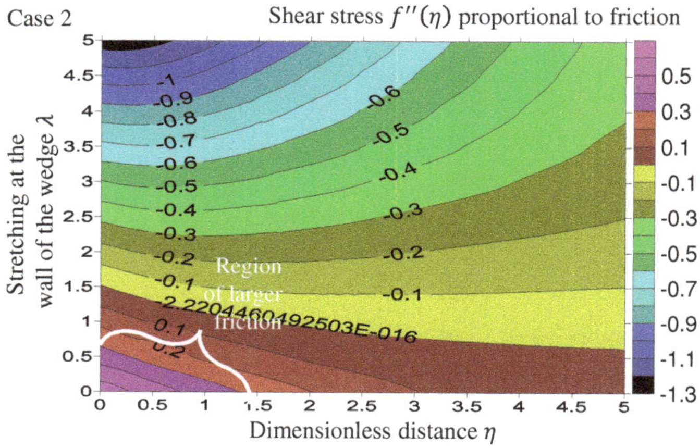

Figure 6. Changes in the shear stress function $f''(\eta)$ with the wall stretching velocity λ—case 2.

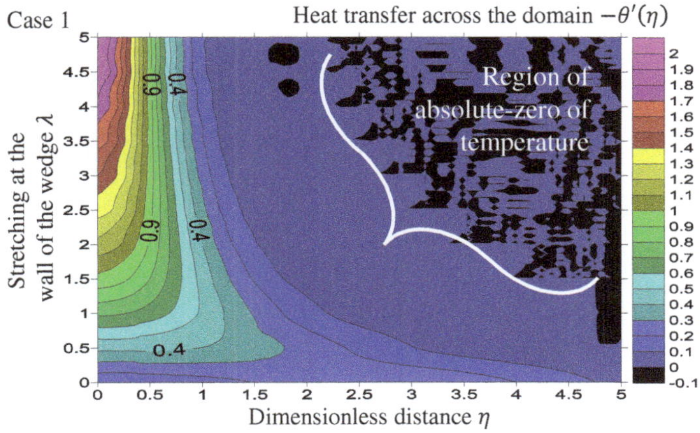

Figure 7. Variation of the heat transfer $-\theta'(\eta)$ with the wall stretching velocity λ—case 1.

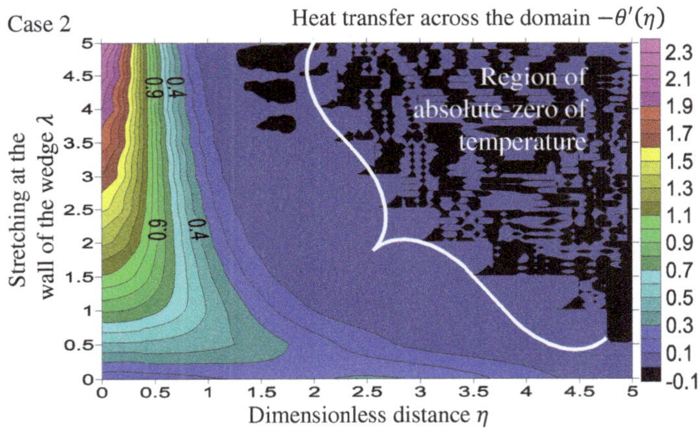

Figure 8. Variation of the heat transfer $-\theta'(\eta)$ with the wall stretching velocity λ—case 2.

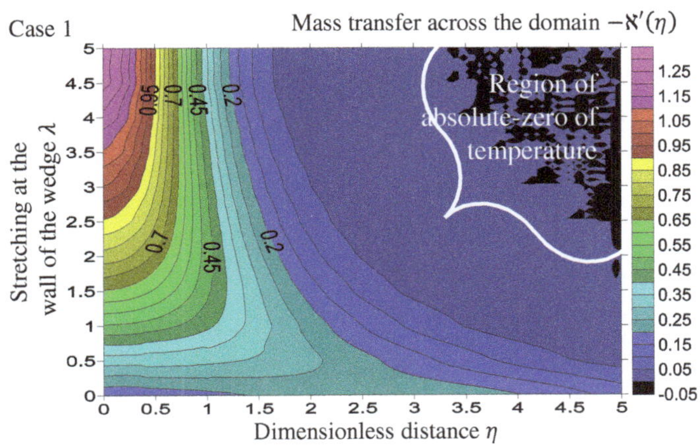

Figure 9. Variation of the mass transfer $-\aleph'(\eta)$ with the wall stretching velocity λ.

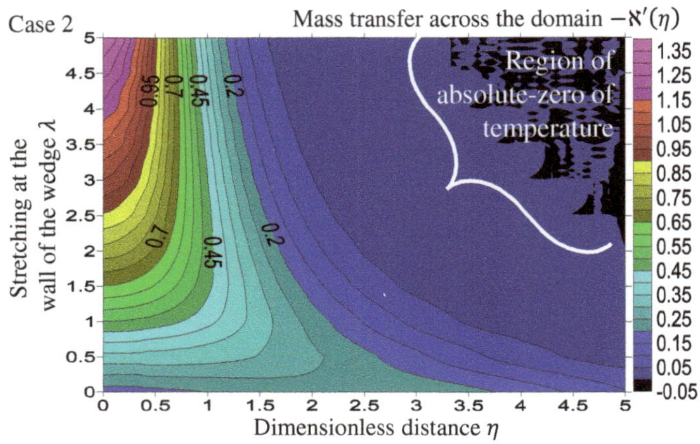

Figure 10. Variation of the gradient of distribution of motile gyrotactic organisms in the entire domain $-\chi'(\eta)$ with the wall stretching velocity λ.

The heat transfer across the case 1 transport phenomenon increases with λ at the rate of 0.521836591. Meanwhile, the heat transfer across the case 2 transport phenomenon increases with λ at the higher rate of 0.591063639; see Tables 4 and 5. The differences between the observed changes in the increase in mass transfer rate for the transport phenomena of cases 1 and 2 are minimal (regarding magnitude). Moreover, the distribution rate of motile gyrotactic organisms was found to be higher as $\lambda \to 2$ in Tables 4 and 5. Figures 11 and 12 depict the gradients of distribution of motile gyrotactic organisms across the domain at different levels of wall stretching velocity λ. Considering the first case of transport phenomenon, the effects of P_e on $-\chi'(\eta)$ when $\lambda = 0$ and $\lambda = 5$ were examined and presented as in Figures 13 and 14. The distribution of motile gyrotactic organisms $\chi(\eta)$ is a decreasing property of P_e. For the static wedge and moving wedge, the higher the Péclet number, the lower the distribution of motile microorganisms within the fluid domain Figures 13 and 14. Moreover, minimal distribution of motile microorganisms within the domain of the ternary hybrid nanofluid is obtainable when the wedge moves faster. Figures 15 and 16 confirm the rise of a gradient of distribution of motile gyrotactic organisms near the wall but reduced far away from the wedge as shown in the 3D plot.

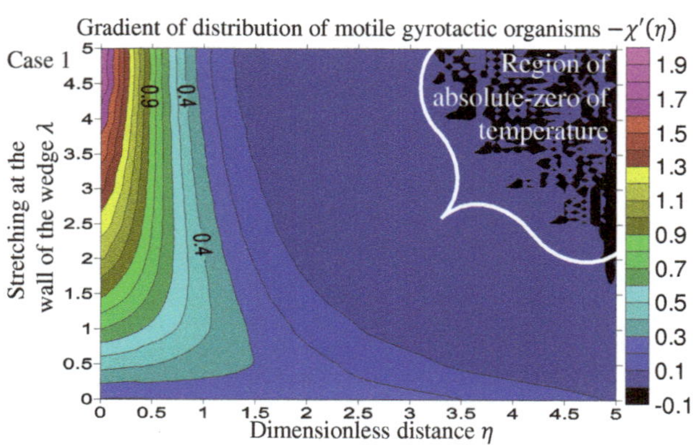

Figure 11. Distribution of motile gyrotactic organisms within the entire domain $\chi(\eta)$ with λ when $\lambda = 0$—case 1.

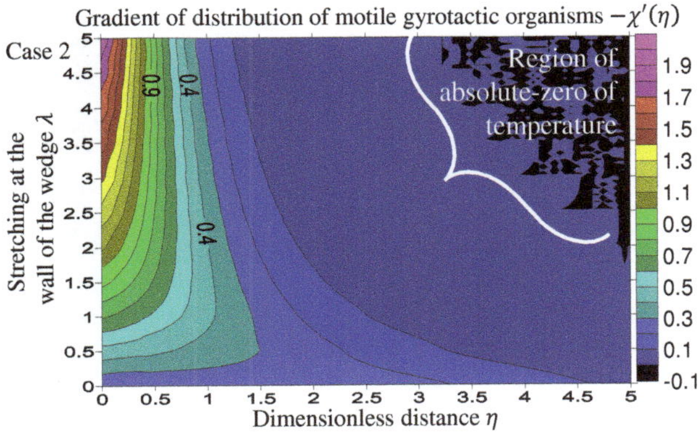

Figure 12. Distribution of motile gyrotactic organisms within the entire domain $\chi(\eta)$ with λ when $\lambda = 0$—case 2.

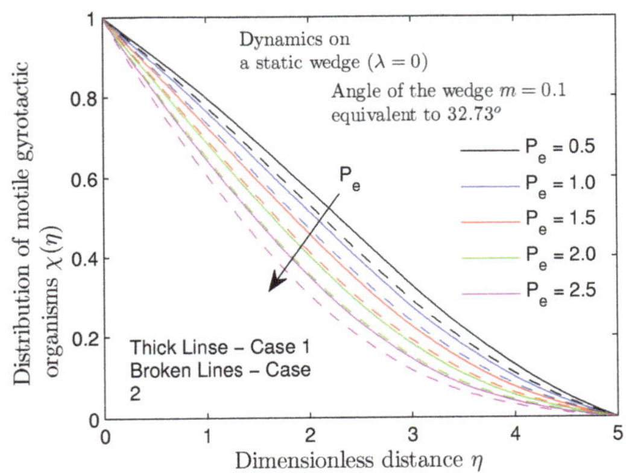

Figure 13. Distribution of motile gyrotactic organisms $\chi(\eta)$ within the entire domain on a static wedge with P_e when $\lambda = 0$.

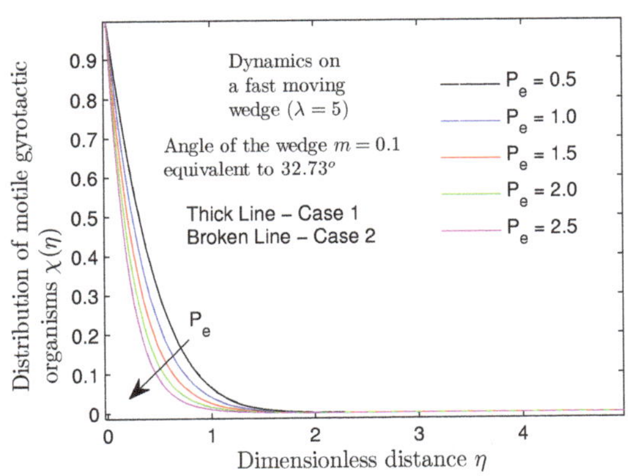

Figure 14. Distribution of motile gyrotactic organisms $\chi(\eta)$ within the entire domain on a fast-moving wedge with P_e when $\lambda = 5$.

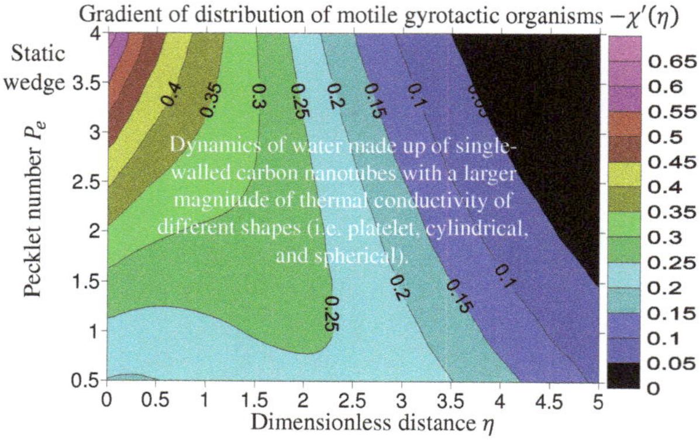

Figure 15. Variation of the gradient of distribution of motile gyrotactic organisms $-\chi'(\eta)$ within the entire domain on a static wedge (i.e., $\lambda = 0$) with P_e and $m = 0.1$.

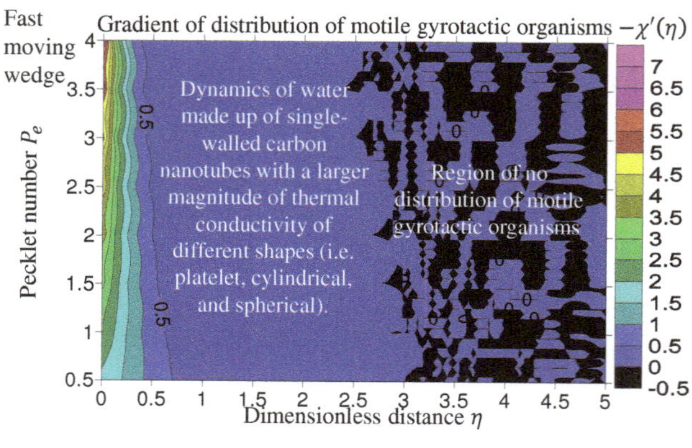

Figure 16. Variation of the gradient of distribution of motile gyrotactic organisms $-\chi'(\eta)$ within the entire domain on a fast-moving wedge (i.e., $\lambda = 5$) with P_e and $m = 0.1$.

The turning points are within the interval $0.5 \leq \eta \leq 2$. The distribution of motile microorganisms is extensive in Figure 15. It was also noticed that the maximum value of $-\chi'(\eta)$ was obtained when $P_e = 4$ at $\eta = 0$ as 6.8992. In Figure 15, when $\lambda = 0$ for a static wedge and $P_e = 0.5$, $\frac{Nm_x}{\sqrt{Re_x}}\sqrt{\frac{2}{m+1}}$ that quantifies the gradient of distribution of motile gyrotactic organisms at the wall decreases with η at the rate -0.020498425. When there exists a higher stretching at the wall of the wedge (i.e., $\lambda = 5$ and $P_e = 0.5$), $-\chi'(\eta)$ decreases with η at the rate of -0.130199045.

4.2. Discussion of the Results

Physically, as the wall velocity $u_w(x,t)$ increases, the layers of the ternary hybrid fluid on the wedge move in the same direction that boosts $v(t,x,y)$ and $u(t,x,y)$. The increment of stretching velocity at the wall of the wedge within the interval $0 \leq \lambda \leq 2$ implies that the wedge moves at an increasing rate with its pointed edge at the front. Higher λ was found capable of enhancing the velocity due to the viscous nature of the ternary hybrid nanofluid that declines when stretching rate b grows; see Figures 2 and 3. The observed increase in the velocity of both water-based ternary hybrid nanofluids parallel and perpendicular to the wedge surface is due to a higher stretching rate b of wall velocity $u_w(x,t)$, which is capable of reducing the viscosity; see Figures 2 and 3. The reduced viscosity of both fluids is used in the industry to facilitate the dynamics between two locations as temperature rises. The outcomes of the result in Figures 2–4 suggest that there exists a significant difference between the velocities of case 1 and case 2 when the wedge is static. A higher level of wall stretching on the wedge (λ) generates greater pressure on the fluid flow; hence the changes in the shear stress are more significant at the wall compared to at the free stream; see Figures 5 and 6. A higher level of friction manifests when the wedge is static ($\lambda = 0$) as shown in Tables 4 and 5. It is worth noting that a higher local skin friction coefficient of 0.562250645 occurs in the case 2 transport phenomenon; see Table 5. When the wedge is static, the shear stress of the domain decreases across the domain at the rate of -0.329789963. However, when there is faster stretching at the wedge (λ), the shear stress across the domain increases at -0.356473325.

When the ternary hybrid nanofluid flows faster, a more significant amount of the substance is bound to move over the wedge surface within a short period. Thus, the average kinetic energy of nanoparticles and molecules forming the ternary hybrid nanofluid decreases. On the distribution of heat energy through the various layers of the atmosphere at different latitudes, McEwen [43] remarked that the temperature typically drops due to the surface's height. There is a possibility for decreasing temperature distribution as $\lambda \to 2$ because of the manifestation of the reduction of total surface energy; see Figure 4. According to Ruiz-Gutierrez (2018), there is an attempt to reduce a liquid droplet's total surface energy when it comes into contact with the inside walls of a channel that fits wedge-shaped. The primary source of energy dissipation near the wedge's apex is friction associated with a viscous force on the fluid flow, which balances the work rate conducted by capillary forces. Just because the higher distribution of temperature across the ternary hybrid nanofluid emerges when ($\lambda = 0$), the concentration and distribution of motile gyrotactic organisms were found to emerge when the wedge was static; see Figure 4. Consequently, stretching on the wedge can reduce the distribution of heat energy, concentration, and distribution of motile gyrotactic organisms across the domain; see Figures 7–12. Wall stretching boosts the pressure inducing the dynamics of each ternary hybrid nanofluid and the flow rate.

An increment in temperature distribution ought to manifest due to a rise in pressure. However, in this case, a larger volume of ternary hybrid nanofluid passes through the wedge surface at each unit of time. The mass transfer, heat transfer, and distribution of motile gyrotactic organisms across the domain become minimal when the nanoparticles' thermal conductivity is considerably large, as in the first case (i.e., water conveying platelet SWCNTs, cylindrical SWCNTs, and spherical SWCNTs). Carbon nanotubes carry heat through the vibrating of the covalent bonds keeping the carbon atoms connected. The carbon nanotubes atoms wig around and convey the heat through the base fluid, unlike metals, which rely on the flow of electrons. Single-walled carbon nanotubes show great promise for enhancing the thermal conductivity of traditional polymers. From Figures 7 and 8, one can see that the transfer of heat energy and mass transfer is predominant at a higher level of wall stretching on the wedge (λ). The net transfer of the species between two locations is referred to as mass transfer. Figures 9 and 10 reveal that the transfer of the species near the wall may be characterized as the mass in transit due to the gradient species concentration in the mixture. The distribution of motile gyrotactic organisms decreases with λ and also from the wall $\eta = 0$ to the free stream $\eta \to 5$, not because the distribution is maximal at the wall of the wedge and minimal at the free stream; Figures 11 and 12.

Flint and Burstein [44] identified the Péclet number as a valuable tool for assessing mixing since it is affected by stochastic particle drift caused by diffusion (i.e., mixing). The observed decrement in the concentration of motile microorganisms within the ternary hybrid fluid due to the rising Péclet number is traced to the associated maximization of temperature distribution across the domain. This is true because when the thermal conductivity of the three kinds of nanoparticles is large in magnitude, the ternary hybrid nanofluid conveys SWCNT nanoparticles, and the distribution of temperature increases with the rising Péclet number. However, this is higher than that of case 2, where the thermal conductivity of the nanoparticles is moderately small (i.e., platelet magnesium oxide nanoparticles, cylindrical aluminum oxide nanoparticles, spherical silicon dioxide nanoparticles); see Figures 13 and 14. Dey and Chutia [45] once noticed that the bioconvection raises the microbe concentration across the flow zone. The collection of microorganisms is higher near the surface and eventually decreases as the perpendicular distance to the wall grows. With higher levels of the Schmidt number associated with bioconvection, the local density number of microorganisms decreases. Figures 15 and 16 reveal the gradient of distribution of motile gyrotactic organisms in a colloidal mixture of water and single-walled carbon nanotubes with a larger magnitude of thermal conductivity of different shapes (i.e., platelet, cylindrical, and spherical) when the wedge is static and moving faster (i.e., $\lambda = 5$). Similar to a moving conveyor belt, more tension is bound to manifest on the carrying side when the layers on the wedge move (i.e., $\lambda = 5$). Consequently, the distribution of motile gyrotactic organisms presented in Figure 16 only occurs near the wall. The observed optimal value of $-\chi'(\eta)$ that was obtained at $\eta = 0$ as 6.8992 can be traced to the heated wall and distribution of motile gyrotactic organisms that spread out quickly due to diffusive processes (i.e., $P_e = 4$).

5. Summary/Conclusions

We studied the dynamics of the unsteady motion of ternary hybrid nanofluid on static and moving wedges. The dynamics of water made up of (i) single-walled carbon nanotubes with a larger magnitude of thermal conductivity of different shapes (i.e., platelet, cylindrical, and spherical) and (ii) moderately small magnitude of thermal conductivity (i.e., platelet magnesium oxide, cylindrical aluminum oxide, spherical silicon dioxide). The following conclusions were drawn based on the findings of the analysis and discussion:

1. Increasing the stretching at the wedge's wall in the same direction as the transport phenomenon is suitable for decreasing the temperature distribution due to the higher velocity of ternary hybrid nanofluids either parallel or perpendicular to the wedge.
2. When the wedge is stationary or travels extremely slowly close to the wall, more friction is achievable at the wall. Although the second scenario of transport phenomena results in the highest friction coefficients along the wall, it is essential to note that increased wall stretching and the dispersion of SWCNTs in water make it possible to achieve the best increase in the same dependent variable.
3. Enhancement of stretching at the wedge wall is a factor responsible for causing a significant transfer of heat energy, species of the nanoparticles, and motile gyrotactic organisms near the wall only.
4. Significant difference exists between (a) the heat transfer of heat energy, (b) the mass transfer of species, and (c) the distribution rate of motile gyrotactic organisms in the dynamics of water made up of (i) single-walled carbon nanotubes with larger magnitudes of thermal conductivity of different shapes (i.e., platelet, cylindrical, and spherical) and (ii) a moderately small magnitude of thermal conductivity (i.e., platelet magnesium oxide, cylindrical aluminum oxide, spherical silicon dioxide).
5. When there is a more significant magnitude of thermal conductivity but smaller densities, and heat capacity of the three types of nanoparticles, as in the case of SWCNTs, the diffusion of motile gyrotactic organisms is significantly influenced in the motion of the bioconvective ternary hybrid nanofluid on a static wedge.

Further study of the dynamics through cavities with or without fins is necessary to grasp the motion outside the boundary layer.

Author Contributions: Conceptualization, I.L.A., Q.M.A.-M. and U.K.; methodology, I.L.A., Q.M.A.-M., U.K. and A.S.A.; software, Q.M.A.-M., U.K.; validation, I.L.A., Q.M.A.-M. and U.K.; formal analysis, I.L.A., Q.M.A.-M. and U.K.; investigation, Q.M.A.-M., U.K.; resources, I.L.A.; data curation, I.L.A., U.K.; writing—original draft preparation, I.L.A., Q.M.A.-M. and U.K.; writing—review and editing, I.L.A., Q.M.A.-M., U.K. and A.S.A.; visualization, Q.M.A.-M.; supervision, Q.M.A.-M. and A.S.A.; project administration, Q.M.A.-M. and A.S.A. All authors have read and agreed to the published version of the manuscript.

Funding: This research received no external funding.

Acknowledgments: This research was funded by Institutional Fund Projects under grant no. (IFPIP: 1284-130-1443). The authors gratefully acknowledge the technical and financial support provided by the Ministry of Education and King Abdulaziz University, DSR, Jeddah, Saudi Arabia.

Conflicts of Interest: On behalf of all authors, the corresponding author states that there is no conflict of interest.

References

1. Baughman, R.H.; Zakhidov, A.A.; De Heer, W.A. Carbon nanotubes—The route toward applications. *Science* **2002**, *297*, 787–792. [CrossRef] [PubMed]
2. Kim, P.; Shi, L.; Majumdar, A.; McEuen, P.L. Thermal transport measurements of individual multiwalled nanotubes. *Phys. Rev. Lett.* **2001**, *87*, 215502. [CrossRef] [PubMed]
3. Prasek, J.; Drbohlavova, J.; Chomoucka, J.; Hubalek, J.; Jasek, O.; Adam, V.; Kizek, R. Methods for carbon nanotubes synthesis. *J. Mater. Chem.* **2011**, *21*, 15872–15884. [CrossRef]
4. Dai, H. Carbon nanotubes: Synthesis, integration, and properties. *Accounts Chem. Res.* **2002**, *35*, 1035–1044. [CrossRef]
5. Li, H.; He, X.; Liu, Y.; Yu, H.; Kang, Z.; Lee, S.T. Synthesis of fluorescent carbon nanoparticles directly from active carbon via a one-step ultrasonic treatment. *Mater. Res. Bull.* **2011**, *46*, 147–151. [CrossRef]
6. Zhang, B.; Liu, C.Y.; Liu, Y. A novel one-step approach to synthesize fluorescent carbon nanoparticles. *Eur. J. Inorg. Chem.* **2010**, *2010*, 4411–4414. [CrossRef]
7. Li, Y.-H.; Qu, W.; Feng, J.-C. Temperature dependence of thermal conductivity of nanofluids. *Chin. Phys. Lett.* **2008**, *25*, 3319. [CrossRef]
8. Maranville, C.; Bollin, S.C.; Sawall, D.; Ohtani, H.; Remillard, J.T.; Ginder, J.M. Thermal conductivity of single-wall carbon nanotube dispersions: Role of interfacial effects. *J. Phys. Chem. C* **2008**, *112*, 654–658. [CrossRef]
9. Sabiha, M.A.; Mostafizur, R.M.; Saidur, R.; Mekhilef, S. Experimental investigation on thermo physical properties of single walled carbon nanotube nanofluids. *Int. J. Heat Mass Transf.* **2016**, *93*, 862–871. [CrossRef]
10. Said, Z. Thermophysical and optical properties of SWCNTs nanofluids. *Int. Commun. Heat Mass Transf.* **2016**, *78*, 207–213. [CrossRef]
11. Namarvari, H.; Razmara, N.; Meneghini, J.R.; Mir, C.R. Effect of SWCNT volume fraction on the viscosity of water-based nanofluids. *J. Mol. Model.* **2021**, *27*, 253. [CrossRef] [PubMed]
12. Alqahtani, A.M.; Adnan, Khan, U.; Ahmed, N.; Mohyud-Din, S.T.; Khan, I. Numerical investigation of heat and mass transport in the flow over a magnetized wedge by incorporating the effects of cross-diffusion gradients: Applications in multiple engineering systems. *Math. Probl. Eng.* **2020**, *2020*, 1–10. [CrossRef]
13. Sarkar, S.; Endalew, M.F. Effects of melting process on the hydromagnetic wedge flow of a Casson nanofluid in a porous medium. *Bound. Value Probl.* **2019**, *2019*, 43. [CrossRef]
14. Jameson, A. Iterative solution of transonic flows over airfoils and wings, including flows at Mach 1. *Commun. Pure Appl. Math.* **1974**, *27*, 283–309. [CrossRef]
15. Stewartson, K. Further solutions of the Falkner-Skan equation. *Math. Proc. Camb. Philos. Soc.* **1954**, *50*, 454–465. [CrossRef]
16. Riccardi, G.; Iafrati, A. Water impact of an asymmetric floating wedge. *J. Eng. Math.* **2004**, *49*, 19–39. [CrossRef]
17. Makinde, O.D.; Animasaun, I.L. Bioconvection in MHD nanofluid flow with nonlinear thermal radiation and quartic autocatalysis chemical reaction past an upper surface of a paraboloid of revolution. *Int. J. Therm. Sci.* **2016**, *109*, 159–171. [CrossRef]
18. Makinde, O.D.; Animasaun, I.L. Thermophoresis and Brownian motion effects on MHD bioconvection of nanofluid with nonlinear thermal radiation and quartic chemical reaction past an upper horizontal surface of a paraboloid of revolution. *J. Mol. Liq.* **2016**, *221*, 733–743. [CrossRef]
19. Schlichting, H. *Boundary-Layer Theory*; McGraw-Hill: New York, NY, USA, 1968.
20. Chelminski, K.; Bulski, W.; Rostkowska, J.; Kania, M. Dynamic wedges-dosimetry and quality control. *Rep. Pract. Oncol. Radiother.* **2006**, *11*, 67–75. [CrossRef]
21. Saminathan, S.; Manickam, R.; Supe, S.S. Comparison of dosimetric characteristics of physical and enhanced dynamic wedges. *Rep. Pract. Oncol. Radiother.* **2012**, *17*, 4–12. [CrossRef]
22. Miksis, M.J.; Vanden-Broeck, J.M. Self-similar dynamics of a viscous wedge of fluid. *Phys. Fluids* **1999**, *11*, 3227–3231. [CrossRef]
23. Ruiz-Gutierrez, E.; Semprebon, C.; McHale, G.; Ledesma-Aguilar, R. Statics and dynamics of liquid barrels in wedge geometries. *J. Fluid Mech.* **2018**, *842*, 26–57. [CrossRef]
24. Xu, L. Accelerating viscous flow past a wedge. *Phys. Fluids* **2020**, *32*, 013604. [CrossRef]
25. Raju, C.S.K.; Hoque, M.M.; Sivasankar, T. Radiative flow of Casson fluid over a moving wedge filled with gyrotactic microorganisms. *Adv. Powder Technol.* **2017**, *28*, 575–583. [CrossRef]
26. Cao, W.; Animasaun, I.L.; Yook, S.-J.; Oladipupo, V.A.; Ji, X. Simulation of the dynamics of colloidal mixture of water with various nanoparticles at different levels of partial slip: Ternary-hybrid nanofluid. *Int. Commun. Heat Mass Transf.* **2022**, *135*, 106069. [CrossRef]

27. Song, Y.-Q.; Obideyi, B.D.; Animasaun, I.L.; Mahrous, Y.M.; Chung, J.D. Significance of haphazard motion and thermal migration of alumina and copper nanoparticles across the dynamics of water and ethylene glycol on a convectively heated surface. *Case Stud. Therm. Eng.* **2021**, *26*, 101050. [CrossRef]
28. Timofeeva, E.V.; Routbort, J.L.; Singh, D. Particle shape effects on thermophysical properties of alumina nanofluids. *J. Appl. Phys.* **2009**, *106*, 014304. [CrossRef]
29. Elnaqeeb, T.; Animasaun, I.L.; Shah, N.A. Ternary-hybrid nanofluids: Significance of suction and dual-stretching on three-dimensional flow of water conveying nanoparticles with various shapes and densities. *Z. Fur Naturforschung A* **2021**, *76*, 231–243. [CrossRef]
30. Ahammed, N.; Asirvatham, L.G.; Wongwises, S. Entropy generation analysis of graphene-alumina hybrid nanofluid in multiport minchannel heat exchanger coupled with thermoelectric cooler. *Int. J. Heat Mass Transf.* **2016**, *103*, 1084–1097. [CrossRef]
31. Sahoo, R.R. Heat transfer and second law characteristics of radiator with dissimilar shape nanoparticle-based ternary hybrid nanofluid. *J. Therm. Anal. Calorim.* **2020**, *146*, 827–839. [CrossRef]
32. Sahoo, R.R. Thermo-hydraulic characteristics of radiator with various shape nanoparticle-based ternary hybrid nanofluid. *Powder Technol.* **2020**, *370*, 19–28. [CrossRef]
33. Hamilton, R.L.; Crosser, O.K. Thermal Conductivity of Heterogeneous Two-Component Systems. *Ind. Eng. Chem. Fundam.* **1962**, *1*, 187–191. [CrossRef]
34. Sahu, M.; Sarkar, J. Steady-State Energetic and Exergetic Performances of Single-Phase Natural Circulation Loop With Hybrid Nanofluids. *J. Heat Transf.* **2019**, *141*, 082401. [CrossRef]
35. Animasaun, I.L.; Shah, N.A.; Wakif, A.; Mahanthesh, B.; Sivaraj, R.; Koriko, O.K. *Ratio of Momentum Diffusivity to Thermal Diffusivity: Introduction, Meta-Analysis, and Scrutinization*; Chapman and Hall: New York, NY, USA; CRC: New York, NY, USA, 2022. ISBN 9781003217374. [CrossRef]
36. Aly, E.H.; Ebaid, A. Exact Analytical Solution for Suction and Injection Flow with Thermal Enhancement of Five Nanofluids over an Isothermal Stretching Sheet with Effect of the Slip Model: A Comparative Study. *Abstr. Appl. Anal.* **2013**, *14*, 721578. [CrossRef]
37. Xiong, X.; Chen, S.; Yang, B. Natural convection of SiO_2-water nanofluid in square cavity with thermal square column. *Appl. Math. Mech.* **2017**, *38*, 585–602. [CrossRef]
38. Kandasamy, R.; Vignesh, V.; Kumar, A.; Hasan, S.H.; Isa, N.M. Thermal radiation energy due to SWCNTs on MHD nanofluid flow in the presence of seawater/water: Lie group transformation. *Ain Shams Eng. J.* **2018**, *9*, 953–963. [CrossRef]
39. Kierzenka, J.; Shampine, L.F. A BVP solver based on residual control and the Maltab PSE. *ACM Trans. Math. Softw.* **2001**, *27*, 299–316. [CrossRef]
40. Al-Mdallal, Q.M.; Syam, M.I.; Anwar, M.N. A collocation-shooting method for solving fractional boundary value problems. *Commun. Nonlinear Sci. Numer. Simul.* **2010**, *15*, 3814–3822. [CrossRef]
41. De Hoog, F.; Weiss, R. The application of Runge-Kutta schemes to singular initial value problems. *Math. Comput.* **1985**, *44*, 93–103. [CrossRef]
42. Animasaun, I.L. Significance of Quartic Autocatalysis on Fluid Flows Conveying Nanoparticles, Dust-Particles and Microstructures over an Object with Non-Uniform Thickness. Ph.D. Thesis, Federal University of Technology, Akure, Nigeria, 2019.
43. McEwen, G.F. A mathematical theory of the vertical distribution of temperature and salinity in water under the action of radiation, conduction, evaporation, and mixing due to the resulting convection. *Trans. Am. Geophys. Union* **1923**, *4*, 139. [CrossRef]
44. Flint, I.M.; Burstein, M.A. FLOTATION | Froth Processes and the Design of Column Flotation Cells. *Encycl. Sep. Sci.* **2000**, 1521–1527. [CrossRef]
45. Dey, D.; Chutia, B. Dusty nanofluid flow with bioconvection past a vertical stretching surface. *J. King Saud Univ.-Eng. Sci.* **2022**, *34*, 375–380. [CrossRef]

applied sciences

MDPI

Article

Numerical Study on the Flow Past Three Cylinders in Equilateral-Triangular Arrangement at $Re = 3 \times 10^6$

Mohan Zhang [1,2], Bo Yin [1,2,*], Dilong Guo [1,2], Zhanling Ji [1,2] and Guowei Yang [1,2]

1 Key Laboratory for Mechanics in Fluid Solid Coupling Systems, Institute of Mechanics, Chinese Academy of Sciences, Beijing 100190, China
2 School of Engineering Science, University of Chinese Academy of Sciences, Beijing 100049, China
* Correspondence: yinbo@imech.ac.cn

Abstract: One of the most common systems in engineering problems is the multi-column system in the form of an equilateral-triangular arrangement. This study used three-dimensional numerical simulations to investigate the flow around three cylinders in this arrangement at the super-critical Reynolds number $Re = 3 \times 10^6$, concentrating on the influence on the spacing ratio (L/D) among cylinders. The instantaneous vortex structures, Strouhal numbers, fluid force coefficients, and pressure distributions are analyzed thoroughly. The present study demonstrated that fluid dynamics is sensitive to L/D, by which five different flow patterns are classified, namely single bluff body flow ($L/D \leq 1.1$), deflected gap flow ($1.2 \leq L/D \leq 1.4$), anti-phase flow ($1.5 \leq L/D \leq 2.3$), in-phase flow ($2.5 \leq L/D < 3.5$), and co-shedding flow ($L/D \geq 3.5$). Critical bounds are identified by significant transitions in the flow structure, discontinuous drop and jump of St, and force coefficients.

Keywords: flow interference; vortex interaction; vortex dynamics; improved delayed-detached eddy simulation (IDDES)

Citation: Zhang, M.; Yin, B.; Guo, D.; Ji, Z.; Yang, G. Numerical Study on the Flow Past Three Cylinders in Equilateral-Triangular Arrangement at $Re = 3 \times 10^6$. *Appl. Sci.* **2022**, *12*, 11835. https://doi.org/10.3390/app122211835

Academic Editors: Vasily Novozhilov and Cunlu Zhao

Received: 29 October 2022
Accepted: 17 November 2022
Published: 21 November 2022

1. Introduction

Manifold cylindrical structures are often used in mechanic practices and industries, such as tower groups, chimneys, and offshore platforms. Flow separation, reattachment, vortex impingement, and flow-induced motion (FIM) may occur in these systems due to the interaction between the flow and cylindrical structures [1]. Particularly, an equilateral-triangular configuration comprising three cylinders is widely encountered in marine and offshore engineering [2–4]. While extensive studies on the flow around two cylinders (with side-by-side or tandem arrangement) have been reported, the flow around three cylinders, particularly at high Reynolds numbers, has received less attention. ($Re = U_0 D/v$, where U_0 is the velocity of the incoming flow, D is the diameter of the cylinder, and v is the kinematic viscosity of the fluid.) In practice, however, turbulent flow is a more frequently encountered situation than laminar flow at low Reynolds numbers, but research on flow past cylinders in the super-critical flow regime is scant. In this study, the fluid forces, spectral characteristics, wake structure and their interactions with three equilateral cylinders are investigated, which support a thorough comprehension of fluid dynamics of flow around cylinders.

The spacing ratio L/D, where L is the separation between the centers of the two cylinders and D is the diameter, can be used to categorize the flow patterns for two cylinders in tandem form [5]. Sumner et al. [6] investigated three flow patterns of the flow around two tandem cylinders for $1.0 \leq L/D \leq 3.0$ at $1200 \leq Re \leq 3800$ by particle image velocimetry (PIV). When L/D = 1.0, the flow field is comparable to a single bluff body. When L/D = 1.5 and 2, the shear-layer reattachment flow features are noticed. When L/D = 2.5 and 3, the flow feature of each cylinder is similar to that in the case of a single cylinder. For two cylinders arranged side by side, Sumner [7] suggested identifying the flow regimes into three types. When $1.0 \leq L/D \leq$ 1.1–1.2, the two cylinders act like one bluff body. When 1.1–1.2 $\leq L/D \leq$ 2–2.2, a biased flow pattern shows up. When $L/D \geq$ 2–2.2, cylinders behave more independent and show parallel vortex streets. More details of the wake flow of two cylinders at variable Re and L/D are summarized by Sumner [7].

For three cylinders in tandem and side-by-side layout, the flow patterns, force coefficients, and pressure distributions have been investigated [8–10]. More complicated than a tandem or side-by-side layout is the flow past through three cylinders arranged in a triangle. Sayers [11] measured the drag and lift force coefficients of one of the three cylinders in triangular arrangement at $Re = 3 \times 10^6$ and $1.25 \leq L/D \leq 5$ with various incidence angles (α). Only one cylinder was measured, so the interference between the cylinders is not clear. The vortex-shedding-frequency data of three cylinders in an equilateral triangle configuration at $Re = 2.1 \times 103$ and 3.5×103 at various incidence angles and L/D were obtained by Lam et al. [12] using a dye-injection approach to visualize the flow. Bi-stable flow characteristics were observed at $L/D < 2.29$, $\alpha = 0°$ and they depend on the starting conditions. The static pressures of each cylinder were individually measured by Tatsuno et al. [13]. The results show that when the spacing ratio is small, the impacts of the flow interference are obvious. By conducting wind tunnel studies, Gu et al. [14] categorized the flow pattern of the three cylinders arranged in an equilateral triangle. It shows that the incidence angle has a huge influence on the pressure distribution on each cylinder and the flow patterns. Four basic levels of interference (small, transition, medium, large spacing) are identified according to the spacing ratio. Furthermore, according to the various incidence angles, the interference type can be identified as proximity effect, shear layer reattachment effect, and wake effect. Pouryoussefi et al. [15] carried out wind tunnel experiments with five subcritical Reynolds numbers at the incidence angle 0° (one cylinder in the upstream and others in the downstream). They showed that as L/D increases, the mean drag coefficient of all cylinders almost increases. When $L/D = 1.5$ and 2, the downstream cylinders' drag coefficients reach their lowest value since there is no vortex shedding from them. Bansal et al. [16] investigated the influence of the incidence angle (0°–60°) when the $L/D = 1.35$ and $Re = 2100$ by PIV and laser Doppler velocimetry. They found that large-scale vortexes shed at about $5D$ downstream from the cylinders for all incidence angles. Yang et al. [17] investigated the effects of L/D (2.8–5.2) and incidence angle (0°–30°) on the flow around an equilateral-triangular-arranged three-cylinder cluster at $Re = 8000$. They classified the flow pattern into two categories, short-spacing ratio shear-layer reattachment regime and big-spacing ratio vortex-shedding regime, respectively. In addition, the influence of the incidence angle is more complex and dependent on the spacing ratio.

Meanwhile, a number of numerical simulations on this three-cylinder configuration have been performed over the past decades. Yan et al. [18] investigated the influence of the spacing ratio (1.5–5) and the incidence angle (0°, 30°, 60°) on the flow patterns at $Re = 100$ by two-dimensional simulation. Yang et al. [19] investigated the characteristic flow regions of three cylinders by the lattice Boltzmann method and fluorescence flow visualization using a laser. They demonstrated that for $Re = 200$ and $S/D = 3$, two different types of flow patterns can be distinguished by T/D, where S is the distance between the centers of the cylinders in the upstream and downstream, and T is the distance between the centers of the two cylinders in the downstream. When $1 \leq T/D \leq 1.2$ and $2.5 \leq T/D \leq 3.1$, typical steady flow occurs, when $1.3 \leq T/D \leq 2.4$ and $3.2 \leq T/D \leq 10$, typical unsteady flow occurs. By using the 2D finite volume approach, Zheng et al. [20] examined the effects of the L/D (1.5–7) and Reynolds number (100–200) on the flow characteristics of three identical cylinders organized in an equilateral-triangular arrangement at incidence angles of 0° and 180°. By using the immersed boundary approach, Chen at al. [21] carefully explored the impact of L/D, Reynolds number, and three-dimensionality on the fluid dynamics of the flow past three circular cylinders arranged in an equilateral-triangular arrangement. Six flow patterns depending on L/D at $Re = 100$ were observed: single bluff-body flow ($1.0 \leq L/D \leq 1.4$), deflected flow ($1.5 \leq L/D \leq 1.9$), flip-flopping flow ($2.0 \leq L/D \leq 2.5$), anti-phase flow ($2.6 \leq L/D \leq 2.8$ and $3.5 \leq L/D \leq 4.1$), in-phase flow ($2.9 \leq L/D \leq 3.4$

and 4.2 $\leq L/D \leq$ 4.5), and fully developed in-phase co-shedding flow (4.6 $\leq L/D \leq$ 6.0). In addition, some research of the flow configuration in the sub-critical regime were studied by the numerical method. Gao et al. [22] simulated the flow past three circular cylinders with the incidence angle of 30° for 200 $\leq Re \leq$ 3900 and 1.25 $\leq L/D \leq$ 4.0 and five flow patterns were identified. The vortex shedding's three-dimensionality grows stronger as the Reynolds number increases.

From the review of the relevant studies, it can be observed that the majority of the numerical and experimental research works were carried out in the laminar and sub-critical flow regime. Nevertheless, the Reynolds number is often between 10^5–10^7 in the majority of real circumstances, such as offshore engineering, which correspond to the super-critical regime and post-critical regime, respectively [23]. In this paper, we focus on the super-critical flow regime, the Reynolds number as high as 3×10^6. In super-critical flow, a rapid transition happens from laminar boundary layer to turbulent boundary layer. It is necessary to conduct a more comprehensive investigation into the flow structure and behavior of the three cylinders when they are arranged in an equilateral triangle at high Reynolds numbers. Schewe [24] carried out wind tunnel tests to achieve the force measurements of single cylinder from the Reynolds number 2.3×10^4 to 7.1×10^6, corresponding to the sub-critical and post-critical regime, respectively. It was observed that the Strouhal number (St) increased while the Cd value decreased at $Re > 3.5 \times 10^5$. Hinsberg [25] measured the unsteady aerodynamic forces and surface pressure of a rough single cylinder at $1.5 \times 10^4 \leq Re \leq 1.2 \times 10^7$. The results showed that the three-dimensional characteristics of the flow in spanwise direction became strong especially in the critical regime. In addition, the wake width became narrower and the drag coefficient decreased, which was observed by Rodriguez et al. [26], too. As for two tandem cylinders, Okajima [27] measured Cd and St by low-speed wind tunnel tests. It was observed that when the Reynolds number came to the super-critical regime, the jump of Cd disappeared at L/D = 3.8 and the effect of L/D on Cd and St became weaker. Hu et al. [23] investigated the tandem cylinders' flow characteristics in the sub-critical and super-critical regimes. When compared to the examples in the sub-critical regime, the vortex shedding frequencies are higher in the super-critical domain. Moreover, the flow separation positions move backward along the cylinder surface, which causes the reattachment position on the back cylinder to shift forward in the super-critical regime.

In summary, few systematic study have been conducted on the flow around three cylinders in equilateral-triangular configuration in the super-critical regime. In the present paper, a three-dimensional numerical simulation of flow past three cylinders in equilateral-triangular arrangement in super-critical regimes ($Re = 3 \times 10^6$) is presented at a spacing ratio of 1.1 $\leq L/D \leq$ 3.5, focusing on how fluctuating forces, flow separation, and vortex shedding frequencies vary with L/D. Five flow patterns are identified by spacing ratio and the characteristics of each pattern are summarized.

2. Numerical Models

2.1. Numerical Method

Detached-eddy simulation (DES) model was first carried out by Spalart [28] to compensate the lack of Reynolds-Averaged Navier–Stokes (RANS) in unsteady turbulence prediction and avoid the heavy demand of grid and computation of Large-Eddy Simulation (LES). Then, for the purpose of solving the modeled stress depletion (MSD), delayed-detached eddy simulation (DDES) was carried out by Spalart [29]. DDES introduces the second length scale of turbulence model in its length scale instead of filter scale of LES. Shur et al. [30] introduced an improved delayed-detached eddy simulation (IDDES), which combines DDES and LES for wall modeling. In the IDDES method, the model stress loss issue from the original DES technique is removed using the DDES length scale, and the boundary layers are predicted using WMLES. In the recent years, this hybrid approach has been widely utilized to examine separated flows which are unstable and geometrically associated [31]. Wang et al. [32] used the $sst\ k - \omega$ IDDES method to perform the reverse flow past an NACA0012 airfoil and had satisfactory results compared to experimental measurements. Moreover, in the field of high-speed train aerodynamics simulations which have complex shapes and usually high Reynolds numbers, the $sst\ k-\omega$ IDDES method is quite popular among scholars due to its capacity of capturing the train wake flow and vortex structures [31,33–35]. Hence, $sst\ k-\omega$ IDDES is adopted in the present study.

The key point of IDDES is to modify the dissipation component in the governing equation for the turbulent kinetic energy equation. The governing equations for the kinetic energy k are written as

$$\frac{\partial(\rho k)}{\partial t}+\frac{\partial(\rho u_i k)}{\partial x_i}=\tau_{ij}S_{ij}-\frac{\rho k^{1.5}}{l_{k-\omega}}+\frac{\partial}{\partial x_i}\left[(\mu_l+\sigma_k\mu_t)\frac{\partial k}{\partial x_i}\right] \tag{1}$$

where $l_{k-\omega}=\sqrt{k}/(\beta^*\omega)$; ρ is the fluid density; t is the time; u_i is the velocity; x_i is the position; τ_{ij} is the Reynolds stress tensor; S_{ij} is the mean strain rate tensor; μ_l is the laminar viscosity coefficient; σ_k is model coefficient; μ_t is the turbulent eddy viscosity. The IDDES model defines the length scale as

$$l_{IDDES}=\tilde{f}_d(1+f_e)l_{k-\omega}+(1-\tilde{f}_d)C_{DES}\Delta \tag{2}$$

where $\Delta=\min[C_w\max(\mathrm{d},\Delta),\Delta]$; d is the distance to the closest wall boundary; $\Delta=\max(\Delta\mathrm{x},\Delta\mathrm{y},\Delta\mathrm{z})$ represents the maximum of the local grid scales; $C_w=0.15$, C_{DES} is a constant calculated by the blending function F_1

$$C_{DES}=(1-F_1)C_{DES}^{k-\varepsilon}+F_1C_{DES}^{k-\omega} \tag{3}$$

with $C_{DES}^{k-\varepsilon}$ is 0.61 and $C_{DES}^{k-\omega}$ is 0.78; $\tilde{f}_d$ is a blending function defined as

$$\tilde{f}_d=\max[(1-f_{dt}),f_b] \tag{4}$$

with

$$f_{dt}=1-\tanh\left[(C_{dt}r_{dt})^3\right] \tag{5}$$

$$r_{dt}=\frac{\mu_t}{\kappa^2d^2[(S^2+\Omega^2)/2]^{0.5}} \tag{6}$$

$$f_b=\min\left[2\exp\left(-9\alpha^2\right),1.0\right] \tag{7}$$

$$\alpha=0.25-d/\Delta \tag{8}$$

where C_{dt} is 20 and κ is the von Karman constant; f_e is elevating function defined as

$$f_e=f_2\cdot\max[(f_1-1.0),0.0] \tag{9}$$

with

$$f_1=\begin{cases}2\exp(-9\alpha^2), & \alpha<0\\ 2\exp(-11.09\alpha^2), & \alpha\geq 0\end{cases} \tag{10}$$

$$f_2=1.0-\max(f_t,f_l) \tag{11}$$

$$f_t=\tanh\left[\left(C_t^2r_{dt}\right)^3\right] \tag{12}$$

$$f_l=\tanh\left[\left(C_l^2r_{dl}\right)^{10}\right] \tag{13}$$

$$r_{dl}=\frac{\mu_l}{\kappa^2d^2[(S^2+\Omega^2)/2]^{0.5}} \tag{14}$$

Please consult Shur et al. [30] for more information.

Numerical simulations are carried out by the STAR-CCM+ 13.06, which is based on the finite-volume method (FVM). For the temporal discretization, a second-order implicit scheme is used. Pressure–velocity coupling is based on the density implicit with the SIMPLE method. A second-order upwind scheme is employed for spatial discretization. In this study, 128 CPUs with 24 h for each case are used for simulation.

2.2. Boundary Conditions and Grid System

The computational domain and the boundary conditions of the present study are displayed in Figure 1. The simulation is 2.5D. The cylinder in the upstream is named Cylinder A and the cylinder in the downstream is named Cylinder B (+y-direction) and C (−y-direction), as shown in Figure 2a. Cylinder A is located in the middle of the domain in the y-direction, and the distance from inlet boundary to the center of the Cylinder A is $10D$. The computation domain is $35D$ (D is the diameter of the cylinder, which is 1 m in the present paper) in the x-direction and $16D$ in the y-direction. The size of the domain in the z-direction is $4D$, which has been proved long enough to capture the relevant flow structure [36,37]. The inlet boundary condition is set as velocity inlet boundary. The outlet boundary is set as a pressure outlet boundary. Periodic boundary conditions are set at spanwise boundaries (z-direction). As for the inlet boundary, turbulence viscosity is 5 times the molecular viscosity. Because when in the super-critical Reynolds regime, the transition in the boundary layer is turbulent. No-slip wall boundary conditions are set at the cylinder surfaces.

The computing mesh of the case when $L/D = 2.5$ (L is the distance between the centers of either two cylinders) is shown in Figure 2. A trimmed mesh is adopted in all the cases. The grids are refined in the region near the cylinder system. The size of each refined mesh region is illustrated in Figure 2a. The finest grid size is 0.025D. Grid size increases by a factor of 2. The grid is the same size in x, y, and z-directions. In the spanwise direction, for the finest grid region, there are 160 cells. The boundary layers are set to be 40 layers to keep $y+ < 1$, as shown in Figure 3. The total number of the cells for $Re = 3 \times 10^6$ is 10 million when $L/D = 2.5$. As the L/D increases, the total number of cells increases, too. As for temporal discretization, the time-step Δt is 5×10^{-4} to ensure the maximum Courant number (defined as $u \times \Delta t/\Delta x$) less than 1. Ten internal iterations exist in each time step. In this paper, the diameter of the cylinder (D) is set to 1 m, the flow velocity (U_0) is 47.02 m/s, fluid density (ρ) is 1.225 kg/m^3, and the dynamic viscosity (μ) of fluid is 0.0000192 Pa · s.

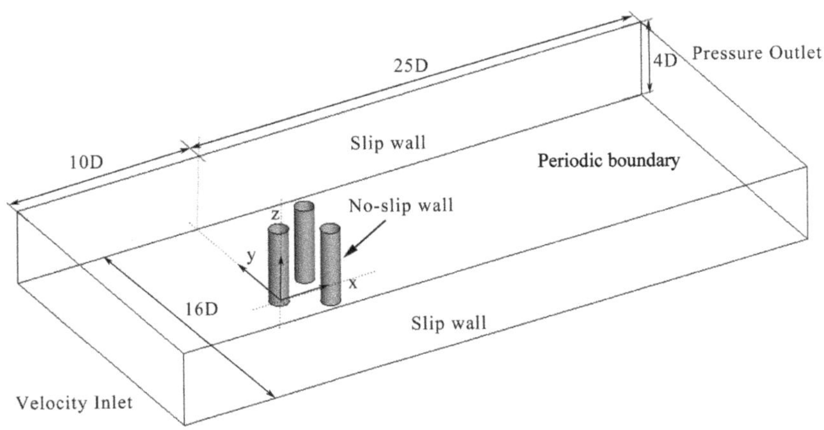

Figure 1. The boundary conditions and the schematics of the computing field.

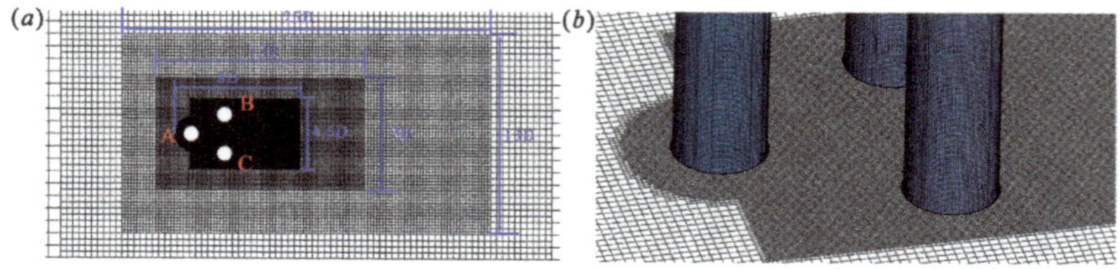

Figure 2. The computational mesh for $L/D = 2.5$: (**a**) mesh refinement using multiple sub-domains; (**b**) mesh near the cylinders surface.

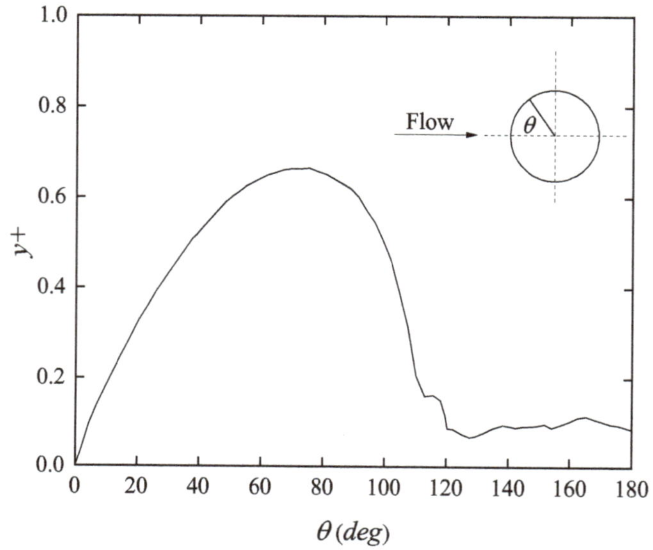

Figure 3. The grid resolution near the cylinder wall.

3. Convergence Study and Calculation Validation

A mesh dependency study of a single cylinder is performed at $Re = 3 \times 10^6$. Three types of meshes (coarse, medium, and fine meshes, respectively) are utilized. The ratio of cell length between coarse mesh and medium mesh, medium mesh and fine mesh is $\sqrt{2}$ in three space directions, which was recommended by Spalart [38]. The boundary layer mesh remains the same. Representative dynamic force coefficients are calculated to evaluate the mesh convergence. The drag and lift coefficients C_d and C_l are defined as

$$C_d = \frac{F_d}{\frac{1}{2}\rho U_0^2 DH} \tag{15}$$

$$C_l = \frac{F_l}{\frac{1}{2}\rho U_0^2 DH} \tag{16}$$

and the Strouhal number S_t is defined as

$$S_t = \frac{f_s D}{U_0} \tag{17}$$

where F_d and F_l are the fluid forces on the cylinder in the x-direction and y-direction, respectively; U_0 is the velocity of the fluid, H is the length of the cylinder in z-direction, f_s is the frequency of the vortex shedding, which is obtained from the power spectrum density (PSD) of the lift force of the cylinder. The pressure coefficient (C_p) is defined as

$$C_p = \frac{P - P_0}{\frac{1}{2}\rho U_0^2} \tag{18}$$

where P_0 represents the reference pressure. C_{pm} is the minimum value of C_p, and C_{pb} is the base pressure coefficient of the cylinder. The overline bar means the time average. The fluctuating lift coefficient C_l' is the root-mean-square (RMS) value of C_l. For statistical accuracy, at least 25 vortex shedding cycles are sampled, which means the flow passes through the computational domain more than 3 times in the statistical process.

As shown in Table 1, the results from the coarse mesh differ from those of the medium mesh and fine mesh. S_t of the three sets of meshes are the same. As can be seen, the medium mesh and fine mesh yield insignificant changes of $\bar{C}_d$, $-\overline{C}_{pm}$ except for $C_l{}'$ and $-\overline{C}_{pb}$, but the difference is in an acceptable range. Thus, the medium mesh is considered to be suitable for the following calculations.

Table 1. Grid density sensitivity study results of a single cylinder at $Re = 3 \times 10^6$.

Case	Cell Count (million)	$\overline{C}_d$	C_l'	S_t	$-\overline{C}_{pm}$	$-\overline{C}_{pb}$
Coarse Mesh	1.9	0.367	0.032	0.383	2.313	0.424
Medium Mesh	3.8	0.377	0.038	0.383	2.318	0.419
Fine Mesh	7.4	0.378	0.046	0.383	2.326	0.437

In order to validate the present numerical model, Table 2 summarizes the results of the present simulation with previous experimental and numerical calculations of other scholars for a single cylinder at $Re = 3 \times 10^6$. Regarding the super-critical regime, the outcomes of the earlier experiments showed significant scattering. This may be primarily caused by a variety of experimental challenges, including surface roughness, turbulent intensity of free flow, cylinder end conditions, wind tunnel blockage ratio, and cylinder aspect ratio, which make it challenging to precisely measure flow at high Reynolds numbers [39]. Considering the aforementioned factors, the present simulation results are acceptable.

Table 2. Comparison of present and previous studies for flow past a single cylinder.

Researchers	Re	Description	$\overline{C}_d$	C_l'	S_t	$-\overline{C}_{pb}$
Present study	3×10^6	k–ω IDDES	0.377	0.033	0.383	0.419
Roshko [40]	1×10^6	Experiments	0.30	–	–	0.37
	1.8×10^6	Experiments	0.42	–	–	0.62
	3.5×10^6	Experiments	0.69	–	–	0.85
Schmidt [41]	$(2.6–3.3) \times 10^6$	Experiments	0.25–0.44	–	–	0.51–0.61
Jones et al. [42]	1×10^6	Experiments	0.21	–	–	0.53
	2.9×10^6	Experiments	0.53	–	–	0.59
	3.7×10^6	Experiments	0.56	–	–	0.61
Schewe [24]	3×10^6	Experiments	0.45	–	–	–
Shih et al. [43]	$(3–3.1) \times 10^6$	Experiments	0.35–0.38	–	–	0.45–0.46
Travin et al. [44]	3×10^6	DES	0.41	0.06	0.35	0.53
Catalanoet al. [45]	1×10^6	LES	0.31	–	0.35	0.32
Ong et al. [46]	3.6×10^6	k–ε URANS	0.457	0.077	0.305	–

In addition, Figure 4 compares the $\bar{C}_p$ distribution on a single cylinder with published numerical studies (Travin et al. [44] at $Re = 3 \times 10^6$ and Catalano et al. [45] at $Re = 1 \times 10^6$) and experimental studies (Warschauer [47] at $Re = 1.2 \times 10^6$ and Zdravkovich [48] at $Re = 6.7 \times 10^5$). As shown in Figure 4, the $\bar{C}_p$ distribution on a single cylinder agrees well with previous results. The skin friction coefficient, $C_f = \tau/(0.5\rho U_0^2)$, where τ is the tangential wall shear stress, is presented in Figure 5. The separation point location $\theta = 116°$

predicted in the present study agrees well with the previous experimental results [49] and other numerical results [44,46]. The overprediction of C_f for $\theta < 90°$ in the present study compared with the experimental values is also observed from other numerical studies which used wall functions [44–46]. This results from the wall function method's assumption of a completely turbulent boundary layer [50].

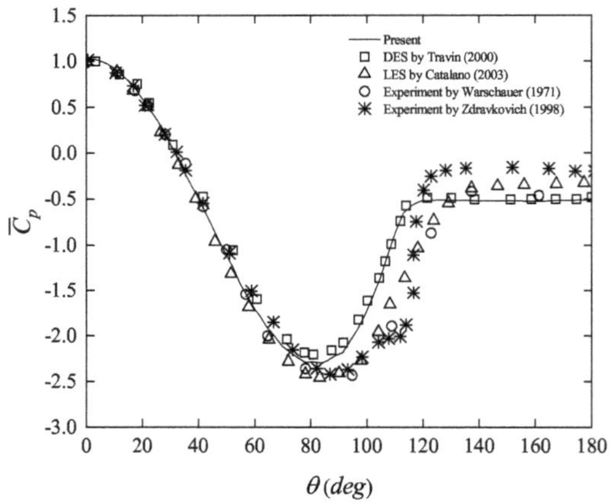

Figure 4. Comparison of pressure distribution on a single cylinder between the present study and published data [44,45,47,48].

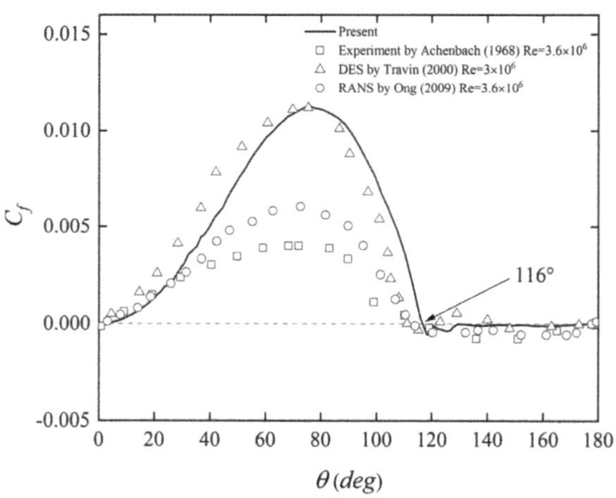

Figure 5. Comparison of skin friction coefficient (C_f) distribution on a single cylinder surface with the published experimental and numerical data [44,46,49].

In general, the present calculation of flow past a single cylinder is in accordance with the outcomes of other published papers. Thus, it is reasonable to assume that the calculation method is also valid for calculating the triangular arrangement of the three cylinders.

4. Results and Discussion

4.1. Flow Patterns

Figures 6–10 show the spanwise normalized vorticities diagrams and the time-averaged streamline diagrams at midspan plane for flow past three cylinders arranged in an equilateral-triangle configuration at different spacing ratio L/D at $Re = 3 \times 10^6$. Clockwise and counterclockwise vorticities are displayed by red and blue colors, respectively. The flow patterns are classified according to L/D based on the careful examination of Figures 6–11 and other parameters such as Cd, Cl, and St. Five different flow patterns were summarized and categorized. The naming of the flow patterns is partially referenced to Chen [21] who found them at $Re = 50$–300. It is worth noting that the specific features and critical spacing ratios are widely divergent according to results from Chen [21], on account of the four-orders-of-magnitude difference in Reynolds number.

The specific characteristics of each flow pattern are as follows.

Single bluff body flow

When $L/D = 1.1$, the shear layers of the upstream cylinder attach to the front surface of the downstream cylinders and pass through the gap between the cylinders, creating an extremely unstable shear flow on the gap-side surface of the downstream cylinders. The free-flow-side of the downstream cylinder has vortex shedding symmetry occurring simultaneously, with the vortex pairs taking a form similar to the typical Karman vortex street. This is unfamiliar with the single bluff body flow at low Reynolds numbers (50–300) [21]. At low Reynolds numbers, the shear layers of the upstream cylinder enclose the downstream cylinders, and when L/D increases, a small amount of fluid will pass through the gap and blend into the shear layers on the free-flow-side of the downstream cylinders and come off together. The critical L/D is 1.4 for low Reynolds numbers [21,51].

Figure 6a,b shows the contours of normalized instantaneous spanwise vorticities and the time-averaged streamlines for single bluff body flow at the middle plane, respectively. The time-averaged streamlines are symmetric relating to the center line of the wake, which indicates that the statistical range used in this paper is acceptable. Figure 6b shows that the recirculation region behind the three cylinders is longish and begins to occur at about one diameter away behind the downstream cylinders. The flow pattern at this L/D is identical to a flow through a single bluff body, as seen in Figure 6b.

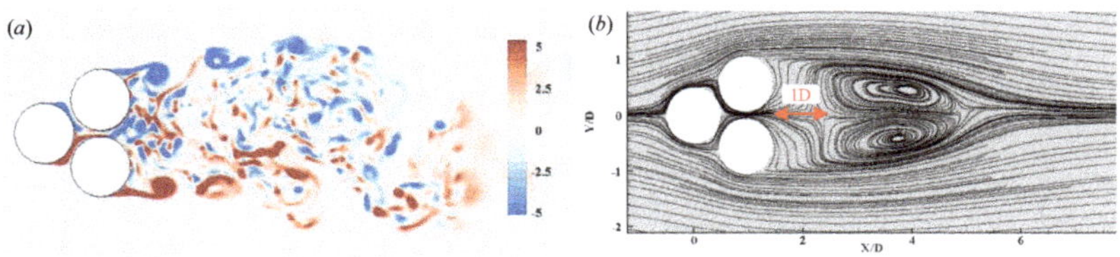

Figure 6. (**a**) Normalized spanwise vorticity $\omega_z D/U$; (**b**) time-averaged streamlines in the plane $z = 2$ D at $L/D = 1.1$.

Deflected gap flow

As L/D increases to 1.2–1.4, the gap flow is deflected towards one of the downstream cylinders and forms a deflected gap flow pattern. At $L/D = 1.2$ and 1.4, gap flow is deflected towards the downstream cylinder C and produces a narrow wake behind it, and the opposite when $L/D = 1.3$. Different computation procedures produce random deflection directions. A switch in the directions of the gap flow is not found throughout the computation process, which is consistent with the findings by Bansal [16] and Chen [21]. As regards double cylinders arranged side by side, nevertheless, Kim [52] and Alam et al. [53] found a shift in the deflection direction of the gap flow. The existence of the upstream cylinder might limit the direction change of the deflection flow.

Figure 7 demonstrates that the vortex shedding of the downstream cylinder with the gap flow bias is suppressed. The suppression lessens with the increase of L/D. From the time-mean streamlines in Figure 7b,d,f, it is observed that the recirculating bubble of the wider wake downstream cylinder is longer and broader. The cylinder with a wide wake has a lower drag coefficient compared to the cylinder with narrow wake, which is consistent with the biased gap flow of the two cylinders arranged side by side [7].

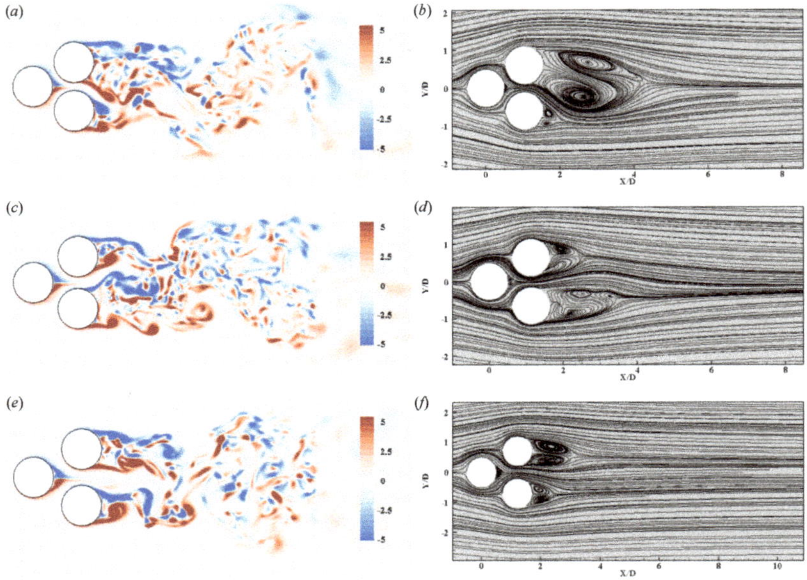

Figure 7. Normalized spanwise vorticity $\omega_z D/U$ and time-averaged streamlines in the plane z = 2 D at: (**a**,**b**) $L/D = 1.2$; (**c**,**d**) $L/D = 1.3$; (**e**,**f**) $L/D = 1.4$.

Anti-phase flow

As L/D increases, the flow pattern becomes anti-phase flow. Anti-phase is referred to the occurrence of 180° phase lag in the time history of lift forces between the downstream cylinders. Figure 11 presents the relationship of the phase lag ϕ of the downstream cylinders with L/D , which is obtained through the FFT analysis of the Cl of two downstream cylinders. Since there is no significant principal frequency of Cl of the downstream cylinder when $L/D < 1.4$, their phase lag data are not shown in the figure.

In the anti-phase flow regime, the vortices on the free-flow-side of the downstream cylinder shed simultaneously and are symmetrical relative to the wake center line. The shear layers of the upstream cylinder appear to roll up and produce vortex shedding, with a weaker intensity compared with the downstream cylinders. They are elongated at the gap, and finally pair with the the vortices produced by gap-side downstream cylinders and blend in with the wake flow. The size of the recirculation zone of the upstream cylinder is way smaller than that of the downstream cylinders, as seen in Figure 8d.

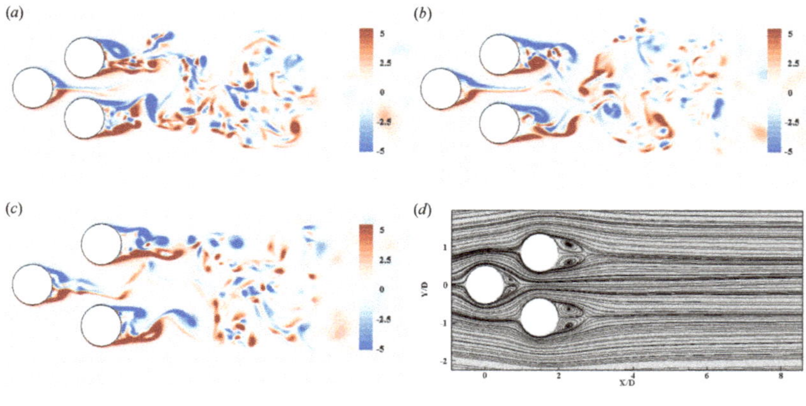

Figure 8. Normalized spanwise vorticity $\omega_z D/U$ in the plane z = 2 D at: (**a**) $L/D = 1.5$; (**b**) $L/D = 1.7$; (**c**) $L/D = 2$; (**d**) time-averaged streamlines at $L/D = 1.7$.

In-phase flow

Figure 9a–c shows the vorticity contours at $L/D = 2.3, 2.5, 3$. The flow pattern belongs to in-phase flow, because the vortex shedding of the downstream cylinder presents an in-phase fashion. As shown in Figure 11, the phase lag of the two downstream cylinders is near 0 when $L/D > 2$. The vortices from free-flow-side shear layers of the cylinder B and the gap-side shear layers of the cylinder C shed synchronously, which is exactly the opposite for anti-phase flow.

The vortex generated from the upstream cylinder is elongated at the gap and interacts with the vortex of the gap-side shear layers of the downstream cylinders. As L/D increases, the vortices shed by the upstream cylinder show an oscillation in the vertical direction in order to pair with the vortex generated from downstream cylinders. These interactions bring on the dissipation of vortices' energy of the upstream cylinder. Figure 9d only shows the time-averaged streamline for $L/D = 2.5$ because the difference between each L/D in in-phase flow is insignificant. Figure 9d indicates that the length of the recirculation zone of the upstream cylinder is larger than that of anti-phase flow. Such a phenomenon results from the increased space at the gap, which allows the shear layers of the upstream cylinder to fully develop.

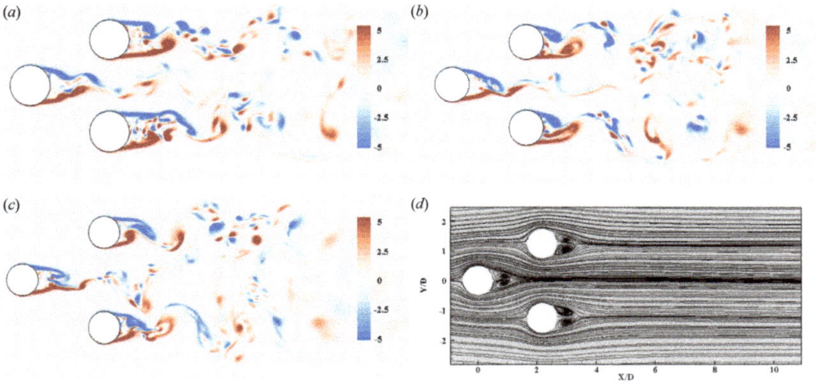

Figure 9. Normalized spanwise vorticity $\omega_z D/U$ in the plane z = 2 D at (**a**) $L/D = 2.3$; (**b**) $L/D = 2.5$; (**c**) $L/D = 3$, (**d**) time-averaged streamlines at $L/D = 2.5$.

Co-shedding flow

When $L/D > 3.5$, the vorticity and time mean streamlines of the three cylinders are shown in Figure 10. Because L/D is large enough, the interaction between the cylinders is relatively small, and the flow pattern around each cylinder approximates the fashion of flow past a single cylinder. However, in fact, it can be seen from the vorticity contour that the vortex shed from the upstream cylinder still has a tendency of pairing with the corresponding downstream vortex when the flow goes through the downstream cylinder. Figure 12 shows the normalized time-averaged x-directional flow velocity U/U_0 in the tail region of the three cylinders, with the x-axis denoting the distance from the center of each cylinder normalized by D. The single cylinder case is also shown for comparison. The wake flow velocity of downstream cylinders is identical to that of a single cylinder. For the upstream cylinder, however, the velocity can be seen to be higher when it passes through the downstream cylinders at about three-quarters of the position, which confirms the vortex-pairing tendency aforementioned. Additionally, the vortex shedding of the downstream cylinders occurs in an in-phase fashion. There is a certain phase lag between the vortex shedding of the upstream cylinder and downstream cylinders, which is determined by L/D.

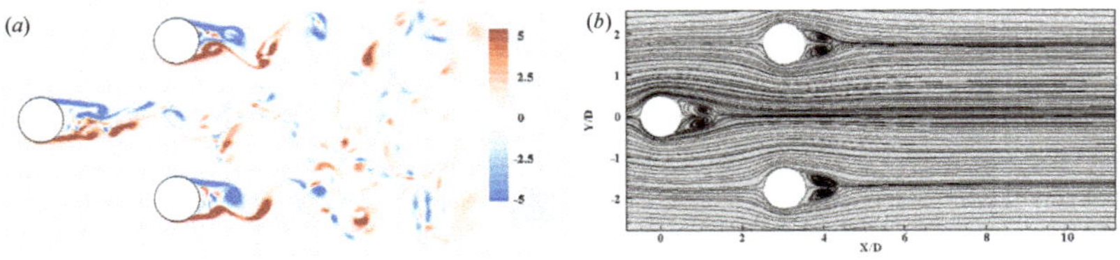

Figure 10. (**a**) Normalized spanwise vorticity $\omega_z D/U$; (**b**) time-averaged streamlines at $L/D = 3.5$.

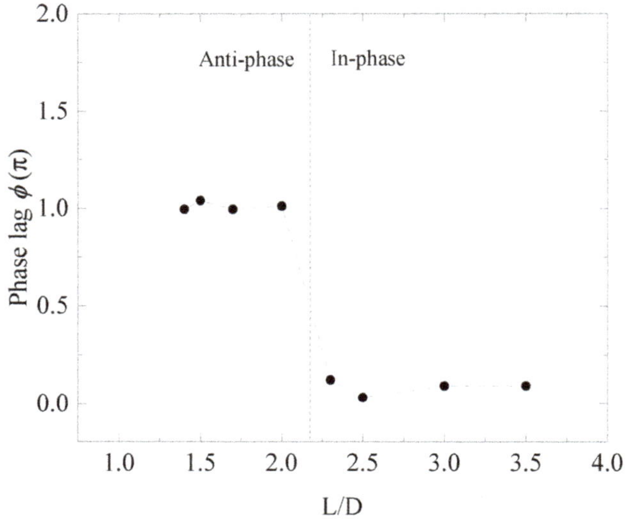

Figure 11. Phase lag of lift force history between two downstream cylinders.

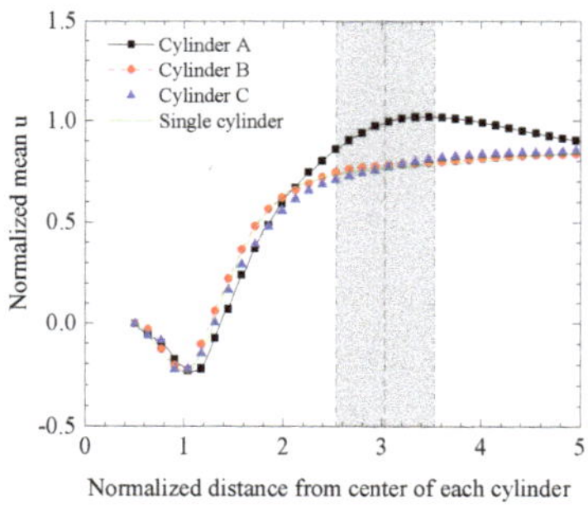

Figure 12. Mean velocity (u_x) at the center line of each cylinder at $L/D = 3.5$. The three gray dashed lines represent the front edge, center line, and end edge of the downstream cylinders, respectively.

With the purpose of understanding the three-dimensional vortex structures, Figure 13 exhibits the instantaneous Q criterion [54] of each flow pattern, for the sake of brevity, only one L/D is shown at each pattern. Q is defined as

$$Q = -\frac{1}{2}\left(\|\mathbf{S}\|^2 - \|\mathbf{\Omega}\|^2\right) \tag{19}$$

where $\mathbf{S}$ and $\mathbf{\Omega}$ represent the strain and rotation tensor, respectively. Significant Kelvin–Helmholtz-type vortex structures are observed. Meantime, instantaneous pressure distributions in the middle section at different flow pattern regimes are shown in Figure 14.

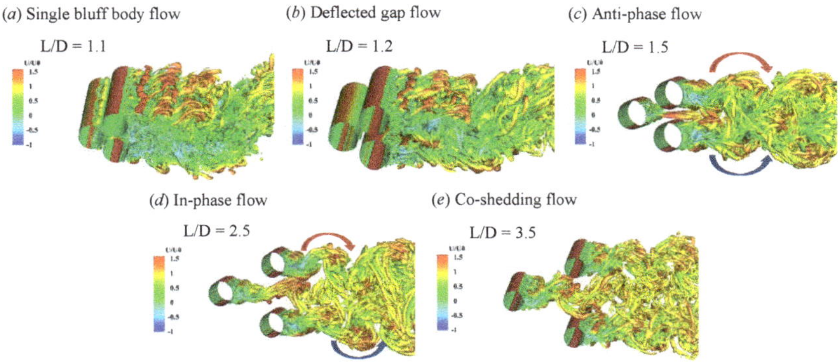

Figure 13. Iso-surface of $Q = 1.5\, U_0{}^2/D$ colored by dimensionless streamwise velocity U/U_0 at different flow pattern regimes.

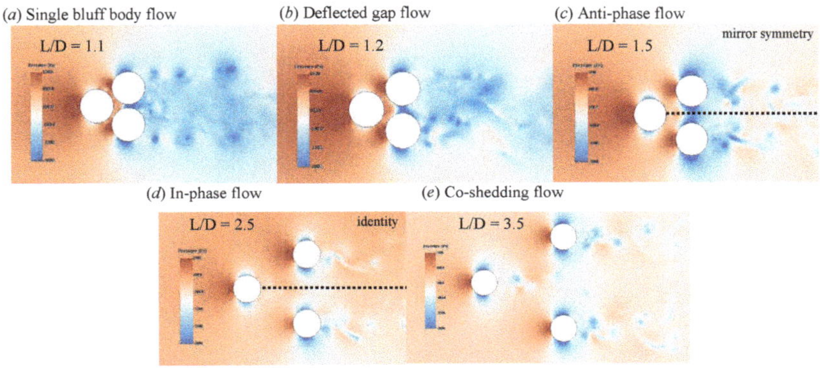

Figure 14. Instantaneous pressure distributions in the middle section (z = 2 m) at different flow pattern regimes.

The dotted line indicates the central line behind the front cylinder. It is worth noting that, in anti-phase and in-phase flow regimes, the flow pattern behind the two downstream cylinders shows obvious different characteristics. A mirror symmetry of the vortex with the central line is observed in anti-phase flow, while an almost identical vortex shedding form exists in in-phase flow.

4.2. Fluid Force Coefficients

Figure 15a,b shows the mean and fluctuating drag force coefficients of the three cylinders, respectively. When $L/D = 1.1$, the drag forces on the downstream cylinders are identical and significantly greater than on the upstream cylinder. When $L/D = 1.2 - 1.4$, two distinct drag force coefficients were obtained for downstream cylinders, which suggests that this L/D belongs to the deflected gap flow regime. As L/D increases, the drag coefficients for the downstream cylinders remain identical and slowly converge to the drag coefficient for the flow past the single cylinder. It is worth noting that at $L/D = 2.5$, which belongs to the early onset of in-phase flow, the drag coefficient suddenly increases. The Cd' shows similar characteristics with Cd as shown in Figure 15b.

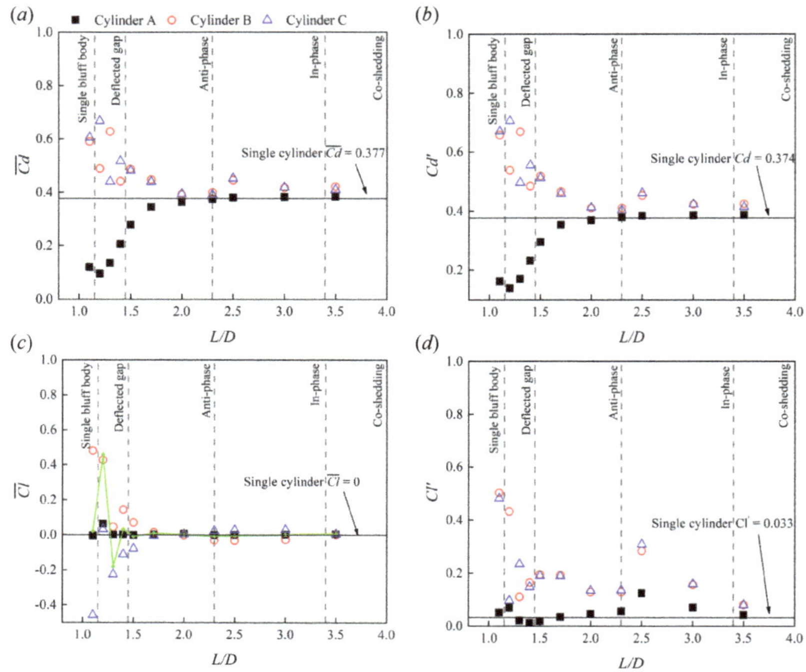

Figure 15. Variations of mean force coefficients and fluctuating force coefficients with different L/D: (**a**) time mean drag force coefficients; (**b**) fluctuating drag force coefficients; (**c**) time mean of lift force coefficients, green line means the sum of two downstream cylinders; (**d**) fluctuating lift force coefficients.

Figure 15c,d presents the mean and fluctuating lift force coefficients for different L/D, respectively. At $L/D = 1.1$, Figure 15c illustrates that there is a strong repulsive force between the two downstream cylinders. It is due to the existence of a pressure difference between the gap side and free-stream side of the downstream cylinders, which can be shown distinctly in pressure distributions in the middle section of Figure 14a. As shown in Figure 14a, in the case of downstream cylinder B, for example, the negative pressure zone on its free-stream side is shifted forward, while the negative pressure zone on its gap side is shifted back, which results in an upward lift force. When L/D = 1.2–1.4, the sum of two downstream cylinders lift force coefficients, which is presented by the green line, is not zero, and the absolute value of the sum lift coefficient decreases with the increase of L/D. The direction of the sum lift coefficient is opposite to the direction the deflected flow is biased towards. When $L/D > 2.3$, there is a weak attractive force between the downstream cylinders. As L/D goes to 3.5, the lift force coefficients of all cylinders is close to zero, which fits the scenario of a single cylinder. As shown in Figure 15d, similarly to the Cd', the downstream cylinders undergo a higher Cl' than that of the upstream cylinder. At $L/D = 1.1$, Cl' of the downstream cylinder reaches a maximum. At L/D = 1.2–1.4, the downstream cylinder with a wider wake has a larger Cl'. As L/D increases, Cl' of the downstream cylinders decreases. Similarly to the Cd and Cd' distributions, Cl' of the three cylinders increases abruptly at $L/D = 2.5$ and then decreases gradually back to be consistent with that of a single cylinder. We regarded that the special phase lag between all three cylinders at $L/D = 2.5$ causes the higher $\bar{Cd}$, Cd', and Cl'. A phase lag of 0π occurring at the early period of the in-phase flow regime enriches the shedding of vortices. The observation of fluid coefficients suggests that the interactions among the three cylinders is apparently strong at single bluff body flow, deflected gap flow, and the start of in-phase flow.

Figures 16 and 17 show the time history curves of Cd and Cl at several critical L/Ds for each flow pattern, respectively.

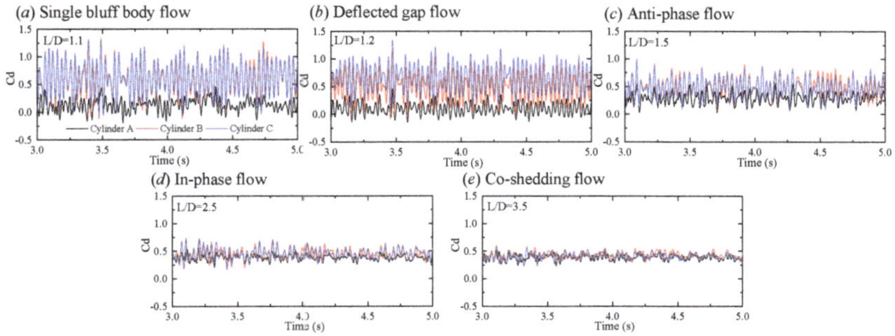

Figure 16. Time histories of Cd of cylinders in different flow pattern regimes.

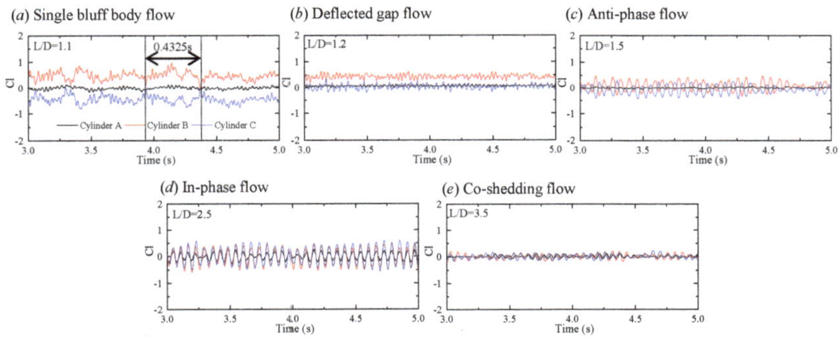

Figure 17. Time histories of Cl of cylinders in different flow pattern regimes.

It can be seen that the values of both drag and lift fluctuations for the upstream cylinder are smaller than those for the downstream cylinders. When $L/D < 2.3$, the reason for the lift fluctuations in the cylinder A is the motion of the shear layers, but for cylinders B and C, it is the vortex shedding. It is worth noting that at $L/D = 1.1$, Cl curve of the downstream cylinders has some certain periodicity. The period ($T = 0.4325s$) illustrated in Figure 17a, which is related to the time interval between the two minimum values of cylinder B and also the two maximum values of cylinder C, corresponding to $St = 0.05$, can be found in Figure 18a in the next section as the main frequency of cylinders B and C at $L/D = 1.1$.

This low frequency is due to the developmentally restricted shear layers of cylinder A.

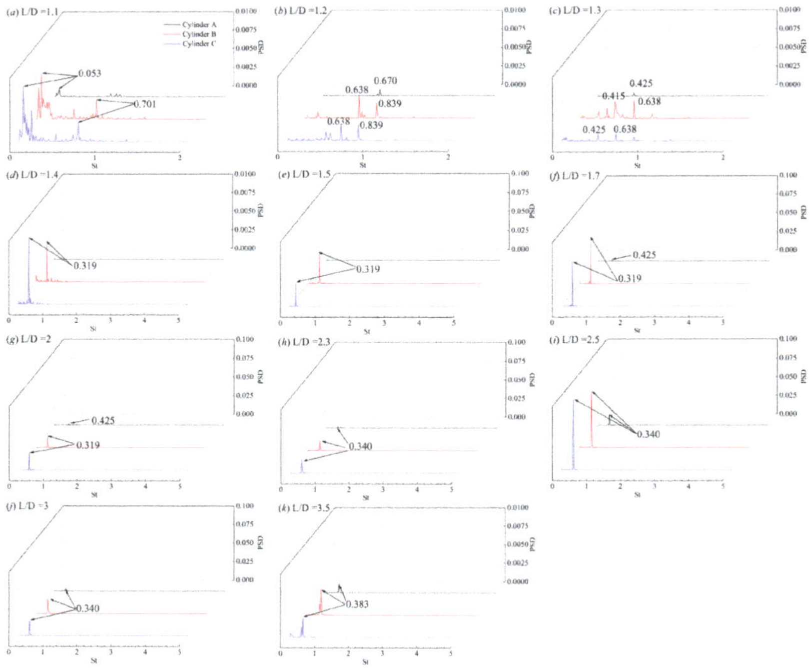

Figure 18. Power spectral density (PSD) of Cl of each cylinder at different L/D.

4.3. Vortex Shedding Frequencies

Figure 19 presents the lift coefficients power spectral density of three cylinders at various L/D, respectively. The Strouhal number St is defined as fD/U_0. At $L/D = 1.1$, the spectra of downstream cylinders are broad-banded. A dominant frequency of $St = 0.053$ can be identified for the three cylinders. In addition, a second harmonic frequency of $St = 0.701$ occurs. It can be attributed to the interactions of shear layers generated at cylinder A with those from the front side of cylinders B and C. The existence of a second harmonic frequency at low L/D at $Re = 3 \times 10^6$ is consistent with that for low Reynolds numbers $(100 - 300)$ by Chen et al. [21]. At $L/D = 1.2, 1.3$, only one spectral peak is identified in the upstream cylinder, $St = 0.67$ and $St = 0.425$, respectively. While downstream cylinders' power spectral density has multiple peaks and the corresponding power of these peaks are at similar values, illustrating that drastic vortex interaction exists. In contrast to the biased-gap flow pattern found for two cylinders placed side by side [7], which led to the observation that a wider wake cylinder has a lower vortex shedding frequency than the narrow wake cylinder, the dominate frequency of the two cylinders with either wide or narrow wake has the same value. This could be because of the presence of cylinder A. As L/D increases to 1.4, dominant frequency, $St = 0.319$, occurs at the downstream cylinders. This implies that the interaction between the shedding processes weakens as the gap space is greater. However, as displayed in Figure 7e,f, the gap flow at $L/D = 1.4$ still has a tendency of bias towards cylinder C. It demonstrates that $L/D = 1.4$ is a critical spacing ratio from deflected gap flow to anti-phase flow. As L/D increases, the St of all cylinders gradually approach that of the flow past a single cylinder. When $L/D = 3.5$, all cylinders have the same frequency ($St = 0.383$), which is equivalent to that of a single cylinder. This indicates that the disturbance becomes very weak and that each vortex shedding behind every cylinder behaves like the flow past a single cylinder. Furthermore, it is notable to highlight that the power density is much higher at the dominant frequency when $L/D = 2.5$. This implies that the vortex shedding at these frequency is pronounced. This observation is in accordance with the discontinuous jump in $\bar{Cd}$, Cd', and Cl' mentioned in the previous section, as shown in Figure 15.

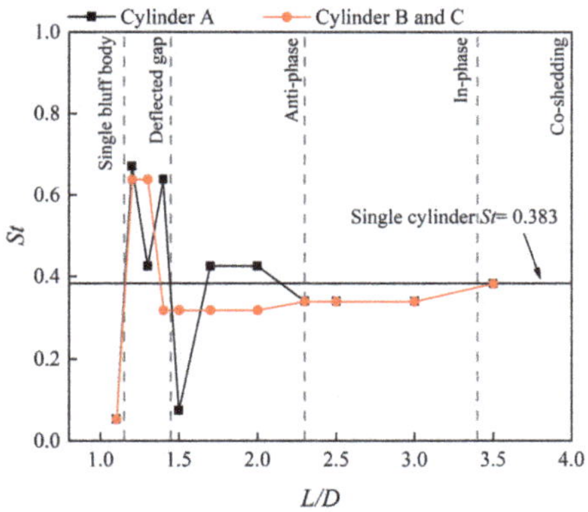

Figure 19. Strouhal numbers of three cylinders at different L/D.

The fluctuation of St with L/D is seen in Figure 13. As for the situations with multi-peak spectral densities, the frequency with the largest power density is collected. It is worth noting that cylinders B and C have the same dominant frequency at all L/D; nevertheless, only one set of data is shown for clarity. The St of three cylinders is equal in single bluff body flow, in-phase flow, and co-shedding flow. It reaches the minimum value at $L/D = 1.1$. In the in-phase flow regime, the value of St is 0.340, which is 88.7% of that for a single cylinder. In the co-shedding flow regime, St equals that for a single cylinder, demonstrating that $L/D = 3.5$ is a crucial spacing ratio of the two flow regimes.

Figure 20 presents the time-averaged pressure distributions of the three cylinders in different flow regimes, where the pressure distribution $\bar{C_p}$ is defined as $\bar{C_p} = (\bar{p} - p_\infty)/(0.5\rho U_0^2)$, in which $\bar{p}$ is the time-averaged local static pressure on the cylinder and p_∞ is the reference pressure. For description convenience, some special parameters, C_{pst1}, C_{pst2}, C_{pb}, C_{pm1}, and C_{pm2} are defined first, as shown in Figure 21. C_{pst1} and C_{pst2} are the pressure coefficients at the upper and lower stagnation points, θ_{pst1} and θ_{pst2}, respectively. C_{pb} is the base pressure coefficient at $\theta = 180°$. C_{pm1} and C_{pm2} are the minimum pressure coefficients at the upper and lower sides, respectively. θ_{pst1} and θ_{pst2} are the corresponding θ of C_{pm1} and C_{pm2}.

As shown in Figure 20a, in the single bluff body flow regime, the base pressure of the upstream cylinder increases compared to that of a single cylinder due to the blockage effect. Stagnation points of downstream cylinders are pretty obvious ($\theta_{pst1} = 6.94°$, $\theta_{pst2} = 353.06°$). The base pressure of upstream cylinder has two minimal values in the local area, located at 150° and 210°. Meanwhile, two local minimum pressure coefficients occur at 30° for cylinder B and 330° for cylinder C. Those arrestive minimal values occur at the positions where cylinder A is closest to cylinders B and C, respectively. The pressure coefficients of downstream cylinders both have a decrease at the gap-side area, and are nearly mirror-symmetric about $\theta = 180°$. For the deflected gap flow shown in Figure 20b, the pressure coefficients of downstream cylinders become asymmetrical because of the jet flow bias towards cylinder C. In addition, stagnation points of downstream cylinders are slightly smaller ($\theta_{pst1} = 4.49°$, $\theta_{pst2} = 355.55°$). As L/D increases into the anti-phase flow regime, the base pressure of the upstream cylinder turns into negative because the blockage effect becomes weak. C_{pm1} of cylinder A is smaller than C_{pm2} of cylinder B, while the opposite condition happens on cylinder C. This is because the flow velocity on the gap side is greater than the flow velocity on the free-flow side. When L/D goes into the in-phase flow regime, stagnation points of downstream cylinders, θ_{pst1} and θ_{pst2}, become zero. In the co-shedding flow regime, the pressure coefficient distributions of all the cylinders are similar with that of a single cylinder.

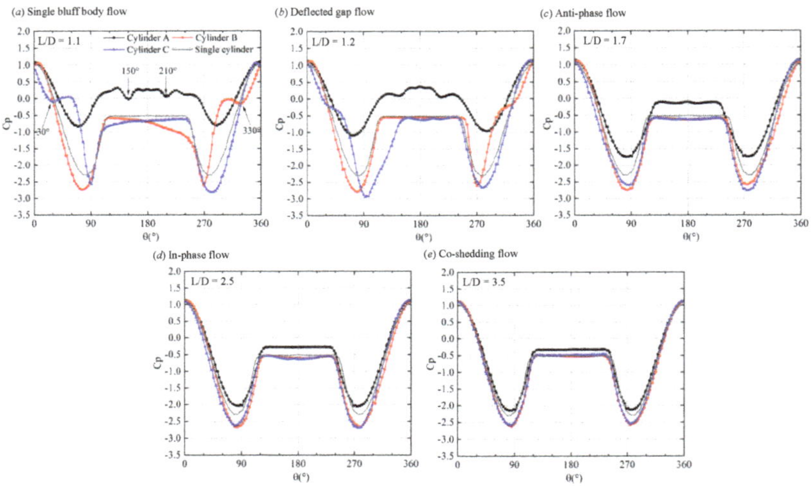

Figure 20. Time-averaged pressure coefficient distributions on surfaces of the cylinders at different L/D.

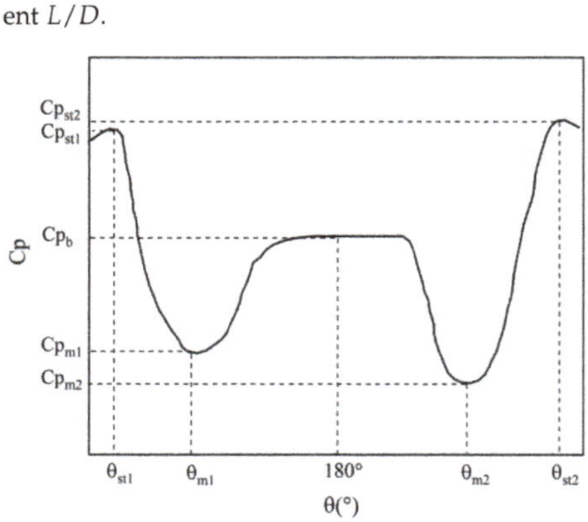

Figure 21. Sketch diagram of special parameters analyzing the pressure coefficient distribution.

4.4. Separation Angles

The separation angles of three cylinders are collected in Figure 22, and they are determined by Cf. Because the flow separation points of cylinder A are symmetrical on the upper and lower sides, only one is shown here for convenience. As shown, when $L/D = 1.1$, the separation points on the upstream cylinder, with black square symbol, is lower than that of a single cylinder. No separation point on the gap-side of the downstream cylinders (red triangle symbol and blue diamond symbol) is identified, because there is no regular vortex shedding at the gap. At $L/D = 1.2$, there is an extremely sharp growth in the separation point of cylinder A, because the increase of gap space leads to the shear layers of the upstream cylinder flow through the gap. The separation point of the upstream cylinder is quite backward because the gap is still small, which limits the separation of shear layers. As L/D goes up, the separation point of the upstream cylinder moves forward rapidly, starting to gradually adhere to the value at the single cylinder when $L/D = 1.7$. For the separation points on the free-stream-side of cylinders B and C, which are shown as red circle symbol and blue star symbol, respectively, they are smaller than in the case of a single cylinder when L/D is small, indicating that the separation position is advanced. As L/D increases, the separation position gradually leans forward, and when $L/D = 1.4$, the separation angle is close to that of single cylinder. For the separation points on the gap-side of downstream cylinders, they are special in the deflected gap flow regime. When

$L/D = 1.2$ and 1.4, the gap-side separation angle of cylinder C is conspicuously large in contrast to that of cylinder B. This is because at these L/D, the gap flow is deflected towards cylinder C, resulting in the delayed separation of the shear layers of cylinder C. The situation is reversed at $L/D = 1.3$.

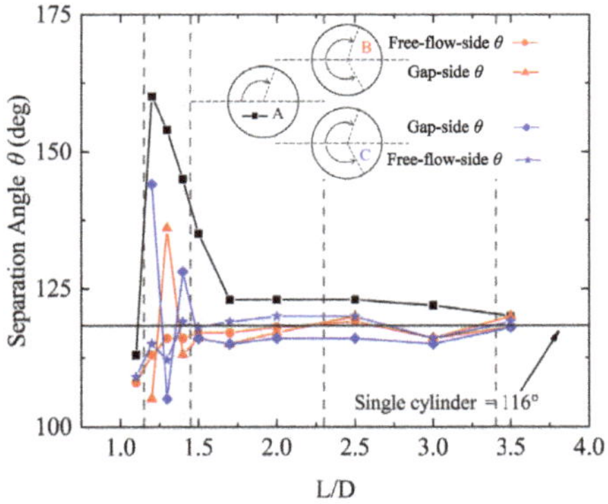

Figure 22. Variation of separation angles of cylinders with L/D.

5. Conclusions

The flow past three equilateral-triangular-arranged cylinders is investigated for spacing ratios L/D = 1.1–3.5 at a super-critical Reynolds number $Re = 3 \times 10^6$. Improved delayed-detached eddy simulation (IDDES) based on SST $k - \omega$ model is utilized to solve the flow field. With the aim to better understand the fluid dynamics of the configuration of three cylinders, the spanwise vorticity (ω_z), force coefficients, and its RMS, Strouhal numbers (S_t), and pressure distribution (C_p) are discussed. The following is a summary of the conclusions:

(i) Five flow regimes are identified depending on the spacing ratios, i.e, single bluff body flow ($L/D = 1.1$), deflected gap flow (L/D = 1.2–1.4), anti-phase flow (L/D = 1.5–2), in-phase flow (L/D = 2–3.5), and co-shedding flow ($L/D > 3.5$).

(ii) When in deflected gap flow, the downstream cylinder with a wide wake experiences lower drag and higher lift (absolute value and RMS), compared to the cylinder with narrow wake. However, their vortex shedding frequencies are identical. In addition, the sum of lift forces of the three cylinders at the deflected gap flow is not zero, either a positive or negative lift may exist.

(iii) When $L/D < 1.5$, the separation points on the free-flow-side of the downstream cylinders are always lower than that of the single cylinder, indicating that the violent flow at the gap also causes the flow separation on the outer side of the cylinder to advance. When $L/D > 3.5$, the separation point of the cylinder in upper stream is close to that of single cylinder, indicating that the minimum L/D for negligible interaction among the cylinders is 3.5.

In conclusion, in this paper, we systematically study the characteristics of different flow regimes of three cylinders with equilateral-triangular arrangement. Fluid forces, flow patterns, vortex shedding frequencies, phase differences, and wake interactions are discussed in detail. Compared with previous studies, this paper focuses on the super-critical Reynolds number regime and a more detailed L/D division is completed. The results of this paper deepen the understanding of the fluid interference of flow past cylinders.

Author Contributions: Methodology, B.Y.; Validation, Z.J.; Data curation, M.Z.; Writing—original draft, M.Z.; Writing–review & editing, B.Y.; Supervision, D.G. and G.Y. All authors have read and agreed to the published version of the manuscript.

Funding: This work was supported by the Strategic Priority Research Program of the Chinese Academy of Sciences (Class A) (Grant No. XDA22040203) and the Strategic Priority Research Program of the Chinese Academy of Sciences (Class B) (Grant No. XDB22020000).

Conflicts of Interest: The authors declare no conflict of interest.

References

1. Han, P.; Pan, G.; Tian, W. Numerical simulation of flow-induced motion of three rigidly coupled cylinders in equilateral-triangle arrangement. *Phys. Fluids* **2018**, *30*, 125107.
2. Liang, Y.; Tao, L.; Xiao, L.; Liu, M. Experimental and numerical study on vortex-induced motions of a deep-draft semi-submersible. *Appl. Ocean Res.* **2017**, *67*, 169–187. [CrossRef]
3. Goldschmidt, M.; Muskulus, M. Coupled mooring systems for floating wind farms. *Energy Procedia* **2015**, *80*, 255–262. [CrossRef]
4. Gonalves, T.R.; Chame, M.E.F.; Hannes, N.H.; de Paula Lopes, P.P.S.; Hirabayashi, S.; Suzuki, H. Experimental study on flow-induced motion of an array of three cylinders with circular, square, and diamond sections. In Proceedings of the 28th International Ocean and Polar Engineering Conference, Sapporo, Japan, 10–15 June 2018.
5. Zdravkovich, M.M. The effects of interference between circular cylinders in cross flow. *J. Fluids Struct.* **1987**, *1*, 239–261. [CrossRef]
6. Sumner, D.; Price, S.J.; Paidoussis, M.P. Tandem cylinders in impulsively started flow. *J. Fluids Struct.* **1999**, *13*, 955–965. [CrossRef]
7. Sumner, D. Two circular cylinders in cross-flow: A review. *J. Fluids Struct.* **2010**, *26*, 849–899. [CrossRef]
8. Harichandan, A.B.; Roy, A. Numerical investigation of low Reynolds number flow past two and three circular cylinders using unstructured grid CFR scheme. *Int. J. Heat Fluid Flow* **2010**, *31*, 154–171. [CrossRef]
9. Igarashi, T.; Suzuki, K. Characteristics of the Flow around Three Circular Cylinders. *Bull. JSME* **1984**, *27*, 2397–2404. [CrossRef]
10. Zhang, H.J.; Zhou, Y. Effect of unequal cylinder spacing on vortex streets behind three side-by-side cylinders. *Phys. Fluids* **2001**, *13*, 3675–3686. [CrossRef]
11. Sayers, A.T. Flow interference between three equispaced cylinders when subjected to a cross flow. *J. Wind Eng. Ind. Aerodyn.* **1987**, *26*, 1–19. [CrossRef]
12. Lam, K.; Cheung, W.C. Phenomena of vortex shedding and flow interference of three cylinders in different equilateral arrangements. *J. Fluid Mech.* **1988**, *196*, 1–26. [CrossRef]
13. Tatsuno, M.; Amamoto, H.; Koji, I.I. Effects of interference among three equidistantly arranged cylinders in a uniform flow. *Fluid Dyn. Res.* **1998**, *22*, 297–315. [CrossRef]
14. Gu, Z.; Sun, T. Classification of flow pattern on three circular cylinders in equilateral-triangular arrangements. *J. Wind Eng. Ind. Aerodyn.* **2001**, *89*, 553–568. [CrossRef]
15. Pouryoussefi, S.G.; Mirzaei, M.; Pouryoussefi, M.H. Force coefficients and Strouhal numbers of three circular cylinders subjected to a cross-flow. *Arch. Appl. Mech.* **2011**, *81*, 1725–1741. [CrossRef]
16. Bansal.; M., S.; Yarusevych.; S.. Experimental study of flow through a cluster of three equally spaced cylinders. *Exp. Therm. Fluid Sci.* **2017**, *80*, 203–217. [CrossRef]
17. Yang, Z.; Wang, X.; Si, J.H.; Li, Y. Flow around three circular cylinders in equilateral-triangular arrangement. *Ocean Eng.* **2020**, *215*, 107838. [CrossRef]
18. Yan, B.; Zhou, D.; Cheng, H. Numerical simulation of flow over three circular cylinders in equilateral arrangements at low Reynolds number by a second-order characteristic-based split finite element method. *Comput. Fluids* **2010**, *39*, 882–899.
19. Yan, W.; Wu, J.; Yang, S.; Wang, Y. Numerical investigation on characteristic flow regions for three staggered stationary circular cylinders. *Eur. J. Mech. B/Fluids* **2016**, *60*, 207–214. [CrossRef]
20. Zheng, S.; Zhang, W.; Lv, X. Numerical simulation of cross-flow around three equal diameter cylinders in an equilateral-triangular configuration at low Reynolds numbers. *Comput. Fluids* **2016**, *130*, 94–108. [CrossRef]
21. Chen, W.; Ji, C.; Alam, M.M.; Williams, J.; Xu, D. Numerical simulations of flow past three circular cylinders in equilateral-triangular arrangements. *J. Fluid Mech.* **2020**, *891, A14*. [CrossRef]
22. Gao, Y.; Qu, X.; Zhao, M.; Wang, L. Three-dimensional numerical simulation on flow past three circular cylinders in an equilateral-triangular arrangement. *Ocean Eng.* **2019**, *189*, 106375. [CrossRef]
23. Hu, X.; Zhang, X.; You, Y. On the flow around two circular cylinders in tandem arrangement at high Reynolds numbers. *Ocean Eng.* **2019**, *189*, 106301. [CrossRef]
24. Schewe, G. On the force fluctuations acting on a circular cylinder in crossflow from subcritical up to transcritical Reynolds numbers. *J. Fluid Mech.* **1983**, *133*, 265–285. [CrossRef]
25. Hinsberg, N.P.V. The Reynolds number dependency of the steady and unsteady loading on a slightly rough circular cylinder: From subcritical up to high transcritical flow state. *J. Fluids Struct.* **2015**, *55*, 526–539. [CrossRef]
26. Rodriguez, I.; Lehmkuhl, O.; Chiva, J.; Borrell, R.; Oliva, A. On the flow past a circular cylinder from critical to super-critical Reynolds numbers: Wake topology and vortex shedding. *Int. J. Heat Fluid Flow* **2015**, *55*, 91–103. [CrossRef]

27. Okajima, A. Flows around Two Tandem Circular Cylinders at Very High Reynolds Numbers. *Bull. JSME Jpn. Soc. Mech. Eng.* **1979**, *22*, 504–511. [CrossRef]
28. Spalart, P.R. Comments on the feasibility of LES for wings, and on a hybrid RANS/LES approach. In Proceedings of the First AFOSR International Conference on DNS/LES, Ruston, LA, USA, 4–8 August 1997.
29. Spalart, P.R.; Deck, S.; Shur, M.L.; Squires, K.D.; Strelets, M.K.; Travin, A. A New Version of Detached-eddy Simulation, Resistant to Ambiguous Grid Densities. *Theor. Comput. Fluid Dyn.* **2006**, *20*, 181. [CrossRef]
30. Shur, M.L.; Spalart, P.R.; Strelets, M.K.; Travin, A.K. A hybrid RANS-LES approach with delayed-DES and wall-modelled LES capabilities. *Int. J. Heat Fluid Flow* **2008**, *29*, 1638–1649. [CrossRef]
31. Liang, X.F.; Chen, G.; Li, X.B.; Zhou, D. Numerical simulation of pressure transients caused by high-speed train passage through a railway station. *Build. Environ.* **2020**, *184*, 107228. [CrossRef]
32. Wang, B.; Liu, J.; Li, Q.B.; Yang, Y.J.; Xiao, Z.X. Numerical studies of reverse flows controlled by undulating leading edge. *Sci.-China-Phys. Mech. Astron.* **2019**, *62*, 974712. [CrossRef]
33. Chen, G.; Li, X.B.; Liu, Z.; Zhou, D.; Wang, Z.; Liang, X.F.; Krajnovic, S. Dynamic analysis of the effect of nose length on train aerodynamic performance. *J. Wind Eng. Ind. Aerodyn.* **2019**, *184*, 198–208. [CrossRef]
34. Tan, C.; Zhou, D.; Chen, G.; Sheridan, J.; Krajnovic, S. Influences of marshalling length on the flow structure of a maglev train. *Int. J. Heat Fluid Flow* **2020**, *85*, 108604. [CrossRef]
35. Dong, T.; Minelli, G.; Wang, J.; Liang, X.; Krajnovi, S. The effect of reducing the underbody clearance on the aerodynamics of a high-speed train. *J. Wind Eng. Ind. Aerodyn.* **2020**, *204*, 104249. [CrossRef]
36. Tamura, T.; Miyagi, T.; Kitagishi, T. Numerical prediction of unsteady pressures on a square cylinder with various corner shapes. *J. Wind Eng. Ind. Aerodyn* **1998**, *74*, 531–542. [CrossRef]
37. Norberg, C. An experimental investigation of the flow around a circular cylinder: Influence of aspect ratio. *J. Fluid Mech.* **1994**, *258*, 287–316. [CrossRef]
38. P.R.; Spalart. Strategies for turbulence modelling and simulations. *Int. J. Heat Fluid Flow* **2000**, *21*, 252–263. [CrossRef]
39. Lehmkuhl, O.; Rodriguez, I.; Borrell, R.; Chiva, J.; Oliva, A. Unsteady forces on a circular cylinder at critical Reynolds numbers. *Phys. Fluids* **2014**, *26*, 125110. [CrossRef]
40. Roshko, A. Experiments on the Flow Past a Circular Cylinder at Very High Re Number. *J. Fluid Mech.* **1961**, *10*, 345–356. [CrossRef]
41. Schmidt, L.V. Flucting force measurements upon a circular cylinder at Reynolds number up to 5×10^6. In Proceedings of the Meeting on Ground Wind Load Problems in Relation to Launch Vehicles, Langley, WA, USA, 7–8 June 1966.
42. Jones, G.W.; Cincotta, J.J.; Walker, R.W. *Aerodynamic Forces on a Stationary and Oscillating Circular Cylinder at High Reynolds Numbers*; Technical report; NASA: Washington, DC, USA, 1969.
43. Shih, W.; Wang, C.; Coles, D.; Roshko, A. Experiments on flow past rough circular cylinders at large Reynolds numbers. *J. Wind Eng. Ind. Aerodyn.* **1993**, *49*, 351–368 [CrossRef]
44. Travin, A.; Shur, M.; Strelets, M.; Spalart, P. Detached-Eddy Simulations Past a Circular Cylinder. *Flow Turbul. Combust.* **2000**, *63*, 293–313. [CrossRef]
45. Catalano, P.; Wang, M.; Iaccarino, G.; Moin, P. Numerical simulation of the flow around a circular cylinder at high Reynolds numbers. *Int. J. Heat Fluid Flow* **2003**, *24*, 463–469. [CrossRef]
46. Ong, M.; Utnes, T.; Holmedal, L.E.; Myrhaug, D.; B, P. Numerical simulation of flow around a smooth circular cylinder at very high Reynolds numbers. *Mar. Struct.* **2009**, *22*, 142–153. [CrossRef]
47. Warschauer, K.A.; Leene, J.A. Experiments on mean and fluctuating pressures of circular cylinders at cross flow at very high Reynolds numbers. In Proceedings of the International Conference on Wind Effects on Buildings and Structures, Tokyo, Japan, 6–11 September 1971.
48. Zdravkovich, M.M.; Bearman, P.W. Flow Around Circular Cylinders -Volume 1: Fundamentals. *ASME J. Fluids Eng.* **1998**, *120*, 216. [CrossRef]
49. Achenbach, E. Distribution of local pressure and skin friction around a circular cylinder in cross-flow up to $Re = 5 \times 10^6$. *J. Fluid Mech.* **1968**, *34*, 625–639. [CrossRef]
50. Janocha, M.J.; Ong, M.C. Vortex-induced vibrations of piggyback pipelines near the horizontal plane wall in the upper transition regime. *Mar. Struct.* **2021**, *75*, 102872. [CrossRef]
51. Yang, X.; Ji, C.; Chen, W.; Zhang, Z. Wake patterns and hydrodynamic forces of flow around circular cylinders in an equilateral triangular arrangement. *Chin. J. Hydrodynom.* **2019**, *34*, 69–76. (In Chinese)
52. Kim, H.J. Investigation of the flow between a pair of circular cylinders in the flopping regime. *J. Fluid Mech.* **1988**, *196*, 431–448. [CrossRef]
53. Alam, M.M.; Moriya, M.; Sakamoto, H. Aerodynamic characteristics of two side-by-side circular cylinders and application of wavelet analysis on the switching phenomenon. *J. Fluids Struct.* **2003**, *18*, 325–346. [CrossRef]
54. Hunt, J.; Wray, A.; Moin, P. Eddies, streams, and convergence zones in turbulent flows. In Proceedings of the 1988 Summer Program, Stanford, CA, USA, 27 June–22 July 1988; pp. 193–208.

applied sciences

MDPI

Article

Thermodynamic and Spectroscopic Studies of SDS in Cinnamaldehyde + Ethanol Mixtures: Influences of Temperature and Composition

Waleed M. Alamier [1], Shadma Tasneem [1,*], Arshid Nabi [2], Nazim Hasan [1] and Firdosa Nabi [2]

[1] Department of Chemistry, Faculty of Science, Jazan University, Jazan P.O. Box 114, Saudi Arabia
[2] Department of Chemistry, Faculty of Science, University of Malaya, Kuala Lumpur 50603, Malaysia
* Correspondence: sthaque@jazanu.edu.sa

Abstract: The study of intermolecular interactions between ethanol (E-OH), cinnamaldehyde (CAD) with anionic surfactant sodium dodecyl sulfate (SDS) in non-aqueous media has been examined by utilizing conductometric and spectroscopic techniques. The critical micelle concentration (CMC) values have been determined. The experimental conductance data were analyzed against temperature and concentration using standard relations. The pseudo phase separation model has been adopted to calculate various thermodynamic parameters like standard free energy, ΔG°_{mic}, enthalpy, ΔH°_{mic}, and entropy, ΔS°_{mic}, of micelle formation. Fourier transforms infrared analysis (FTIR), and Fluorescence spectra were taken out to assess the possible interactions prevailing in the micellar systems. The findings demonstrated that the presence of SDS, and the composition of CAD + ethanol might affect the thermodynamic parameters. The discrepancy in these parameters with the surfactant concentration or with the temperature change indicates the manifestation of different interactions prevailing in the studied systems.

Keywords: ethanol; CAD; thermodynamic parameters; CMC; hydrophobic–hydrophobic interactions; FTIR analysis

Citation: Alamier, W.M.; Tasneem, S.; Nabi, A.; Hasan, N.; Nabi, F. Thermodynamic and Spectroscopic Studies of SDS in Cinnamaldehyde + Ethanol Mixtures: Influences of Temperature and Composition. *Appl. Sci.* **2022**, *12*, 12020. https://doi.org/10.3390/app122312020

Academic Editors: Antony C. Calokerinos and Qi-Huang Zheng

Received: 20 September 2022
Accepted: 14 November 2022
Published: 24 November 2022

1. Introduction

The studies of interactions between surfactants and bioactive molecules are of enormous importance because of their applications in biological systems, pharmaceuticals, and biotechnological processes [1,2]. Surfactants are fascinating compounds because of their amphiphilic nature, hence they possess both hydrophilic and hydrophobic characteristics in the same mixtures and have soluble properties in polar and non-polar solvents [3]. Surfactants have the ability to form aggregates at a particular concentration in both aqueous and non-aqueous media. The concentration, at which the aggregation originates, is known as critical micelle concentration [CMC], and the aggregate is called a micelle. The phenomenon of aggregation in molecules incorporates both attractive and repulsive interactions. The nature of hydrophilic and hydrophobic moieties determines how bio-active compounds interact during formulation, agrregation and biological point of view in drug –surfacatnt interactions. Because of the presence of substituents on the hydrophobic core or variations of hydrocarbon chain length, it is evident that in solution systems, the behavior of such molecules can be varied [4].

To understand the intermolecular interactions of SDS in CAD + E-OH mixtures, we determined several physicochemical characteristics. The prevailing molecular interactions and micelle growth at particular concentrations are interpreted using different techniques such as conductivity, surface tension, refractive index, Uv-Visible spectroscopy and fluorescence [5]. However, we use conductivity and spectroscopic methods in this study; SDS was chosen because it is widely known for its ability to create micellar solutions. SDS is amphiphilic and features a 12-carbon tail connected to the sulfate group. It is said to be more effective than urea and guanidine hydrochloride at denaturing proteins. Because of its high foaming properties, ease of accessibility, and low production cost, SDS is considered the best option. SDS has numerous biological functions, including antibacterial and skin cleaning effects [6]. It is well-documented that the interaction due to the presence of an oxygen group acts as both an acceptor and a donator [7]. In this work, SDS is chosen to quantify the impact of various solution environments on physical characteristics in the continued development of innovative drug delivery systems, including micellar solution.

On the other hand, cinnamaldehyde [CAD] is a liquid aldehyde derived from the bark of Cinnamomum trees, a naturally occurring flavonoid often used as a natural flavoring and fragrance agent in the kitchen and industry [8]. In addition, CAD is believed to have various beneficial properties, for instance antimicrobial [9], antioxidative [10], and inhibiting antiapoptotic [11] properties. CAD is found to protect against gram-positive/negative infection [1], diabetes [12], gastric ulcer [13], cardiomyopathy hypertrophy, [14] brain I/R injuries [15]. Cinnamaldehyde (CAD) is a representative substrate containing the carbonyl and vinyl groups simultaneously. It is worth mentioning, that the proposed work was designed to increase the solubility of hydrophobic CAD and to encapsulate the CAD molecules via micelles to offer protection against degradation. However, here we study the possible molecular interaction and micelle formation between CAD molecules in SDS ethanol mixed media. This study has further applications in drug delivery vehicles as an encapsulated micelle formulation, and is expected to improve cellular uptake with reduced side effects. However, the slight solubility of CAD in water restricts the study of interactions in aqueous media. Meanwhile, ethanol, a polar solvent self-associated through hydrogen bonding, is expected to interact strongly with other fluids by hydrogen-bonding, and was preferred in the present study. Thus, it would be interesting to make a comparison of the micellar properties of SDS in pure ethanol mixtures and in ethanol–CAD mixtures.

CAD is usually used as a flavoring agent, as well as a fragrance agent in global market like cosmetics, soaps, detergents, deodorants, shampoos, etc. [10–15]. Thus, it is essential to develop a simple, sensitive and selective analytical approach to detect the interaction properties of CAD. So, overwhelmed with its multiple applications in the diverse field, this work proposed a simple technique to interpret possible molecular interactions of CAD with surfactant in ethanol media using critical micelle concentration of SDS and spectroscopic techniques. Conductivity and thermodynamic empirical equations at different temperatures were used for molecular interaction parameters analysis.

In the present study, we evaluate the micelle formation of sodium dodecyl sulfate (SDS) with CAD molecules in ethanol media by conductometric, and spectroscopic methods. It is worth observing the effect of aggregation in bioactive molecules. Interestingly both SDS and CAD have different polarity regions with hydrophilic and hydrophobic substitutional groups of corresponding characteristics. Thus, it is meaningful to inspect the process of surfactant aggregation in a non-aqueous media.

2. Experiment

2.1. Materials

Ethanol E-OH 99% (Sigma–Aldrich, St. Louis, MO, USA) was used without further purification. Cinnamaldehyde (CAD) with preeminent quality was purchased from Sigma Life Sciences (Sigma-Aldrich) and kept under low temperature before use. The sodium dodecyl sulfate (SDS) was purchased from Friedmann Schmidt chemicals, purity 95% as reported by the vendor. The chemicals were dried in vacuum over P_2O_5 for about 72 h at room temperature.

2.2. Methods

2.2.1. Stock Solution Preparations

The stock solutions of CAD 0.05 m (mol kg^{-1}) in ethanol were prepared by stirring the solution mixture for about 8 h at room temperature. Samples of 0.001, 0.002, 0.003, 0.004, 0.005, 0.006, 0.007, 0.008, 0.009, 0.010, 0.011, 0.012, 0.013, 0.014, 0.015, 0.016, 0.017,

0.018 and 0.019 m SDS solutions were prepared for the experimental investigations. An electronic balance Shimadzu AY220, Japan, with precision of ±0.0001 g, was used for weighing the freshly prepared samples. The appropriate arrangements were made for storage for all concentrations to avoid evaporation. Moreover, the chemicals used were systematically described by their chemical structures: (a) sodium dodecyl sulfate (SDS), (b) Cinnamaldehyde (CAD), and (c) ethanol, shown in Figure 1.

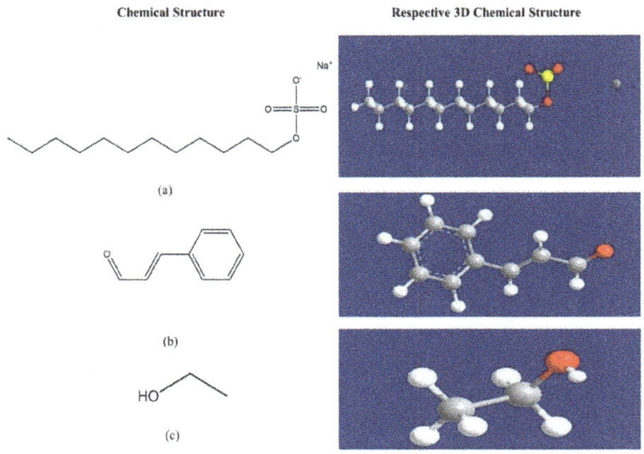

Figure 1. Chemicals and their respective 3D chemical structures of (**a**) sodium dodecyl sulfate, (**b**) cinnamaldehyde and (**c**) ethanol, respectively.

2.2.2. Conductivity Measurements

A digital conductivity meter (PC 510 Bench/Conductivity Meter, EUTECH Instruments) was used to determine the conductance of each sample. The conductivity meter was calibrated before measuring the conductance of the solutions using standard solutions of 0.01 and 0.10 N KCl (purity > 99%) prepared in doubly-distilled water, with a conductivity of 1413 μS/cm at 298.15 °K. The temperature of the water bath was maintained at a constant level for each measurement. To achieve thermal equilibrium and record conductivity data, the electrode assembly with glass tube was submerged in each sample solution. The conductivity measurements have a measurement error of 0.5%.

2.2.3. Fluorescence Measurement

To evaluate the interaction of cinnamaldehyde (CAD) and SDS, a fixed volume of ethanol solvent was separately mixed to dilute the solution. Each sample was incubated for five minutes at room temperature and then fluorescence spectra (Shimadzu RF-6000, Spectrofluorophotometer, Kyoto, Japan) were recorded at the excitation wavelength of 295 nm. For different detection range assessments, various sets of solutions of fixed volume were prepared by mixing ethanol. The fluorescence spectra were measured in a range from 200 nm to 400 nm after using an excitation wavelength of 295 nm. Every set of sample solution fluorescence emission spectra was recorded three times and evaluated. Cinnamaldehyde (CAD) quenched the maximum fluorescence intensity of SDS-ethanol that is further finely fitted in the Stern–Volmer equation, hence it is used in clinical science as an anticancer agent, stomachic, antipyretic and antiallergic drug [16]. The cinnamaldehyde (CAD) acts as a quencher to determine the sample mixture's binding constant (K_b). Furthermore, the interaction between the micellar solution of SDS and CAD with ethanol was also studied.

2.2.4. FTIR Measurement

A liquid-sample-based Fourier transform infrared (FTIR) spectrometer (Nicolet iS10, Thermo Fisher Scientific, Waltham, Mass, USA) was used to evaluate the interaction of CAD and SDS during pre- and post-micelle formation in an ethanol system. All measurements were carried out using an attenuated total reflectance (ATR) accessory with a resolution of 4 cm^{-1}, and 15 scans of each sample were performed. A hardware specification of a flow-rate top plate fitted with a 421 ZnSe reflection crystal and a depth of penetration of 2 mm was used for spectral analysis.

2.3. Calculation of the Thermodynamic Properties of Micellization

The pseudo-phase separation model was used to interpret the thermodynamic parameters. As a result, the standard free energy of micellization can be calculated using the following Equations (1)–(6):

$$\Delta G_m^{\circ} = (2 - \beta) RT \ln X_{cmc} \tag{1}$$

where, X_{cmc} is the *cmc* values expressed in mole fraction, β, is the degree of ionization of the micelles obtained by the ratio of two linear segments of conductivity (SDS + -E-OH) and (SDS + E-OH + CAD) plots above and below cmc values [16]. The other thermodynamic properties for instance enthalpy, ΔH_m° and entropy, ΔS_m° of micellization can be evaluated from the fundamental thermodynamic equations as:

$$\Delta H_m^{\circ} = RT^2(2 - \beta)\frac{d \ln X_{cmc}}{dT} \tag{2}$$

$$\Delta S_m^{\circ} = \frac{\Delta H_m^{\circ} - \Delta G_m^{\circ}}{T} \tag{3}$$

where, $\frac{d \ln X_{cmc}}{dT}$ values fitted to a polynomial function with the values of $d \ln X_{cmc} \Delta T$ as:

$$\ln X_{cmc} = a + b(T/K) + c(T/K)^2 \tag{4}$$

where, a, b and c are the polynomial constants, and the above equation can also be expressed as:

$$\frac{\ln X_{cmc}}{dT} = b + 2c(T/K) \tag{5}$$

Further, the results obtained from the above thermodynamic Equations (1)–(5) and the values presented in Table 1, provides the values of cmc, $\ln X_{cmc}$, β, ΔG_m°, ΔH_m°, ΔS_m° and $T\Delta S_m^{\circ}$, respectively.

Table 1. The thermodynamic parameters of micellization at different temperatures (298.15 K to 318.15 K) for SDS + E-OH and SDS + E-OH + CAD, respectively.

T (K)	CMC (M)	α	ΔG_m° (kJ mol^{-1})	ΔH_m° (kJ mol^{-1})	ΔS_m° (J mol^{-1} K^{-1})	$T\Delta S_m^{\circ}$ (J mol^{-1} K^{-1})
			SDS + Ethanol			
298.15	0.00822	0.65	−29.51	−13.06	0.055	16.9051
303.15	0.00881	0.64	−29.88	−13.56	0.054	16.3701
308.15	0.00939	0.60	−31.16	−14.47	0.054	16.6401
313.15	0.00995	0.59	−31.68	−15.05	0.053	16.5969
318.15	0.01073	0.56	−32.50	−15.83	0.052	16.5438
			SDS + Ethanol + Cinnamaldehyde			
298.15	0.0120	0.76	−22.99	−1.72	0.071	21.1686
303.15	0.0122	0.73	−23.74	−2.68	0.069	20.9173
308.15	0.0125	0.71	−24.49	−3.74	0.067	20.6460
313.15	0.0127	0.69	−25.32	−4.90	0.065	20.3547
318.15	0.0129	0.66	−26.15	−6.17	0.063	20.0434

3. Results and Discussion

3.1. Conductometric Study

Electrical conductivity techniques determine the interaction and association of molecules in diverse fluid systems [17,18]. In the current study, two sets of conductivity tests were performed to evaluate the interaction behavior and affiliation of different molecules in the component systems. The conductance was measured as a function of SDS in the presence of ethanol (Figure 2) and SDS in presence of CAD + ethanol in the concentration range of 0.001 to 0.020 mol kg^{-1} at 298.15, 303.15, 308.15 and 313.15 K and are presented in Figure 3. The intersection of straight lines above and below the breaking points of conductance (κ) against surfactant concentration at different temperatures acquired the CMC values. The values of CMC for SDS in pure ethanol solutions are presented in Table 1. The data given in Table 1 presents the minimum value of CMC verses temperature curve for SDS + ethanol system at lower temperature (298.15 K). The CMC of SDS rises from 298.15 to 313.15 K as the temperature rises. The influence of temperature on CMC has been shown to be system dependent [19]. Depending on the type of solvent solution employed, the micellization process might either be instantaneous or sluggish [20]. In the instance of the SDS in CAD + ethanol solution, the specific conductivity values gave a distinct breakpoint suggesting instantaneous micellization.

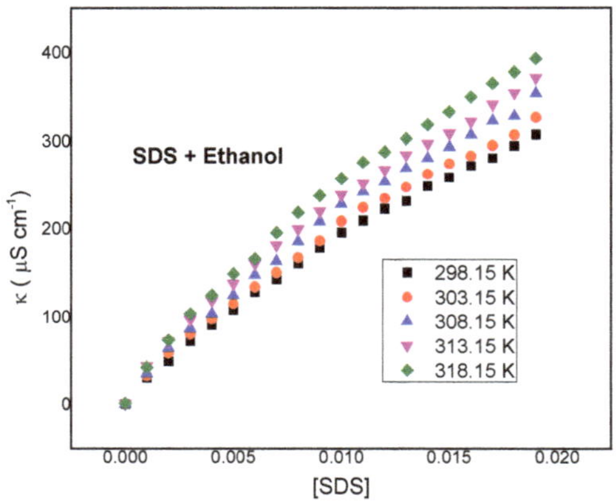

Figure 2. Specific conductance of SDS in ethanol at different temperatures.

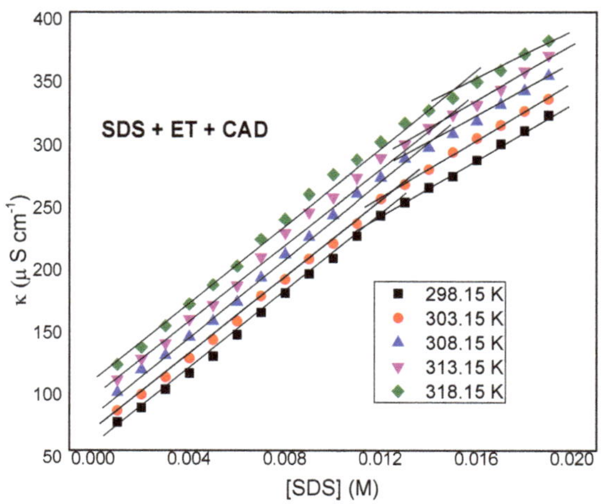

Figure 3. Specific conductance of SDS in cinnamaldehyde + ethanol at different temperatures.

However, In Figure 3, CMC values increase when CAD is introduced into the system, and the micellization process observed is different. CAD molecules are insoluble in an aqueous medium so we conducted the experiment using an ethanol medium. Ethanol acts as a structure-breaking solvent which causes the destruction of the inherent H-bond association of CAD and SDS. The incorporation of CAD molecules with SDS molecules resulted in slower micellization. This slight change in CMC values in the (SDS in CAD + Ethanol) system in Table 1 reveals that an addition of CAD molecules led to an effect on the types of interactions occurring in the system. However, micellization is delayed when CAD is added, which can be explained by the presence of an aromatic ring in the CAD's structure, which may establish a fine balance of hydrophobic interactions between SDS's long hydrocarbon tails and repulsive interactions between the ionic head groups. In the presence of CAD, a delay in CMC values was reported. The values of CMC rise as the temperature rises. Due to the existence of various components in the mixture, the effect of temperature on the magnitude of the CMC values of surfactant was frequently examined. The disruption of the ordered molecules surrounds the surfactant's hydrophobic groups at higher temperatures. Furthermore, it is advantageous that at higher temperatures, thermal motion increases, resulting in de-micellization due to the distraction of the micelle's palisade layer, which increases the CMC of the surfactants [21]. This is due to the fact that when the temperature rises, the high solubility of hydrocarbons stabilizes the surfactant monomers, preventing micelle formation, resulting in a larger SDS CMC [22].

The evaluations of the thermodynamic parameters of the pseudo-phase separation model were deduced from various thermodynamic micellization equations as described above. The Gibbs free energy of micellization, ΔG_m°, entropy of micellization, ΔS_m° and enthalpy of micellization, ΔH_m° are worth applying to understand molecular interaction and are also the core concepts behind the process of micellization [23]. Furthermore, the data obtained from the thermodynamic parameters are useful for determining the influence of environmental and structural contributions to CMC values, as well as observing the effects of novel structures and environmental deviations in the presence of various components.

Meanwhile, in an ionic surfactant it has been stated that the values of ΔG_m° lie in the range from -23 to -42 kJ mol^{-1} at 298.15 K [24]. It was noted that the values -29.51 kJ mol^{-1} for SDS in ethanol and -22.99 kJ mol^{-1} for SDS in cinnamaldehyde + ethanol at the corresponding temperature lie in the reported range. There were more negative values of free energy of micellization, ΔG_m° in SDS with -E-OH mixtures than SDS + E-OH + CAD mixtures at all temperature from 298.15 to 313.15 K, observed experimentally. Moreover, the free energy of micellization ΔG_m° examined in Table 1 is negative for both systems at all temperature from 298.15 to 313.15 K. Therefore, the values of ΔG_m° of micellization are more for the SDS +E-OH system than in the SDS + E-OH + CAD system. This signifies the readiness and thermodynamically more spontaneous micellization of SDS with the E-OH mixture than the SDS + E-OH + CAD mixtures. It was noticed that the values of ΔG_m° increase with an increase in temperature, suggesting that the process of micellizations in both studied systems was thermodynamically spontaneous. It was recognized that the desolvation of the hydrophilic groups of the surfactants took place if the ΔG_m^{o} values declined with increasing temperature [25].

Moreover, the observed increased negative values of ΔG_m° with elevated temperatures are because of the mutual effects of the β and ln X_{cmc}values. The β values increased while the ln X_{cmc} values decreased at higher temperatures and vice versa. The reason for the increased values of β with rising temperature is desolvation of the hydrogen bond which persisted among the H atoms of –OH groups of ethanol and the polar head groups of the SDS molecules. Sequentially, the surface charge density and electrostatic repulsions between head groups of SDS + E-OH + CAD molecules increased more with the elevated temperatures. Thus, it disfavors the process of micellization with increased temperatures. Similar to ΔG_m° values, the values of ΔH_m° are negative for all the systems at all temperatures. It is worth mentioning that the observed values of ΔG_m°, ΔH_m° and ΔS_m° in E-OH + SDS were found to be -29.51 kJ mol^{-1}, -13.06 kJ mol^{-1} and 0.055 kJ mol^{-1} K^{-1} at 298.15 K, respectively, and for SDS + E-OH + CAD were found to be −22.99 kJ mol^{-1}, −1.72 kJ mol^{-1} and 0.071 kJ mol^{-1} K^{-1}, which are much lower than the literature values −39.70 kJ mol^{-1}, −22.98 kJ mol^{-1} and 0.06 kJ mol^{-1} K^{-1} [26], respectively, of SDS in pure water at the corresponding temperatures. Lower values of SDS in ethanol than in pure water may be explained by considering the nature of the solvent (ethanol). The marked difference in the thermodynamic parameters may be because of the relative permittivity of water (e = 79.99 at 293.15 K) and ethanol (e = 25.02 at 293 K) [27] is expected to affect the thermodynamic parameters of SDS in these media. The values of standard free energy of micellization becomes less negative with the addition of CAD which indicates that the micellization is less favored with the addition of CAD in the SDS ethanolic media. Similar results have also been reported for the drug chloroquine with SDS in polar aqueous solution, where ΔG_m°, ΔH_m° and ΔS_m° values were found to be −21.74 kJ mol^{-1}, −3.45 kJ mol^{-1} and 62.42 kJ mol^{-1} K^{-1} at 298.15 K, respectively [28]. The increased negative values of ΔH_m° and the decreased positive values of ΔS_m° (Table 1) weaken the hydrophobic interactions between the SDS and CAD molecules. In contrast, the electrostatic interactions become stronger with rising temperatures. The values of ΔH_m° are believed to derive from the interactions of electrostatic, hydrophobic, or polar head group hydrations and because of the counter ion micelle bindings. Moreover, the investigated $T\Delta S_m^\circ$ values are much higher in magnitude as compared to the ΔH_m° values. Hence, this indicates micellization is entropy driven, having the potential for hydrophobic groups of surfactants to orient themselves to the interior of micelle core from the bulk solvent [29]. Behind the possibility of micellization is the tendency of structure-breaking E-OH to act as a solvent on the surfactant (SDS) molecules. This may be possible because of the higher freedom of hydrophobic chains in the vicinity of the nonpolar interior of micelles than in the solvent environment [18,30].

3.2. Fluorescence Spectra Behaviour of Cinnamaldehyde (CAD) in SDS-Ethanol Media

Cinnamaldehyde (CAD + ethanol + SDS) exhibits the maximum fluorescence emission at 310 nm. The maximum emission was shown between 300–400 nm for SDS in ethanol + CAD. The study was further inferred by utilizing specified sets of parameters, such as to investigate the fluorescence intensities in a region before and after micellization [30]. The pre-micellar region of SDS with varying concentration in CAD + ethanol was inferred as depicted in Figure 4. In addition, it was observed that after a certain point i.e., 0.012 m concentration of SDS with CAD in ethanol, a linear decrease in fluorescence intensity with decreasing concentrations was attained. The observed results are believed to be due to the presence of more SDS monomers in the surfactant micellar system which further modifies monomer interactions in aggregates, thereby alter the fluorescence intensities in the premicellar and postmicellar region. In premicellar region, CAD molecules possibly interact onto the surface of SDS via hydrogen bonding between the head group of SDS and substitute (H-C=O) groups of CAD molecules resulting in aggregates rather than micelles. In other words, with an increase in the concentration of SDS molecules to CAD in an ethanol solution, more aggregates are possibly formed leading to increase in fluorescence intensities up to a concentration i.e., critical micellar concentrations (CMC). In addition, at the micellar region a sudden jump was observed in fluorescence intensity, leading to a constant decrease in fluorescence intensities, with an increase in the concentration of SDS with CAD in ethanol media. However, in the post-micellar concentration another rearrangement occurs between the SDS monomers and drug molecules in ethanol media. The maximum fluorescence intensity was observed at 0.019 m concentration in the post-micellar region as depicted in Figure 5. Micellar systems consisting of a surfactant and an additive such as an organic compound usually self-organize as a series of worm-like micelles that ultimately form a micellar network. The nature of the additive influences micellar structure and properties such as aggregate lifetime. CAD molecules show π–π stacking interaction that somewhat segregates into SDS micelle at higher micellar concentrations [29,30]. It is believed that at higher micellar concentrations, the hydrogen bonding forces onto the surface between CAD and SDS monomers become more favorable and organize as micelle assemblies in comparison to the sole structural forces of the solvent i.e., ethanol molecules. In addition, the hydrophilic counterpart of the CAD molecules does not intervene in the micellar growth. Therefore, the fluorescence quenching of SDS with the cinnamaldehyde is mainly achieved by photo-excited electron transfer between the micelles of SDS and cinnamaldehyde molecules, and the process is thermodynamically feasible with a ΔG value of -4.30×10^4 kcal mol^{-1}. The maximum fluorescence intensity of SDS is quenched gradually with the cinnamaldehyde in ethanol and the quenching efficiency of CAD can be finely fitted with the Stern–Volmer equation. Fluorescence measurement offers the information about the binding process during the self-aggregation process of SDS in the CAD-ethanol system. The fluorescence spectra of the observed pre- and post-micellar regions are shown in Figures 4 and 5 at maximum fluorescence excitation (λ_{ex} 295 nm). The fluorescence intensities increase in the pre-micellar region whereas with the increase of SDS concentration the intensities decrease steeply resulting in the fluorescence quenching. Quenching can appear as a result of micellization depicting the role of hydrophobic interaction in the binding of molecules with the micelles. Fluorescence quenching is the decrease of the quantum yield of fluorescence from a fluorophore induced by a variety of molecular interactions with quencher molecules [31]. Therefore, the fluorescence data below the CMC and above the CMC are well fitted in the Stern–Volmer equation as given below:

$$\frac{F^\circ}{F} = 1 + K_q\tau^\circ\,[Q] = 1 + K_{SV}\,[Q], \tag{6}$$

where, F° and F are the fluorescence intensities in the absence and presence of quencher, respectively. A linear regression plot is drawn between the F°/F against the molarity of varying concentration of the quencher [Q] (Figures 6 and 7). K_{sv} is the Stern–Volmer constant, and its values are listed in Table 2. Below the CMC, the plot is linear, showing lower value of K_{sv} than in the post-micellar region, i.e., above the CMC, suggesting that the micellization is controlled by the dynamic quenching mechanism rather than the static mechanism. K_q is the quenching rate constant, and the values are evaluated by using Equation (6) [32] in the following form:

$$K_q = \frac{K_{SV}}{\tau^\circ} \tag{7}$$

where, τ° is the average lifetime of the fluorescent molecule with the value (10^{-8} s^{-1}), and the Kq values estimated from Equation (7) are presented in Table 2. Hydrophobicity of the SDS molecule in the pre- and post-micelle regions has a significant influence on the binding constant, and the equilibrium between free and bound molecules is given by the Equation (8) [32].

$$\mathrm{Log}_{10}\frac{F^\circ - F}{F} = \log_{10} K_b + n\,\log_{10}\,[Q] \tag{8}$$

where, K_b is the binding constant and n is the binding sites, and the value of K_b and n is obtained from the intercept and slope of the linear regression plot between $\log_{10}$ ((F° − F)/F) versus $\log_{10}$ [Q] below and above the CMC. The values are shown in Table 2. The higher value of K_b above the CMC than the binding constant below the CMC is positive evidence of the hydrophobic interaction. The observed comparatively lower fluorescence intensities in pre-micellar concentrations occurred after the repulsive electrostatic forces and dipolar interactions between CAD and SDS molecules in the ethanol medium, in turn resulting in a reduction of the fluorescence intensities. In addition, at pre-micellar concentrations, the partitioning of drug molecules has maximum probabilities as more forces are prevailing to form fewer aggregates in comparison with the micellar region hence, hydrophobic forces are predominating major driving force for micellization in the micellar region.

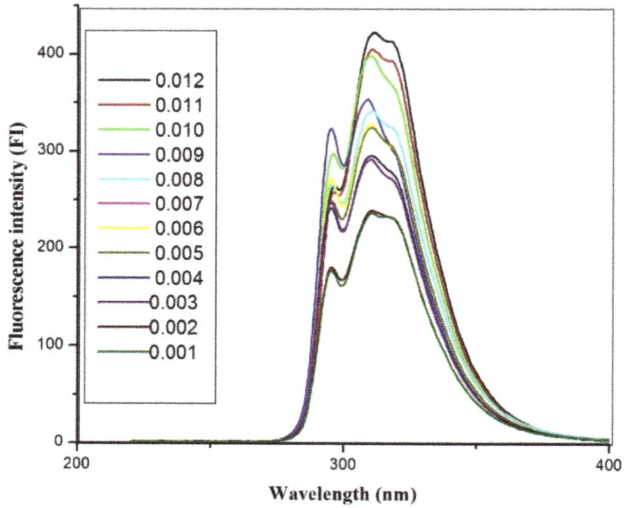

Figure 4. Fluorescence intensities (FI) versus wavelength for pre-micellar concentrations of SDS in CAD + E-OH.

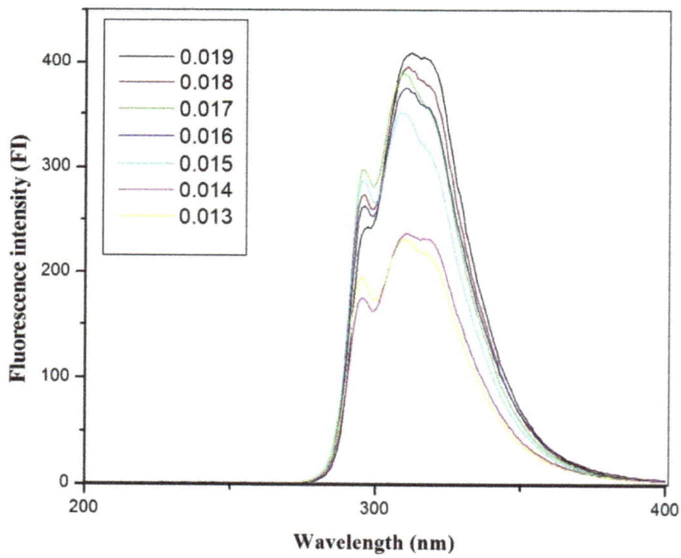

Figure 5. Fluorescence intensities (FI) versus wavelength for post-micellar concentrations of SDS in CAD + E-OH.

By knowing the binding constant of SDS with the CAD-ethanol system below and above the CMC, thermodynamic parameter free energy change (ΔG°) can also be evaluated easily in the surfactant transition process of monomer to micellar phase by the following equation [32]:

$$\Delta G^{\circ} = -RTln\ K_b \tag{9}$$

The obtained negative values of free energy change (ΔG°) (listed in Table 2) signify the SDS thermodynamic stability and governs the binding process favoring the micellization process.

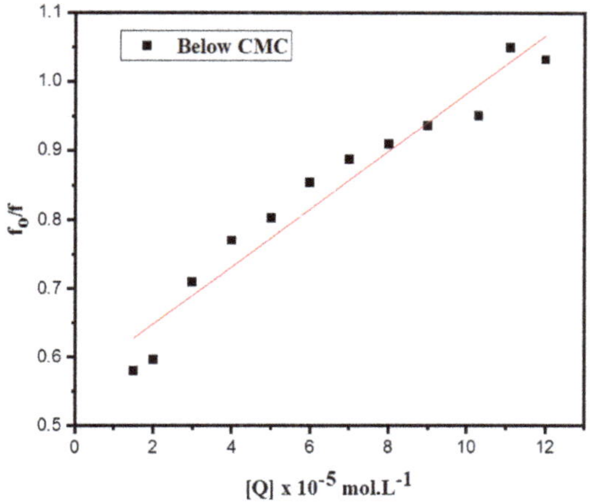

Figure 6. Stern–Volmer plot of interaction of SDS in the CAD-ethanol system at 298.15 K below CMC.

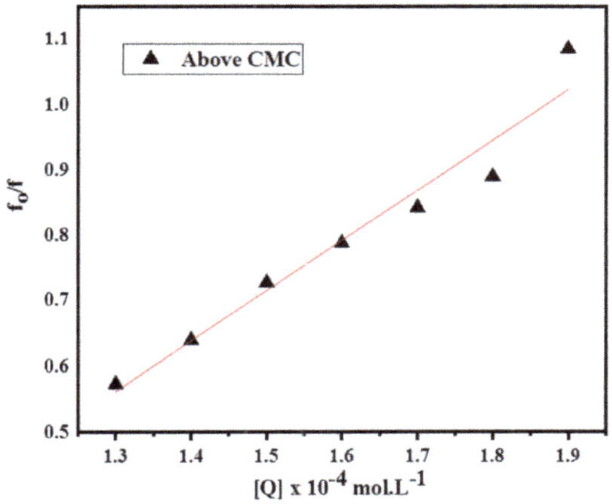

Figure 7. Stern–Volmer plot of interaction of SDS in CAD-ethanol system at 298.15 K above CMC.

Table 2. Stern–Volmer quenching constants (K_{sv}), quenching rate constant (K_q), binding constants (K_b), number of binding sites (n), and Gibbs free energy ($\Delta G°$) for the interaction of cinnamaldehyde (CAD) in SDS-ethanol media at 298.15 K temperature.

(Measured at λ_{ex} 295 nm)	Below CMC	Above CMC
K_{sv} (10^3 M^{-1})	4.19	7.68
K_q	4.5×10^{10}	3.3×10^{15}
K_b (M^{-1})	4.51×10^2	3.34×10^7
n	1.22	4.38
$\Delta G°$ (10^4 kJ mol^{-1})	−1.51	−4.30

3.3. FTIR Analysis of CAD + SDS + E-OH Media

The FTIR analysis was inferred to interpret the possible interactions in an ethanol medium between the cinnamaldehyde (CAD) and sodium dodecyl sulphate (SDS) molecules. The presence of different substitutions with their available surrounding environment between different molecules influences molecular interactions in their aqueous media [33]. Hereby, we determined the possible existing molecular interactions between the CAD and SDS as deduced from both the structural information of pure CAD, pure SDS, and varying concentrations of CAD + SDS in an ethanol medium as depicted in Figure 8. Some of the observed essential peak intensities were lower, which confirms the existence of major functional groups of pure cinnamaldehyde. The intensities at 3062.6 cm^{-1} and 3026.7 cm^{-1} are attributed to presence of -C-H structuring vibrations and aromatic C–H bond structuring vibrations, 2814.5 cm^{-1} and 2742.6 cm^{-1} are because of –CH_2 structuring vibrations, and the presence of peak intensities within the range of 1648 cm^{-1} to 1746 cm^{-1}, such as 1713.0 cm^{-1}, 1667.8 cm^{-1}, and 1619.7 cm^{-1} are corresponding -C=O vibrations; the presence of peak intensities between 1463 cm^{-1} and 1627 cm^{-1}, such as 1495.7 cm^{-1} and 1574.5 cm^{-1}, are attributed to aromatic ring –C=C structural structuring vibrations. Meanwhile, C=C of aromatic ring vibrations were observed at 687.5 cm^{-1}; the presence of these peak intensities is attributed to pure cinnamaldehyde (Figure 6 (b; black line)) [34]. In addition, the major peak intensities as observed at 3324.6 cm^{-1}, 2971.6 cm^{-1}, 2926.4 cm^{-1}, 2881.2 cm^{-1}, 1085.8 cm^{-1}, and 1015.2 cm^{1}, respectively were observed in the pure spectrum of SDS molecules as depicted in Figure 6 (c; red line). Observed peak intensity at 3324.6 cm^{-1} was assigned to the -O-H structuring vibrations modes, 2971.6, 2926.4, and 2881.2 cm^{-1}, which are attributed to the presence of asymmetric and symmetric structuring vibration modes of C-H aliphatic carbon chains [35]. In addition, the observed peak intensities at 1085.8 and 1015.2 cm^{-1} are assigned to the corresponding S-O and S=O structuring vibrational modes respectively [36,37]. Based on the pre- and post-micellization concentrations of the FTIR analysis of combined CAD + SDS in ethanol solution as depicted in Figure 6 (F; cyan line and G; pink line), we observed the disappearance of CAD peak intensity at 1667.6 cm^{-1} (i.e., presence of -C=O vibrations), and also the observed lengthening of -O-H structuring vibrations at 3324.6 cm^{-1} towards the higher frequency, i.e., 3327.0 cm^{-1} in mixed CAD + SDS, emphasizing the possible hydrogen bond interaction in the mixed ethanol medium. The observed results of the FTIR analysis further established the aggregation of CAD with SDS molecules in the pre-micellar region and the formation of micelle assemblies. Additionally, in the pre- and post-micellar regions the possible hydrophobic and hydrogen bonding interaction between CAD and SDS are established after their function group interactions in an ethanol environment.

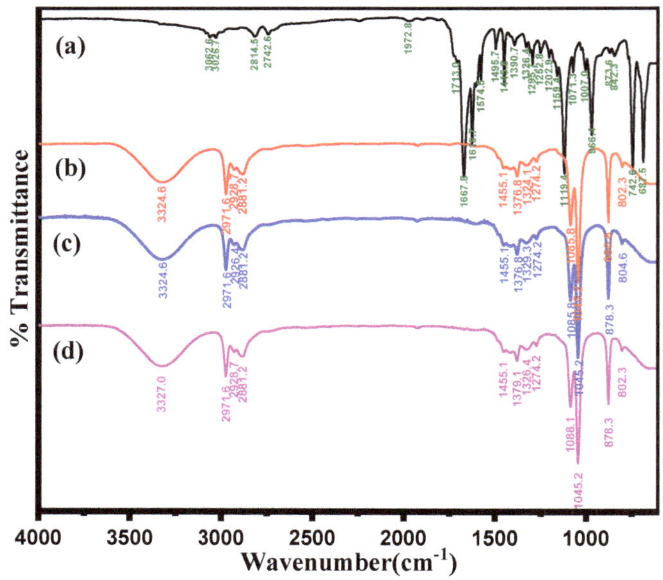

Figure 8. FTIR analysis of CAD + SDS at varying concentrations: (**a**) FTIR analysis of CAD, (**b**) SDS alone in ethanol, (**c**) Pre–micellization SDS + E-OH + CAD (**d**) Post-micellization SDS + E-OH + CAD.

4. Conclusions

The present study was conducted to deduce the possible molecular interaction between SDS and CAD molecules in an ethanol medium. The result obtained from the conductivity measurements implies delayed micellization after the molecular interactions in SDS + E-OH + CAD. The observations established the fine balancing hydrophobic force between the aromatic ring of CAD molecules and SDS's long hydrocarbon tails. whereas repulsive interactions also encounter the ionic head groups. Meanwhile, in SDS + E-OH and SDS + E-OH + CAD with elevated temperatures, increased thermal motions are quite possible with the distraction of the micelle's palisade layer, and in turn results in de-micellization, as observed with the increased CMC values of surfactant molecules. The values of ΔG_m° of micellization were more for the SDS + E-OH system than in the SDS + E-OH + CAD system. This signifies the readiness and thermodynamically more spontaneous micellization of SDS with E-OH mixture than SDS + E-OH + CAD mixtures. In addition, increased numerical negative values of ΔH_m° and the decreased numerical positive values of ΔS_m° prompt the hydrophobic interactions between the SDS and CAD molecules weaken, while the electrostatic interactions become stronger with rising temperatures.

Moreover, from fluorescence analysis we believe the hydrogen bonding forces onto the surface between CAD and SDS monomers become more favorable and organize as micelle assemblies in comparison to the sole structural forces of the solvent i.e., ethanol molecules. Furthermore, lower fluorescence intensities were observed at lower micellar concentrations after the repulsive electrostatic forces and dipolar interactions between CAD and SDS molecules in the ethanol medium. Our observation of the molecular interaction and micelle growth was further supported upon an FTIR functional group analysis, from different peak intensities at varying concentration, in the pre- and post-micellar region. The possible hydrophobic and hydrogen bonding interactions between CAD and SDS were established as a regard of consequence of function group interactions in an ethanol medium.

Author Contributions: Conceptualization, W.M.A. and N.H.; methodology, W.M.A. and N.H.; software, S.T.; validation, S.T. and N.H.; formal analysis, S.T. and N.H.; investigation, S.T.; resources, W.M.A.; data curation, S.T.; writing—original draft preparation, A.N.; writing—review and editing, A.N., S.T. and F.N.; visualization, S.T.; supervision, S.T. and W.M.A.; project administration, N.H.; funding acquisition, W.M.A. All authors have read and agreed to the published version of the manuscript.

Funding: This research was funded by Deanship of Scientific Research, Jazan University (Grant No. RUP2-02).

Institutional Review Board Statement: Not applicable.

Informed Consent Statement: Not applicable.

Acknowledgments: The authors extend their appreciation to Deanship of Scientific Research Work, Jazan University, for supporting this research through the Research Units Support Program, Support Number: RUP2-02.

Conflicts of Interest: The authors declare no conflict of interest.

References

1. Doyle, A.A.; Stephens, J.C. A review of cinnamaldehyde and its derivatives as antibacterial agents. *Fitoterapia* **2019**, *139*, 104405–104454. [CrossRef]
2. Tan, Y.; Liu, X.; Zhang, L.; Liu, F.; Wang, A.; Zhang, T. Producing of cinnamyl alcohol from cinnamaldehyde over supported gold nanocatalyst. *Chin. J. Catal.* **2021**, *42*, 470–481. [CrossRef]
3. Abbot, V.; Bhardwaj, V.; Sharma, P. Investigation of intermolecular interactions of anionic surfactant SDS and rutin: A physicochemical approach for pharmaceutical application. *J. Mol. Liq.* **2021**, *337*, 116352–116371. [CrossRef]
4. Sharma, P.; Bhardwaj, V.; Chaudhary, T.; Sharma, I.; Kumar, P.; Chauhan, S. Thermodynamics and micellization of cetyltrimethyl ammonium bromide in the presence of lysozyme. *J. Mol. Liq.* **2013**, *187*, 287–293. [CrossRef]
5. Nabi, A.; Tasneem, S.; Jesudason, C.G.; Lee, V.S.; Md Zain, S.B. Study of interaction between cationic surfactant (CTAB) and paracetamol by electrical conductivity, tensiometric and spectroscopic methods. *J. Mol. Liq.* **2018**, *256*, 100–107. [CrossRef]

6. Roy, S.; Sengupta, P.S.; Guin, P.S. Electrochemical and UV-Vis spectroscopic studies on the interaction of sodium 1,4- dihydroxy-9,10-anthraquinone-2-sulphonate with cetyltrimethylammonium bromide micelles. *Chem. Phys. Lett.* **2018**, *694*, 7–13. [CrossRef]
7. Tanwar, L.K.S.; Banjare, M.K.; Sharma, S.; Ghosh, K.K. Physicochemical studies on the micellization of anionic surfactants in the presence of long alkyl chain ionic liquid. *Chem. Phys. Lett.* **2021**, *769*, 138399. [CrossRef]
8. Autelitano, A.; Minassi, A.; Pagani, A.; Taglialatela-Scafati, O.; Appendino, G. The reaction of cinnamaldehyde and cinnam(o)yl derivatives with thiols. *Acta Pharm. Sin. B* **2017**, *7*, 523–526. [CrossRef]
9. Sawicki, R.; Golus, J.; Przekora, A.; Ludwiczuk, A.; Sieniawska, E.; Ginalska, G. Antimycobacterial Activity of Cinnamaldehyde in a Mycobacterium tuberculosis(H37Ra) Model. *Molecules* **2018**, *23*, 2381. [CrossRef]
10. Tanaka, Y.; Uchi, H.; Furue, M. Antioxidant cinnamaldehyde attenuates UVB-induced photoaging. *J. Dermatol. Sci.* **2019**, *96*, 151–158. [CrossRef]
11. El-Ezz, A.; Maher, A.; Sallam, N.; El-Brairy, A.; Kenawy, S. Trans-cinnamaldehyde Modulates Hippocampal Nrf2 Factor and Inhibits Amyloid Beta Aggregation in LPS-Induced Neuroinflammation Mouse Model. *Neurochem. Res.* **2018**, *43*, 2333–2342. [CrossRef] [PubMed]
12. El-Baroty, G.S.; Abd El-Baky, H.H.; Farag, R.S.; Saleh, M.A. Characterization of antioxidant and antimicrobial compounds of cinnamon and ginger essential oils. *Afr. J. Biochem. Res.* **2010**, *4*, 167–174.
13. Tankam, J.M.; Sawada, Y.; Ito, M. Regular ingestion of cinnamomi cortex pulveratus offers gastroprotective activity in mice. *J. Nat. Med.* **2013**, *67*, 289–295. [CrossRef] [PubMed]
14. Yang, L.; Wu, Q.-Q.; Liu, Y.; Hu, Z.-F.; Bian, Z.-Y.; Tang, Q.-Z. Cinnamaldehyde attenuates pressure overload-induced cardiac hypertrophy. *Int. J. Clin. Exp. Pathol.* **2015**, *8*, 14345–14354. [PubMed]
15. Qi, X.; Zhou, R.; Liu, Y.; Wang, J.; Zhang, W.-N.; Tan, H.-R.; Niu, Y.; Sun, T.; Li, Y.-X.; Yu, J.-Q. Trans-cinnamaldehyde protected PC12 cells against oxygen and glucose deprivation/reperfusion (OGD/R)-induced injury via anti-apoptosis and anti-oxidative stress. *Mol. Cell. Biochem.* **2016**, *421*, 67–74. [CrossRef]
16. Ray, G.B.; Chakraborty, I.; Ghosh, S.; Moulik, S.P.; Palepu, R. Self-Aggregation of Alkyltrimethylammonium Bromides (C10-, C12-, C14-, and C16TAB) and Their Binary Mixtures in Aqueous Medium: A Critical and Comprehensive Assessment of Interfacial Behavior and Bulk Properties with Reference to Two Types of Micelle Formation. *Langmuir* **2005**, *21*, 10958–10967. [CrossRef]
17. Muhammad, M.T.; Khan, M.N. Study of electrolytic effect on the interaction between anionic surfactant and methylene blue using spectrophotometric and conductivity methods. *J. Mol. Liq.* **2017**, *234*, 309–314. [CrossRef]
18. Rosen, M.J.; Kunjappu, J.T. *Surfactants and Interfacial Phenomena*, 4th ed.; John Wiley & Sons: New York, NY, USA, 2012. [CrossRef]
19. Ray, G.B.; Ghosh, S.; Moulik, S.P. Physicochemical Studies on the Interfacial and Bulk Behaviors of Sodium N-Dodecanoyl Sarcosinate (SDDS). *J. Surfactants Deterg.* **2009**, *12*, 131–143. [CrossRef]
20. Patist, A.; Oh, S.G.; Leung, R.; Shah, D.O. Kinetics of micellization: Its significance to technological processes. *Colloids Surf. A Physicochem. Eng. Asp.* **2001**, *176*, 3–16. [CrossRef]
21. Shah, S.S.; Jamroz, N.U.; Sharif, Q.M. Micellization parameters and electrostatic interactions in micellar solution of sodium dodecyl sulfate (SDS) at different temperatures. *Colloids Surf. A Physicochem. Eng. Asp.* **2001**, *178*, 199–206. [CrossRef]
22. Bakshi, M.S. Micelle Formation by Sodium Dodecyl Sulfate in Water–Additive Systems. *Bull. Chem. Soc. Jpn.* **1996**, *69*, 2723–2729. [CrossRef]
23. Alhmoud, H.A. The effect of surfactant above and below the critical micelle concentration (CMC) and the mathematical models used to determine the kinetics of drug release from the matrix system. *Afr. J. Pharm. Pharmacol.* **2016**, *10*, 88–94. [CrossRef]
24. Mehta, S.K.; Bhasin, K.K.; Chauhan, R.; Dham, S. Effect of temperature on critical micelle concentration and thermodynamic behavior of dodecyldimethylethyl ammonium bromide and dodecyltrimethylammonium chloride in aqueous media. *Colloids Surf. A Physicochem. Eng. Asp.* **2005**, *255*, 153–157. [CrossRef]
25. Sharma, V.K.; Yadav, O.P.; Singh, J. Physicochemical studies of aqueous sodium dodecylsulphate solutions in pyridine and isomeric picolines. *Colloids Surf. A Physicochem. Eng. Asp.* **1996**, *110*, 23–25. [CrossRef]
26. Sheppard, V.B.; Figueiredo, M.; Canar, J.; Goodman, M.; Caicedo, L.; Kaufman, A.; Norling, G.; Mandelblatt, J. Latina a LatinaSM†: Developing a breast cancer decision support intervention. *Psycho-Oncol. J. Psychol. Soc. Behav. Dimens. Cancer* **2008**, *17*, 383–391. [CrossRef] [PubMed]
27. Mohsen-Nia, M.; Amiri, H.; Jazi, B. Dielectric Constants of Water, Methanol, Ethanol, Butanol and Acetone: Measurement and Computational Study. *J. Solut. Chem.* **2010**, *39*, 701–708. [CrossRef]
28. Usman, M.; Siddiq, M. Surface and micellar properties of Chloroquine Diphosphate and its interactions with surfactants and Human Serum Albumin. *J. Chem. Thermodyn.* **2013**, *58*, 359–366. [CrossRef]
29. Balasubramanian, D.; Srinivas, V.; Gaikar, V.G.; Sharma, M.M. Aggregation behavior of hydrotropic compounds in aqueous solution. *J. Phys. Chem.* **1989**, *93*, 3865–3870. [CrossRef]
30. Vinarov, Z.; Katev, V.; Radeva, D.; Tcholakova, S.; Denkov, N.D. Micellar solubilization of poorly water-soluble drugs: Effect of surfactant and solubilizate molecular structure. *Drug Dev. Ind. Pharm.* **2018**, *44*, 677–686. [CrossRef]
31. Valeur, B. *Molecular Fluorescence. Principles and Applications*, 2nd ed.; Wiley-VCH Verlag GmbH: New York, NY, USA, 2001; ISBN 978-3-527-32837-6.
32. Feng, X.Z.; Jin, R.X.; Qu, Y.; He, X.W. Study on the ion effect on the binding interaction between HP and BSA. *Chem. J. Chin. Univ.* **1996**, *17*, 866–869.

33. Salanci, E.; Malík, I.; Šandrik, R.; Pecher, D.; Andriamainty, F. Determination of the critical micelle concentration and thermodynamic parameters of phenylcarbamic acid derivatives using a fluorescence method. *Chem. Pap.* **2021**, *75*, 3081–3090. [CrossRef]
34. Bissantz, C.; Kuhn, B.; Stahl, M. Medicinal Chemist's Guide to Molecular Interactions. *J. Med. Chem.* **2010**, *53*, 5061–5084. [CrossRef] [PubMed]
35. Sastrohamidjojo, H. *Dasar-Dasar Spektroskopi*; UGM Press: Yogyakarta, Indonesia, 2018; ISBN 978-979-420-817-5.
36. Ma, C.-L.; Sun, X.-D. Preparation and characterization of SnO2 nanoparticles with a surfactant-mediated method. *Nanotechnology* **2002**, *13*, 565–569. [CrossRef]
37. Tao, Q.; He, H.; Frost, R.L.; Yuan, P.; Zhu, J. Nanomaterials based upon silylated layered double hydroxides. *Appl. Surf. Sci.* **2009**, *255*, 4334–4340. [CrossRef]

energies

Article

Self-Starting Characteristics and Flow-Induced Rotation of Single- and Dual-Stage Vertical-Axis Wind Turbines

Muhammad Saif Ullah Khalid [1], David Wood [2] and Arman Hemmati [1,*]

[1] Department of Mechanical Engineering, University of Alberta, Edmonton, AB T6G 1H9, Canada
[2] Department of Mechanical and Manufacturing Engineering, University of Calgary, 2500 University Dr NW, Calgary, AB T2N 1N4, Canada
* Correspondence: arman.hemmati@ualberta.ca

Abstract: Despite offering promising opportunities for wind energy harvesting in urban environments, vertical axis wind turbines face limitations in terms of poor starting characteristics. In this study, we focus on analyzing improvements offered by dual-stage turbines for a range of wind velocities. Numerical simulations are performed for different phase angles between the rotors (a measure of relative angular positions of the blades in the two rotors) to quantify the response time for their starting behavior. These simulations rely on a through sliding mesh technique coupled with flow-induced rotations. We find that for $U_\infty = 4$ m/s, the phase angles of 30° and 90° substantially reduce starting time in comparison to a single-stage turbine. Dual-stage turbines with a phase angle of 90° exhibit similar or better starting behavior for other wind speeds. The phase angle of 0° in double-rotor turbines shows the poorest starting response. Moreover, it is revealed that stabilization of shear layers generated by the blades passing through the windward side of the turbine, vortex-entrapment by these rotating blades, and suppressing of flow structures in the middle of the wake enhance the capacity of VAWTs to achieve faster steady angular speed.

Keywords: wind energy; vertical axis wind turbines; wake dynamics; computational fluid dynamics

Citation: Khalid, M.S.U.; Wood, D.; Hemmati, A. Self-Starting Characteristics and Flow-Induced Rotation of Single- and Dual-Stage Vertical-Axis Wind Turbines. *Energies* **2022**, *15*, 9365. https://doi.org/10.3390/en15249365

Academic Editor: Vasily Novozhilov

Received: 3 October 2022
Accepted: 18 November 2022
Published: 10 December 2022

1. Introduction

Intensified effects of climate change and implications of greenhouse gas (GHG) emissions have recently accelerated research in the field of alternate energy systems. These resources are advantageous to existing and conventional energy harvesting and power generation technologies due to their limited GHG emissions. In this context, enormous energy can be harnessed from wind through turbines in urban and off-shore environments for power generation. To this effect, horizontal-axis wind turbines (HAWTs) have gained popularity due to their capacity for greater power generation. However, vertical axis wind turbines (VAWTs) provide more promising alternatives in urban environments due to their small size and better aerodynamic performance at lower wind speeds. VAWTs are quieter and their omnidirectional performance is suited to high turbulence levels in the urban setting. There are some design concerns over the performance of these turbines that require further research. Mainly, Darrieus-type or straight-bladed VAWTs require a faster self-starting mechanism that does not rely on any external support or excitation. There exist several definitions in the literature for defining the self-starting phenomenon in VAWTs [1–4]. For a horizontal-axis turbine, Ebert and Wood [5] suggested that it was simply necessary to minimize the time taken for blades to reach a reasonable rotational speed to start loading the generator in order to produce power. According to Kirke [1], a turbine must generate significant power for the self-starting process to commence adequately. Lunt [2] defined this process for a turbine when it accelerates from a stationary position to produce usable power output. Moreover, Takao et al. [3] presented a more appropriate definition for a self-starting turbine without involving vague terms, such as usable outputs or significant power. They defined a turbine as self-starting if it accelerated from a static state and attains a steady-state speed that is greater than the wind velocity. It means that a turbine should have tip-speed ratio more than one to complete the self-starting process. Usually, improvements in designs of VAWTs to address this particular issue drastically impact their power production [6]. Hence, novel techniques are needed for VAWTs to enhance their self-starting process adequately towards a desired power outcome.

In the literature, various efforts were reported to analyze and hasten the self-starting of VAWTs. For example, Dominy et al. [7] developed a numerical model with symmetric NACA-0012 blade profiles to demonstrate that a three-bladed turbine always had a potential to self-start under steady wind flow. Their work also showed that the self-starting process of a two-bladed VAWT depended on the initial orientations of the blades. Later, a theoretical framework was developed by [8] using momentum models, and experiments were also performed to examine the effects of airfoil shape on a vertical axis rotor. Such momentum-based models were also used to determine virtual camber in VAWTs to correctly predict their power on very low tip-speed ratios [9]. Worasinchai et al. [10] explained similarities in the aerodynamics of flapping wings and VAWTs. They suggested that utilizing unsteadiness in the upstream wind by the rotor geometry could enhance thrust generation by the blades. They also concluded that this exploitation of unsteadiness was essential for the VAWT to self-start, because the turbine was locked in a certain "dead" regime when thrust was not continuously generated by the blades. To this effect, many studies [11–14] have recently focused on modifying blade profiles and turbine geometries to increase the starting time of VAWTs. Sun et al. [15] examined the effects of symmetric and asymmetric cross sections of blades and their inertial properties on the self-starting capacity of VAWTs. They reported that variations in the geometry of turbine blades could help enhance the strength of the initially formed vortex in terms of vorticity and size over a blade's surface to reduce the starting time. Based on this finding, blades with EN0005-based cross-sectional profiles showed better self-starting characteristics. However, they also observed that other asymmetric blade profiles (e.g., NACA-1425 or NACA-4425) with selected camber and thicker symmetric profiles (e.g., DU06-W-200) showed superior power production at wind speeds below 6 m/s. Moreover, blades with thinner cross sections generated more power at higher wind speeds. Celik et al. [6] performed numerical simulations to examine how the number of blades and moment-of-inertia impact the self-starting of H-type straight-bladed VAWTs. Their findings indicated that this process remained insensitive to inertial characteristics of the blades. Another important observation in their work involved reducing the starting time of a turbine by increasing the number of blades, which resulted in reduced power output. More recently, Sun et al. [16] also studied the impact of variations in the number of blades and offsets in their pitching angle on the power output. Their findings revealed that an optimum pitching angle of -4° significantly increased the power output as well as minimized the time for a turbine to self-start at wind speeds over 9 m/s. A very recent work by [17] presented an interesting approach to compute localized flow speeds near blades and their dependence on tip-speed ratios of VAWTs during different revolutions when they are self-starting. They concluded that drag force along with lift could be a substantial contributor in driving these turbines initially when the blades would be likely to experience high angles-of-attack.

Another very promising modification in designs of VAWTs involves the addition of another rotor inside the primary rotor of H-type VAWTs. A few studies have reported the benefits of increase in power generation and improved static torque by such configurations [18–28]. Particularly, Torabi Asr et al. [21] performed two-dimensional numerical simulations to show faster starting and more power extraction of double-stage VAWTs than a single-rotor counterpart by offering smaller cut-in wind speeds and lower starting time. Later, Chen et al. [29] employed computational simulations to investigate the role of distance between the two rotors for the aerodynamics and starting of dual-stage VAWTs. In their cases, these turbines produced less power compared to a single-rotor configuration, but significantly reduced the starting time. With this background, it is important to further analyze effects of the phase angle between rotors of dual-stage turbines for a range of wind speeds to identify adequate locking mechanisms under distinct flow conditions to attain the desired results. Another motivation for choosing such configurations for aerodynamic designs comes from schooling behavior of fish, which can move in circular formations. It demonstrates prospects to enhance performance metrics as also recently shown by [30] for a Savonius turbine. More nature-inspired mechanisms to improve energy harvesting from VAWTs were described by [31]. Hence, we focus on this particular subject in our present work and employ computational simulations to quantify the performance of both single- and dual-stage turbines and qualitatively analyze their self-starting behavior and power generation capacity. The primary novelty of our present work lies in explaining physical mechanisms around the rotating blades in VAWTs, which play a critical role in quickly reaching steady-state. Another aspect is explaining the impact of different geometric configurations on their self-starting behavior. The simulations were made without loading the rotor because any load would increase the starting time. It is emphasized that the simulations do not give information on the power extraction capacity of dual-rotor VAWTs. This is an important topic, but one that should be explored only if the dual-rotor turbine starts more rapidly than a single-stage one.

The manuscript is organized as follows. Section 2 explains the geometric and kinematic configurations of vertical axis wind turbines as well as our numerical methodology for flow-induced rotations of vertical axis wind turbines. It also presents detailed validations and verification of our computational strategy. In section sec:esults, we first elucidate the quantitative performance metrics of the single- and dual-rotors VAWTs supplemented later by the governing unsteady physical mechanisms that lead the turbines to their steady-state rotations.

2. Numerical Methodology

In this section, we describe the computational methodology employed to perform simulations for single-stage and dual-stage VAWTs. This analysis includes simulating their flow-induced rotation at various operating conditions. We also provide details on the verification and validation studies associated with the current setup.

2.1. Geometry and Kinematics

Two-dimensional single-stage and dual-stage configurations of H-type Darrieus vertical axis wind turbines with rigid symmetric NACA0018 airfoils were employed for this study. It is important to note that the flow-induced motion of the rotors in the present work does not include structural deformations of the blades. Schematics of the single- and double-rotor configurations of the VAWTs presently under consideration and a schematic of the flow domain are shown in Figures 1 and 2, respectively. The key details about kinematics and geometries are outlined in Table 1 along with the ranges of governing parameters that quantify the performance of VAWTs.

Table 1. Geometric and flow parameters for VAWTs, where subscript "1" denotes the primary, outer rotor and "2" the inner, secondary one.

Parameters	Value
Cross section of blades	NACA0018
No. of blades in single-stage turbines	3
No. of blades in dual-stage turbines	6
Free-stream velocity (U_∞)	4–10 m/s
Chord length of blades in primary rotors (c_1)	0.06 m
Radius of the primary (outer) rotor (R_1)	0.5 m
Ratios of radii for primary and secondary (inner) rotors	$R_1/R_2 = 0.85$
Geometric ratios for blades in primary and secondary rotors	$c_1/D_1 = c_2/D_2 = 0.06$
Angle between the blades of primary and secondary rotors (ϕ)	0°–90°
Reference area (A)	1 m^2

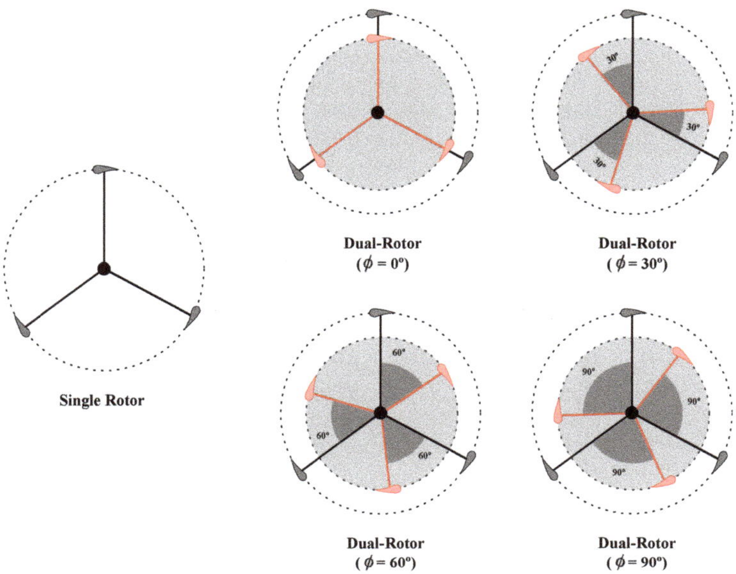

Figure 1. Schematics of single- and dual-stage turbines.

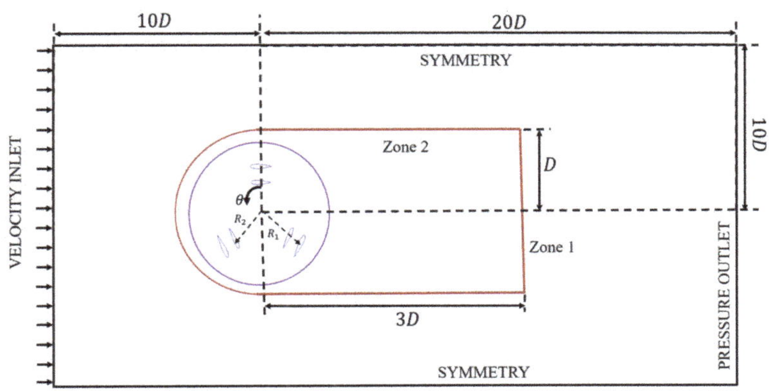

Figure 2. A schematic (not to scale) of the computational domain with the description of boundary conditions.

2.2. Flow Solver

The simulations were performed using ANSYS Fluent 2020R2 [32], a commercial finite-volume based computational solver, which has gained popularity among researchers for simulations of flow over wind turbines [33–36]. Here, incompressible continuity and unsteady Reynolds-averaged Navier–Stokes (URANS) equations are solved in Cartesian coordinates through the pressure-based algorithm. Although a PISO (pressure implicit with splitting operators) scheme is recommended for unsteady flows [37], it is usually advantageous when a large time step (Δt) was adopted to computationally march in time. Hence, the Semi-Implicit Method for Pressure Linked Equation (SIMPLE) algorithm is adopted for our current simulations to improve the computational efficiency.

The least square cell-based technique was utilized for the computation of gradient terms, second-order scheme for convective pressure terms, and second-order upwind technique for diffusion terms in the momentum equation. Although third-order algorithms may also be used for the terms with Laplacian operator, those are computationally expensive. The advantage of upwind schemes is the provision of greater stability in numerical simulations. The unsteady term was numerically approximated by the second-order implicit scheme, as also recommended by [38]. These unsteady simulations were performed through the sliding-mesh technique, which allows physical rotation of the turbine without disturbing the original mesh. Rezaeiha et al. [39] suggested that Shear Stress Transport (SST) turbulence models performed well in capturing the flow features for VAWTs and results obtained through them closely matched with experiments. Therefore, a four-equation transition SST turbulence model was employed to predict the turbulent flow features and accurately capture the laminar-to-turbulence transition. The convergence criterion for the iterative solution at each time step was set to 10^{-4}. All the simulations ran for sufficient time, so that the VAWTs attain their steady-state tip-speed ratios.

2.3. Computational Domain and Boundary Conditions

An H-grid method with a rectangular computational domain was used in this study, shown in Figure 2. A uniform flow-velocity was prescribed at the inlet boundary located at a distance of $10D$ from the central axis of the turbine. Gauge pressure was set as zero on the pressure outlet boundary located $20D$ away from this axis. Top and bottom boundaries were set as symmetry, and each one was located at a distance of $10D$ from the center of the turbine. All domain boundaries were adjusted following the recommendations of Rezaeiha et al. [40]. The main objective of performing sensitivity studies on domain-size independence is to ensure accuracy in capturing the physical phenomena around rotating blades and computing their performance parameters. Rezaeiha et al. [40] concluded through extensive $2D$ and $2.5D$ simulations that inlet, outlet, and side boundaries should be kept at a minimum distance of $10D$ from the center of the turbine to minimize their impact on numerical accuracy, let the wake fully resolve, and avoid blockage effects.

To incorporate the sliding mesh techniques, the computational domain was divided into three zones for both single- and dual-stage VAWTs. Zones 1 and 2 are shown in Figure 3. Both these zones remain stationary and are connected by an interface, which enabled communication of flow information between neighboring domains through a non-conformal meshing algorithm. Zone 2 was used to capture complex details of the wake of the rotating turbine. Figure 3a presents the meshing features inside all three zones. Here, Zone 3 contains blades for the outer and inner stages of the turbine. This domain rotates around the central axis of the VAWT. Figure 3b,c exhibit the very fine grid near the leading and trailing edges of the foils with $y^+ = 1$, respectively, which is sufficient to accurately capture the boundary layers around these rotating structures. It is evident from Figure 3d that we control these mesh settings in order to slowly vary the mesh size and avoid large gradients. Numerical errors are avoided by keeping the same cell size around the interface boundary between the rotating and static domains. The current setup follows the recommendations of Rezaeiha et al. [40], whose study reveals that the radius of a rotating domain in such simulations does not have significant effects on the aerodynamics of wind turbines. Although the Reynolds number Re for flows around a rotating machinery involves velocity scales that are dependent on both free-stream/local air speed and the angular velocity of the machines, the *"averaged" Re can be defined as:*

$$\mathrm{Re} = \frac{({U_\infty}^2 + \Omega^2)^{0.5} D_1}{\nu} \tag{1}$$

where Ω denotes the angular speed of the turbine, and ν is the kinematic viscosity of air. Considering the speeds of air in this work, the initial values of Re range from 2.7×10^5 to 6.75×10^5. As a turbine accelerates to gain its steady rotational speed, Re increases substantially.

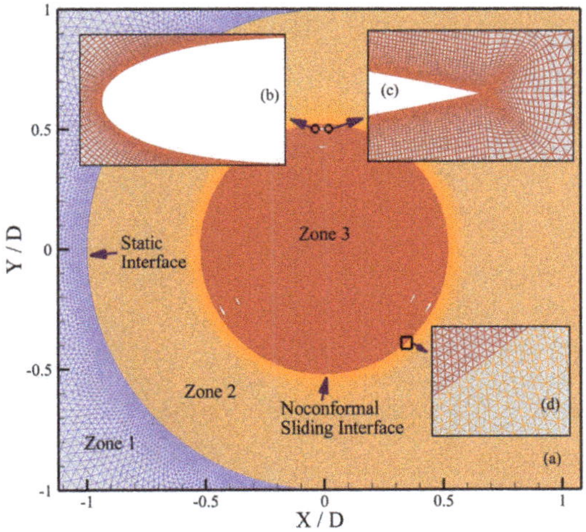

Figure 3. Mesh settings in difference zones of the flow domain and near the blades.

2.4. Flow-Induced Rotation of Turbines

The simulations for flow-induced rotation of single-stage and dual-stage VAWTs provide insights to their self-starting with different geometric configurations. In such cases, turbines are allowed to rotate passively as a result of flow-induced forces and moments on blades instead of prescribing the angular velocity of rotors. For this purpose, the six degrees-of-freedom (6DoF) solver in ANSYS Fluent [32] was employed to solve the following equation for dynamics of the rotational structure.

$$I\ddot{\theta} = \sum M_{\circ} \tag{2}$$

where I represents the moment of inertia of the VAWT, $\ddot{\theta}$ denotes its angular acceleration, and $M_{\circ}$ is the total moment produced by the blades of the turbines around the central axis, which makes it equal to τ. Next, the angular velocity ($\Omega = \dot{\theta}$) of the turbine is computed by numerically integrating Equation (2) through a trapezoidal rule. It is important to mention that coupled rotation of the two rotors for dual-stage turbines is considered in the present study, where both outer and inner rotors rotate have the same angular velocity. Here, the summation of aerodynamic moments represents the total torque induced by the wind on rotors. The mass of each blade in the outer and inner rotors is considered as 0.06 kg and 0.051 kg, respectively. Their respective moments-of-inertia are then computed as:

$$\begin{aligned} I_{outer} &= 3m_1{R_1}^2 \\ I_{inner} &= 3m_2{R_2}^2 \end{aligned} \tag{3}$$

where m_1 and m_2 show masses of each blade in the outer and inner stages, respectively. Hence, the single- and dual-stage turbines have $I = 0.045$ kgm^2 and 0.0726 kgm^2, respectively. It is important to note that the values of masses for blades in these turbines are chosen to keep moments-of-inertia in the range of those provided by others in the literature [6].

2.5. Key Performance Parameters

The following parameters are generally used to describe the flow as well as the geometric and kinematic features of wind turbines. First is the tip-speed ratio defined as:

$$\lambda = \frac{\Omega R_1}{U_\infty} \tag{4}$$

where R_1 is the radius of the outer rotor of a turbine, and U_∞ represents the free-stream flow velocity.

In order to measure the aerodynamic performance of wind turbines, non-dimensional torque (moment) and power coefficients, denoted as C_m and C_P, respectively, are computed for all cases, which are defined as:

$$C_m = \frac{\tau}{qR_1A}, \quad C_P = \frac{\tau\Omega}{qU_\infty A} \tag{5}$$

where τ denotes torque of the VAWT, A is the swept area of the turbine, and $q = \rho U^2{}_\infty/2$ is the dynamic pressure. The swept area is a factor computed through multiplying the turbine height (1 m for the current 2D cases) with its outer diameter.

2.6. Validation and Verification

The accuracy of computational results depends on mesh topology, which also has a strong impact on boundary layer flows. Hence, we performed detailed grid convergence tests to ensure that the results were not significantly impacted by an increase in mesh sizes. For this purpose, the computational domain around a dual-stage VAWT was decomposed in non-homogeneous grids with unstructured triangular cells in the fluid domain and 26 layers of quadrilateral elements around each blade to accurately resolve boundary layers. We controlled the grid size by changing the maximum sizes of the grid in different zones while keeping y^+ value of the order of 1. This parameter helps compute the height of the first cell from the blade's surface. Its accuracy plays a critical part in resolving the viscous sublayer in turbulent boundary layers. Here, three mesh sizes were considered for a constant tip-speed-ratio of $\lambda = 4.0$, the details of which are provided in Table 2. In these simulations, we set the time-step size according to a change of 0.2° in the azimuthal angle per time step. The maximum cell size for the outermost static domain, identified as Zone 1 in Figure 3 is 0.5 for all the cases considered here, because the far-wake region does not influence the dynamics of turbine significantly.

Table 2. Details of mesh topology in different zones for grid-independence tests, where the sizes are nondimensionalized by the outer diameter ($2R1$) of the VAWTs

Details of Grid	G1 Coarse	G2 Medium	G3 Fine
Mesh Nodes on Each Blade	400	400	400
Maximum Size in Zone 2	0.02	0.015	0.01
Maximum Size in Zone 3	0.002	0.00135	0.001
Total Number of Cells	271,223	510,530	1,199,418

Figure 4 presents profiles of moment coefficient of one outer blade in a dual-stage rotor, undergoing rotation with $\lambda = 4.0$. It is apparent that all three grids produce the same C_m, and small differences arose only for $240° < \theta < 350°$. Because the grid configuration G_2 matches more closely to G_3, the remaining simulations were run with the mesh settings of G_2. In order to further demonstrate the effectiveness of our grid-convergence study, we employed another parameter called the determination coefficient (R^2) and used by [17,41] to quantify variations in the final solution for different grids. The expression is,

$$R^2 = 1 - \frac{\sum_{\theta=0°}^{360°}[C_m(\theta) - C_{m,final}(\theta)]^2}{\sum_{\theta=0°}^{360°}[C_m(\theta) - C_{m,average}]^2} \tag{6}$$

where $C_m(\theta)$ and $C_{m,average}$ are the instantaneous and cycle-averaged moment coefficients for the grid under consideration, In addition, $C_{m,final}(\theta)$ is the moment coefficient for the finest mesh. This determination coefficient takes values of 0.9829, 0.9844, and 1 for the coarse, medium, and finest grids, respectively, in the present study.

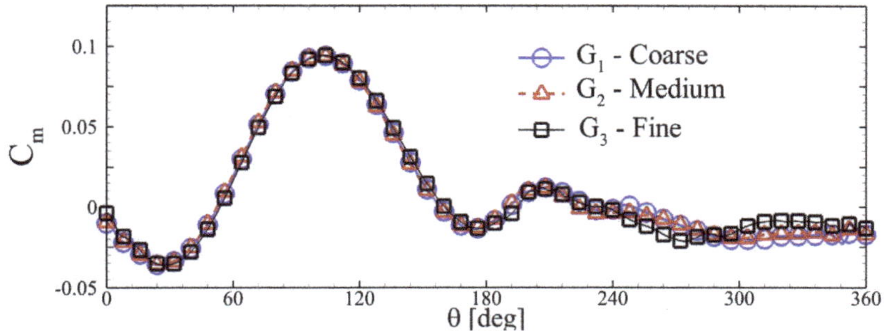

Figure 4. C_m of a single blade in a dual-stage turbine ($\phi = 0°$) rotating with $\lambda = 4.0$ for different grid configurations.

The rotation of blades in close vicinity could influence flow separation from their surfaces. There are two important aspects for modeling such phenomena: (i) fine resolution of mesh in the vicinity of the rotating blades so that unsteadiness and important length scales of flow features may be well captured and (ii) accurate modeling of turbulent flows around the blades, which may be subjected to transitional effects. Figure 3 shows the fine resolution of mesh in the regions between and around the blades. As demonstrated by [40], SST-based models provide the best agreements with experimental results for flow features around blades in VAWTs. These features include dynamic stall along with the related growth, shedding, and traversing of vortices in the wake. The four-equation transition SST model proves to be a superior choice to predict circulation and timings of vortex shedding process from the rotating blades.

We validate our computational setup using previously published experimental [42] and numerical studies [15,43]. The specifications of the VAWT for the validation study are: NACA0018 represents the cross section of a blade with chord-length of 0.083 m, the radius of the rotor in this case is 0.375 m, moment-of-inertia of the turbine is 0.018 kgm^2, and the wind speed is 6 m/s. Figure 5 shows variations in tip-speed ratios of a vertical axis wind turbine, where the present simulation methodology performs better in comparison to the previously reported numerical investigations with respect to experiments of Rainbird [42]. Here, T denotes the time when the turbine attains its steady-state tip-speed ratio. When the turbine begins its rotation, it undergoes a slow acceleration stage up to $t/T = 0.7$, after which a rapid increase in its rotational velocity is observed.

Rainbird [42] reported the steady-state tip-speed ratio of ~3.40. Asr et al. [43] and Sun et al. [15] found this value to be 4.3 and 3.63, respectively. The present methodology not only predicts the rotation rate during the acceleration stages with comparative accuracy, but also gives the final value of the tip-speed ratio as 3.70, which is close to the one determined experimentally. It is important to mention that Asr et al. [43] and Sun et al. [15] employed first-order implicit time-marching numerical schemes in their studies, which could cause overprediction of performance parameters for the VAWT. A plausible reason for more accurately capturing the temporal variations in the angular velocity in our simulation is the utilization of the second-order implicit algorithm to march in time. Using the sliding-mesh technique may also help achieve better results because dynamic meshing techniques introduce additional diffusion and constant variations in the grid quality. The reason for a slight overprediction of the steady-state λ is the two-dimensional modeling of this VAWT. The span of each blade in the experiments of Rainbird [42] was 0.60 m. This feature introduces span-wise flow over a blade and consequent reduction in aerodynamic torque. Nevertheless, our two-dimensional simulations capture the relevant quantitative and qualitative details with reasonable accuracy. The time-step size (dt) used for these simulations is 0.0005 s, which is sufficiently small to capture highly unsteady flow characteristics around the turbines, and it is also consistent with the recommendations of [43].

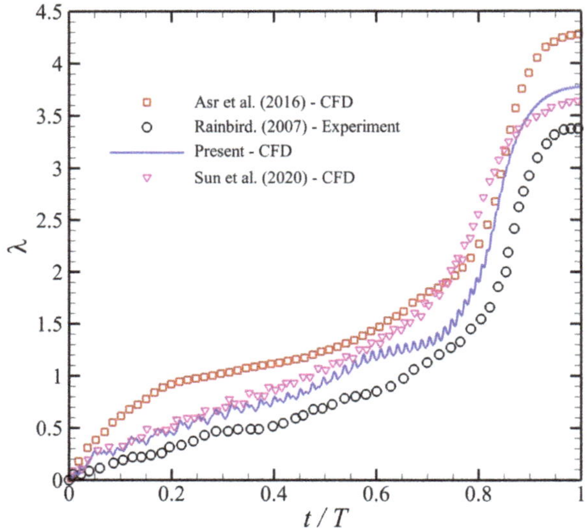

Figure 5. Results to validate the present computational methodology using experimental and numerical results from literature [15,42,43].

3. Results and Discussion

In this section, we discuss effects of the wind velocity on self-starting characteristics and passive rotation of VAWTs. In this context, the performance of single-stage and dual-stage VAWTs are compared in detail. To explore the dynamic response of multi-stage turbines, the phase angle between inner and outer rotors is varied from 0° to 90°.

There are different perspectives within the wind turbine community on the definition of the self-starting process of VAWTs [7,10,44], as well as how best to illustrate it. Generally, a turbine is considered as self-starting if it can accelerate and attain a steady-state angular velocity in response to the aerodynamic forces and moments on its own without any external excitation [8,15]. Its ability to self-start can be determined through the time taken to reach a steady-state rotational speed. Figure 6 shows variations in λ versus time, non-dimensionalized by the factor U_∞/R_1 for single- and dual-stage turbines. For the lowest wind speed, i.e., $U_\infty = 4$ m/s in Figure 6a, the dual-stage turbines are observed to reach steady-state λ more quickly compared to their single-stage counterpart. Here, time is nondimensionalized by U_∞/R_1, and it is denoted with the symbol $t^\star$. The single-stage VAWT experiences slow acceleration until $t^\star = 80$ from which point onward it rapidly increases its rotation rate and attains the steady-state condition of $\lambda = 5.06$ at $t^\star = 114$. It is interesting to notice that the dual-stage VAWT with $\phi = 0°$ cannot attain high λ even after rotating for a longer period of time. Its maximum λ remains at 0.63, which is achieved at $t^\star = 100$. Its dynamic response seems to become locked in a state, which is unrecoverable, and the underlying reasons are unclear. This kind of behavior was previously observed by Liu et al. [44] for a vertical axis tidal turbine, and they termed it as the "*unstable equilibrium mode*". For $\phi = 30°$ and 90°, the dual-stage turbine accelerates more quickly compared to the single-rotor VAWT. The dual-stage versions, except for the case of $\phi = 0°$, attain steady-state $\lambda = 3.83$. However, the turbine with $\phi = 30°$ reaches its steady-state condition faster than the other cases. Hence, dual-stage turbines outperform the one with a single rotor at very low wind speeds in terms of better self-starting characteristics.

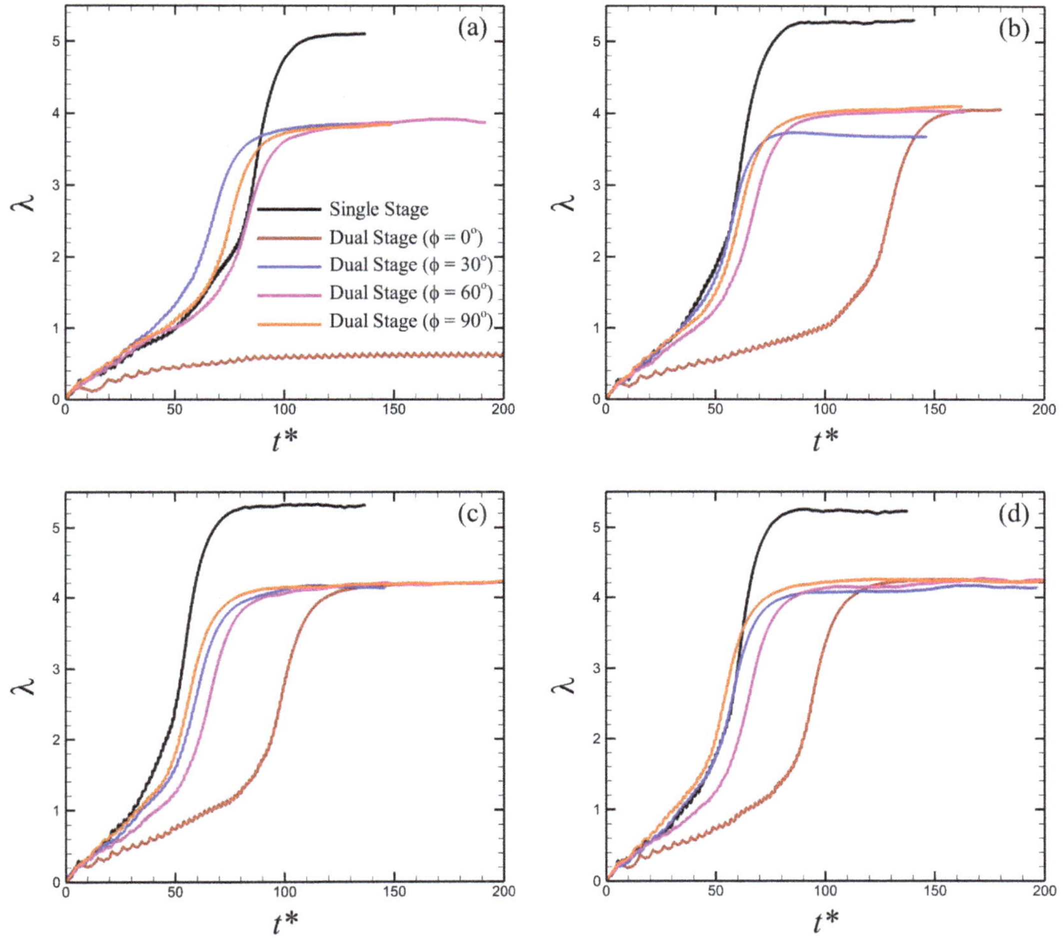

Figure 6. Temporal variations in tip-speed ratios (λ versus nondimensional time t^*) of single- and dual-stage turbines for U_∞ = (**a**) 4.0; (**b**) 6.0; (**c**) 8.0; and (**d**) 10.0 m/s.

For the case of a higher wind speed at $U_\infty = 6.0$ m/s, the single-stage VAWT accelerates very rapidly and achieves $\lambda = 5.26$ at $t^\star = 88.6$. Performance of the dual-rotor VAWT with $\phi = 0°$ remains the lowest of all configurations. Although it attains the same steady-state condition with $\lambda = 4.04$, as was observed for turbines with $\phi = 60°$ and 90°, it takes more time. The configuration with $\phi = 30°$ is the fastest amongst the configurations considered here; however, it achieves the maximum λ of 3.7. As shown in Figure 6c, the self-starting capability of the single-stage VAWT is superior to that of other dual-stage turbines at $U_\infty = 8.0$ m/s. The VAWTs with two rotors spend more time to reach their steady-state with $\lambda = 4.15$, whereas those with $\phi = 30°$, 60°, and 90° take $t^\star = 107$, 115, and 95 to reach steady-state, respectively. Looking at the trends of λ for $U_\infty = 10$ m/s in Figure 6d, it can be assessed that $\phi = 30°$ or 90° is advantageous for dual-stage VAWTs. In addition, a reasonable argument for the rapidly accelerating single-stage turbine at higher wind speed is related to its lower moment-of-inertia. It is evident that the multi-rotor VAWTs with phase angles between 30° and 90° offer more promising self-starting characteristics even with 38% greater inertia. Simultaneously, configurations with $\phi = 0°$ should be avoided.

In order to have more insight into the aerodynamic performances of these turbines, the total torque experienced by the single-stage turbine and four different configurations of dual-rotor VAWTs are plotted in Figure 7. Here, all turbines initially experience small torque with more intense random fluctuations. Nevertheless, all the cases exhibit the production of positive time-averaged τ, which enhances acceleration of turbines. As time progresses, these rotating structures, except the one with $\phi = 0°$, experience sharp increments in their torque profiles, which we call the "overshoot" phenomenon. After passing through this stage, the single-stage turbine and dual-rotor VAWTs with $\phi = 30°, 60°$, and $90°$ reach their periodic steady-state response in terms of aerodynamic torque. After attaining their steady-state operations stages, all turbines produce zero time-averaged torque, which reflects the attainment of steady angular speed. Although the dual-stage VAWT with $\phi = 0°$ does not experience an "overshoot", it exhibits periodic oscillations in its torque profile as shown in Figure 7b. It also elucidates the reason for the VAWT not undergoing any rapid acceleration and maintaining a very low steady-state value of λ.

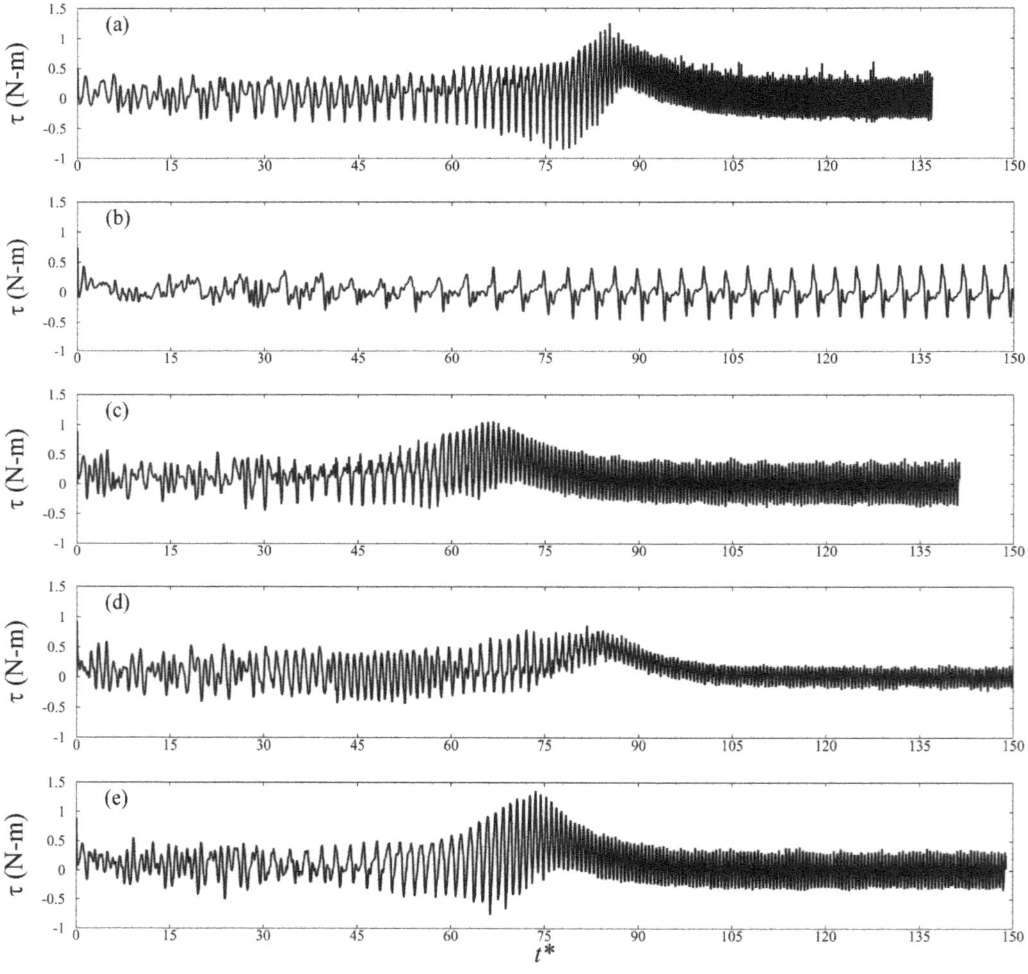

Figure 7. Variations in τ for the (**a**) single-stage turbine and dual-stage turbines with (**b**) $\phi = 0°$; (**c**) $\phi = 30°$; (**d**) $\phi = 60°$; and (**e**) $\phi = 90°$ at $U_\infty = 4$ m/s.

The quantifiable aerodynamic performance of turbines depends on the flow physics in their vicinity and its interaction with blades. Instantaneous locations of blades determine their effective angles-of-attack (α_{eff}), which can be defined as:

$$\alpha_{eff} = \tan^{-1}\frac{\sin\theta}{\cos\theta + \lambda} \tag{7}$$

Figure 8 demonstrates how α_{eff} varies as a function of the azimuthal position of blades, denoted as θ. Figure 8a explains why a dual-rotor VAWT with $\phi = 0°$ does not perform well at $U_\infty = 4$ m/s. This configuration experiences very large effective angles-of-attack. The remaining VAWTs show identical sinusoidal profiles for their angles-of-attack.

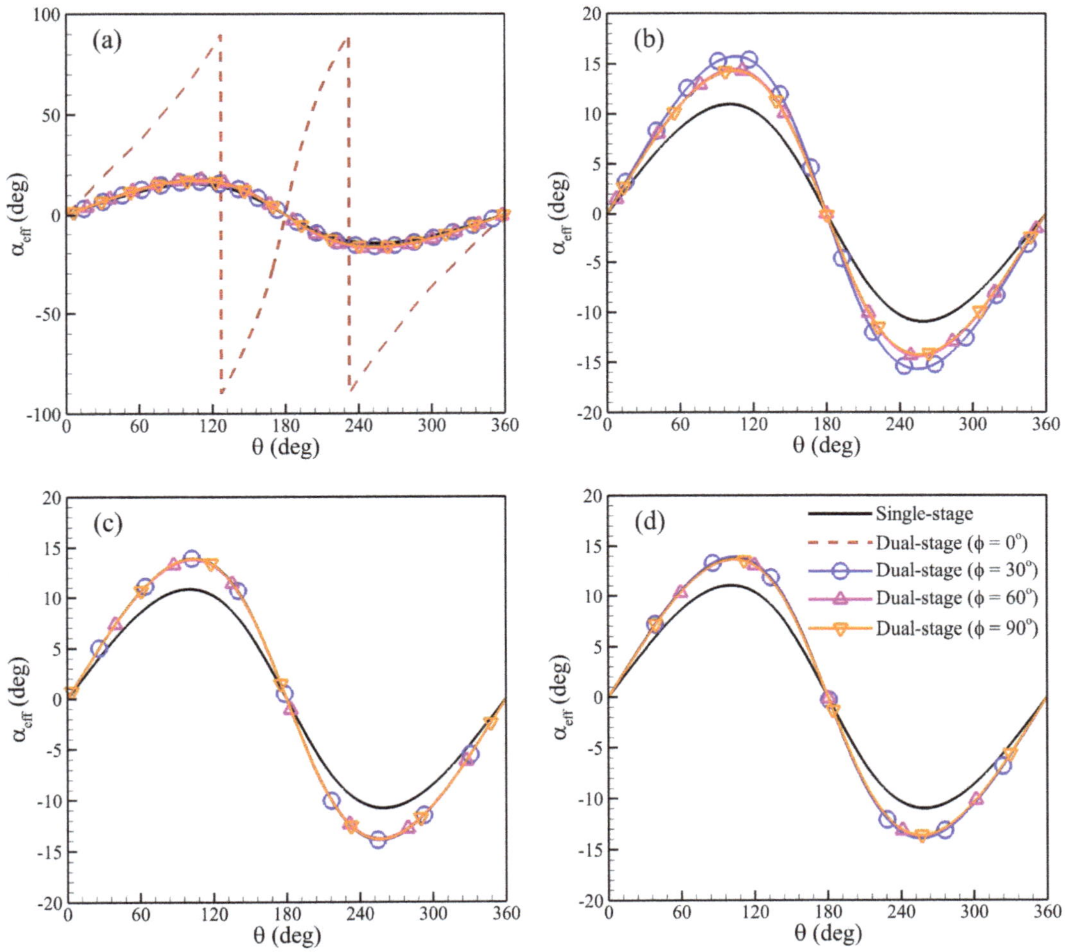

Figure 8. Variations in α_{eff} as the function of the azimuthal position (θ) of blades in the outer rotor for U_∞ = (**a**) 4 m/s; (**b**) 6 m/s; (**c**) 8 m/s; and (**d**) 10 m/s.

Comparing this information with unsteady aerodynamic forces presented in Figure 9a1,a2, the large effective angles-of-attack (α_{eff}) are responsible for the production of only negative lift and lower drag over the blades of a dual-stage turbine with $\phi = 0°$. It is also noteworthy that the dual-stage VAWT with $\phi = 90°$ produces the maximum amplitudes for lift and drag at very low wind speeds. For all the remaining wind speeds in Figure 9b–d, α_{eff} remains lower than 16°. Under these flow conditions at large Reynolds numbers, dynamic stall may occur at relatively lower angles-of-attack, usually in the range of 12°–17° for NACA-0012 and NACA-0018 airfoils [45,46]. We also notice in Figure 9 that blades of the dual-stage VAWT with $\phi = 60°$ produce smaller amplitudes of unsteady lift and drag forces, which further impedes the production of greater torque.

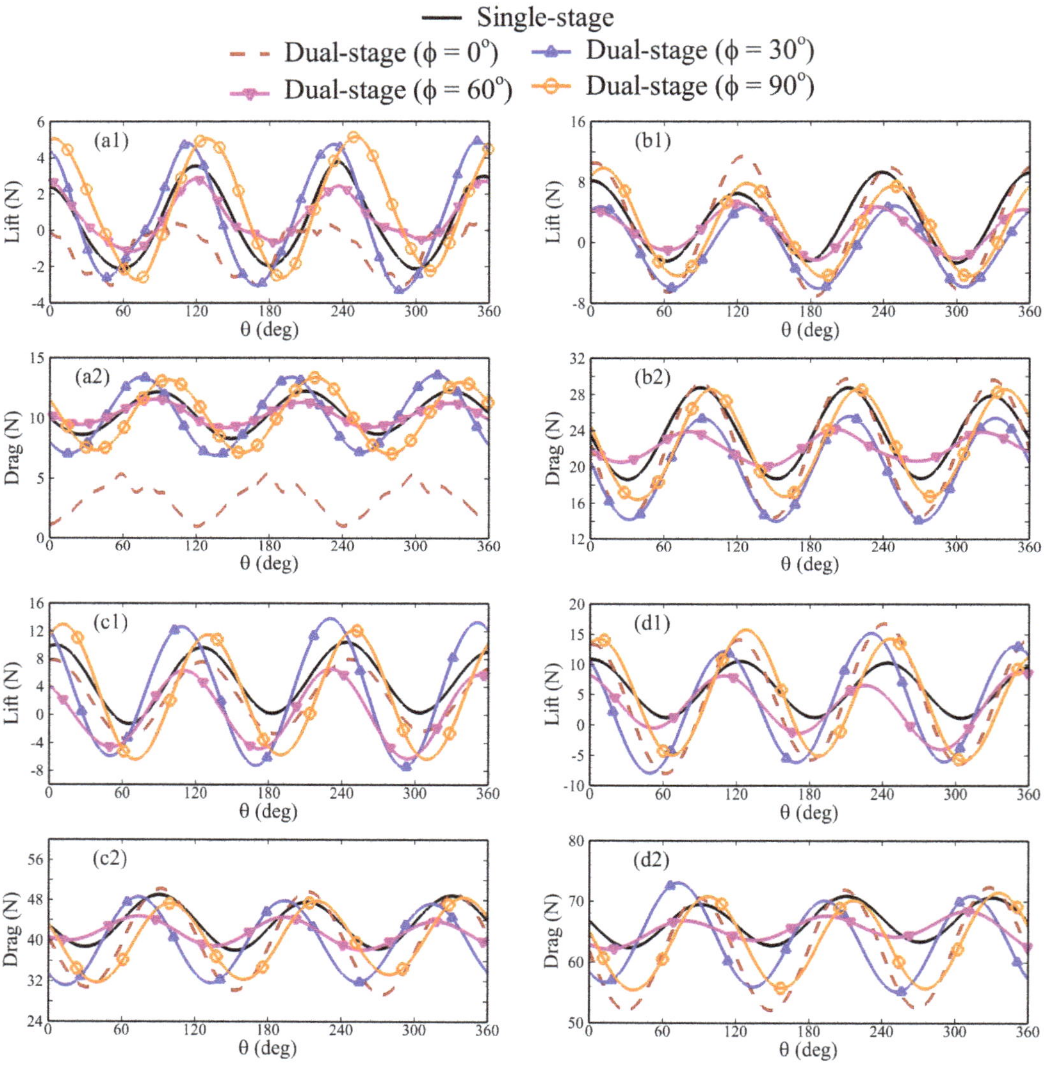

Figure 9. Variations in aerodynamic lift and drag forces versus azimuth angle (θ) for (**a1**,**a2**) $U_\infty = 4$ m/s (**b1**,**b2**) $U_\infty = 6$ m/s; (**c1**,**c2**) $U_\infty = 8$ m/s; and (**d1**,**d2**) $U_\infty = 10$ m/s.

At the end of the Introduction, it was stated that determining the power generation by dual-stage turbines should be performed only after establishing their self-starting characteristics. If they do not start faster, we believe there is no point in pursuing multi-stage rotors for VAWTs. The 6DoF starting calculations do not provide the extracted power as they only consider the flow-induced motion of rotors, whereas computations of power extraction require different solver settings to apply the equivalent of a generator load, as explained by [44]. It is important to note that other simulations, such as performed by [6], using 6DoF solvers assumed that the instantaneous torque and omega provided an equivalent metric to the steady values of power extraction, which is fundamentally wrong.

This measure may only provide the rate of work performed by the passing fluid on the blades for their rotation.

Next, we elucidate the starting process and the underlying governing flow physics for the VAWTs. When a single-stage VAWT begins rotating at $U_\infty = 4$ m/s, the blade on the windward side ($\theta = 0^\circ$) of the flow domain produces two shear layers of opposite vorticity in the first revolution. Figure 10a illustrates this scenario. These shear layers are rolled into vortices when the blade is at $\theta > 45^\circ$. Because the other two blades experience larger angles-of-attack at this instant, they produce large strong vortices traversing in the downstream direction. During initial revolutions of the turbine, blades start capturing vortices shed by other blades when they pass through the windward side. These interactions between blades and coherent structures cause intermittent vortex dynamics in the wake as shown in Figure 10b. Primary flow activity remains confined towards the middle of the flow domain. These flow patterns explain the production of random-looking fluctuations in temporal profiles of unsteady forces and torque. At this stage, λ remains below 0.50, while there are oscillations of low amplitudes in plots of λ (see Figure 6a). With increasing acceleration at $t^\star = 32$, the blades passing through the windward side produce distinct vortices for $\theta \sim 80^\circ$ (see Figure 10c). It allows the blades to maintain their shear layers for a longer circumferential distance which they begin to form at $\theta = 0^\circ$. Each following blade interacts with these shear layers, and more coherent flow structures traverse along the windward side of the wake. The vortices travelling on the leeward end of the turbines are shed from the blades passing through $180^\circ < \theta < 270^\circ$. Nevertheless, this flow activity expanded over the whole plane of the flow field as the turbine picks up more speed. Intense interactions of each blade with vortices and shear layers produced by the other two blades are observed in Figure 10d. We also notice the mitigation of randomness in the fluctuations of unsteady loads on blades with an increase in their amplitudes.

For $t^\star > 80$, the turbine undergoes rapid acceleration, and we see an overshoot in the torque profile (see Figure 7a). Now, the wake is clearly bounded by two vortex streets with negative and positive vortices traversing on the windward and leeward sides, respectively, as presented in Figure 10e. Vortices in the middle are not strong with smaller length scales and quickly diffuse, which means that convection of secondary coherent flow structures is suppressed by the turbine in order to reach its steady-state. The primary reason for the disappearance of flow activity from the middle of the wake is the entrapment of vortices between the rotating blades with high angular velocities on the windward side. As the turbine approaches its steady-state rotation, blades stop producing distinct vortices during their entire rotation cycle. Thus, distinct shear layers are formed and maintained, which is due to the blades experiencing angles-of-attack lower than their stall angles such that they do not undergo dynamic stall in this operational phase with increasing speed of the turbine. Stabilization of these shear layers then causes the production of helical vortices in three-dimensional flows around rotating machines [47]. In this stage of their flow-induced rotation, the shear layers of blades are only destabilized when the blades pass through the regions of entrapped vorticity on windward and leeward sides. Due to this phenomenon, more distinct coherent structures are shed on the windward side and an elongated shear layer of positive vorticity is seen to traverse in the downstream direction on the leeward side. It means that entrapment of vorticity by the rotating blades plays an important role in setting up flow dynamics in the wake to a steady-state and help the turbine attain its final λ. Although the wake is clearly asymmetric in this state, it resembles a classical von Karman vortex street behind an oscillating/rotating cylinder [48,49]. We observe similar flow phenomena for the whole range of U_∞ considered in this study. However, three important phenomena occur with increasing U_∞: (1) the region of entrapment for vortices expands between the rotating blades, (2) the vortices shed by the turbines on windward and leeward sides become stronger and bigger in size, and (3) the turbines are able to attain their steady-state operational phase in less time.

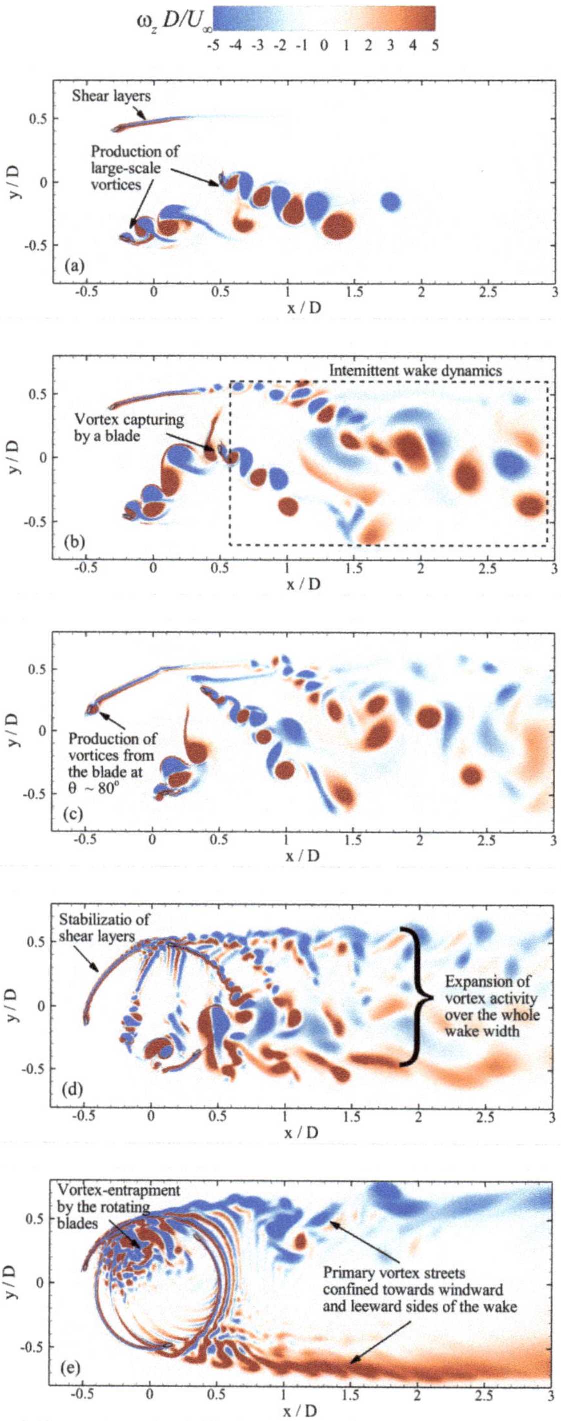

Figure 10. Vorticity contours in the flow field of a single-stage turbine at $U_\infty = 4$ m/s at (**a**) $t^\star = 6.1$; (**b**) $t^\star = 24$; (**c**) $t^\star = 32$; (**d**) $t^\star = 80.8$; and (**e**) $t^\star = 104.8$.

Similar flow physics also hold for dual-stage turbines. The readers are encouraged to watch the flow animations provided with this paper, which elucidate vorticity dynamics around the single-stage and dual-stage VAWTs for $U_\infty = 6$ m/s. Now, we describe the governing flow mechanism for the dual-stage turbine with a zero phase angle between its two stages, which appears as an exception from the abovementioned reasoning. Figure 11 explains the governing flow physics of this dual-stage VAWT. It is evident that shear layers formed in the wake of the rotating outer blades at $\theta = 0°$ are destablized due to their continuous interactions with the closely rotating inner blades. This phenomenon does not allow the blades to delay the stall process over them even after a number of rotations. Consequently, vortices are continuously formed and shed into the wake. These vortices cannot be trapped inside the rotating turbine and keep traversing in the wake. Moreover, the proximity of the inner blades and its impact on the destabilization of the shear layers could be one of the main contributing factors for the poorest starting performance of dual-stage VAWTs with $\phi = 0°$ as presented in Figure 6. Another important point here is about the dual-rotor turbines with $\phi = 60°$, reaching their steady-state rotational speeds little later than those with $\phi = 30°$ and $90°$. A reasonable argument for this behavior is the stabilization of the shear layers around the rotors before their shedding in the wake. The blades arranged in tandem configurations perform better for self-starting. Their immediate presence behind their counterparts helps the turbines attain that state of stabilization. On this ground, an arrangement with $\phi = 60°$ seems to be less advantageous compared to $\phi = 30°$ and $90°$.

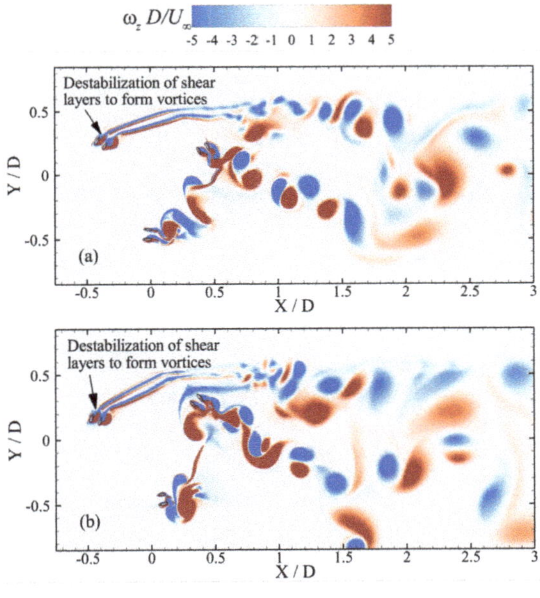

Figure 11. Vorticity contours in the flow field of a dual-stage turbine with $\phi = 0°$ at $U_\infty = 4$ m/s at (**a**) $t^\star = 47.7$ and (**b**) $t^\star = 122.5$.

4. Conclusions

In this study, numerical simulations are carried out to examine the self-starting characteristics of dual-rotor turbines in comparison to those with a single rotor. We find that, for very low wind speed of 4 m/s, dual-stage turbines with $\phi = 30°$ and $90°$ significantly reduce the response time to attain a steady-state tip-speed ratio, which demonstrates improvements in their self-starting process. Having $\phi = 0°$ in double-rotor turbines is not recommended because they take longer time to undergo rapid acceleration stage and reach steady-state. It experiences very large angles-of-attack during its rotation, which locks it in a dead band. Moreover, the dual-stage VAWT with $\phi = 90°$ is the best option to perform well in all wind conditions. The following three physical mechanisms play a critical role to help a turbine attain a steady angular speed: (i) stabilization of the two shear layers comprising the wake of blades passing through the windward side of a turbine, (ii) vortex-entrapment due to rotating blades and shear layers attached to them, and (iii) suppression of secondary vortex structures causing intermittent flow dynamics in the middle section of the wake of a turbine. It is important to highlight that a constant ratio of radii for the two stages are considered in our present study. Nevertheless, how variations in this important geometric parameter influence the aerodynamic performance and power generation capacity of multi-stage turbines should be the subject of a future research study.

It is also important to mention that this work ignores any load due to a generator or other drivetrain components with these VAWTs which may cause the turbine to experience large resistive moments. Their effects on the performance of dual-stage turbines are still unknown and need to be further investigated. The impact of other geometric features of blades, such as camber or chord-to-diameter ratio, should also be explored and analyzed to mitigate limitation of vertical axis wind turbines. In order to better understand the flow-induced motion of turbines and their reliance on complex vortical flow structures, three-dimensional simulations for flows around VAWTs with the addition of geometric features, such as connecting rods between the hub and blades etc., will also be carried out and reported in near future.

Author Contributions: Conceptualization, M.S.U.K. and A.H.; methodology, M.S.U.K.; software, M.S.U.K.; validation, M.S.U.K.; formal analysis, M.S.U.K., A.H. and D.W.; investigation, M.S.U.K.; writing—original draft preparation, M.S.U.K. and A.H.; writing—review and editing, M.S.U.K., A.H. and D.W.; visualization, M.S.U.K. and A.H.; funding acquisition, M.S.U.K. All authors have read and agreed to the published version of the manuscript.

Funding: This research received no external funding.

Data Availability Statement: Not applicable.

Acknowledgments: M.S.U. Khalid acknowledges the support from Natural Sciences and Engineering Research Council, Canada, through Discovery Grant.

Conflicts of Interest: The authors declare no conflict of interest.

References

1. Kirke, B.K. Evaluation of Self-Starting Vertical Axis Wind Turbines for Stand-Alone Applications. Ph.D. Thesis, Griffith University Gold Coast, Southport, Australia, 1998.
2. Lunt, P. *An Aerodynamic Model for a Vertical-Axis Wind Turbine*; MEng Project Report; School of Engineering, University of Durham: Durham, UK, 2005.
3. Takao, M.; Kuma, H.; Maeda, T.; Kamada, Y.; Oki, M.; Minoda, A. A straight-bladed vertical axis wind turbine with a directed guide vane row-Effect of guide vane geometry on the performance-. *J. Therm. Sci.* **2009**, *18*, 54–57. [CrossRef]
4. Douak, M.; Aouachria, Z.; Rabehi, R.; Allam, N. Wind energy systems: Analysis of the self-starting physics of vertical axis wind turbine. *Renew. Sustain. Energy Rev.* **2018**, *81*, 1602–1610. [CrossRef]
5. Ebert, P.; Wood, D. Observations of the starting behaviour of a small horizontal-axis wind turbine. *Renew. Energy* **1997**, *12*, 245–257. [CrossRef]
6. Celik, Y.; Ma, L.; Ingham, D.; Pourkashanian, M. Aerodynamic investigation of the start-up process of H-type vertical axis wind turbines using CFD. *J. Wind. Eng. Ind. Aerodyn.* **2020**, *204*, 104252. [CrossRef]
7. Dominy, R.; Lunt, P.; Bickerdyke, A.; Dominy, J. Self-starting capability of a Darrieus turbine. *Proc. Inst. Mech. Eng. Part A J. Power Energy* **2007**, *221*, 111–120. [CrossRef]
8. Bianchini, A.; Ferrari, L.; Magnani, S. Start-up behavior of a three-bladed H-Darrieus VAWT: Experimental and numerical analysis. *Turbo Expo Power Land Sea Air* **2011**, *54617*, 811–820.
9. Bianchini, A.; Carnevale, E.A.; Ferrari, L. A model to account for the Virtual Camber Effect in the Performance Prediction of an H-Darrieus VAWT Using the Momentum Models. *Wind. Eng.* **2011**, *35*, 465–482. [CrossRef]
10. Worasinchai, S.; Ingram, G.L.; Dominy, R.G. The physics of H-Darrieus turbine starting behavior. *J. Eng. Gas Turbines Power* **2016**, *138*, 062605. [CrossRef]
11. Batista, N.; Melício, R.; Mendes, V.; Calderón, M.; Ramiro, A. On a self-start Darrieus wind turbine: Blade design and field tests. *Renew. Sustain. Energy Rev.* **2015**, *52*, 508–522. [CrossRef]

12. Singh, M.; Biswas, A.; Misra, R. Investigation of self-starting and high rotor solidity on the performance of a three S1210 blade H-type Darrieus rotor. *Renew. Energy* **2015**, *76*, 381–387. [CrossRef]
13. Sengupta, A.; Biswas, A.; Gupta, R. Studies of some high solidity symmetrical and unsymmetrical blade H-Darrieus rotors with respect to starting characteristics, dynamic performances and flow physics in low wind streams. *Renew. Energy* **2016**, *93*, 536–547. [CrossRef]
14. Du, L.; Ingram, G.; Dominy, R.G. Experimental study of the effects of turbine solidity, blade profile, pitch angle, surface roughness, and aspect ratio on the H-Darrieus wind turbine self-starting and overall performance. *Energy Sci. Eng.* **2019**, *7*, 2421–2436. [CrossRef]
15. Sun, X.; Zhu, J.; Hanif, A.; Li, Z.; Sun, G. Effects of blade shape and its corresponding moment of inertia on self-starting and power extraction performance of the novel bowl-shaped floating straight-bladed vertical axis wind turbine. *Sustain. Energy Technol. Assess.* **2020**, *38*, 100648. [CrossRef]
16. Sun, X.; Zhu, J.; Li, Z.; Sun, G. Rotation improvement of vertical axis wind turbine by offsetting pitching angles and changing blade numbers. *Energy* **2021**, *215*, 119177. [CrossRef]
17. Mohamed, O.S.; Elbaz, A.M.; Bianchini, A. A better insight on physics involved in the self-starting of a straight-blade Darrieus wind turbine by means of two-dimensional Computational Fluid Dynamics. *J. Wind. Eng. Ind. Aerodyn.* **2021**, *218*, 104793. [CrossRef]
18. Didane, D.H.; Rosly, N.; Zulkafli, M.F.; Shamsudin, S.S. Performance evaluation of a novel vertical axis wind turbine with coaxial contra-rotating concept. *Renew. Energy* **2018**, *115*, 353–361. [CrossRef]
19. Didane, D.H.; Rosly, N.; Zulkafli, M.F.; Shamsudin, S.S. Numerical investigation of a novel contra-rotating vertical axis wind turbine. *Sustain. Energy Technol. Assess.* **2019**, *31*, 43–53. [CrossRef]
20. ZSB-AB. Doppelter Darrieus-rotor. DE202011002702U, May 2011.
21. Torabi Asr, M.; Osloob, R.; Mustapha, F. Double-Stage H-Darrieus Wind Turbine-Rotor Aerodynamics. In *Applied Mechanics and Materials*; Trans Tech Publications Ltd.: Wollerau, Switzerland, 2016; Volume 829, pp. 21–26.
22. Scungio, M.; Arpino, F.; Focanti, V.; Profili, M.; Rotondi, M. Wind tunnel testing of scaled models of a newly developed Darrieus-style vertical axis wind turbine with auxiliary straight blades. *Energy Convers. Manag.* **2016**, *130*, 60–70. [CrossRef]
23. Arpino, F.; Cortellessa, G.; Dell'Isola, M.; Scungio, M.; Focanti, V.; Profili, M.; Rotondi, M. CFD simulations of power coefficients for an innovative Darrieus style vertical axis wind turbine with auxiliary straight blades. *J. Phys. Conf. Ser.* **2017**, *923*, 012036. [CrossRef]
24. Arpino, F.; Scungio, M.; Cortellessa, G. Numerical performance assessment of an innovative Darrieus-style vertical axis wind turbine with auxiliary straight blades. *Energy Convers. Manag.* **2018**, *171*, 769–777. [CrossRef]
25. Arpino, F.; Cortellessa, G.; Massarotti, N.; Mauro, A.; Scungio, M. Numerical performance assessment of a novel Darrieus-style VAWT with auxiliary straight blades. *J. Phys. Conf. Ser.* **2020**, *1589*, 012020. [CrossRef]
26. Su, H.; Dou, B.; Qu, T.; Zeng, P.; Lei, L. Experimental investigation of a novel vertical axis wind turbine with pitching and self-starting function. *Energy Convers. Manag.* **2020**, *217*, 113012. [CrossRef]
27. Malael, I.; Dragan, V. Numerical and Experimental Efficiency Evaluation of a Counter-Rotating Vertical Axis Wind Turbine. *Eng. Technol. Appl. Sci. Res.* **2018**, *8*, 3282–3286. [CrossRef]
28. Tahani, M.; Razavi, M.; Mirhosseini, M.; Razi Astaraei, F. Unsteady aerodynamic performance of Dual-Row H-Darrieus vertical axis wind turbine. *Energy Equip. Syst.* **2020**, *8*, 55–80.
29. Chen, J.; Liu, P.; Xu, H.; Chen, L.; Yang, M.; Yang, L. A detailed investigation of a novel vertical axis Darrieus wind rotor with two sets of blades. *J. Renew. Sustain. Energy* **2017**, *9*, 013307. [CrossRef]
30. Hashem, I.; Zhu, B. Metamodeling-based parametric optimization of a bio-inspired Savonius-type hydrokinetic turbine. *Renew. Energy* **2021**, *180*, 560–576. [CrossRef]
31. Islam, M.R.; Mekhilef, S.; Saidur, R. Progress and recent trends of wind energy technology. *Renew. Sustain. Energy Rev.* **2013**, *21*, 456–468. [CrossRef]
32. Manual. ANSYS FLUENT 2020R2. In *Theory Guide*; ANSYS, Inc.: Canonsburg, PA, USA, 2020.
33. Mohamed, M.H. Aero-acoustics noise evaluation of H-rotor Darrieus wind turbines. *Energy* **2014**, *65*, 596–604. [CrossRef]
34. Siddiqui, M.S.; Durrani, N.; Akhtar, I. Quantification of the effects of geometric approximations on the performance of a vertical axis wind turbine. *Renew. Energy* **2015**, *74*, 661–670. [CrossRef]
35. Ghasemian, M.; Nejat, A. Aero-acoustics prediction of a vertical axis wind turbine using Large Eddy Simulation and acoustic analogy. *Energy* **2015**, *88*, 711–717. [CrossRef]
36. Mohamed, M.H. Reduction of the generated aero-acoustic noise of a vertical axis wind turbine using CFD (Computational Fluid Dynamics) techniques. *Energy* **2016**, *96*, 531–544. [CrossRef]
37. The, J.; Yu, H. A critical review on the simulations of wind turbine aerodynamics focusin on hybrid RANS-LES methods. *Energy* **2017**, *138*, 257–289. [CrossRef]
38. Balduzzi, F.; Bianchini, A.; Maleci, R.; Ferrara, G.; Ferrari, L. Critical issues in the CFD simulation of Darrieus wind turbines. *Renew. Energy* **2016**, *85*, 419–435. [CrossRef]
39. Rezaeiha, A.; Montazeri, H.; Blocken, B. On the accuracy of turbulence models for CFD simulations of vertical axis wind turbines. *Energy* **2019**, *180*, 838–857. [CrossRef]

40. Rezaeiha, A.; Kalkman, I.; Blocken, B. CFD simulation of a vertical axis wind turbine operating at a moderate tip speed ratio: Guidelines for minimum domain size and azimuthal increment. *Energy* **2017**, *107*, 373–385. [CrossRef]
41. Balduzzi, F.; Bianchini, A.; Ferrara, G.; Ferrari, L. Dimensionless numbers for the assessment of mesh and timestep requirements in CFD simulations of Darrieus wind turbines. *Energy* **2016**, *97*, 246–261. [CrossRef]
42. Rainbird, J. The Aerodynamic Development of a Vertical Axis Wind Turbine. Master's Thesis, University of Durham, Durham, UK, 2007.
43. Asr, M.T.; Nezhad, E.Z.; Mustapha, F.; Wiriadidjaja, S. Study on start-up characteristics of H-Darrieus vertical axis wind turbines comprising NACA 4-digit series blade airfoils. *Energy* **2016**, *112*, 528–537. [CrossRef]
44. Liu, Z.; Qu, H.; Shi, H. Numerical study on self-starting performance of Darrieus vertical axis turbine for tidal stream energy conversion. *Energies* **2016**, *9*, 789. [CrossRef]
45. Worasinchai, S.; Ingram, G.; Dominy, R. A low-Reynolds-number, high-angle-of-attack investigation of wind turbine aerofoils. *Proc. Inst. Mech. Eng. Part A J. Power Energy* **2011**, *225*, 748–763. [CrossRef]
46. Timmer, W. Two-dimensional low-Reynolds number wind tunnel results for airfoil NACA 0018. *Wind Eng.* **2008**, *32*, 525–537. [CrossRef]
47. Posa, A.; Balaras, E. Large Eddy Simulation of an isolated vertical axis wind turbine. *J. Wind. Eng. Ind. Aerodyn.* **2018**, *172*, 139–151. [CrossRef]
48. Nobari, M.; Ghazanfarian, J. A numerical investigation of fluid flow over a rotating cylinder with cross flow oscillation. *Comput. Fluids* **2009**, *38*, 2026–2036. [CrossRef]
49. Mehmood, A.; Abdelkefi, A.; Hajj, M.R.; Akhtar, I. On the onset of bifurcation and nonlinear characterization of vortex-induced vibrations under varying initial conditions. *Nonlinear Dyn.* **2020**, *99*, 575–592. [CrossRef]

MDPI AG
Grosspeteranlage 5
4052 Basel
Switzerland
Tel.: +41 61 683 77 34

MDPI Books Editorial Office
E-mail: books@mdpi.com
www.mdpi.com/books

www.ingramcontent.com/pod-product-compliance
Lightning Source LLC
Chambersburg PA
CBHW061501220926
51850CB00004B/144
9783725814633